Wavelets, Images, and Surface Fitting

Wavelets, Images, and Surface Fitting

edited by

Pierre-Jean Laurent
Université Joseph Fourier
Grenoble, France

Alain Le Méhauté
Ecole Nationale Supérieure
Télécommunications
de Bretagne, France

Larry L. Schumaker
Vanderbilt University
Nashville, Tennessee, USA

A K Peters
Wellesley, Massachusetts

Editorial, Sales, and Customer Service Office

A K Peters, Ltd.
289 Linden Street
Wellesley, MA 02181

Library of Congress Cataloging-in-Publication Data

Wavelets, images, and surface fitting / [edited by] Pierre-Jean Laurent, Alain Le Méhauté, Larry L. Schumaker.
p. cm.
Papers from the Second International Conference on Curves and Surfaces, held in Chamonix-Mont-Blanc, France, June 10–16, 1993.
Includes bibliographical references.
ISBN 1-56881-040-7 :
1. Wavelets—Congresses. 2. Approximation theory—Congresses. 3. Curves—Computer simulation—Congresses. 4. Surfaces—Computer simulation—Congresses. I. Laurent, Pierre-Jean, 1937– . II. Le Méhauté, Alain. III. Schumaker, Larry L., 1939– .
IV. International Conference on Curves and Surfaces (2nd : 1993 : Chamonix-Mont-Blanc, France
QA403.3.W394 1994
511'.42—dc20 94-11330
CIP

Printed in the United States of America
98 97 96 95 94 10 9 8 7 6 5 4 3 2 1

CONTENTS

PREFACE

During the week of June 10–16, 1993, an International Conference on *Curves and Surfaces* was held in Chamonix-Mont-Blanc (France). It was organized by the *Association Française d'Approximation* (A. F. A.). The organizing committee consisted of D. Apprato (Pau), M. Attéia (Toulouse), J.-M. Chassery (Grenoble), P. Chenin (Grenoble), J.-P. Dedieu (Toulouse), J.-C. Fiorot (Valenciennes), J. Gaches (Toulouse), B. Lacolle (Grenoble), Y. Lafranche (Rennes), P.-J. Laurent (Grenoble), A. Le Méhauté (Brest), M.-L. Mazure (Grenoble), C. Rabut (Toulouse), P. Sablonnière (Rennes), L. L. Schumaker (Nashville, USA), and C. Vercken (Paris).

The Conference was attended by 260 mathematicians from 22 different countries, and the program included 10 invited one-hour lectures and 140 research talks. The survey lectures dealt with several particularly active subareas of *Approximation Theory*, including *approximation of images, computer-aided geometric design, constrained approximation, finite elements, multivariate splines, Padé approximation, subdivision, surface compression,* and *wavelets.* A number of the research talks were presented in minisymposia on *Spline Conversion*, *Software Infrastructure for CAGD*, *Rational Approximation*, *Wavelets*, and *Constrained Approximation*, organized by J. Hoschek, R. Bartels, J.-C. Fiorot, A. Cohen, and F. Utreras, respectively.

The proceedings of this conference consist of a total of 106 papers. This volume contains 48 papers on wavelets, images, surface fitting, and related topics. The volume *Curves and Surfaces in Geometric Design*, also published by A K PETERS, contains 58 papers.

We would like to thank the following institutions and companies for their financial support: Universités J. Fourier, Grenoble and P. Sabatier, Toulouse, and Vanderbilt University, Nashville; Inst. Nat. Sciences Appliquées de Toulouse; Institut IMAG; Laboratoire de Modélisation et de Calcul (Grenoble); Centre National de la Recherche Scientifique (C.N.R.S.); Direction des Recherches et Etudes Doctorales (DRED); Direction des Recherches et Etudes Techniques (DRET, Contract 93-1110/A000/DRET/DS/SR); US Air Force European Office of Aerospace Research and Development (EOARD); US Army Research, Development and Standardization Group, UK (USARDSG); US Navy, Office of Naval Research, European Office; and Elf-Aquitaine.

Local organization of the Conference was carried out by the LMC-IMAG Institute (Grenoble). We would like to thank D. Iglesias and his colleagues at the printing service. In addition to the members of the organizing committee, several of our colleagues and students helped with the organization of the Conference, including O. Daoudi, M. Duc-Jacquet, and A. Eberhard. Finally, we want to express our special thanks to our wives Guizou Laurent, Michelle Le Méhauté, and Gerda Schumaker.

Nashville, Tennessee

February 28, 1994

CONTRIBUTORS

Numbers in parentheses indicate pages on which authors' contributions begin.

K. N. AGBÉVÉ (335), *3-135, place de Venise, 59650 Villeneuve D'Ascq, FRANCE*

L. AMODEI (1), *Météo-France, Centre National de Recherches Météorologiques, 42, Av. G. Coriolis, 31057 - Toulouse cedex, FRANCE* [luca.amodei@meteo.fr]

I. J. ANDERSON (11), *DITC, NPL, Teddington, TW11 0LW, ENGLAND* [iga@uk.co.npl.newton]

ANESTIS ANTONIADIS (21), *Université Joseph Fourier, IMAG-LMC, BP 53 X, 38041 Grenoble cedex FRANCE* [antonia@imag.fr]

M. ATTEIA (29), *Laboratoire d'Analyse Numérique, 118, Route de Narbonne, 31062 Toulouse cedex, FRANCE*

GÜNTER BASZENSKI (37), *Fachbereich Nachrichtentechnik, Fachhochschule Dortmund, Postfach 10 50 18, D – 44047 Dortmund, GERMANY* [baszenski@ruba.rz.ruhr-uni-bochum.d400.de]

M. N. BENBOURHIM (1), *Laboratoire d'Analyse Numérique, Université Paul Sabatier, 118, Rte de Narbonne, 31062 - Toulouse cedex, FRANCE*

ABDERRAHMAN BOUHAMIDI (51), *Dept. Math. Appliquées, Univ. du Littoral, 50 rue Ferdinand Buisson, BP 699, F-62228 Calais, FRANCE*

M. BOZZINI (59), *Dipartimento di Matematica, via C. Saldini 50, 20133 Milano, ITALY* [bozzini@vmimat.mat.unimi.it]

JEFFREY L. BROWN (67), *Mathematical Sciences Department, University of North Carolina at Wilmington, 601 South College Road, Wilmington, North Carolina 28403* [brownj@seq.cms.uncwil.edu]

GISELA BRUMME (75), *Fachbereich Mathematik, Universität Rostock, D-18051 Rostock, GERMANY* [brumme@mathematik.uni-rostock.d400.de]

LIONEL BRUNIE (83), *Laboratoire d'informatique médicale, Faculté de médecine, 38700 La Tronche, FRANCE* [lbrunie@lip.ens-lyon.fr]

MARTIN D. BUHMANN (93), *D–Mathematik, ETH–Zentrum, CH–8092 Zürich, SWITZERLAND* [mdb@math.ethz.ch]

VALÉRIE BURDIN (269), *Groupe Traitement d'Images, Telecom Bretagne, BP 832, Brest cedex, FRANCE* [burdin@gti.enst-bretagne.fr]

M. Morandi Cecchi (189), *Dip. di Matematica Pura ed Applicata, Università di Padova, Via Belzoni 7, I-35131, Padova, ITALY* [mcecchi@pdmat1.math.unipd.it]

Bernard Chalmond (113), *Cergy-Pontoise University, 8, Le Campus, 95033 Cergy-Pontoise cedex, FRANCE* [chalmond@diam1.ens.cachan.fr]

Charles K. Chui (121), *Center for Approximation Theory, Department of Mathematics, Texas A&M University, College Station, TX 77843-3368* [cchui@tamu.edu]

Philippe Cinquin (83), *Laboratoire d'informatique médicale, Faculté de médecine, 38700 La Tronche, FRANCE* [cinquin@timb.imag.fr]

F. Coldefy (113), *SUDIMAGE group, Cachan, FRANCE*

M. G. Cox (11), *DITC, NPL, Teddington, TW11 0LW, ENGLAND* [mgc@uk.co.npl.newton]

Stephan Dahlke (141), *Institut für Geometrie und Praktische Mathematik, RWTH Aachen, Templergraben 55, 52062 Aachen, GERMANY* [dahlke@igpm.rwth-aachen.de]

Wolfgang Dahmen (157), *Institut für Geometrie und Praktische Mathematik, RWTH Aachen, Templergraben 55, 52056 Aachen, GERMANY* [dahmen@igpm.rwth-aachen.de]

S. De Marchi (189), *Dip. di Matematica Pura ed Applicata, Università di Padova, Via Belzoni 7, I-35131, Padova, ITALY* [demarchi@pdmat1.math.unipd.it]

F. De Tisi (59), *Dipartimento di Matematica, via C. Saldini 50, 20133 Milano, ITALY*

Ronald A. DeVore (197), *Department of Mathematics, University of South Carolina, Columbia, SC 29208* [devore@math.scarolina.edu]

Harvey Diamond (221), *Mathematics Department, West Virginia University, Morgantown, WV 26506* [diamond@math.wvu.edu]

Paul Dierckx (511), *Katholieke Universiteit Leuven, Department of Computer Science, Celestijnenlaan 200 A, B-3001 Heverlee, BELGIUM*

Nira Dyn (229), *School of Mathematical Sciences, Sackler Faculty of Exact Sciences, Tel Aviv University, Tel Aviv 69978, ISRAEL* [niradyn@taurus.bitnet]

Mariano Gasca (245), *Departamento de Matemática Aplicada, Universidad de Zaragoza, 50009 Zaragoza, SPAIN* [gasca@cc.unizar.es]

Faouzi Ghorbel (269), *Groupe Image INT-ENIC, ENIC – Cité scientifique, 59658 Villeneuve d'Ascq cedex, FRANCE*

G. GREINER (277), *Universität Erlangen, IMMD IX- Graphische Datenverarbeitung, Am Weichselgarten 9, D-91058 Erlangen, GERMANY* [greiner@informatik.uni-erlangen.de]

S. HARBOUR (11), *AMOR group, Cranfield (RMCS), Shrivenham, SN6 8LA, ENGLAND* [harbour@uk.ac.cran.rmcs]

WALTER HESS (287), *Institute of Numerical Mathematics, Technical University of Dresden, D – 01062 Dresden, GERMANY* [numath@urzdfn.mathematik.tu-dresden.dbp.de]

KURT JETTER (295), *FB Mathematik, Universität Duisburg, D–47048 Duisburg, GERMANY* [hn277je@unidui.uni-duisburg.de]

JOHANNES KAASA (303), *Dept. of Industrial Mathematics, SINTEF, P. O. Box 124 Blindern, N-0314 Oslo, NORWAY* [Johannes.Kasa@si.sintef.no]

Y. L. KERGOSIEN (311), *Dépt. de Mathématiques, Bât 425, Université de Paris-Sud, 91405 Orsay, FRANCE* [kergos@psisun.u-psud.fr]

OGNYAN I. KOUNCHEV (319), *Dept. of Mathematics, University of Duisburg, Lotharstr. 65, Duisburg, 47058 GERMANY* [hn314ko@unidui.uni-duisburg.de]

ANGELA KUNOTH (327), *Institut für Informatik, Fachbereich Mathematik und Informatik, Freie Universität Berlin, Takustr. 9, D-14195 Berlin, GERMANY* [kunoth@inf.fu-berlin.de]

B. LAVAYSSIÈRE (113), *DER-SDM Department, EDF, Chatou, FRANCE*

ALAIN LE MÉHAUTÉ (51, 335), *Département de Mathématiques et système de communication, ENST–Bretagne, B.P 832, 29285 Brest cedex, FRANCE* [Alain.LeMehaute@enst-bretagne.fr]

J. LEVESLEY (343), *Department of Mathematics and Computer Science, University of Leicester, University Road, Leicester LE1 7RH, UK* [jl1@mathematics-and-computer-science.leicester.ac.uk]

DAVID LEVIN (229), *School of Mathematical Sciences, Sackler Faculty of Exact Sciences, Tel Aviv University, Tel Aviv 69978, ISRAEL* [dlevin@taurus.bitnet]

WILL LIGHT (351), *Department of Mathematics and Computer Science, University of Leicester, Leicester LE1 7RH, ENGLAND* [pwl@mcs.le.ac.uk]

J. C. MASON (11), *AMOR group, Cranfield (RMCS), Shrivenham, SN6 8LA, ENGLAND* [mason@uk.ac.cran.rmcs]

CHARLES A. MICCHELLI (245), *Department of Mathematical Sciences, IBM T. J. Watson Research Center, P. O. Box 218, Yorktown Heights, New York 10598* [cam@yktvmx.bitnet]

ROGER MOHR (521), *LIFIA-INRIA, 46, avenue Félix Viallet, 38031 Grenoble cedex, FRANCE* [mohr@lifia.imag.fr]

PIERRE MONTÈS (371), *Ecole Polytechnique de Montréal, Département de Génie Civil, 5182 Snowdon, Montréal Québec, CANADA H3W 2G1*

BERND MULANSKY (379), *Institut für Numerische Mathematik, Technische Universität, D–01062 Dresden, GERMANY* [mulansky@math.tu-dresden.de]

J. M. PEÑA (245), *Departamento de Matemática Aplicada, Universidad de Zaragoza, Edificio de Matemáticas, Planta 1a, 50009 Zaragoza, SPAIN* [jmpena@cc.unizar.es]

GERLIND PLONKA (387), *FB Mathematik, Universität Rostock, D-18051 Rostock, GERMANY* [nfa016@nfa.uni-rostock.d400.de]

CHRISTINE POTIER (399), *Telecom Paris, 46 rue Barrault, 74634, Paris cedex 13, FRANCE* [Christine.Potier@inf.enst.fr]

JÜRGEN PRESTIN (407), *FB Mathematik, Universität Rostock, Universitätsplatz 1, 18051 Rostock, GERMANY* [prestin@mathematik.uni-rostock.d400.de]

EWALD QUAK (407, 419), *Center for Approximation Theory, Department of Mathematics, Texas A&M University, College Station, TX 77843-3368* [quak@euler.math.tamu.edu]

CHRISTOPHE RABUT (429), *Centre de Mathématiques, Institut National des Sciences Appliquées, Complexe Scientifique de Rangueil, F-31077 Toulouse cedex, FRANCE* [Rabut@cict.fr]

LOUISE ARAKELIAN RAPHAEL (221), *Mathematics Department, Howard University, Washington, D.C. 20059* [alama38@humain.howard.edu]

O. RIOUL (437), *CNET Paris B, 38-40, rue du Général Leclerc, 92 131 Issy-Les-Moulineaux, FRANCE* [rioul@balsa.issy.cnet.fr]

AMOS RON (93), *University of Wisconsin–Madison, Department of Computer Sciences, 1210 West Dayton Street, Madison, WI 53706* [amos@cs.wisc.edu]

M. ROSSINI (59), *Dipartimento di Matematica, via C. Saldini 50, 20133 Milano, ITALY*

PAUL SABLONNIÈRE (445), *INSA 20, Avenue des Buttes de Coësmes, 35042 Rennes cedex, FRANCE*

C. SBERT (453), *Dept. of Mathematics and Computer Science, Univ. de les Illes Balears, Ctra Valldemossa Km. 7,5, 07071 Palma de Mallorca, SPAIN* [dmicsj0@ps.uib.es]

ROBERT SCHABACK (459), *Institut für Numerische und Angewandte Mathematik, Georg-August-Universität Göttingen, Lotzestr. 16–18, D-37083 Göttingen, GERMANY* [schaback@namu01.gwdg.de]

JOCHEN W. SCHMIDT (287, 379), *Institute of Numerical Mathematics, Technical University of Dresden, D – 01062 Dresden, GERMANY* [numath@urzdfn.mathematik.tu-dresden.dbp.de]

JÖRN SPRINGER (467), *GeoForschungsZentrum Potsdam, Telegrafenberg A51, D-14473 Potsdam, GERMANY* [pr@gfz-potsdam.dbp.de]

MANFRED TASCHE (37, 475), *Fachbereich Mathematik, Universität Rostock, D–18051 Rostock, GERMANY* [tasche@mathematik.uni-rostock.d400.de]

JUAN JOSÉ TORRENS (485), *Departamento de Matemática e Informática, Universidad Pública de Navarra, Campus de Arrosadía s/n, 31006 Pamplona, SPAIN* [jjtorrens@upna.es]

CHRISTINE VERCKEN (399), *Telecom Paris, 46 rue Barrault, 74634, Paris cedex 13, FRANCE* [Christine.Vercken@inf.enst.fr]

LARS F. VILLEMOES (493), *Department of Mathematics, Royal Institute of Technology, S-10044 Stockholm, SWEDEN* [larsv@math.kth.se]

MARSHALL WALKER (501), *Department of Computer Science and Mathematics, Atkinson College, York University, North York M3J 1P3, Ontario, CANADA* [mwalker@sol.yorku.ca]

JIANZHONG WANG (121), *Department of Mathematics, University of North Carolina, Charlotte, NC 28223* [jiawang@mosaic.uncc.edu]

GEIR WESTGAARD (303), *Dept. of Industrial Mathematics, SINTEF, P. O. Box 124 Blindern, N-0314 Oslo, NORWAY* [Geir.Westgaard@si.sintef.no]

NORMAN WEYRICH (419), *Department of Mathematics, AFIT/ENC, 2950 P street. Wright-Patterson AFB, OH 45433-7765* [nweyrich@afit.af.mil]

KARIN WILLEMANS (511), *Katholieke Universiteit Leuven, Department of Computer Science, Celestijnenlaan 200 A, B-3001 Heverlee, BELGIUM* [karinw@cs.kuleuven.ac.be]

DANIEL A. WILLIAMS (221), *Mathematics Department, Howard University, Washington, D.C. 20059* [daw@scs.howard.edu]

FATIMA JEEAWOCK-ZEDEK (445), *Université de Lille, Département de Mathé matiques, 59655 Villeneuve d'Ascq cedex, FRANCE*

CHANGSHENG ZHAO (521), *LIFIA-INRIA, 46, avenue Félix Viallet, 38031 Grenoble cedex, FRANCE* [zhao@lifia.imag.fr]

A Vector Spline Quasi-Interpolation

L. Amodei and M. N. Benbourhim

Abstract. We introduce a new vector quasi-interpolant based on a vector spline function given by the authors [1,2]. Following an established technique [6], the elementary matrix approximant is obtained by convolution of the spline semi-kernel associated to the particular semi-norm considered and a discretized version of the partial differential operator for which the semi-kernel is a fundamental solution. However, in order to get a symmetric and positive definite elementary matrix approximant, a factorization of the differential operator has to be considered. Some general regularity properties of the elementary matrix are given along with results on the convergence of the quasi-interpolant. Finally we present some numerical examples of the elementary matrix.

1. Introduction: the Vector Spline Minimization Problem and its Solution.

The authors [1,2] introduced a vector spline function as the solution of the following minimization problem (P_ρ) depending on a fixed coefficient $\rho > 0$:

$$\begin{cases} \text{Min}\Big[\rho \int_{\mathbb{R}^2} \mid \nabla \operatorname{div} u \mid^2 \, dxdy \quad + \quad \int_{\mathbb{R}^2} \mid \nabla \operatorname{rot} u \mid^2 \, dxdy\Big] \\ \qquad \text{for} \quad u = (u_1, u_2) \in \mathcal{V}^2 = V^2 \times V^2 \quad \text{and} \\ \qquad\qquad u(t^i) = u^i,\ i = 1, \ldots, k. \end{cases}$$

Here V^2 is the classical Beppo-Levi space of order 2 over $\mathbb{R}^2$:

$$V^2 = \{ v \in \mathcal{D}'(\mathbb{R}^2) : D^\gamma v \in L^2(\mathbb{R}^2), \mid \gamma \mid = 2 \},$$

$\operatorname{div} u = \partial_1 u_1 + \partial_2 u_2$ is the 2D divergence operator,

$\operatorname{rot} u = \partial_1 u_2 - \partial_2 u_1$ is the 2D rotational operator,

$\nabla = (\partial_1 , \partial_2)$ is the 2D nabla operator (formal transpose of div),

$\text{Rot} = (-\partial_2 , \partial_1)$ is the formal transpose of rot,

$| \quad |$ is the classical Euclidean norm of $\mathbb{R}^2$,

$t^i = (t^i_1, t^i_2) \in \mathbb{R}^2$ are the interpolation points,

Wavelets, Images and Surface Fitting
P. J. Laurent, A. Le Méhauté, and L. L. Schumaker (eds.), pp. 1–10.

ISBN 1-56881-040-7.

and $u^i = (u_1^i, u_2^i) \in \mathbb{R}^2$, $i = 1, \ldots, k$, is the vector of data.

In [1,2], it is shown that the problem (P_ρ) admits a unique solution u_ρ. It is also interesting to mention that for $\rho = 1$, the problem splits into two separate problems, namely two thin-plate spline problems (of order 2), one for each of the components u_1 and u_2. On the contrary, for $\rho \neq 1$, the problem is effectively coupled in the two components. The coefficient ρ allows to control the relative magnitude of the two terms of the functional. Moreover, when $\rho \to 0$ and $\rho \to +\infty$, the solution u_ρ tends towards finite limits.

The expression of the functional is related to the fact that each vector field $u \in \mathcal{V}^2$ may be decomposed into the sum

$$u = \nabla\chi + \operatorname{Rot}\psi,$$

with a divergent part $\nabla\chi$ (∇ satisfies $\operatorname{rot}\nabla = 0$) and a rotational part $\operatorname{Rot}\psi$ (Rot satisfies $\nabla \operatorname{Rot} = 0$). In fluid mechanics this equality is known as the *Helmholtz decomposition.* The function χ is called a *velocity potential* and ψ is called a *stream function.*

This vector spline may be used to interpolate 2D velocity fields (fluid flows, horizontal wind field in meteorology, etc.). The solution u_ρ is calculated by solving the linear system

$$\begin{cases} \Delta(\rho\,\partial_1 \operatorname{div} u - \partial_2 \operatorname{rot} u) &= \displaystyle\sum_{i=1}^{k} a_i \delta_{t^i} \\ \Delta(\rho\,\partial_2 \operatorname{div} u + \partial_1 \operatorname{rot} u) &= \displaystyle\sum_{i=1}^{k} b_i \delta_{t^i}, \end{cases} \tag{1}$$

with δ_{t^i} the Dirac distribution at point t^i and $a_i, b_i \in \mathbb{R}$.

Equation (1) involves the operators

$$P_\rho(D) = \Delta\Big[\rho\begin{pmatrix} \partial_{11}^2 & \partial_{12}^2 \\ \partial_{12}^2 & \partial_{22}^2 \end{pmatrix} + \begin{pmatrix} \partial_{22}^2 & -\partial_{12}^2 \\ -\partial_{12}^2 & \partial_{11}^2 \end{pmatrix}\Big] = \Delta R_\rho(D), \tag{2}$$

and the solution is obtained via a fundamental solution F_ρ of the system

$$P_\rho(D) F_\rho = \delta\, I_2. \tag{3}$$

We observe that

$$\det(P_\rho(D)) = \rho\Delta^4,$$

so that a solution F_ρ of (3) is given by

$$F_\rho = \widetilde{P_\rho(D)}\frac{1}{\rho}K_4\, I_2,$$

where K_4 is a fundamental solution of $\Delta^4 K_4 = \delta$ and $\widetilde{P_\rho(D)}$ is the adjoint of $P_\rho(D)$, i. e. the transpose of the cofactor matrix of $P_\rho(D)$ which satisfies

$$\widetilde{P_\rho(D)}\, P_\rho(D) = P_\rho(D)\, \widetilde{P_\rho(D)} = \det(P_\rho(D))\, I_2.$$

The equality (2) allows us to rewrite the solution as

$$F_\rho = \widetilde{R_\rho(D)}\frac{1}{\rho}\Delta K_4\, I_2 = \widetilde{R_\rho(D)}\frac{1}{\rho} K_3\, I_2.$$

Specifically, $\widetilde{R_\rho(D)}$ is given by

$$\widetilde{R_\rho(D)} = \begin{pmatrix} \partial_{11}^2 & \partial_{12}^2 \\ \partial_{12}^2 & \partial_{22}^2 \end{pmatrix} + \rho \begin{pmatrix} \partial_{22}^2 & -\partial_{12}^2 \\ -\partial_{12}^2 & \partial_{11}^2 \end{pmatrix}.$$

Finally, we get the solution $u_\rho(t)$ of (P_ρ):

$$u_\rho(t) = \mu * F_\rho(t) + q(t),$$

with $\mu = (\sum_{i=1}^{k} a_i \delta_{t^i}, \sum_{i=1}^{k} b_i \delta_{t^i})$ and $q(t) \in (\mathcal{P}_1)^2$.

Remark. a) The theory can be generalized for any order $m \geq 2$ by considering $u \in \mathcal{V}^m = V^m \times V^m$ and the semi-norm

$$\rho \mid \operatorname{div} u \mid_{m-1}^2 + \mid \operatorname{rot} u \mid_{m-1}^2 .$$

Here $|\ \ |_{m-1}$ is the classical semi-norm of V^{m-1}.

b) The elements of the matrix F_ρ are no longer radial functions like K_3, but this fact has no bearing on the general properties of the quasi-interpolant.

§2. Elementary Matrix Approximant

To simplify the notations, from now on we will omit the dependency on the coefficient ρ for the solution and the differential operators involved by problem (P_ρ).

The elementary 2-harmonic cardinal B-spline (see [6]) is obtained by applying a discretized bilaplacian to a fundamental solution of the continuous operator (namely the radial function $\psi(r) = c\, r^2 \log(r)$). In a natural way, we are led to define the elementary vector approximant based on F and (3) by

$$P(D_h)F,$$

where $P(D_h)$ is a discretization of the differential operator. However, we can easily verify that $P(D_h)F$ is neither a symmetric nor a positive definite matrix.

To get a symmetric expression, we have to consider first the formal positive square root of the operator $P(D)$. This decomposition may also be obtained from an alternate formulation of the semi-norm which defines the spline problem (P_ρ). For this, we observe that the operators ∇ and Rot are orthogonal in the space V^1, i.e.,

$$\int_{\mathbf{R}^2} (\nabla\phi, \operatorname{Rot}\varphi)\, dxdy = 0, \qquad \forall\, \phi, \varphi \in V^1,$$

where (,) is the Euclidean scalar product. This is obtained from the density of $\mathcal{D}+\mathcal{P}_0$ in the space V^1 (see [1,2]) and a double integration by parts which is free of boundary terms.

For $u, v \in \mathcal{V}^2$, we have $\operatorname{div} u \in V^1$ and $\operatorname{rot} v \in V^1$, so that

$$\int_{\mathbb{R}^2} (\nabla \operatorname{div} u, \operatorname{Rot}\operatorname{rot} v)\, dxdy = 0.$$

Observing that $\displaystyle\int_{\mathbb{R}^2} |\nabla \operatorname{rot} u|^2\, dxdy = \int_{\mathbb{R}^2} |\operatorname{Rot}\operatorname{rot} u|^2\, dxdy$, we get

$$\rho \int_{\mathbb{R}^2} |\nabla \operatorname{div} u|^2\, dxdy \quad + \quad \int_{\mathbb{R}^2} |\nabla \operatorname{rot} u|^2\, dxdy =$$
$$= \int_{\mathbb{R}^2} |\sqrt{\rho}\nabla \operatorname{div} u - \operatorname{Rot}\operatorname{rot} u|^2\, dxdy = \int_{\mathbb{R}^2} |Q(D)\, u|^2\, dxdy,$$

where $Q(D)$ is defined by

$$Q(D) = \sqrt{\rho}\begin{pmatrix} \partial_{11}^2 & \partial_{12}^2 \\ \partial_{12}^2 & \partial_{22}^2 \end{pmatrix} + \begin{pmatrix} \partial_{22}^2 & -\partial_{12}^2 \\ -\partial_{12}^2 & \partial_{11}^2 \end{pmatrix}.$$

The optimality condition (1) and the symmetry of $Q(D)$ lead to the equality

$$P(D) = Q(D)^t\, Q(D) = Q(D)\, Q(D).$$

Proposition 1. $Q(D)\, F\, Q(D) = \delta I_2$.

Proof: The notation $Q(D)\, F\, Q(D)$ stands for the convolution product

$$(Q(D)\, \delta\, I_2) * F * (Q(D)\, \delta\, I_2).$$

We have $F = \widetilde{Q(D)}\, \widetilde{Q(D)} \frac{1}{\rho} K_4\, I_2$. Premultiplying by $Q(D)$, we get

$$Q(D)\, F = \det(Q(D))\, \widetilde{Q(D)} \frac{1}{\rho} K_4\, I_2 = \sqrt{\rho}\, \Delta^2\, \widetilde{Q(D)} \frac{1}{\rho} K_4\, I_2.$$

Postmultiplying by $Q(D)$, we get

$$Q(D)\, F * (Q(D)\, \delta\, I_2) = \sqrt{\rho}\, \Delta^2 \sqrt{\rho}\, \Delta^2 \frac{1}{\rho} K_4\, \delta\, I_2 = \delta\, I_2. \quad \blacksquare$$

For the discretization, we use the following notations. The centered difference operator is defined for $h \in \mathbb{R}$ by

$$\partial_{h,j} = h^{-1}(\delta_{-\frac{h}{2}e_j} - \delta_{+\frac{h}{2}e_j}),$$

with e_j the jth basis vector of $\mathbb{R}^2$, and the iterated operator by

$$D_h^\alpha = \underbrace{(\partial_{h,1} * \cdots * \partial_{h,1})}_{\alpha_1 \text{ times}} * \underbrace{(\partial_{h,2} * \cdots * \partial_{h,2})}_{\alpha_2 \text{ times}},$$

for $\alpha = (\alpha_1, \alpha_2) \in \mathbb{N}^2$. We denote by $P(D_h)$ the discretization of a general constant coefficient operator $P(D)$.

The Fourier transform $\widehat{\partial_{h,j}}$ of $\partial_{h,j}$ is given by

$$\widehat{\partial_{h,j}}(\xi) = h^{-1}\, 2i \sin(\pi h \xi_j),$$

with $\xi = (\xi_1, \xi_2) \in \mathbb{R}^2$. Prop. 1 suggests the following definition.

Definition 1. *The elementary matrix approximant is defined by*

$$F_h = h^2\, Q(D_h) * F * Q(D_h). \tag{4}$$

It is easy to see that the elements of F_h are of the form $D^\alpha(\mu * K_3)$, with $|\alpha| = 2$ and $\mu \in \mathcal{E}'$ orthogonal to $\mathcal{P}_3$.

Proposition 2. *The 2×2 matrix F_h is symmetric and positive definite.*

Proof: From Definition 1, and using the fact that the 2×2 matrix $\widehat{F}$ is positive, we verify that the symmetric matrix $\widehat{F_h}(\xi) = \begin{pmatrix} \widehat{F_{h,11}}(\xi) & \widehat{F_{h,12}}(\xi) \\ \widehat{F_{h,21}}(\xi) & \widehat{F_{h,22}}(\xi) \end{pmatrix}$ is positive. ■

§3. Regularity of the Elementary Matrix Approximant

From the Fourier transform of $D^\beta F_h$, we can obtain some regularity results on F_h.

Proposition 3. *1) For $1 - \frac{1}{p} < \frac{|\alpha|}{2} < 3 - \frac{1}{p}$ and for any $\mu \in \mathcal{E}'$ such that $\mu(q) = 0$, $\forall q \in \mathcal{P}_3$,*

$$\mathcal{F}(D^\alpha(\mu * K_3))(\xi) \in \mathrm{L}^p(\mathbb{R}^2).$$

2) For $k, l = 1, 2$, the coefficients $F_{h,kl}$ of the matrix F_h belong to $\mathrm{H}^2(\mathbb{R}^2)$, and for $|\beta| \le 1$, $D^\beta F_{h,kl}$ is bounded and continuous with $\lim\limits_{|t|\to+\infty} |D^\beta F_{h,kl}(t)| = 0$.

Proof: 1) Following [5], we have

$$\widehat{K_3}(\xi) = \xi^\alpha\, \widehat{\mu}(\xi)(c_1 \mathrm{Fp}\frac{1}{|\xi|^6} + c_2 \Delta^2 \delta). \tag{5}$$

The hypothesis $\mu(q) = 0$, $\forall q \in \mathcal{P}_3$, shows that $(D^\beta \widehat{\mu})(0) = 0$ for $|\beta| \le 3$. We easily deduce that $\xi^\alpha \widehat{\mu}(\xi) \Delta^2 \delta = 0$ for $|\alpha| \ge 1$ and the term $\Delta^2\delta$ may be dropped from the expression (5). We consider two cases:

a) For $\xi \in \mathrm{B}(0,1)$ (the unit ball), we have $|\widehat{\mu}(\xi)| \le c_3 |\xi|^4$ since $(D^\beta \widehat{\mu})(0) = 0$ for $|\beta| \le 3$. Hence $|\widehat{K_3}(\xi)| \le c_1\, c_3 \dfrac{1}{|\xi|^{2-|\alpha|}}$ which is in $\mathrm{L}^p(\mathrm{B}(0,1))$ if and only if $2 - |\alpha| < \dfrac{2}{p}$.

b) For ξ in the complement $\widetilde{\mathrm{B}}$ of $\mathrm{B}(0,1)$, we get $|\widehat{K_3}(\zeta)| \le c_1\, c_4 \dfrac{1}{|\xi|^{6-|\alpha|}}$ since $|\widehat{\mu}(\xi)| \le c_4$ and $\widehat{K_3}(\xi) \in \mathrm{L}^p(\widetilde{\mathrm{B}})$ for $6 - |\alpha| > \dfrac{2}{p}$.

2) By taking $p = 2$, the preceding result shows that $F_{h,kl} \in \mathrm{H}^2(\mathbb{R}^2)$. With $p = 1$, we get $(\widehat{D^\beta F_{h,kl}})(\xi) \in \mathrm{L}^1(\mathbb{R}^2)$ for $|\beta| \le 1$ and the Riemann-Lebesgue theorem gives the result. ■

If $D^\gamma(\widehat{F_{h,kl}}) \in \mathrm{L}^1(\mathbb{R}^2)$, for $|\gamma| \le a$, the same theorem (RL) gives $F_{h,kl}(t) = o(|t|^{-a})$ near infinity. We denote by δ_{kl} the Kronecker symbol.

Proposition 4. *For $k, l = 1, 2$, we have*

$$F_{h,kl}(t) = o(|t|^{-3}),$$

for $|t| \to +\infty$, and

$$\widehat{F_{h,kl}}(0) = h^2\, \delta_{kl},$$

$$D^{\beta_0}(\widehat{F_{h,kl}})(0) = 0, \quad D^{\beta}(\widehat{F_{h,kl}})(\frac{j}{h}) = 0, \forall j \in \mathbb{Z}^2 - \{0\},$$

respectively with $|\beta_0| = 1$ and $|\beta| \leq 1$.

Proof: Eq. (4) shows that the general form of $\widehat{F_{h,kl}}$ is

$$\widehat{F_{h,kl}}(\xi) = h^2 \sum_{\substack{r,s \\ \gamma,\, |\gamma|=4}} \pi^2 \frac{\xi_r \xi_s}{|\pi\xi|^6} h^{-4} \left(\sin(\pi h\, \xi_1)\right)^{\gamma_1} \left(\sin(\pi h\, \xi_2)\right)^{\gamma_2}. \tag{6}$$

Using the power series $\sin(x) = x - \frac{x^3}{3!} + \cdots$ near 0, we get

$$\widehat{F_{h,kk}}(\xi) = h^2 + \frac{p_8(\xi)}{|\xi|^6} + \frac{p_{10}(\xi)}{|\xi|^6} + \cdots \tag{7}$$

and for $k \neq l$

$$\widehat{F_{h,kl}}(\xi) = \frac{q_8(\xi)}{|\xi|^6} + \frac{q_{10}(\xi)}{|\xi|^6} + \cdots, \tag{8}$$

where $p_8, p_{10}, \cdots, q_8, q_{10}, \cdots$ are homogeneous polynomials of degree given by their index. Observing that each derivation decreases the degree of the rationnal fractions by one, we deduce that $D^\gamma(\widehat{F_{h,kl}})(\xi)$ is integrable near 0 for $|\gamma| \leq 3$. Near infinity, (6) shows that $\widehat{F_{h,kl}}(\xi) \leq c\frac{1}{|\xi|^4}$ and $D^\gamma(\widehat{F_{h,kl}})(\xi)$ is integrable for all γ. We can conclude that $D^\gamma(\widehat{F_{h,kl}})(\xi) \in \mathrm{L}^1(\mathbb{R}^2)$ for each $|\gamma| \leq 3$. The second part of the proposition is easily deduced from (7), (8) and (6). ■

Remark. Using the same arguments as in [4] (Corollary 4.3), expression $F_{h,kl}(t) = o(|t|^{-3})$ may be strengthened into $F_{h,kl}(t) = \mathcal{O}(|t|^{-4})$.

§4. Quasi-Interpolation and Convergence

The following property which is verified as well by the elementary 2-harmonic cardinal B-spline suggests our definition of the quasi-interpolant.

Proposition 5. *The equality*

$$F_h(t) = F_1(\frac{t}{h}), \forall t \in \mathbb{R}^2,$$

is satisfied for each h.

Proof: Equality (6) shows that

$$\widehat{F_h}(\xi) = h^2 \widehat{F_1}(h\xi).$$

A general property of the Fourier transform gives $\widehat{\psi(\frac{\cdot}{h})}(\xi) = h^2 \widehat{\psi}(h\xi)$. By taking $\psi = F_1$ and using the inverse Fourier transform, we obtain the result. ■

Definition 2. *For $u = (u_1, u_2) \in (\mathcal{C}^0(\mathbb{R}^2))^2$, the vector quasi-interpolant, denoted by $F_h *_h u$, is defined by*

$$\begin{cases} (F_h *_h u)_1(t) = \sum_{j\in\mathbb{Z}^2} F_{h,11}(t-hj)u_1(hj) + \sum_{j\in\mathbb{Z}^2} F_{h,12}(t-hj)u_2(hj) \\ (F_h *_h u)_2(t) = \sum_{j\in\mathbb{Z}^2} F_{h,21}(t-hj)u_1(hj) + \sum_{j\in\mathbb{Z}^2} F_{h,22}(t-hj)u_2(hj). \end{cases}$$

Prop. 5 allows to replace $F_{h,kl}(t-hj)$ by $F_{1,kl}(\frac{t}{h}-j)$, which is the classical expression given in the literature of quasi-interpolation (see [3,4,6]). The natural expected property of polynomial reproduction is shown in the following proposition.

Proposition 6. *For all $p \in (\mathcal{P}_1)^2$, we have*

$$F_h *_h p = p.$$

Proof: The proof follows from the Poisson formula

$$\sum_{j\in\mathbb{Z}^2} \psi(t-hj) = \frac{1}{h^2} \sum_{j\in\mathbb{Z}^2} \widehat{\psi}(\frac{j}{h}) e^{2i\pi(j,\frac{t}{h})}.$$

If we suppose that $\psi \in \mathcal{C}^0$, $|t|^3\,|\psi(t)| < +\infty$ and $\sum_{j\in\mathbb{Z}^2} |\widehat{\psi}(\frac{j}{h})|$ finite, this identity is pointwise (see [4]).

From Prop. 3, 2) for the continuity and Prop. 4, we obtain

$$\sum_{j\in\mathbb{Z}^2} F_{h,kl}(t-hj) = \delta_{kl}. \tag{9}$$

By taking $\psi(t) = p(t)F_{h,kl}(t)$, with $p(t)$ homogeneous of degree 1, Prop. 4 shows that $\widehat{(pF_{h,kl})}(\frac{j}{h}) = 0$, $\forall j \in \mathbb{Z}^2$. Using the fact that $|t|^4|F_{h,kl}(t)|$ is bounded (see the remark after Prop. 4), we conclude that

$$\sum_{j\in\mathbb{Z}^2} p(t-hj)F_{h,kl}(t-hj) = 0. \tag{10}$$

But $p(t-hj)=p(t)-p(hj)$, and from (9) and (10), we obtain

$$\sum_{j\in\mathbb{Z}^2} p(hj)\, F_{h,kl}(t-hj)=p(t)\,\delta_{kl}, \tag{11}$$

and the proof is complete. ■

As for the scalar case, the convergence may be stated for different orders, according to the regularity of the function u (see for example [4]). The following proposition is intended to show how the identities (9) lead to convergence results, in the case of a vector interpolation.

Proposition 7. *Let $F_h(t)$ be a 2×2 continuous matrix verifying Prop. 5 and $|t|^3|F_{h,kl}(t)|\le M$, $\forall t\in\mathbb{R}^2$ (see Prop. 4). Let $u=(u_1,u_2)$ be a bounded and uniform continuous function. Then, we have*

$$\lim_{h\to+\infty} |u-F_h *_h u|_\infty = 0,$$

if and only if F_h satisfies (9).

Proof: The specific point comes from the equalities

$$\begin{aligned} u_1(t)-(F_h *_h u)_1(t) &= \sum_{j\in\mathbb{Z}^2}(u_1(t)-u_1(hj))\,F_{h,11}(t-hj)\\ &+\sum_{j\in\mathbb{Z}^2}(u_2(t)-u_2(hj))\,F_{h,12}(t-hj)\end{aligned}$$

and

$$\begin{aligned} u_2(t)-(F_h *_h u)_2(t) &= \sum_{j\in\mathbb{Z}^2}(u_1(t)-u_1(hj))\,F_{h,21}(t-hj)\\ &+\sum_{j\in\mathbb{Z}^2}(u_2(t)-u_2(hj))\,F_{h,22}(t-hj),\end{aligned}$$

deduced from (9). In order to get the same approach as in [4] (Th. 8.1), $F_{h,kl}(t-hj)$ is replaced by $F_{1,kl}(\frac{t}{h}-j)$ using Prop. 5; the bounds on each sum are then obtained in the same way. ■

§5. Numerical Examples

The normalized coefficients of the first column of the matrix F_h are shown in Figures 1 and 2. The vector field representing the first column of F_h is shown in Figures 3 and 4. It corresponds to the quasi-interpolation obtained from a value (1,0) at point (0,0) (center of the figures) and (0,0) on other gridpoints.

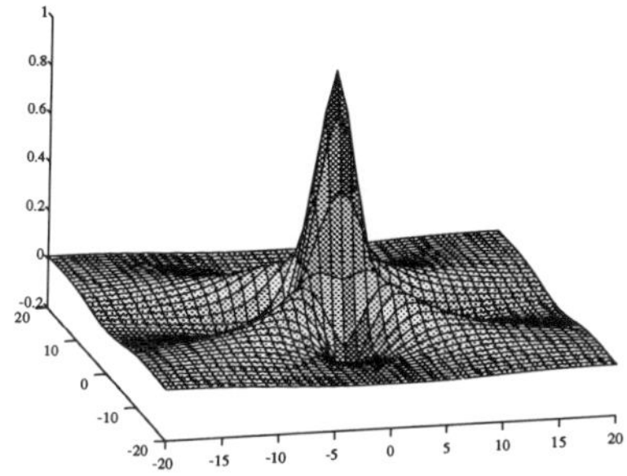

Fig. 1. Coefficient $F_{h,11}$ for parameter $\rho = 0.1$.

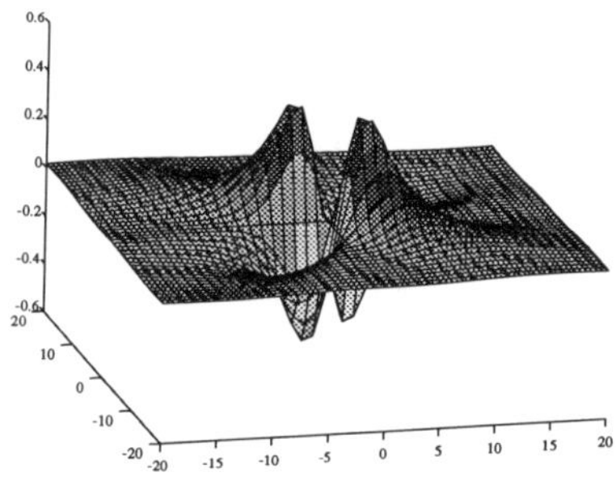

Fig. 2. Coefficient $F_{h,21}$ for parameter $\rho = 0.1$.

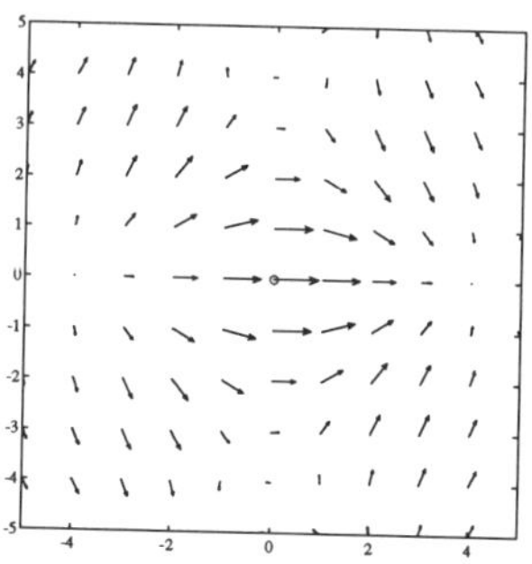

Fig. 3. Vector field for parameter $\rho = 0.1$; divergence/convergence is amplified.

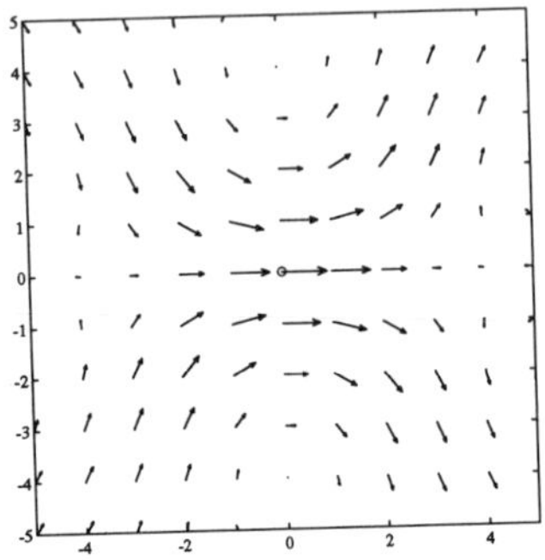

Fig. 4. Vector field for parameter $\rho = 10$; rotation is amplified.

References

1. Amodei, L. and M. N. Benbourhim, A vector spline approximation. J. Approx. Th. **67** (1991).
2. Amodei, L., Etude d'une classe de fonctions splines vectorielles en vue de l'approximation d'un champ de vitesse. Applications à la météorologie. Thèse, Université de Toulouse, France, 1993.
3. Jackson, I. R. H., Radial basis function methods of multivariate approximation. Ph. D. Dissertation, University of Cambridge, 1988.
4. Powell, M. J. D., The theory of radial basis function approximation in 1990. DAMTP report NA 11, University of Cambridge, 1990.
5. Schwartz, L., *Théorie des Distributions*, Hermann, Paris, 1966.
6. Rabut, C, B - splines polyharmoniques cardinales: interpolation, quasi-interpolation, filtrage. Thèse d' Etat, Université de Toulouse, France, 1990.

L. Amodei
Météo-France
Centre National de Recherches Météorologiques
42, Av. G. Coriolis
31057 - Toulouse Cedex, FRANCE
luca.amodei@ meteo.fr

M. N. Benbourhim
Laboratoire d'Analyse Numérique
Université Paul Sabatier
118, Rte de Narbonne
31062 - Toulouse cedex, FRANCE

Implementation on a Shared Memory Parallel Computer of Algorithms for Approximating Data on a Family of Parallel Lines

I. J. Anderson, M. G. Cox, S. Harbour, and J. C. Mason

Abstract. We review serial algorithms for fitting structured data, collected on a family of parallel lines, by appropriate tensor-product functions. We consider specifically algorithms based on the method of least-squares which exploit this structure, namely a well-established polynomial algorithm and a new variant of a recent spline algorithm. Parallel processing procedures are then investigated for further exploiting the structure of each algorithm, and implementations made for the KSR1, a virtual shared memory parallel machine, are discussed. Early evidence suggests that high (parallel) speed-up may be achieved, at least in parts of the computation.

§1. Introduction

Structured data such as ordinates specified at points on or near a set of parallel lines or a family of curves can advantageously be fitted by exploiting the data structure — even when the points are irregularly spaced and are in different positions on each line or curve. Existing algorithms of this type, based on separation of variables, include an algorithm for polynomial approximation of lines of data [6], an algorithm which extends this approach to curves of data [15], and an algorithm which uses an iterative method to construct a spline surface that interpolates data on or near lines [1]. These algorithms are now extended in two principal ways: first, the interpolation algorithm [1] is generalised to allow splines to be fitted in a least-squares sense to (overspecified) data on lines and, second, parallel processing procedures are investigated and implemented for exploiting the structure of the algorithms. It is relatively straightforward to envisage the principles of parallel algorithms for these problems, but it is more difficult to carry this out in practice on a specific parallel platform in a way which avoids excessive communication and organisational overheads.

Wavelets, Images and Surface Fitting
P. J. Laurent, A. Le Méhauté, and L. L. Schumaker (eds.), pp. 11–20.

ISBN 1-56881-040-7.

In Section 2 we review the established theory of fitting tensor-product functions to data whose ordinates are specified at the vertices of a rectangular mesh, so as to set the scene for later developments involving lines of data. In Sections 3 and 4 we describe an existing algorithm and a new algorithm, respectively, for polynomials and splines. The architecture of the KSR1 computer is outlined and the (parallel) implementation of the algorithms of Sections 3 and 4 are discussed in Section 5. The results of these implementations are given in Section 6 and conclusions are drawn in Section 7.

§2. Fitting Data on a Rectangular Mesh

Consider the problem of fitting a rectangular mesh of data using a tensor-product approximating function. Given a set of $m_x \times m_y$ data values, $\{f_{k,\ell}\}$, whose abscissae lie at the vertices of a rectangular mesh, (x_k, y_ℓ), with $k = 1, \ldots, m_x$ and $\ell = 1, \ldots, m_y$, find the least-squares approximation $F(x, y)$ that minimises the sum of the squares of the residuals,

$$\sum_{k=1}^{m_x} \sum_{\ell=1}^{m_y} e_{k,\ell}^2, \tag{1}$$

where

$$e_{k,\ell} = f_{k,\ell} - F(x_k, y_\ell),$$

and $F(x, y)$ is a tensor-product combination of suitable basis functions $\{\phi_i(x)\}$ and $\{\psi_j(y)\}$:

$$F(x, y) = \sum_{i=1}^{r} \sum_{j=1}^{s} c_{i,j} \phi_i(x) \psi_j(y).$$

The structure of the approximating function and data allows the problem to be *separated* into two univariate problems [12]. Thus, for each $\ell = 1, \ldots, m_y$, we fit in the least-squares sense the univariate function $F_\ell(x) = \sum_{i=1}^{r} b_{i,\ell} \phi_i(x)$ to the data $\{(x_k, f_{k,\ell})\}_{k=1}^{m_x}$ on the line $y = y_\ell$ to obtain *intermediate* coefficients $\{b_{i,\ell}\}_{i=1}^{r}$. Then, for each $i = 1, \ldots, r$, we fit, also in the least-squares sense, the univariate function $b_i(y) = \sum_{j=1}^{s} c_{i,j} \psi_j(y)$ to the intermediate coefficients $\{(y_\ell, b_{i,\ell})\}_{\ell=1}^{m_y}$. The resulting coefficients $c_{i,j}$ minimise the residual sum of squares (1) [12].

Two significant advantages accrue from separability. First, a vast decrease in computation time can be achieved: the speed-up is a factor $\mathcal{O}(r^3)$ for general basis functions, in the case $r = s$. For basis functions of compact support the speed-up may be less if structure is exploited in the full bivariate and separate univariate formulations. This represents a significant saving even for problems of modest size. Second, properties such as uniqueness and rank deficiency can be tested independently for each variable. Such testing is generally simpler than in the bivariate case; indeed, a bivariate test may not be readily available.

§3. Polynomial Fitting to Data on Parallel Lines

A particularly useful class of approximating functions is the bivariate polynomials of degree r in x and degree s in y. Each function in this class has the tensor-product representation

$$P(x,y)=\sum_{i=0}^{r}\sum_{j=0}^{s}c_{i,j}p_i(x)q_j(y),$$

where p_i is the ith polynomial basis function in x and similarly for q_j in y. Data is assumed to be specified on a set of lines $y=y_\ell$, for $\ell=1,\ldots,m_y$, parallel to the x-axis, and it is required to fit such a function to data $f_{k,\ell}$ given at points $(x_{k,\ell},y_\ell)$, for $k=1,2,\ldots,m_\ell$, where m_ℓ represents the number of data values on the ℓth line.

A method [6] for constructing such an approximation is an extension of that for the rectangular mesh. The basis functions used are Chebyshev polynomials and the form of the approximation is

$$P(x,y)=\sum_{i=0}^{r}\sum_{j=0}^{s}c_{i,j}T_i(x)T_j(y),$$

with $T_i(x)$ representing the Chebyshev polynomial of degree i in x and similarly for $T_j(y)$. This use of Chebyshev polynomials requires that the data points are normalised to the square $[-1,1]\times[-1,1]$. Such normalisation is readily achieved by constructing the smallest rectangle with sides parallel to the x- and y-axes that contains the points and transforming linearly their x- and y-coordinates. The Chebyshev form has many advantages including concise storage and ease of interpretation [6].

The first stage of fitting determines a least-squares polynomial curve $p_r^{(\ell)}(x)$ of degree r to the function values on each line $y=y_\ell$. This is done by the construction of polynomials that are orthogonal with respect to the data abscissae for that line. These polynomials are generated by a recurrence relation, based on the x-values [10]. The polynomial curves produced are then converted to Chebyshev form [5]. So the least-squares polynomial curve to the data on the line $y=y_\ell$ is expressed as

$$p_r^{(\ell)}(x)=\sum_{i=0}^{r}a_{i,\ell}T_i(x),$$

where $\{a_{i,\ell}\}_{i=0}^{r}$ are the coefficients obtained from those of the fitted orthogonal polynomials. The work involved in producing this polynomial is $\mathcal{O}(m_\ell r^2)$ and since the data for each line are different the total time required for the first stage is $\mathcal{O}(mr^2)$, where $m=\sum_{j=1}^{m_y}m_\ell$, the total number of data points. Note that for polynomials of low degree the work involved is dominated by lower order terms, principally $\mathcal{O}(mr)$.

The second stage involves fitting the coefficients of the Chebyshev polynomials in the y-direction to form a tensor-product polynomial surface. Such fitting is only possible since the coefficients are related to the same basis polynomials along each line. This second stage can be solved in $\mathcal{O}(rm_y s^2)$ time. Had the polynomial curves $p_r^{(\ell)}$ been left in their orthogonal polynomial form, we would be unable to fit the coefficients in the y-direction since the orthogonal polynomials would be different for each line due to the different distribution of x-values along each line.

The resulting coefficients form a tensor-product Chebyshev polynomial approximation to the data. This approximation is *not* the true least-squares solution to the fitting problem, but nevertheless it can generally be expected to be close to this [6].

§4. Spline Fitting of Data on Parallel Lines

Using one of the existing standard notations [8], consider finite intervals $I = [x_{min}, x_{max}]$ and $J = [y_{min}, y_{max}]$, and the knot sets

$$\begin{aligned} x_{min} = \lambda_{1-n_x} &= \cdots = \lambda_0 < \lambda_1 \le \cdots \\ &\cdots \le \lambda_{q_x - n_x} < \lambda_{q_x - n_x + 1} = \cdots = \lambda_{q_x} = x_{max}, \\ y_{min} = \mu_{1-n_y} &= \cdots = \mu_0 < \mu_1 \le \cdots \\ &\cdots \le \mu_{q_y - n_y} < \mu_{q_y - n_y + 1} = \cdots = \mu_{q_y} = y_{max}. \end{aligned}$$

This chosen set of knots allows us to define a complete set of B-spline basis functions in terms of which a bivariate spline can be represented. Let $\boldsymbol{\lambda} = \{\lambda_i\}$ and $\boldsymbol{\mu} = \{\mu_i\}$.

A bivariate tensor-product polynomial spline is represented over the rectangular region $I \times J$ as

$$s(x,y) = \sum_{i=1}^{q_x} \sum_{j=1}^{q_y} c_{ij} N_{n_x i}(\boldsymbol{\lambda}; x) N_{n_y j}(\boldsymbol{\mu}; y).$$

Here $N_{n_x i}(\boldsymbol{\lambda}; x)$ denotes the ith normalised B-spline of order n_x in x [3,7], which is non-zero only over its *support*, $[\lambda_{i-n_x}, \lambda_i]$. Clearly, there are q_x B-spline basis functions in x. An analogous definition applies to $N_{n_y j}(\boldsymbol{\mu}; y)$. The (i,j)th B-spline coefficient is represented by c_{ij}.

We now consider the spline counterpart of Section 3, namely, spline fitting to data on parallel lines. As before, we wish to fit, in a least-squares sense, data $f_{k,\ell}$ at $(x_{k,\ell}, y_\ell)$ by the spline $s(x,y)$ (with given knot sets $\boldsymbol{\lambda}$ and $\boldsymbol{\mu}$) by minimising the residual sum of squares

$$\sum_{\ell=1}^{m_y} \sum_{k=1}^{m_\ell} e_{k,\ell}^2,$$

where

$$e_{k,\ell} = f_{k,\ell} - S(x_{k,\ell}, y_\ell).$$

Let $\mathbf{c}$ contain the coefficients $c_{i,j}$ expressed as a vector, and $\mathbf{f}$ the ordinates $f_{k,\ell}$ similarly. Let A be a matrix of order $m \times q$, with $m = \sum_{\ell=1}^{m_y} m_\ell$, the total number of data points as before, and $q = q_x q_y$. Let the elements of $A = \{a_{g,h}\}$ be the basis functions evaluated at the data points. Thus

$$a_{g,h} = N_i(x_g)N_j(y_g),$$

where $h = i + (j-1)q_x$ and x_g is the g-th element in the order sequence $\{\{x_{k,\ell}\}_{k=1}^{m_\ell}\}_{\ell=1}^{m_y}$. (This corresponds to a lexicographic ordering of the basis functions and data points). The spline-fitting problem can be expressed as the least-squares solution of the overdetermined linear algebraic equations $A\mathbf{c} = \mathbf{f}$.

This system of equations could be solved, for the coefficients $\mathbf{c}$, using a method based on orthogonal transformations to ensure stability, in $\mathcal{O}(mq_x^2q_y^2)$ time if the system is considered as full and in $\mathcal{O}(mq_x^2n_y^2)$ time if the coarse structure of the system is taken into account [13].

However, instead of solving this large system we can, as in the polynomial case, separate the algorithm into two stages. The first stage involves treating each line independently and carrying out a univariate fit to the appropriate data, i.e., fitting a spline curve to the section of the surface defined by the data on that line. An overdetermined system of linear equations of the form $A_x^{(\ell)}\mathbf{b}^{(\ell)} = \mathbf{f}^{(\ell)}$ is produced, where $A_x^{(\ell)}$ contains the basis functions in x evaluated at the points on the ℓth line. This univariate system is solved in the least-squares sense for spline curve coefficients $\mathbf{b}^{(\ell)}$ relating to the ℓth line. The total time required to solve the univariate systems for all lines is $\mathcal{O}(mn_x^2)$.

The second stage involves fitting the spline-curve coefficients from each line in the y-direction to produce a spline surface fit. An overdetermined system, $A_yC = B$, is solved, also in the least-squares sense, to produce the required surface coefficients $\mathbf{c}$. The elements of A_y are the basis functions in y evaluated at the values $y = y_\ell$, C represents the surface coefficients $\mathbf{c}$ in matrix form and B is a matrix whose ℓth row contains the intermediate coefficient values $\mathbf{b}^{(\ell)}$. This stage requires $\mathcal{O}(m_yn_yq_x)$ time and is therefore faster than the first stage by a factor of $\mathcal{O}(m_xn_x/q_x)$, where m_x is the mean number of data points on a line. It is possible to calculate all coefficients C by a single factorisation of the matrix A_y, and these timings are based on the assumption that this approach has been employed.

There is one important detail that we have so far omitted. Clearly, each univariate spline in x must be compatible with all the other univariate splines if fitting in the y-direction is to be successful. We need a procedure for choosing a knot set in x that will allow adequate fitting along each line while ensuring that there is not an excessive number of knots leading to non-uniqueness or spurious oscillations of the spline fit. An approach has been developed [1] which generates and uses a *common knot set* in x that applies to the whole surface being generated, and only a "suitable" subset of this common knot set is used for fitting along each line. When the univariate fit is complete, knot insertion [4] is used as required so that all line fits are defined with respect to the common knot set. The approach is designed specifically for spline interpolation, but we have successfully extended it to the current least-squares fitting procedure.

§5. Implementation on the KSR1

The Kendall Square Research machine (KSR1) is a large *virtual* shared memory parallel computer (with 32 processors) of the multiple instruction, multiple data stream (MIMD) class. Its (ALLCACHE) memory [14] is distributed amongst large (32 Mb) local caches which communicate via a search engine. This machine offers a powerful, but flexible and user-friendly, environment in which to develop efficient parallel code

A key feature of the KSR1, which we exploit in our implementation, is a process known as *tiling* [14], by which the work contained within a `do` loop is distributed among the processors. It is necessary that the operations carried out in each partition of the `do` loop can be performed independently of the other partitions.

5.1. Exploiting parallelism

The algorithms of Sections 4 and 5 are implemented on the KSR1 by making each processor responsible for a group of consecutive data lines. The size of the groups is determined by sharing the total number of lines 'equally' between the processors available. We consider $m_y = kp$ lines of data, where p is the number of processors available. We wish to assign the first k lines of data to the first processor, the second k lines to the second processor, and so on. The following fragment of FORTRAN code demonstrates the implementation of a tile for this purpose.

```
c*ksr* user tile(j)
       do j = 1, my
          call fit_on_j_th_line_of_data(...)
       end do
c*ksr* end tile
```

The comment statements inform the KSR1 compiler that segments of the code are to be tackled in parallel. Specifically, the directive `c*ksr* user tile(j)` requires the compiler to share the execution of the `do` loop in the variable `j` between the various processors (as described above).

Here we note that since all of the parallel instructions appear as comment statements the algorithm may be compiled and executed serially on any standard FORTRAN machine. To first order, the computational load is balanced across the processors. The second order load-balancing effects can be taken into account, although doing so is pointless if the gain achieved is negated by the pre-computation required. One consideration in this regard is the unevenness in the load if m_y/p is not integral (some processors will be allocated more data lines than others). Another consideration is the fact that the value of m_ℓ, the number of points on the ℓth data line, is not in general a constant. In fact, a small amount of pre-computation can address such second-order load-balancing issues. Our results in the following section do not take these issues into account. Considerable variability in the values of m_ℓ could, of course, have a highly significant effect.

5.2. Coding and implementation issues

In order to capitalise on existing software, access was obtained to the relevant DASL [2] and NAG [9] library source codes (subroutines `T2FLE` and `E02CAF` respectively) for the problem of data fitting on parallel lines.

In parallelising the codes, the problems encountered were as follows.

- Since the initial code had obviously been optimised for a serial processor, a parallel code had to be developed from first principles, which inevitably performed less well than the initial code on a single processor.
- The original serial code contained a thorough set of data-acceptability tests which were executed in-situ. A parallel approach might have led to one processor terminating through a failure condition while the remaining processors were unaffected, but this possibility was avoided by organising the code such that all tests were carried out in serial at the beginning of the algorithm.

§6. Results

The main aim of this section is to report on the performance aspects of the parallel algorithm, and consequently little emphasis is placed on the algorithm's ability to fit data representing functions of different degrees of complexity. The data sets chosen were all well-behaved and generated as follows.

A suitable value for m_y was chosen and the y values y_ℓ were taken to be uniformly spaced in $[-1, 1]$. The values m_ℓ were chosen randomly to be the closest integers to values sampled from a normal distribution of mean $2m_y$ and standard deviation $m_y/2$. The x-values along each line were also taken to be uniformly spaced in $[-1, 1]$. The ordinates were taken from an underlying function of tensor-product form. For the first algorithm, the parallel version of bivariate polynomial fitting, the underlying function was a linear combination of tensor-product Chebyshev polynomials of degree 12 in x and degree 10 in y. The (i, j)th coefficient of the function was taken as a value sampled from a normal distribution with mean zero and standard deviation $(i + j - 1)^{-2}$. For the second algorithm, the parallel version of bivariate spline fitting, the underlying function was a linear combination of tensor-product B-splines of order four in both x and y and with a coefficient matrix of order 13×11. The (i, j)th coefficient of the function was taken as a value sampled from a normal distribution with zero mean and unit standard deviation.

The form of the fitting functions for the two algorithms was chosen to be identical to that of the underlying function from which the ordinates were

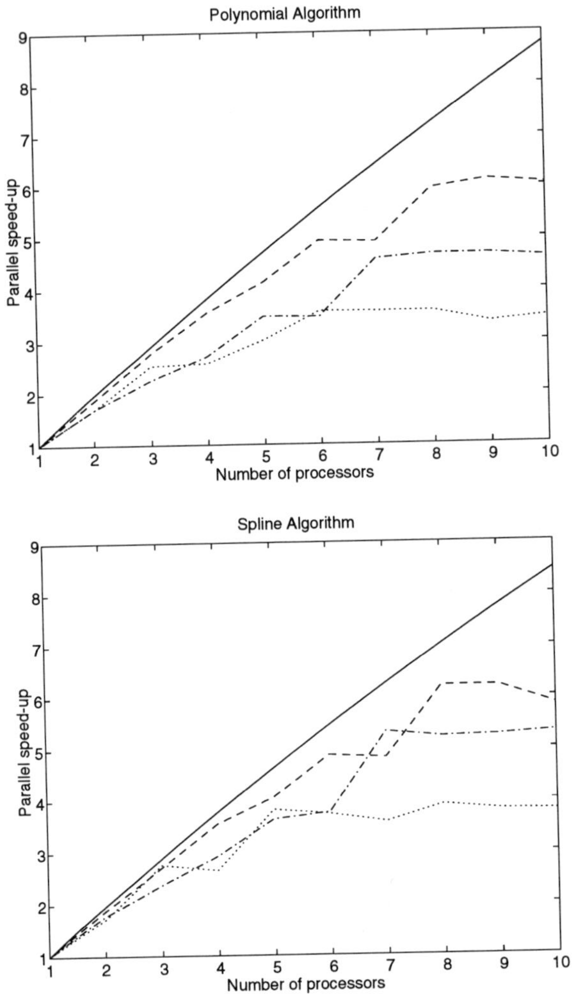

Fig. 1. Speed-up graphs of implemented algorithms for problems with 360 lines of data (broken line), 200 lines (dash-dot line) and 140 lines (dotted line).

taken. From the magnitude of the generated ordinates, the well-conditioned nature of the data-fitting problems for these data sets and the inherent stability of the underlying algorithms, the difference between the fitted coefficients and the true coefficients could be expected to be at most a small multiple of the machine precision. As a consequence, we could be confident in testing the effectiveness of the parallel algorithms, without being unduly influenced by other considerations.

Figure 1 shows the speed-up ratios for three data sets using various numbers of processors. The broken, dash-dot and dotted lines represent the data sets generated, as described above, with the number of data lines $m_y = 360$, 200 and 140, respectively. The smooth curve shows the theoretical bound on the speed-up ratio given by Amdahl's law [11]

$$\text{Speed-up} \le \frac{1}{(1-\tau)+\tau/p}, \tag{2}$$

where p is the number of processors and τ is the fraction of the algorithm that is parallelisable.

The achieved speed-up would appear to be significant in the sense that, for the largest value of m_y considered, it comes close to attaining the theoretical bounds given by Amdahl's law (2) for both the polynomial and spline algorithms. The best values obtained were for $p = 9$ processors, where speed-ups of 6.1 for the polynomial algorithm and 6.2 for the spline algorithm were achieved. Both of the graphs demonstrate a fall-off in performance as the number of processors grows — this is believed to be caused by the cost of overheads and the increase in communication out-weighing the parallel gain.

§7. Conclusions

We have achieved a significant speed-up in the performance of two algorithms for the least-squares fitting to lines of data by adopting a parallel-processing approach — indeed the parallel efficiency, measured as the ratio of speed-up to theoretically maximum speed-up (as measured by the Amdahl bound (2)), was at least 70% for the large data set and at least 40% for the small data set. Moreover we have developed flexible and potentially portable parallel software for a polynomial algorithm that is currently available only in a strictly serial software form in DASL and the NAG library.

Overhead costs led to some deterioration in performance for large numbers of processors. We have not exploited every special feature of KSR `FORTRAN` but have concentrated on features that allow the software to be readily implemented on alternative architectures. As a result we might achieve further improvements in performance if portability requirements were relaxed.

References

1. I. J. Anderson, M. G. Cox and J. C. Mason, Tensor-product spline interpolation to data on or near a family of lines, in *Algorithms for Approximation III*, M. G. Cox and J. C. Mason (eds.), J. C. Baltzer AG, Basel, Switzerland, 1993, 193–204.
2. G. T. Anthony and M. G. Cox, The National Physical Laboratory's Data Approximation Subroutine Library, in *Algorithms for Approximation*, J. C. Mason and M. G. Cox (eds.), Clarendon Press, Oxford, 1987, 669–687.
3. C. de Boor, On calculating with B-splines, J. Approx. Theory **6** (1972) 50–62.

4. W. Boehm, Inserting new knots into B-spline curves, Comput. Aided Design **12** (1980) 199–201.
5. C. W. Clenshaw, Curve fitting with a digital computer, Computer. J. **2** (1960) 170–173.
6. C. W. Clenshaw and J. G. Hayes, Curve and surface fitting, J. Inst. Math. Appl. **1** (1965) 164–183.
7. M. G. Cox, The numerical evaluation of B-splines, J. Inst. Math. Appl. **10** (1972) 134–149.
8. M. G. Cox, Data approximation by splines in one and two independent variables, in *The State of the Art in Numerical Analysis*, A. Iserles and M. J. D. Powell (eds.), Oxford University Press, Oxford, 1987, 111-138.
9. B. Ford and J. Bentley and J. J. du Croz and S. J. Hague, The NAG Library 'Machine', Software - Practice and Experience **9** (1979) 56–72.
10. G. E. Forsythe, Generation and use of orthogonal polynomials for data fitting with a digital computer, J. Soc. Ind. Appl. Math. **5** (1957) 74.
11. T. L. Freeman and C. Phillips, *Parallel Numerical Algorithms*, Prentice Hall, 1992.
12. T. N. E. Greville, Note on fitting of functions of several independent variables, J. Soc. Ind. Appl. Math. **9** (1961) 109–115.
13. J. G. Hayes and J. Halliday, The least-squares fitting of cubic spline surfaces to general data sets, J. Inst. Math. Appl. **14** (1974) 89–103.
14. Kendall Square Research Corporation, *KSR Parallel Programming*, Waltham, MA, 1991.
15. J. C. Mason and R. P. Bennell, Efficient multivariate approximation methods for lines and curves of data, in *Mathematics of Surfaces IV*, Oxford University Press, London, 1994, in press.

I. J. Anderson and M. G. Cox
DITC, NPL
Teddington
TW11 0LW, U. K.
iga@uk.co.npl.newton
mgc@uk.co.npl.newton

I. J. Anderson, S. Harbour, and J. C. Mason
AMOR group
Cranfield (RMCS)
Shrivenham
SN6 8LA, U. K.
andersij@uk.ac.cran.rmcs
harbour@uk.ac.cran.rmcs
mason@uk.ac.cran.rmcs

Wavelet Methods for Smoothing Noisy Data

Anestis Antoniadis

Abstract. Wavelet theory has the potential to provide statisticians with powerful new techniques for nonparametric inference. It combines recent advances in approximation theory with insights gained from applied signal analysis. This paper connects wavelets with some problems of nonparametric statistical inference in the setting of some simple nonparametric curve estimation problems. Wavelet analogues of some familiar kernel and orthogonal series estimators are introduced, and their finite sample and asymptotic properties are studied.

§1. Introduction

The purpose of this paper is to contribute to the methodology available for estimating smooth regression functions from noisy data.

Let $Y_0, Y_1, \ldots, Y_{n-1}$ denote data generated by the fixed-design regression model

$$Y_i = g(x_i) + \epsilon_i, \quad 0 \le i \le n-1,$$

where $n = 2^J$ and the design points $x_0, x_1, \ldots, x_{n-1}$ all belong to an interval $[0, T]$ and are assumed to be equally spaced. The random variables (Y_i) are measurements of the unknown regression function $g : [0, T] \to \mathbb{R}$, contaminated with errors ϵ_i which are assumed to be independent identically distributed (i.i.d.) random variables with mean 0 and variance σ^2.

There are two main approaches to estimate the regression function g: the parametric approach which assumes that g follows a parametric (linear or nonlinear) model and the nonparametric approach which only assumes that g is "smooth". In recent years several methods of nonparametric curve fitting have been proposed and investigated. Recent references with extensive bibliographies and interesting discussions on the applicability of such smoothing methods are Eubank [4], and Wahba [7].

We shall introduce wavelet versions of the most frequently used kernel and orthogonal series estimators for these models. Our estimators are either delta sequence smoothers based on wavelet kernels, or orthogonal projection estimates. By means of the properties of the wavelet transform and the wavelets bases that we are using, we will obtain precise rates of convergence.

Wavelets, Images and Surface Fitting
P. J. Laurent, A. Le Méhauté, and L. L. Schumaker (eds.), pp. 21–28.

ISBN 1-56881-040-7.

§2. Wavelet Kernel Smoothers

Wavelet kernel smoothers are delta sequence smoothers based on wavelet kernels $E_m(\cdot,\cdot)$ as defined in [6]. These kernels represent integral operators E_m that project onto closed subspaces V_m of $L^2(\mathbb{R})$. The increasing sequence of subspaces V_m form a so-called *multiresolution analysis* of $L^2(\mathbb{R})$. The basic idea is that the V_m provide successive approximations, with details being added as m increases. Thus m acts as a tuning parameter, much as the bandwidth does for standard kernel smoothers. A key aspect of wavelet estimators is that the tuning parameter ranges over a much more limited set of values than is common with other nonparametric regression techniques. In practice only a small number of values of m (say three or four) need to be considered.

For the fixed design model the proposed estimator is

$$\hat{g}(t) = \sum_{i=0}^{n-1} Y_i \int_{A_i} E_m(t,s)\,ds,$$

where the $A_i = [s_{i-1}, s_i)$ are intervals that partition $[0,1]$ with $s_{i-1} \le x_i < s_i$. The estimator $\hat{g}$ is a wavelet version of Gasser and Müller's [5] (convolution) kernel estimator, and was first introduced in [2].

The delta sequence E_m used in $\hat{g}$ can be written as

$$E_m(t,s) = 2^m \sum_{k\in\mathbb{Z}} \varphi(2^m t - k)\varphi(2^m s - k),$$

where φ is a compactly supported scaling function associated to a multiresolution analysis of $L^2(\mathbb{R})$. Since φ has compact support, this is a finite sum, each term of which can be evaluated by the cascade algorithm (see [3]). To evaluate the weights $\int_{A_i} E_m(t,s)\,ds$ in $\hat{g}(t)$, we use

$$\int_u^v \varphi_j(x)\,dx = \sum_{k\in\mathbb{Z}} a_k^j \int_u^v \chi(2^j x - k)\,dx,$$

where χ is the indicator function of the interval $[-\frac{1}{2}, \frac{1}{2}[$. As j goes to infinity, the sequence $\int_u^v \varphi_j(x)\,dx$ converges to $\int_u^v \varphi(x)\,dx$ for each $u < v$.

Our first result gives consistency of $\hat{r}$.

Theorem 1. *If g is continuous at t, and $m \to \infty$ as $n \to \infty$, then $\hat{g}(t)$ is mean square consistent.*

For a proof of such a result the reader is referred to [2]. Strong consistency of $\hat{g}(t)$ can be obtained under a more refined condition on the rate of increase of m. In order to obtain deeper results we need the regression function g to satisfy

(1) $g \in H^\nu$, for some $\nu > \frac{1}{2}$;

(2) g is Lipschitz of order $\gamma > 0$;

Functions belonging to H^ν for $\nu > \frac{3}{2}$ are continuously differentiable, so condition (2) is redundant when $\nu > \frac{3}{2}$. We also need some additional assumptions on the scaling function φ:

(3) φ has compact support;

(4) φ is Lipschitz;

(5) $|\hat{\varphi}(\xi) - 1| = \mathcal{O}(|\xi|)$ as $\xi \to 0$.

Here $\hat{\varphi}$ denotes the Fourier transform of φ. The compactly supported scaling functions ${}_N\varphi$, $N \geq 3$, of Daubechies satisfy all of these conditions. For our asymptotic normality results we will need φ to be regular of order $q \geq 1$. However, to obtain good rates of convergence for the mean square error of $\hat{g}$ we need to adapt the regularity of φ to the smoothness of g. Thus we assume

(6) φ is regular of order $q \geq \nu$.

We shall also assume that for some Lipschitz function $\kappa(\cdot)$,

$$\rho(n) \equiv \max_i \left| s_i - s_{i-1} - \frac{\kappa(s_i)}{n} \right| = o(n^{-1}).$$

This is a standard assumption for the fixed design model, but somewhat weaker than the "asymptotic equidistance" assumption of Gasser and Müller [5] in which $\kappa(t) \equiv 1$ and $\rho(n) = \mathcal{O}(n^{-\delta})$ for some $\delta > 1$. Denoting by $\mathbb{E}$ the expectation operator, the next results, whose proofs can be found in [2], give an asymptotic bound for the bias and the variance of $\hat{g}$.

Theorem 2.

$$\mathbb{E}\hat{g}(t) - g(t) = \mathcal{O}(n^{-\gamma}) + \mathcal{O}(\eta_m)$$

where

$$\eta_m = \begin{cases} \left(1/2^m\right)^{\nu - 1/2} & \text{if } \frac{1}{2} < \nu < \frac{3}{2}, \\ \sqrt{m}/2^m & \text{if } \nu = \frac{3}{2}, \\ 1/2^m & \text{if } \nu > \frac{3}{2}. \end{cases}$$

$$\mathrm{Var}(\hat{g}(t)) = \frac{\sigma^2 2^m}{n} \kappa(t)(w^2 + o(1)) + \mathcal{O}(2^m \rho(n)) + \mathcal{O}\left(\frac{2^{2m}}{n^2}\right),$$

where $w^2 = \sum_{k \in \mathbb{Z}} \varphi^2(k)$.

The leading term of the variance of $\hat{g}(t)$ is $\sigma^2 2^m n^{-1} \kappa(t) w^2(t_m)$, where $t_m = 2^m t - [2^m t]$ and w^2 is the function defined by

$$w^2(u) = \int_{\mathbb{R}} E_0^2(u, v)\, dv.$$

Notice that for dyadic t and m sufficiently large, $t_m = 0$, so the variance of $\hat{g}(t)$ is asymptotically stable. But if t is non-dyadic then the sequence t_m wanders around the unit interval and fails to converge. For example, at $t = \frac{1}{3}$, it oscillates between $\frac{1}{3}$ (m even) and $\frac{2}{3}$ (m odd), so the variance oscillates between $w^2(\frac{1}{3})$ and $w^2(\frac{2}{3})$. The sequence t_m belongs to the class of exponential lacunary sequences studied in ergodic theory. It is known that except for at most countably many t, the sequence t_m has infinitely many accumulation points. It is also interesting to note that for irrational t's, the sequence is eventually confined to the interval $[\frac{1}{3}, \frac{2}{3}]$.

Generally, higher regularity of the wavelet basis reduces instability in the asymptotic variance of $\hat{g}(t)$, although this comes at the expense of larger bias (the support of the scaling function increases with the regularity).

Optimal rates.

In order to give a rate of convergence for the mean squared error of their estimates, Gasser and Müller [5] assume that g is k-times continuously differentiable and use a kernel of order $k \geq 2$. They find that the best rate of convergence for the mean squared error is $n^{-2k/(2k+1)}$. An analogous result holds in our case. Assume that g is $k = q+1$ times continuously differentiable, where q is the regularity of the scaling function. Since polynomials of degree $\leq q$ are invariant under $E_m(t,s)$, we get by using a Taylor expansion of g that the best rate of convergence for the mean squared error of $\hat{g}$ at dyadic points is the same as for the kernel estimator, and is attained for $m = \log_2 n/(2k+1)$.

It is worth stressing that the wavelet approach allows us to obtain rates under much weaker assumptions on g than second order differentiability.

Our next result concerns asymptotic normality of $\hat{g}$. It can be applied to the unmodified estimator $\hat{g}$ at dyadic points.

Theorem 3. *For any dyadic point $t \in [0,1]$, if $n2^{-m} \to \infty$ and $n2^{-2m\nu^*} \to 0$, then $\sqrt{n2^{-m}}(\hat{g}(t) - g(t))$ is asymptotically normal with zero mean and variance $\sigma^2 w^2 \kappa(t)$.*

§3. Orthogonal Wavelet Series Estimators

In the fixed design regression model considered in the introduction, the function g is viewed as a member of a class $\mathcal{F}$ of possible regressions over $[0,1]$. Hereafter, we shall assume that $\mathcal{F}$ is the space of functions $\lfloor\beta\rfloor$-times continuously differentiable in $\mathbb{R}$, and such that their $\lfloor\beta\rfloor$th derivative satisfy a lipschitz condition of order $\beta - \lfloor\beta\rfloor$.

From now on, ϕ and ψ will denote the scaling function and the wavelet associated to an r-regular multiresolution analysis of $L^2(\mathbb{R})$. We will further assume that the scaling function ϕ is a coiflet (see [3]) of order $L = 2K$ with $L > \lfloor\beta\rfloor + 1$.

Now g does not necessarily belong to $L^2(\mathbb{R})$, and we replace g by a function f such that

$$f(t) = g(t) \quad \text{for all } t \text{ in } [0,1],$$

with f in $C^\beta(\mathbb{R})$, compactly supported in an interval $[-\epsilon, 1+\epsilon]$ for some small ϵ. This substitution will have little effect since the properties of the estimator are only evaluated on $[0,1]$, and the observed model coincides with

$$Y_i = f(t_i) + \epsilon_i, \quad 0 \leq i \leq n-1.$$

It follows that $f \in H^\beta(\mathbb{R})$, where $H^\beta(\mathbb{R})$ denotes the Sobolev space of order β. Since

$$L^2(\mathbb{R}) = V_J \oplus (\oplus_{j \geq J} W_j),$$

the function f admits the following expansion in H^β (and in L^2):

$$f(t) = \sum_{k \in \mathbb{Z}} \langle f, \phi_{J,k} \rangle \phi_{J,k}(t) + \sum_{j \geq J} \sum_{\ell \in \mathbb{Z}} \langle f, \psi_{j,\ell} \rangle \psi_{j,\ell}(t).$$

A first step is to consider the wavelet approximation for a function using its sampled values over the interval. Since the coiflets have L vanishing moments, one can approximate the projection P_J on V_J by some operator Π_J in terms of the sampled values $f(\frac{k}{2^J})$ and to derive then a reasonable estimator of the approximation $\Pi_J f$. Such an estimator of $\Pi_J f$ is given by

$$\hat{f}_J(t) = \widehat{\Pi}_J f(t) = 2^{-J/2} \sum_{k \in \mathbb{Z}} Y_k \phi_{J,k}(t) = 2^{-J/2} \sum_{k=0}^{2^J-1} Y_k \phi_{J,k}(t). \qquad (1)$$

This choice can be justified by the following lemma, whose proof can be found in [1].

Lemma 1. *The set of nonzero coefficients $g^{\{J\}}(k) = 2^{J/2}\langle g, \phi_{J,k}\rangle$ has a cardinal equivalent to $\mathcal{O}(2^J)$. Moreover, with $L > [\beta]+1$, the following uniform bound (in k) holds:*

$$|g^{\{J\}}(k) - g(\frac{k}{2^J})| \leq C_1 \, 2^{-J\beta},$$

where C_1 is a constant only depending on the coiflet ϕ.

By Lemma 1, one is therefore able to approximate the coefficients $\langle g, \phi_{J,k}\rangle$ with an error $\mathcal{O}(2^{-J/2}2^{-J\beta})$, and it is then natural to approximate $P_J g(t)$ for $t \in [a,b]$, by $(\Pi_J f)(t)$. Observing that $\mathbb{E}(Y_k) = g(\frac{k}{2^J})$ justifies completely our choice of the estimator $\hat{f}_J$, at least within the interval $[0,1]$.

The above calculations suggest that given our observed sampled values of g, the observations are equivalent to the estimator $\hat{f}_J$. This estimator, while presenting a very small bias, leads to a highly oscillatory solution being to fidel to the data. In order to smooth correctly the data, we will associate to each sample size $n = 2^J$ a resolution $m(n)$, and we shall estimate the unknown function g by the orthogonal projection of $\hat{f}_J$ onto $V_{m(n)}$.

We will use in the sequel a weighted integrated mean squared error (WIMSE) of the form

$$R_n = R(\hat{g}) = \int_0^1 \mathbb{E}\left[\hat{g}(t) - g(t)\right]^2 u(t)\, dt$$

to assess the performance of an estimator $\hat{g}$ of g on $[0,1]$. Here $u > 0$ denotes a continuously differentiable weight function on $[0,1]$, supported in an interval $[a,b] \subset (0,1)$ and included to reduce the impact of boundary effects and sampling approximation.

The following theorem, whose proof may be found in [1], addresses the appropriate decay rate of the weighted integrated mean squared error defined above.

Theorem 4. *Under the smoothness assumptions imposed on the function g and the weight function u, the WIMSE satisfies*

$$R_n \le \mathcal{O}(2^{-2m(n)J\min([\beta],q)}) + \mathcal{O}(2^{m(n)-J}) + \mathcal{O}(2^{m(n)}2^{-2(J-m(n))(\alpha+1/2)}),$$

where α denotes the Hölder exponent of the coiflet ϕ.

If one takes $m(n) = J/(1+2[\beta]) = \log_2(n)/(1+2[\beta])$, it is clear that, with regular enough coiflets, our estimator attains the best possible convergence rate as in the previous section.

Regarding the pointwise consistency of the coiflet estimator, we have (see [1])

Theorem 5. *If the errors ϵ_i, $i = 0,\ldots,n-1$, are independent identically distributed with zero mean and finite variance σ^2, then, for any dyadic $t \in (0,1)$,*
(a) *$\hat{g}_J(t) \to g(t)$ almost surely.*
(b) *$\sqrt{n}\,(\hat{g}_J(t) - g(t))$ converges in distribution to a centered Gaussian random variable with variance $\sigma^2 w^2(t)$, where*

$$w^2(t) = \lim_{n\to\infty} \frac{1}{2^n} \sum_{k=0}^{2^n-1} \phi^2_{n,k}(t).$$

The interesting feature of this orthogonal series estimator compared to the kernel estimator studied in Section 2, is that it takes full advantage of the nested structure of a multiresolution analysis and its computation can be very fast using the efficient tree-structured algorithm of the discrete wavelet transform.

§5. Example

To illustrate the techniques given so far, we now consider a real example. The example presented in the following figures is based on a data set discussed in Example 3.4.5 of Eubank [4], and represents the voltage drop in the battery of a guided missile motor during flight. In this example the assumptions of the fixed design model are reasonable. We find that there is an undesirable boundary effect in the kernel estimator $\hat{g}$ at time 0. The time reversed $\hat{g}$ has

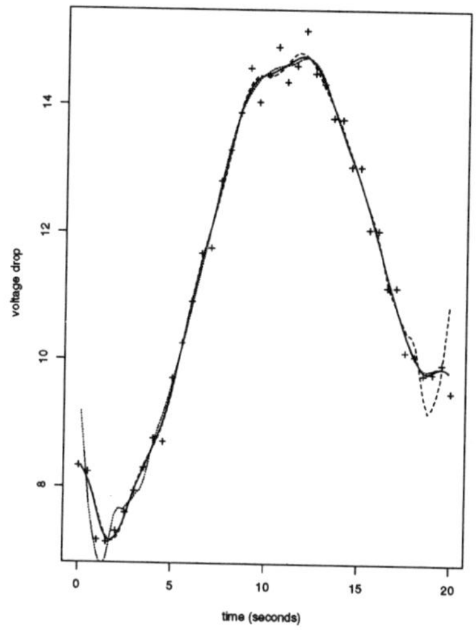

Fig. 1. Plot of the voltage drop data and the symmetrized wavelet regression estimate (solid line) for $m = 4$ and scaling function ${}_6\varphi$.

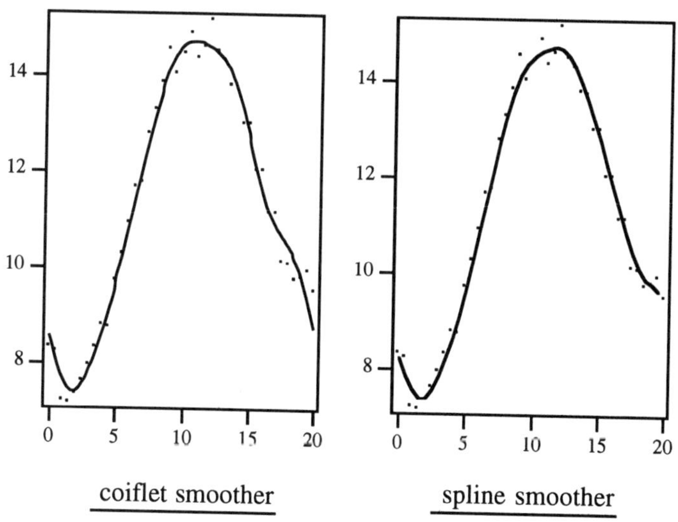

Fig. 2. Plot of the voltage drop data. The coiflet wavelet estimator is on the left, the cubic spline fitted by cross-validation, on the right.

a similar problem at the right-hand end of the design interval. However, a symmetrized estimator $\hat{g}_{sym}$ produces an acceptable fit (see Figure 1). In fact, considering that $\hat{g}_{sym}$ uses a tuning parameter setting chosen from among only three different values ($m = 3, 4$ and 5), it gives a very good result compared with other nonparametric regression estimates.

For the coiflet series smoother the data was interpolated by a piecewise polynomial and the interpolated values on an equispaced grid of 128 points were fitted. The resulting coiflet smoother obtained by projection onto the space V_3 is displayed in the left plot of Figure 2. For comparison purposes, the right plot displays the cross-validated spline smoother fitted to the original data with a cross-validation parameter $\lambda = 0.094$. Considering that the wavelet estimator choses a tunning parameter from among only two different values ($m(n) = 2$ or 3), it gives an outstanding result.

As an overall conclusion, the coiflet smoother gives good performance, does not rely on a data driven method for choosing its smoothing parameters, and is very fast to compute when compared to the wavelet kernel smoother.

References

1. Antoniadis, A, Smoothing noisy data with Coiflets, *Rapport technique*, LMC-IMAG, (1993).
2. Antoniadis, A., G. Grégoire, and I. W. McKeague, Wavelet methods for curve estimation, *Rapport de Recherche* 890-M, LMC-IMAG, (1992).
3. Daubechies, I., *Ten Lectures on Wavelets.* CBMS-NSF regional conferences series in applied mathematics, SIAM, Philadelphia, Pennsylvania, (1992).
4. Eubank, R. L., *Spline Smoothing and Nonparametric Regression.* Marcel Dekker, New York, 1988.
5. Gasser, T. and H. M. Müller, Estimating regression functions and their derivatives by the kernel method. *Scand. J. Statist.* **11** (1984), 171–185.
6. Meyer, Y., *Ondelettes et opérateurs, Vol. I*, Hermann. Paris, 1990.
7. Wahba, G., *Spline Models for Observational Data*, CBMS-NSF regional conferences series in applied mathematics, SIAM, Philadelphia, Pennsylvania, 1990.

Anestis Antoniadis
Université Joseph Fourier
IMAG-LMC
BP 53 X
38041 Grenoble cedex FRANCE
antonia@imag.fr

Quasi-interpolants and (quasi-) Wavelets P(D) Manifold

M. Atteia

Abstract. Let $\zeta = (\zeta_1, \ldots, \zeta_n) \in \mathbb{R}^n$ and $P_{\ell_j}(\zeta)$, $1 \le \ell \le m$, $1 \le j \le q$, be a real polynomial in the variables $\zeta_1, \ldots, \zeta_n$ with constant coefficients. We set $P(\zeta) = (P_{\ell_j}(\zeta))$. In [1], with convenient assumptions, we have proved the existence of a spline manifold σ relating to $P(D)$ (where $P(D) = \left(\frac{\partial}{\partial x_n}, \ldots, \frac{\partial}{\partial x_n}\right))$ and we proved that σ is characterized by a Hilbertian (semi-)kernel $\mathcal{L}$ relating to ${}^tPP$. In this lecture, we define and study quasi-interpolant (resp. box-spline) manifolds and quasi-wavelet manifolds, when P is homogeneous. They are obtained by approximating in a convenient way the formula ${}^tPP\mathcal{L} = \delta$ (δ is the Dirac functional) on a sequence of regular meshes of $\mathbb{R}^n$. We also give some examples.

§1. Introduction

Three years ago, I presented a lecture on spline manifolds. The idea was to approach a curve or a surface using spline manifolds taking into account its physical nature. In particular, M. N. Benbourhim and I have studied spline elastic manifolds. Now, I would consider, in the same spirit, quasi-interpolants and wavelets.

§2. Basic notions

2.1. The differential operator $P(D)$

We consider a $P(D)$-differential operator with constant coefficients:

$$P(D) = (P_\ell(D))_{1 \le \ell \le q} \quad \text{such that for} \quad u = (u_j)_{1 \le j \le m},$$

$$P_\ell(D)u = \sum_{j=1}^{m} \Big(\sum_{k=1}^{n} \gamma_{\ell j}^k D_k u_j \Big), \quad D_k = \frac{\partial}{\partial x_k}.$$

Wavelets, Images and Surface Fitting
P. J. Laurent, A. Le Méhauté, and L. L. Schumaker (eds.), pp. 29–36.

ISBN 1-56881-040-7.

So, in matrix form, $P(D) = \Gamma Du$, i.e.,

$$\underbrace{\begin{pmatrix} P_1(D)u \\ \cdots\cdots \\ P_\ell(D)u \\ \cdots\cdots \\ P_q(D)u \end{pmatrix}}_{P(D)} = \underbrace{\begin{pmatrix} \gamma_{11}^1 & \cdots & \gamma_{11}^n & \cdots & \gamma_{1m}^1 & \cdots & \gamma_{1m}^n \\ \cdots & \cdots & \cdots & \cdots & \cdots & \cdots & \cdots \\ \gamma_{\ell 1}^1 & \cdots & \gamma_{\ell 1}^n & \cdots & \gamma_{\ell m}^1 & \cdots & \gamma_{\ell m}^n \\ \cdots & \cdots & \cdots & \cdots & \cdots & \cdots & \cdots \\ \gamma_{q1}^1 & \cdots & \gamma_{q1}^n & \cdots & \gamma_{qm}^1 & \cdots & \gamma_{qm}^n \end{pmatrix}}_{\Gamma} \underbrace{\begin{pmatrix} D_1u_1 \\ \cdots\cdots \\ D_nu_1 \\ \cdots\cdots \\ D_1u_m \\ \cdots\cdots \\ D_nu_m \end{pmatrix}}_{Du}$$

Here Γ is a (q, mn) matrix. With $P(D)$ we can associate $P(\zeta) = (P_\ell(\zeta))_{1\leq\ell\leq q}$ $= (\sum_{j=1}^m \sum_{k=1}^n \gamma_{\ell j}^k \zeta_k)_{1\leq\ell\leq q}$.

Examples 1.1.

Maxwell equations :

$$P(\zeta) = \begin{pmatrix} 0 & -\zeta_3 & \zeta_2 \\ \zeta_3 & 0 & -\zeta_1 \\ -\zeta_2 & \zeta_1 & 0 \end{pmatrix}$$

Acoustics :

$$P(\zeta) = \begin{pmatrix} 0 & 0 & \zeta_1 \\ 0 & 0 & \zeta_2 \\ \zeta_1 & \zeta_2 & 0 \end{pmatrix}$$

Divergence-Curl :

$$P(\zeta) = \begin{pmatrix} \zeta_1 & \zeta_2 & \zeta_3 & 0 & 0 & 0 \\ 0 & 0 & 0 & 0 & -\zeta_3 & \zeta_2 \\ 0 & 0 & 0 & \zeta_3 & 0 & -\zeta_1 \\ 0 & 0 & 0 & -\zeta_2 & \zeta_1 & 0 \end{pmatrix}$$

Elasticity :

$$P(\zeta) = \begin{pmatrix} \zeta_1 & 0 & 0 \\ 0 & \zeta_2 & 0 \\ 0 & 0 & \zeta_3 \\ \zeta_2 & \zeta_1 & 0 \\ \zeta_3 & 0 & \zeta_1 \\ 0 & \zeta_3 & \zeta_1 \end{pmatrix}$$

2.2. The Hilbert space L.

It can be proved (cf. [3,4]), that there exists a Hilbert space $(L, \langle \cdot \mid \cdot \rangle)$ such that :

i) $\forall u, v \in L$,

$$\langle u \mid v \rangle = \int_{\mathbb{R}^n} P(D)u \cdot P(D)v \, dt, \quad P(D)u, P(D)v \in L^2(\mathbb{R}^n)$$

ii) $P(D)$ is a closed image operator from L into $L^2(\mathbb{R}^n)$ and :

$$\dim(\mathrm{Ker}\ P(D)) < +\infty.$$

iii) $\forall u \in L$,

$$\rho_i(u) = \int_{\omega_i} u(s)\, ds = 0$$

where ω_i is a "small" ball in $\mathbb{R}^n$, $i \in I$, Card I $= \dim(\mathrm{Ker}\ P(D))$.

iv) L is a Hilbert subspace of a linear space of distributions E

$$(E = \mathcal{D}'(\mathbb{R}^n),\ \mathcal{S}'(\mathbb{R}^n), \ldots)$$

v) L is contained in $C^0(\mathbb{R}^n)$ or in $L^2_{loc}(\mathbb{R}^n)$.

Proposition 2.1. *Let $\mathcal{L}$ be the Hilbertian Kernel of L.*

α) $\mathcal{L}$ is a vectorial Kernel $\mathcal{L} = (\mathcal{L}_{pj})_{1 \le p,j \le m}$

β) $\mathcal{L}$ is shift-invariant

γ) $\forall \psi \in E,\ (\mathcal{L}\psi)(t) = \int \mathcal{L}(t-s)\psi(s)\, ds$

δ) $\mathcal{L}$ is characterized by the following relation :

$$\forall \psi \in E, \forall u \in L, \quad \langle \mathcal{L}\psi \mid u \rangle = \langle \psi \mid u \rangle = \int \sum_{j=1}^{m} \psi_j u_j \, dt.$$

We deduce that for all $\psi \in E$,

$$^tP(D) \cdot P(D)\ \mathcal{L}\,\psi \;=\; \psi \;-\; \sum_{i \in I} \left(\underbrace{\int \psi(s)\ v^i(s)\ ds}_{P_i^*(\psi)} \right) 1_{\omega_i}$$

with $P_i(v^j) = \delta_{ij}$, $i, j \in I$. Moreover, we notice that

$$Q(D) = (Q_{rj}(D))_{1 \le r,j \le m}$$

$$Q_{rj}(D) = -\sum_{\ell=1}^{q} P_{\ell r}(D) \cdot P_{\ell j}(D).$$

As $Q(D)$ admits a fundamental solution $\mathcal{Q}$ such that

$$\mathcal{Q} * Q(D) = Q(D) \cdot \mathcal{Q} = \delta \cdot I_m,$$

we can prove that

$$\mathcal{L}(t-s) = \mathcal{Q}(t-s) - \sum_i v^i(t)\rho_i(\mathcal{Q}(\cdot - s)) - \sum_j v^j(s)\rho_j(\mathcal{Q}(t-\cdot)) + \sum_{i,j} v^i(t)v^j(s)\rho_{i,t}(\rho_{j,s}(\mathcal{Q}(t-s))).$$

Moreover, it is easy to prove that

$$\forall \phi \in E, \quad \langle \mathcal{Q} \cdot ({}^tP(D)\phi), {}^tP(D)\phi \rangle = \langle \mathcal{L} \cdot ({}^tP(D)\phi), {}^tP(D)\phi \rangle \geq 0.$$

So, $\mathcal{Q}$ is a semi-Hilbertian Kernel. Finally, we notice that for all $\phi \in E$,

$$Q(D) \cdot \left(\mathcal{L} \underbrace{({}^tP(D)\phi)}_{\psi} \right) = Q(D) \cdot \left(\mathcal{Q} \underbrace{({}^tP(D)\phi)}_{\psi} \right) = \underbrace{{}^tP(D)\phi}_{\psi},$$

$\psi \in (\operatorname{Ker} P(D))^\circ$.

§3. Quasi-$P(D)$ interpolants

Let $\mathbb{R}^n$ be nested by a regular grid $\mathcal{M}(t)$, with t as origin and mesh size $h = (h_1, \ldots, h_n) \in \left(\mathbb{R}_+^*\right)^n$. When $L \subset L^2_{loc}(\mathbb{R}^n)$ ($L \not\subset C^0(\mathbb{R}^n)$), we associate to each $a \in \mathbb{R}^n$, a ball $\omega(a)$ such that

$$\alpha \text{ meas}(\omega(a)) << \|h\|_2.$$

We set (for sake of simplicity!): $\forall v \in L$,

$$v(a) = \frac{1}{\alpha} \int_{\omega(a)} v(s)ds.$$

We denote by E^{h_j} the shift operator such that

$$E^{h_j} v = v(\cdot + h_j e_j), \quad e_j = (0, \ldots, 0, \underbrace{1}_{j}, 0, \ldots, 0)$$

and

$$D_{h_j} v = \frac{E^{h_j} - I}{h_j} v$$

$$\text{(resp. } D_{h_j}v = \frac{E^{2h_j} - E^{-2h_j}}{4h_j}v, \ldots)$$

For $v = (v_1, \ldots, v_m)$ we set:

$$^t(D_h v) = (D_{h_1}v_1, \ldots, D_{h_n}v_1, \ldots, D_{h_1}v_m, \ldots, D_{h_n}v_m)$$

and

$$P(D_h)v = \Gamma \cdot (D_h v)\,.$$

Let $Q_h = Q(D_h) = {}^tP(D_h) \cdot P(D_h)$ (resp. $Q_h = {}^tP(D_h) \cdot P(D)$, $Q_h = {}^tP(D) \cdot P(D_h)$) and:

$$\mathcal{Q}_h = Q_h \mathcal{Q}.$$

So, $\mathcal{Q}_h(\cdot - s)$ is an approximant of the Dirac functional at the point s. Its regularity is conditionned by $(P(D)$ and its precision by that of D_h, which is an approximant of D.

Remark. In the above, we can replace rectangular grid by "oblique" grid.

Proposition 3.1.

i) *Ker* $P(D) \subset$ *Ker* $P(D_h)$

ii) $\mathcal{Q}_h \cdot (Q(D)\phi) = Q_h\phi$

iii) $\langle \mathcal{Q}_h \cdot (Q_h\phi)\,, \phi\rangle = \langle \mathcal{L} \cdot (Q_h\phi)\,, Q_h\phi\rangle \geq 0$

iv) *If* $p \in$ *Ker* $P(D)$,

$$\mathcal{Q}_h p = \int Q_h \cdot \mathcal{Q}(\cdot - s)ds = \textit{constant} \cdot p$$

v) $\int \mathcal{Q}(t - s)ds =$ *constant (independent of* t*)*

vi) $\mathcal{Q}_h$ *is a "bell-shaped" function.*

§4. Wavelets

4.1. Relation between D_{2h} and D_h.

Let us assume that

$$D_{h_j} = \frac{E^{h_j} - I}{h_j}.$$

Then,

$$D_{2h_j} = \frac{E^{2h_j} - I}{2h_j} = \frac{E^{h_j} + I}{2}D_{h_j}$$

$$\text{(resp. } D_{h_j} = \frac{E^{h_j} - E^{-h_j}}{2h_j} \Rightarrow D_{2h_j} = \frac{E^{h_j} + E^{-h_j}}{2}D_{h_j}, \ldots)$$

Notice that

$$D_{2h_j} - D_{h_j} = \left(\frac{E^{h_j} + I}{2} - I\right)D_{h_j} = h_j(D_{h_j})^2$$

$$(\text{resp. } D_{2h_j} - D_{h_j} = h_j \left(\frac{D_{h_j} - D_{-h_j}}{2} \right) D_{h_j}, \ldots, \text{ etc}).$$

Therefore, $D_{2h}v = A(h)D_h v$ where $A(h)$ is a diagonal matrix.

Remark. We can consider the case where $A(h)$ would be a full matrix which means that the approximants of the derivatives

$$D_k v_j, \quad 1 \le k \le n, 1 \le j \le m,$$

are all connected.

First case: Γ is invertible : $(\Rightarrow q = mn)$

$$P(D_{2h}v) = \Gamma D_{2h} v = \Gamma\left(A(h)\right) D_h v \Rightarrow P(D_{2h}v) = \left(\Gamma A(h) \Gamma^{-1}\right) P(D_h v).$$

Second case: $q > mn$, $\text{Rank}(\Gamma) = mn$
Assume (which is possible by changing the numbering) that

$$\widetilde{\Gamma} = \begin{pmatrix} \gamma_{11}^1 & \cdots & \gamma_{1m}^n \\ \cdots & \cdots & \cdots \\ \gamma_{mn,1} & \cdots & \gamma_{mn,m} \end{pmatrix}$$

is invertible. Set $\tilde{P}(D)v = \tilde{\Gamma}(Dv)$. Then,

$$\tilde{P}(D_{2h}v) = \left(\tilde{\Gamma}\tilde{A}(h)\tilde{\Gamma}^{-1}\right) \tilde{P}(D_h v).$$

Third case: $q < mn$, $\text{Rank}(\Gamma) = q$:

$$\widetilde{\Gamma} = \begin{pmatrix} & & & \Gamma & & & \\ 0 & \cdots & 0 & & \times & \cdots & \times \\ \cdots & \cdots & \cdots & & \cdots & \cdots & \cdots \\ 0 & \cdots & 0 & & \times & \cdots & \times \end{pmatrix}$$

such that $\tilde{\Gamma}$ is invertible. Thus,

$$\text{if } \tilde{P}(D)v = \tilde{\Gamma}(Dv), \quad \text{then} \quad \tilde{P}(D_{2h})v = (\tilde{\Gamma}\tilde{A}(h)\tilde{\Gamma}^{-1})\tilde{P}(D_h v).$$

Example 4.1. Linear elasticity

$$\Gamma = \begin{pmatrix} 1 & 0 & 0 & 0 \\ 0 & 1/2 & 1/2 & 0 \\ 0 & 0 & 0 & 1 \end{pmatrix}, \quad P(D)v = \begin{pmatrix} D_1 v_1 \\ \frac{1}{2}(D_2 v_1 + D_1 v_2) \\ D_2 v_2 \end{pmatrix}$$

$$\tilde{\Gamma} = \begin{pmatrix} 1 & 0 & 0 & 0 \\ 0 & 1/2 & 1/2 & 0 \\ 0 & 0 & 0 & 1 \\ 0 & 0 & 1 & 0 \end{pmatrix}, \quad \tilde{P}(D)v = \begin{pmatrix} D_1 v_1 \\ \frac{1}{2}(D_2 v_1 + D_1 v_2) \\ D_2 v_2 \\ D_1 v_2 \end{pmatrix}$$

4.2. (Quasi-) wavelets.

Proposition 4.1.

i) $\tilde{\mathcal{Q}}_{2h} = B(h)\tilde{\mathcal{Q}}_h$

ii) $\tilde{\mathcal{Q}}_{2h}(2\cdot) = 2^{\alpha(P)}\tilde{\mathcal{Q}}_h(\cdot), \quad \alpha(P) \in \mathbb{Z}$

iii) *If* $\tilde{Q}_h = {}^t\tilde{P}(D_h)\tilde{P}(D_h)$, *then* $\tilde{\mathcal{Q}}_h$ *is a semi-Hilbertian Kernel.*

Definition 4.1. *We set*

$$\phi_j = \tilde{\mathcal{Q}}_{2^j h}, \quad \phi_{j,k} = \phi_j(\cdot - kh), \quad j,k \in \mathbb{Z}.$$

Proposition 4.2. *Let V_j be the closed linear space of $L^2(\mathbb{R}^n)$ generated by the family $\{\phi_{j,k}, k \in \mathbb{Z}\}$. Then $\{V_j, j \in \mathbb{Z}\}$ is a multiresolution analysis related to $P(D)$ for the space $P(D)[\mathrm{Ker}(P(D))^{\perp}]$.*

Definition 4.2. $\psi_0 = \phi_{-1} - \phi_0$ *will be called quasi-wavelet related to* $P(D)$.

Proposition 4.3. *Let* $\psi_{j,k} = 2^j\psi(2^j \cdot - kh)$, $j,k \in \mathbb{Z}$.

i) $\psi_{j,k}$ *has the same properties as* $\phi_{j,k}$.

ii) $\{\psi_{j,k}; k \text{ odd}\}$ *is a basis of* W_j *such that* $V_{j+1} = V_j + W_j$.

As above, we can construct wavelets from

$$\tilde{\mathcal{L}}_h = \tilde{Q}_h\mathcal{L} = {}^t\tilde{P}_t(D_h)\tilde{P}_s(D_h)$$

§5. Extensions

With the same method, it is now easy to study quasi-interpolants and quasi-wavelets relating to

$$P^{\mu}(D) = \underbrace{P(D) \circ \cdots \circ P(D)}_{\mu}, \quad \mu \in \mathbb{N}^*, \mu > 1,$$

or

$$P^{\mu}(D) = P(D^{\mu}), \quad \mu \in \mathbb{N}^*, \mu > 1,$$

and so on.

Quasi-interpolants and wavelets on bounded sets can also be studied by that method.

References

1. Atteia M., Spline manifolds, in *Curves and Surfaces*, P. J. Laurent, A. Le Méhauté, and L. L. Schumaker (eds), Academic Press, New York, 1991, 1–4.
2. Atteia M., *Hilbertian Kernels and Spline Functions*, Studies in Computational Mathematics, Elsevier North-Holland, 1992.
3. Atteia M., Benbourhim M. N., and Gonzalez-Casanova P., *Quasi-interpolants elastic manifolds*, to appear.
4. Atteia M. and Benbourhim M. N., A vector spline approximation with application to meteorology, in *Curves and Surfaces*, P. J. Laurent, A. Le Méhauté, and L. L. Schumaker (eds), Academic Press, New York, 1991, 5–10.

M. Atteia
Laboratoire d'Analyse Numérique
118, Route de Narbonne
31062 Toulouse cedex, FRANCE

Fast DCT–Algorithms, Interpolating Wavelets, and Hierarchical Bases

G. Baszenski and M. Tasche

Abstract. Cosine and Chebyshev interpolation are based on discrete cosine transforms (DCT). The aim of this paper is to describe the close connection between these interpolation schemes, fast DCT–algorithms, interpolating wavelets and hierarchical bases. We obtain error and stability estimates for the considered interpolation methods. Further, we present efficient decomposition and reconstruction algorithms for the corresponding interpolating wavelets.

§0. Introduction

Considering cosine and Chebyshev interpolation, we describe a new, close relationship between these interpolation schemes, fast DCT–algorithms, interpolating wavelets and hierarchical bases. Related results concerning complex trigonometric interpolation can be found in [1, 4].

First we sketch fast algorithms for various DCTs, where we extend earlier results in [2,7,8]. In Section 2 we explain a multiscale method for approximation. Starting with a chain of interpolation projectors, we give a constructive approach to interpolating wavelets [3] and hierarchical bases [4,9]. In Section 3, we apply the results of Section 2 to cosine and Chebyshev interpolation. In Section 4, we derive new error and stability estimates for cosine and Chebyshev interpolation. Finally, we present decomposition and reconstruction algorithms. Both procedures based on DCT–algorithms of Section 1 are fast and numerically stable. This method leads to a new insight into efficient interpolation schemes.

§1. Fast DCT–Algorithms

The interpolation schemes considered here are mainly based on DCT. In the following we define the variants of transforms that we use.

Wavelets, Images and Surface Fitting
P. J. Laurent, A. Le Méhauté, and L. L. Schumaker (eds.), pp. 37–50.

ISBN 1-56881-040-7.

The *type* I–DCT *of length* $2^{j+1}+1$ (DCT-I $(2^{j+1}+1)$) is a mapping of $\mathbb{R}^{2^{j+1}+1}$ into itself defined by $(a_{j+1,k}(\mathbf{f}_{j+1}))_{k=0}^{2^{j+1}} :=$ DCT-I $(2^{j+1}+1)\,\mathbf{f}_{j+1}$, where

$$a_{j+1,k}(\mathbf{f}_{j+1}) := \sum_{l=0}^{2^{j+1}} \varepsilon_{j+1,l}\, f_{j+1,l} \cos\frac{kl\pi}{2^{j+1}} \quad (k=0,\ldots,2^{j+1}) \tag{1.1}$$

with $\mathbf{f}_{j+1} := (f_{j+1,l})_{l=0}^{2^{j+1}} \in \mathbb{R}^{2^{j+1}+1}$ and $\varepsilon_{j+1,0} = \varepsilon_{j+1,2^{j+1}} = 1/2$, $\varepsilon_{j+1,k} := 1$ $(k=1,\ldots,2^{j+1}-1)$. Its inverse mapping is given by 2^{-j} DCT-I $(2^{j+1}+1)$.

The *type* II–DCT *of length* 2^j (DCT-II (2^j)) is a mapping of $\mathbb{R}^{2^j}$ into itself defined by $(b_{j,k}(\mathbf{g}_j))_{k=0}^{2^j-1} :=$ DCT-II $(2^j)\,\mathbf{g}_j$, where

$$b_{j,k}(\mathbf{g}_j) := \sum_{l=0}^{2^j-1} g_{j,l} \cos\frac{k(2l+1)\pi}{2^{j+1}} \quad (k=0,\ldots,2^j-1) \tag{1.2}$$

with $\mathbf{g}_j := (g_{j,l})_{l=0}^{2^j-1} \in \mathbb{R}^{2^j}$.

The *type* III–DCT *of length* 2^j (DCT-III (2^j)) is a mapping of $\mathbb{R}^{2^j}$ into itself defined by $(c_{j,k}(\mathbf{g}_j))_{k=0}^{2^j-1} :=$ DCT-III $(2^j)\,\mathbf{g}_j$,

$$c_{j,k}(\mathbf{g}_j) := \sum_{l=0}^{2^j-1} \varepsilon_{j,l}\, g_{j,l} \cos\frac{(2k+1)l\pi}{2^{j+1}} \quad (k=0,\ldots,2^j-1). \tag{1.3}$$

The inverse of DCT-II (2^j) is 2^{1-j} DCT-III (2^j).

Fast algorithms for DCT-II (2^j) and DCT-III (2^j) are described for instance in [2,7,8].

In the following we give two procedures for computing DCT-I $(2^{j+1}+1)$. Applying the divide-and-conquer technique, we obtain:

Theorem 1.1. *The* DCT-I $(2^{j+1}+1)$ *of the data vector* $\mathbf{f}_{j+1} \in \mathbb{R}^{2^{j+1}+1}$ *can be computed recursively by* DCT-I (2^j+1) *and* DCT-II (2^j) *or by* DCT-I (2^j+1) *and* DCT-III (2^j) *as follows:*
(i) *Decimation in time algorithm. If* $0 \le k < 2^j$, *then*

$$a_{j+1,k}(\mathbf{f}_{j+1}) := a_{j,k}(\mathbf{f}_{j+1}^{(0)}) + b_{j,k}(\mathbf{f}_{j+1}^{(1)}), \tag{1.4}$$

$$a_{j+1,2^{j+1}-k}(\mathbf{f}_{j+1}) := a_{j,k}(\mathbf{f}_{j+1}^{(0)}) - b_{j,k}(\mathbf{f}_{j+1}^{(1)}) \tag{1.5}$$

and if $k = 2^j$, *then*

$$a_{j+1,2^j}(\mathbf{f}_{j+1}) := a_{j,2^j}(\mathbf{f}_{j+1}^{(0)}) \tag{1.6}$$

with

$$\mathbf{f}_{j+1}^{(0)} := (f_{j+1,2l})_{l=0}^{2^j} \in \mathbb{R}^{2^j+1},$$
$$\mathbf{f}_{j+1}^{(1)} := (f_{j+1,2l+1})_{l=0}^{2^j-1} \in \mathbb{R}^{2^j}$$

(ii) *Decimation in frequency algorithm. If* $0 \le k \le 2^j$, *then*

$$a_{j+1,2k}(\mathbf{f}_{j+1}) := a_{j,k}(\mathbf{g}_j) \tag{1.7}$$

and if $0 \le k < 2^j$, *then*

$$a_{j+1,2k+1}(\mathbf{f}_{j+1}) := c_{j,k}(\mathbf{h}_j) \tag{1.8}$$

with

$$\begin{aligned} \mathbf{g}_j &:= (f_{j+1,l} + f_{j+1,2^{j+1}-l})_{l=0}^{2^j} \in \mathbb{R}^{2^j+1}, \\ \mathbf{h}_j &:= (f_{j+1,l} - f_{j+1,2^{j+1}-l})_{l=0}^{2^j-1} \in \mathbb{R}^{2^j}. \end{aligned}$$

Proof: (i) For $k = 0, \ldots, 2^{j+1}$ we have by definition (1.1)

$$a_{j+1,k}(\mathbf{f}_{j+1}) = \sum_{l=0}^{2^j} \varepsilon_{j,l}\, f_{j+1,2l} \cos\frac{kl\pi}{2^j} + \sum_{l=0}^{2^j-1} f_{j+1,2l+1} \cos\frac{k(2l+1)\pi}{2^{j+1}}. \tag{1.9}$$

In the case $0 \le k < 2^j$ we obtain (1.4). Replacing k in (1.9) by $2^{j+1} - k$ and observing

$$\cos\frac{(2^{j+1}-k)l\pi}{2^j} = \cos\frac{kl\pi}{2^j}, \tag{1.10}$$

$$\cos\frac{(2^{j+1}-k)(2l+1)\pi}{2^{j+1}} = -\cos\frac{k(2l+1)\pi}{2^{j+1}}, \tag{1.11}$$

we get (1.5). If $k = 2^j$, then (1.6) follows again from (1.9).

(ii) For $k = 0, \ldots, 2^j$ we obtain from (1.1) and (1.10) that

$$\begin{aligned} a_{j+1,2k}(\mathbf{f}_{j+1}) &= \sum_{l=0}^{2^j-1} \varepsilon_{j,l}\, f_{j+1,l} \cos\frac{kl\pi}{2^j} + (-1)^l f_{j+1,2^j} \\ &\quad + \sum_{l=0}^{2^j-1} \varepsilon_{j,l}\, f_{j+1,2^{j+1}-l} \cos\frac{k(2^{j+1}-l)\pi}{2^j} \\ &= \sum_{l=0}^{2^j} \varepsilon_{j,l}(f_{j+1,l} + f_{j+1,2^{j+1}-l}) \cos\frac{kl\pi}{2^j} = a_{j,k}(\mathbf{g}_j). \end{aligned}$$

For $k = 0, \ldots, 2^j - 1$ we conclude similarly using (1.1) and (1.11) that

$$\begin{aligned} a_{j+1,2k+1}(\mathbf{f}_{j+1}) &= \sum_{l=0}^{2^j-1} \varepsilon_{j,l}\, (f_{j+1,l} - f_{j+1,2^{j+1}-l}) \cos\frac{(2k+1)l\pi}{2^{j+1}} \\ &= c_{j,k}(\mathbf{h}_j). \quad \blacksquare \end{aligned}$$

Summarizing we can say that all fast DCT–algorithms described here require only $O(j\,2^j)$ arithmetical operations.

§2. Interpolating Wavelets and Hierarchical Bases

In this section we explain a multiscale method for approximation [3,4,9,10]. Starting with a chain of interpolation projectors, we give a constructive approach to define interpolating wavelets [3] and hierarchical bases [4,9].

Let $L^2(I)$ denote the Hilbert space of square integrable functions on a compact interval $I \subset \mathbb{R}$. Let $\{\phi_k : k \in \mathbb{N}_0\}$ be a complete orthogonal basis of $L^2(I)$. For $N_j \in \mathbb{N}_0$ with $N_j < N_{j+1}$ $(j \in \mathbb{N}_0)$ we set

$$V_j := \text{span}\,\{\phi_0, \dots, \phi_{N_j}\}.$$

Let $K_j := \{t_{j,l} \in I : l = 0, \dots, N_j\}$ be sets of interpolation nodes with $K_j \subset K_{j+1}$ $(j \in \mathbb{N}_0)$. Assume that for every continuous function $f \in C(I)$ and every refinement level j there exists a unique function $v \in V_j$ with

$$f(t_{j,l}) = v(t_{j,l}) \quad (l = 0, \dots, N_j).$$

Hence we have in particular that the nodes of one level are pairwise different. We set $P_j f := v$. Note that $\operatorname{Im} P_j = V_j$. We denote the coefficients of the interpolant $P_j f$ as follows:

$$P_j f = \sum_{k=0}^{N_j} \alpha_{j,k}(f)\, \phi_k. \tag{2.1}$$

These interpolation projectors P_j fulfill the property

$$P_j P_{j+1} = P_{j+1} P_j = P_j \Longleftrightarrow: P_j \le P_{j+1} \quad (j \in \mathbb{N}_0). \tag{2.2}$$

Note that $\le$ defined in this way is an order relation. Next we introduce

$$Q_j := P_{j+1} - P_j \quad (j \in \mathbb{N}_0). \tag{2.3}$$

The operator Q_j is a projector, since by (2.2)

$$\begin{aligned} Q_j^2 &= (P_{j+1} - P_j)(P_{j+1} - P_j) \\ &= P_{j+1}^2 - P_{j+1}P_j - P_j P_{j+1} + P_j^2 \\ &= P_{j+1} - 2P_j + P_j = Q_j. \end{aligned}$$

Then $W_j := \operatorname{Im} Q_j$ is called *interpolating wavelet space of level* j. We have $W_j \subset V_{j+1}$. In view of

$$P_{j+1} = P_j + Q_j, \quad P_j Q_j = 0 \tag{2.4}$$

we have the direct sum decomposition

$$V_{j+1} = V_j \dotplus W_j \quad (j \in \mathbb{N}_0). \tag{2.5}$$

Iterating this equation, we get

$$V_{j+1} = V_0 \dot{+} W_0 \dot{+} W_1 \dot{+} \cdots \dot{+} W_j. \tag{2.6}$$

Therefore, every $v \in V_{j+1}$ has the representation

$$v = P_0 v + \sum_{k=0}^{j} Q_k v. \tag{2.7}$$

Note that V_j and W_j are not orthogonal with respect to the $L^2(I)$-scalar product. Instead, we have for $v \in V_j$ and $w \in W_j$ the relation

$$\sum_{l=0}^{N_j} v(t_{j,l})\,\overline{w(t_{j,l})} = 0.$$

For every level j we want to construct a basis of W_j. Note that $w \in W_j$ satisfies $P_j w = 0$. Defining *interpolating wavelets of level* j (see [3]) by

$$\psi_r := \phi_r - P_j \phi_r \quad (r = N_j + 1, \ldots, N_{j+1}), \tag{2.8}$$

we obtain for $r = N_j + 1, \ldots, N_{j+1}$

$$P_j \psi_r = 0, \quad Q_j \psi_r = \psi_r \in W_j,$$

since

$$\begin{aligned} Q_j \psi_r &= (P_{j+1} - P_j)\psi_r = P_{j+1}\psi_r = P_{j+1}(\phi_r - P_j \phi_r) \\ &= P_{j+1}\phi_r - P_j \phi_r = \phi_r - P_j \phi_r = \psi_r. \end{aligned}$$

The number of interpolating wavelets $\psi_r \in W_j$ is $N_{j+1} - N_j$, which is also the dimension of W_j, because by (2.5) we have

$$\dim W_j = \dim V_{j+1} - \dim V_j = N_{j+1} - N_j.$$

Further ψ_r $(r = N_j + 1, \ldots, N_{j+1})$ are linearly independent, since

$$P_j \phi_r \in V_j = \operatorname{span}\{\phi_0, \ldots, \phi_{N_j}\}$$

and since ϕ_k $(k = 0, \ldots, N_{j+1})$ are linearly independent by assumption. Thus

$$W_j = \operatorname{span}\{\psi_{N_j+1}, \ldots, \psi_{N_{j+1}}\}.$$

The projector Q_j can be represented as

$$Q_j f = \sum_{r=N_j+1}^{N_{j+1}} \beta_{j,r}(f)\,\psi_r \quad (f \in C(I)). \tag{2.9}$$

The union

$$B_{j+1} = \{\phi_0, \ldots, \phi_{N_0}, \psi_{N_0+1}, \ldots, \psi_{N_1}, \ldots, \psi_{N_j+1}, \ldots, \psi_{N_{j+1}}\} \tag{2.10}$$

of the bases of $V_0, W_0, W_1, \ldots, W_j$ is a basis of V_{j+1} by (2.6). In addition we set $B_0 := \{\phi_0, \ldots, \phi_{N_0}\}$. Then B_j $(j \in \mathbb{N}_0)$ is called *hierarchical basis of* V_j $(j \in \mathbb{N}_0)$ (see [9]). Obviously, a hierarchical basis has the following properties for $j \in \mathbb{N}_0$:

$$\begin{aligned} &B_j \subset B_{j+1}, \\ &V_j = \operatorname{span} B_j, \\ &W_j = \operatorname{span}(B_{j+1} \setminus B_j). \end{aligned}$$

§3. Cosine and Chebyshev Interpolation

In the following we consider two special cases. First we discuss a trigonometric interpolation scheme. Let $I := [0,1]$ and $\phi_k := \cos(k\pi\,\cdot\,)$ $(k \in \mathbb{N}_0)$. With $N_j := 2^j$ $(j \in \mathbb{N}_0)$ we form V_j. As interpolation nodes we choose $t_{j,l} := 2^{-j}l$ $(l = 0, \ldots, 2^j)$. Then the interpolation projector P_j is given for $f \in C(I)$ by (2.1) with

$$P_j f = \sum_{k=0}^{2^j} \alpha_{j,k}(f)\, \cos(k\pi\,\cdot\,) \tag{3.1}$$

with

$$\alpha_{j,k}(f) := 2^{1-j} \varepsilon_{j,k}\, a_{j,k}(\mathbf{f}_j), \tag{3.2}$$

where $a_{j,k}(\mathbf{f}_j)$ is explained by (1.1) and where

$$\mathbf{f}_j := \big(f(t_{j,l})\big)_{l=0}^{2^j}. \tag{3.3}$$

In order to construct the hierarchical basis, we need the following

Lemma 3.1. *For $j, r \in \mathbb{N}_0$ and $k = 0, \ldots, 2^j$ we have*

$$\begin{aligned}\alpha_{j,k}(\cos(r\pi\,\cdot\,)) &= 2^{1-j} \varepsilon_{j,k} \sum_{l=0}^{2^j} \varepsilon_{j,l} \cos\frac{rl\pi}{2^j} \cos\frac{kl\pi}{2^j} \\ &= \begin{cases} 1 & \text{for } r \equiv \pm k \mod 2^{j+1}, \\ 0 & \text{otherwise.}\end{cases}\end{aligned}$$

The proof is by straightforward calculations.

For every level j we form the functions (2.8), i.e., for $r = 2^j + 1, \ldots, 2^{j+1}$ we set

$$\begin{aligned}\psi_r &= \cos(r\pi\,\cdot\,) - P_j(\cos(r\pi\,\cdot\,)) \\ &= \cos(r\pi\,\cdot\,) - \sum_{k=0}^{2^j} \alpha_{j,k}\,(\cos(r\pi\,\cdot\,)) \cos(k\pi\,\cdot\,) \\ &= \cos(r\pi\,\cdot\,) - \cos((2^{j+1} - r)\pi\,\cdot\,).\end{aligned} \tag{3.4}$$

We summarize:

Lemma 3.2. *For $j \in \mathbb{N}_0$, the set*

$$B_{j+1} = B_0 \cup \bigcup_{i=0}^{j} \{\cos(r\pi\,\cdot\,) - \cos((2^{i+1} - r)\pi\,\cdot\,) : r = 2^i + 1, \ldots, 2^{i+1}\} \tag{3.5}$$

with $B_0 := \{1, \cos(\pi\,\cdot\,)\}$ is a hierarchical basis of V_{j+1}.

Next we want to represent $Q_j f$ as a linear combination of hierarchical basis elements in the form (2.9).

Lemma 3.3. *For $f \in C[0,1]$ and $j \in \mathbb{N}_0$ we have*

$$Q_j f = \sum_{r=2^j+1}^{2^{j+1}} \alpha_{j+1,r}(f) \left(\cos(r\pi\,\cdot) - \cos((2^{j+1}-r)\pi\,\cdot)\right) \tag{3.6}$$

where $\alpha_{j+1,r}(f)$ is given by (3.2).

Proof: Inserting (1.4) – (1.6) into (3.2), we obtain for $k = 0, \ldots, 2^j - 1$:

$$\alpha_{j+1,k}(f) = 2^{-j}\varepsilon_{j,k}\, a_{j,k}(\mathbf{f}_j) + 2^{-j}\varepsilon_{j,k}\, b_{j,k}(\mathbf{f}_{j+1}^{(1)}),$$
$$\alpha_{j+1,2^{j+1}-k}(f) = 2^{-j}\varepsilon_{j,k}\, a_{j,k}(\mathbf{f}_j) - 2^{-j}\varepsilon_{j,k}\, b_{j,k}(\mathbf{f}_{j+1}^{(1)})$$

and $\alpha_{j+1,2^j}(f) = \alpha_{j,2^j}(f)$.

For $f \in C[0,1]$ and $j \in \mathbb{N}_0$ we have by (2.3) and (2.1)

$$\begin{aligned} Q_j f = P_{j+1}f - P_j f = &\sum_{k=0}^{2^j-1} \alpha_{j+1,k}(f)\cos(k\pi\,\cdot) + \alpha_{j+1,2^j}(f)\cos(2^j\pi\,\cdot) \\ &+ \sum_{k=0}^{2^j-1} \alpha_{j+1,2^{j+1}-k}(f)\cos((2^{j+1}-k)\pi\,\cdot) \\ &- \sum_{k=0}^{2^j-1} \alpha_{j,k}(f)\cos(k\pi\,\cdot) - \alpha_{j,2^j}(f)\cos(2^j\pi\,\cdot). \end{aligned}$$

Using the above equations, we obtain

$$\begin{aligned} Q_j f &= \sum_{k=0}^{2^j-1} 2^{-j}\varepsilon_{j,k}\left(-a_{j,k}(\mathbf{f}_j) + b_{j,k}(\mathbf{f}_{j+1}^{(1)})\right)\cos(k\pi\,\cdot) \\ &\quad + \sum_{k=0}^{2^j-1} \alpha_{j+1,2^{j+1}-k}(f)\cos((2^{j+1}-k)\pi\,\cdot) \\ &= \sum_{k=0}^{2^j-1} \alpha_{j+1,2^{j+1}-k}(f)\left(\cos((2^{j+1}-k)\pi\,\cdot) - \cos(k\pi\,\cdot)\right). \quad \blacksquare \end{aligned}$$

Note that the coefficients $\alpha_{j+1,r}(f)$ $(r = 2^j+1, \ldots, 2^{j+1})$ of $Q_j f$ in the hierarchical basis coincide with the last coefficients (3.2) of $P_{j+1}f$ in the original cosine basis.

Performing a change of variables

$$x := \frac{1}{\pi}\arccos t \quad (t \in [-1,1])$$

the previous interpolation scheme can be transformed into Chebyshev interpolation. The details of this polynomial interpolation are as follows. Let

$I := [-1,1]$ and $\phi_k = T_k := \cos(k \arccos \cdot)$ $(k \in \mathbb{N}_0)$. We set $N_j := 2^j$ $(j \in \mathbb{N}_0)$ and $V_j := \operatorname{span}\{1, T_1, \ldots, T_{2^j}\}$. The interpolation knots are $t_{j,l} := \cos(2^{-j} l\pi)$ $(l = 0, \ldots, 2^j)$. For $f \in C[-1,1]$ we have

$$\tilde{P}_j f := \sum_{k=0}^{2^j} \alpha_{j,k}(f)\, T_k$$

with $\alpha_{j,k}(f)$ given by (3.2).

Lemma 3.4. *For $j, r \in \mathbb{N}_0$ and $k = 0, \ldots, 2^j$ we have*

$$\alpha_{j,k}(T_r) = \begin{cases} 1 & \text{if } r \equiv \pm k \mod 2^{j+1}, \\ 0 & \text{otherwise.} \end{cases}$$

For every level j we have by (2.8) and Lemma 3.4 for $r = 2^j + 1, \ldots, 2^{j+1}$

$$\psi_r = T_r - \tilde{P}_j T_r = T_r - \sum_{k=0}^{2^j} \alpha_{j,k}(T_r)\, T_k = T_r - T_{2^{j+1}-r}. \tag{3.7}$$

Thus we obtain the following hierarchical basis:

Lemma 3.5. *For $j \in \mathbb{N}_0$, the set*

$$\tilde{B}_{j+1} := \tilde{B}_0 \cup \bigcup_{i=0}^{j} \{T_r - T_{2^{i+1}-r} : r = 2^i + 1, \ldots, 2^{i+1}\} \tag{3.8}$$

with $\tilde{B}_0 := \{1, T_1\}$ is a hierarchical basis of V_{j+1} $(j \in \mathbb{N}_0)$.

By the same computations as in Lemma 3.3 we obtain a representation of the projector $\tilde{Q}_j := \tilde{P}_{j+1} - \tilde{P}_j$:

Lemma 3.6. *For $f \in C[-1,1]$ and $j \in \mathbb{N}_0$ we have*

$$\tilde{Q}_j f = \sum_{r=2^j+1}^{2^{j+1}} \alpha_{j+1,r}(f)\,(T_r - T_{2^{j+1}-r}),$$

where $\alpha_{j+1,r}(f)$ is given by (3.2).

§4. Error and Stability Estimates

First we derive an error estimate for cosine interpolation. Let $f \in C^{2m+2}[0,1]$ $(m \in \mathbb{N}_0)$ be a function with $f^{(2\mu+1)}(0) = f^{(2\mu+1)}(1) = 0$ $(\mu = 0, \ldots, m-1)$. Let

$$\alpha_r(f) := 2\int_0^1 f(x)\cos(r\pi x)\, dx \quad (r \in \mathbb{N}),$$

$$\alpha_0(f) := \int_0^1 f(x)\, dx.$$

Then the Fourier-cosine series of f converges uniformly:

$$f = \sum_{r=0}^{\infty} \alpha_r(f) \cos(r\pi \cdot). \tag{4.1}$$

Let

$$S_j f := \sum_{r=0}^{2^j} \alpha_r(f) \cos(r\pi \cdot).$$

By $(2m+2)$-fold integration by parts we get for $r \in \mathbb{N}$:

$$\begin{aligned}\alpha_r(f) = \frac{2}{(r\pi)^{2m+2}} \Big((-1)^{r+1} f^{(2m+1)}(1) + f^{(2m+1)}(0) \\ + \int_0^1 f^{(2m+2)}(x) \cos(r\pi x)\, \mathrm{d}x \Big),\end{aligned}$$

which implies

$$|\alpha_r(f)| \le C r^{-2m-2}$$

with

$$C := \frac{2}{\pi^{2m+2}} \Big(|f^{(2m+1)}(1)| + |f^{(2m+1)}(0)| + \int_0^1 |f^{(2m+2)}(x)|\, \mathrm{d}x \Big). \tag{4.2}$$

Then we obtain for $x \in [0,1]$ the estimate

$$\begin{aligned}|f(x) - S_j f(x)| &\le \sum_{r=2^j+1}^{\infty} |\alpha_r(f)| \le C \sum_{r=2^j+1}^{\infty} r^{-2m-2} \\ &\le \frac{C}{2m+1}\, 2^{-j(2m+1)}.\end{aligned} \tag{4.3}$$

The coefficients of the cosine interpolation projector P_j given by (3.1) satisfy the following aliasing formula

$$\alpha_{j,k}(f) = \alpha_k(f) + \sum_{\nu=1}^{\infty} \alpha_{2^{j+1}\nu+k}(f) \qquad (k = 0,\, 2^j),$$

$$\alpha_{j,k}(f) = \alpha_k(f) + \sum_{\nu=1}^{\infty} (\alpha_{2^{j+1}\nu+k}(f) + \alpha_{2^{j+1}\nu-k}(f)) \qquad (k = 1, \ldots, 2^j - 1).$$

This follows from the uniform convergence of (4.1), which implies

$$\alpha_{j,k}(f) = \sum_{r=0}^{\infty} \alpha_r(f)\, \alpha_{j,k}(\cos(r\pi \cdot)) \quad (k = 0, \ldots, 2^j),$$

and from Lemma 3.1.

In order to appraise the interpolation error, we consider

$$P_j f - S_j f = \sum_{k=0}^{2^j} \big(\alpha_{j,k}(f) - \alpha_k(f)\big) \cos(k\pi\,\cdot).$$

Using the above aliasing formula and (4.3), we conclude

$$\begin{aligned}
\big|P_j f(x) - S_j f(x)\big| &\le \sum_{k=0}^{2^j} \big|\alpha_{j,k}(f) - \alpha_k(f)\big| \\
&\le \sum_{\nu=1}^{\infty} \big|\alpha_{2^{j+1}\nu}(f)\big| + \sum_{k=0}^{2^j-1} \sum_{\nu=1}^{\infty} \big(\big|\alpha_{2^{j+1}\nu+k}(f)\big| + \big|\alpha_{2^{j+1}\nu-k}(f)\big|\big) \\
&\quad + \sum_{\nu=1}^{\infty} \big|\alpha_{2^{j+1}\nu+2^j}(f)\big| \\
&= \sum_{r=2^j+1}^{\infty} \big|\alpha_r(f)\big| \le \frac{C}{2m+1}\, 2^{-j(2m+1)}.
\end{aligned}$$

Thus we obtain the error estimate:

Theorem 4.1. *For given* $f \in C^{2m+2}[0,1]$ $(m \in \mathbb{N}_0)$ *with* $f^{(2\mu+1)}(0) = f^{(2\mu+1)}(1) = 0$ $(\mu = 0, \ldots, m-1)$ *we have*

$$\big\|f - P_j f\big\|_\infty \le \big\|f - S_j f\big\|_\infty + \big\|S_j f - P_j f\big\|_\infty \le \frac{2C}{2m+1}\, 2^{-j(2m+1)}$$

with a constant C *given by* (4.2).

As a corollary of Theorem 4.1 we obtain an error estimate for Chebyshev interpolation in the following way: The change of variables $t := \cos(\pi x)$ $(x \in [0,1])$ transforms $\tilde{f} \in C^{2m+2}[-1,1]$ into $f := \tilde{f}(\cos(\pi\,\cdot)) \in C^{2m+2}[0,1]$ with $f^{(2\mu+1)}(0) = f^{(2\mu+1)}(1) = 0$ $(\mu = 0, \ldots, m)$ and the Chebyshev interpolant $\tilde{P}_j \tilde{f}$ into the cosine interpolant $P_j f$. Therefore, we get

Corollary 4.2. *Let* $j, m \in \mathbb{N}_0$. *For* $\tilde{f} \in C^{2m+2}[-1,1]$ *we have*

$$\big\|\tilde{f} - \tilde{P}_j \tilde{f}\big\|_\infty = \big\|f - P_j f\big\|_\infty \le \frac{2C}{2m+1}\, 2^{-j(2m+1)}.$$

Estimates for derivatives can be found in [5,6]. In the following we consider only cosine interpolation. First we derive estimates for the norm of $Q_j f$ $(f \in C[-1,1])$ given by (3.6).

Theorem 4.3. *For* $f \in C[-1,1]$ *and* $j \in \mathbb{N}_0$ *we have*

$$\big\|Q_j f\big\|_2^2 = \sum_{r=2^j+1}^{2^{j+1}-1} \big|\alpha_{j+1,r}(f)\big|^2 + \frac{3}{2}\, \big|\alpha_{j+1,2^{j+1}}(f)\big|^2, \tag{4.4}$$

$$\big\|Q_j f\big\|_\infty \le 2 \sum_{r=2^j+1}^{2^{j+1}} \big|\alpha_{j+1,r}(f)\big|. \tag{4.5}$$

Proof: Equation (4.4) follows from the orthogonality of the interpolating wavelets ψ_r $(r = 2^j + 1, \ldots, 2^{j+1})$ given by (3.4) and from

$$\|\psi_r\|_2^2 = \int_0^1 \psi_r(x)^2 \, dx = \begin{cases} 1 & \text{for } r = 2^j + 1, \ldots, 2^{j+1} - 1, \\ \frac{3}{2} & \text{for } r = 2^{j+1}. \end{cases}$$

Since $\|\psi_r\|_\infty \le 2$ $(r = 2^j + 1, \ldots, 2^{j+1})$, the estimate (4.5) is obvious. ■

Remark 4.4. The asymptotic behaviour of $\|Q_j f\|_\infty$, where a function $f \in C^{2m+2}[-1,1]$ with $f^{(2\mu+1)}(0) = f^{(2\mu+1)}(1) = 0$ $(\mu = 0, \ldots, m-1)$ is given, follows from Theorem 4.1 and can be described by

$$\|Q_j f\|_\infty \le \|P_{j+1} f - f\|_\infty + \|f - P_j f\|_\infty \le \frac{2C\,(2^{-2m-1} + 1)}{2m+1}\, 2^{-j(2m+1)}.$$

The constructed hierarchical basis is not orthogonal, but on one decomposition level we have the following stability property.

Theorem 4.5. *Any element of* $V_{j+1} = V_j \dot{+} W_j$ *has the form* $v + w$ *with*

$$v := \sum_{k=0}^{2^j} \lambda_k \, \phi_k \in V_j, \quad w := \sum_{k=0}^{2^{j+1}} \lambda_k \, \psi_k \in W_j \quad (\lambda_k \in \mathbf{C})$$

and we have

$$\frac{1}{6} \sum_{k=0}^{2^j - 1} |\lambda_k|^2 \le \|v + w\|_2^2 \le \frac{3}{2} \sum_{k=0}^{2^j - 1} |\lambda_k|^2.$$

Proof: Writing $v + w$ in the original orthogonal basis $\{\phi_k : k = 0, \ldots, 2^{j+1}\}$, we obtain

$$v + w = (\lambda_0 - \lambda_{2^{j+1}}) + \sum_{k=1}^{2^j - 1} (\lambda_k - \lambda_{2^{j+1}-k}) \cos(k\pi \,\cdot\,) + \sum_{k=2^j}^{2^{j+1}} \lambda_k \cos(k\pi \,\cdot\,).$$

Therefore,

$$\begin{aligned} \|v + w\|_2^2 = & \Big(|\lambda_0 - \lambda_{2^{j+1}}|^2 + \frac{1}{2}|\lambda_{2^{j+1}}|^2\Big) \\ & + \frac{1}{2} \sum_{k=1}^{2^j - 1} \Big(|\lambda_k - \lambda_{2^{j+1}-k}|^2 + |\lambda_{2^{j+1}-k}|^2\Big) + \frac{1}{2} |\lambda_{2^j}|^2. \end{aligned}$$

Since

$$\begin{aligned} \frac{1}{5} \Big(|\lambda_0|^2 + |\lambda_{2^{j+1}}|^2\Big) &\le |\lambda_0 - \lambda_{2^{j+1}}|^2 + \frac{1}{2} |\lambda_{2^{j+1}}|^2 \\ &\le 2 |\lambda_0|^2 + \frac{5}{2} |\lambda_{2^{j+1}}|^2, \\ \frac{1}{3} \Big(|\lambda_k|^2 + |\lambda_{2^{j+1}-k}|^2\Big) &\le |\lambda_k - \lambda_{2^{j+1}-k}|^2 + |\lambda_{2^{j+1}-k}|^2 \\ &\le 2 |\lambda_k|^2 + 3 |\lambda_{2^{j+1}-k}|^2 \quad (k = 1, \ldots, 2^j - 1), \end{aligned}$$

the result follows. ■

§5. Decomposition and Reconstruction Algorithms

The following results hold for cosine and Chebyshev interpolation. For sake of simplicity we use only the notation of cosine interpolation. First we consider the following decomposition problem: Let $P_{j+1}f \in V_{j+1}$ for a continuous function f be given in the form

$$P_{j+1}f = \sum_{k=0}^{2^{j+1}} \alpha_{j+1,k}(f)\,\phi_k. \tag{5.1}$$

We assume that the coefficients $\alpha_{j+1,k}(f)$ are known, for instance precomputed by trigonometric interpolation (see (3.2)). We want to know the coefficients of $P_{j+1}f$ with respect to the hierarchical basis of V_{j+1}. For that reason we decompose $P_{j+1}f$ into the sum $P_jf + Q_jf$ with

$$P_jf = \sum_{k=0}^{2^j} \alpha_{j,k}(f)\,\phi_k, \quad Q_jf = \sum_{k=2^j+1}^{2^{j+1}} \beta_{j,k}(f)\,\psi_k \tag{5.2}$$

and we compute the unknown coefficients $\alpha_{j,k}(f)$, $\beta_{j,k}(f)$. If we iterate this decomposition, we obtain the coefficients of $P_{j+1}f$ with respect to the hierarchical bases (3.5), (3.8).

Theorem 5.1. *The coefficients of P_jf and Q_jf can be computed by*

$$\begin{aligned}\alpha_{j,k}(f) &:= \alpha_{j+1,k}(f) + \alpha_{j+1,2^{j+1}-k}(f) \quad (k = 0,\ldots,2^j-1),\\ \alpha_{j,2^j}(f) &:= \alpha_{j+1,2^j}(f),\\ \beta_{j,k}(f) &:= \alpha_{j+1,k}(f) \quad (k = 2^j+1,\ldots,2^{j+1}).\end{aligned}$$

Proof: We obtain by (3.4), (3.7) that

$$\begin{aligned}P_{j+1}f &= \sum_{k=0}^{2^j-1} \alpha_{j+1,k}(f)\,\phi_k + \alpha_{j+1,2^j}(f)\,\phi_{2^j}\\ &\quad + \sum_{k=2^j+1}^{2^{j+1}} \alpha_{j+1,k}(f)\,(\phi_k - \phi_{2^{j+1}-k}) + \sum_{k=0}^{2^j-1} \alpha_{j+1,2^{j+1}-k}(f)\,\phi_k\\ &= \sum_{k=0}^{2^j-1} \big(\alpha_{j+1,k}(f) + \alpha_{j+1,2^{j+1}-k}(f)\big)\,\phi_k\\ &\quad + \alpha_{j+1,2^j}(f)\,\phi_{2^j} + \sum_{k=2^j+1}^{2^{j+1}} \alpha_{j+1,k}(f)\,\psi_k.\end{aligned}$$

Since $\{\phi_0,\ldots,\phi_{2^j},\psi_{2^j+1},\ldots,\psi_{2^{j+1}}\}$ is a basis of $V_{j+1} = V_j \dot{+} W_j$, this representation of $P_{j+1}f$ is unique. ■

The iteration for j yields:

Corollary 5.2. *For $j \in \mathbb{N}_0$ we have the hierarchical basis representation*

$$P_{j+1}f = P_0 f + \sum_{i=0}^{j} Q_i f$$

$$= \alpha_{0,0}(f)\,\phi_0 + \alpha_{0,1}(f)\,\phi_1 + \sum_{i=0}^{j} \sum_{k=2^i+1}^{2^{i+1}} \alpha_{i+1,k}(f)\,\psi_k. \tag{5.3}$$

Note that the coefficients in the above representation (5.3) are all generated (apart from the power of two scaling factors) by the decimation in time algorithm for DCT-I $(2^{j+1}+1))$ with input $\mathbf{f}_{j+1} := \big(f(t_{j+1,l})\big)_{l=0}^{2^{j+1}}$. Since the coefficients $\big(\alpha_{j+1,k}(\mathbf{f}_{j+1})\big)_{k=0}^{2^j}$ are not used in the above representation we conclude that $P_{j+1}f$ in the form (5.3) can be computed from $\mathbf{f}_{j+1}$ somewhat cheaper than the standard representation (3.1) of $P_{j+1}f$. Similarly, it is possible to compute $\mathbf{f}_{j+1}$ when $P_{j+1}f$ is given in the form (5.3) by performing the decimation in frequency algorithm, again with some operations saved.

Next we consider the so-called reconstruction problem: Let $P_j f \in V_j$ and $Q_j f \in W_j$ with continuous f be given in the form (5.2). We want to know the coefficients of $P_{j+1}f = P_j f + Q_j f$ with respect to the original basis of V_{j+1}. From Theorem 5.1 it follows immediately:

Theorem 5.3. *The coefficients of $P_{j+1}f$ can be computed by*

$$\alpha_{j+1,k}(f) := \alpha_{j,k}(f) - \beta_{j,2^{j+1}-k}(f) \quad (k = 0, \ldots, 2^j - 1),$$
$$\alpha_{j+1,2^j}(f) := \alpha_{j,2^j}(f),$$
$$\alpha_{j+1,k}(f) := \beta_{j,k}(f) \quad (k = 2^j + 1, \ldots, 2^{j+1}).$$

Hence our decomposition and reconstruction algorithms require for the j–th level only 2^j additions, if the corresponding coefficients are given or precomputed by a fast DCT–algorithm (see Section 1).

Acknowledgements

The research of the second author was supported by the Deutsche Forschungsgemeinschaft.

References

1. Baszenski, G., and Delvos, F.-J., A discrete Fourier transform scheme for Boolean sums of trigonometric operators, in *Multivariate Approximation Theory* IV, C. K. Chui, W. Schempp and K. Zeller (eds.), Birkhäuser, Basel, 1989, 15–24.
2. Baszenski, G., and Tasche, M., Fast algorithms for simultaneous polynomial approximation, in *Multivariate Approximation and Wavelets*, K. Jetter and F. I. Utreras (eds.), World Scientific, Singapore, in print.

3. Donoho, D. L., Interpolating wavelet transforms, Preprint, Stanford University, Berkeley, 1992.
4. Hallatschek, K., Fouriertransformation auf dünnnen Gittern mit hierarchischen Basen, Numer. Math. **63** (1992), 83–97.
5. Havelkamp, R., Approximationsfehler der Ableitungen von Interpolationspolynomen, J. Approx. Th. **30** (1980), 180–196.
6. Havelkamp, R., Zur Konvergenz der Ableitungen von Interpolationspolynomen, Computing **32** (1984), 343–355.
7. Steidl, G., Fast radix-p discrete cosine transform, Appl. Algebra Engrg. Comm. Comput. **3** (1992), 39–46.
8. Tasche, M., Fast algorithms for discrete Chebyshev–Vandermonde transforms and applications, Numer. Algorithms **5** (1993), 453–464.
9. Yserentant, H., On the multi-level splitting of finite element spaces, Numer. Math. **49** (1986), 379–412.
10. Zenger, C., Sparse grids, in *Notes on Numerical Fluid Mechanics*, Bd. 31, W. Hackbusch (ed.), Vieweg, Braunschweig, 1991, 241–251.

Günter Baszenski
Fachbereich Nachrichtentechnik
Fachhochschule Dortmund
Postfach 10 50 18
D – 44047 Dortmund, GERMANY
baszenski@ ruba.rz.ruhr-uni-bochum.d400.de

Manfred Tasche
Fachbereich Mathematik
Universität Rostock
D – 18051 Rostock, GERMANY
tasche@ mathematik.uni-rostock.d400.de

Spline Curves and Surfaces with Tension

A. Bouhamidi and A. Le Méhauté

Abstract. Using a differential operator and the associated reproducing kernel, we construct a surface spline under tension as a particular case of an $L^{m,\ell,s}$–spline, a generalization of thin plate splines. We recall briefly some properties of these functions and show how it works on some examples. The same idea, restricting to the univariate case, gives a theoretical framework (Hilbert spaces and reproducing kernels) for spline curves under tension.

§1. $\mathbf{L^{m,\ell,s}}$–splines for Interpolation of Scattered Data

Given a real number s, let $\widetilde{H}^s(\mathbb{R}^d)$ be the space of tempered distributions u whose Fourier transform $\mathcal{F}u$ (or $\hat{u}$) satisfies $\int_{\mathbb{R}^d} |2\pi\xi|^{2s} |\mathcal{F}u(\xi)|^2 \, d\xi < +\infty$. For $s < \frac{d}{2}$, the space $\widetilde{H}^s(\mathbb{R}^d)$ endowed with the scalar product $(u,v)_s = \int_{\mathbb{R}^d} |2\pi\xi|^{2s} \mathcal{F}u(\xi) \overline{\mathcal{F}v(\xi)} \, d\xi$ is a Hilbert subspace of $\mathcal{S}'(\mathbb{R}^d)$ [4] .

Let $m \geq 1$ be an integer such that

$$-m + \frac{d}{2} < s < \frac{d}{2}. \tag{1.1}$$

For any positive integer $\ell \in \mathbb{N}$, let $D_\ell^{-m}(\widetilde{H}^s)$ be the space of distributions whose k–th derivatives, with $m \leq k \leq m + \ell$, are in $\widetilde{H}^s(\mathbb{R}^d)$, namely

$$D_\ell^{-m}(\widetilde{H}^s) = \left\{ v \in \mathcal{D}'(\mathbb{R}^d) : \ \forall \alpha \in \mathbb{N}^d, m \leq |\alpha| \leq m + \ell, D^\alpha v \in \widetilde{H}^s(\mathbb{R}^d) \right\}.$$

In this space, we consider the semi–inner product defined by

$$(u,v)_{m,\ell,s} = \sum_{m \leq |\alpha| \leq m+\ell} C_\alpha \int_{\mathbb{R}^d} |2\pi\xi|^{2s} \mathcal{F}D^\alpha u(\xi) \ \ \overline{\mathcal{F}D^\alpha v(\xi)} \, d\xi, \tag{1.2}$$

where the real positive constants C_α are subject to the following conditions

$$\begin{cases} \forall k \in \mathbb{N} \ ; \ m \leq k \leq m + \ell \ ; \ \exists \, C_k \, , C'_k > 0 \quad \text{such that :} \\ C_k \, |\xi|^{2k} \leq \Big| \sum_{|\alpha|=k} C_\alpha \xi^{2\alpha} \Big| \leq C'_k \, |\xi|^{2k} \quad \forall \xi \in \mathbb{R}^d \, . \end{cases} \tag{1.3}$$

Wavelets, Images and Surface Fitting
P. J. Laurent, A. Le Méhauté, and L. L. Schumaker (eds.), pp. 51–58.

ISBN 1-56881-040-7.

With $(.,.)_{m,\ell,s}$ and the associated semi–norm $|.|_{m,\ell,s}$, the space $D_\ell^{-m}(\widetilde{H}^s)$ is a semi–Hilbert space of tempered distributions which are in fact continuous functions, and separate points in $\mathbb{R}^d$. The assumption (1.1) implies that $-m-\frac{d}{2}<s<\frac{d}{2}$, which is a necessary condition for $\mathcal{S}(\mathbb{R}^d)$ to be algebraically enclosed in $D_\ell^{-m}(\widetilde{H}^s)$ (see [2,7]).

The null space of $D_\ell^{-m}(\widetilde{H}^s)$ is exactly the space $\mathbb{P}_{m-1}(\mathbb{R}^d)$, the space of all polynomials of degree $\leq m-1$ on $\mathbb{R}^d$. Let $d(m)$ denotes its dimension .

Now, let us define $L^{m,\ell,s}$–splines and for simplicity, let us assume that $L=(\ell_1,\cdots,\ell_p)$ $(p>d(m))$ is a collection of Dirac measures $\ell_k=\delta_{a_k}$ defined on a set $A=\{a_1,\cdots,a_p\}$ of scattered data points in $\mathbb{R}^d$ which contains a $\mathbb{P}_{m-1}$–unisolvent subset $\widehat{A}$ (for more general cases, see [2,7]) .

From Urysohn's lemma, there exists a family $(\varphi_j)_{1\leq j\leq p}$ of functions in $\mathcal{D}(\mathbb{R}^d)$ such that

$$\langle \ell_i,\varphi_j\rangle=\delta_{ij} \quad i,j=1,\cdots,p, \tag{1.4}$$

where δ_{ij} is the Kronecker symbol.

For any $z=(z_1,\cdots,z_p)\in\mathbb{C}^p$, let $G(z)=\{u\in D_\ell^{-m}(\widetilde{H}^s);\ u(a_i)=z_i\ ;\ i=1,\cdots,p\}$. Since $\mathcal{D}(\mathbb{R}^d)\subset\mathcal{S}(\mathbb{R}^d)\subset D_\ell^{-m}(\widetilde{H}^s)$, by (1.4), $G(z)$ is a nonempty subset of $D_\ell^{-m}(\widetilde{H}^s)$.

Definition 1. *Given $z\in\mathbb{C}^p$, a $L^{m,\ell,s}$–spline (or interpolating spline) is an element σ of $D_\ell^{-m}(\widetilde{H}^s)$ which is a solution of the following problem:*

$$(\mathcal{P})\begin{cases} \qquad\qquad \sigma\in G(z) \\ \text{and}\quad |\sigma|_{m,\ell,s}=\inf_{u\in G(z)}|u|_{m,\ell,s}. \end{cases}$$

Proposition 1. *[2,7] For any $z\in\mathbb{C}^p$, there exists one and only one $L^{m,\ell,s}$–spline σ interpolating z.*

Let r be a real number . Following Rabut [8] , we denote by Δ^r the iterated Laplacean operator of real order r, defined for any tempered distribution T on $\mathbb{R}^d$ as the tempered distribution whose Fourier transform is given by $\mathcal{F}(\Delta^r T)=e^{i\pi r}\,|2\,\pi\,\xi\,|^{2r}\,\mathcal{F}T$. Let us denote by $\Delta_{m,\ell,s}$ the linear differential operator defined for any tempered distribution T on $\mathbb{R}^d$ by

$$\Delta_{m,\ell,s}T=e^{-i\pi s}\,\Delta^s\big[\Delta_{m,\ell}T\big] \text{ where } \Delta_{m,\ell}T=\sum_{m\leq|\alpha|\leq m+\ell}(-1)^{|\alpha|}\,C_\alpha\,D^{2\alpha}\,T$$

and the constants C_α are defined in (1.2), (1.3).

From Malgrange (see[9]), there exists a fundamental solution $K_{m,\ell,s}$ of the differential operator $\Delta_{m,\ell,s}$, i.e., such that $\Delta_{m,\ell,s}K_{m,\ell,s}=\delta$ and we assume that $K_{m,\ell,s}$ is a tempered distribution associated to a function defined on $\mathbb{R}^d$. (In the particular case when s is a positive integer, it is not even necessary to assume that $K_{m,\ell,s}$ is a tempered distribution).

Let us consider the quotient space $\overset{\bullet}{D}_\ell{}^{-m}(\widetilde{H}^s) = D_\ell^{-m}(\widetilde{H}^s)/\mathbb{P}_{m-1}(\mathbb{R}^d)$. with the scalar product $[\overset{\bullet}{u}, \overset{\bullet}{v}]_{m,\ell,s} = (u,v)_{m,\ell,s}, \quad \forall\, \overset{\bullet}{u}, \overset{\bullet}{v} \in \overset{\bullet}{D}_\ell{}^{-m}(\widetilde{H}^s)$. The associated norm is denoted by $\| \overset{\bullet}{u} \|_{m,\ell,s}$.

It is obvious that $\overset{\bullet}{D}_\ell{}^{-m}(\widetilde{H}^s)$ is an Hilbert space, and by the closed graph theorem, we obtain that $\overset{\bullet}{D}_\ell{}^{-m}(\widetilde{H}^s)$ is continuously imbedded in $E_m = \mathcal{D}'(\mathbb{R}^d)/\mathbb{P}_{m-1}(\mathbb{R}^d)$. The antidual space $\overline{E'}_m = \mathcal{D}(\mathbb{R}^d)/\mathbb{P}_{m-1}(\mathbb{R}^d)$ of E_m is isomorphic to the space of compactly supported and $\mathcal{C}^\infty$ functions orthogonal to $\mathbb{P}_{m-1}(\mathbb{R}^d)$. We can identify these two spaces and write

$$\overline{E'}_m = \left\{ \varphi \in \mathcal{D}(\mathbb{R}^d) \mid \int_{\mathbb{R}^d} \xi^\alpha \varphi(\xi) d\xi = 0, \quad \forall \alpha \in \mathbb{N}^d;\ |\alpha| \le m-1 \right\}$$

From Schwartz [9] and the topological properties of $\overline{E'}_m$ and $\overset{\bullet}{D}_\ell{}^{-m}(\widetilde{H}^s)$, the Hilbert space $\overset{\bullet}{D}_\ell{}^{-m}(\widetilde{H}^s)$ has a reproducing kernel related to E_m which is the unique mapping H from $\overline{E'}_m$ into $\overset{\bullet}{D}_\ell{}^{-m}(\widetilde{H}^s)$ such that

$$[\overset{\bullet}{u}, H\varphi]_{m,\ell,s} = \langle u, \overline{\varphi} \rangle, \qquad \forall\, \overset{\bullet}{u} \in \overset{\bullet}{D}_\ell{}^{-m}(\widetilde{H}^s), \quad \forall \varphi \in \overline{E'}_m, \tag{1.5}$$

and we can prove the following characterization:

Proposition 2 . *The reproducing kernel of* $\overset{\bullet}{D}_\ell{}^{-m}(\widetilde{H}^s)$ *related to* E_m *is the mapping* H *defined from* $\overline{E'}_m$ *to* $\overset{\bullet}{D}_\ell{}^{-m}(\widetilde{H}^s)$ *by* $H : \varphi \longmapsto \varphi \star K_{m,\ell,s} + \mathbb{P}_{m-1}$.

Proof: The assumption (1.1) implies that for all φ element of $\overline{E'}_m$ the convolution product $\varphi \star K_{m,\ell,s}$ is in $D_\ell^{-m}(\widetilde{H}^s)$ [2] . Then, H is a mapping defined from $\overline{E'}_m$ to $\overset{\bullet}{D}_\ell{}^{-m}(\widetilde{H}^s)$, which satisfies (1.5) and thus is the reproducing kernel. ∎

Let $G_m = \mathcal{C}(\mathbb{R}^d)/\mathbb{P}_{m-1}(\mathbb{R}^d)$ (quotient space of $\mathcal{C}(\mathbb{R}^d)$ by $\mathbb{P}_{m-1}(\mathbb{R}^d)$). Its antidual space $\overline{G'}_m = \mathcal{C}'(\mathbb{R}^d)/\mathbb{P}_{m-1}(\mathbb{R}^d)$ is the quotient of the space of compactly supported Radon measures by the subspace of polynomials $\mathbb{P}_{m-1}(\mathbb{R}^d)$. G'_m is isomorphic to the space of compactly supported Radon measures orthogonal to $\mathbb{P}_{m-1}(\mathbb{R}^d)$. We can identify these spaces and write

$$\overline{G'}_m = \left\{ \mu \in \mathcal{C}'(\mathbb{R}^d) \mid \int_{\mathbb{R}^d} \xi^\alpha \, d\mu(\xi) = 0, \quad \forall \alpha \in \mathbb{N}^d;\ |\alpha| \le m-1 \right\}.$$

Hereafter, we assume that $K_{m,\ell,s} \in \mathcal{C}(\mathbb{R}^d)$. Then we get

Proposition 3. *The space* $\overset{\bullet}{D}_\ell{}^{-m}(\widetilde{H}^s)$ *is a Hilbert subspace of* G_m *whose reproducing kernel is the mapping* $\widehat{H}$ *defined from* $\overline{G'}_m$ *to* $\overset{\bullet}{D}_\ell{}^{-m}(\widetilde{H}^s)$ *by*

$$\widehat{H} : \mu \longmapsto \mu \star K_{m,\ell,s} + \mathbb{P}_{m-1}(\mathbb{R}^d), \qquad \forall \mu \in \overline{G'}_m$$

and the space $D_\ell^{-m}(\widetilde{H}^s)$ *is continuously imbedded in* $\mathcal{C}(\mathbb{R}^d)$.

Proof: The proof is a consequence of Corollary 3 in Schwartz [9] (page 179). We first prove that the kernel H of $\dot{D_\ell}^{-m}(\widetilde{H}^s)$ related to E_m can be extended to $\widehat{H}$ and then that $\widehat{H}$ is weakly continuous from $\overline{G'}_m$ to G_m. ■

The next proposition provides a characterization of the $L^{m,\ell,s}$–spline σ.

Proposition 4. *([2.7]) The $L^{m,\ell,s}$–spline σ has the following properties:*

i) $\overline{\Delta_{m,\ell,s}\,\sigma}$ is orthogonal to $\mathbb{P}_{m-1}(\mathbb{R}^d)$ and is a linear combination of the ℓ_i's i.e., $\overline{\Delta_{m,\ell,s}\,\sigma} = \sum_{i=1}^{p} \lambda_i\,\ell_i$ with $\lambda_i = (\varphi_i, \sigma)_{m,\ell,s}$.

ii) σ is the unique element of $\dot{\sigma} \bigcap G(z)$.

iii) The reproducing kernel $\widehat{H}$ of the Hilbert space $\dot{D_\ell}^{-m}(\widetilde{H}^s)$ relative to G_m maps $S = \sum_{i=1}^{p} \lambda_i\,\ell_i$ into $\dot{\sigma}$, i.e., $\dot{\sigma} = \widehat{H}\,S = S \star K_{m,\ell,s} + \mathbb{P}_{m-1}(\mathbb{R}^d)$.

It is now possible to define σ as an interpolating function by :

Corollary 1. *The $L^{m,\ell,s}$–spline σ solution of $(\mathcal{P})$ is explicitly given by*

$$\sigma = \sum_{i=1}^{p} \lambda_i\,K_{m,\ell,s}(\bullet - a_i) + \sum_{j=1}^{d(m)} \alpha_j \pi_j,$$

where $(\pi_j)_{j=1,\ldots,d(m)}$ is a basis of $\mathbb{P}_{m-1}(\mathbb{R}^d)$ and the real coefficients λ_i, $i = 1, \ldots, p$ and α_j, $j = 1, \ldots, d(m)$ are solutions of the linear system :

$$\begin{cases} \text{interpolation conditions :} \\ \qquad \begin{cases} \sigma(a_k) = z_k, \\ k = 1, \ldots, p. \end{cases} \\ \text{orthogonality conditions :} \\ \qquad \begin{cases} \sum_{i=1}^{p} \lambda_i\,\pi_j(a_i) = 0, \\ j = 1, \ldots, d(m). \end{cases} \end{cases} \qquad (S)$$

§2. Practical Examples

2.1. Spline curves under tension

In [10], Schweikert introduces the idea of spline curves under tension. Let $a = x_1 < x_2 < \ldots < x_p = b$ be a set of points in $[a,b]$. Given $r = (r_1,\ldots,r_p) \in \mathbb{R}^p$, the spline curve under tension σ_τ is the unique element in the space $\mathcal{H} = \{f : f' \text{ absolutely continuous on } [a,b] \text{ and } f'' \in L^2([a,b])\}$ with $\sigma_\tau(x_i) = r_i$, $i = 1,\ldots,p$, minimizing $\int_a^b [f''(x)]^2\,dx + \tau^2 \int_a^b [f'(x)]^2\,dx$ (where τ is the tension parameter). The spline σ_τ is defined locally on each subinterval. For $x \in [x_i, x_{i+1}]$, $i = 1,\ldots,p-1$, we have $\sigma_\tau(x) = a_i + b_i x + c_i sh(\tau x) + d_i ch(\tau x)$ and the coefficient a_i, b_i, c_i et d_i are defined by interpolation and continuity conditions.

Another way to consider splines under tension is to characterize σ_τ as a $L^{m,\ell,s}$ spline. Let the interpolation scheme be $L = (\delta_{x_1},\ldots,\delta_{x_p})$. As soon as we consider more than 2 distinct points x_i, L is $\mathbb{P}_0(\mathbb{R})$–unisolvent. Then, we choose $s = 0$ and $m = \ell = d = 1$, and the assumption (1.1) is fulfilled. Let us consider the space $D_0^{-1}(L^2(\mathbb{R})) = \{u \in \mathcal{D}'(\mathbb{R}); u', u" \in L^2(\mathbb{R})\}$ with the semi–scalar product $(u,v)_{1,1,0} = \int_{-\infty}^{+\infty} u"(x)\,v"(x)\,dx + \tau^2 \int_{-\infty}^{+\infty} u'(x)\,v'(x)\,dx$ associated to the differential operator $\Delta_{1,1,0}u = \Delta^2 u - \tau^2 \Delta u = \frac{d^4u}{dx^4} - \tau^2 \frac{d^2u}{dx^2}$. A fundamental solution [6] is the function $E_\tau(x)$ of class $\mathcal{C}^2(\mathbb{R})$, $E_\tau(x) = -\frac{1}{2\tau^3}(e^{-\tau|x|} + \tau|x|)$. The spline under tension now has the following expression:

$$\sigma_\tau(x) = \sum_{i=1}^{p} \lambda_i\, E_\tau(x - x_i) + C,$$

where the coefficients λ_i, $i = 1,\ldots,p$ and the constant C are solutions of

$$\begin{cases} \sum_{i=1}^{p} \lambda_i\, E_\tau(x_j - x_i) + C = r_j \quad ; \quad j = 1,\ldots,p \\ \sum_{i=1}^{p} \lambda_i = 0. \end{cases}$$

In order to illustrate the effect of the tension parameter in the univariate case, we present an example (see Fig. 1). The data are taken from Akima [1].

2.2. Surface splines under tension

In [5], Franke defined thin plate splines under tension as surfaces minimizing

$$\sum_{|\alpha|=2} \frac{2!}{\alpha!} \int_{\mathbb{R}^2} |D^\alpha u(\xi)|^2\, d\xi + \tau^2 \sum_{|\alpha|=1} \int_{\mathbb{R}^2} |D^\alpha u(\xi)|^2\, d\xi$$

with Lagrange interpolation contraints defined on a set of p scattered data points in $\mathbb{R}^2$ and τ as a tension parameter. With the notation of the previous

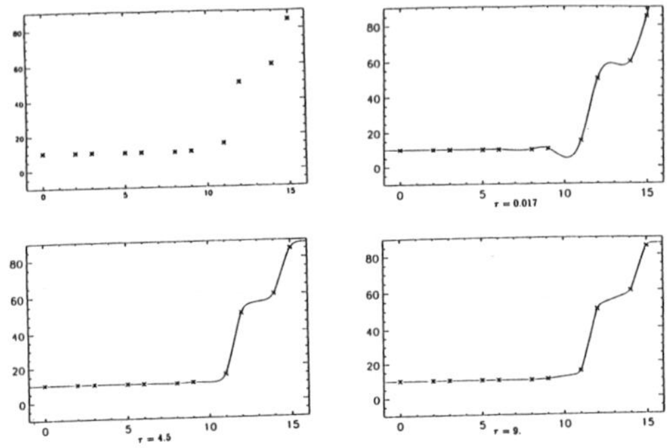

Fig. 1. Spline curves under various tension parameters.

section, we can characterize these thin plate splines under tension as a $L^{1,1,0}$–spline in $\mathbb{R}^2$, where L is a Lagrange interpolation scheme associated to a set of p distinct points of $\mathbb{R}^2$. They will be in the space

$$D_1^{-1}(L^2(\mathbb{R}^2)) = \left\{u \in \mathcal{D}'(\mathbb{R}^2),\ \forall\alpha \in \mathbb{N}^2\ |\alpha| = 1, 2;\ D^\alpha u \in L^2(\mathbb{R}^2)\right\},$$

with the semi–inner product

$$(u, v)_{1,1,0} = \sum_{|\alpha|=2} \frac{2!}{\alpha!} \int_{\mathbb{R}^2} D^\alpha u(\xi)\, D^\alpha v(\xi)\, d\xi + \tau^2 \sum_{|\alpha|=1} \int_{\mathbb{R}^2} D^\alpha u(\xi)\, D^\alpha v(\xi) d\xi.$$

The coefficients C_α are $C_\alpha = \frac{2!}{\alpha!}$ for $|\alpha| = 2, C_\alpha = \tau^2$ for $|\alpha| = 1$ and the associated differential operator is $\Delta_{1,1,0}u = \Delta^2 u - \tau^2 \Delta u$. The null space of $D_1^{-1}(L^2(\mathbb{R}^2))$ is $\mathbb{P}_0(\mathbb{R}^2)$, the space of constant polynomials.

Unfortunately, for such a representation, the assumption (1.1) is not fulfilled (with $m = 1$, $d = 2$ and $s = 0$), and the existence of the spline is not even guaranteed! Another trouble is of a numerical matter, since a fundamental solution of the differential operator $\Delta_{1,1,0}$ involves $K_0(\tau r)$, the first Bessel function of order zero , where $r = \sqrt{x^2 + y^2}$, and the Bessel function needs to be approximated $p \times p$ times in order to solve (S).

In order to enforce the theoretical requirement (1.1) and give an elementary representation of the basis function, we propose another way to consider surface spline under tension. We consider $\mathbb{R}^2$ as a subspace of $\mathbb{R}^3$ and consider splines under tension for $d = 3$. The behavior of the resulting surfaces illustrates the effect for various values of the tension parameter, see Fig. 2.

Let us consider the space

$$D_1^{-2}(L^2(\mathbb{R}^3)) = \left\{u \in \mathcal{D}'(\mathbb{R}^3), D^\alpha u \in L^2(\mathbb{R}^3);\ |\alpha| = 2, 3 \forall\, \alpha \in \mathbb{N}^3\right\}$$

with the semi–scalar product

$$(u, v)_{2,1,0} = \sum_{|\alpha|=3} \frac{3!}{\alpha!} \int_{\mathbb{R}^3} D^\alpha u(\xi) D^\alpha v(\xi)\, d\xi + \tau^2 \sum_{|\alpha|=2} \frac{2!}{\alpha!} \int_{\mathbb{R}^3} D^\alpha u(\xi) D^\alpha v(\xi) d\xi$$

and the associated semi-norm $|u|_{2,1,0}$. The null space of $D_1^{-2}(L^2(\mathbb{R}^3))$ is $\mathbb{P}_1(\mathbb{R}^3)$. We have now $m = 2$, $\ell = 1$, $s = 0$ and $d = 3$. The associated coefficients C_α are $C_\alpha = \frac{3!}{\alpha!}$ for $|\alpha| = 3$, $C_\alpha = \tau^2 \frac{2!}{\alpha!}$ for $|\alpha| = 2$.

In this case, the assumption (1.1) is satisfied, and in order to get an explicit representation of the interpolation function as a $L^{2,1,0}$–spline type we have to find a fundamental solution E_τ of the differential operator $\Delta_{2,1,0}\, u = -\Delta^3 u + \tau^2 \Delta^2 u$. Let $v = \Delta u$, such that $-\Delta^2 v + \tau^2 \Delta v = \delta$. A fundamental solution for v is given [6] by $v_\tau(r) = \frac{1}{4\pi r\,\tau^2}\,(e^{-\tau\pi} - 1)$. We need to integrate twice in order to have for u a fundamental solution of $\Delta_{2,1,0}$, which is

$$E_\tau(x,y,z) = \frac{1}{8\pi\tau^4 r}\left\{2e^{-\tau r} - (1-\tau r)^2 + C\right\},$$

where C is an arbitrary constant. In order to have a fundamental solution which can be extended continously in $\mathbb{R}^3$, we choose $C = -1$. Thus,

$$E_\tau(x,y,z) = \begin{cases} \dfrac{1}{8\pi\tau^4 r}\left\{2e^{-\tau r} - (1-\tau r)^2 - 1\right\} & \text{if} \quad r \neq 0 \\ 0 \quad \text{if} \quad (x,y,z) = (0,0,0). \end{cases}$$

For Lagrange interpolation at p scattered data points $A_i = (x_i, y_i, z_i)$, $i = 1, \ldots, p$ in $\mathbb{R}^3$, interpolating $V = (r_1, \ldots, r_p)$, the $L^{2,1,0}$–spline σ_τ is given by $\sigma_\tau(X) = \sum_{i=1}^p \lambda_i\, E_\tau(X - A_i) + \alpha_1\, x + \alpha_2\, y + \alpha_3\, z + \alpha_4$ where the coefficients λ_i, $i = 1, \ldots, p$ and α_1,α_2,α_3,α_4, are solutions of

$$\begin{cases} \begin{cases} \displaystyle\sum_{i=1}^p \lambda_i E_\tau(A_j - A_i) + \alpha_1 x_j + \alpha_2 y_j + \alpha_3 z_j + \alpha_4 = r_j \\ \qquad \text{with} \quad j = 1, \ldots, p \quad \text{and} \end{cases} \\ \displaystyle\sum_{i=1}^p \lambda_i x_i = 0, \quad \sum_{i=1}^p \lambda_i y_i = 0, \quad \sum_{i=1}^p \lambda_i z_i = 0, \quad \sum_{i=1}^p \lambda_i = 0. \end{cases}$$

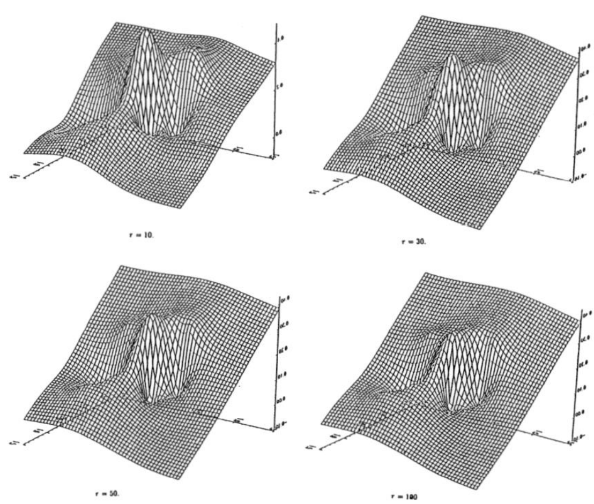

Fig. 2. Spline surfaces under various tension parameters.

Suppose we are given a set $A = \{a_i = (x_i, y_i) \in \mathbb{R}^2, i = 1, \ldots, p\}$ of p distinct points in the plane. Also, suppose we are given a set of real values $V = (r_1, \ldots, r_p)$, and we wish to use the $L^{2,1,0}$–spline interpolation method. We associate a third component z_i to each point a_i of A. In order to satisfy the $\mathbb{P}_1(\mathbb{R}^3)$–unisolvence of $\mathcal{A} = \{A_i = (a_i, z_i) \in \mathbb{R}^3, i = 1, \ldots, p\}$ we choose $z_i = 0$ except for one of them. The $L^{2,1,0}$–spline $\sigma(x, y, z)$ is computed by solving the linear system (S) and is plotted as a bivariate function $\sigma(x, y, 0)$.

The data for Figure 2 are taken from Franke [5], Table 3 (ROCKY MTN JRNL FNCTN), with 23 points. We chose all the $z_i = 0$ with the exception of $z_5 = 1$. The influence of the value and the choice of the non zero third component is under investigation. The four surfaces interpolate the same data . As one can see, this example illustrates the behavior of the surfaces for various values of the tension parameter.

References

1. Akima. H., A new method of interpolation and smooth curve fitting based on local procedures, J. Assoc. Comut. Mach **17** (1970), 589–602.
2. Bouhamidi. A., Interpolation et approximation par des fonctions splines radiales à plusieurs variables, Thèse, Université de Nantes, 1992.
3. Duchon. J., Interpolation des fonctions de deux variables suivant le principe de la flexion des plaques minces, R.A.I.R.O, Analyse numérique **10**, n°12 (1975), 5–12.
4. Duchon. J., Fonctions–spline à énergie invariante par rotation, Rapport de Recherche n°27, Grenoble, 1976.
5. Franke. R. Scattered data interpolation using the thin plate spline with tension. Report #NPS53-85-0005. Naval Postgraduate School, Monterey, CA, 1984.
6. Garnir. H. G., Sur les distributions résolvantes des opérateurs de la physique mathématique, Bull. Société Royale des Sciences de Liège, 1951, 174–195 and 271–296.
7. Le Méhauté. A, and Bouhamidi. A., $L^{m,\ell,s}$–splines in $\mathbb{R}^d$, Numerical Methods of Approximation Theory, Vol 9, ISNM 105, Birkhauser Verlag, 1992, 135–154.
8. Rabut. C., B–splines polyharmoniques cardinales : Interpolation, Quasi-interpolation, Filtrage. Thèse d'Etat, Toulouse, 1990.
9. Schwartz. L., Sous–espaces hilbertiens d'espaces vectoriels topologiques et noyaux associés (noyaux reproduisants), J. Anal. Math., Paris, 1964.
10. Schweikert. D. G., An interpolation curve using a spline in tension. J. Math. and Phys. **45** (1966), 312–317.

A. Bouhamidi
Département de Mathématiques Appliquées, Université du Littoral
B.P 699, 62228 Calais Cedex–France
A. Le Méhauté
ENST–Bretagne
B.P 832, 29285 Brest Cedex–France

Irregularity Detection from Noisy Data with Wavelets

M. Bozzini, F. De Tisi, and M. Rossini

Abstract. In this paper we present a method based on the wavelet transform to detect from noisy data the location and the order of the irregularities of unknown functions.

§1. Introduction

The problem of detection, classification, and measurement of irregularities of a function $f(t)$, $x = t \in [c, d]$, arises in many applications. Its importance is well-known in computer vision, computer graphics, geology, tomography, etc. For this reason many authors have studied this problem. Girard [3], Lee [5] and Mallat [4] tackled this problem from different points of view.

In this note we present a method to detect and in some cases to measure the function irregularities using a test based on the wavelet transform.

§2. Detection of Irregularities of Functions

It is well-known that the more suitable representation for a periodic phenomena $f(t)$ is the spectral one. When the frequencies vary in time, and high frequencies can be present in short duration phenomena, it is useful to consider the wavelet transform

$$W_f(a,b) = \frac{1}{a} \int_{-\infty}^{+\infty} f(t)\overline{\psi(\frac{t-b}{a})}dt \tag{1}$$

in order to get better information on the details. This is also the case for function irregularities. For this reason, we recall the following definition:

Wavelets, Images and Surface Fitting
P. J. Laurent, A. Le Méhauté, and L. L. Schumaker (eds.), pp. 59–66.

ISBN 1-56881-040-7.

Definition 1. *Given a positive integer n and a real number α such that $n < \alpha \le n+1$, a function $f(t)$ is Lipschitz-α in $t = t_0$ iff there exist two constants $A, h_0 > 0$ and a polynomial $P_n(t)$ of degree n that for all $h < h_0$,*

$$|f(t_0 + h) - P_n(h)| < Ah^\alpha.$$

Thus, if $f(t)$ is not Lipschitz-$(n+1)$ in t_0,

A) $f^{(n+1)}(t_0)$ does not exist or $f^{(n+1)}(t_0)$ is not bounded.

On the other hand $f(t)$ could be Lipschitz-$(n+1)$ in t_0, but

B) $f^{(n+1)}(t_0^+) \neq f^{(n+1)}(t_0^-)$.

In both of these cases, we say that $f(t)$ has an *irregularity point of order* $n+1$ at t_0. Consider a family of wavelets

$$\psi_{a,b}(t) = \psi(\frac{t-b}{a}), \qquad a, b \in \mathbb{R}, \quad a \neq 0,$$

such that the base wavelet satisfies the admissibility condition

$$C_\psi = \int_{-\infty}^{+\infty} \frac{|\hat{\psi}(\omega)|^2}{|\omega|} d\omega < \infty. \tag{2}$$

We have the following theorem:

Theorem 1. *Let $f(t) \in L^2[c,d]$ and $t_0 \in [c,d]$ be an irregularity point of type B, that is $f^{(i)}(t_0^+) = f^{(i)}(t_0^-)$, $\forall i \le n$, $f^{(n+1)}(t_0^+) \neq f^{(n+1)}(t_0^-)$, and let*

i) *$f(t)$ be identical to a polynomial $P(t)$ of degree $n+1$ in $[t_0 - L, t_0)$ and to a different polynomial $\tilde{P}(t)$ of degree $n+1$ in $(t_0, t_0 + L]$, $L > 0$.*

Consider a family of wavelets such that

ii) *$\psi(t)$ has compact support $[-A, A]$ with $A \le \frac{L}{2}$;*

iii) *$\psi(t)$ has at least $n+1$ null moments;*

iv) *$\psi(t)$ is the $n+2$ derivative of a function $\Psi(t)$ with compact support $[-A, A]$.*

Then

$$W_f(a, t_0) = (-1)^{n+2} a^{n+1} \overline{\Psi(0)} \big(f^{(n+1)}(t_0^+) - f^{(n+1)}(t_0^-)\big) \tag{3}$$

$$W_f(a, b + t_0) = \frac{W_f(a, t_0)\overline{\Psi(-\frac{b}{a})}}{\overline{\Psi(0)}} \quad \forall b \in [-aA, aA]. \tag{4}$$

Proof: Setting $y = \frac{t-b}{a}$, the wavelet transform (1) becomes

$$W_f(a,b) = \int_{-A}^{+A} f(b + ay)\overline{\psi(y)} dy.$$

Assume $b = t_0$. We can rewrite $W_f(a, t_0)$ as

$$\int_{-A}^{0} f(t_0 + ay)\overline{\psi(y)}dy + \int_{0}^{+A} f(t_0 + ay)\overline{\psi(y)}dy. \tag{5}$$

Considering $P(t)$ and $\tilde{P}(t)$ equal to the first $n+1$ terms of a Taylor series, (5) becomes

$$\begin{aligned} W_f(a, t_0) = &\sum_{i=0}^{n} \frac{f^{(i)}(t_0)}{i!} a^i \int_{-A}^{A} y^i \overline{\psi(y)}dy \\ &+ \frac{f^{(n+1)}(t_0^-)}{(n+1)!} a^{n+1} \int_{-A}^{0} y^{n+1}\overline{\psi(y)}dy \\ &+ \frac{f^{(n+1)}(t_0^+)}{(n+1)!} a^{n+1} \int_{0}^{A} y^{n+1}\overline{\psi(y)}dy. \end{aligned} \tag{6}$$

Since $\psi(t)$ and $\Psi(t)$ have compact support and $\psi(t)$ has $n+1$ null moments, from (6) we obtain

$$\begin{aligned} W_f(a, t_0) &= \frac{f^{(n+1)}(t_0^+) - f^{(n+1)}(t_0^-)}{(n+1)!} a^{n+1} \int_0^A y^{n+1}\overline{\psi(y)}dy \\ &= (-1)^{n+2} a^{n+1} \overline{\Psi(0)} \big(f^{(n+1)}(t_0^+) - f^{(n+1)}(t_0^-)\big). \end{aligned}$$

Now we consider $W_f(a, b + t_0)$. Making the substitution $y = \frac{t-b-t_0}{a}$ and proceeding as in the previous case, we get

$$\begin{aligned} W_f(a, b+t_0) =&(-1)^{n+2}\overline{\Psi(-\frac{b}{a})} a^{n+1} \big(f^{(n+1)}(t_0^+) - f^{(n+1)}(t_0^-)\big) \\ =&\frac{W_f(a, t_0)\overline{\Psi(-\frac{b}{a})}}{\overline{\Psi(0)}} \quad \forall b \in [-aA, aA]. \ \blacksquare \end{aligned}$$

Theorem 1 can be extended also to the cases where hypothesis i) doesn't hold, or t_0 is an irregularity point of type A. In fact, we have

Corollary 1. *Let $f \in L^2[c, d]$ and $t_0 \in [c, d]$ be such that $f^{(i)}(t_0^+) = f^{(i)}(t_0^-)$ for all $i \leq n$. Suppose the hypotheses ii) – iv) of Theorem 1 hold, and that 1) $f^{(n+1)}(t_0^+) \neq f^{(n+1)}(t_0^-)$ or 2) t_0 is an irregularity point of type A. Then for every $\epsilon > 0$,*

$$\begin{aligned} W_f(a, t_0) \approx& \big(f^{(n+1)}(t_0 + \epsilon) - f^{(n+1)}(t_0 - \epsilon)\big) a^{n+1} \overline{\Psi(0)} \\ W_f(a, b + t_0) \approx& \frac{W(a, t_0)\overline{\Psi(-\frac{b}{a})}}{\overline{\Psi}(0)}. \end{aligned}$$

Proof: The proof follows from the existence of $f^{(i)}(t_0)$ $i = 0, \ldots, n$, and from the arbitrariness of ϵ. ■

Remark 1. Theorem 1 gives informations on the irregularity measure. In fact, from (3) we get

$$|f^{(n+1)}(t_0^+) - f^{(n+1)}(t_0^-)| = \frac{|W_f(a, t_0)|}{a^{n+1}|\overline{\Psi(0)}|}.$$

§3. Description of the Method

3.1. Irregularity detection from noisy data

We consider a stochastic process

$$y(t) = f(t) + e(t)$$

and its wavelet transform

$$\begin{aligned} W_y(a,b) &= \int_{-\infty}^{+\infty} (f(t)+e(t))\overline{\psi(\frac{t-b}{a})}dt \\ &= W_f(a,b) + W_e(a,b). \end{aligned}$$

It is well-know that if ϕ is square integrable and $e(t)$ is ergodic

$$E_{t\in\mathbb{R}}(e*\phi) = E\{e\}\int_{\mathbb{R}}\phi(t)dt \tag{7}$$

$$E_{t\in\mathbb{R}}(\{e*\phi\}^2) = \int_{\mathbb{R}}|\hat{\phi}(\omega)|^2 S(\omega)d\omega. \tag{8}$$

If t_0 is an irregularity point of order k for the trend, from Theorem 1 and from (8) we get

$$\begin{aligned} E_{b\in[-aA,aA]}\Big(\{W_y(a,b+t_0) - \frac{W_f(a,t_0)\overline{\Psi(-\frac{b}{a})}}{\overline{\Psi(0)}}\}^2\Big) &= \\ &= \int_{\mathbb{R}}|\hat{\psi}(\omega)|^2 S(\omega)d\omega, \end{aligned} \tag{10}$$

Relation (10) gives us the variance of the wavelet transform of the process over time interval $[t_0 - aA, t_0 + aA]$.

On the other hand it is well-known that the irregularity point belongs to the set of wavelet modulus maxima, [4,2]. In this set there are also the irregularities due to the noise. Therefore it is necessary to minimize the spurious response.

To this end, we recall [1] the expression of the signal to noise ratio in t_0:

$$\rho = \frac{E(\{W_f(a,b_0)\}^2)}{E(\{W_e(a,b)\}^2)}, \tag{11}$$

and the expression of the expected number of extrema in the unit interval [1]:

$$\lambda = \frac{1}{\pi}\sqrt{(\frac{E(\{W_e''(a,b)\}^2)}{E(\{W'_e(a,b)\}^2)})}. \tag{12}$$

In our case, with simple calculations, (11), (12) become

$$\rho = \frac{\alpha_1 a^{2k+1}}{||\hat{\psi}(\omega)||_2^2} \tag{13}$$

$$\lambda = \frac{\alpha_2 ||\hat{\psi}''(\omega)||_2}{a||\hat{\psi}'(\omega)||_2}. \tag{14}$$

Therefore, when the scale a increases, the expected number of maxima due to the noise decreases, and in t_0 the signal prevails over the noise. So for suitable values of a, we can suppose

$$W_y(a, t_0) \approx W_f(a, t_0).$$

3.2. Description of the algorithm

Suppose we have a sample of dimension N:

$$y(t_i) = f(t_i) + e(t_i), \qquad i = 1, \ldots, N,$$

obtained by observing an ergodic normal process with expected value equal to zero and power spectra $S(\omega)$.

If the trend of the process has a discontinuity of order k in t_0, a suitable approximation of the process $y(x)$ can be obtained using perfect splines $g(x)$ of order k, and the corresponding wavelet transform $W_g(a, b)$ will approximate $W_y(a, b)$:

$$W_g(a, b) \approx W_f(a, b) + W_e(a, b)$$

Such approximations become better as N becomes large.

Once we have chosen a suitable values of a, we look for the points x_i $i = 1, \ldots, m$, where $|W_g(a, b)|$ is maximum, and $\forall x_i$ we compute the ratio

$$r(x_i) = \frac{E\{W_g(a, b + x_i) - W_g(a, x_i)\}^2}{E\{W_e\}^2}.$$

From (10) it follows that, if x_i is an irregularity point for $f(t)$, $r(x_i)$ has a value near one.

So it is possible to construct the following algorithm for the detection and, in some cases, the measure of the function irregularities up to a fixed order j. On the basis of the sample, the steps are:

For $k = 0, \ldots, j$

1) consider a base wavelet $\psi(t)$ such that $\psi(t) = \Psi^{(k+1)}(t)$ and compute $\int_{\mathbb{R}} |\hat{\psi}(\omega)|^2 S(\omega) d\omega$;
2) construct a perfect spline $g(t)$ of order k;
3) compute the $|W_g(a, b)|$ for a suitable value of a;
4) compute the modulus maxima abscissae x_i $i = 1, \ldots, m$;

5) for $i = 1, \ldots, m$ compute $r(x_i)$;

$$\text{if} \quad 0.8 < r(x_i) < 1.2, \tag{15}$$

x_i is an irregularity point for the function of order k.

Inequality (15) indicates that the estimated values of

$$E\{W_g(a, b + x_i) - W_g(a, x_i)\}^2$$

has the correct value of the true value $E\{W_e\}^2$.

The constants in (15) can be chosen differently depending on the sample dimension and on computational efficiency. For example, in some case it may be sufficient for $r(x_i) \in [0.1, 10]$, which furnishes the correct value of magnitude, as well as $r(x_i) \in [0.5, 2]$ which furnishes the correct size.

§4. Numerical Results

The algorithm described in (3.2) has been successfully tested on several functions, including

1) $f(t) = \sqrt{|t|}$

2) $f(t) = \begin{cases} \exp \frac{-1}{2}(t - 4.5)^2 & 0 \leq t < 5 \\ \exp \frac{-1}{2}(t - 5.5)^2 & 5 < t \leq 10 \end{cases}$

3) $f(t) = \dfrac{\sin(2\pi t) + |\sin(\frac{8}{3}\pi t)|}{12}$

4) $f(t) = \begin{cases} -t^2 + 0.5 & t < 0 \\ -2t^2 + 0.5 & t \geq 0 \end{cases}$

4) $f(t) = \sqrt[3]{x}$

5) $f(t) = \begin{cases} -0.5 & \text{if } t \leq -2 \\ -0.25 & -2 < t \leq 0 \\ 0.25 & 0 < t \leq 2 \\ 0 & t > 2 \end{cases}$

6) $f(t) = \log(1 + |t - 4.5|)$

We have considered samples of dimension $N = 129$ coming from different Gaussian processes whose data have been chosen

i) equispaced,

ii) scattered

iii) data with jitter, i.e.,

$$\tilde{x}_i = x_i + \epsilon_i, \quad \epsilon_i \in N(0, \sigma), \quad \sigma \in (0, \frac{1}{6}), \quad \epsilon_i \in (x_i - \frac{1}{2}, x_i + \frac{1}{2}).$$

We have used the wavelet

$$\psi(t) = (1 - t^2) exp(-\frac{t^2}{2}).$$

For sake of brevity we present only some of the results obtained in the case of white noise in order make a comparison with those obtained by Mallat in [4].

The figures show the modulus of wavelets transforms for different values of a. Figure 1 and Figure 2 refer to the function 5 which has three discontinuities of order 0 in $-2, 0, 2$ with size $0.25, 0.5, -0.25$, respectively. Figure 3 refers to function 2 which has a discontinuities of order 1 in $t = 5$ with size 0.882.

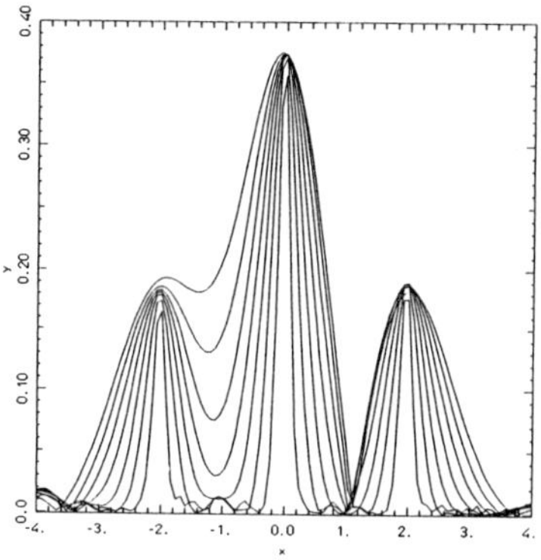

$\frac{\sigma}{f} \approx \frac{1}{50}$, $a = 0.5$, $E\{W_e\}^2 = 0.02531$		
x_i	$E\{\}^2$	*size*
-2	0.02355	0.24333
0	0.01296	0.49801
2	0.02262	-0.25192

Fig. 1. Moduli of the wavelet transforms for Function 5.

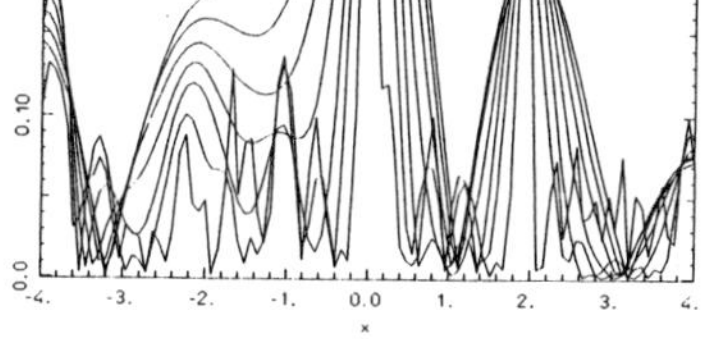

$\frac{\sigma}{f} \approx \frac{1}{5}$, $a = 0.5$ $E\{W_e\}^2 = 0.02388$		
x_i	$E\{\}^2$	*size*
-2	0.02543	0.19143
0	0.01313	0.49713
2	0.02326	-0.27865

Fig. 2. Moduli of the Wavelet Transforms for Function 5.

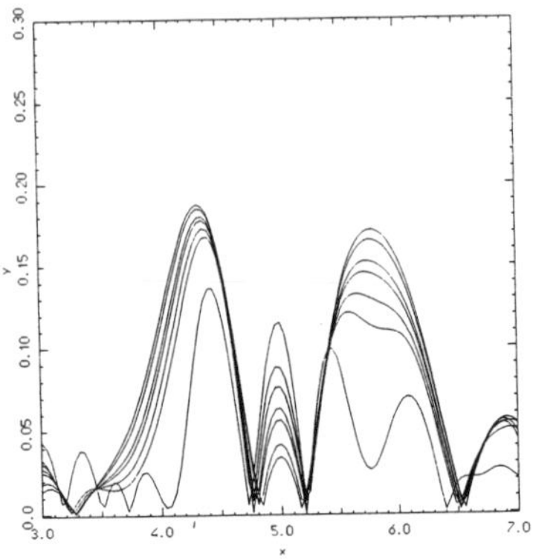

$\frac{\sigma}{f} \approx \frac{1}{50}$, $a = 0.3$ $E\{W_e\}^2 = 0.01$		
x_i	$E\{\}^2$	*size*
4.3	0.0038	-0.7189
5.0	0.0097	0.3159
5.6	0.0056	-0.5504
6.9	0.0042	0.2264

Fig. 3. Moduli of the Wavelet Transforms for Function 2.

References

1. Bendat J. S., *Principles and Applications of Random Noise Theory*, New York: Wiley, 1958.
2. Di Nunzio P., Applicazioni delle ondine all'analisi delle irregolarità delle funzioni, Thesis, 90/91.
3. Girard D., From Template matching to optimal approximation by piecewise smooth curves, in *Curves and Surfaces in Computer Vision and Graphics*, February 1990, Santa Clara, California, 174–182.
4. Mallat S. and Hwang W. L., Singularity Detection and Processing with Wavelets, IEEE Transactions on Information Theory, **38** (1992).
5. Lee D., Coping with Discontinuities in Computer Vision: Their Detection, Classification and Measurement, IEEE Transaction on Pattern Analysis and Machine Intelligence **12** (1990), 321–344.
6. Papoulis A., *Probability, Random Variables and Stochastic Processes*, New York, McGraw-Hill, 1984.

M. Bozzini, F. De Tisi, M. Rossini
Dipartimento di Matematica
via C. Saldini 50, 20133 Milano, ITALY
bozzini@vmimat.mat.unimi.it

Natural Neighbor Interpolation on the Sphere

J. L. Brown

Abstract. Natural neighbor coordinates have been used as building blocks for interpolation schemes by Sibson and by Farin and others. In this paper we will define coordinate systems for points scattered on the sphere which are analogous to natural neighbor coordinates in the plane. These coordinates provide an interpolant over the sphere which has local support and is C^1, except at the data sites, where it is C^0. As in the planar case, the local support property makes these coordinate functions good candidates for use in other interpolation schemes.

§1. Introduction

In 1980, Robin Sibson [2] defined natural neighbor coordinate systems associated with scattered points in $\mathbb{R}^m$. These coordinates have proved to be useful in a variety of interpolation and surface building schemes (see [2] and [1]). Given a set of points $\{\mathbf{p}_1, \mathbf{p}_2, ..., \mathbf{p}_n\}$ scattered in $\mathbb{R}^m$, Sibson defined coordinate functions $b_i(\mathbf{p})$ for points $\mathbf{p}$ in the convex hull of the data sites $\{\mathbf{p}_i\}$which satisfy the following identities:

$$b_i(\mathbf{p}) \geq 0 \tag{1}$$

$$b_i(\mathbf{p}_j) = \delta_{ij} \tag{2}$$

$$1 = b_1(\mathbf{p}) + \cdots + b_n(\mathbf{p}) \tag{3}$$

$$\mathbf{p} = b_1(\mathbf{p})\mathbf{p}_1 + \cdots + b_n(\mathbf{p})\mathbf{p}_n. \tag{4}$$

The coordinate functions b_i are C^1, except at the data sites, where they are C^0, and they have *local support.*

Wavelets, Images and Surface Fitting
P. J. Laurent, A. Le Méhauté, and L. L. Schumaker (eds.), pp. 67–74.

ISBN 1-56881-040-7.

We will define, in an analogous manner, coordinate systems associated with points scattered on the sphere and prove that the coordinate functions are C^1 except at the data sites, where they are C^0. We will also introduce a natural way to smooth these coordinate functions, making them globally C^1.

Our coordinate functions will satisfy (1)–(3), but there is no possibility of satisfying (4) over the sphere. Consider the case where all the data sites are in the first octant. A convex combination of the data sites will be a point in the first octant of $\mathbb{R}^3$, so an arbitrary point $\mathbf{p}$ cannot be recovered as such.

However, interpolation schemes based on natural neighbor coordinates do not depend on (4), so we will be able to define a C^1 interpolant with local support over S^2 based on our coordinate functions.

§2. Definitions and Results

We are given a set of points $\{\mathbf{x}_1, ..., \mathbf{x}_n\}$ scattered on the sphere S^2.

Definition 1. *For $i = 1, ..., n$ define $A_i \subseteq S^2$ by*

$$A_i = \{\mathbf{z} \in S^2 \ : \ d(\mathbf{z}, \mathbf{x}_i) \leq d(\mathbf{z}, \mathbf{x}_j) \text{ for } j = 1, \ldots, n\}.$$

The sets A_i are spherical polygons with geodesic edges, and they comprise a tesselation of the sphere which is analogous to the Dirichlet tesselation of the plane. The tile A_i consists of the points on the sphere which are at least as close to $\mathbf{x}_i$ as they are to any of the other data sites.

Let $\mathbf{x}$ be any point on the sphere. Insert $\mathbf{x}$ into the tesselation, and find it's tile. Let T_i be the area of the portion of $\mathbf{x}$'s tile which lies in the original tile of $\mathbf{x}_i$. See Figure 1.

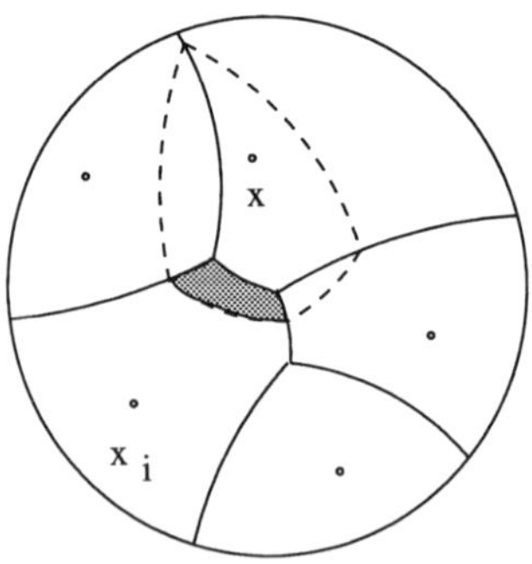

Fig. 1. The area of the shaded region is T_1.

Definition 2. *For $i = 1, ..., n$ the coordinate of $\mathbf{x}$ with respect to $\mathbf{x}_i$ is defined to be*

$$b_i(\mathbf{x}) = \frac{T_i}{T_1 + \cdots + T_n}.$$

Discussion: Note on implementation. Just as in the case of $\mathbb{R}^n$, there is a triangulation (Delaunay) of the sphere which is dual to the tesselation described above. The triangulation is defined as follows: two data sites are connected by an edge if and only if their tiles share a facet. An insertion algorithm analogous to the one commonly used to compute Delaunay triangulations in the plane is an efficient way to compute the spherical triangulation. The vertices of the tile corresponding to a data site are the circumcenters of the triangles containing the site. ■

Theorem 3. *The coordinate functions b_i are C^1 everywhere on S^2 except at the data sites $\mathbf{x}_j$.*

Proof: Let $\mathbf{x}$ be a point on S^2, and let T_i be the area of the portion of $\mathbf{x}$'s tile which lies in $\mathbf{x}_i$'s tile, as above. By a slight abuse of notation, we also let T_i denote the polygon whose area is T_i.

Choose i, $1 \le i \le n$. We will show that T_i is a C^1 function of $\mathbf{x}$, provided $\mathbf{x}$ is not one of the data sites. Since b_i is a rational function of the T_j's, this will complete the proof of the theorem.

The data site $\mathbf{x}_i$ is a vertex of several triangles in the Delaunay triangulation. The circumcircles of these triangles will be called the Dirichlet circles associated with $\mathbf{x}_i$. Each of the Dirichlet circles bounds two disks on the sphere, and the one whose center is the circumcenter of one of the triangles will be called a Dirichlet disk.

As the point $\mathbf{x}$ varies on the sphere, the number of sides of the polygon T_i varies. The points at which the number of sides changes are exactly the points on the Dirichlet circles associated with $\mathbf{x}_i$, and the support of T_i is the union of the Dirichlet disks.

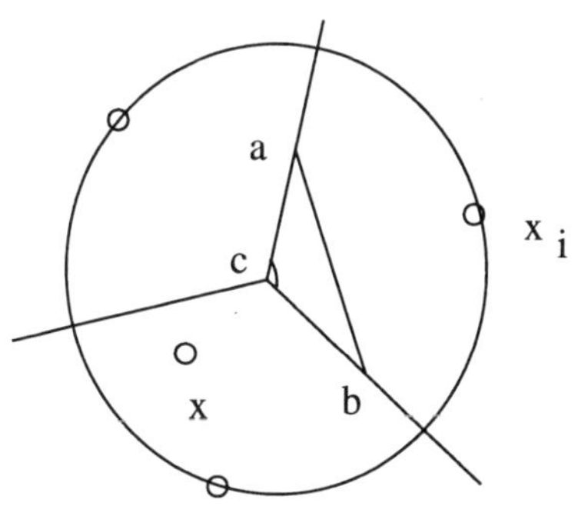

Fig. 2. One of the Dirichlet circles of the data site x_i.

Let D be a portion of a Dirichlet disk of $\mathbf{x}_i$ which is not in any of $\mathbf{x}_i$'s other Dirichlet disks, and let $\mathbf{x} \in \mathbf{D}$. Then the region T_i is a triangle. Let a, b, c be the angles of T_i, so

$$T_i = a + b + c - \pi. \tag{5}$$

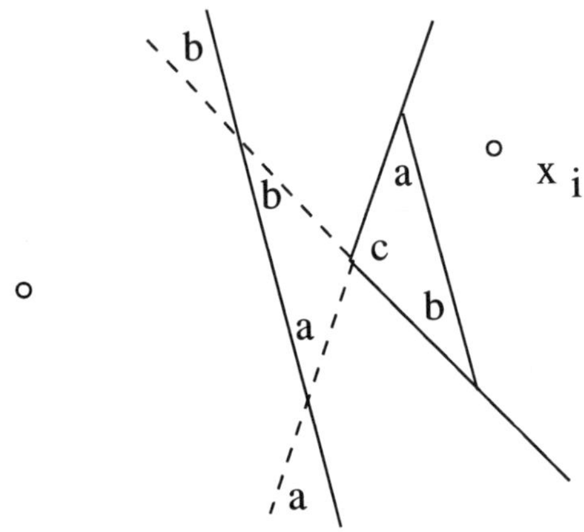

Fig. 3. $a+b+c-\pi$ is positive when x is outside the Dirichlet disks.

The angle c is fixed (see Figure 2) and the angles a and b are found by a calculation of the form *arccos(dot product)*. Hence, as long as the *dot product* $\neq 1$ ($a \neq 0$, $b \neq 0$), the angle functions are C^∞. The only way that a or b could be zero is for $\mathbf{x}$ to be one of the data sites. Therefore, at all points of D other than the data sites on the boundary, T_i is C^∞.

Outside the Dirichlet disks $T_i = 0$, so it suffices to show that on the boundary of D, all the directional derivatives of T_i vanish. Let $\mathbf{x}$ be a point on the boundary of D, other than a data site, and choose a direction in which to take the derivative of T_i (with respect to arc length).

Notice that $a+b+c-\pi$ is defined (and smooth) at points outside the Dirichlet disks. Furthermore, the value of $a+b+c-\pi$ is *positive* at outside points (still the area of a triangle), while $T_i = 0$ (see Figure 3). This means that along the chosen direction line (geodesic), $a+b+c-\pi$ has a relative minimum at $\mathbf{x}$, so the directional derivative of T_i is 0, as desired.

Now let $\mathbf{x}$ be a point such that T_i has more than three sides. It is clear that T_i is C^1 (in fact C^∞) at points where the number of sides is not changing, i.e. away from the Dirichlet circles.

Let T_i be a quadrilateral, and extend two edges to form a triangular region S, as in Figure 4. Our previous argument shows that the triangles S and $T_i \bigcup S$ are smooth at the point where S degenerates, so $T_i = (T_i + S) - S$ is also smooth. This argument may be applied to any region T_i. ■

Let $\mathbf{x}$ be a data site. Consider the *one sided* directional derivatives of T_i at $\mathbf{x}$. These exist, but they are *positive*, so the directional derivative does not exist. There is a sharp point on the graph of b_i over the data site $\mathbf{x}_i$, and there are also singularities at some of the other data sites.

Notice that the functions T_i^p for $p > 1$ do have directional derivatives at the data sites, because the one-sided directional derivatives have a factor of T_i^{p-1} which is zero at data sites other than $\mathbf{x}_i$.

Therefore we can build a C^1 interpolant by using the following coordinate functions. Choose a number $p \geq 1$, and define the coordinate $s_i^{(p)}$ of $\mathbf{x}$ with respect to $\mathbf{x}_i$ by

$$s_i^{(p)} = \frac{T_i^p}{T_1^p + \cdots + T_n^p}.$$

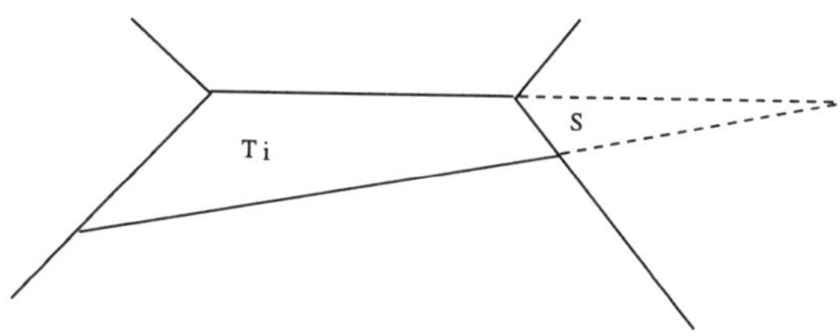

Fig. 4. $T_i \bigcup S$ is a triangle.

Theorem 4. *If* $p > 1$, *then* $s_i^{(p)}$ *is* C^1 *over the entire sphere.*

Note that the same technique can be used to smooth planar natural neighbor coordinate functions, but then the barycentric property (4) will no longer hold.

§3. Examples

In this section we show examples of the interpolants defined above. We first used 30 points scattered at random in the first octant. All but three of the function values were set to 0.1 to illustrate the local support property. Then we calculated the interpolant based on s_i at about 3200 points in the first octant. The graph of the result is shown in Figure 5.

Figure 6 shows the result of using the same data set as was used in Figure 5, with the interpolant based on $s_i^{(p)}$ with $p = 1.8$.

Next we used a set of 46 points spread uniformly over the sphere, with all but three function values set to 0.1. Figure 7 shows the result of interpolation at about 8000 points using the coordinate functions s_i. Figure 8 shows the result of using the coordinate functions $s_i^{(2)}$.

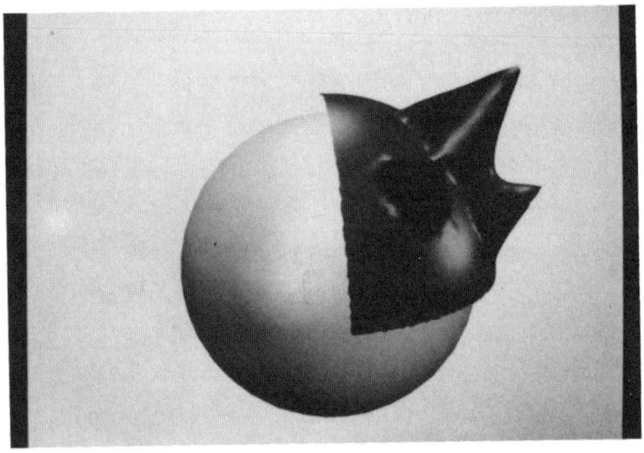

Fig. 5. Interpolant based on s_i.

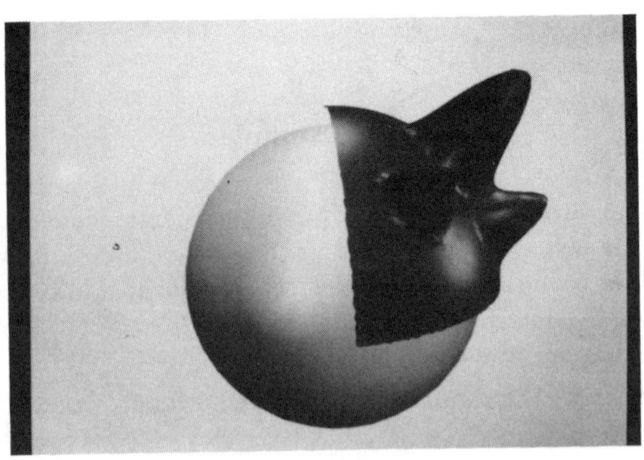

Fig. 6. Interpolant based on $s_i^{(p)}$, with $p = 1.8$.

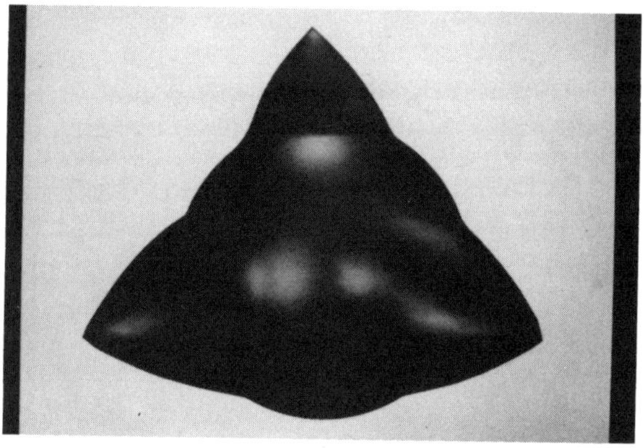

Fig. 7. Interpolant based on s_i.

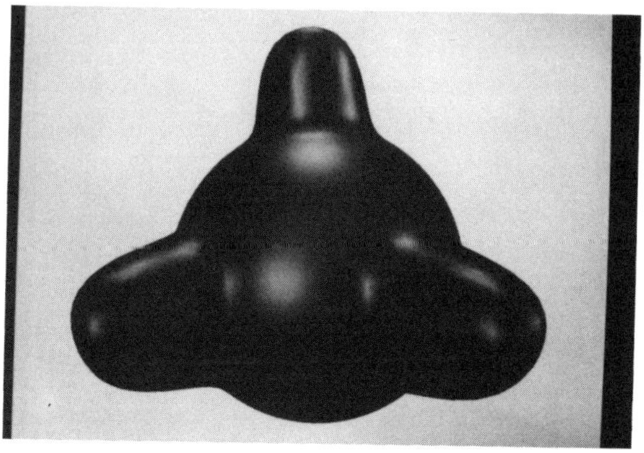

Fig. 8. Interpolant based on $s_i^{(p)}$ with $p = 2.0$.

References

1. Farin, G., Surface over Dirichlet tesselations, Comp. Aided Geom. Design **7** (1990), 281–292.
2. Sibson, R., A vector identify for the Dirichlet tesselation, Math. Proc. Camb. Phil. Soc. **87** (1980), 151–155.

Jeffrey L. Brown
Mathematical Sciences Department
University of North Carolina at Wilmington
601 South College Road
Wilmington, North Carolina 28403
brownj@ seq.cms.uncwil.edu

Error Estimates for Periodic Interpolation by Translates

Gisela Brumme

Abstract. In this paper we describe an error analysis for periodic interpolation by translates in the $L_2(\mathbb{T}_N^d)$–norm. For this purpose we introduce Strang–Fix conditions for the periodic case. Finally, an application to the method of attenuation factors for Fourier coefficients is presented.

§0. Introduction

Periodic interpolation by translates and the method of attenuation factors have been treated by Gautschi [3], Locher [6] and Delvos [2] in the univariate case, while Gutknecht [4] and ter Morsche [7] investigated the multivariate case. Only ter Morsche has given any error estimates, but his arguments are based on periodization of a non–periodic function.

In Section 2 we present an error estimate in the $L_2(\mathbb{T}_N^d)$–norm which works for periodic as well as for periodized functions. Our goal is the introduction of Strang–Fix conditions for the periodic case and the investigation of the Fourier coefficients of the interpolant. An application of periodic interpolation by translates to the method of attenuation factors for Fourier coefficients is the topic of Section 3. Moreover, we give a new version of ter Morsche's result which does not depend on periodization. In our paper we use some new ideas of Jetter [5] applying Fourier techniques in a unified approach to cardinal interpolation by translates.

§1. Interpolation of Periodic Data

First we recall some basic notations for d–variate N–periodic functions ($N \in \mathbb{N}$). Let B_N denote the following set of multi–indices

$$B_N := \{\alpha \in \mathbb{Z}^d : -N/2 \leq \alpha_i < N/2;\ i = 1, \ldots, d\}$$

Wavelets, Images and Surface Fitting
P. J. Laurent, A. Le Méhauté, and L. L. Schumaker (eds.), pp. 75–82.

ISBN 1-56881-040-7.

and let $\mathbb{T}_a^d$ be the d–dimensional torus represented by the cube

$$\mathbb{T}_a^d := \{x \in \mathbb{R}^d : |x_i| \le a/2;\ i = 1, \ldots, d\},$$

whose opposite sides are identified. The norm of a function f in $L_2(\mathbb{T}_N^d)$ is given by $\|f\|_2^2 := N^{-d} \int_{\mathbb{T}_N^d} |f(x)|^2\, dx$.
The *Fourier coefficients* $c_k(f)$ for a function f in $L_1(\mathbb{T}_N^d)$ are defined by

$$c_k(f) := \frac{1}{N^d} \int_{\mathbb{T}_N^d} f(x) e^{-i\frac{2\pi}{N}xk}\, dx, \qquad k \in \mathbb{Z}^d.$$

The *discrete Fourier coefficients* $c_k^N(f)$ are given by

$$c_k^N(f) := \frac{1}{N^d} \sum_{l \in B_N} f(l) e^{-i\frac{2\pi}{N}lk}, \qquad k \in B_N.$$

Obviously, this expression can be obtained by discretization of $c_k(f)$.

The fundamental relation between $c_k(f)$ and $c_k^N(f)$ for a function f in $C(\mathbb{T}_N^d)$ possessing an absolutely convergent Fourier series is described by the *aliasing formula*

$$c_k^N(f) = \sum_{l \in \mathbb{Z}^d} c_{k+Nl}(f) \quad , \ k \in B_N.$$

Now, we turn to the problem of periodic interpolation by translates. Let $\varphi \in C(\mathbb{T}_N^d)$ be an N–periodic "basis function" and let $\{c_k(\varphi)\}_{k \in \mathbb{Z}^d}$ belong to $l_1(\mathbb{Z}^d)$. We first want to construct the corresponding *N–periodic fundamental interpolant* Λ^N, *i.e.*,

$$\Lambda^N(x) := \sum_{\beta \in B_N} \lambda_\beta \varphi(x - \beta), \qquad \lambda_\beta \in \mathbb{R},$$

where Λ^N interpolates the N–periodic Kronecker–data $\delta_{0\alpha}^N$ $(\alpha \in \mathbb{Z}^d)$

$$\Lambda^N(\alpha) = \delta_{0\alpha}^N := \begin{cases} 1 & \text{for } \alpha \equiv 0 \pmod N, \\ 0 & \text{for } \alpha \not\equiv 0 \pmod N. \end{cases}$$

Here we understand the relation $\alpha \equiv \beta \pmod N$ for $\alpha,\ \beta \in \mathbb{Z}^d$ as $\alpha_i \equiv \beta_i \pmod N$ $(i = 1, \ldots, d)$. Inserting the interpolation conditions and computing the discrete Fourier coefficients with $\lambda := \{\lambda_\alpha\}_{\alpha \in B_N}$ leads to $N^{-d} = N^d c_k^N(\lambda) c_k^N(\varphi)$, $(k \in B_N)$. Furthermore, the Fourier coefficients of Λ^N are $c_k(\Lambda^N) = N^d c_k^N(\lambda) c_k(\varphi)$, $(k \in \mathbb{Z}^d)$. Combining both results yields

$$c_k(\Lambda^N) = \frac{1}{N^d} \frac{c_k(\varphi)}{c_k^N(\varphi)} = \frac{1}{N^d} \frac{c_k(\varphi)}{\sum_{l \in \mathbb{Z}^d} c_{k+Nl}(\varphi)}, \qquad k \in \mathbb{Z}^d.$$

From this expression it is obvious that the condition $c_k^N(\varphi) \neq 0$ $(k \in B_N)$ is necessary and sufficient for the existence and uniqueness of Λ^N (see Locher [6], Delvos [2] for $d = 1$ and Gutknecht [4], ter Morsche [7] for $d \ge 1$).

For the interpolation of arbitrary N–periodic data we consider a given function $f \in C(\mathbb{T}_N^d)$. The N–periodic interpolant $\mathcal{I}(f)$ with the property $\mathcal{I}(f)(\alpha) = f(\alpha)$, $(\alpha \in B_N)$ can be constructed by setting

$$\mathcal{I}(f)(x) := \sum_{\beta \in B_N} f(\beta)\Lambda^N(x-\beta) \, .$$

Thus, we obtain for the corresponding Fourier coefficients

$$c_k(\mathcal{I}(f)) = N^d c_k^N(f) c_k(\Lambda^N) = c_k^N(f)\frac{c_k(\varphi)}{c_k^N(\varphi)} \, , \qquad k \in \mathbb{Z}^d.$$

For a 1–periodic function $g \in C(\mathbb{T}_1^d)$ we find the 1–periodic interpolant $\mathcal{I}^h(g)$ by scaling

$$\mathcal{I}^h(g)(x) := \sum_{\beta \in B_N} g(h\beta)\Lambda^N(x/h - \beta), \qquad h = 1/N$$

with $\mathcal{I}^h(g)(h\alpha) = g(h\alpha)$, $(\alpha \in B_N)$. Note that $c_k(\mathcal{I}(f)) = c_k(\mathcal{I}^h(g))$ for $k \in \mathbb{Z}^d$, if $f(\alpha) = g(h\alpha)$ for $\alpha \in B_N$.

§2. Main Results

Our aim is an error estimate of the form

$$\|f - \mathcal{I}(f)\|_2 \le const |f|_{m,2,\mathbb{T}_N^d},$$

with $|f|_{m,2,\mathbb{T}_N^d}$ denoting the *Sobolev seminorm of order* m for $f \in W_2^m(\mathbb{T}_N^d)$ defined by

$$|f|^2_{m,2,\mathbb{T}_N^d} := \sum_{k \in \mathbb{Z}^d} |2\pi k/N|_2^{2m} \, |c_k(f)|^2,$$

$|\cdot|_2$ being the usual euclidean norm in $\mathbb{R}^d$. Such an estimate immediately implies the approximation order h^m for scaled interpolants in the following sense

$$\|f - \mathcal{I}^h(f)\|_{L_2(\mathbb{T}_1^d)} \le const \, h^m |f|_{m,2,\mathbb{T}_1^d},$$

where $|f|^2_{m,2,\mathbb{T}_1^d} := \sum_{k \in \mathbb{Z}^d} |2\pi k|_2^{2m} |c_k(f)|^2$ for $f \in W_2^m(\mathbb{T}_1^d)$.

Let $S_n(f)$ denote the truncated Fourier sum of $f \in L_2(\mathbb{T}_N^d)$ given by

$$S_n(f)(x) := \sum_{k \in B_n} c_k(f) e^{i\frac{2\pi}{N}kx},$$

and let T_N be the set of trigonometric polynomials of the form

$$T_N := span\{e^{i\frac{2\pi}{N}k\cdot}; \; k \in B_N\}.$$

Now, we estimate the interpolation error in more detail as

$$\|f-\mathcal{I}(f)\|_2 \le \|f-S_N(f)\|_2+\|\mathcal{I}(f-S_N(f))\|_2+\|S_N(f)-\mathcal{I}(S_N(f))\|_2. \quad (*)$$

Clearly, the smoothness of the original function f mainly influences the first and second expression on the right hand side. We are mostly interested in the third term which depends on the approximation properties of our "basis function" φ.

In order to describe approximation properties of φ, we introduce a periodic version of the strong Strang–Fix conditions (Jetter [5]).

We say that $\varphi \in C(\mathbb{T}_N^d)$ with $c_k^N(\varphi) \neq 0$ for $k \in B_N$ satisfies the *periodic Strang–Fix conditions (SFC)* if for some $m \in \mathbb{R}_+$ and all $k \in B_N$ the inequalities

$$\begin{array}{lll} (i) & \left|1-\frac{c_k(\varphi)}{c_k^N(\varphi)}\right| \le c_0 \left|\frac{2\pi k}{N}\right|_2^m, & \\ (ii) & \left|\frac{c_{k+N\alpha}(\varphi)}{c_k^N(\varphi)}\right| \le c_\alpha \left|\frac{2\pi k}{N}\right|_2^m, & \alpha \in \mathbb{Z}^d\backslash\{0\} \end{array} \qquad (SFC)$$

hold for some sequence of positive constants $\{c_\alpha\}_{\alpha\in\mathbb{Z}^d} \in l_2(\mathbb{Z}^d)$.

Examples

1. The simplest example of an N–periodic "basis function" would be the choice of $\varphi \in T_N$. If we put $\varphi(x) := \sum_{\beta\in B_N} t_\beta e^{i\frac{2\pi}{N}\beta x}$, $(t_\beta \in \mathbb{R})$, we immediately get for $t_\beta \neq 0$ $(\beta \in B_N)$

$$\frac{c_k(\varphi)}{c_k^N(\varphi)} = \begin{cases} 1 & \text{for } k \in B_N, \\ 0 & \text{for } k \notin B_N . \end{cases}$$

Inserting this result in the periodic Strang–Fix conditions, the right hand sides of the inequalities vanish. Thus, a function $\varphi \in T_N$ satisfies the periodic Strang–Fix conditions for arbitrary $m \in \mathbb{R}_+$.

2. Another important type of "basis functions" for periodic interpolation by translates are periodized functions. We start with a non–periodic function $\psi \in C(\mathbb{R}^d)$ with $\{\psi(\alpha)\}_{\alpha\in\mathbb{Z}^d} \in l_1(\mathbb{Z}^d)$ and obtain an N–periodic function by the usual periodization $\varphi(x) := \sum_{\alpha\in\mathbb{Z}^d} \psi(x+N\alpha)$. Furthermore, we denote by $\hat{\psi} := \int_{\mathbb{R}^d} \psi(x)e^{-ix\cdot}\,dx$ the Fourier transform of ψ and by $\tilde{\psi}$ the series $\tilde{\psi} := \sum_{\alpha\in\mathbb{Z}^d} \psi(\alpha)e^{-i\alpha\cdot}$. Then we obtain

$$\frac{c_k(\varphi)}{c_k^N(\varphi)} = \frac{\hat{\psi}(2\pi k/N)}{\tilde{\psi}(2\pi k/N)}, \qquad k \in \mathbb{Z}^d.$$

Using this expression, we establish the periodic version of the strong Strang–Fix conditions Jetter has given for the cardinal case [5]. It follows that the number m for the periodized function φ is the same as for ψ, if ψ satisfies the strong Strang–Fix conditions of order m for the cardinal case.

The periodic Strang–Fix conditions are very useful for the proof of the following result. It concerns the leading part of the interpolation error influenced by the "basis function" φ.

Theorem 1. *If $f \in T_N$ and if $\varphi \in C(\mathbb{T}_N^d)$ satisfies (SFC) and $c_k^N(\varphi) \neq 0$ for $k \in B_N$, then it holds that*

$$\|f - \mathcal{I}(f)\|_2 \leq \gamma_1 |f|_{m,2,\mathbb{T}_N^d},$$

where $\gamma_1^2 = \sum_{\alpha \in \mathbb{Z}^d} c_\alpha^2$ with $\{c_\alpha\}_{\alpha \in \mathbb{Z}^d}$ as in (SFC) .

Proof: Applying Parseval's identity and the aliasing formula, we get

$$\begin{aligned}
\|f - \mathcal{I}(f)\|_2 &= \sum_{k \in \mathbb{Z}^d} |c_k(f) - c_k(\mathcal{I}(f))|^2 \\
&= \sum_{k \in \mathbb{Z}^d} \left| \left(1 - \frac{c_k(\varphi)}{c_k^N(\varphi)}\right) c_k(f) - \sum_{l \in \mathbb{Z}^d \setminus \{0\}} c_{k+Nl}(f) \frac{c_k(\varphi)}{c_k^N(\varphi)} \right|^2 \\
&= \sum_{k \in B_N} \left| 1 - \frac{c_k(\varphi)}{c_k^N(\varphi)} \right|^2 |c_k(f)|^2 + \sum_{k \in B_N} \sum_{l \in \mathbb{Z}^d \setminus \{0\}} \left| \frac{c_{k+Nl}(\varphi)}{c_k^N(\varphi)} \right|^2 |c_k(f)|^2.
\end{aligned}$$

The last equality holds by the orthogonality of the summands. An estimation by (SFC) proves the theorem. ■

Remark. Theorem 1 gives the result for the periodic case which is analogous to the cardinal case for band–limited functions [5]. Furthermore, Theorem 1 and the estimates of the other two terms of $(*)$ show that it is sufficient to consider the interpolation of trigonometric polynomials in order to describe the interpolation error depending on the approximation properties of the "basis function" φ.

We want to continue our considerations by an estimation of $\|f - S_N(f)\|_2$ in terms of $|f|_{m,2,\mathbb{T}_N^d}$. A similar result was given by de Boor, DeVore and Ron [1] for the cardinal case.

Lemma 1. *If $f \in W_2^m(\mathbb{T}_N^d)$, then*

$$\|f - S_N(f)\|_2 \leq \pi^{-m} |f|_{m,2,\mathbb{T}_N^d}.$$

Proof: By standard arguments we get

$$\begin{aligned}
\|f - S_N(f)\|_2^2 &= \sum_{k \in \mathbb{Z}^d \setminus B_N} |c_k(f)|^2 = \sum_{k \in \mathbb{Z}^d \setminus B_N} \frac{|2\pi k/N|_2^{2m} |c_k(f)|^2}{|2\pi k/N|_2^{2m}} \\
&\leq \frac{1}{\pi^{2m}} \sum_{k \in \mathbb{Z}^d \setminus B_N} |2\pi k/N|_2^{2m} |c_k(f)|^2 \\
&= \frac{1}{\pi^{2m}} \frac{\sum_{k \in \mathbb{Z}^d \setminus B_N} |2\pi k/N|_2^{2m} |c_k(f)|^2}{\sum_{k \in \mathbb{Z}^d} |2\pi k/N|_2^{2m} |c_k(f)|^2} |f|_{m,2,\mathbb{T}_N^d}^2. \qquad ■
\end{aligned}$$

To deal with the second term on the right hand side of $(*)$, we consider the following theorem. We introduce Φ as the *N–periodic autocorrelation function* of φ given by

$$\Phi(x) := \frac{1}{N^d} \int_{\mathbb{T}_N^d} \varphi(t) \varphi(x+t)\, dt,$$

with Fourier coefficients $c_k(\Phi) = |c_k(\varphi)|^2$, $(k \in \mathbb{Z}^d)$.

Theorem 2. *Let $f \in C(\mathbb{T}_N^d) \cap W_2^m(\mathbb{T}_N^d)$, and let $\varphi \in C(\mathbb{T}_N^d)$ satisfy $c_k^N(\varphi) \neq 0$ for $k \in B_N$ and $\{c_k(\Phi)\}_{k\in\mathbb{Z}^d} \in l_1(\mathbb{Z}^d)$. Then with $m > d/2$,*

$$\|\mathcal{I}(f - S_N(f))\|_2 \leq \pi^{-m}\gamma_2\gamma_3|f|_{m,2,\mathbb{T}_N^d},$$

where the constants are given by $\gamma_2^2 = \max\limits_{k\in B_N} c_k^N(\Phi)/|c_k^N(\varphi)|^2$ and

$\gamma_3^2 = \max\limits_{k\in B_N} \sum\limits_{l\in\mathbb{Z}^d\setminus\{0\}} |2(k/N + l)|_2^{-2m} < \infty.$

Proof: With Parseval's identity and the aliasing formula for Φ we establish

$$\begin{aligned}\|\mathcal{I}(f - S_N(f))\|_2^2 &= \sum_{k\in\mathbb{Z}^d}\left|c_k^N(f - S_N(f))\frac{c_k(\varphi)}{c_k^N(\varphi)}\right|^2\\ &= \sum_{k\in B_N}\sum_{l\in\mathbb{Z}^d}|c_k^N(f - S_N(f))|^2\frac{|c_{k+Nl}(\varphi)|^2}{|c_k^N(\varphi)|^2} = \sum_{k\in B_N}|c_k^N(f - S_N(f))|^2\frac{c_k^N(\Phi)}{|c_k^N(\varphi)|^2}\\ &\leq \max_{k\in B_N}\frac{c_k^N(\Phi)}{|c_k^N(\varphi)|^2}\sum_{k\in B_N}|c_k^N(f - S_N(f))|^2.\end{aligned}$$

Applying the Cauchy–Schwarz inequality and using the same argument as in the proof of Lemma 1 yields for the second factor

$$\begin{aligned}&\sum_{k\in B_N}|c_k^N(f - S_N(f))|^2\\ &\leq \sum_{k\in B_N}\left(\sum_{l\in\mathbb{Z}^d\setminus\{0\}}\left|\frac{2\pi}{N}(k + Nl)\right|_2^{-2m}\right)\left(\sum_{l\in\mathbb{Z}^d\setminus\{0\}}\left|\frac{2\pi}{N}(k + Nl)\right|_2^{2m}|c_{k+Nl}(f)|^2\right)\\ &\leq \max_{k\in B_N}\sum_{l\in\mathbb{Z}^d\setminus\{0\}}\frac{1}{|2(k/N + l)|_2^{2m}}\frac{\sum_{k\in\mathbb{Z}^d\setminus B_N}|2\pi k/N|_2^{2m}|c_k(f)|^2}{\pi^{2m}\sum_{k\in\mathbb{Z}^d}|2\pi k/N|_2^{2m}|c_k(f)|^2}|f|_{m,2,\mathbb{T}_N^d}^2\end{aligned}$$

Since the first factor converges for $m > d/2$, the proof is complete. ■

Combining the previous theorems and Lemma 1 we obtain

Theorem 3. *Let $f \in C(\mathbb{T}_N^d) \cap W_2^m(\mathbb{T}_N^d)$. Further let $\varphi \in C(\mathbb{T}_N^d)$ satisfy (SFC), $c_k^N(\varphi) \neq 0$ for $k \in B_N$ and $\{c_k(\Phi)\}_{k\in\mathbb{Z}^d} \in l_1(\mathbb{Z}^d)$. Then the following estimate holds for $m > d/2$*

$$\|f - \mathcal{I}(f)\|_2 \leq const|f|_{m,2,\mathbb{T}_N^d},$$

where $const = \pi^{-m}(1 + \gamma_2\gamma_3) + \gamma_1$ with γ_1, γ_2 and γ_3 from Theorem 1 and Theorem 2, respectively.

§3. Application to the Method of Attenuation Factors

The method of attenuation factors for Fourier coefficients as considered by Gautschi [3], Locher [6], Gutknecht [4] and ter Morsche [7] can be briefly described as follows: we look for approximate values of the Fourier coefficients $c_k(f)$, where f is a given 1–periodic function. A first idea is the discretization of the integral which leads to the discrete Fourier coefficients. Clearly, the discrete Fourier coefficients $c_k^N(f)$ are N–periodic, while the Fourier coefficients $c_k(f)$ tend to zero for $|k| \to \infty$. In order to simulate the decreasing behaviour of the Fourier coefficients, the discrete Fourier coefficients are multiplied by so–called attenuation factors.

These *attenuation factors* are generated by a given "basis function" φ in $C(\mathbb{T}_N^d)$ and have the form

$$\frac{c_k(\varphi)}{c_k^N(\varphi)}, \qquad k \in \mathbb{Z}^d.$$

The new approximation of the Fourier coefficients $c_k(f)$ is

$$c_k^N(f)\frac{c_k(\varphi)}{c_k^N(\varphi)} = c_k(\mathcal{I}^h(f)), \qquad h = 1/N,\ k \in \mathbb{Z}^d.$$

In other words, the method of attenuation factors describes nothing else but the Fourier coefficients of the interpolant $\mathcal{I}^h(f)$ generated by φ.

Theorem 3 immediately gives the error for this method in the $\|\cdot\|_{l_2(\mathbb{Z}^d)}$–norm via Parseval's identity. More precisely, we define the $\|\cdot\|_{l_2(\mathbb{Z}^d)}$–norm for some sequence $\mathbf{d} := \{d_k\}_{k\in\mathbb{Z}^d} \in l_2(\mathbb{Z}^d)$ by $\|\mathbf{d}\|^2_{l_2(\mathbb{Z}^d)} := \sum_{k\in\mathbb{Z}^d} |d_k|^2$.

Now let $\mathbf{c}(f) := \{c_k(f)\}_{k\in\mathbb{Z}^d}$ be the sequence of the Fourier coefficients of f. Then we can state

Theorem 4. *Let $f \in C(\mathbb{T}_1^d) \cap W_2^m(\mathbb{T}_1^d)$. Further let $\varphi \in C(\mathbb{T}_1^d)$ satisfy (SFC), $c_k^N(\varphi) \neq 0$ for $k \in B_N$ and $\{c_k(\Phi)\}_{k\in\mathbb{Z}^d} \in l_1(\mathbb{Z}^d)$. Then for $m > d/2$,*

$$\|\mathbf{c}(f - \mathcal{I}^h(f))\|_{l_2(\mathbb{Z}^d)} \leq const\ h^m |f|_{2,m,\mathbb{T}_1^d},$$

and in particular,

$$|c_k(f) - c_k(\mathcal{I}^h(f))| \leq const\ h^m |f|_{2,m,\mathbb{T}_1^d}, \qquad k \in \mathbb{Z}^d$$

with the same constant as in Theorem 3.

Proof: The first proposition is obvious from Theorem 3. The second inequality can be established by using the Cauchy–Schwarz inequality and Theorem 3. ■

Theorem 4 improves the result of ter Morsche [7], because our estimate works for periodic as well as for periodized functions.

References

1. de Boor, C., DeVore, R. A. and A. Ron, Approximation from shift-invariant subspaces of $L_2(\mathbb{R}^d)$, CMS Techn. Report #92–2, University of Wisconsin, Madison (1991).
2. Delvos, F.–J., Periodic interpolation on uniform meshes, J. Approx. Th. **51** (1987), 71–80.
3. Gautschi, W., Attenuation factors in practical Fourier analysis, Numer. Math. **18** (1972), 373–400.
4. Gutknecht, M. H., Attenuation factors in multivariate approximation, Numer. Math. **51** (1987), 615–629.
5. Jetter, K., Multivariate approximation from the cardinal interpolation point of view, in *Approximation Theory VII*, E. W. Cheney, C. K. Chui, and L. L. Schumaker, eds., Academic Press, New York, 1992, 131–161.
6. Locher, F., Interpolation on uniform meshes by translates of one function and related attenuation factors, Math. Comp. **37** (1981), 404–416.
7. ter Morsche, H. G., Attenuation factors and multivariate periodic spline interpolation, in *Topics in Multivariate Approximation*, C. K. Chui, L. L. Schumaker, and F. I. Utreras, eds., Academic Press, New York, 1987, 165–174.

Gisela Brumme
Fachbereich Mathematik
Universität Rostock
D-18051 Rostock, GERMANY
brumme@ mathematik.uni-rostock.d400.de

An Energy-based Paradigm for Multimodal Images Fusion

L. Brunie and P. Cinquin

Abstract. This paper deals with the problem of the estimation, from 2D images, of the pose of a rigid 3D object of which a representation of the surface is known. Based on a mechanical model, a new iterative approach is proposed which solves that problem when the contours of the object in the images are known. The general case of non-segmented images is studied in the second part of the paper. Based on a similarity function integrating data provided by various procedures of image analysis and by the knowledge of the context, various strategies are analysed.

§0. Introduction

One of the most basic abilities of any human or artificial intelligence is the inference of knowledge by matching various pieces of information [20]. Thus, in computer vision, the analysis and the interpretation of a scene can rarely be performed using a single source of information. In order to remove remaining uncertainties and ambiguities, additional information is required. Two main approaches can be used: either reconstruct the scene from a large set of data [11,19]; or, when only a few data are available, introduce *a priori* knowledge to compensate for the lack of information and match it with the data. Such is the frame of model-based computer vision.

Model-based vision allows *a priori* knowledge of the shape and appearance of the objects of the scene described in the image(s). These shape informations are commonly called the *"models"* of the objects. Powerful inferences about the scene can generally be made by matching, putting into relation such 3D models with the information from thc image(s).

A good example of this procedure is the fusion of scanner pre-operative 3D images and X-ray intra-operative 2D images in order to estimate, during the intervention, the pose (i.e., the location and orientation) of target anatomical or tumoral structures.

Wavelets, Images and Surface Fitting
P. J. Laurent, A. Le Méhauté, and L. L. Schumaker (eds.), pp. 83–91.

ISBN 1-56881-040-7.

We begin the paper with a review of related work in the field of 3D/2D matching (Section 1). We then present in Section 2 the new formulation of the problem we propose. Section 3 displays some experimental results, with both synthetic data and real data. Section 4 deals with the problem of non-segmented images.

§1. Related Work

Techniques for recovering pose from 2D images (the 3D/2D matching problem) can be classified into three general categories, based on the matching primitives used. The first class of techniques is based on matching features identified on the surface of the object [7]. The correspondence between features can be computed using either direct or coarse-to-fine procedures. These approaches resemble stereo matching algorithms and are particularly well-suited to a graph-theoretic formulation [17,15,12]. Reducing the space of the possible poses [9] and using contextual knowledge [4] can be both be used to speed up the matching.

A second, more global, approach estimates the pose by aligning the extremal contours of the object with the available projections. For objects modeled as a set of polygonal surface patches, Lowe [16] describes an iterative approach for aligning the projected extremal contours with edges found in the image. His approach is similar to ours in that it uses non-linear least squares. It differs in measuring errors in the image plane and assuming a polyhedral model (which may not be suitable for complex anatomical surfaces).

For smooth surfaces, inferring the contour generators (i.e., the curves that belong to the surface and that form the image contours) can be quite difficult [14]. A traditional approach is to restrict the shape of the object, e.g., to objects of revolution [10] or generalized cones [5]. Ponce and Kriegman [14] have used rational parametric surface patches, implicit algebraic equations, and elimination theory to obtain analytic expressions for the projected contours. However, their method is restricted at present to objects modeled by a few patches.

By comparison, our approach can deal with objects of arbitrary shape and complexity, since it does not need to explicitly compute the projected contours. In fact, our approach is quite close to classical chamfer matching [3] which maximizes the correlation between 2D features in the image and the projections of 3D object features. However, our distance measure is computed in 3D, which, as we will see, removes the need to identify 3D object features.

A third approach to 3D/2D matching uses a topological analysis of the object based on catastrophe theory [1,13] and differential geometry [8]. Unfortunately, no one has yet been able to develop an algorithm using these concepts. Recent analysis by Ponce of aspect graphs for smooth objects further demonstrates the high complexity of this approach.

In that context, the method described below attempts to solve a still largely open problem : the pose estimation of a free-form smooth object.

§2. An Energy-based Paradigm for Multimodal Images Fusion

Mechanical approach

Let Ref_{3D} be the reference system associated with the surface of the object and Ref_{2D} the reference system of the 2D projections. The problem is to estimate the transformation $\mathbf{T}$ between Ref_{2D} and Ref_{3D}. $\mathbf{T}$ is called the *final pose* of the object and can be defined by a translation vector $\mathbf{t}$ and a 3×3 rotation matrix $\mathbf{R}$ represented by its Euler angles.

Given N_P projection contours of an object S, we first select a set of M_P pixels $\{\mathcal{P}_i, i = 1 \ldots M_P\}$, that belong to the N_P contours. This means that we do not need to perfectly extract the projection contours of S. Instead, only some pixels $\mathcal{P}_i$ that lie on the contours with a high certainty need to be known.

Each pixel $\mathcal{P}_i$ corresponds to a 3D *matching line* L_i joining $\mathcal{P}_i$ to the focal point of the projection (Fig. 1).

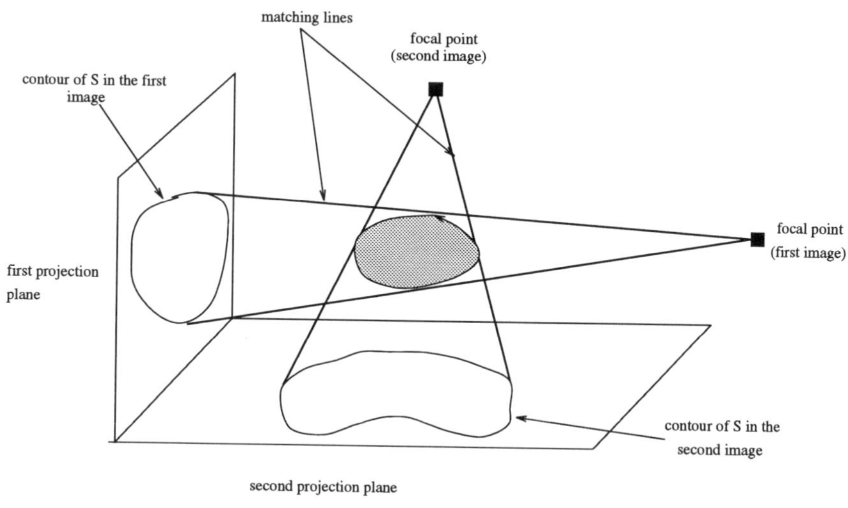

Fig. 1. 3D scene: in the final pose, all the matching lines are tangent to the surface.

In the final pose $\mathbf{T}$, every matching line is tangent to the surface S of the object (Fig. 1). We can reasonably suppose that for complex objects, in a large neighborhood of an initial estimate $\mathbf{T}^{(0)}$, $\mathbf{T}$ is the only 3D pose of S leading to such a geometric configuration (this will later be confirmed in our experiments). Based on this important assumption, we design an algorithm which moves the surface S towards the matching lines (or equivalently, moves the rigid set of matching lines towards the surface) until they are all tangent to S.

An intuitive idea of how to perform this task is to imagine a mechanical device in which springs (of null minimal length and of elasticity k_i), sliding along each matching line L_i), is attached between the line L_i and the point of S the closest to it (remark: if L_i is piercing the surface S, the spring will be, in fact, attached from the point of L_i interior of S which is furthest from S). Clearly the surface will globally be pulled towards the matching lines, i.e., the distance from S to the set of lines will decrease. So, the object is placed in a force field in which a spring-type force is applied on the object from every matching line. Classically, the mechanical motion in the 3D space of such an isolated system is defined by Newtonian dynamics. During this motion, the potential energy of the system is continuously minimized until the system reaches a local minimum where the resultant forces cancel each other. At any instant, the energy $E(\mathbf{p})$ depends on the current pose $\mathbf{p}$ of S and is defined by

$$E(\mathbf{p}) = \sum_{i=1}^{M_p} \frac{k_i}{2}[d(L_i(p), S)]^2, \tag{1}$$

where $d(L_i(p), S)$ is the Euclidean distance from L_i to the surface of S (i.e., the length of the i'th spring) when the object is in attitude p .

The main drawback of using mechanical modeling is that the system may take a long time to converge to its minimum energy configuration. That is why we propose to implement a dual approach, based on a least squares modelling of the above problem.

Least squares modelling

To formulate the least squares problem, we need a function which estimates the distance from any matching line to the surface of S. We first define the 3D unsigned distance between a point r and the surface S, $d_E(r)$, as the minimum Euclidean distance between r and all the points of S. If we define the distance between a line $L_i(p)$ and the surface S as the minimum of $d_E(r)$ over all the points that belong to $L_i(p)$, then the distance associated with a line piercing the surface would be zero (i.e. the "best" possible distance) even though the line would not be tangent to S. To avoid this problem, we define the signed distance, $d(r)$, between a point r and a closed surface S as $d_E(r)$ if r is exterior to S, and as the negative of this distance if r is interior to S. We then define the signed distance between a line $Li(p)$ and the surface S, $dl(L_i(p))$, as the minimum of $d(r)$ over all the points that belong to $L_i(p)$.

Therefore, while the minimum unsigned distance along a line piercing the surface may be zero, the signed distance will be negative. Thus, whether the line $L_i(p)$ passes through the surface or not, since we look for the minimum value of the signed distance along $L_i(p)$, the absolute value of $dl(L_i(p))$ indicates whether the line $L_i(p)$ is close and tangent to S When the final pose is reached, every line $L_i(p)$ is tangent to S, and every signed distance $dl(L_i(p))$ is therefore zero. Thus, we have to deal with a mere problem of data modelling: estimate the parameters $\mathbf{p}$ such that $dl(L_i(p))$ is zero for any line L_i.

A criterion of goodness of fit minimized in a least squares model, $E(\mathbf{p})$, is

$$E(\mathbf{p}) = \sum_{i=1}^{M_P} \frac{1}{\sigma_i^2}[dl(L_i(p), S)]^2, \tag{2}$$

where σ_i^2 is the variance of the noise of the measurement $dl(L_i(p))$ [18], which is homomorphic to the potential energy of the system of springs (eq. 1).

Our problem of pose estimation can therefore be modelled as either a mechanical problem or a least squares problem. Experimentally the paths followed by S in both estimations are very close to each other. For robustness and efficiency reasons, we will use a least squares method. Complementarily, the mechanical model gives a much more intuitive idea of the estimation process which can be very valuable to understand the matching process.

We implemented the least squares method using the Levenberg-Marquardt algorithm [18]. It should be noticed that first derivatives of $E(\mathbf{p})$ have an analytical expression (the computation is very simple (see [6] for full details)).

In the experiments we ran, local minima have always been very few and very far from the final pose (convergence domain: $\pm 20°$ on each rotation axis; no limitation in tranlation).

§3. Experimental Results

The validity of the model described above has been tested on both simulated and real anatomical surfaces. Tests have shown that the object transformation converges towards the final pose when the potential energy converges towards the global minimum [6] which justifies the use of the potential energy as goodness of fit.

In practise, the algorithm allows to estimate the final pose with an accuracy of 1° in rotation and 1mm in translation as soon as the initial pose is less than 20° (on each of the three rotation axes) far from the final pose.

Figures 2 and 3 display an example of convergence of the algorithm. Average estimation time on a DEC workstation 5000 is 1.5 seconds.

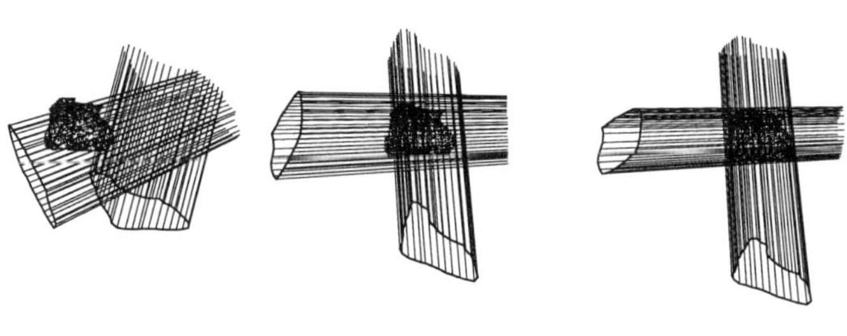

Fig. 2. Convergence of the algorithm for an anatomic structure (brain grey nucleous), observed from a general point of view. At the end of the process, all the matching lines are tangent to the surface. Right: start of the process; center: after 2 iterations; left: after 6 iterations (end).

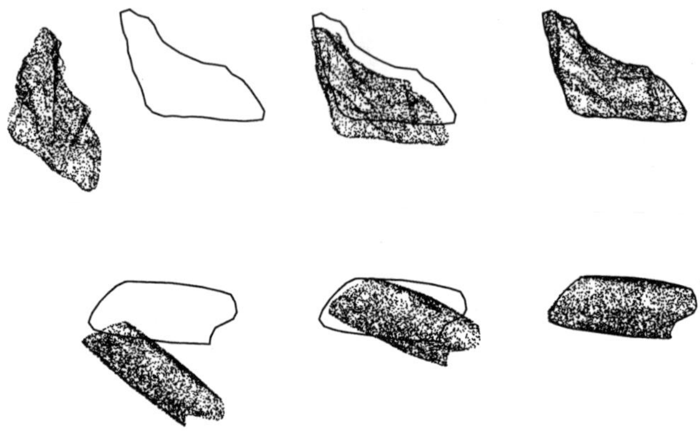

Fig. 3. same experiment, observed from the two projection axes.
Right: start of the process; center: after 2 iterations;
left: after 6 iterations (end).

§4. From Energy to Credibility: Dealing with the Non-segmented Images Problem

In medical informatics, X-rays are the most commonly used 2D imaging device. However X-rays are not mere 2D projections . A pixel is not the projection of a 3D point: it integrates the whole density information along the X-ray line joining the X source to the the image plane. This generally implies very serious segmentation difficulties. To our knowledge, no general efficient technique has been proposed to segment X-rays.

Similar difficulties of segmentation appear also in classical video vision, especially when the scene includes reflectant surfaces.

Notion of credibility

Though segmentation techniques often fail to extract the contour of the biomedical structures visible in an X-ray, nevertheless they give some useable information. Let us consider for example a gradient image of an X-ray. Due to the superposition, high gradient pixels can be found inside some structures and not only on the contours. In the same way, contour pixels may show low gradient levels if other structures project over the contour.

Nevertheless, low gradient based (with extra contour pixels) or high gradient based (with many missing contour pixels) segmentations define some kind of potentiality, of credibility for a pixel to belong to the contour. Basically, high gradient pixels are more likely to belong to a contour.

Similar statements can be made for any segmentation technique (region-based, snakes, Canny-Deriche gradient, etc.). Performing several of these methods and analysing their results allows us to define many *credibility fields* defining for one pixel its chances to be a contour element *according to each of these techniques.* At last, merging these fields together leads to the definition of a *global credibility field.*

Similarly, *a priori* knowledge of the context can also be very useful. Regions of important superposition (or reflectance), leading to low gradient contour pixels, are very often known. In the same way, in biomedicine, structures projecting inside the contour of a bigger structure (thus providing false high gradient pixels) are not secret. Global credibilities can therefore be weighted using such knowledge in order to avoid misclassifying of ambiguous pixels. This knowledge may be either dynamic or static:

1) (static) If the weighting is directly made on the X-rays, without reference to the model. This only consists in creating a new context-based credibility field which can be merged with the others.
2) (dynamic) If the weighting is made with respect to the model and its current attitude. For example, 3D structures which project without ambiguities are *a priori* known; so it can be decided to weight by an interesting factor the pixels close to the projection of such structures.

Toward the notion of attitude credibility

Based on the previous notion of credibility of a pixel (or more precisely of a projection line), various strategies may be imagined to estimate the final dynamic attitude of the object. First, this notion reminds us of the weighting of data in the least-squares algorithms (Section 2). A first idea for an algorithm consists in selecting the potentially interesting projection lines (by thresholding the credibility) and then running the previous energetic procedure using these lines as matching features and their credibilities as weights on their contribution.

In contrast to the "normal" case, some of these projection lines do not correspond to "contour lines" i.e. are not based on contour pixels. So in the dynamic attitude, they are not tangent to the surface. Consequently, the dynamic attitude is not an attitude of null energy. However, if these "false matching lines" are uncertainly distributed around the real contour, this attitude realizes a minimum of energy, which allows to use our minimization procedure. Note that dynamic contextual credibilities cannot be included into the ponderations but can be taken into consideration for the computation of the energies. However such a process implies distortions in the Jacobian.

Another idea consists in selecting the projection lines of very high credibility (i.e. for which we are almost sure they must be tangent to the surface) and to run the energetic algorithm with them (with or without ponderation). Most of the time, these features are not numerous enough to allow a good accuracy in the estimation but this process allows to bring the object in the neighborhood of the final dynamic attitude. Then projections lines of lower credibility can be added to the previous ones and the first procedure performed to find a better estimation.

A last strategy consists in defining the notion of *credibility of an attitude* as the sum of the credibilities of the projection lines tangent to the surface.

If the credibilities give a reliable information, the dynamic attitude can be expected to have a maximum credibility which allows to estimate it using classical function maximization procedures. Kalman filtering could also be envisaged. As previously, the use of a first estimation based on high gradient lines can also be interesting when the starting attitude is far from the solution.

All of these algorithms are presently under development.

§5. Conclusion

In this paper, a new method was developed to estimate the pose of a 3D object from 2D projections or X-rays. Based on a mechanical model, this very fast and accurate technique has been largely validated on both simulations and experiments

That method shows two main advantages. First, it works for *any free-form smooth surface*, since no *a priori* assumptions (symmetry hypotheses, algebraic forms, or special curves) are used. Second, *only a few pixels on the contours are needed* which allows to estimate the attitude of the object even if it is *partially occluded.*

A potential weakness of our method lies in the fact that our algorithm does not use discontinuities in the projected contours. However, in X-ray images, such points always coincide with some event in the projected 3D scene [13], which could be interpreted to infer additional information.

In the second part of the paper, various approaches have been studied to deal with non- segmented X-rays. To our knowledge, no algorithm had been proposed until now to tackle that problem. The methods proposed in this paper may constitute a first starting point to study that class of problems.

Acknowledgements. The researche described in this paper was partially supported by Safir-Groupe SEM and Digital Equipment Corporation.

References

1. Arnol'd, V. I., *Catastrophe theory*, Springer Verlag, 1984.
2. Ayache, N. *Artificial Vision for Mobile Robots: Stereo Vision and Multisensory Perception*, MIT Press, Cambridge, Massachusetts, 1991.
3. Barrow, H. G., Tenenbaum, J. M., Bolles, R. C., and Wolf, H. C., Parameteric Correspondence and Chamfer Matching: Two New Techniques for Image Matching, in *Fifth International Joint Conference on Artificial Intelligence (IJCAI-77)*, Cambridge, Massachusetts, 1977, 659–663.
4. Beauchamp, A., Dansereau, J., De Guise, J., and Labelle,JH., Computer-assisted digitization system for scoliotic spinal stereo-radiographs, in *International Symposium on 3D Scoliotic Deformities*, Dansereau, J. (ed.), Gustav Fischer Verlag, Montreal, Canada, 1992, 18–25.
5. Brooks, R., Symbolic reasoning among 3D models and 2D images, Artificial intelligence **17** (1981), 285–348.
6. Brunie, L., Fusion d'images médicales multimodales, PhD dissertation, Joseph Fourier University, Grenoble, France, 1992.

7. Demazeau, Y., Niveaux de repésentation pour la vision par ordinateur; indices d'image et indices de scène, PhD dissertation, INPG, Grenoble, France, 1986.
8. Demazure, M., *Catastrophes et bifurcations*, Ellipses, Paris, 1989.
9. Dhome, M., Richetin, M., Lapreste, J. T., and Rives, G., Determination of the attitude of 3D objects from a single perspective view, IEEE PAMI **11**(12) (1989), 1265–1278.
10. Glachet, R., Dhome, M., and Lapreste, J. T., Finding the pose of an object of revolution, in *Second European Conference on Computer Vision (ECCV'92)*, Sandini, G. (ed.), Springer Verlag, LNCS Series Vol. 588, Santa Margherita, Italy, 1992, 681–686.
11. Herman, M., Kanade, T. and Kuroe, S., Incremental Acquisition of a Three-Dimensional Scene Model from Images, IEEE Transactions on Pattern Analysis and Machine Intelligence **6**(3) (1984), 331–340.
12. Horaud, R., New Methods for Matching 3D objects with Single Perspective Views, IEEE Transactions on Pattern Analysis and Machine Intelligence **9**(3) (1987), 401–412.
13. Kergosien, Y. L., Projection of smooth surfaces: stable primitives, in *MARI- COGNITIVA*, Paris, France, 1987, 447–454.
14. Kriegman, D. J. and Ponce, J., On Recognizing and Positioning Curved 3-D Objects from Image Contours, IEEE PAMI, **12**(12) (1990), 1127–1137.
15. Liu, Y., and Huang, T., Determination of camera location from 2D to 3D line and point correspondences, IEEE Transactions on Pattern Analysis and Machine Intelligence, **12**(january 1990), 28–37.
16. Lowe, D. G., Fitting parameterized three-dimensional models to images, IEEE Transactions on Pattern Analysis and Machine Intelligence, **13**(5) (1991), 441– 450.
17. Lowe, D. G., *Perceptual Organization and Visual Recognition*, Kluwer Academic Publishers, Boston, Massachusetts, 1985.
18. Press, W. H., Flannery, B. P., Teukolsky, S. A., and Vetterling, W. T., *Numerical Recipes: The Art of Scientific Computing*, Cambridge University Press, Cambridge, England, 1986.
19. Szeliski, R., Shape from rotation, in *IEEE Computer Society Conference on Computer Vision and Pattern Recognition (CVPR'91)*, Maui, Hawai, 1991.
20. Wackenheim, A., *Perception, commentaire et interpretation dc l'image par les intelligences naturelle et artificielle*, Springer Verlag, 1987.

Lionel Brunie, Philippe Cinquin
Laboratoire d'informatique médicale
Faculté de médecine
38700 La Tronche, FRANCE
cinquin@timb.imag.fr

Radial Basis Functions: L^p-Approximation Orders with Scattered Centres

Martin D. Buhmann and Amos Ron

Abstract. In this paper we generalize several results on uniform approximation orders with radial basis functions in (Buhmann, Dyn and Levin, 1993) and (Dyn and Ron, 1993) to L^p-approximation orders. These results apply, in particular, to approximants from spaces spanned by translates of radial basis functions by *scattered centres.* Examples to which our results apply include quasi-interpolation and least-squares approximation from radial function spaces.

1. Introduction

Radial basis function methods are tools for multivariable approximation where functions $f : \mathbb{R}^n \to \mathbb{R}$ are approximated from spaces spanned by translates of normally just one function $\varphi : \mathbb{R}^n \to \mathbb{R}$. The points by which φ is translated are usually called "centres". The function φ may or may not be a radially symmetric function, i.e., $\varphi = \tilde{\varphi}(\|\cdot\|_2)$, but for the most common and best studied examples it is, such as the ubiquitous multiquadric function, where $\tilde{\varphi} = \sqrt{(\cdot)^2 + c^2}$, c being a parameter, or the equally well-known thin plate spline, where $\tilde{\varphi} = (\cdot)^2 \log(\cdot)$. This is why these methods are still called radial basis function methods, although their performance depends much more on φ's smoothness and growth properties than on its spherical symmetry.

There are several recent reviews of the state-of-the-art of radial basis function methods, but we mention here only the one due to the first author [1993]. It is apparent from this survey that, in most articles about radial basis functions and their approximation properties so far, only translates along multiintegers have been studied in detail. Quasi-interpolation and least-squares approximation from the radial basis function spaces, and interpolation at the centres, have been investigated thoroughly for multiinteger centres, but, to the best of our knowledge, extensions of this work to infinitely many scattered centres have not been provided until recently, when Buhmann, Dyn

Wavelets, Images and Surface Fitting
P. J. Laurent, A. Le Méhauté, and L. L. Schumaker (eds.), pp. 93–112.

ISBN 1-56881-040-7.

& Levin [1993], and Dyn & Ron [1993] were completed. In both papers, quasi-interpolation from radial basis function spaces with infinitely many scattered centres is analyzed. In the former paper, quasi-interpolants for sets of scattered centres that enjoy certain regularity conditions are explicitly constructed. It is shown that they provide the same approximation orders as the quasi-interpolants on gridded centres (which they agree with in the special case when the scattered centres happen to be gridded) if the centres become dense in the underlying $\mathbb{R}^n$. The class of radial basis functions for which this approach applies is restricted by various conditions, but none of the usually considered φs is excluded.

In Dyn & Ron [1993] a different approach is proposed. Rather than constructing approximation schemes that employ scattered translates of φ directly, that paper suggests and analyzes a method for converting known "gridded centre approximation schemes" to the scattered case. This is done in such a way that the difference of the latter and the former is at least of the same order as the approximation order anticipated (when the gridded centres become dense in the underlying space). The approach to these results rests on the idea that the gridded centre approximant can itself be approximated by a function from the space spanned by scattered centre translates. For this, one replaces each multiinteger translate of the radial basis function, $\varphi(\cdot - j)$ for a $j \in \mathbb{Z}^n$ say, by an approximation φ_j from the space spanned by the scattered translates. The difference $\varphi(\cdot - j) - \varphi_j$ then has to satisfy a certain decay condition, and there are other conditions on the gridded centre scheme. This work applies to the class of radial basis functions as defined in Buhmann, Dyn & Levin [1993]. It is, in fact, more general because not just one quasi-interpolant for scattered centres is manufactured. Instead, a general conversion method is established that allows many approximation schemes to be converted from gridded to scattered centres.

All the error estimates in the literature mentioned so far are in $L^\infty(\mathbb{R}^n)$, i.e., they are uniform estimates. Error analyses of quasi-interpolation schemes in general $L^p(\mathbb{R}^n)$-norms are usually more subtle, and are less common than their $L^\infty(\mathbb{R}^n)$-counterparts. We mention here the work of Lei & Jia [1991], where $L^p(\mathbb{R}^n)$-approximation orders are established from spaces spanned by gridded translates of a compactly supported piecewise-exponential function; in this regard we also mention arguments employed in the book on box splines by de Boor, Höllig & Riemenschneider [1993] for a related theorem on L^p-approximation orders. Later, Jia & Lei [1993] offered approximation schemes and a complementary error analysis for spaces generated by the gridded translates of several non-compactly supported functions. This applies to quasi-interpolation with basis functions of global support, such as is required here for radial basis function approximants, in contrast to the L^p-approximation order result by de Boor, Höllig & Riemenschneider [1993] which only applies to compactly supported basis functions φ. On the other hand, the scheme proposed by Jia and Lei is more specific in nature.

In our work, we amalgamate these approaches and provide L^p-error estimates between the gridded and the scattered centre quasi-interpolants by basis functions of global support. Therefore, they apply tó quasi-interpolation schemes using radial basis functions. We complement these results by simplified approaches to estimating the L^p-approximation orders of quasi-interpolants even for gridded centres. Examples that show the usefulness of our work are provided in the paper too.

2. L^p-approximation Orders on Uniform Grids

We always assume that the given approximand is a function $f \in W_p^k(\mathbb{R}^n)$, $p \in [1,\infty)$, where $W_p^k(\mathbb{R}^n)$ is the usual L^p-Sobolev space of order k of functions in n variables. We restrict attention to $p \in [1,\infty)$ because the results for $p=\infty$ are readily available, as pointed out in the introduction, and they are of the same nature as our L^p-results, but simpler in various respects. In fact, all our theorems remain true *verbatim* if every occurrence of p is replaced by ∞. The positive integer k is associated with the polynomial recovery of the semi-discrete convolution operator

$$\psi *' : f \mapsto \sum_j \psi(\cdot - j)\, f(j)$$

(all sums are over $\mathbb{Z}^n$ unless otherwise indicated) with a given basis function ψ, i.e., we assume that $\psi *'$ is the identity on Π_{k-1} which is the space of polynomials in n variables of total degree less than k. Such ψ's that are linear combinations of radial basis functions exist in abundance, for instance $k=n+1$ is permissible for multiquadrics and $k=n+3$ for thin plate splines when n is odd or even respectively, see Buhmann [1989], for instance. In connection with the L^p-Sobolev spaces, we shall shortly require the following further notation. By $|\cdot|_{k,p}$ we shall denote the homogeneous kth order L^p-Sobolev semi-norm on $W_p^k(\mathbb{R}^n)$, i.e.,

$$|f|_{k,p} = \sum_{\substack{\alpha\in\mathbb{Z}_+^n \\ |\alpha|=k}} \|D^\alpha f\|_p,$$

where $D = \left(\frac{\partial}{\partial x_1}, \frac{\partial}{\partial x_2}, \ldots, \frac{\partial}{\partial x_n}\right)$ is the vector of partial derivatives and $\|D^\alpha f\|_p$ is the L^p-norm of $D^\alpha f$. Moreover, $D^\alpha = D_1^{\alpha_1} D_2^{\alpha_2} \cdots D_n^{\alpha_n}$, $|\alpha| = \alpha_1 + \alpha_2 + \cdots + \alpha_n$ and, as we shall require in the sequel, $\alpha! = \alpha_1!\alpha_2!\cdots\alpha_n!$, all for $\alpha \in \mathbb{Z}_+^n$ and $\alpha = (\alpha_1, \alpha_2, \ldots, \alpha_n)$. Finally, $\|\cdot\|_{p,\Omega}$ and $|\cdot|_{k,p,\Omega}$ denote the L^p-norm and the homogeneous kth order L^p-Sobolev semi-norm, restricted to a prescribed subset $\Omega \subseteq \mathbb{R}^n$, respectively, and $\|\cdot\|_{p,\mathbb{Z}^n}$ stands for the $\ell^p(\mathbb{Z}^n)$ norm.

The approximation to f is provided in the form $\psi *' g$, for a suitably chosen g. In case f is smooth enough (say, $f \in W_p^k(\mathbb{R}^n) \cap C^k(\mathbb{R}^n)$), we may take $g = f$. Otherwise, a smoothing procedure is required, and g is defined as $g := \lambda * f$, where $\lambda \in C^k(\mathbb{R}^n)$ satisfies the two decay conditions

$$\| \|\cdot\|_\infty^k \lambda\|_1 < \infty \quad \text{and} \quad \sum_j \|\lambda\|_{\infty, j+\mathrm{U}} < \infty. \tag{2.1}$$

Here $\mathrm{U} := (-\frac{1}{2}, \frac{1}{2})^n$. We also demand that λ satisfy the moment conditions

$$\int \lambda(x) x^\alpha \, dx = \delta_{0\alpha}, \qquad \forall\, |\alpha| < k. \tag{2.2}$$

All integrals are over $\mathbb{R}^n$ unless noted otherwise. We note that functions λ that satisfy the above conditions do exist, and any such function will do.

The approximation scheme $f \approx \psi *' (\lambda * f)$ can be refined by dilation, that is, with σ_h the scaling operator

$$\sigma_h : f \mapsto f(\cdot/h),$$

and finer approximations to f are provided by

$$f \approx \sigma_h(\psi *' (\lambda * (\sigma_{1/h} f))).$$

Here and throughout, $*$ denotes convolution, either discrete or continuous, but never semi-discrete, as we use the $*'$ notation for the latter. The exact meaning of $*$ will be clear from the context. Note that this latter approximant is a linear combination of the $h\mathbb{Z}^n$-translates of $\sigma_h \psi$. One should note that in order to derive approximation orders, it suffices to establish a bound of the form

$$\|f - \psi *' (\lambda * f)\|_p \le C |f|_{k,p}, \tag{2.3}$$

(where here and everywhere in the paper C denotes a universal positive constant), since then it is straightforward to derive the estimate

$$\|f - \sigma_h(\psi *' (\lambda * (\sigma_{1/h} f)))\|_p \le C h^k |f|_{k,p}, \qquad h \to 0. \tag{2.4}$$

Our principal result in this direction is as follows:

Theorem 1. *Let $\psi : \mathbb{R}^n \to \mathbb{R}$ be a function that satisfies the following conditions:*

(i) Decay condition: *for all $x \in \mathbb{R}^n$,*

$$|\psi(x)| \le C(1 + \|x\|_\infty)^{-\nu_\psi - k},$$

where k is a given positive integer, $\nu_\psi > n$ and C does not depend on x.

(ii) Polynomial reproduction: *$\psi *'$ is the identity on Π_{k-1}.*

Let $f \in W_p^k(\mathbb{R}^n)$, and let $g \in W_p^k(\mathbb{R}^n)$ satisfy the following three conditions:

(a) *g is k-times continuously differentiable.*

(b) *f, g satisfy*

$$\left(\sum_j |g|^p_{k,\infty,j+\mathrm{U}} \right)^{\frac{1}{p}} \le C |f|_{k,p}. \tag{2.5}$$

(c) *f, g satisfy*

$$\|f - g\|_p \le C |f|_{k,p}. \tag{2.6}$$

Then

$$\|f - \psi *' g\|_p \le C|f|_{k,p}. \tag{2.7}$$

Thus, according to Theorem 1, we choose $g = f$, as long as f satisfies the first two conditions (a)–(b) required of g. Otherwise, the above detailed smoothing process is suitable, i.e., $g := \lambda * f$ satisfies the three required properties. Indeed, we note first that $g = \lambda * f \in C^k(\mathbb{R}^n) \cap W_p^k(\mathbb{R}^n)$ due to standard properties of the convolution, see e.g., Stein & Weiss [1971], so property (a) holds always. The fact that the two other required properties, i.e., (b) and (c), also hold for $g := \lambda * f$, is the content of the following two propositions.

Proposition 1. *Let $g := \lambda * f$, with λ satisfying the moment conditions* (2.2) *as well as the decay conditions* (2.1). *Then,*

$$\|f - g\|_p \le C|f|_{k,p}, \tag{2.8}$$

that is, g satisfies condition (c) *of* Theorem 1.

Proof. According to the integral remainder form of the multivariable Taylor theorem and according to (2.2),

$$(g - f)(x) = \int \lambda(u) \sum_{\substack{\gamma \in \mathbb{Z}_+^n \\ |\gamma| = k}} \frac{(-u)^\gamma}{\gamma!} \int_0^1 (1-t)^{k-1} D^\gamma f(x - ut)\, dt\, du, \quad x \in \mathbb{R}^n. \tag{2.9}$$

Therefore, using the generalized Minkowski inequality, (2.9) implies that the left-hand side of (2.8) is at most a constant multiple of

$$\int \int_0^1 |\lambda(u)||u|^k (1-t)^{k-1}\, dt\, du\, |f|_{k,p}. \tag{2.10}$$

By virtue of (2.1), expression (2.10) is a constant multiple of $|f|_{k,p}$. ■

The proof shows that the constant C in (2.8) may depend on λ and k, but is independent of p and f.

Proposition 2. *Let g be as in* Proposition 1. *Then* (2.5) *holds, and, further,*

$$|g|_{k,p} \le C|f|_{k,p}. \tag{2.11}$$

Proof. We use Hölder's inequality and obtain, for *any* $\widetilde{f} \in L^p(\mathbb{R}^n)$, for any $j, i \in \mathbb{Z}^n$, and for every $x \in j + \mathrm{U}$,

$$\left| \int_{i+\mathrm{U}} \lambda(u) \widetilde{f}(x - u)\, du \right| \le \|\widetilde{f}\|_{p, j-i+2\mathrm{U}} \|\lambda\|_{p', i+\mathrm{U}}, \tag{2.12}$$

where $\frac{1}{p} + \frac{1}{p'} = 1$. Introducing the two sequences

$$a(i) := \|\lambda\|_{p',i+\mathrm{U}}, \quad b(i) := \|\widetilde{f}\|_{p,i+2\mathrm{U}}, \qquad i \in \mathbb{Z}^n,$$

and summing (2.12) over all integer i, we obtain the bound

$$\|\lambda * \widetilde{f}\|_{\infty,j+\mathrm{U}} \leq (a * b)(j), \qquad j \in \mathbb{Z}^n.$$

Therefore, by Young's inequality (cf., e.g., Stein & Weiss [1971], p. 178),

$$\Big(\sum_j \|\lambda * \widetilde{f}\|^p_{\infty,j+\mathrm{U}}\Big)^{\frac{1}{p}} \leq \|a\|_{1,\mathbb{Z}^n}\|b\|_{p,\mathbb{Z}^n}.$$

Because of the second assumption in (2.1), $\|a\|_{1,\mathbb{Z}^n} < \infty$. Also, $\|b\|_{p,\mathbb{Z}^n} \leq 2^{n/p}\|\widetilde{f}\|_p$, and we thus have proved that

$$\Big(\sum_j \|\lambda * \widetilde{f}\|^p_{\infty,j+\mathrm{U}}\Big)^{\frac{1}{p}} \leq C\|\widetilde{f}\|_p.$$

Substituting $\widetilde{f} := D^\gamma f$, for all possible γ with $|\gamma| = k$, we obtain the desired claim. The second inequality (2.11) is an easy consequence of Young's inequality and the fact that, as a result of (2.1), $\lambda \in L^1(\mathbb{R}^n)$. ■

Propositions 1 and 2 assure us of the existence of a suitable g for the sake of constructing the approximation scheme $f \approx \psi *' g$. We can, therefore, turn our attention to the proof of Theorem 1.

The condition (c) on g (in Theorem 1) allows us to bound the error by

$$\|f - \psi *' g\|_p \leq \|f - g\|_p + \|g - \psi *' g\|_p \leq C|f|_{k,p} + \|g - \psi *' g\|_p, \quad (2.13)$$

and therefore leads to the desired estimate (2.7), as soon as one has an estimate of the form

$$\|g - \psi *' g\|_p \leq C|f|_{k,p} \qquad (2.14)$$

in hand. One can in fact prove more generally for *any* $g \in C^k(\mathbb{R}^n) \cap W^k_p(\mathbb{R}^n)$, and $p \in [1, \infty)$, that

$$\|g - \psi *' g\|_p \leq C|g|_{k,p} \qquad (2.15)$$

so long as g satisfies

$$\Big(\sum_j |g|^p_{k,\infty,j+\mathrm{U}}\Big)^{\frac{1}{p}} \leq C|g|_{k,p}. \qquad (2.16)$$

The arguments below that lead to (2.16) also support the derivation of (2.14) from our condition (2.5): one simply needs to replace in the proofs all applications of (2.16) by (2.5).

We first establish our approximation order result for the (relatively simple case) when ψ is of compact support. Thereafter we will provide a proof for Theorem 1 in its full generality.

Theorem 2. *The conclusions of* Theorem 1 *hold in case* ψ *is bounded and of compact support. More generally, if* $g \in C^k(\mathbb{R}^n) \cap W_p^k(\mathbb{R}^n)$, *and* ψ *is bounded and of compact support and satisfies* (ii), *then* (2.16) *implies* (2.15).

Proof. Let $x \in \mathbb{R}^n$ be arbitrary. If T denotes the Taylor polynomial to $g \in C^k(\mathbb{R}^n)$ about x of degree $k-1$, we get (since $\psi *' T = T$)

$$|(\psi *' g - g)(x)| \le \sum_j |\psi(x-j)||g(j) - T(j)|.$$

Let m be a positive integer that satisfies $\operatorname{supp}\psi \subset (2m-1)\mathrm{U}$. Therefore

$$\|\psi *' g - g\|_p \le \left[\int \Big(\sum_{j\in Z_x} |\psi(x-j)||g(j)-T(j)|\Big)^p dx\right]^{\frac{1}{p}}, \tag{2.17}$$

where $Z_x := \{j \in \mathbb{Z}^n \mid \|x-j\|_\infty < \frac{1}{2}(2m-1)\}$. By the integral form of the Taylor remainder, the right-hand side of (2.17) is at most $\|\psi\|_\infty$-times

$$\begin{aligned}
&\left[\int \Big(\sum_{j\in Z_x} |g(j)-T(j)|\Big)^p dx\right]^{\frac{1}{p}} \\
&=\left[\int \Big(\sum_{j\in Z_x} \Big| \sum_{\substack{\gamma\in\mathbb{Z}_+^n\\|\gamma|=k}} \frac{(j-x)^\gamma}{\gamma!} \int_0^1 (1-t)^{k-1} D^\gamma g((1-t)x+tj)\,dt\Big|\Big)^p dx\right]^{\frac{1}{p}} \\
&\le \sum_{\substack{\gamma\in\mathbb{Z}_+^n\\|\gamma|=k}} \frac{m^k}{\gamma!} \left[\int \Big(\int_0^1 \sum_{j\in Z_x} (1-t)^{k-1} |D^\gamma g((1-t)x+tj)|\,dt\Big)^p dx\right]^{\frac{1}{p}} \\
&\le \sum_{\substack{\gamma\in\mathbb{Z}_+^n\\|\gamma|=k}} \frac{m^k}{\gamma!} \int_0^1 (1-t)^{k-1} \Big\| \sum_{j\in Z_{(\cdot)}} |D^\gamma g((1-t)(\cdot)+tj)| \Big\|_p dt,
\end{aligned}$$

where we have used the Minkowski inequality and the generalized Minkowski inequality in the penultimate and in the last inequality, respectively.

We estimate the above expression term-by-term in γ and for each $t \in (0,1)$ separately. Hence we consider

$$\Big\| \sum_{j\in Z_{(\cdot)}} |D^\gamma g((1-t)(\cdot)+tj)| \Big\|_p, \tag{2.18}$$

leaving out the factor of $(1-t)^{k-1}$. Note that $Z_x = Z_\ell$ if $x \in \mathrm{U}+\ell$. Hence we have for each $t \in (0,1)$ that (2.18) is at most the p-th root of

$$\begin{aligned}
&\int \Big(\sum_{j\in Z_x} |D^\gamma g(tj+(1-t)x)|\Big)^p\,dx \\
&= \sum_\ell \int_{\mathrm{U}+\ell} \Big(\sum_{j\in Z_\ell} |D^\gamma g(tj+(1-t)x)|\Big)^p\,dx \\
&= \sum_\ell \int_{\mathrm{U}+\ell} \Big(\sum_{j\in Z_\ell} |D^\gamma g(j+(1-t)(x-j))|\Big)^p\,dx \\
&\le \sum_\ell (2m)^{np} \|D^\gamma g\|^p_{\infty, Z_\ell+2m\mathrm{U}} \\
&\le C \sum_\ell |g|^p_{k,\infty,\ell+\mathrm{U}},
\end{aligned}$$

the last inequality being a simple consequence of the fact that $\{\ell+\mathrm{U}\}_{\ell\in\mathbb{Z}^n}$ form a tiling of $\mathbb{R}^n$.

Now, if assumption (2.16) on g holds, we obtain that the last expression is bounded by a constant multiple of $|g|^p_{k,p}$. We then take the p-th root to obtain (2.15). If, instead, we apply condition (b), i.e., replace the usage of (2.16) by using (2.5), we arrive directly at the desired result (2.7). ■

We remark that the various constants that appear in the proof of Theorem 2 depend on k, m, $\|\psi\|_\infty$, n and p. These constants, however, do not depend on f.

Proof of Theorem 1. Before we begin the proof, we point out once again that there are many ψs that satisfy the assumptions of this theorem which are linear combinations of radial basis functions (see, for instance, Buhmann [1989], and see in particular our example at the end of this note).

To embark on the proof, we first follow the argument in the proof of Theorem 1, until (2.18) is reached. Here, due to the global support of ψ, there is no need to introduce the set Z_x. Instead, we use the decay assumption (i) on ψ, and obtain the bound

$$\|\psi *' g - g\|_p \le \Big[\int \Big(\sum_j |\psi(x-j)||g(j)-T(j)|\Big)^p\,dx\Big]^{\frac{1}{p}}$$

$$\le C \sum_{\substack{\gamma\in\mathbb{Z}^n_+\\ |\gamma|=k}} \frac{1}{\gamma!} \int_0^1 (1-t)^{k-1} \Big\|\sum_j (1+\|\cdot - j\|_\infty)^{-\nu_\psi} |D^\gamma g(tj+(1-t)(\cdot))|\Big\|_p\,dt,$$

where we have used Minkowski's inequality and its generalization as in the previous proof. We observe that for each $t \in (0,1)$, the integrand of this expression is less than or equal to

$$\Big(\sum_i \Big\|\sum_j (1+\|\cdot-j\|_\infty)^{-\nu_\psi}|D^\gamma g(tj+(1-t)(\cdot))|\Big\|_{p,\mathrm{U}+i}^p\Big)^{\frac{1}{p}}. \tag{2.19}$$

Expression (2.19) equals

$$\begin{aligned}&\Big(\sum_i \Big\|\sum_j (1+\|\cdot-j\|_\infty)^{-\nu_\psi}|D^\gamma g(i+tj+(1-t)(\cdot))|\Big\|_{p,\mathrm{U}}^p\Big)^{\frac{1}{p}}\\ &\le \Big(\sum_i \Big\|\sum_j (\tfrac{1}{2}+\|j\|_\infty)^{-\nu_\psi}|D^\gamma g(i+tj+(1-t)(\cdot))|\Big\|_{p,\mathrm{U}}^p\Big)^{\frac{1}{p}}\\ &\le \Big(\sum_i \Big(\sum_j (\tfrac{1}{2}+\|j\|_\infty)^{-\nu_\psi}\big\|D^\gamma g(i+tj+\cdot)\big\|_{\infty,\mathrm{U}}\Big)^p\Big)^{\frac{1}{p}},\end{aligned} \tag{2.20}$$

where the last estimate depends on the triangle inequality for the uniform norm and on $t \in (0,1)$.

For $t=1$, the last expression can be identified as $\|B*G\|_{p,\mathbb{Z}^n}$, with $B(j) := (\frac{1}{2}+\|j\|_\infty)^{-\nu_\psi}$, $j\in\mathbb{Z}^n$, and $G(j) := \big\|D^\gamma g(j+\cdot)\big\|_{\infty,\mathrm{U}}$, $j \in \mathbb{Z}^n$. For that case, Young's inequality can be invoked to yield

$$\|B*G\|_{p,\mathbb{Z}^n} \le \|B\|_{1,\mathbb{Z}^n}\|G\|_{p,\mathbb{Z}^n}.$$

While for $t<1$ the expression in question is *not* a convolution product, it can still be majorised by an appropriate convolution, as the following argument shows. For each $t\in(0,1)$ fixed, we define the map $N(j) := \lfloor(1-t)j\rfloor$, $j\in\mathbb{Z}^n$, with $\lfloor\cdot\rfloor$ denoting (componentwise) the greatest integer less than $(\cdot+\frac{1}{2})$. We can then write the above sequence B as the sum $\sum_\ell B_\ell$ of sequences of disjoint supports defined as follows:

$$B_\ell(j) := \begin{cases} B(j), & \ell = N(j),\\ 0, & \text{otherwise},\end{cases} \qquad j,\ell\in\mathbb{Z}^n.$$

By using the bound

$$\big\|D^\gamma g(i+tj+\cdot)\big\|_{\infty,\mathrm{U}} \le \big\|D^\gamma g(i+j-N(j)+\cdot)\big\|_{\infty,2\mathrm{U}},$$

defining $\widetilde{G}(j) := \big\|D^\gamma g(j+\cdot)\big\|_{\infty,2\mathrm{U}}$, $j\in\mathbb{Z}^n$, and using the nonnegativity of the entries of the sequences B and $\widetilde{G}$, we can bound the last expression in the chain of inequalities (2.20) from above by

$$\Big\|\sum_\ell B_\ell * E^\ell\widetilde{G}\Big\|_{p,\mathbb{Z}^n} = \Big\|\Big(\sum_\ell E^\ell B_\ell\Big)*\widetilde{G}\Big\|_{p,\mathbb{Z}^n},$$

where $E^\ell : f \mapsto f(\cdot - \ell)$. Therefore, Young's inequality applies once again to yield the upper bound

$$\Big\|\sum_\ell E^\ell B_\ell\Big\|_{1,\mathbb{Z}^n} \|\widetilde{G}\|_{p,\mathbb{Z}^n} = \|B\|_{1,\mathbb{Z}^n}\|\widetilde{G}\|_{p,\mathbb{Z}^n} =$$

$$\sum_j \big(\tfrac{1}{2} + \|j\|_\infty\big)^{-\nu_\psi}\Big(\sum_j \widetilde{G}(j)^p\Big)^{\frac{1}{p}}.$$

Finally, the sum $\big(\sum_j \widetilde{G}(j)^p\big)^{1/p}$ can certainly be bounded by a constant multiple of $\big(\sum_j G(j)^p\big)^{1/p}$, and this latter expression is bounded by $C|f|_{k,p}$, by our assumption (b) (or by $C|g|_{k,p}$, if, instead of assuming (b), we adopt (2.16)). The theorem is proved. ■

If we apply Propositions 1 and 2, we arrive at the following

Corollary 1. *If $f \in W_p^k(\mathbb{R}^n)$, if the assumptions* (i)–(ii) *of* Theorem 1 *hold and if $\lambda \in C^k(\mathbb{R}^n)$ satisfies* (2.2) *and* (2.1), *then* (2.3), *and thus* (2.4) *too, hold. Furthermore, if f is k-times continuously differentiable and satisfies*

$$\Big(\sum_j |f|^p_{k,\infty,j+\mathrm{U}}\Big)^{\frac{1}{p}} \le C|f|_{k,p},$$

then

$$\|f - \sigma_h(\psi *' \sigma_{1/h} f)\|_p \le Ch^k |f|_{k,p}$$

for h tending to zero.

The second assertion in the above theorem says in particular that for sufficiently smooth f one needs to know only the values $f_{|h\mathbb{Z}^n}$ in order to compute its L^p-approximant. This is especially important from a practical point of view, since computing the values on $h\mathbb{Z}^n$ of $\lambda * f$ can be hard.

Our final note in this section concerns the conditions imposed on ψ in Theorem 1. The decay condition (i) is essential to the approximation orders derived. Though approximation orders can be derived under weaker decay conditions, such derivations either involve some complementary information of the exact decay rates of ψ (as in Buhmann [1989] or in Dyn, Jackson, Levin & Ron [1992]), or require a substantial modification of the approximation schemes as well as their error analyses (as in de Boor & Ron [1992a] and de Boor, DeVore & Ron [1993]). However, the second condition, i.e., that $\psi *'$ be the identity on Π_{k-1} can be relaxed to (the more standard one) that $\psi *'$ induces a linear isomorphism on Π_{k-1}. An appropriate approximation scheme in this case has the form

$$f \approx \psi *' (\rho * \lambda * f),$$

with λ, as before, satisfying (2.2) and (2.1), and ρ, say, compactly supported and satisfying

$$\psi *' (\rho * p) = p, \quad p \in \Pi_{k-1}.$$

Such constructions are discussed in detail in the literature, and we refer the reader to de Boor & Ron [1992b] where such schemes were thoroughly analysed for *compactly supported* ψ. The modifications required in our arguments in order to cover that more general setup are mostly straightforward. Moreover, one should notice that, if ρ is selected to have its support lying in $\mathbb{Z}^n$, then $\psi *' (\rho * \cdot) = (\psi * \rho) *'$, and, for the new function $\Psi := \psi * \rho$, $\Psi *'$ is the identity on Π_{k-1}. Therefore, our results apply directly to this important case.

3. L^p-approximation Orders using Scattered Centres

We use the results of the previous section in the derivation of approximation schemes that employ scattered translates of the basis function φ. The approach as well as the error analysis follow the work of Dyn & Ron [1993], where L^∞-approximation orders were obtained.

We let ψ be a linear combination

$$\psi = \varphi *' \mu, \tag{3.1}$$

where φ is a given function in $C(\mathbb{R}^n)$ such as one of our radial basis functions and where $\mu : \mathbb{Z}^n \to \mathbb{R}$ is such that the infinite sum in (3.1) converges uniformly on compacta. We shall as before consider the approximation operator $f \mapsto Lg := \psi *' g$, where $g = \lambda * f$ with $\lambda \in C^k(\mathbb{R}^n)$ satisfying (2.2) and (2.1). We make assumptions according to those in Dyn & Ron [1993]: we always assume from now on that the thus constructed $L : C(\mathbb{R}^n) \cap L^\infty(\mathbb{R}^n) \to C(\mathbb{R}^n) \cap L^\infty(\mathbb{R}^n)$ is a bounded operator without further mention. As in the previous section, we use here *stationary schemes*, i.e., define the h-refinement L_h by dilation:

$$L_h := \sigma_h L \sigma_{1/h}.$$

We shall further assume that

$$\left.\begin{array}{ll} |\mu(\ell)| \le C(1+\|\ell\|_\infty)^{-\nu_\mu}, & \ell \in \mathbb{Z}^n, \\ |\psi(x)| \le C(1+\|x\|_\infty)^{-\nu_\psi}, & x \in \mathbb{R}^n, \end{array} \quad \nu_\mu, \nu_\psi > n. \right\} \tag{3.2}$$

We take, for a given set of (scattered) centres $\Xi \subset \mathbb{R}^n$, approximations to each translate $\varphi(\cdot - j)$ of the form

$$\varphi_j := \sum_{\xi \in \Xi} A_{j,\xi} \varphi(\cdot - \xi),$$

where $\{A_{j,\xi}\}_{j \in \mathbb{Z}^n, \xi \in \Xi}$ are suitable coefficients which provide that

$$|\varphi_j(x) - \varphi(x-j)| \le C(1+\|x-j\|_\infty)^{-\nu_A} \begin{cases} x \in \mathbb{R}^n, \\ j \in \mathbb{Z}^n, \end{cases} \tag{3.3}$$

holds for a $\nu_A > n$ and a positive constant C which is uniform in x and j. We call the functions $\{\varphi_j\}_j$ *pseudo-shifts* (as they are approximants to the "true" shifts $\{\varphi(\cdot - j)\}_j$ from our space), and use these pseudo-shifts to construct another set of approximations, now to the localized functions $\{\psi(\cdot - j)\}_j$. We carry out this second task by imitating the localization process of (3.1). Precisely, we define

$$\psi_j := \sum_\ell \varphi_\ell \, \mu(\ell - j). \tag{3.4}$$

Finally, the approximation scheme L_A, that employs the scattered translates $\{\varphi(\cdot - \xi)\}_{\xi\in\Xi}$, is obtained from L by replacing each $\psi(\cdot - j)$ by ψ_j:

$$L_A : f \mapsto \sum_j f(j)\psi_j.$$

As before, there are functions f for which convolution with λ need be performed:

$$f \approx L_A(\lambda * f).$$

The h-refinement here is the same as before for L, that is

$$L_{h,A} := \sigma_h L_A \sigma_{1/h}.$$

One verifies that $L_{h,A}$ maps into the span of the $h\Xi$-translates of φ. We remark that other, more flexible refinements may be allowed, too, and refer the interested reader to the discussion of that point in §2.6 of Dyn & Ron [1993] and to the specific strategy suggested for the purpose of treating sets of scattered centres becoming dense in the underlying space in the paper of Buhmann, Dyn & Levin [1993].

Theorem 3. *Let φ, μ, ψ, ψ_j, L_h and $L_{h,A}$ be as above. Assume that (3.2) and (3.3) hold. Suppose also that*

$$\nu_\mu \geq n + k + \varepsilon - 1 \quad \text{for an } \varepsilon \in (0,1] \tag{3.5}$$

and

$$\mu : f \mapsto \sum_\ell \mu(-\ell) f(\ell) \quad \text{satisfies} \quad \mu\Pi_{k-1} = \{0\}. \tag{3.6}$$

Then, for $h \in (0,1)$ and $f \in W_p^k(\mathbb{R}^n)$,

$$\|(L_h - L_{h,A})(\lambda * f)\|_p \leq Ch^k(|f|_{k-1,p} + |f|_{k,p}) \begin{cases} h^{\varepsilon-1} & \text{if } \varepsilon < 1, \\ -\log h & \text{if } \varepsilon = 1. \end{cases} \tag{3.7}$$

Moreover, if $\varepsilon = 1$ and $\sum_j |\mu(j)||j|^k < \infty$, then

$$\|(L_h - L_{h,A})(\lambda * f)\|_p \leq Ch^k |f|_{k,p}. \tag{3.8}$$

Implicit in the statement of theorem is the assertion that $L_A(\lambda * f)$ is well-defined, i.e., that the series $L_A(\lambda * f)$ converges absolutely for every, say, bounded f. This well-definedness has been established in Dyn & Ron [1993]; cf. Lemma 2.2.6 there.

Proof. We begin with establishing (3.8). We may consider L and L_A instead of L_h and $L_{h,A}$, because, as we have pointed out earlier, the powers of h are introduced by a suitable scaling. Set $g := \lambda * f$. Since the series that define L and L_A converge absolutely, we may sum $(L - L_A)g$ by parts to obtain

$$\|(L - L_A)g\|_p \leq \Big\|\sum_j |\varphi(\cdot - j) - \varphi_j| \Big|\sum_\ell \mu(-\ell) g(j + \ell)\Big|\Big\|_p.$$

Since μ annihilates Π_{k-1} by (3.6), we have, for each $j \in \mathbb{Z}^n$,

$$\sum_\ell \mu(-\ell) g(j + \ell) = \sum_\ell \mu(-\ell)(g(j + \ell) - T_j(j + \ell)),$$

with T_j the $(k - 1)$-degree Taylor polynomial of g about j. Using, as in the previous section, the integral remainder formula for the Taylor expansion, followed by the triangle inequality and summation by parts, we obtain the following upper bound on $\|(L - L_A)g\|_p$:

$$\Big\|\sum_\ell \sum_{\substack{\gamma \in \mathbb{Z}_+^n \\ |\gamma| = k}} \int_0^1 t^{k-1} |\mu(-\ell) \frac{\ell^\gamma}{\gamma!}| \sum_j |\varphi(\cdot - j) - \varphi_j| \Big|D^\gamma g(j + (1 - t)\ell)\Big| \, dt\Big\|_p.$$

Using the Minkowski inequality we can take the summations over ℓ and γ out of the p-norm expression. Furthermore, using the generalized Minkowski inequality we can also interchange integration over t and the L^p-norm. This leaves us with the task of estimating the expression

$$\Big\|\sum_j |\varphi(\cdot - j) - \varphi_j| \Big|D^\gamma g(j + (1 - t)\ell)\Big|\Big\|_p. \tag{3.9}$$

Because of (3.3), we further can bound expression (3.9) by a constant multiple of

$$\Big\|\sum_j (1 + \|\cdot - j\|_\infty)^{-\nu_A} |D^\gamma g(j + (1 - t)\ell)|\Big\|_p. \tag{3.10}$$

The estimation of expression (3.10) follows a similar estimation done in the proof of Theorem 1, only the present case is in fact slightly simpler. Indeed,

$$\begin{aligned}
&\Big\|\sum_j (1+\|\cdot-j\|_\infty)^{-\nu_A}|D^\gamma g(j+(1-t)\ell)|\Big\|_p^p \\
&= \sum_i \Big\|\sum_j (1+\|\cdot-j\|_\infty)^{-\nu_A}|D^\gamma g(j+(1-t)\ell)|\Big\|_{p,\mathrm{U}+i}^p \\
&\le \sum_i \Big(\sum_j (\tfrac{1}{2}+\|j\|_\infty)^{-\nu_A}|D^\gamma g(i-j+(1-t)\ell)|\Big)^p \\
&\le C\sum_i |D^\gamma g(i+(1-t)\ell)|^p \\
&\le C\sum_i |g|_{k,\infty,\mathrm{U}+i}^p \\
&\le C|f|_{k,p}^p,
\end{aligned}$$

where we have used the Young inequality in the second estimate and Proposition 2 in the last. Now we may use the fact that $\sum_\ell |\mu(-\ell)||\ell|^k$ is finite and take p-th roots to deduce the estimate (3.8) from the above chain of inequalities.

The estimate (3.7) is proved similarly, but we can no longer rely on the convergence of $\sum_\ell |\mu(-\ell)||\frac{\ell^\gamma}{\gamma!}|$, and need to split the sum over ℓ into two partial sums: $\sum_{\ell\in\mathbb{Z}^n\cap h^{-1}\mathrm{U}}$ and the remaining sum. The first sum is estimated exactly as above, and the bound obtained then takes the form

$$C|f|_{k,p}\sum_{\ell\in\mathbb{Z}^n\cap h^{-1}\mathrm{U}} |\mu(-\ell)|\Big|\frac{\ell^\gamma}{\gamma!}\Big|$$

for each γ with $|\gamma|=k$. In the remaining sum, the zero at j of $g(j+\ell)-T_j(j+\ell)$, though being of order k, is treated as being of order $k-1$, thus leading to a Taylor remainder formula of the form

$$\sum_{\substack{\gamma\in\mathbb{Z}_+^n\\ |\gamma|=k-1}} \int_0^1 t^{k-2}\Big|\mu(-\ell)\frac{\ell^\gamma}{\gamma!}\Big|\sum_j |\varphi(\cdot-j)-\varphi_j|\Big|D^\gamma(g-T_j)(j+(1-t)\ell)\Big|\,dt.$$

Since $D^\gamma T_j = D^\gamma g(j)$ for all admissible γ, we can follow the argument that was used to prove (3.8), to obtain a bound of the form

$$C|f|_{k-1,p}\sum_{\ell\in\mathbb{Z}^n,\,\ell\notin h^{-1}\mathrm{U}} |\mu(-\ell)|\Big|\frac{\ell^\gamma}{\gamma!}\Big|$$

for each γ with $|\gamma|=k-1$. (In the derivation of this latter estimate, one invokes Proposition 2, with k there replaced by $k-1$.) Now, we can use the fact, that, whenever the required decay estimate on μ holds for $\varepsilon\in(0,1)$,

$$\sum_{\ell\in\mathbb{Z}^n\cap h^{-1}\mathrm{U}} |\mu(-\ell)|\Big|\frac{\ell^\gamma}{\gamma!}\Big| \le Ch^{\varepsilon-1}, \tag{3.11}$$

$$\sum_{\ell\in\mathbb{Z}^n,\,\ell\notin h^{-1}\mathrm{U}} |\mu(-\ell)|\left|\frac{\ell^\gamma}{\gamma!}\right| \le Ch^\varepsilon, \tag{3.12}$$

whereas if $\varepsilon = 1$, the right-hand side of (3.11) has to be replaced by $-C\log h$. All these estimates were proved for instance in Buhmann, Dyn & Levin [1993], and are also well-documented in Dyn & Ron [1993]. In summary, the total error bound for, say, $\varepsilon < 1$ takes the form

$$\|(L - L_A)(\lambda * f)\|_p \le C(h^{\varepsilon-1}|f|_{k,p} + h^\varepsilon |f|_{k-1,p}). \tag{3.13}$$

It is straightforward to insert the $\log h$-term into this expression when $\varepsilon = 1$. When finally estimating $(L_h - L_{h,A})(\lambda * f)$, we simply invoke (3.13) with f replaced by $\sigma_{1/h} f$ for the given value of h, and then scale to obtain (3.7). ■

In analogy to Theorem 2.3.1 in Dyn & Ron [1993] (where the case $p = \infty$ was discussed) we establish the following result. Its purpose is to relax the decay conditions on μ in the previous result which requires μ to decay fast enough so as to be summable against polynomials. It is now replaced by a condition on the Fourier series of μ, a condition that may be valid even when μ decays in a milder fashion. Incidentally, we will see that the rates derived in that result may be superior to those of Theorem 3.

In order to present the next theorem, we have to introduce some additional notation. Given $1 \le p < \infty$, let $\widetilde{W}_p^k(\mathbb{R}^n)$ be the space of all functions f whose Fourier transform $\widehat{f}$ satisfies $(1+\|\cdot\|_2)^k \widehat{f} \in L^{p'}(\mathbb{R}^n)$, where $\frac{1}{p}+\frac{1}{p'} = 1$. The $L^{p'}$-norm of $\|\cdot\|_2^k \widehat{f}$ is denoted by $|f|'_{k,p}$. Note that, for a nonnegative integer k and $p \ge 2$, the Hausdorff–Young Theorem shows that the space $\widetilde{W}_p^k(\mathbb{R}^n)$ equipped with the norm $\|f\|'_{k,p} := \|(1+\|\cdot\|_2)^k \widehat{f}\|_{p'}$ is continuously embedded into $W_p^k(\mathbb{R}^n)$. Further, for $p = 2$ and any nonnegative integer k, the two spaces coincide. We have the following

Theorem 4. *Assume $1 \le p < \infty$. Let all the assumptions of the previous theorem be in force (including in particular (3.2)) except for (3.5) and (3.6). Let $f \in \widetilde{W}_p^k(\mathbb{R}^n)$ be bounded, and assume that the inequality*

$$|\mu f| \le C|f|'_{k,p} \tag{3.14}$$

holds. Then

$$\|(L - L_A)f\|_p \le C|f|'_{k,p}. \tag{3.15}$$

The estimate (3.15) remains valid even if we replace f on its left-hand side by $(\lambda * f)$, provided that, in addition to all other requirements on λ, $\widehat{\lambda} \in L^\infty(\mathbb{R}^n)$.

Note that the dash-semi-norm $|f|'_{k,p}$ reacts to scaling in the correct manner: upon scaling of approximand and approximant suitably by h, one does get in (3.15) the power h^k into the bound, as required. Note also that the smoothing process $f \approx \lambda * f$ is permitted but not required in Theorem 4. It may, however, actually be required for guaranteeing that the gridded centre approximant to the function f approximates it to the best order established in Theorem 1.

We will discuss the nature of condition (3.14) after proving this theorem. In particular, we will compare that condition to the polynomial annihilation condition (3.6) assumed in Theorem 3.

Proof of Theorem 4. The proof follows the argument used in Theorem 3, but is much simpler. In Theorem 3, we bounded the error by

$$\sum_j |\varphi(\cdot - j) - \varphi_j| \Big|\sum_\ell \mu(-\ell) f(j+\ell)\Big|. \tag{3.16}$$

Since no smoothing is required, we replaced g by f in (3.16). We bound the error in the same fashion in the present proof, and remark that, since f is bounded and $\mu \in \ell_1(\mathbb{Z}^n)$ by (3.2), the sum $\mu(f(\cdot + j)) = \sum_\ell \mu(-\ell) f(j+\ell)$ converges absolutely. By our assumption (3.14) in the theorem and the fact that the semi-norm $|\cdot|'_{k,p}$ is translation-invariant, we have $|\mu(f(\cdot + j))| \le C|f|'_{k,p}$. We thus obtain the bound

$$C|f|'_{k,p} \sum_j |\varphi(\cdot - j) - \varphi_j|.$$

The desired result is now a consequence of the assumption (3.3). The proof so far also provides that

$$\|(L - L_A)(\lambda * f)\|_p \le C|\lambda * f|'_{k,p} \le C\|\widehat{\lambda}\|_\infty |f|'_{k,p}, \tag{3.17}$$

thus showing that the difference $(L - L_A)(\lambda * f)$ is bounded to the asserted order too. The first inequality in (3.17) is a consequence of (3.15) and the second one is a consequence of the definition of our dash-semi-norm. The theorem is proved. ■

Of course, Theorem 4 is useful only if one can establish verifiable conditions on f which ensure (3.14). In this regard, we remind the reader that, if f fails to satisfy the conditions of the theorem, for instance because f is not bounded, we may still invoke the result with respect to a mollified $g := \lambda * f$, and combine (3.15) with an inequality

$$|g|'_{k,p} \le C|f|'_{k,p} \tag{3.18}$$

to obtain the desired estimate. This is of course what we already did at the end of the above proof, where (3.18) was a consequence of the boundedness of λ's Fourier transform.

The crucial bound (3.14) holds under a variety of possible conditions on f, k and μ, and it is beyond the scope of this article to discuss all possible versions. The following seems to be the most useful variant:

Proposition 3. *Assume f is bounded and $\widehat{f} \in L^1(\mathbb{R}^n) \cap \widetilde{W}_p^k(\mathbb{R}^n)$. If*

$$\|\cdot\|_2^{-k}\widehat{\mu} \in L^p(\mathbb{R}^n), \tag{3.19}$$

$\widehat{\mu}$ being the Fourier series of μ, then (3.14) holds.

Proof. Recall that $\mu \in \ell_1(\mathbb{Z}^n)$, which renders μf well-defined. The crux in the proof is the identity

$$\mu f = \int \widehat{\mu}\widehat{f}. \tag{3.20}$$

This can be derived as in Dyn & Ron [1993] (cf. the proof of the lemma in Theorem 2.3.1 in that article), where the Poisson Summation Formula (Stein & Weiss [1971], p. 252) is the salient ingredient to the proof and which else requires only $\widehat{f} \in L^1(\mathbb{R}^n)$ and boundedness of f. Now it follows from an application of Hölder's inequality that

$$|\mu f| \leq \left(\int (\|\cdot\|_2^{-k}\widehat{\mu})^p\right)^{\frac{1}{p}} \left(\int (\|\cdot\|_2^{k}\widehat{f})^{p'}\right)^{\frac{1}{p'}}. \tag{3.21}$$

Combining (3.19) and (3.21) leads to (3.14). ∎

The assumption $\widehat{f} \in L^1(\mathbb{R}^n)$ is not very restrictive. Indeed, for $k > n/p$ that condition is implied by the requirement $f \in \widetilde{W}_p^k(\mathbb{R}^n)$. When k is too small and $\widehat{f} \notin L^1(\mathbb{R}^n)$, we still have $\widehat{\lambda * f} \in L^1(\mathbb{R}^n)$ if λ is sufficiently smooth. This smoothness requirement of λ can be shown to be mild, and it decreases with the increase of p. For example, if $p \geq 2$, we only need λ to be in $L^2(\mathbb{R}^n)$ which we assume anyway. In short, for an appropriately chosen λ and under the conditions assumed in Theorem 4, save the boundedness of f and the inequality (3.14), the bound

$$\|(L - L_A)(\lambda * f)\|_p \leq C|f|'_{k,p} \tag{3.22}$$

is valid, as soon as (3.19) holds.

We see that the important assumption in the proposition is the one with respect to μ, and we would like to compare it to the polynomial annihilation condition (3.6) assumed in Theorem 3. If μ decays fast enough to satisfy, say, the condition $\sum_j |\mu(j)||j|^k < \infty$ which was assumed in Theorem 3, then $\widehat{\mu}$ is k-times continuously differentiable everywhere and in particular at the origin. It follows easily that (3.6) implies that $\widehat{\mu}$ has a zero of order k at the origin. If $k > n/p$, this certainly implies (3.19). Moreover, this latter condition is valid for $\|\cdot\|_2^{-k'}\widehat{\mu}$, with $k' < k + n/p$. By choosing $n/p < k' < k + n/p$, we therefore obtain from Theorem 4 the bound (3.22), which yields that, upon scaling, the difference between the scattered centre approximant and the gridded approximant decays to zero with order $h^{k'}$. This is very satisfactory, since the approximation order we expect L_h to provide (in the best circumstances) is only k (not to mention the fact that we had relaxed the decay condition on μ). On the other hand, we should mention that, for example, when $p > 2$, the space of approximands to which Theorem 4 applies (viz., $\widetilde{W}_p^k(\mathbb{R}^n)$) is only a proper subspace of the Sobolev space $W_p^k(\mathbb{R}^n)$ to which Theorem 3 applies.

As an application of our results, we consider the orthogonal projection on a space spanned by translates of a radial basis function. A comprehensive analysis of the best least squares approximations onto radial functions spaces and their approximation orders can be found in Ron [1992], but we will use our theorems in this paper to deduce L^p-convergence orders of best least squares approximations. We take φ to belong to a class specified in Buhmann & Micchelli [1992] which contains all the functions that are mentioned in the current literature, see, e.g., the survey by Buhmann [1993]. We do not state here the full set of assumptions required for φ to be in that class, but mention the salient condition that $\widehat{\varphi}$ must have a singularity at the origin, viz., $\widehat{\varphi}(t) \sim \|t\|_2^{-k+1-\delta}$ for small $\|t\|_2$, where $\delta \in (0,1]$ and k is a positive integer. It must also satisfy a decay condition $|\widehat{\varphi}(t)| = O(\|t\|_2^{-n-\varepsilon})$ for $\varepsilon > 0$ and $\|t\|_2 \to \infty$. The space of approximants is then defined as

$$S_2(\varphi) := \Big\{ \sum_j d_j \varphi(\cdot - j) \mid |d_j| \le C\|j\|_\infty^{-k+1-\delta-\delta'} \text{ for some positive } \delta' \Big\} \cap L^2(\mathbb{R}^n).$$

In fact, spaces other than the above $S_2(\varphi)$ can be used as the underlying approximation space but we restrict attention to $S_2(\varphi)$ in this example for the sake of simplicity. We have set the notations in such a way that k here can be identified with the k of Theorem 1. The approximation scheme we use is the *orthogonal projector* $P : L^2(\mathbb{R}^n) \to S_2(\varphi)$. It follows from the required properties of φ that P can be written as

$$Pf = \sum_j (f, \psi(\cdot - j))\psi(\cdot - j), \qquad f \in L^2(\mathbb{R}^n), \tag{3.23}$$

as we shall explain in the sequel. The inner product in (3.23) is the standard inner product

$$(f_1, f_2) = \int f_1 \overline{f}_2, \qquad f_1, f_2 \in L^2(\mathbb{R}^n). \tag{3.24}$$

Further, the function ψ in (3.23) is a function in $S_2(\varphi)$ whose multiinteger translates are mutually orthonormal with respect to (3.24) and which can be defined by its Fourier transform

$$\widehat{\psi} = \frac{\widehat{\varphi}}{\sqrt{\sum_j |\widehat{\varphi}(\cdot - 2\pi j)|^2}}. \tag{3.25}$$

The orthonormality property we claim ψ to have is easily verified. Further, it is established in Theorem 3.1 on p. 330 and Theorem 3.3 on p. 333 of Buhmann & Micchelli [1993] that for φ from the said class of radial basis functions, ψ satisfies the decay assumption on ψ imposed in the statement of Theorem 1 for $\nu_\psi = n + \delta$. It therefore decays fast enough to admit actually any $f \in L^p(\mathbb{R}^n)$, $p \in [1, \infty]$, in (3.23), namely, for every $f \in L^p(\mathbb{R}^n)$, the series in (3.23) converges unconditionally and uniformly on compacta to an $L^p(\mathbb{R}^n)$ function. The sum in (3.23) can be rearranged so that the coefficients associated with the multiinteger translates of the radial function φ decay fast enough to satisfy the requirement in the definition of our space $S_2(\varphi)$, cf. Theorem 3.2 on p. 332 in the mentioned paper. Therefore it is settled that (3.23) is indeed the desired projection operator.

We need to verify that the remaining conditions on ψ of Theorem 1 and 3 hold and that (3.23) is actually of the form of the approximation operators we study in this work. To begin with, we observe that the coefficients μ that form ψ as the semi-discrete convolution $\psi = \varphi *' \mu$ satisfy the decay assumptions of Theorem 3 for $\varepsilon = \delta$, and the singularity of $\widehat{\varphi}$ at the origin implies (3.6) because

$$\widehat{\mu} = \frac{1}{\sqrt{\sum_j |\widehat{\varphi}(\cdot - 2\pi j)|^2}}.$$

Next, we need to know that the coefficients of $\psi(\cdot - j)$, $j \in \mathbb{Z}^n$, in the present approximation scheme (3.23) are derived from a convolution process with an appropriate convolutor λ. Indeed, taking $\lambda := \psi(-\cdot)$, we certainly have $(f, \psi(\cdot - j)) = (\lambda * f)(j)$. Also, as proved in Theorem 3.1 on p. 330 of the mentioned paper, ψ decays in such a way to provide (2.1). It satisfies the moment conditions (2.2) for the present k due to $\widehat{\varphi}$'s singularity at the origin and due to $\widehat{\psi}$'s form (3.25). Consequently, Theorems 1 and 3 are applicable to the present case as soon as we show that $\psi *'$ is the identity on Π_{k-1}. That final requirement follows, by a standard argument, from the fact that ψ satisfies the so-called Strang-Fix conditions, namely

$$D^\alpha \widehat{\psi}(2\pi\beta) = \delta_{\beta 0}\delta_{\alpha 0}, \quad \beta \in \mathbb{Z}^n, \quad |\alpha| < k.$$

These conditions can be verified directly from the definition of $\widehat{\psi}$ in (3.25) and from the singularity of the radial function's Fourier transform at the origin. Therefore, our Theorems 1 and 3 apply to this example and provide the desired L^p-approximation orders.

References

1. de Boor, C., K. Höllig and S. Riemenschneider (1993) *Box Splines*, Springer (New York).
2. de Boor, C., R. A. DeVore and A. Ron (1993) Approximation from shift-invariant subspaces of $L_2(\mathbb{R}^d)$, to appear in Trans. Amer. Math. Soc.
3. de Boor, C., and A. Ron (1992a) Fourier analysis of approximation power of principal shift-invariant spaces, Constr. Approx. **8** (1992), 427–462.

4. de Boor, C., and A. Ron (1992b) The exponentials in the span of the integer translates of a compactly supported function: approximation orders and quasi-interpolation, J. London Math. Soc. **45** (1992), 519–535.
5. Buhmann, M. D. (1989) *Multivariable interpolating using radial basis functions*, Ph.D. Dissertation, University of Cambridge.
6. Buhmann, M. D. (1993) New developments in the theory of radial basis function interpolation, in *Multivariate Approximation: From CAGD to Wavelets*, K. Jetter and F. I. Utreras, eds., World Scientific (Singapore), 35–75.
7. Buhmann, M. D., N. Dyn and D. Levin (1993) On quasi-interpolation with radial basis functions on non-regular grids, to appear in Constr. Approx.
8. Buhmann, M. D. and C. A. Micchelli (1992) On radial basis approximations on periodic grids, Mathematical Proceedings of the Cambridge Philosophical Society **112**, 317–334.
9. Dyn, N., I. R. H. Jackson, D. Levin and A. Ron (1992) On multivariate approximation by the integer translates of a basis function, Israel J. Math. **78**, 95–130.
10. Dyn, N. and A. Ron (1993) Radial basis function approximation: from gridded centres to scattered centres, CMS Technical Summary Report #94–3, University of Wisconsin–Madison.
11. Jia, R.-Q. and J.-J. Lei (1993) Approximation by multiinteger translates of functions having global support, J. Approx. Th. **72**, 2–23.
12. Lei, J.-J., and R.-Q. Jia (1991) Approximation by piecewise exponentials, SIAM J. Math. Anal. **22**, 1776–1789.
13. Ron, A. (1992) The L_2-approximation orders of principal shift-invariant spaces generated by a radial basis function, in *Numerical Methods of Approximation Theory*, Vol. 9, D. Braess and L. L. Schumaker, eds., Birkhäuser (Basel), 245–268.
14. Stein, E. M. and G. Weiss (1971) *Introduction to Fourier Analysis in Euclidean Spaces*, Princeton University Press (Princeton).

Acknowledgements. The first author's research was supported by a BASF Research Grant to the University of Wisconsin–Madison. This work was also sponsored by the National Science Foundation (grant DMS-9224748), and by the U.S. Army Research Office (contract DAAL03-90-G-0090).

University of Wisconsin–Madison
Department of Computer Sciences
1210 West Dayton Street
Madison, WI 53706

E-mail: $\left.\begin{matrix}\text{mdb}\\ \text{amos}\end{matrix}\right\}$@cs.wisc.edu

3D Curve Reconstruction from Degraded Projections

B. Chalmond, F. Coldefy, and B. Lavayssière

Abstract. Our purpose is the reconstruction of a 3D curve from a limited number of projections. These projections are highly blurred. The curve is represented by a parameterized spline function. The estimation of the spline parameters is done in the context of hidden data. In this framewok we build an iterative algorithm which gives a new viewpoint of the "active curve" methods developed in image analysis.

§1. Introduction

We assume the data are composed of a small number L of projections of a 3D unknown curve, say $\mathcal{C}^*$, ($L < 10$ in our application). The projection images are very degraded: each observed projection of the curve $\mathcal{C}^*$ is not a binary image of a 2D curve, but is a grey level image digitized on a square grid S^ℓ. Such an image can be seen as a blurred projection, or in other words as a plane having discontinuities represented by an uneven valley in which the theoretical curve projection of $\mathcal{C}^*$ lies approximately in the bottom, [1]. Furthermore, due to the degradation, the valley may be the union of several valleys. For the images $g = \{g^\ell, \ell = 1, ..., L\}$ we assume that $E(g_s^\ell) < 0$ if the site $s \in S^\ell$ belongs to a valley, and $E(g_s^\ell) = 0$ otherwise, E stands for the mathematical expectation. Note that we do not assume any knowledge about the physical degradation system. We denote by $\{\mathcal{P}^\ell, \ell = 1, ..., L\}$ the geometrical projection functions on the image planes. For instance, in our application each $\mathcal{P}^\ell$ is defined from a point source O^ℓ which illuminates the curves $\mathcal{C}^*$ with a 3D fan-beam of γ-rays. Denoting $\tilde{S}^\ell$ the projection plane associated to the source O^ℓ, for every point $c \in \mathcal{C}^*$, the non-degraded projection is simply $\mathcal{P}^\ell(c) = \tilde{S}^\ell \cap \overline{O^\ell c}$, where $\overline{O^\ell c}$ is the ray emitted by O^ℓ and passing through c.

Our goal is to determine a 3D curve, say $\mathcal{C}$, whose projections $\mathcal{P}^\ell(\mathcal{C})$ are coherent with the degraded projections of $\mathcal{C}^*$, that is with the valleys. This is an attempt to reconstruct $\mathcal{C}^*$, but L being too small, it is illusory to hope for an accurate reconstruction. Hence, we shall only seek a curve which presents

Wavelets, Images and Surface Fitting
P. J. Laurent, A. Le Méhauté, and L. L. Schumaker (eds.), pp. 113–119.

ISBN 1-56881-040-7.

the following characteristics of $\mathcal{C}^*$: its situation in the space, its orientation, and its general shape. In our presentation, we shall not deal with the technical difficulties arising from the ordering of the discretized curve points.

§2. The Curve Model

Our approach is related to the "active curve" method, mainly developed in image analysis research. In the two dimensional case, a model curve is placed on the image and is subject to the action of "external forces" which move and deform it, to best fit it to the desired features in the image. Formulations of problems of this kind are mostly deterministic and involve variational principles [2,3]. Stochastic formulations based on probabilistic models have also been proposed [4]. If ω denotes the generic curve, these approaches consist in constructing an energy function $U(\omega, g)$ which tells us how plausible any curve ω is, given the data g, and then to compute iteratively $\min_\omega U(\omega, g)$, starting with an initial curve ω^0. In fact in most of the cases, the active curve ω^k obtained at each iteration k converges to a local minimum which is highly dependent of ω^0: the curve is attracted by the nearest features around ω^0 in g. Then, the practical strategy consists in choosing ω^0 in the vicinity of the desired features. Here, we want to alleviate this drawback since we know there is only one curve to reconstruct.

We search for a curve represented by $\mathcal{C} = \omega + \delta$, where ω is a smooth curve and $\omega + \delta$ is a deformed version of ω by the deformation δ. ω is a deterministic global representation. As in [5], each coordinate x, y, z of ω is a cubic spline function on $[0,1]$:

$$\omega(u) = (x(u), y(u), z(u)) = \sum_{j=1}^{m} B_j(u)\theta_j,$$

where $u \in [0,1]$. Here $\theta_j = (\theta_j^x, \theta_j^y, \theta_j^z)$ are the spline parameters, and m is the number of knots defining the spline function on $[0,1]$, while B_j are the associated B-spline basis functions. Pratically, the curve ω is discretized into $\{\omega_i = \omega(u_i); i = 1, ..., n\}$, $m < n$. The fraction n/m defines the smoothing power of the spline. Denoting by $B_j^i = B_j(u_i)$ the generic element of the B-spline matrix for this discretization, we have $\omega_i = \sum_{j=1}^m B_j^i \theta_j$, and finally the matrix form of the discrete curve

$$\omega = B\theta,$$

where now ω is a matrix of size $n \times 3$. δ is a stochastic local representation. $\omega_i + \delta_i$ is seen as a random deformation of ω_i by δ_i, according to a Gibbs distribution, that will be described below. The global representation captures the general shape, whereas the local one reconstructs details that the global abstraction misses.

§3. The Energy without δ

Let $\omega^* = B\theta^*$ be the *true unknown* global curve. θ^* is unknown and is the main parameter to estimate. If "noisy" discretized curve versions of $\mathcal{P}^\ell(\omega^*)$ were given, say ξ^ℓ, then an estimation of θ^* could be

$$\hat{\theta} = \arg\min_\theta \sum_\ell \| \xi^\ell - \mathcal{P}^\ell(B\theta) \|^2 . \tag{1}$$

Here, the difficulty arises from the fact that ξ cannot be directly observed.

For every n-polygonal line ω, (not necessary of spline type), we define the log-likelihood of θ with respect to (ω, g) by

$$\log P_\theta(\omega \mid g) \propto -U_\theta(\omega, g)$$
$$U_\theta(\omega, g) = \sum_{\ell=1}^{L} U_\theta^\ell(\mathcal{P}^\ell(\omega), g^\ell),$$

where P_θ denotes a probability. The energies $\{U_\theta^\ell\}$ are dependent because of the projections $\mathcal{P}^\ell$ which are applied to the same ω. They are defined by

$$U_\theta^\ell(\mathcal{P}^\ell(\omega), g^\ell) = \sum_{i=1}^{n} \| \mathcal{P}^\ell(\omega_i) - \mathcal{P}^\ell(B_i\theta) \|^2 + \sum_{i=1}^{n} \| s_i^\ell(\omega, g) - \mathcal{P}^\ell(B_i\theta) \|^2,$$

where $s_i^\ell(\omega, g) \in S^\ell$ is called the *"global attractor site"* of $\mathcal{P}^\ell(\omega_i)$ and is computed from g^ℓ. Its definition is application-dependent, and is rather difficult to design. At this point, any prior information concerning the characteristics of the degradation can be taken into account. For example, consider the simplest definition, that is the barycenter

$$s_i^\ell(\omega, g) \propto \sum_{j \in A_i^\ell} g_j^\ell \, j , \tag{2}$$

where A_i^ℓ is the set of sites in the projection plane for which $\mathcal{P}^\ell(\omega_i)$ is the nearest point of the projected curve: $j \in A_i^\ell$ if $i = \arg\min_t \| j - \mathcal{P}^\ell(\omega_t) \|$. Note that every site in g^ℓ contributes to the definition of these attractors and this energy favours the bottom valleys.

§4. The Algorithm without δ

The question now is how to estimate θ^* ? We place this estimation problem in the context of hidden data, taking g as the observed data and ω as the hidden data. This terminology is meaningful when one tries to estimate θ^* by the maximum likelihood principle. The interested reader can refer to [6] for a general presentation of this topic. Since ω is hidden, the maximun likelihood principle is not directly applicable, and thus we propose the following iterative algorithm. Let Ω denote the random curve whose ω is the occurence. At step $(k+1)$, the estimate of θ^* is

$$\theta^{k+1} = \arg\min_\theta E[U_\theta(\Omega, g) \mid \theta^k] ,$$

where the mathematical expectation $E[\cdot]$ is defined by

$$E[U_\theta(\Omega, g) \mid \theta^k] = \sum_\omega U_\theta(\omega, g) P_{\theta^k}(\omega) .$$

We choose the simplest expression for the probability distribution P_{θ^k}:

$$P_{\theta^k}(\omega) = I(\omega = B\theta^k), \tag{3}$$

where I is the indicator function. In this case, it follows $E[U_\theta(\Omega, g) \mid \theta^k] = U_\theta(\omega^k, g)$. Thus, at the step $(k+1)$, the solution is

$$\begin{aligned} \theta^{k+1} &= \arg\min_\theta U_\theta(\omega^k, g) , \\ \omega^{k+1} &= B\theta^{k+1} . \end{aligned} \tag{4}$$

At each step k, note that the "observations" ξ^ℓ introduced in (1), can be indentified with $(\mathcal{P}^\ell(\omega^k), s^\ell(\omega^k, g))$. In (4), we have to solve a non-linear least-square problem. In our application, if (x, y, z) denotes the coordinate system with the source $\mathcal{O}^\ell$ at position $(\mathcal{O}^\ell_x, \mathcal{O}^\ell_y, \mathcal{O}^\ell_z)$, and if the projection plane $\tilde{S}^\ell$ is horizontal at $z = \kappa$, it is straigtforward to prove that

$$\mathcal{P}^\ell_x(B_i\theta) = \frac{B_i\theta^x - \mathcal{O}^\ell_x}{B_i\theta^z - \mathcal{O}^\ell_z}(\kappa - \mathcal{O}^\ell_z) + \mathcal{O}^\ell_x,$$

and with a similar expression for the y coordinate. The convergence of this deterministic algorithm is guaranted since at each iteration the log-likelihood decreases, and it is very fast. Let us emphasize that a serious weakness of this kind of approach is the use of the attractor sites, a difficulty which is also present in the other active curve methods when we have to express the "external forces" energy.

§5. The Algorithm with δ

The role of the above energy is to determine a global representation of the curve taking account of the whole data set g. But such a representation must also be stable with respect to the local data surrounding the curve. To this end, we rewrite the energy as $U_\theta(\omega + \delta, g)$, where δ is included in order to determine a curve $\mathcal{C} = \omega + \delta$ which satisfies this stability. On δ, we define a probability distribution $P(\delta \mid \omega, g)$ which characterises the local environment, and then the energy is iteratively minimized according to a Bayesian approach based on this distribution.

The hidden data now are composed of ω and δ. Let Δ denote the random deformation whose δ is the occurrence. At the step $(k+1)$, the estimate of θ^* is $\theta^{k+1} = \arg\min_\theta E[U_\theta(\Omega+\Delta, g) \mid \theta^k]$. The mathematical expectation $E[\cdot]$ is defined by

$$\begin{aligned} E[U_\theta(\Omega+\Delta, g) \mid \theta^k] &= \sum_{\omega,\delta} U_\theta(\omega+\delta, g) P(\delta \mid \omega, g) P_{\theta^k}(\omega) \\ &= \sum_{\delta} U_\theta(\omega^k+\delta, g) P(\delta \mid \omega^k, g), \end{aligned} \tag{5}$$

where again the distribution (3) has been considered for P_{θ^k}. In this new expression, note that the model $B\theta$ concerns $\omega^k+\delta$. We define the following Gibbs distribution on δ:

$$\begin{aligned} P(\delta \mid \omega^k, g) &\propto \; P(\delta)\; P(g \mid \omega^k+\delta) \\ &\propto \; \exp - \sum_i \| \delta_i - \delta_{i-1} \|^2 \\ &\cdot \; \exp \; -\lambda \sum_\ell \sum_i \| \bar{s}_i^\ell(\omega^k, g) - \mathcal{P}^\ell(\omega_i^k) \|^2, \end{aligned}$$

where $\bar{s}_i^\ell(\omega^k, g) \in S^\ell$ is called the *"local attractor site"* of $\mathcal{P}^\ell(\omega_i^k)$. The global attractor sites are computed by visiting all the sites in S^ℓ, and looking for the nearest point on ω^k for any such site. Conversely, for the local attractor sites, one moves along the curve ω^k, and for every site $i \in \omega^k$ one looks for the nearest valley sites in a neighborhood $\mathcal{N}_i^\ell$ of this site. As in (2), the simplest definition is the barycenter

$$s_i^\ell(\omega, g) \;\propto\; \sum_{j \in \mathcal{N}_i^\ell} g_j^\ell \, j \, .$$

λ is a weighting parameter which balances the influence of the two terms in the Gibbs distribution, and is determined by heuristic considerations. The sum (5) is computed using a Monte Carlo method that simulates a finite set of configurations δ according to the Gibbs distribution, [4,6]. It remains to give a scheme for the chaining of the iterations. Unfortunately, we have no result concerning the convergence of this second algorithm. We think it is dependent on the scheme type. In our application, the reconstruction is a two-stage procedure. First it is started by using the first algorithm, and then is finished by the second one, but with ω fixed to the configuration obtained at the first stage.

§6. Experimental Results

In the context of our application, we now illustrate the method on artificial data. Figure 1 presents the degraded projections respectively obtained from

$L = 3$ aligned γ-sources. If G denotes the barycenter of $\mathcal{C}^*$, then the sources are placed in such a way that their angles satisfy $(\overline{O^1G}, \overline{O^2G}) = (\overline{O^2G}, \overline{O^3G})$ $= 20^o$. Furthermore, the projection images belong to the same plane. On these images, the noise is real, but the valleys of crescent shape have been artificially added, in order to control the reconstruction results. We have run our two-stage algorithm, starting with an initial curve ω^0 randomly drawn in the 3D space. The first stage needs ten iterations. Figure 1 shows also how the projections of the reconstructed curve at different iterations, are coherent with the degraded projections. The first row of Figure 1 shows the projections $\{\mathcal{P}^\ell(\omega^0)\}$. The second row depicts $\{\mathcal{P}^\ell(\omega^1)\}$, and the third one gives a $\{\mathcal{P}^\ell(\mathcal{C})\}$ set obtained after convergence.

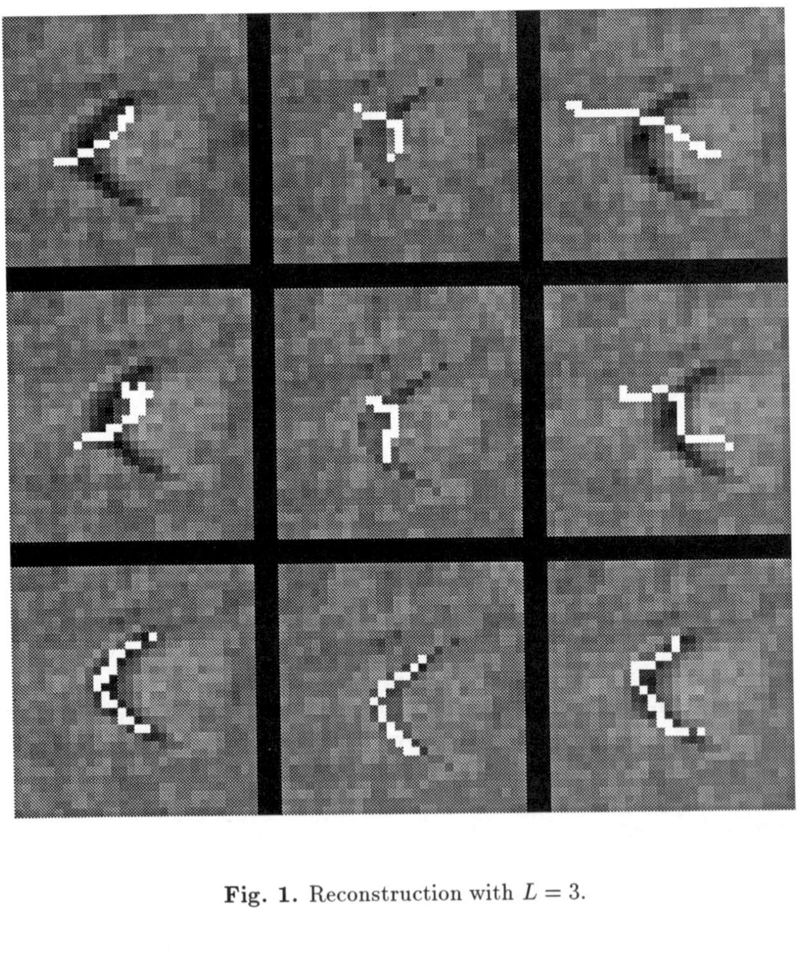

Fig. 1. Reconstruction with $L = 3$.

Acknowledgments. This work was supported by Electricité De France (DER - SDM) and SUDIMAGE (Automatic Imaging).

References

1. Azencott, R., B. Chalmond and F. Coldefy, Association of adaptative smoothing and Markovian models for detection of valley bottoms on strongly noisy images, in *Proc. of Int. Conf. on Patt. Recogn.*, The Hague, 1992.
2. Munford, D. and Shah J., Boundary detection by minimizing functionals, I., in *Proc. of Comp. Vision Patt. Recogn.*, San Francisco, 1985.
3. Terzopoulos, D., A. Witkin and M. Kass, Constraints deformable models: Recovery 3D shape and nonrigid motion, Artificial Intelligence **36** (1988), 91–123.
4. Grenander, U. and Keenan D., Towards automated image understanding, J. Appl. Statis., **16** (1989), 207–221.
5. Leitner, F., I. Marque, S. Lavallée and P. Cinquin, Dynamic segmentation: Finding the edges with snake splines, in *Curves and Surfaces*, P.-J. Laurent, A. LeMéhauté, and L.L. Schumaker (eds.), Academic Press, New-York, (1991), 279-284.
6. Dempster, A. P., N. M. Laird and Rubin D. B., Maximum likelihood from incomplete data via the EM algorithm. J. R. Statist. Soc. B. **39** (1977), 1–38.
7. Chalmond, B., An iterative Gibbssian technique for reconstruction of m-ary images. Pattern Recognition **22** (1989), 747–761.

Bernard Chalmond
Cergy-Pontoise University
8, Le Campus
95033 Cergy-Pontoise cedex, FRANCE
chalmond@ diam1.ens.cachan.fr

F. Coldefy is with SUDIMAGE group, Cachan, FRANCE

B. Lavayssière is with the DER-SDM Department,
EDF, Chatou, FRANCE

A Study of Compactly Supported Scaling Functions and Wavelets

Charles K. Chui and Jianzhong Wang

Abstract. This is a somewhat comprehensive study of compactly supported scaling functions and wavelets. While many papers have already been written recently on various problems in this subject, this paper is intended to give a systematic treatment as well as to report on our recent results on asymptotically optimal time-frequency localization by symmetric totally positive scaling functions (STOPLETS) and their corresponding completely oscillating wavelets (COWLETS).

§1. Introduction

This paper is devoted to the study of compactly supported scaling functions and their corresponding semi-orthogonal wavelets (also known as pre-wavelets) with minimum support. As usual, we only consider real-valued functions and restrict our attention to the $L^2 := L^2(\mathbb{R})$ setting. We will also use the notion of multiresolution analysis (MRA) of L^2, of Meyer [21] and Mallat [20], generated by any such scaling functions.

Recall that corresponding to any scaling function ϕ, there is a sequence $\mathbf{p} = \{p_k\} \in \ell^2 := \ell^2(\mathbb{Z})$ that uniquely determines ϕ (under a certain normalization condition, say $\widehat{\phi}(0) = 1$ provided that $\widehat{\phi}$ is continuous) as follows:

$$\phi(x) = \sum_{k \in \mathbb{Z}} p_k \phi(2x - k). \tag{1.1}$$

This sequence $\mathbf{p}$ is called the *two-scale sequence* (or *mask*) of the scaling function ϕ, while the Laurent series

$$P(z) := \frac{1}{2} \sum_{k \in \mathbb{Z}} p_k z^k, \tag{1.2}$$

is called the *two-scale symbol* of ϕ. Note that this notation differs from the usual symbol notation by a multiplicative factor of $\frac{1}{2}$.

Wavelets, Images and Surface Fitting
P. J. Laurent, A. Le Méhauté, and L. L. Schumaker (eds.), pp. 121–140.

ISBN 1-56881-040-7.

In general, corresponding to a sequence $\mathbf{p} \in \ell^2$ that satisfies certain mild conditions, the functional equation (1.1) always has a solution ϕ which, however, may only be a distribution. Nevertheless, even if ϕ were a function in L^2, such a solution may not generate an MRA of L^2, and hence, is not necessarily a scaling function. For this reason, it is better to call ϕ a *refinable distribution* in general. In this paper, we will be concerned mainly with refinable distributions, and particularly scaling functions that are compactly supported. Our goal, however, is not to discuss the issues of existence, uniqueness, and regularity (either global or local) of the solution of the functional equation (1.1), since these topics are well documented in the two fundamental papers listed in [13] and the well-known book [12] as well.

Our first task, to be discussed in Section 2, is to classify all compactly supported scaling functions that generate the same MRA. Among all such functions, we will show that there is one with minimum support which possesses two very nice algebraic properties: the collection of all of its integer translates is linearly independent, and any compactly supported scaling function that generates the same MRA is a finite linear combination of its integer translates. These results were first mentioned by our earlier work [5], but a more complete and systematical treatment will be given in Section 2 of this paper.

In Section 3, we will devote our attention to the study of symmetric scaling functions and symmetric or anti-symmetric wavelets. One new, and perhaps interesting, result reported in this section is that if the compactly supported scaling function in an MRA with minimum support is asymmetric, then any other compactly supported scaling function in this MRA is also asymmetric.

In Section 4, we will restrict our attention to those scaling functions that are totally positive (TP). The importance of such functions is that they give rise to asymptotically optimal time-frequency localization as governed by the uncertainty principle. As to their corresponding semi-orthogonal wavelets, we will modify the notion of frequency localization to the consideration of positive values only and also demonstrate that these wavelets yield asymptotically optimal time-frequency localization as governed by the corresponding modified uncertainty principle for bandpass filters.

On the other hand, when we study scaling functions that are either interpolating or orthonormal in Section 5, we will see that such scaling functions and their corresponding wavelets do not give good time-frequency localization if we desire very high orders.

In spite of the many fruitful results in the literature, however, a few problems remain open. We will discuss some of them in Section 6, where we will also briefly comment on some of the references concerning this present work.

To facilitate our presentation, we will need the following notations:

$$\left.\begin{array}{l} \pi := \text{the class of all Laurent polynomials.} \\ \pi^0 := \{P \in \pi : P(1) = 1\}. \\ \text{For any finite sequence } \mathbf{p},\ |\mathbf{p}| := \text{ length of } \mathbf{p}. \\ \text{For any } P \in \pi \text{ with coefficient sequence } \mathbf{p},\ \deg P := |\mathbf{p}| - 1. \end{array}\right\} \tag{1.3}$$

In addition, when we write

$$A(z) = \sum_{n=k_1}^{k_2} a_n z^n \tag{1.4}$$

we will always assume that $k_1 \leq k_2$ and $a_{k_1} a_{k_2} \neq 0$.

A complex number z_0 is called a symmetric zero of a Laurent polynomial $A(z)$ if $z_0 \neq 0$ and $A(z_0) = A(-z_0) = 0$. Later when we consider the zeros of a Laurent polynomial we always assume those different from 0; i.e. by letting $\mathbb{C}^0 := \mathbb{C}\backslash\{0\}$, we will only count the zeros of a Laurent polynomial in the set $\mathbb{C}^0$. Throughout, the unit circle will be denoted by

$$\Gamma = \{z\colon\ |z| = 1, \qquad z \in \mathbb{C}\}.$$

Let g be a distribution. Then if $\int g\varphi = 0$ holds for any test function φ that vanishes on a compact interval $[a, b]$, we say that g has *compact support* and the notation supp $g \subset [a, b]$ is used. If supp $g \subset [a, b]$ but supp $g \not\subset [\alpha, \beta]$ for any proper subinterval $[\alpha, \beta]$ of (a, b), we write supp $g = [a, b]$ and say that the *support* of g is $[a, b]$. Note that the definition of the support of a distribution here is slightly different from that in Harmonic Analysis. If supp $g = [a, b]$ the support length of g is $|\text{supp } g| = b - a$.

Let f be a function in L^2, or more weakly, an increasing tempered distribution. Then its Fourier transform is denoted by

$$\widehat{f}(\omega) = \int_{-\infty}^{\infty} f(x) e^{-ix\omega} dx.$$

Furthermore, the autocorrelation of any $\phi \in L^2$ will be denoted by

$$\Phi(x) = \int_{-\infty}^{\infty} \phi(x + y)\overline{\phi(y)} dy \tag{1.5}$$

(with lower case for the function and upper case for its autocorrelation) and the *Euler-Frobenius Laurent series* (EFL series) corresponding to any $\phi \in L^2$ is denoted by

$$B_\phi(z) = \sum_{n\in\mathbb{Z}} \Phi(n) z^n. \tag{1.6}$$

Observe that since $\phi \in L^2$, we have $\Phi \in C_0$, where

$$C_0 := \{f \in C(R)\colon\ f(x) \to 0 \text{ as } x \to \pm\infty\}.$$

Of course, if ϕ is compactly supported, so is Φ, so that $B_\phi \in \pi$ is an EFL polynomial.

Finally, N_m will denote the m-th order polynomial B-spline with knot sequence $\mathbb{Z}$, defined inductively by

$$N_m(x) := (N_{m-1} * N_1)(x) = \int_0^1 N_{m-1}(x-t)dt,$$

where $N_1 = \chi_{[0,1)}$.

§2. Minimally Supported Scaling Functions and Classification of Scaling Functions

In this section, we will discuss how compactly supported scaling functions are classified. The motivation is based upon the fact that different scaling functions can be used to generate the same MRA. A trivial observation is that if φ is a scaling function that generates some MRA, then $c\varphi(x-j)$, for any $j \in \mathbb{Z}$ and $c \neq 0$, will generate the same MRA. Let $P(z)$ be the two-scale symbol of φ. Then the two-scale symbol of $c\varphi(x-j)$ turns out to be $z^j P(z)$. Recall that $\varphi(0) \neq 0$ for any compactly supported scaling function φ. Hence, when we classify scaling functions of the same MRA, we will first normalize a compactly supported scaling function φ by $\widehat{\varphi}(0) = 1$, and then identify φ with any of its integer translates. That is, the two-scale symbols $P(z)$ and $z^j P(z)$ are considered to be equivalent.

In general, since the Fourier transform of a compactly supported refinable distribution ϕ is an entire function, the function $\widehat{\phi}(\omega)$ has at most a finite multiple zero at $\omega = 0$. Hence, by dividing $\widehat{\phi}$ by a suitable power of ω, say ω^k, we get $\widehat{\varphi}(\omega) = \frac{\widehat{\phi}(\omega)}{\omega^k}$ with $\widehat{\varphi}(0) \neq 0$. It is easy to check that φ is still a refinable distribution. For convenience, we will always assume, in this section, that a compactly supported refinable distribution ϕ is normalized by $\widehat{\phi}(0) = 1$. Then the two-scale symbol $P(z)$ of ϕ always satisfies $P(1) = 1$.

Now we begin our discussion by considering the relation between the support lengths of refinable distributions and the polynomial orders of their two-scale symbols.

Lemma 2.1. *Let φ be a refinable distribution with two-scale symbol $Q(z) \in \pi$. Then there is a sufficiently large integer m such that $\varphi * N_m \in C(\mathbb{R})$.*

Proof: Let m be an integer so chosen that

$$\sup_{|z|=1} |Q(z)| < 2^{m-1}.$$

Then the symbol of $\varphi * N_m$ is $\left(\frac{1+z}{2}\right)^m Q(z)$, and it follows from [12], Lemma 7.1.7, that $\varphi * N_m \in C(\mathbb{R})$. ■

Lemma 2.2. *Let φ be a refinable distribution with two-scale symbol $Q(z) = \frac{1}{2}\sum_{j=k_1}^{k_2} q_j z^j$. Then supp $\varphi = [k_1, k_2]$.*

Proof: That supp $\varphi \subset [k_1, k_2]$ has been proved in [14]. We will now prove that supp $\varphi \supset [k_1, k_2]$. Without loss of generality, we assume that $k_1 = 0$ and $k_2 = k$, and consider

$$\varphi(x) = \sum_{j=0}^{k} q_j \varphi(2x - j), \qquad q_0 q_k \neq 0. \tag{2.1}$$

Let us first assume that $\varphi \in C(\mathbb{R})$. If supp $\varphi \not\supset [0, k]$, then there exists a positive integer n such that φ vanishes either on $\left[0, \frac{1}{2^n}\right]$ or on $\left[k - \frac{1}{2^n}, k\right]$. Now we assume that

$$\varphi(x) = 0, \quad x \in \left[0, \frac{1}{2^n}\right]. \tag{2.2}$$

Writing $\varphi_j(x) = \varphi(x - j + 1)$, $j = 1, \ldots, k$, $x \in [0, 1]$, we obtain

$$\varphi_j(x) = \begin{cases} \displaystyle\sum_{\ell=1}^{2j-1} q_{2j-\ell-1} \varphi_\ell(2x), & x \in \left[0, \frac{1}{2}\right] \\ \displaystyle\sum_{\ell=1}^{2j} q_{2j-\ell} \varphi_\ell(2x - 1), & x \in \left[\frac{1}{2}, 1\right], \end{cases} \tag{2.3}$$

$j = 1, \ldots, k$. Choosing $j = 1$ in (2.3), we get, by using (2.2) first, that $\varphi_1(x) = 0$, $x \in \left[0, \frac{1}{2^n}\right]$, and then $\varphi_1(x) = 0$, $x \in \left[\frac{1}{2^n}, \frac{1}{2^{n-1}}\right]$. By a repeated argument, we arrive at $\varphi_1(x) = 0$, $x \in [0, 1]$, and similarly, $\varphi_2(x) = 0$ by the second equation in (2.3) with $j = 1$. By successively choosing $j = 2, 3, \ldots, k$ in (2.3) and using the same trick as above, we finally obtain $\varphi_j(x) = 0$, $j = 1, \ldots, k$, $x \in [0, 1]$. That is, $\varphi(x) = 0$. This contradiction means that (2.2) is false. Similarly φ does not vanish on $\left[k - \frac{1}{2^n}, k\right]$. This establishes the lemma for $\varphi \in C(\mathbb{R})$.

Now we turn to the case that φ is only a refinable distribution but not necessarily in $C(\mathbb{R})$. By Lemma 1, we can choose m such that $\varphi * N_m \in C(\mathbb{R})$. It is well-known that since the two-scale symbol of $\varphi * N_m$ is $\left(\frac{1+z}{2}\right)^m Q(z)$, we have supp $\varphi * N_m = [0, k + m]$. On the other hand, we also have supp $N_m = [0, m]$. Hence, it follows that supp $\varphi = [0, k]$. ■

In general, the converse of Lemma 2.2 is not true. However, we still have the following result.

Lemma 2.3. *The two-scale symbol of a compactly supported refinable distribution is a rational function.*

Proof: Let φ be a refinable distribution with two-scale relation

$$\varphi(x) = \sum_{j \in Z} q_j \phi(2x - j)$$

and two-scale symbol $Q(z)$. If φ is a continuous function, then

$$Q(z) := \frac{1}{2}\sum_{j\in Z} q_j z^j = \frac{\sum\limits_{j\in Z}\varphi(\frac{j}{2})z^j}{\sum\limits_{j\in Z}\varphi(j)z^j}.$$

So, since φ has compact support, $Q(z)$ is a rational function.

In general, by Lemma 2.1, we may choose a positive integer m such that $\varphi * N_m \in C(\mathbb{R})$. It is obvious that $\varphi * N_m$ is compactly supported. Then the two-scale symbol $Q(z)\left(\frac{1+z}{2}\right)^m$ of $\varphi * N_m$ is rational, and so is the two-scale symbol of φ. ■

Now we are ready to prove the following.

Theorem 2.1. *A rational function $P(z)$ is a two-scale symbol of a compactly supported refinable distribution $\varphi(x)$ with $\widehat{\varphi}(0) = 1$ if and only if there are two polynomials $r(z)$, $\ell(z) \in \pi^0$ such that $P(z) = r(z)\frac{\ell(z^2)}{\ell(z)}$.*

Proof: If $P(z)$ is the two-scale symbol of a compactly supported refinable distribution φ with $\widehat{\varphi}(0) = 1$, then

$$\widehat{\varphi}(\omega) = P(z)\widehat{\varphi}\left(\frac{\omega}{2}\right), \qquad z = e^{-i\frac{\omega}{2}}. \tag{2.4}$$

Since φ is compactly supported, $\widehat{\varphi}(\omega)$ is an entire function on the complex plane $\mathbb{C}$. Hence, $\widehat{\varphi}$ has at most a finite number of 2π-periodic zeros in $\mathbb{C}$. That is, $\widehat{\varphi}(\omega) = \ell(e^{-i\omega})\widehat{\phi}(\omega)$ with $\ell \in \pi^0$, where $\widehat{\phi}$ is an entire function with no 2π-periodic zeros and $\widehat{\phi}(0) = 1$. From (2.4), we have

$$\widehat{\phi}(\omega) = r(z)\widehat{\phi}\left(\frac{\omega}{2}\right), \qquad z = e^{-i\frac{\omega}{2}}, \tag{2.5}$$

where $r(z) = P(z)\frac{\ell(z)}{\ell(z^2)}$ is a rational function and $r(1) = 1$. We rewrite $r(z) = \frac{r_1(z)}{r_2(z)}$, where $r_1, r_2 \in \pi^0$ and have no common factors. Recall that $\widehat{\phi}(\omega)$ has no 2π-periodic zeros. Therefore $\widehat{\phi}(\frac{\omega}{2})$ has no 4π-periodic zeros. From the fact that $\widehat{\phi}(\omega)$ and $\widehat{\phi}(\frac{\omega}{2})$ are entire functions, we conclude that $\deg r_2 = 0$ and $r \in \pi^0$. Hence, we have

$$P(z) = r(z)\frac{\ell(z^2)}{\ell(z)}, \qquad r, \ell \in \pi^0.$$

Conversely, let φ be a refinable distribution with symbol $P(z) = r(z)\frac{\ell(z^2)}{\ell(z)}$, $r, \ell \in \pi^0$. If ℓ is a trivial polynomial, then $P(z) = r(z) \in \pi^0$, and by Lemma 2.2, φ is compactly supported. So, let us assume that $\deg \ell \geq 1$, and let ϕ be a refinable distribution defined by

$$\widehat{\phi}(\omega) = r(z)\widehat{\phi}\left(\frac{\omega}{2}\right), \qquad z = e^{-i\frac{\omega}{2}}.$$

By Lemma 2.2, ϕ is compactly supported. It is easy to see that $\widehat{\varphi}(\omega) = \ell(e^{-i\omega})\widehat{\phi}(\omega)$, $\ell \in \pi^0$. This means that φ is a finite linear combination of integer translates of ϕ, so that φ itself is compactly supported. ■

Later we will consider

$$\pi^1 = \{P(z)\colon\ P(z) = r(z)\frac{\ell(z^2)}{\ell(z)}, \quad r, \ell \in \pi^0\}.$$

Then $\pi^0 \subset \pi^1$. For any $P(z) = r(z)\frac{\ell(z^2)}{\ell(z)}$, we write $\deg P = \deg r + \deg \ell$. From Theorem 2.1, we obtain

Corollary 2.1. *For a compactly supported refinable distribution φ with two-scale symbol $P(z)$,*

$$|supp\ \varphi| = \deg P. \tag{2.6}$$

Proof: Let $P(z) = r(z)\frac{\ell(z^2)}{\ell(z)}$ and ϕ be the compactly supported distribution with two-scale symbol given by $r(z)$, $r \in \pi^0$. By Lemma 2.2, we have $|\text{supp}\ \phi| = \deg r$. Writing $\ell(z) = \sum_{j=k_1}^{k_2} \ell_j z^j$, we also have

$$\varphi(x) = \sum_{j=k_1}^{k_2} \ell_j \phi(x-j), \qquad \ell_{k_1}\ell_{k_2} \neq 0,$$

which implies that $|\text{supp}\ \varphi| = |\text{supp}\ \phi| + k_2 - k_1 = \deg r + \deg \ell = \deg P$. ■

Now we consider all the scaling functions that generate the same MRA. Recall that a function ϕ is a scaling function if and only if $\phi \in L^2$ and it generates an MRA. Then a refinable distribution, even if it is in L^2 (and in this case we call it a refinable function), it is not necessarily a scaling function. It is well-known that a compactly supported refinable function ϕ is a scaling function if and only if there are two positive numbers A and B such that

$$A \leq \sum_{k\in\mathbb{Z}} |\widehat{\phi}(\omega + 2k\pi)|^2 \leq B. \tag{2.7}$$

The condition in (2.7) is called the "stability" condition of ϕ. It is a necessary condition for a noncompactly supported function in L^2 to be a scaling function (cf. [20,21]).

A direct consequence of (2.7) is that if ϕ is a compactly supported refinable function but is not a scaling function, then any finite linear combination of integer translates of ϕ is not a scaling function either. Later we will again refer to this fact.

For any compactly supported scaling function ϕ, we will use the notation MRA(ϕ) to denote the MRA generated by ϕ and use $V(\phi)$ to denote the L^2-closure of span$\{\phi(\cdot - j)\}_{j\in\mathbb{Z}}$. Then any function in $V(\phi)$ is refinable. (Again, to avoid unnecessary statements, we will identify a function $f \in V(\phi)$ with $cf(x-j)$, $c \neq 0$, $j \in \mathbb{Z}$, as mentioned at the beginning in this section.) It is evident that among all compactly supported functions in $V(\phi)$ there is one with minimum support. This minimally supported function in $V(\phi)$ has certain significant properties as shown in the following.

Theorem 2.2. *Let ϕ be a compactly supported scaling function and φ a minimally supported function in $V(\phi)$. Then the two-scale symbol of φ has minimum degree and $\widehat{\varphi}$ has no 2π-periodic zeros in $\mathbf{C}^0$. Furthermore, the converse also holds. In addition, any compactly supported function in $V(\phi)$ is a finite combination of some integer translates of φ. Finally, φ is unique.*

Proof: The fact that the two-scale symbol of φ has minimum degree is a direct consequence of Corollary 2.1. Using the same trick as in the proof of Theorem 2.1, we see that $\widehat{\varphi}$ has no 2π-periodic zeros. The converse is evident. The proof of the other statements can be found in our earlier work [5]. ■

We point out that the minimally supported scaling function is linear independent. This means that if $\sum_{j \in Z} c_j \varphi(x-j) = 0$ for an arbitrary complex sequence $\{c_j\}$, then $c_j = 0$ for all j. The following result was established in [11,22].

Lemma 2.4. *A compactly supported distribution ϕ is linearly independent if and only if $\widehat{\phi}$ has no 2π-periodic zeros.*

As a consequence of this result, the first statement in Theorem 2.2 can also be stated as follows.

Corollary 2.2. *φ is minimally supported in $V(\phi)$ if and only if φ is linearly independent.*

Since the minimally supported scaling function is unique (in an MRA), we can classify all compactly supported scaling functions in the same MRA by using the minimally supported one. That is, we have the following result.

Corollary 2.3. *Let $\mathcal{A}$ be the collection of all compactly supported scaling functions that generate the same MRA. Then there is a unique function φ in $\mathcal{A}$ with minimum support. Furthermore, any function ϕ in $\mathcal{A}$ is a finite linear combination of some integer translates of φ. In addition, if $P_\phi(z)$ and $P_\varphi(z)$ are the two-scale symbols of ϕ and φ, respectively, then*

$$P_\phi(z) = P_\varphi(z)\frac{\ell(z^2)}{\ell(z)},$$

where $\ell \in \pi^0$ has no root on Γ.

§3. Symmetric Scaling Functions and Wavelets

In many applications, the property of symmetry or anti-symmetry of scaling functions and wavelets is important. In general, a distribution φ is called *symmetric* (*anti-symmetric*) with respect some $a \in R$ if $\varphi(x) = \varphi(a-x)$ ($\varphi(x) = -\varphi(a-x)$), while a sequence $\mathbf{a} = (a_j)_{j\in Z}$ is called symmetric (anti-symmetric) with respect to an integer k, if $a_j = a_{k-j}$ ($a_j = -a_{k-j}$).

If the Fourier transform of a distribution φ exists, (for example, φ is an increasing tempered distribution), then φ is symmetric (anti-symmetric) with respect to some a if and only if $\widehat{\varphi}(\omega) = e^{-ia\omega}\widehat{\varphi}(-\omega)$ ($\widehat{\varphi}(\omega) = -e^{-ia\omega}\widehat{\varphi}(-\omega)$). Similarly a sequence $\mathbf{a}$ is symmetric (anti-symmetric) with respect to some k if and only if its symbol $A(z)$ satisfies $A(z) = z^k A(\frac{1}{z})$ ($A(z) = -z^k A(\frac{1}{z})$). There is a relation between the symmetry of refinable distributions and that of their masks. We have the following result.

Theorem 3.1. *Let ϕ be a refinable increasing tempered distribution with $\widehat{\phi}(\omega) \neq 0$ a.e. in $(2\pi, 2\pi+\epsilon)$ for some $\epsilon > 0$ and let $P(z)$ be its two-scale symbol. If ϕ is symmetric with respect to $d \in R$, then d must be an integer. Furthermore, ϕ is symmetric with respect to d if and only if its mask is symmetric with respect to d.*

Proof: We have

$$\widehat{\phi}(2\omega) = P(e^{-i\omega})\widehat{\phi}(\omega) \tag{3.1}$$

and

$$\widehat{\phi}(\omega) \neq 0 \qquad \text{a.e. in} \quad (2\pi, 2\pi+\epsilon).$$

By (3.1), $\widehat{\phi}(\omega) \neq 0$ a.e. in $\left(\frac{2\pi}{2^n}, \frac{2\pi+\epsilon}{2^n}\right)$, for all $n \in \mathbb{Z}^+$.

We now choose n_0 such that $\frac{2\pi+\epsilon}{2^{n_0}} \leq \epsilon$, and let $\beta = \frac{2\pi}{2^{n_0}}$, $\gamma = \frac{2\pi+\epsilon}{2^{n_0}}$. Then $\widehat{\phi}(2\omega) \neq 0$, a.e. in (β, γ); and it follows that $P(e^{-i\omega}) \neq 0$, a.e. in (β,γ). Since $(2\pi+\beta, 2\pi+\gamma) \subset (2\pi, 2\pi+\epsilon)$, we have

$$P(e^{-i\omega}) = \frac{\widehat{\phi}(2\omega)}{\widehat{\phi}(\omega)} = \frac{\widehat{\phi}(2(\omega+2\pi))}{\widehat{\phi}(\omega+2\pi)} \qquad \text{a.e. in } (\beta,\gamma).$$

By using the fact that $\widehat{\phi}(\omega) = e^{-id\omega}\widehat{\phi}(-\omega)$, we have $e^{-id\omega} = e^{-i(\omega+2\pi)d}$, a.e. in (β, γ). It follows that $e^{-2id\pi} = 1$. Hence, $d \in \mathbb{Z}$, and $P(e^{-i\omega}) = e^{-id\omega}P(e^{i\omega})$. This means that ϕ is symmetric with respect to d which implies that the mask is also symmetric with respect to d. Now if the coefficient sequence of $P(z)$ is symmetric with respect to d, we have $P(z) = z^d P(\frac{1}{z})$. Hence

$$\widehat{\phi}(\omega) = \prod_{j=1}^{\infty} P(e^{i\omega/2^j}) = \prod_{j=1}^{\infty}[P(e^{i\omega/2^j})e^{-d\omega i/2^j}] = e^{-d\omega i}\widehat{\phi}(-\omega),$$

so that ϕ is symmetric with respect to d. ■

In Theorem 3.1, the condition $\widehat{\phi}(\omega) \neq 0$ a.e. in $(2\pi, 2\pi+\epsilon)$ does not seem to be quite natural. However, this condition cannot be omitted for the theorem to hold. The following counterexample is borrowed from Zhou [24].

Let ϕ be a Meyer scaling function defined by

$$\widehat{\phi}(\omega)=\begin{cases}\frac{1}{\sqrt{2\pi}}, & |\omega|\le\frac{2\pi}{3}\\ \frac{1}{\sqrt{2\pi}}\cos\left[\frac{\pi}{2}\nu\left(\frac{3}{2\pi}|\omega|-1\right)\right], & \frac{2\pi}{3}\le|\omega|\le\frac{4\pi}{3}\\ 0 & \text{otherwise,}\end{cases}$$

where ν is a corner smooth function, satisfying

$$\nu(x)=\begin{cases}0 & \text{if} \quad x\le 0\\ 1 & \text{if} \quad x>1,\end{cases}$$

and

$$\nu(x)+\nu(1-x)=1, \quad \text{all} \quad x.$$

Now let $\phi_a(x) := \phi(x-a)$. It is easy to see that $\phi_a(x)$ is symmetric with respect to $2a$, and $\phi_a(x)$ remains to be a scaling function when $|a|<2/3$, but $2a$ is not necessarily an integer. When ϕ has compact support, $\widehat{\phi}$ is an entire function. Hence, $\widehat{\phi}$ satisfies the all conditions in Theorem 3.1. On the other hand, we have the following.

Theorem 3.2. *A compactly supported refinable distribution is symmetric if and only if its mask is symmetric.*

We now turn to consider the property of symmetry of wavelets. Without loss of generality, we assume that the two-scale symbols of compactly supported scaling functions are always of the form $P(z)=\frac{1}{2}\sum_{j=0}^{n}p_jz^j$, with $p_0p_n\neq 0$.

Definition 3.1. *Let ϕ be a compactly supported scaling function with two-scale symbol $P(z)=\frac{1}{2}\sum_{j=0}^{n}p_jz^j$. Let $B_\phi(z)$ be the EFL polynomial of ϕ. Then the B-wavelet ψ corresponding to ϕ is defined by*

$$\widehat{\psi}(\omega)=e^{i\frac{\omega}{2}}\overline{P(-e^{-i\frac{\omega}{2}})}B_\phi(-e^{-i\frac{\omega}{2}})\widehat{\phi}\left(\frac{\omega}{2}\right). \tag{3.2}$$

Theorem 3.3. *If a compactly supported scaling function ϕ is symmetric, then the corresponding B-wavelet ψ is symmetric for even n and anti-symmetric for odd n.*

The proof of this result can be found in our earlier work [5].

In Section 2, we have already proved that in any MRA generated by a compactly supported scaling function there is a unique minimally supported scaling function. The following theorem shows that the lack of symmetry of this minimally supported function characterizes the "asymmetry" of the MRA it generates.

Theorem 3.4. *If the minimally supported function φ in the MRA (φ) is not symmetric, then any compactly supported refinable function in $V(\varphi)$ cannot be symmetric.*

Proof: Let ϕ be a compactly supported refinable function in $V(\varphi)$. By Theorem 2.2, ϕ is a finite linear combination of some integer translates of φ. Hence, there is polynomial $A(z)$, such that

$$\widehat{\phi}(\omega) = A(e^{-i\omega})\widehat{\varphi}(\omega).$$

Now suppose that ϕ is symmetric, then there is an integer d such that $\widehat{\phi}(\omega) = e^{-i\omega d}\widehat{\phi}(-\omega)$. It follows that

$$A(e^{-i\omega})\widehat{\varphi}(\omega) = e^{-i\omega d}A(e^{i\omega})\widehat{\varphi}(-\omega). \tag{3.3}$$

Let $z_0 = e^{-i\omega_0}$ be a zero of $A(z)$. Here, $z_0 \in \mathbb{C}^0$ and ω_0 is not necessarily a real number. Since $\widehat{\varphi}$ has no 2π-periodic zeros, there is an integer $k_0 \in \mathbb{Z}$ such that $\widehat{\varphi}(-\omega_0 + 2k_0\pi) \neq 0$. From (3.3), we have

$$\begin{aligned} 0 &= A(e^{-i(\omega_0 - 2k_0\pi)})\widehat{\varphi}(\omega_0 - 2k_0\pi) \\ &= e^{-i\omega_0 d}A(e^{i\omega_0})\widehat{\varphi}(-\omega_0 + 2k_0\pi). \end{aligned}$$

It follows that $A(e^{-i\omega_0}) = 0$, or $A(\frac{1}{z_0}) = 0$. Recall that $A(z)$ is a polynomial. Hence, we may conclude that $A(z) = z^k A(\frac{1}{z})$ for some integer k. That is, $A(e^{-i\omega}) = e^{-ik\omega}A(e^{i\omega})$.

Finally we obtain $\widehat{\varphi}(\omega) = e^{-i(d-k)\omega}\widehat{\varphi}(-\omega)$. This means that φ is symmetric, which is a contradiction. ■

Write $V_0 = V(\varphi)$ and $V_n = \{f\colon\ f(2^{-n}x) \in V_0\}$. Then the following corollary is obvious.

Corollary 3.1. *If φ satisfies the conditions in Theorem 3.4, then any compactly supported refinable function in V_n, $n \in \mathbb{Z}$, is asymmetric. In particular, the corresponding B-wavelet cannot be symmetric or anti-symmetric.*

§4. Time-frequency Localization of Stoplets and Cowlets

If the mask of a compactly supported scaling function is a Pólya frequency sequence, then the scaling function has properties linked to total positivity. Therefore, scaling functions of this type and their corresponding wavelets may have special properties that are useful for time-frequency localizations. This section is devoted to study this type of scaling functions and wavelets.

Recall that a real sequence **a** is called a *Pólya frequency sequence* if all the minors of the bi-infinite matrix

$$A = [A_{ij}] = [a_{j-i}]$$

are non-negative; that is

$$A\begin{pmatrix} i_1, \ldots, i_p \\ j_1, \ldots, j_p \end{pmatrix} := \det_{k,\ell=1,\ldots,p} [A_{i_k j_\ell}] \geq 0,$$

for all integers $i_1 < \cdots < i_p$ and $j_1 < \cdots < j_p$. The properties of Pólya frequency sequences are documented in [17]. Recently, Goodman and Micchelli [15] introduced the notion of "*ripplets*" and the following result was essentially proved in their paper (see [15; Corollary 3.1] and also [8]).

Lemma 4.1. *Let ϕ be a scaling function with a Pólya frequency two-scale sequence. Then the coefficient sequence of its Euler-Frobenius Laurent series $B_\phi(z)$ is also a Pólya frequency sequence.*

Now we have the following.

Theorem 4.1. *If ϕ is a compactly supported scaling function with a Pólya frequency two-scale sequence, then ϕ and its corresponding B-wavelet ψ are linearly independent in the sense that the family of integer translates of each of ϕ and ψ is a linearly independent family.*

Proof: It is well-known that the symbol $P(z)$ of the two-scale sequence of ϕ has only non-positive real zeros. This in turn implies that the scaling function ϕ is linear independent. Now let us consider ψ. Observe that since

$$\widehat{\psi}(\omega) = e^{i\omega/2}\overline{P(-e^{-i\omega/2})}B_\phi(-e^{-i\omega/2})\widehat{\phi}\left(\frac{\omega}{2}\right),$$

we have

$$\overline{P(-e^{-i\omega_0/2})}B_\phi(-e^{-i\omega_0/2})\widehat{\phi}\left(\frac{\omega_0}{2} + 2k\pi\right) = 0, \qquad k \in Z,$$

and

$$\overline{P(e^{-i\omega_0/2})}B_\phi(e^{-i\omega_0/2})\widehat{\phi}\left(\frac{\omega_0}{2} + \pi + 2k\pi\right) = 0, \qquad k \in \mathbb{Z},$$

whenever ω_0 is 2π-periodic (and possibly complex-valued) zero of $\widehat{\psi}$. Since $\widehat{\phi}$ has no 2π-periodic zeros, we obtain

$$\overline{P(-e^{-i\omega_0/2})}B_\phi(-e^{-i\omega_0/2}) = 0 \tag{4.1}$$

and

$$\overline{P(e^{-i\omega_0/2})}B_\phi(e^{-i\omega_0/2}) = 0 \tag{4.2}$$

provided $\widehat{\psi}(\omega_0 + 2k\pi) = 0$, $k \in \mathbb{Z}$.

On the other hand, since P and B_ϕ only have non-negative real zeros, (4.1) and (4.2) cannot hold simultaneously. Hence, ψ is linearly independent. ■

Later, we will always assume that the two-scale symbol of a compactly supported scaling function is of the form $P(z) = \frac{1}{2}\sum_{j=0}^{m} p_j z^j$. It is known that if $P(z)$ is "symmetric" then ϕ is symmetric, and its corresponding B-wavelet ψ is symmetric for even m and anti-symmetric for odd m.

A scaling function with a symmetric Pólya frequency two-scaled sequence will be called a "*stoplet*" and its B-wavelet is called "*cowlet*". Here "stop" stands for symmetric total positivity and "co" for complete oscillation (see Chui [3, Section 6.6]). Stoplets and cowlets will be shown to provide good time-frequency localizations. As in [3], we define the time-frequency windows as follows.

Let $\phi \in L^2$. If

$$t_\phi = \lim_{N\to\infty} \int_{-N}^{N} x|\phi(x)|^2 dx \Big/ \int_{-\infty}^{\infty} |\phi(x)|^2 dx$$

and

$$w_\phi = \lim_{N\to\infty} \int_{-N}^{N} \omega|\widehat{\phi}(\omega)|^2 d\omega \Big/ \int_{-\infty}^{\infty} |\widehat{\phi}(\omega)|^2 d\omega$$

exist, then they will be called the time-center and the frequency-center of ϕ respectively. In addition, let

$$\Delta_\phi = \frac{\left(\int_{-\infty}^{\infty} (x - t_\phi)^2 |\phi(x)|^2 dx\right)^{\frac{1}{2}}}{\left(\int_{-\infty}^{\infty} |\phi(x)|^2 dx\right)^{\frac{1}{2}}}$$

and

$$\Delta_{\widehat{\phi}} = \frac{\left(\int_{-\infty}^{\infty} (\omega - \omega_\phi)^2 |\widehat{\phi}(\omega)|^2 d\omega\right)^{\frac{1}{2}}}{\left(\int_{-\infty}^{\infty} |\widehat{\phi}(\omega)|^2 d\omega\right)^{\frac{1}{2}}}.$$

The values of $2\Delta_\phi$ and $2\Delta_{\widehat{\phi}}$ (if finite) are used as widths of the time and frequency localization windows of ϕ, respectively. A function ϕ with $\Delta_\phi \Delta_{\widehat{\phi}} < \infty$ is called a window function. If a window function ψ satisfies $\widehat{\psi}(0) = 0$, it is called a *bandpass filter*; and in general, $\widehat{\psi}$ has two "peaks" in the frequency domain, one in the positive half and the other in the negative half. Hence, the window of a bandpass filter ψ needs some modification in order to match its time-frequency character. Usually we modify this window as follows. We preserve the definition for time localization, such as the time center and window width, but introduce two frequency centers instead. That is, we define, for a bandpass filter ψ, its positive and negative frequency centers, respectively, by

$$\omega_{\widehat{\psi}}^{+} = \frac{\int_0^{\infty} \omega|\widehat{\psi}(\omega)|^2 d\omega}{\int_0^{\infty} |\widehat{\psi}(\omega)|^2 d\omega}$$

and

$$\omega_{\widehat{\psi}}^{-} = \frac{\int_{-\infty}^{0} \omega |\widehat{\psi}(\omega)|^2 d\omega}{\int_{-\infty}^{0} |\widehat{\psi}(\omega)|^2 d\omega}.$$

Then we define its positive and negative frequency localization window widths, respectively, by $2\Delta_{\widehat{\psi}}^{+}$ and $2\Delta_{\widehat{\psi}}^{-}$, where

$$\Delta_{\widehat{\psi}}^{+} = \frac{\left(\int_{0}^{\infty} (\omega - \omega_{\widehat{\psi}}^{+})^2 |\widehat{\psi}(\omega)|^2 d\omega\right)^{\frac{1}{2}}}{\left(\int_{0}^{\infty} |\widehat{\psi}(\omega)|^2 d\omega\right)^{\frac{1}{2}}}$$

and

$$\Delta_{\widehat{\psi}}^{-} = \frac{\left(\int_{-\infty}^{0} (\omega - \omega_{\widehat{\psi}}^{-})^2 |\widehat{\psi}(\omega)|^2 d\omega\right)^{\frac{1}{2}}}{\left(\int_{-\infty}^{0} |\widehat{\psi}(\omega)|^2 d\omega\right)^{\frac{1}{2}}}.$$

When ψ is a real-valued function, it is clear that $|\widehat{\psi}|$ is an even function, so that $\omega_{\widehat{\psi}}^{-} = -\omega_{\widehat{\psi}}^{+}$ and $\Delta_{\widehat{\psi}}^{+} = \Delta_{\widehat{\psi}}^{-}$. Since we are only concerned with real-valued functions in this paper, the positive frequency behavior of ψ is sufficient in characterizing all of its frequency localization properties. Therefore, for bandpass filters, we can ignore $\omega_{\widehat{\psi}}^{-}$ and $\Delta_{\widehat{\psi}}^{-}$.

The quantity $\Delta_{\phi}\Delta_{\widehat{\phi}}$ and $\Delta_{\psi}\Delta_{\widehat{\psi}}^{+}$ measure the precision of time-frequency localizations by the lowpass filter ϕ and the bandpass filter ψ, respectively. For stoplets ϕ and cowlets ψ, we have the following result, the proof of which is given in [8] in detail.

Theorem 4.2. *For each natural number n, let $\mathbf{a}^n = (a_j^n)_{j=0}^{k_n}$, $a_0^n a_{k_n}^n \neq 0$, be a finite symmetric Pólya frequency sequence and with two-scale symbol given by*

$$A_n(z) = \left(\frac{1+z}{2}\right)^n P_n(z)$$

where the algebraic polynomial $P_n(z)$ satisfies $P_n(1) = 1$, $P_n(-1) \neq 0$ and $\deg P_n \leq Cn$ for some constant C independent of n. Let ϕ_n be the scaling function with two-scale symbol $A_n(z)$. Then the quantity

$$\sigma_n := \left(\int_{-\infty}^{\infty} \left(x - \frac{k_n}{2}\right)^2 \phi_n(x) dx\right)^{1/2}$$

is a positive (real) number, and

$$\lim_{n\to\infty} \left\| \widehat{\phi}_n\left(\frac{\omega}{\sigma_n}\right) e^{i\frac{k_n \omega}{2\sigma_n}} - e^{-\frac{\omega^2}{2}} \right\|_{L^p} = 0, \qquad 1 \leq p < \infty, \tag{4.3}$$

or equivalently,

$$\lim_{n\to\infty}\left\|\sigma_n\phi_n\left(\sigma_n x+\frac{k_n}{2}\right)-\frac{1}{2\pi}e^{-\frac{x^2}{2}}\right\|_{L^q}=0,\qquad 2\le q<\infty. \tag{4.4}$$

Furthermore,

$$\lim_{n\to\infty}\frac{1}{\sigma_n}\Delta_{\phi_n}=\lim_{n\to\infty}\sigma_n\Delta_{\widehat{\phi}_n}=\frac{1}{\sqrt{2}}\,, \tag{4.5}$$

so that

$$\lim_{n\to\infty}\Delta_{\phi_n}\Delta_{\widehat{\phi}_n}=\frac{1}{2}\,. \tag{4.6}$$

For the corresponding cowlets, the following result was also established in our work [8].

Theorem 4.3. *Let ψ_n be the B-wavelets corresponding to the scaling functions ϕ_n considered in Theorem 4.2; i.e.*

$$\widehat{\psi}_n(\omega)=c_n e^{i\omega/2}\overline{A_n(-e^{-i\frac{\omega}{2}})}B_{\phi_n}(-e^{-i\omega/2})\widehat{\phi}_n\left(\frac{\omega}{2}\right), \tag{4.7}$$

where c_n is a constant so chosen that $\|\widehat{\psi}_n\|_\infty=1$. Then

$$\lim_{n\to\infty}\left\|e^{i\omega/2\tau_n}\widehat{\psi}_n\left(\frac{\omega}{\tau_n}\right)-e^{-\frac{(\omega-\omega_n\tau_n)^2}{2}}\right\|_{L^p(0,+\infty)}=0,\qquad 1\le p<\infty, \tag{4.8}$$

so that for even n,

$$\lim_{n\to\infty}\left\|\tau_n\psi_n\left(\tau_n x-\frac{1}{2}\right)-\frac{1}{2\pi}\cos(\tau_n\omega_n x)e^{-\frac{x^2}{2}}\right\|_{L^q}=0,\quad 2\le q<\infty, \tag{4.9}$$

and for odd n,

$$\lim_{n\to\infty}\left\|\tau_n\psi\left(\tau_n x-\frac{1}{2}\right)-\frac{1}{2\pi}\sin(\tau_n\omega_n x)e^{-\frac{x^2}{2}}\right\|_{L^q}=0,\quad 2\le q<\infty. \tag{4.10}$$

Furthermore,

$$\lim_{n\to\infty}\frac{1}{\tau_n}\Delta_{\psi_n}=\lim_{n\to\infty}\tau_n\Delta^{+}_{\widehat{\psi}_n}=\frac{1}{\sqrt{2}}\,, \tag{4.11}$$

so that

$$\lim_{n\to\infty}\Delta_{\psi_n}\Delta^{+}_{\widehat{\psi}_n}=\frac{1}{2}\,, \tag{4.12}$$

where $\tau_n=\sqrt{|\widehat{\psi}''_n(\omega_n)|}$ and ω_n is the (unique) maximum of $|\widehat{\psi}_n|$ on $[0,+\infty)$.

§5. Interpolating Scaling Functions and Wavelets

In many applications, approximating functions by using an interpolating scheme is usually very efficient and is certainly a very popular approach. However, in general, wavelet bases are constructed in L^2. In order to consider interpolation, wavelets in the space C of continuous functions on $\mathbb{R}$ are required. In this section, we assume that the function, say f, to be approximated is bounded and continuous, that is in $L^\infty \cap C$. Hence, multiresolution analyses of $L^\infty \cap C$ as introduced in Chui and Li [4] are needed. Any function in $L^\infty \cap C$ that generates such an MRA is also called a *scaling function.* If a scaling function ϕ is fundamental in the sense that $\phi(j) = \delta_{i,j}$, we call it an *interpolating scaling function.* It is obvious that an interpolating scaling function is always linearly independent. It is well-known that the two-scale symbol $P(z)$ of an interpolating scaling function satisfies

$$P(z) + P(-z) = 1, \qquad |z| = 1. \tag{5.1}$$

Under certain very mild conditions, it is also well-known that (5.1) is a sufficient condition for a scaling function ϕ to be an interpolating one.

An interpolating scaling function φ that satisfies $\widehat{\varphi} \geq 0$ will be said to be *H-interpolating.* Such a function φ is in the Hardy space H, which is properly contained in L^2. Many standard interpolating scaling functions are H-interpolating. Typical examples are the fundamental cardinal splines.

There is an obvious relation between H-interpolating scaling functions and orthonormal scaling functions: If φ is an orthonormal scaling function with exponential decay, then $\varphi * \overline{\varphi(-\cdot)}$ is an H-interpolating scaling function.

For orthonormal scaling functions as well as H-interpolating scaling functions, there is a very interesting phenomenon: Roughly speaking, under certain conditions, the smoother the scaling function, the worse its time-frequency localization. We will describe it more precisely in the following.

We say that a family functions f_n is uniformly Lip α ($\alpha > 0$) at x_0, if there are constants $c > 0$ and $\delta > 0$, independent on n, such that

$$|f_n(x) - f_n(x_0)| \leq c|x - x_0|^\alpha, \qquad n \in \mathbb{N}, \quad |x - x_0| < \delta. \tag{5.2}$$

The following result was recently established in our work [7].

Lemma 5.1. *If $\{\phi_n\}$ is a sequence of H-interpolating scaling functions with corresponding two-scale symbols $\{P_n(z)\}$ satisfying*

$$\{P_n(e^{-i\omega})\} \text{ is uniformly Lip } \alpha \text{ at } \omega = 0, \tag{5.3}$$

and

$$\lim_{n\to\infty} P_n(e^{-i\omega}) = \chi_{\left[-\frac{\pi}{2},\frac{\pi}{2}\right]}, \qquad \omega \in [-\pi, \pi]. \tag{5.4}$$

Then

$$\lim_{n\to\infty} \|\widehat{\phi}_n(\omega) - \chi_{[-\pi,\pi]}\|_{L^p} = 0, \qquad p \in [1, +\infty), \tag{5.5}$$

$$\liminf_{n\to\infty} \Delta_{\widehat{\phi}_n} \geq \frac{\pi}{\sqrt{3}}, \tag{5.6}$$

and

$$\lim_{n\to\infty} \Delta_{\phi_n} = +\infty. \tag{5.7}$$

The following result, also established in [7], is a consequence of the above lemma.

Theorem 5.2. *Let $_N\phi$ and $_N\psi$ be the Daubechies compactly supported orthonormal scaling functions and wavelets of order N, respectively. Then*

$$\lim_{N\to\infty} \| |_N\widehat{\phi}| - \chi_{[-\pi,\pi]} \|_{L^2} = 0, \tag{5.8}$$

$$\lim_{N\to\infty} \| |_N\widehat{\psi}| - \chi_{[-2\pi,-\pi]\cup[\pi,2\pi]} \|_{L^2} = 0; \tag{5.9}$$

$$\lim_{N\to\infty} \Delta_{_N\widehat{\phi}} = \frac{\pi}{\sqrt{3}}, \tag{5.10}$$

$$\lim_{N\to\infty} \Delta^+_{_N\widehat{\psi}} = \frac{\pi}{2\sqrt{3}}; \tag{5.11}$$

and

$$\lim_{N\to\infty} \int_{-\infty}^{\infty} x^2 \; _N\Phi(x) dx = +\infty, \tag{5.12}$$

$$\lim_{N\to\infty} \int_{-\infty}^{\infty} x^2 \; _N\Psi(x) dx = +\infty, \tag{5.13}$$

where $_N\Phi$ and $_N\Psi$ are the autocorrelations of $_N\phi$ and $_N\psi$, respectively.

Examples of non-compactly supported orthonormal scaling functions and wavelets that have the same behaviour as the Daubechies functions are the Battle-Lemarié orthonormal scaling functions and wavelets [1,19].

More precisely, let ϕ_n be the nth order cardinal B-spline and $\phi_n^\perp$ and $\psi_n^\perp$ be the corresponding orthonormal scaling functions and (Battle-Lemarié) wavelets. Then we have the following.

Theorem 5.3. *The orthonormal cardinal splines and wavelets satisfy the following.*

$$\lim_{n\to\infty} \|\widehat{\phi}_n^\perp - \chi_{[-\pi,\pi]}\|_{L^2} = 0, \tag{5.14}$$

$$\lim_{n\to\infty} \|[\text{sgn}(\cdot)]^n \widehat{\psi}_n^\perp - \chi_{[-2\pi,\pi]\cup[\pi,2\pi]}\|_{L^2} = 0; \tag{5.15}$$

$$\lim_{n\to\infty} \Delta_{\widehat{\phi}_n^\perp} = \frac{\pi}{\sqrt{3}} \tag{5.16}$$

$$\lim_{n\to\infty} \Delta^+_{\widehat{\psi}_n^\perp} = \frac{\pi}{2\sqrt{3}} \tag{5.17}$$

and

$$\lim_{n\to\infty} \Delta_{\phi_n^\perp} = +\infty, \tag{5.18}$$

$$\lim_{n\to\infty} \Delta^+_{\psi_n^\perp} = +\infty. \tag{5.19}$$

§6. Remarks and Open Problems

1. The topic discussed in Section 2 was first studied in our earlier work [5] and the results in Section 2 may be considered as an improvement. There are several versions of the necessary and sufficient conditions for a compactly supported scaling function to be linearly independent. We quote them now for comparison.

In [16], a compactly supported refinable distribution φ (with $\widehat{\varphi}(0) = 1$) is linearly independent if and only if (1) its two-scale symbol $P(z)$ has no symmetric zeros in $\mathbb{C}\backslash\{0\}$, and (2) for any odd integer $m > 1$ and a primitive m-th root of unity, say ξ_m, there exists a positive integer d such that $p(-\xi_m^{2^d}) \neq 0$.

In [9], a compactly supported refinable distribution φ (with $\widehat{\varphi}(0) = 1$) is linearly independent if and only if there exists a compact set K congruent to $[-\pi, \pi]$ and containing a neighborhood of 0 so that $\inf\limits_{k>0} \inf\limits_{\omega\in K} |p(e^{-i\frac{\omega}{2^k}})| > 0$, where $P(z)$ is the two-scale symbol of φ.

In [23], a compactly supported refinable distribution φ (with $\widehat{\varphi}(0) = 1$) is linearly independent if and only if its two-scale symbol $P(z)$ cannot be written in the form of $P(z) = r(z)\frac{\ell(z^2)}{\ell(z)}$ with r and ℓ in π^0, and $\deg \ell \geq 1$.

In [18], a necessary and sufficient condition for the orthonormality of φ was given. This condition has a close link to linear independence.

2. The relation between the symmetry of a refinable distribution and its mask was first observed in our work [5]. When this result was stated as Theorem 5.3.2 in [3], the condition $\widehat{\phi}(\omega) \neq 0$ on $(2\pi, 2\pi + \epsilon)$ was not imposed on ϕ. (Recall that this condition is automatically satisfied by a refinable distribution with exponential decay.) The first author would like to thank Charles Micchelli for pointing this out, and the helpful conversation with him lead to the proof of Theorem 3.1.

3. For complete decomposition and reconstruction of functions in L^2 by using a wavelet basis, one needs to construct the dual scaling functions and dual wavelets. The interested reader is referred to [3,6,10].

4. The following questions still seem to be unanswered.

(1) If a compactly supported scaling function ϕ is linearly independent, is its corresponding B-wavelet minimally supported and linearly independent?

(2) If a compactly supported scaling function ϕ is linearly independent and asymmetric, does there exist any nontrivial (noncompactly supported) symmetric function that generates the same MRA?

(3) Let ${}_N\widetilde{\phi}$ and ${}_N\widetilde{\psi}$ denote the Daubechies compactly supported scaling function and wavelet of order N with the minimum phase (cf. [12; Ch. 8]). Then do we have

$$\lim_{N\to\infty} \|\widehat{{}_N\widetilde{\phi}} - \chi_{[-\pi,\pi]}\|_{L^2} = 0,$$

and

$$\lim_{N\to\infty} \|\widehat{{}_N\widetilde{\psi}} - \chi_{[-2\pi,\pi]\cup[\pi,2\pi]}\|_{L^2} = 0?$$

Acknowledgements. The research of C.K. Chui was supported by NSF Grant # DMS-92-06928 and ARO Contract # DAAH 04-93-G-0047.

References

1. Battle, G., A block spin construction of ondelettes, Part I: Lemarié functions, Comm. Math. Phys. **110** (1987), 601-615.
2. Cavaretta, A. S., W. Dahmen, C. Micchelli, *Stationary subdivision*, Mem. Amer. Math. Soc., Providence. RI, 1993.
3. Chui, C. K., *An introduction to wavelets*, Academic Press, Boston, 1992.
4. Chui, C. K., and C. Li, Dyadic affine decompositions and functional wavelet transforms, CAT report #295, Texas A&M University, 1993.
5. Chui, C. K., and J. Z. Wang, A general framework for compactly splines and wavelets, J. Approx. Theory **71** (1992), 263-304.
6. Chui, C. K., and J. Z. Wang, On compactly supported spline-wavelets and a duality principle, Trans. Amer. Math. Soc. **330** (1992), 903-915.
7. Chui, C. K., and J. Z. Wang, Poor time-frequency localization by orthonormal scaling functions and wavelets of high orders, CAT Report #322, Texas A&M University, 1993.
8. Chui, C. K., and J. Z. Wang, A study of asymptotically optimal time-frequency windows of scaling functions and wavelets, CAT Report #323, Texas A&M University, 1993.
9. Cohen, A., Ondelettes, analyses multirésolutions et filtres miroir en quadrature, Ann Inst. H. Poincaré, Anal. non linéaire **7** (1990), 439-459.
10. Cohen, A., I. Daubechies, and J. C. Feauveau, Biorthogonal bases of compactly supported wavelets, Comm. Pure Appl. Math. **45** (1992), 485-560.
11. Dahmen, W., and C. A. Micchelli, Translates of multivariate splines, Linear Algebra and Appl. **54** (1983), 217-234.
12. Daubechies, I., *Ten Lectures on Wavelets*, CBMS-NSF Regional Conference Series in Appl. Math. #61, SIAM Publ., Philadelphia, 1992.
13. Daubechies, I., and J. Lagarias, Two-scale difference equations I. Existence and global regularity of solutions SIAM J. Math. Anal. **22** (1991), 1388-1410. II. Local regularity, infinite products of matrices and fractals, SIAM J. Math. Anal. **23** (1992), 1031-1079.
14. Deslauriers, G., and S. Dubuc, Interpolation dyadique, in *Fractals, Dimensions non entières et applications*, G. Cherbit, ed., Masson, Paris, 1987, pp. 44-55.
15. Goodman, T. N. T., and C. A. Micchelli, On refinement equations determined by Pólya frequency sequence, SIAM J. Math. Anal. **23** (1992), 776-784.
16. Jia, R. Q., and J. Z. Wang, Stability and linear independence associated with wavelet decompositions, Proc. Amer. Math. Soc. **117** (1993), 1115-1124.
17. Karlin, S., *Total Positivity*, Stanford University Press, Stanford, California, 1968.

18. Lawton, W., Tight frames of compactly supported wavelets, J. Math. Phys. **31** (1990), 1898-1901.
19. Lemarié, P. G., Ondelettes à localisation exponentielles, J. Math. Pure et Appl. **67** (1988), 227-236.
20. Mallat, S., Multiresolution approximation and wavelets, Trans. Amer. Math. Soc. **315** (1989), 69-88.
21. Meyer, Y., Ondelettes et opérateurs, I: Ondelettes, Hermann, Paris, 1991.
22. Ron, A., A necessary and sufficient condition for the linear independence of the integer translates of a compactly supported distribution, Constructive Approx. **5** (1989), 297-308.
23. Ron, A., Characterizations of linear independence and stability of the shifts of a univariate refinable function in terms of its refinement mask, preprint.
24. Zhou, Ding xue, A remark on the symmetricity of scaling functions, preprint.

Charles K. Chui
Center for Approximation Theory
Department of Mathematics
Texas A&M University
College Station, TX 77843-3368
cchui@tamu.edu

Jianzhong Wang
Department of Mathematics
University of North Carolina
Charlotte, NC 28223
jiawang@mosaic.uncc.edu

Multiresolution Analysis and Wavelets on Locally Compact Abelian Groups

Stephan Dahlke

Abstract. In this paper, the concept of a multiresolution analysis approximation of functions is generalized to locally compact Abelian groups. An existence theorem based on generalized B-splines is proved and several examples are given. Furthermore, the associated wavelet basis is constructed.

§1. Introduction

In recent years, the concept of a *multiresolution analysis* introduced by Mallat [11] has become an important mathematical tool. For instance, a quite natural way to construct a *wavelet basis* of $L^2(\mathbb{R}^n)$, *i.e.*, a basis that consists of the scaled and integer translated versions of a finite number of functions, is to use a multiresolution analysis.

On $\mathbb{R}^n$, a multiresolution analysis consists of a nested sequence

$$\cdots \subset V_{j-1} \subset V_j \subset V_{j+1} \subset \cdots$$

of closed subspaces of $L^2(\mathbb{R}^n)$ such that the closure of their union is $L^2(\mathbb{R}^n)$ while their intersection is $\{0\}$. The space V_0 is assumed to be shift-invariant, i.e.,

$$f \in V_0 \iff f(\cdot\, - k) \in V_0, \quad k \in \mathbb{Z}^n\,,$$

and to be spanned by the integer translates of a single function ϕ, called the *generator* of the multiresolution analysis. Furthermore, one assumes that the space V_j can be generated from the space V_{j-1} by a simple scaling process, i.e.,

$$f \in V_{j-1} \iff f(2\cdot) \in V_j\,.$$

It is also possible to scale with certain integer matrices. This case was studied by Cohen and Daubechies [2]. The shift-invariance of V_0 can be interpreted as an invariance property with respect to the action of a discrete subgroup, and the scaling process can be viewed as the action of a certain group automorphism. Therefore, on a general locally compact Abelian group we can define the following:

Wavelets, Images and Surface Fitting
P. J. Laurent, A. Le Méhauté, and L. L. Schumaker (eds.), pp. 141–156.

ISBN 1-56881-040-7.

Definition 1.1. *Let G be a locally compact Abelian group, and suppose that G contains a discrete countable subgroup K such that the quotient group G/K is compact. Furthermore, let us assume that there exists an automorphism A of G such that $A(K) \subset K$. Then a* multiresolution analysis *of $L^2(G, m_G)$, where m_G denotes the Haar measure on G, is a sequence $\{V_j\}_{j\in\mathbb{Z}}$ of closed subspaces of $L^2(G, m_G)$ such that*

$$\cdots \subset V_{j-1} \subset V_j \subset V_{j+1} \subset \cdots \tag{1.1}$$

$$f(g) \in V_j \iff f(A(g)) \in V_{j+1} \tag{1.2}$$

$$f(g) \in V_0 \iff f(g \circ k) \in V_0, \quad k \in K \tag{1.3}$$

$$\overline{\bigcup_{j=-\infty}^{\infty} V_j} = L^2(G, m_G) \tag{1.4}$$

$$\bigcap_{j=-\infty}^{\infty} V_j = \{0\}. \tag{1.5}$$

There exists a function ϕ and constants $c_1, c_2 \in \mathbb{R}^+$ such that V_0 is the closed linear span of $\phi(\cdot \circ k)$ and

$$c_1||\lambda||_{l_2,K} \leq ||\sum_{k\in K} \lambda_k \phi(\cdot \circ k)||_2 \leq c_2||\lambda||_{l_2,K} \quad \text{for all } \lambda \in l_2(K), \tag{1.6}$$

where $||\lambda||_{l_2,K}$ is defined by

$$||\lambda||_{l_2,K} = \left(\sum_{k\in K} |\lambda_k|^2\right)^{\frac{1}{2}}. \tag{1.7}$$

Remark 1.2. *It is a well-known fact that a Haar measure, i.e., a translation–invariant Borel measure, exists on a locally compact Abelian group. The Haar measure is unique up to multiplication by constants, see [3] for details.*

Remark 1.3. *A discrete subgroup that yields a compact coset space is sometimes called a* uniform lattice. *Such a uniform lattice often exists, see [13] for details.*

Remark 1.4. *We will henceforth assume that the automorphism A is continuous and that G is second countable and metrizable. Without loss of generality (see [8], § 8), the metric can be chosen to be* translation-invariant, *i.e.,*

$$d(a \circ g, a \circ h) = d(g, h) \qquad \text{for all } a,\ g,\ h \in G.$$

This paper is concerned with the following natural question arising from our setting: under which conditions does there exist a multiresolution analysis, and is it generated by a compactly supported, continuous generator?

A quite natural way to construct a multiresolution analysis of $L^2(\mathbb{R})$ is to use cardinal B-splines. This case was intensively studied by Chui and Wang [4,5]. Therefore, we wish to define generalized B-splines on locally compact Abelian groups. This is done in Section 2. It turns out that most of the useful properties of cardinal B-splines on $\mathbb{R}$ carry over to the general situation. In Section 3 we state and prove our main result which says that under certain conditions the generalized B-splines generate a multiresolution analysis. It turns out that our approach is connected with the theory of self-similar lattice tilings. On $\mathbb{R}^n$, this theory was studied by Gröchenig and Madych [7], see also [6]. In Section 4, we give a straightforward generalization of one of their main results which is useful for our purpose. In Section 5, we study some examples to which the general results of the preceeding sections apply. Finally, in Section 6, we give an outline how the resulting generalized B–spline wavelets can be constructed.

By applying a quite different method, similar things were done before by Lemarié [10] for certain Lie groups. Furthermore, this work is connected with the recent investigations of Strichartz for nilpotent Lie groups [14].

§2. Generalized Cardinal B-Splines

One way to introduce cardinal B-splines on $\mathbb{R}$ is to define them recursively as a convolution product, *i.e.*,

$$\begin{aligned} N_1(x) &= \chi_{[0,1)} \\ N_m(x) &= (N_{m-1} * N_1)(x) \,. \end{aligned} \tag{2.1}$$

But for two integrable functions H, L on a locally compact group G the convolution

$$(H * L)(g) = \int_G H(h)\, L(h^{-1} \circ g)\, dm_G(h) \tag{2.2}$$

is well-defined. Therefore, it seems natural to carry over the recursive Definition 2.1 to a more general situation. For that purpose we want to use a *fundamental region* for K, *i.e.*, a set $F \subset G$ such that

$$K \circ F = G\,, \qquad k \circ F \cap F = \emptyset, \quad k \neq e, \tag{2.3}$$

where e denotes the identity element of the group. In the following, we will always assume that $e \in F$. Furthermore, since $G/_K$ is compact, $\overline{F}$ is compact. Using the characteristic function χ_F of the fundamental region, we set:

Definition 2.1. *The function $N_m : G \to \mathbb{R}$ defined recursively by*

$$N_m(g) = (N_{m-1} * N_1)(g)\,, \tag{2.4}$$

where

$$N_1(g) = \chi_F(g)$$

is called generalized cardinal B-spline of order m.

In the sequel, we will assume that the Haar measure is normalized in such a way that the fundamental region F has measure one. Then some of the useful properties of cardinal B-splines on $\mathbb{R}$ carry over to the general situation considered here. One has

Lemma 2.2. *The generalized B-spline of order m in Definition 2.1 satisfies:*

i) supp N_m *is compact*

ii) $N_m(g) \geq 0$ *(positivity)*

iii) $\displaystyle\int_G N_m(g)\, dm_G = 1$ *(normalization)*

iv) $\displaystyle\sum_{k\in K} N_m(g \circ k) = 1$ *(partition of unity)*

v) there exist constants $B_1,\, B_2 \in \mathbb{R}^+$ *such that*

$$B_1||\lambda||_{l_2,K} \leq \Big|\Big| \sum_{k\in K} \lambda_k\, N_m(g\circ k)\Big|\Big|_2 \leq B_2||\lambda||_{l_2,K} \qquad \textit{(stability).}$$

Proof: i) and ii) follow immediately by definition, while iii) and iv) can be easily proved by induction on m. The proof of stability is a little bit more involved. First we show the upper estimate. Using ii), the Cauchy-Schwarz-inequality, iv), and iii) we get

$$\begin{aligned}
&\Big|\Big| \sum_{k\in K} \lambda_k\, N_m(\cdot\, \circ\, k)\Big|\Big|_2^2 \\
&= \int_G \Big| \sum_{k\in K} \lambda_k\, N_m(g\circ k)\Big|^2 dm_G(g) \\
&\leq \int_G \left(\sum_{k\in K} |\lambda_k|\, N_m(g\circ k)\right)^2 dm_G(g) \\
&\leq \int_G \left(\sum_{k\in K} |\lambda_k|^2\, N_m(g\circ k)\right) \left(\sum_{k\in K} N_m(g\circ k)\right) dm_G(g) \\
&= \sum_{k\in K} |\lambda_k|^2 \int_G N_m(g\circ k)\, dm_G(g) = \sum_{k\in K} |\lambda_k|^2 \ .
\end{aligned}$$

To show the lower bound, we will make use of Lemma 2.4 proved below. It states a sufficient condition for the validity of the lower bound in terms of the Fourier transform of integrable functions, which is defined by

$$\widehat{H}(\gamma) = \int_G H(g)\, \gamma(g^{-1})\, dm_G(g)\,, \tag{2.5}$$

where γ is a character of G, *i.e.*, a continuous homomorphism of G into the multiplicative group of complex numbers of modulus 1.

Using the fact that N_m is defined by a convolution, we obtain

$$\sum_{\tilde\gamma\in\Gamma} \big|\widehat{N_m}(\cdot\, \circ\, \tilde\gamma)\big|^2 = \sum_{\tilde\gamma\in\Gamma} \big|\widehat{N_1 * \cdots * N_1}(\cdot\, \circ\, \tilde\gamma)\big|^2 = \sum_{\tilde\gamma\in\Gamma} \big|\widehat{N_1}(\cdot\, \circ\, \tilde\gamma)\big|^{2m}.$$

Assume that there exists some character γ such that

$$\sum_{\tilde\gamma\in\Gamma} \big|\widehat{N_1}(\gamma\circ\tilde\gamma\big|^{2m} = 0.$$

This immediately implies

$$\sum_{\tilde{\gamma}\in\Gamma} \left|\widehat{N_1}(\gamma \circ \tilde{\gamma})\right|^2 = 0\,.$$

But from formula (2.10) proved below and the definition of N_1, we can conclude that

$$\sum_{\tilde{\gamma}\in\Gamma} \left|\widehat{N_1}(\gamma \circ \tilde{\gamma})\right|^2 = 1,$$

which implies the lower estimate in v) by Lemma 2.4. ■

Remark 2.3. *It is a consequence of the strong continuity of the left regular representation that the generalized B-splines are continuous functions for $m \geq 2$, see [15], Proposition 3.4, for details. However, if G admits a differentiable structure, i.e., if G is a Lie group, then one would expect that the regularity of the splines increases with m, see Example 5.3 below.*

Lemma 2.4. *Let G be a locally compact Abelian group with uniform lattice K. Let ϕ be a compactly supported function that satisfies for some $B_1 > 0$*

$$B_1 \leq \sum_{\tilde{\gamma}\in\Gamma} \left|\hat{\phi}(\cdot \circ \tilde{\gamma})\right|^2, \tag{2.6}$$

where Γ denotes the set of all characters $\tilde{\gamma}$ that satisfy

$$\tilde{\gamma}(k) = 1\,, \qquad k \in K\,. \tag{2.7}$$

Then

$$B_1 ||\lambda||^2_{l_{2,K}} \leq \Big|\Big| \sum_{k\in K} \lambda_k\, \phi(\cdot \circ k)\Big|\Big|_2^2. \tag{2.8}$$

Proof: Let $\widehat{G}$ denote the dual group of G, i.e., the set of all characters of G. First we check that Γ is a uniform lattice for $\widehat{G}$. In general, if H is a subgroup of G, and $\check{H}$ is the set of all characters that are identically equal to 1 on H, one has the canonical identifications (see [12] for details).

$$(\widehat{G/H}) = \check{H}\,, \qquad \widehat{H} = \widehat{G}/\check{H}\,. \tag{2.9}$$

On the other hand, we can make use of the fact that a locally compact Abelian group is compact if and only if its dual group is discrete. Therefore, since G/K is compact, it follows from the first identity in (2.9) that $\check{K} = \Gamma$ is discrete. Furthermore, since K is discrete, it follows from the second identity in (2.9) that $\widehat{G}/\Gamma = \widehat{K}$ is compact.

Next we sketch the proof of the following straightforward generalization of the Poisson summation formula:

$$\sum_{k\in K} \langle \phi(\cdot), \phi(\cdot \circ k)\rangle\, \gamma(k^{-1}) = \sum_{\tilde{\gamma}\in\Gamma} \left|\hat{\phi}(\gamma^{-1} \circ \tilde{\gamma})\right|^2, \quad \gamma \in \widehat{G}\,. \tag{2.10}$$

Since ϕ is assumed to be compactly supported, (2.10) implies that the right-hand side of (2.10) is a bounded continuous function.

(2.10) can be proved by following the lines of the proof of Theorem 3.2 in [9]. A simple calculation yields

$$\sum_{k\in K} \langle \phi(\cdot), \phi(\cdot \circ k)\rangle \, \gamma(k^{-1}) = \int_{G/K} \Big|\sum_{l\in K} \phi(g\circ l)\,\gamma(g\circ l)\Big|^2 \, dm_{G/K} \, .$$

Since ${}^{G}/_{K}$ is compact, it follows from the Theorem of Peter and Weyl that the discrete set $(\widehat{G/K})$ forms an orthonormal basis for $L^2({}^{G}/_{K})$, see [15], Theorem 3.3. Consequently, the periodic function $\sum_{l\in K} \phi(g\circ l)\,\gamma(g\circ l)$ can be expanded into a Fourier series where the Fourier coefficients $\alpha(\tilde{\gamma})$ are determined by

$$\alpha(\tilde{\gamma}) = \hat{\phi}(\gamma^{-1}\circ\tilde{\gamma}) \, , \tag{2.11}$$

since, by (2.9), $(\widehat{G/K})$ can be identified with Γ. A simple application of Parseval's identity yields the result.

After these preliminaries we are now ready to prove (2.8). By Plancherel's identity,

$$\begin{aligned}
\Big|\Big|\sum_{k\in K} \lambda_k\,\phi(\cdot\circ k)\Big|\Big|_2^2 &= \Big|\Big|\sum_{k\in K} \lambda_k\,\widehat{\phi(\cdot\circ k)}\Big|\Big|_2^2 \\
&= \int_{\hat{G}} \Big|\sum_{k\in K} \lambda_k\,\gamma(k^{-1})\Big|^2 \, |\hat{\phi}(\gamma)|^2 \, dm_{\hat{G}} \\
&= \int_{\hat{G}/\Gamma} \Big|\sum_{k\in K} \lambda_k\,\gamma(k^{-1})\Big|^2 \sum_{\tilde{\gamma}\in\Gamma} |\hat{\phi}(\gamma\circ\tilde{\gamma})|^2 \, dm_{\tilde{G}/\Gamma}
\end{aligned}$$

where we used the fact that $\tilde{\gamma}\in\Gamma$ is equal to 1 on K, so that, employing condition (2.6), we finally obtain

$$\Big|\Big|\sum_{k\in K} \lambda_k\,\phi(\cdot\circ k)\Big|\Big|_2^2 \geq B_1 \int_{\hat{G}/\Gamma} \Big|\sum_{k\in K} \lambda_k\,\gamma(k^{-1})\Big|^2 \, dm_{\hat{G}/\Gamma} \, .$$

The complex-valued mapping that sends γ to $\gamma(k^{-1})$ can be interpreted as a character on $\hat{G}$, and, by Pontryagin's duality theorem, every character can be represented in this way for a certain $g\in G$, see [8], Section 24 for details. Once again, by (2.9), $\left(\widehat{\hat{G}/\Gamma}\right)$ can be identified with the characters that are constant on Γ, *i.e.*, with the mappings $\gamma \longrightarrow \gamma(k)$. Therefore, employing the theorem of Peter and Weyl and Parseval's identity yields

$$\Big|\Big|\sum_{k\in K} \lambda_k\,\phi(\cdot\circ k)\Big|\Big|_2^2 \geq B_1 \int_{\hat{G}/\Gamma} \Big|\sum_{k\in K} \lambda_k\,\gamma(k^{-1})\Big|^2 \, dm_{\hat{G}/\Gamma} = B_1\, ||\lambda||_{l_2,K}^2$$

which finishes the proof. ■

§3. Main Result

In this section we will prove the main result of this paper which says that under certain conditions the generalized B-splines introduced in Section 2 generate a multiresolution analysis of $L^2(G, m_G)$.

The fundamental region according to a discrete subgroup is by no means unique. It turns out that we have a look for a special kind of region that guarantees that the B-splines are refinable functions. Furthermore, the inverse of the automorphism according to Definition 1.1. has to be a contraction. More precisely, one has

Theorem 3.1. *Let G be a locally compact, metrizable and second countable Abelian group with uniform lattice K. Furthermore, let A be an automorphism of G such that $A(K) \subset K$ and*

i.) there exists a fundamental region F such that

$$F = \bigcup_{r \in R} A^{-1}(F) \circ A^{-1}(r),$$

where $R \subset K$ is a finite set,

ii.) A^{-1} is a contraction, i.e.,

$$d\left(A^{-1}(g), A^{-1}(h)\right) \leq c\, d(g,h)$$

for some $c < 1$ and for all g, $h \in G$.

Let N_m be a generalized B-spline with respect to F and

$$V_0 := \overline{\operatorname{span}\{N_m(\cdot \circ k) | k \in K\}}\,.$$

Then the sequence $\{V_j\}_{j \in \mathbb{Z}}$ of closed subspaces defined by

$$V_j \ni f(\cdot) \iff f(A^{-j}\,\cdot) \in V_0$$

is a multiresolution analysis of $L^2(G, m_G)$.

Remark 3.2. *Subsets that satisfy condition i.) are sometimes called self-similar with respect to A. Therefore, since F is a fundamental region, it gives rise to a self-similar tiling of G. As mentioned above, for the case $G = (\mathbb{R}^n, +)$ the theory of tilings was already studied by K. H. Gröchenig and W. Madych, see [7]. In Section 4, we will present a method how such a tiling can be constructed.*

Proof of Theorem 3.1. First we show that the generalized B-splines are refinable functions in the sense that they satisfy for some finite set $H \subset K$

$$N_m(g) = \sum_{k \in H} a_k\, N_m(A(g) \circ k)\,. \tag{3.1}$$

(3.1) can be proved by induction on m. For $m = 1$, using i.), one has

$$N_1(g) = \sum_{r\in R} \chi_{A^{-1}(F)\circ A^{-1}(r)}(g) = \sum_{r\in R} \chi_F\left(A(g)\circ r^{-1}\right) = \sum_{r\in R} N_1(A(g)\circ r^{-1}) .$$

Now, let us suppose that (3.1) holds for $m-1$. Then we get

$$\begin{aligned} N_m(g) &= \int_G \left(\sum_{k\in H} a_k\, N_{m-1}\left(A(h)\circ k\right)\right)\left(\sum_{r\in R} N_1\left(A(h^{-1}\circ g)\circ r^{-1}\right)\right) dm_G(h) \\ &= \sum_{k\in H}\sum_{r\in R} a_k \int_G N_{m-1}\left(A(h)\circ k\right)\, N_1\left((A\,h)^{-1}\circ A(g)\circ r^{-1}\right)\, dm_G(h) . \end{aligned}$$

It is easy to check that the distribution μ of an automorphism with respect to m_G is a translation-invariant measure. Therefore, μ differs from m_G only by multiplication with a constant c_1. This yields

$$\begin{aligned} N_m(g) &= c_1 \sum_{k\in H}\sum_{r\in R} a_k \int_G N_{m-1}(h\circ k)\, N_1\left(h^{-1}\circ A(g)\circ r^{-1}\right)\, dm_G(h) \\ &= c_1 \sum_{k\in H}\sum_{r\in R} a_l \int_G N_{m-1}(h)\, N_1\left((h\circ k^{-1})^{-1}\circ A(g)\circ r^{-1}\right)\, dm_G(h) \\ &= c_1 \sum_{k\in H}\sum_{r\in R} a_l\, (N_{m-1} * N_1)\left(A(g)\circ r^{-1}\circ k\right) \end{aligned}$$

and (3.1) is proved.

Now we have to check that $\{V_j\}_{j\in\mathbb{Z}}$ is a multiresolution analysis of $L^2(G, m_G)$. (1.2) and (1.3) follow immediately from the definition of the spaces V_j. (1.1) is a consequence of (3.1), and (1.6) was already shown in Lemma 2.2 v.). It remains to prove (1.4) and (1.5). (1.5) can be shown by following the lines of the proof of Theorem 2.3 in [9]. If

$$f \in \bigcap_{-\infty}^{\infty} V_j ,$$

then

$$f(A^j\,\cdot) \in V_0 \qquad \text{for all } j \in \mathbb{Z}.$$

Therefore, there exist sequences $\{a_k^j\}_{k\in K}$ such that

$$f\left(A^j(g)\right) = \sum_{k\in K} a_k^j\, N_m(g\circ k) .$$

The function value of f can be estimated by

$$\left|f\left(A^j(g)\right)\right| \le \Big|\sum_{k\in K} a_k^j\, N_m(g\circ k)\Big| \le \sup_{k\in K} |a_k^j| \sum_{k\in K} \left|N_m(g\circ k)\right| \le \|a_k^j\|_{l_{2,K}},$$

and therefore

$$\int_{B_r(e)} |f(g)|^2 \, dm_G \leq ||a^j||_{l_2,K}^2 c_1^{-j} m_g(A^{-j} B_r(e))$$

with c_1 the constant from above, and $B_r(e)$ denotes the set

$$B_r(e) := \{g \in G | \, d(g,e) \leq r\} .$$

From Lemma 2.2. v.) we know that N_m has stable translates, *i.e.*, there exists some constant B_1 such that

$$||a^j||_{l_2,K} \leq \frac{1}{B_1} \left(\int_G |f(A^j g)|^2 \, dm_G \right)^{\frac{1}{2}} \leq \frac{1}{B_1} c_1^{\frac{j}{2}} \left(\int_G |f(g)|^2 \, dm_G \right)^{\frac{1}{2}} .$$

This yields

$$\int_{B_r(e)} |f(g)|^2 \, dm_G \leq \frac{1}{B_1^2} ||f||_2^2 \, m_G(A^{-j} B_r(e)).$$

If we let j tend to infinity and use condition ii.), we obtain

$$\int_{B_r(e)} |f(g)|^2 \, dm_G = 0 \qquad \text{for all } r \in \mathbb{R}$$

and therefore

$$f \equiv 0 ,$$

so that (1.5) is proved.
It remains to show (1.4). We will make use of the fact that the space of continuous functions with compact support is dense in $L^2(G, m_G)$. This is true for L^P-spaces on locally compact, second countable topological spaces with respect to a Borel measure, see [1], § 43 and § 44 for details. Let f be a continuous function with compact support. Defining

$$f_j(g) := \sum_{k \in K} f\left(A^{-j}(k^{-1}\right) N_m \left(A^j(g) \circ k\right) ,$$

we notice that supp f and supp f_j are contained in a compact set B. Therefore, to prove (1.4), it is sufficient to check that f_j converges uniformly to f on B. Employing Lemma 2.2. ii.) and iv.) yields

$$\begin{aligned} |f_j(g) - f(g)| &= \Big| \sum_{k \in K} f\left(A^{-j}(k^{-1})\right) N_m \left(A^j(g) \circ k\right) - f(g) \Big| \\ &= \Big| \sum_{k \in K} \left(f\left(A^{-j}(k^{-1})\right) - f(g)\right) N_m \left(A^j(g) \circ k\right) \Big| \\ &\leq \sum_{k \in K} \left|f\left(A^{-j}(k^{-1})\right) - f(g)\right| N_m \left(A^j(g) \circ k\right) . \end{aligned}$$

The sum on the right-hand side differs from zero only if

$$g \in A^{-j}\left(\text{supp } N_m \circ k^{-1}\right) .$$

We also have the inclusion

$$F \circ k^{-1} \subset \text{supp } N_m \circ k^{-1}$$

and therefore, since $k^{-1} \in F \circ k^{-1}$,

$$A^{-j}(k^{-1}) \in A^{-j}\left(\text{supp } N_m \circ k^{-1}\right) .$$

Combining the two expressions we obtain

$$\begin{aligned} d\left(g, A^{-j}(k^{-1})\right) &\leq \text{diam}\left(A^{-j}(\text{supp } N_m \circ k^{-1})\right) \\ &\leq c^j \text{diam}\left(\text{supp } N_m \circ k^{-1}\right) \\ &\leq c^j \text{diam}\left(\text{supp } N_m\right) \end{aligned}$$

since we required that the metric on G is translation-invariant. The function f is continuous and compactly supported and therefore uniformly continuous, so we can find, for a given $\epsilon > 0$, a number $\delta > 0$ such that

$$d\left(g, A^{-j}(k^{-1})\right) \leq \delta$$

implies

$$\left|f\left(A^{-j}(k^{-1})\right) - f(g)\right| \leq \epsilon .$$

But since $c < 1$, there exists $j_0 \in \mathbb{N}$ such that

$$d\left(g, A^{-j}(k^{-1})\right) \leq \delta , \quad j \geq j_0 .$$

Using Lemma 2.2. iv.) again, we finally get

$$\begin{aligned} \left|f_j(g) - f(g)\right| &\leq \sum_{k \in K} \left|f\left(A^{-j}(k^{-1})\right) - f(g)\right| N_m\left(A^j(g) \circ k\right) \\ &\leq \epsilon \sum_{k \in K} N_m\left(A^j(g) \circ k\right) \leq \epsilon , \end{aligned}$$

i.e., f_j converges to f uniformly. This finishes the proof of Theorem 3.1. ■

Remark 3.3. *If G is a connected Lie group and A a C^1-diffeomorphism, then condition ii.) of Theorem 3.1 holds if*

$$||T_e A^{-1}|| \leq c < 1. \tag{3.2}$$

§4. Self-Similar Tilings on Locally Compact Abelian Groups

In this section, we present an algorithm to construct a self-similar tiling in the sense of Theorem 3.1. For that purpose, suppose that $r_1 \circ A\,K, \ldots, r_m \circ A\,K$ is a complete enumeration of the cosets in $K/_{AK}$, i.e.,

$$K = \bigcup_{r \in R} r \circ A\,K\,, \qquad R = \{r_1, \ldots, r_m\}\,.$$

To construct a self-similar tiling, we have to find a fundamental region F that satisfies

$$F = \bigcup_{r \in R} A^{-1}\,F \circ A^{-1}\,r\,.$$

Consider the mapping

$$\begin{aligned} \mathcal{A}: \quad P_c(G) &\longrightarrow P_c(G) \\ Q &\longmapsto \bigcup_{r \in R} A^{-1}(Q) \circ A^{-1}(r)\,, \end{aligned}$$

where $P_c(G)$ denotes the set of all compact subsets of G. With the metric

$$\rho(P, Q) = \max\left(r(P, Q), r(Q, P)\right)\,,$$

where

$$r(P, Q) = \sup_{h \in P}\,\inf_{g \in Q} d(h, g)\,,$$

$P_c(G)$ is a complete metric space. To construct a self-similar tiling, we have to find a fixed point of $\mathcal{A}$. This means that we have to check that the fix point iteration

$$Q_{n+1} = \mathcal{A}(Q_n) = \bigcup_{r \in R} A^{-1}(Q_n) \circ A^{-1}(r)$$

converges to a compact set in the metric ρ. Following the lines of the proof of Lemma 3 in [7], it is easy to show the following result.

Lemma 4.1. *Let $\{r_1, \ldots, r_m\}$ be some enumeration of cosets in $K/_{AK}$. If Q_0 is a compact set, then the sequence*

$$Q_{n+1} = \mathcal{A}(Q_n) = \bigcup_{r \in R} A^{-1}(Q_n) \circ A^{-1}(r)$$

converges in the metric ρ to the set

$$Q = \left\{ g \in G \,\middle|\, g = \lim_{j \to \infty} \left(A^{-j}(\epsilon_j) \circ A^{-(j-1)}(\epsilon_{j-1}) \circ \ldots \circ A^{-1}(\epsilon_1)\right)\,,\ \epsilon_j \in R \right\}.$$

Furthermore, if the limit set Q has measure one, then it gives rise to a self-similar tiling.

For further information concerning self-similar tilings and related topics, the reader is referred to [7] and [6].

§5. Examples

In this section we will present some examples to illustrate the above construction.

Example 5.1. Let $G = \mathbb{R}^n$, $K = \mathbb{Z}^n$, $A(x) = 2\,x$. Then G/K is the n-Torus T^n, and $[0,1)^n$ is a self-similar fundamental region in the sense of Theorem 3.1i. In this case, the generalized B-splines are just the tensor product cardinal B-splines.

Example 5.2. Let $G = \mathbb{R}^2$, $K = \mathbb{Z}^2$ and

$$A\begin{pmatrix} x_1 \\ x_2 \end{pmatrix} = \begin{pmatrix} 1 & -1 \\ 1 & 1 \end{pmatrix} \begin{pmatrix} x_1 \\ x_2 \end{pmatrix}.$$

The self-similar fundamental region constructed by Lemma 4.1 is the set

$$\Delta = \left\{ \sum_{n=1}^{\infty} \epsilon_n \left(\frac{1-i}{2} \right)^n \,|\{\epsilon_n\}_{n\in\mathbb{N}} \in \{0,1\}^{\mathbb{N}} \right\},$$

which is a well-known fractal object called the *twin dragon set.* This example was independently studied by Cohen and Daubechies [2] and by Gröchenig and Madych [7]. In [7], further examples using different scaling matrices are presented.

Example 5.3. Let $G = \mathbb{R}^+$, viewed as a multiplicative group. A Haar measure on this group can be defined by

$$m_{\mathbb{R}^+}(A) = \int_A \frac{dx}{x}\,.$$

Let the set $\{2^n\}_{n\in\mathbb{Z}}$ be the discrete subgroup K. An automorphism A with $A(K) \subset K$ can, for example, be defined by

$$A(x) = x^2\,.$$

By using formula (3.2), it is easy to check that A^{-1} is a contraction in the sense of Theorem 3.1ii. We can choose the intervall $[1,2)$ as a self-similar fundamental region since it can be decomposed as

$$[1,2) = [1,\sqrt{2}) \cup [\sqrt{2},2) = \left(A^{-1}[1,2) \circ A^{-1}(1)\right) \cup \left(A^{-1}[1,2) \circ A^{-1}(2)\right).$$

It follows from Theorem 3.1 that the generalized B-splines with respect to $[1,2)$ generate a multiresolution analysis of $L^2\left(\mathbb{R}^+, \dfrac{dx}{\ln(2)\,x}\right)$. (We use the normalized Haar measure). In this case

$$N_2(x) = \frac{1}{\ln(2)} \int_0^\infty N_1(y)\,N_1(y^{-1}\,x)\,\frac{dy}{y} = \begin{cases} \dfrac{\ln(x)}{\ln(2)}, & \text{if } x \in [1,2) \\[2ex] \dfrac{\ln(\frac{4}{x})}{\ln(2)}, & \text{if } x \in [2,4) \\[2ex] 0, & \text{elsewhere} \end{cases}$$

so that N_2 is a continuous, piecewise logarithmic function which can be viewed as a generalized hat function. Consequently, N_3 is a C^1 function. In fact, one has

$$N_3(x) = (N_2 * N_1)(x) = \begin{cases} \dfrac{(\ln(x))^2}{2\,(\ln(2))^2}, & \text{if } x \in [1,2) \\ 1 - \dfrac{(\ln(\frac{x}{2}))^2}{2\,(\ln(2))^2} - \dfrac{(\ln(\frac{x}{4}))^2}{2\,(\ln(2))^2}, & \text{if } x \in [2,4) \\ \dfrac{(\ln(\frac{x}{8}))^2}{2\,(\ln(2))^2}, & \text{if } x \in [4,8) \\ 0, & \text{elsewhere.} \end{cases}$$

Example 5.4. Consider the subgroup of the upper triangular 3×3-matrices of the form

$$M = \begin{pmatrix} 1 & a & b \\ 0 & 1 & -a \\ 0 & 0 & 1 \end{pmatrix},$$

where $a,\ b \in \mathbb{R}$. This group can be identified with $\mathbb{R}^2$ equipped with the product

$$\{a_1, b_1\} \circ \{a_2, b_2\} = \{a_1 + a_2, b_1 + b_2 - a_1 a_2\}.$$

Then $G = (\mathbb{R}^2, \circ)$ is a locally compact Abelian group with

$$e = \{0,0\} \qquad \text{and} \qquad \{a,b\}^{-1} = \{-a, -b - a^2\}.$$

A Haar measure on this group is the Lebesgue-measure. Clearly, $K = \mathbb{Z}^2$ is a discrete subgroup and $G/_K$ is compact. The mapping

$$A: \quad \begin{aligned} G &\longrightarrow G \\ \{a,b\} &\longmapsto \{2a, 4b\} \end{aligned}$$

is an automorphism, A^{-1} is a contraction, and $A(K) \subset K$. A complete enumeration of distinct cosets of $\mathbb{Z}^2/_{A\,\mathbb{Z}^2}$ is the set

$$R = \{\{0,0\}, \{0,1\}, \{1,1\}, \{1,0\}, \{0,2\}, \{1,2\}, \{0,3\}, \{1,3\}\}.$$

In Figure 1, we plotted the resulting self-similar fundamental region computed with the aid of Lemma 4.1. It follows from Theorem 3.1 that the generalized B-splines with respect to this set generate a multiresolution analysis of $L^2(\mathbb{R}^2)$.

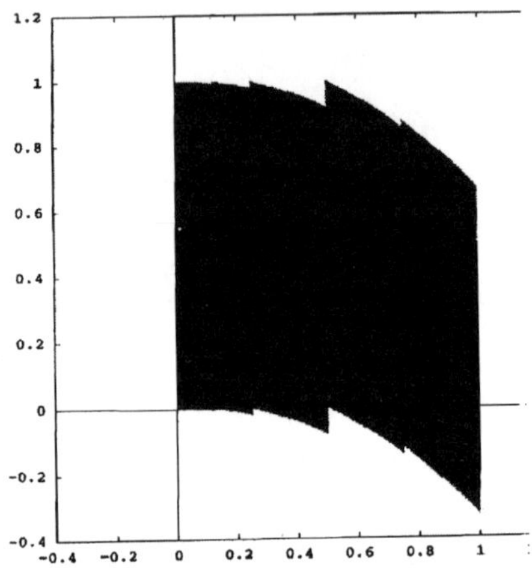

Fig. 1. Self-similar fundamental region.

§6. Generalized B–Spline Wavelets

In our setting, a system of functions $\{\psi^r\}_{r\in R\backslash\{e\}}$, is called a family of (mother) *wavelets* if the translates of $\{\psi^r\}_{r\in R\backslash\{e\}}$ span the orthogonal complement of V_0 in V_1, i.e.,

$$V_1 = V_0 \oplus W_0, \quad V_0 \perp W_0, \quad W_0 = \overline{span\{\psi^r(\cdot \circ k) | k \in K, r \in R\backslash\{e\}\}}. \tag{6.1}$$

Since the union of the spaces $\{V_j\}_{j\in Z}$ is dense in $L^2(G, dm_G)$ while their intersection is zero, it is easy to check that $L^2(G, dm_G)$ is spanned by all scaled and translated versions of the functions $\{\psi^r\}_{r\in R\backslash\{e\}}$. To construct such a wavelet basis, we introduce the *symbol* of a sequence $\{c_k\}_{k\in K} \in l_1(K)$ as the function

$$c(\gamma) := \sum_{k\in K} c_k \gamma(A^{-1}(k)), \qquad \gamma \in \hat{G}. \tag{6.2}$$

Furthermore, we define the *subsymbols* $c_r(\gamma), r \in R$ by

$$c_r(\gamma) = \sum_{k\in K} c_{r\circ A(k)} \gamma(k). \tag{6.3}$$

A straightforward calculation shows that

$$c(\gamma) = \sum_{r\in R} \gamma(A^{-1}(r)) \cdot \sum_{k\in K} c_{r\circ A(k)} \gamma(k) = \sum_{r\in R} \gamma(A^{-1}(r)) c_r(\gamma). \tag{6.4}$$

Theorem 6.1. *Let $\{V_j\}_{j\in Z}$ denote the multiresolution analysis constructed in Section 3. Let $a^e(\gamma)$ be defined by (6.2) by using the coefficients of the* *refinement equation (3.1). Suppose that there exist finitely supported subsymbols $a^r_{\tilde{r}}(\gamma), r \in R\backslash\{e\}, \tilde{r} \in R$ such that the matrix*

$$C(\gamma) := (a^r_{\tilde{r}}(\gamma))_{r,\tilde{r}\in R} \tag{6.5}$$

is nonsingular for all $\gamma \in \hat{G}$ and possesses an inverse $C(\gamma)^{-1}$ whose entries $(b^r_{\tilde{r}}(\gamma))_{r,\tilde{r}\in R}$ have coefficients in $l_1(K)$. Furthermore, suppose that

$$\sum_{\delta\in\tilde{R}} a^e(\gamma\circ\delta)\overline{a^r(\gamma\circ\delta)}\sum_{\tilde{\gamma}\in\Gamma}\left|\tilde{\phi}(\tilde{A}(\gamma)\circ\tilde{\gamma}\circ\delta)\right|^2 = 0 \quad \text{if } r \neq e, \tag{6.6}$$

where $\tilde{R}$ is a complete set of representers of $\Gamma/\tilde{A}\Gamma$ for $\tilde{A}$ defined by

$$\tilde{A}(\gamma)(\cdot) := \gamma(A(\cdot)). \tag{6.7}$$

Then the functions $\{\psi^r\}_{r\in R\backslash\{e\}}$ defined by

$$\psi^r(g) = \sum_{k\in K} a^r_k \phi(Ag\circ k) \tag{6.8}$$

form a wavelet basis.

Remark 6.2. *If $C(\gamma)$ is nonsingular, then the entries of $C(\gamma)^{-1}$ are continuous Γ–periodic functions, so that, once again by the theorem of Peter and Weyl and Pontryagin's duality theorem, each entry in $C(\gamma)^{-1}$ can be written in the form of (6.3) where the coefficients are in $l_2(K)$. We need here the slightly stronger condition that they are in $l_1(K)$.*

Proof of Theorem 6.1. It is sufficient to check that there exist sequences $\{b^r_{\tilde{r}\circ A(k)}\}_{k\in K}$ such that

$$\phi(A(g)\circ\tilde{r}) = \sum_{k\in K} b^e_{\tilde{r}\circ A(k)}\phi(g\circ k) + \sum_{r\in R\backslash\{e\}}\sum_{m\in K} b^r_{\tilde{r}\circ A(m)}\psi^r(g\circ m) \tag{6.9}$$

for all $\tilde{r} \in R$. Assume that (6.9) is true. Employing formula (6.8) and the refinement equation (3.1), we obtain

$$\phi(A(g)\circ\tilde{r}) = \sum_{r\in R}\sum_{k\in K}\sum_{l\in K} b^r_{\tilde{r}\circ A(k)} a^r_l \phi(A(g)\circ A(k)\circ l). \tag{6.10}$$

Taking the Fourier transform of formula (6.10) and using formula (6.4) yields

$$\begin{aligned} c_1\gamma(A^{-1}(\tilde{r}))\hat{\phi}(\tilde{A}^{-1}(\gamma)) &= c_1\sum_{r\in R}\sum_{k\in K}\sum_{l\in K} b^r_{\tilde{r}\circ A(k)} a^r_l \gamma(k)\gamma(A^{-1}(l))\hat{\phi}(\tilde{A}^{-1}(\gamma)) \\ &= c_1\sum_{r\in R} b^r_{\tilde{r}}(\gamma)a^r(\gamma)\hat{\phi}(\tilde{A}^{-1}(\gamma)) \\ &= c_1\sum_{r\in R}\sum_{\hat{r}\in R} b^r_{\tilde{r}}(\gamma)a^r_{\hat{r}}(\gamma)\gamma(A^{-1}(\hat{r}))\hat{\phi}(\tilde{A}^{-1}\gamma), \end{aligned} \tag{6.11}$$

for $\gamma \in \hat{G}$. Formula (6.11) holds if

$$\sum_{r\in R} b^r_{\tilde{r}}(\gamma)a^r_{\hat{r}}(\gamma) = \delta_{\tilde{r}\hat{r}}. \tag{6.12}$$

This means that (6.11) and therefore (6.9) holds if $(b^r_{\tilde{r}}(\gamma))_{r,\tilde{r}} = C(\gamma)^{-1}$, which makes sense from our assumptions. Equation (6.6) clearly implies that $\langle \phi(\cdot), \psi^r(\cdot \circ k)\rangle = 0$ for all $r \in R\backslash\{e\}, k \in K$. ■

Acknowledgements. Thanks to Angela Kunoth for many helpful remarks, and to Vera Latour who helped produce the figure.

References

1. Bauer, H., *Wahrscheinlichkeitstheorie und Grundzüge der Masstheorie*, Walter de Gruyter, Berlin New York, 1978.
2. Cohen, A., and I. Daubechies, Non-separable bidimensional wavelet bases, preprint 1991.
3. Cohn, D., *Measure Theory*, Birkhäuser, Basel, 1980.
4. Chui, C., and I. Wang, A cardinal spline approach to wavelets, CAT Report 211, Texas A&M University, 1990.
5. Chui, C., and I. Wang, A general framework of compactly supported splines and wavelets, CAT Report, Texas A&M University, 1990.
6. Gröchenig, K., and A. Haas, Self-similar lattice tilings, preprint 1992.
7. Gröchenig, K., and W. R. Madych, Multiresolution analysis, Haar bases, and self-similar tilings of $\mathbb{R}^n$, IEEE Transactions on information theory, **38** (1992), 556–568.
8. Hewitt, E., and K. A. Ross, *Abstract Harmonic Analysis*, Volume I, Springer Verlag, Berlin, 1963.
9. Jia, R. Q., and C. A. Micchelli, Using the refinement equations for the construction of Pre-wavelets II: Powers of two, in: *Curves and Surfaces*, P. J. Laurents, A. Le Méhauté and L. L. Schumaker (eds.), Academic Press, New York, 1991.
10. Lemarié, P. G., Bases d'ondelettes sur les groupes de Lie stratifiés, Bulletin de la Société Mathématique de France **117** (1989), 211–232.
11. Mallat, S., Multiresolution approximation and wavelet orthonormal bases of L^2, Trans. Amer. Math. Soc., **315** (1989), 69–88.
12. Pontryagin, L. S., *Topological Groups*, Princeton University Press, 1939.
13. Raghunathan, M. S., *Discrete Subgroups of Lie Groups*, Springer Verlag, Berlin, 1972.
14. Strichartz, R. S., Self-similarity on nilpotent Lie groups, preprint 1992.
15. Sugiura, K., *Unitary Representation and Harmonic Analysis*, Halsted Press, 1975.

Stephan Dahlke
Institut für Geometrie und Praktische Mathematik
RWTH Aachen
Templergraben 55
52062 Aachen, GERMANY

dahlke@igpm.rwth-aachen.de

Some Remarks on Multiscale Transformations, Stability, and Biorthogonality

Wolfgang Dahmen

Abstract. This paper is concerned with the concepts of stability and biorthogonality for a general framework of multiscale transformations. In particular, stability criteria are derived which do not make use of Fourier transform techniques but rather hinge upon classical Bernstein and Jackson estimates. Therefore they might be useful when dealing with possibly nonuniform discretizations or with bounded domains.

§1. Introduction

Let $\mathbf{c}$ be some string of data $c_k, k \in I$, where I is some (finite or possibly infinite) index set. These data could represent grey scale values of a digital image, statistical noisy data, or control points in some curve or surface representation, or approximate solutions of some discretized operator equation. The common ground for these rather different interpretations is that these data could be viewed as coefficients of some expansion

$$f = \sum_{k \in I} c_k \varphi_k, \tag{1.1}$$

where the φ_k are (typically scalar-valued shape) functions defined on some domain (or manifold) Ω (which is topologically equivalent to some bounded or unbounded domain) in $\mathbb{R}^s$. As a familiar simple example one could take φ_k as B-splines relative to some knot sequence in an interval Ω. When each c_k is a point in $\mathbb{R}^3$ say, f represents a space curve. The c_k then already convey explicit geometrical information on the curve or, more precisely, on the location of the points $f(x), x \in \Omega$. It is well known that this kind of information can be drawn from the data c_k under much more general circumstances, namely when the φ_k have good localization properties and sum to one. Note that this localization represents the resolution of the underlying object only with respect to a *single scale.*

Wavelets, Images and Surface Fitting
P. J. Laurent, A. Le Méhauté, and L. L. Schumaker (eds.), pp. 157–188.

ISBN 1-56881-040-7.

However, in many applications it is important to extract or exploit information on the data that could be associated with different scales ranging from a very coarse to a very fine level. For instance, when dealing with images one would want to separate local from global features. The coefficients corresponding to higher scales represent successive updates of increasingly local nature. Moreover, the magnitude of such coefficients indicates the significance of such updates and therefore forms the basis for a primary objective of multiscale representations, namely *data compression*. In a similar way they facilitate smoothing noisy data (cf. [19,32]). In fact, these ideas carry much further. It is not only possible to reduce the complexity of images and surfaces or to smooth noisy data but similar principles apply to multiscale representations of operators with global Schwartz kernel. Here compression strategies lead to sparse approximations to such operators which in turn can be further processed by efficient iterative solvers (see e.g. [2,13,14,15,16]).

When dealing with elliptic operators, multiscale bases usually form much better approximations to the eigenfunctions of the operator than the *nodal* bases corresponding to the finest scale only. As a consequence, corresponding stiffness matrices can be efficiently *preconditioned* by diagonal scalings [11,20,33].

Aside from compression and preconditioning multiscale representations of data usually convey accurate information about the smoothness or regularity of the underlying continuous object. This fact is for instance exploited (in combination with the above mentioned effects) in [23,24,25] to speed up numerical schemes for hyperbolic conservation laws by significantly reducing flux calculations which in this context form the bulk of computations.

The central ingredient of any of the above multiscale applications is an appropriate transformation $\mathbf{T}$ relating the *fine scale data* $\mathbf{c}$ to their *multiscale representation* $\mathbf{d}$

$$\mathbf{c} = \mathbf{T}\mathbf{d}. \tag{1.2}$$

The success of such a scheme then relies mainly on the following two requirements. Firstly, the transformation (1.2) should be efficient. This means it should only require the order of $\#I$ floating point operations. Secondly, such transformations should not entail any significant loss of accuracy in the data, i.e., the condition numbers should remain uniformly bounded

$$\|\mathbf{T}\| \, \|\mathbf{T}^{-1}\| = \mathcal{O}(1), \qquad \#I \to \infty, \tag{1.3}$$

independent of the size of the problem. This latter requirement is trivially satisfied when $\mathbf{T}$ is an *orthogonal transformation*. This is the case when dealing with *wavelet representations*, i.e., when $\mathbf{d}$ are the coefficients of an expansion in terms of (orthogonal) wavelets. However, depending on the setting at hand, the construction of orthogonal multiscale bases by itself could be a very difficult task and computationally as expensive as the whole problem. This is the case when dealing with unstructured grids or discretizations of bounded domains. Even for uniform grids on all of $\mathbb{R}^s$ it is by no means simple to construct compactly supported wavelets. In many applications one therefore has to dispense with orthogonality and be content with *stability* in the sense of (1.3). This is also the point of view taken in [23,24,25]. There a very general framework is developed which is formulated almost entirely in discrete terms deferring any reference to an underlying system of functions as far as possible. However, when it comes to ascertaining stability (1.3), it seems that one ultimately cannot avoid associating the discrete data with expansions in terms of appropriate basis functions. Nevertheless, the only setting for which this line of arguments has been carried through in a concrete way is the one of classical *multiresolution analysis* for *shift-invariant* function spaces on $\mathbb{R}^s$ or on the torus [5,17,27,28,29,30]. While stability in the above sense is a trivial attribute of orthogonal wavelets, it takes more effort to be ascertained when dealing with so called *biorthogonal wavelets.* Sharp stability conditions in connection with biorthogonal wavelets are established in [7]. However, these results rely crucially on Fourier transform techniques which do not apply in many cases of practical interest such as classical finite elements or unstructured grids for bounded domains.

The objective of this paper is therefore to discuss some aspects of the issue of stability for multiscale transformations in a sufficiently general setting covering the above mentioned cases of interest. Several equivalent ways of viewing stability will be pointed out. We are mainly interested in criteria which do not make use of Fourier transforms. In particular, classical *Bernstein* and *Jackson estimates* turn out to play a crucial role in this regard.

The layout of this paper is as follows. In Section 2 we describe a general format of multiscale transformations and discuss some important requirements imposed by two types of applications. We take the opportunity of reviewing some basic ideas and arguments relating stability to *Riesz-bases.* The essence of this material is certainly not new. Rephrasing it in the present general context is merely to help focusing on some aspects which might be useful for further progress and to prepare the ground for the development in Section 3. Using standard facts from functional analysis, these concepts are reformulated there in terms of linear projectors associated with nested closed subspaces of a Banach space. This leads to stability estimates for a variety of norms. The main result establishes the equivalence between such norms and multiscale sequence norms expressed in terms of differences of *biorthogonal* projectors, assuming only the validity of certain Bernstein and Jackson inequalities. It is interesting to note that such relations are needed for establishing the equivalence between Besov-norms of positive and negative order and certain sequence norms [26] (see Theorem 3.11 below). This is used in [26] for developing multilevel preconditioners in connection by saddle point problems which arise when appending boundary conditions for elliptic problems by means of Lagrange multipliers. We will conclude Section 3 with briefly pointing out how these observations relate to the characterization of biorthogonal Riesz-bases in the context of classical multiresolution for shift-invariant spaces (cf. [7,8]).

We conclude in Section 4 with some comments on the construction of stable biorthogonal bases by means of generalized subdivision schemes.

§2. Multiscale Transformations

As in the previous section, $\mathbf{c} = \mathbf{c}^j$ will denote a sequence of data $c_k^j, k \in I_j$, associated with a single discretization level. However, we will deal with a whole hierarchy of discretization levels (of increasing fineness) represented by (possibly infinite) index sets $I_j, j \in \mathbb{N}_0$. $\mathbf{c}^j$ should be viewed as a way of encoding information by means of a single scale of resolution, in brief, *single scale data.* We wish to reexpress the information carried by such single scale data in terms of *multiscale data.* Roughly speaking $\mathbf{c}^j$ is to be split into a blurred version $\mathbf{c}^{j-1}$ and the detail $\mathbf{d}^{j-1}$ lost by this compression. We proceed now describing the form of transformations interrelating both data types. To this end, we will always assume that

$$I_{j+1} \simeq I_j \cup J_j, \quad I_j \cap J_j = \emptyset. \tag{2.1}$$

The data $\mathbf{c}^j$ are supposed to belong to some (finite or infinite dimensional) sequence space $l(I_j)$ endowed with a norm $\|\cdot\|_{l(I_j)}$. Standard examples are the l_p-norms

$$\|\mathbf{c}\|_{l_p(I_j)} := \left(\sum_{k \in I_j} |c_k|^p \right)^{1/p},$$

for $1 \le p \le \infty$. $\mathcal{L}(l(I), l(I'))$ denotes the space of bounded linear operators from $l(I)$ to $l(I')$ represented by matrices $\mathbf{A} = (a_{k,m})_{k \in I', m \in I}$ (with possibly complex) entries $a_{k,m}$. $l^*(I)$ is the dual space of $l(I)$ relative to the standard form $\langle \mathbf{c}, \mathbf{c}' \rangle_{l(I) \times l^*(I)} := \sum_{k \in I} c_k \overline{c'_k}$ and $\mathbf{A}^*$ is the corresponding adjoint of $\mathbf{A}$. For each level j let

$$\mathbf{B}_{j,0} \in \mathcal{L}(l(I_j), l(I_{j+1})), \quad \mathbf{B}_{j,1} \in \mathcal{L}(l(J_j), l(I_{j+1})). \tag{2.2}$$

In view of (2.1), the composed mapping

$$\mathbf{B}_j := (\mathbf{B}_{j,0}, \mathbf{B}_{j,1}) \tag{2.3}$$

is to be viewed as an element of $\mathcal{L}(l(I_j) \oplus l(J_j), l(I_{j+1}))$, and in all relevant examples one may identify $l(I_j) \oplus l(J_j)$ with $l(I_{j+1})$ so that $\mathbf{B}_j$ actually can be interpreted as a change of bases in $l(I_{j+1})$. Apparently this corresponds to a step of a typical *pyramid type scheme* arising in connection with wavelet decompositions, in that $\mathbf{B}_{j,0}^* \mathbf{c}^{j+1}$ stands for a *blurred* version of $\mathbf{c}^{j+1}$ while $\mathbf{B}_{j,1}^* \mathbf{c}^{j+1}$ represents the detail lost by this compression. The only difference is that here the transformations are allowed to depend on the level j in order to deal with bounded domains or unstructured grids. Thus setting

$$\mathbf{c}^{j-1} := \mathbf{B}_{j-1,0}^* \mathbf{c}^j, \quad \mathbf{d}^{j-1} := \mathbf{B}_{j-1,1}^* \mathbf{c}^j, \quad j \in \mathbb{N}, \tag{2.4}$$

leads to the following scheme

$$\begin{array}{ccccccccc}
 & \mathbf{B}^*_{j-1,0} & & \mathbf{B}^*_{j-2,0} & & \mathbf{B}^*_{j-3,0} & & & \\
\mathbf{c}^j & \rightarrow & \mathbf{c}^{j-1} & \rightarrow & \mathbf{c}^{j-2} & \rightarrow & & \cdots & \mathbf{c}^0 \\
 & \mathbf{B}^*_{j-1,1} & & \mathbf{B}^*_{j-2,1} & & \mathbf{B}^*_{j-3,1} & & & \\
 & \searrow & & \searrow & & \searrow & & \cdots & \\
 & & \mathbf{d}^{j-1} & & \mathbf{d}^{j-2} & & & \cdots & \mathbf{d}^0
\end{array} \tag{2.5}$$

which transforms the single scale data $\mathbf{c}^j$ into multiscale data

$$\mathbf{d}^{(j)} := (\mathbf{c}^0, \mathbf{d}^0, \mathbf{d}^1, \ldots, \mathbf{d}^{j-1})^T. \tag{2.6}$$

To reverse the transformation (2.5), i.e., to recover $\mathbf{c}^j$ from $\mathbf{d}^{(j)}$, let

$$\mathbf{A}_{j,0} \in \mathcal{L}(l(I_j), l(I_{j+1})), \quad \mathbf{A}_{j,1} \in \mathcal{L}(l(J_j), l(I_{j+1})) \tag{2.7}$$

satisfy

$$\mathbf{A}_{j,0}\mathbf{B}^*_{j,0} + \mathbf{A}_{j,1}\mathbf{B}^*_{j,1} = I, \tag{2.8}$$

or equivalently

$$\mathbf{B}^*_j \mathbf{A}_j = I. \tag{2.9}$$

Here $\mathbf{A}_j = (\mathbf{A}_{j,0}, \mathbf{A}_{j,1})$ and I denotes the identity in the space $l(I_{j+1}), l(I_j)$, respectively. The inverse operation to (2.5) can then be described as follows:

$$\begin{array}{ccccccccc}
 & \mathbf{A}_{0,0} & & \mathbf{A}_{1,0} & & \mathbf{A}_{2,0} & & \mathbf{A}_{j-1,0} & \\
\mathbf{c}^0 & \rightarrow & \mathbf{c}^1 & \rightarrow & \mathbf{c}^2 & \rightarrow & \cdots & \rightarrow & \mathbf{c}^j \\
 & \mathbf{A}_{0,1} & & \mathbf{A}_{1,1} & & \mathbf{A}_{2,1} & & \mathbf{A}_{j-1,1} & \\
 & \nearrow & & \nearrow & & \nearrow & \cdots & \nearrow & \\
\mathbf{d}^0 & & \mathbf{d}^1 & & \mathbf{d}^2 & & & &
\end{array} \tag{2.10}$$

Alternatively, defining

$$\hat{\mathbf{A}}_j := \begin{pmatrix} \mathbf{A}_j & 0 \\ 0 & I \end{pmatrix} \in \mathcal{L}(l(I_j) \oplus_{m=j}^{n-1} l(J_m), l(I_{j+1}) \oplus_{m=j+1}^{n-1} l(J_m)), \tag{2.11}$$

we can write (cf. (2.6))

$$\mathbf{c}^n = \mathbf{T}_n \mathbf{d}^{(n)}, \tag{2.12}$$

where

$$\mathbf{T}_n := \hat{\mathbf{A}}_{n-1} \cdots \hat{\mathbf{A}}_0. \tag{2.13}$$

Of course, the particular structure of the transformations (2.5) and (2.10) is motivated by viewing $\mathbf{c}^j$ as the coefficients of some expansion

$$f_j = \sum_{k \in I_j} c^j_k \varphi^j_k$$

relative to some basis $\Phi_j = \{\varphi_k^j : k \in I_j\}$ of a space S_j while $\mathbf{d}^{(j)}$ are the coefficients of the same function f_j relative to another basis. Such an underlying system of bases may never enter the computations explicitly but it turns out to be crucial for at least two reasons. It should be relevant for the type of information to be extracted from both representations. It enters when trying to prove various important properties of the transformation detailed below.

To be more specific, let $\mathcal{S} = \{S_j\}_{j \in \mathbb{N}_0}$ be a sequence of strictly nested closed subspaces S_j of $L_2(\Omega)$. Suppose that each S_j is generated by an l_2-*stable* collection of functions . Here l_2-stable means that there exist positive constants c_1, c_2 such that

$$c_1 \|\mathbf{c}\|_{l_2(I_j)} \leq \left\| \sum_{k \in I_j} c_k \varphi_k^j \right\|_{L_2(\Omega)} \leq c_2 \|\mathbf{c}\|_{l_2(I_j)} \tag{2.14}$$

holds uniformly in $\mathbf{c} \in l_2(I_j)$ and $j \in \mathbb{N}_0$. We will also use the notation

$$\|f\|_{L_p(\Omega)} := \left(\int_\Omega |f(x)|^p dx \right)^{1/p}$$

for $1 \leq p \leq \infty$ (with the standard interpretation for $p = \infty$), and

$$(f, g)_\Omega := \int_\Omega f(x)\overline{g(x)}dx.$$

Here and in the sequel we will assume that

$$\left\| \varphi_k^j \right\|_{L_2(\Omega)} \sim 1, \tag{2.15}$$

where '$\sim$' is to express a relation of the type (2.14) to hold uniformly with respect to all the involved parameters.

Remark 2.1. *The nestedness of the S_j combined with l_2-stability implies the existence of some matrix $\mathbf{A}_{j,0} = (a_{k,m}^j)_{k \in I_{j+1}, m \in I_j} \in \mathcal{L}(l_2(I_j), l_2(I_{j+1}))$ such that*

$$\varphi_k^j = \sum_{m \in I_{j+1}} a_{m,k}^j \varphi_m^{j+1}. \tag{2.16}$$

$\mathbf{A}_{j,0}$ is sometimes referred to as *refinement* or *subdivision matrix*. It is perhaps worth describing two typical examples.

Example 2.2. *Suppose Ω is some polygonal domain in $\mathbb{R}^2$ and let $\mathcal{T}_j$ denote the triangulation of Ω obtained by subdividing each triangle in $\mathcal{T}_{j-1}$ into four congruent triangles introducing the midpoints of the edges in $\mathcal{T}_{j-1}$ as new vertices. Here one can take I_j as the set of vertices in $\mathcal{T}_j$ and φ_k^j as the unique continuous piecewise linear functions relative to $\mathcal{T}_j$ satisfying*

$$\varphi_k^j(m) = 2^j \delta_{m,k}, \quad m, k \in I_j.$$

It is easy to see that the Φ_j are uniformly stable in the sense of (2.14) and that (2.15) is valid (see e.g. [11]). Moreover, one readily checks that

$$\varphi_k^j = \sum_{m \in I_{j+1}} 2^{-j-1} \varphi_k^j(m) \varphi_m^{j+1},$$

i.e., $a_{m,k}^j := 2^{-j-1} \varphi_k^j(m)$.

Example 2.3. *Let $\Omega = \mathbb{R}^s$ and let M be some fixed $s \times s$ matrix with integer entries whose eigenvalues are all strictly larger than one (cf. e.g. [12,21]). Suppose $\varphi \in L_2(\mathbb{R}^s)$ is $(\mathbf{a}^0, M)$-refinable, i.e., there exists some mask $\mathbf{a}^0 = \{a_k^0\}_{k \in \mathbb{Z}^s}$ such that*

$$\varphi(x) = \sum_{k \in \mathbb{Z}^s} a_k^0 \varphi(Mx - k), \quad x \in \mathbb{R}^s, \text{ a.e..} \tag{2.17}$$

Thus here one has $I_j = M^{-j}\mathbb{Z}^s$ and

$$\mathbf{A}_{j,0} = \mathbf{A}_0 = (a_{m-Mk}^0)_{m,k \in \mathbb{Z}^s}. \tag{2.18}$$

Now suppose one wants to reexpress an element in S_j by successively updating lower-level information. To this end, one has to build a *multiscale basis* by successively completing lower level bases. More precisely, given $\mathbf{A}_{j,0}$, i.e., Φ_j satisfying (2.16), one has to find

$$\begin{aligned} &\mathbf{A}_{j,1} \in \mathcal{L}(l_2(J_j), l_2(I_{j+1})), \\ &\mathbf{B}_{j,0} \in \mathcal{L}(l_2(I_j), l_2(I_{j+1})), \\ &\mathbf{B}_{j,1} \in \mathcal{L}(l_2(J_j), l_2(I_{j+1})), \end{aligned} \tag{2.19}$$

such that the functions

$$\psi_k^j = \sum_{m \in I_{j+1}} a_{m,k}^j \varphi_m^{j+1}, \quad k \in J_j, \tag{2.20}$$

satisfy

$$\varphi_k^{j+1} = \sum_{m \in I_j} b_{k,m}^j \varphi_m^j + \sum_{m \in J_j} b_{k,m}^j \psi_m^j. \tag{2.21}$$

Thus

$$\Psi^j := \{\psi_k^j : k \in J_j\} \tag{2.22}$$

completes Φ_j to a basis for S_{j+1}. Following [3] one can state

Proposition 2.4. *Let the matrices* $\mathbf{A}_{j,e}, \mathbf{B}_{j,e}, e \in \{0,1\}$ *satisfy the assumptions of Remark 2.1 and (2.19). Assume that* Φ_j *and* Φ_{j+1} *are* l_2*-stable. Then the following statements are equivalent:*

(i) Ψ^j, *defined by (2.20) and (2.22) is* l_2*-stable and satisfies (2.21).*
(ii) *The matrices* $\mathbf{A}_j, \mathbf{B}_j$, *defined as in (2.3), satisfy*

$$\mathbf{A}_j \mathbf{B}_j^* = \mathbf{B}_j^* \mathbf{A}_j = I. \tag{2.23}$$

Thus under either condition one has

$$S_{j+1} = S_j \bigoplus W_{j+1}, \tag{2.24}$$

where W_{j+1} is the (L_2-closure of the) linear span of Ψ^j. Noting that (2.23) is equivalent to

$$\mathbf{B}_{j,e'}^* \mathbf{A}_{j,e} = \delta_{e,e'} I, \quad e, e' \in \{0,1\}, \quad \mathbf{A}_{j,0}\mathbf{B}_{j,0}^* + \mathbf{A}_{j,1}\mathbf{B}_{j,1}^* = I, \tag{2.25}$$

and recalling (2.8), Proposition 2.4 yields

Remark 2.5. *One has*

$$\sum_{k\in I_n} c_k^n \varphi_k^n = \sum_{k\in I_0} c_k^0 \varphi_k^0 + \sum_{j=0}^{n-1} \sum_{k\in J_j} d_k^j \psi_k^j \tag{2.26}$$

if and only if $\mathbf{c}^n = \mathbf{T}_n \mathbf{d}^{(n)}$, *where* $\mathbf{d}^{(n)}$ *and* $\mathbf{T}_n$ *are defined by (2.6) and (2.13), respectively. Thus* $\mathbf{T}_n$ *is the mapping that takes the coefficients relative to the multiscale basis*

$$\Psi_n = \Phi_0 \cup_{j=0}^{n-1} \Psi^j \tag{2.27}$$

into the coefficients relative to the single scale basis Φ_n.

There are two issues of central practical importance that immediately come to mind.

Complexity: One would prefer the basis functions in Φ_j and Ψ_j to be all compactly supported. This leaves the chance of finding $\mathbf{A}_j$ that are uniformly *sparse* by which we mean that all rows and columns contain only a uniformly bounded finite number of nonzero entries. If $\#I_j$ behaves essentially like a fixed fraction of $\#I_{j+1}$ for all j this would ensure the transformation (2.10) tc be *efficient* in that it requires only $\mathcal{O}(\#I_j)$ operations. Likewise, one would want (2.5) to be efficient in the same sense. In this context the following should be kept in mind.

Remark 2.6. *Usually the matrices* $\mathbf{A}_{j,0}$ *are given along with the bases* Φ_j. *One then has to find suitable (sparse) matrices* $\mathbf{A}_{j,1}$ *such that* $\mathbf{A}_j := (\mathbf{A}_{j,0}, \mathbf{A}_{j,1})$ *is invertible. A characterization of all possible such completions*

$\mathbf{A}_{j,1}$ is given in [3]. Since the inverse of a sparse matrix is generally not sparse it is then usually hard to find such $\mathbf{A}_{j,1}$ that $\mathbf{B}_j$ is sparse as well.

Stability: Aside from computational complexity, the *stability* of the transformation $\mathbf{T}_n$ is of central importance. More precisely, setting

$$\mathcal{J}_n := I_0 \cup_{j=0}^{n-1} J_j,$$

one has to ensure that for $\mathbf{c}^n = \mathbf{T}_n \mathbf{d}^{(n)}$

$$\|\mathbf{c}^n\|_{l_2(I_n)} \sim \left\|\mathbf{d}^{(n)}\right\|_{l_2(\mathcal{J}_n)},$$

uniformly in $n \in \mathbb{N}$, which is equivalent to

$$\|\mathbf{T}_n\|, \ \left\|\mathbf{T}_n^{-1}\right\| = \mathcal{O}(1), \quad n \to \infty. \tag{2.28}$$

Here $\|\cdot\|$ denotes the operator norm induced by the underlying sequence or vector norm.

The following well-known fact indicates how to go about the stability question.

Remark 2.7. *Suppose the matrices $\mathbf{A}_j$ and $\mathbf{B}_j$ are associated with Φ_j and Ψ^j through (2.16), (2.20), (2.21), and let $\mathbf{T}_n$ be defined by (2.13). If the Φ_j and Ψ^j are orthonormal bases, then the $\mathbf{T}_n$ are uniformly stable, i.e., (2.28) holds.*

In fact, when dealing with orthonormal bases the transformations $\mathbf{T}_n$ are clearly unitary matrices so that $\|\mathbf{T}_n\| = \|\mathbf{T}_n^{-1}\| = 1$. Since in this case the inverse is the adjoint matrix, sparsity of the $\mathbf{A}_j$ already implies sparsity of the $\mathbf{B}_j$. Thus if one could construct an orthonormal multiscale basis consisting of compactly supported functions, one would simultaneously solve the complexity and stability problem. However, the concrete construction of such bases suitable for practical purposes seems to be confined so far to the classical concept of *orthonormal wavelets* on $\mathbb{R}^s$ [17,18,27], i.e., the underlying sequence $\mathcal{S}$ consists of shift-invariant spaces generated by dilates and integer translates of a single refinable function φ as described in Example 2.3. Even then one seems to be essentially confined to tensor products of univariate wavelets. On the other hand, it is also well-known that one needs actually much less. In fact, suppose $\mathbf{A}_j$ have been found such that (2.21) holds, and the ψ_k^j are only orthogonal between levels, i.e.,

$$(\psi_k^j, \psi_m^l)_\Omega = 0 \quad \text{for } j \neq l \ \text{ and } \ \text{all } k \in J_j, m \in J_l. \tag{2.29}$$

Since by Proposition 2.4 the Ψ^j are also l_2-stable, one gets for $\mathbf{c}^n = \mathbf{T}_n \mathbf{d}^{(n)}$

and $f_n = \sum_{k\in I_n} c_k^n \varphi_k^n$

$$\begin{aligned}
\|\mathbf{c}^n\|^2_{l_2(I_n)} &\sim \|f_n\|^2_{L_2(\Omega)} = \Big\| \sum_{k\in I_0} c_k^0\varphi_k^0 + \sum_{j=0}^{n-1}\sum_{k\in J_j} d_k^j\psi_k^j \Big\|^2_{L_2(\Omega)} \\
&= \Big\| \sum_{k\in I_0} c_k^0\varphi_k^0 \Big\|^2_{L_2(\Omega)} + \sum_{j=0}^{n-1} \Big\| \sum_{k\in J_j} d_k^j\psi_k^j \Big\|^2_{L_2(\Omega)} \\
&\sim \|\mathbf{c}^0\|^2_{l_2(I_0)} + \sum_{j=0}^{n-1} \|\mathbf{d}^j\|^2_{l_2(J_j)} = \Big\|\mathbf{d}^{(n)}\Big\|^2_{l_2(\mathcal{J}_n)} \qquad (2.30)
\end{aligned}$$

Orthogonality between levels, which is much easier to realize than full orthogonality (see e.g. [1,5,22,29,30]), clearly suffices to ensure stability, but the matrices $\mathbf{B}_j$ controlling the scheme (2.5) are then generally *not* sparse. Nevertheless, we have repeated the standard reasoning (2.30) only to emphasize that the key to stability estimates lies in switching to function norms, i.e., in making explicit use of the underlying bases Φ_j, Ψ_j. In fact, replacing the equal sign in (2.30) (which is due to the Pythagorean Theorem) by another '$\sim$' yields a bit more flexibility which can actually help to remedy the above drawback. To describe this, let

$$\Psi := \Phi_0 \bigcup_{j=0}^{\infty} \Psi^j. \qquad (2.31)$$

Moreover, in the following it will be convenient to set

$$J_{-1} := I_0, \quad \Psi^{-1} := \Phi_0, \quad \mathbf{d}^{-1} := \mathbf{c}^0,$$

and

$$\mathcal{J} := \bigcup_{j=-1}^{\infty} J_j.$$

The following proposition summarizes essentially familiar facts but will serve as a motivation for the subsequent discussion.

Proposition 2.8. *Suppose the $\Phi_j, j\in \mathbb{N}_0$, are uniformly stable and that S is dense in $L_2(\Omega)$, i.e., $\overline{\bigcup_{j\in\mathbb{N}_0} S_j} = L_2(\Omega)$. Then $\mathbf{T}_n$ are uniformly stable, i.e., (2.28) holds, if and only if Ψ is a Riesz-basis of $L_2(\Omega)$, i.e., every $f\in L_2(\Omega)$ has a unique expansion*

$$f = \sum_{j=-1}^{\infty}\sum_{k\in J_j} d_k^j(f)\psi_k^j \qquad (2.32)$$

which converges in $L_2(\Omega)$ such that

$$\|f\|_{L_2(\Omega)} \sim \left(\sum_{j=-1}^{\infty} \sum_{k \in J_j} |d_k^j(f)|^2 \right)^{1/2} . \tag{2.33}$$

Proof: Suppose first that (2.28) holds, and let $f_n = \sum_{k \in I_n} c_k^n \varphi_k^n$ denote the orthogonal projection of $f \in L_2(\Omega)$ to S_n, where we assume that f does not vanish identically. Thus, since $\mathcal{S}$ is dense, one has for n sufficiently large that

$$2\|f_n\| \geq \|f\|_{L_2(\Omega)} \geq \|f_n\|_{L_2(\Omega)} - \|f - f_n\|_{L_2(\Omega)} \geq \frac{1}{2}\|f_n\|_{L_2(\Omega)} . \tag{2.34}$$

Now let

$$\mathbf{c}^n = \mathbf{T}_n \mathbf{d}^{(n)}. \tag{2.35}$$

By the uniform stability of the Φ_n, we infer from (2.34) and (2.35) that there exists some constant c independent of n such that for n sufficiently large

$$\|f\|_{L_2(\Omega)} \geq \frac{c}{\left\|\mathbf{T}_n^{-1}\right\|} \left\|\mathbf{d}^{(n)}\right\|_{l_2(\mathcal{J}_n)} .$$

Thus when (2.28) holds, one concludes that

$$\limsup_{n \to \infty} \left\|\mathbf{d}^{(n)}\right\|_{l_2(\mathcal{J}_n)} \leq c\|f\|_{L_2(\Omega)} , \tag{2.36}$$

where c is independent of n and f. Since therefore

$$\|f_n - f_{n+m}\|_{L_2(\Omega)} \geq c \left\|\mathbf{d}^{(n)} - \mathbf{d}^{(n+m)}\right\|_{l_2(\mathcal{J})} ,$$

it is easy to derive from (2.36) the existence of an expansion (2.32) as well as the estimate

$$c\|f\|_{L_2(\Omega)} \geq \left(\sum_{j=-1}^{\infty} \sum_{k \in J_j} |d_k^j(f)|^2 \right)^{1/2} .$$

Likewise we deduce from (2.34) and (2.35) that

$$\|f\|_{L_2(\Omega)} \leq c\|\mathbf{T}_n\| \left\|\mathbf{d}^{(n)}\right\|_{l_2(\mathcal{J}_n)}$$

which proves (2.33). The converse is trivial. ■

We conclude this section by briefly reviewing two types of applications of the above transformations. The transformation from single to multiscale data (2.5) is needed explicitly when decomposing a signal or image in order to smooth or compress the $\mathbf{d}^{(n)}$. In addition, (2.10) is used for reconstruction from the compressed multiscale data. There is another type of application where the object to be analysed is not given explicitly, but is sought for instance as the solution of an operator equation

$$\mathcal{A}u = f \tag{2.37}$$

(with suitable side conditions incorporated in (2.37)) on some domain Ω.

Now given such a sequence $\mathcal{S}$ of trial spaces, a Galerkin scheme for determining an approximate solution

$$u^n = \sum_{k\in I_n} c_k^n \varphi_k^n \in S_n$$

of (2.37) requires solving the linear system

$$\mathcal{A}_{\Phi_n}\mathbf{c}^n = \mathbf{f}^n, \tag{2.38}$$

where $f_k^n := (f, \varphi_k^n)_\Omega$ and

$$\mathcal{A}_{\Phi_n} = ((\mathcal{A}\varphi_k^n, \varphi_m^n)_\Omega)_{m,k\in I_n}$$

is the *nodal* or *single-scale* stiffness matrix. When Φ_n consists of compactly supported functions and $\mathcal{A}$ is an elliptic differential operator of order r the matrix $\mathcal{A}_{\Phi_n}$ is sparse but its spectral condition numbers typically grow like $\mathcal{O}(\#I_n^{r/s})$. This makes iterative solvers prohibitively slow. On the other hand, direct solvers cannot be applied because the size of such problems is typically very large, and direct solvers usually do not preserve sparsity. In contrast, stiffness matrices with respect to appropriate *multiscale bases* are much better behaved in that when subjected to a simple diagonal scaling resulting condition numbers remain uniformly bounded independent of the size of the problem [11,26]. As a consequence, iterative schemes such as the conjugate gradient method produce a solution with desired accuracy at the expense of $\mathcal{O}(\#I_n)$ operations.

Note that (2.38) is equivalent to

$$\mathcal{A}_{\Psi_n}\mathbf{d}^{(n)} = \mathbf{T}_n^*\mathbf{f},$$

where

$$\mathcal{A}_{\Psi_n} := \mathbf{T}_n^*\mathcal{A}_{\Phi_n}\mathbf{T}_n$$

is the stiffness matrix relative to the multiscale basis Ψ_n. As mentioned before, for suitable choices of W_{j+1} or equivalently of $\mathbf{A}_{j,1}$, the matrix $\mathcal{A}_{\Psi_n}$ can easily be preconditioned by a diagonal scaling [11].

When $\mathcal{A}$ is an operator with global Schwartz kernel, one can show that in contrast to $\mathcal{A}_{\Phi_n}$ the matrix $\mathcal{A}_{\Psi_n}$ can be approximated by a sparse matrix [2,14,15]. Both tasks as well as the transformation to the single scale representation (which is needed at the end) require *only* the process (2.10) involving the matrices $\mathbf{A}_j$.

The above observations show that uniform stability of multiscale transformations is intimately tied to identifying an underlying Riesz-basis. In [8] Riesz-bases are constructed through so called *biorthogonal* wavelets. Moreover, it is shown there that both the matrices $\mathbf{A}_j$ and $\mathbf{B}_j$ can be chosen to be sparse. Unfortunately, the approach in [7,8] relies crucially on Fourier transform techniques which are not applicable in the present general setting. The remainder of this paper is therefore devoted to a discussion of various aspects of biorthogonality and to appropriate reformulations. This will ultimately lead us to a different type of criteria that ensure certain norm equivalences covering (2.33) as a special case.

§3. Riesz-Bases and Biorthogonality

After interrelating stability and Riesz-bases, we will now tie this into the concept of biorthogonality (see e.g. [8] for the case of shif-invariant spaces). We adhere to the notation of the previous section and suppose that Ψ defined in (2.31) forms a Riesz-basis of $L_2(\Omega)$. Thus every $f \in L_2(\Omega)$ has an expansion (2.32) satisfying (2.33). Hence the mapping $f \to d_k^j(f)$ is a bounded linear functional on $L_2(\Omega)$. Therefore there exists a $\tilde{\psi}_k^j \in L_2(\Omega)$ such that

$$d_k^j(f) = (f, \tilde{\psi}_k^j)_\Omega \tag{3.1}$$

and therefore

$$(\psi_k^j, \tilde{\psi}_m^l)_\Omega = \delta_{j,l}\delta_{k,m}, \quad (j,k),(l,m) \in \mathcal{J}, \tag{3.2}$$

where

$$\mathcal{J} := \{(j,k) : k \in J_j, j = -1, \ldots, \infty\}.$$

Thus the Riesz-basis leads in a natural way to a *biorthogonal system* $(\Psi, \tilde{\Psi})$. Before exploring the properties of such a system in more detail we observe that the above reasoning immediately carries over to a more general setting which may be described as follows. Let $\mathcal{F}$ be a Banach space endowed with a norm $\|\cdot\|_{\mathcal{F}}$. Let $\mathcal{F}'$ be its *dual* (or conjugate space), i.e., the space of all bounded linear functionals on $\mathcal{F}$. Denoting the action of $f' \in \mathcal{F}'$ on $f \in \mathcal{F}$ by $f'(f) = \langle f, f'\rangle$, the space $\mathcal{F}'$ becomes of course also a Banach space under the norm

$$\|f'\|_{\mathcal{F}'} = \sup_{\|f\|_{\mathcal{F}}=1} |\langle f, f'\rangle|. \tag{3.3}$$

It will be convenient to assume $(\lambda f')(f) = \overline{\lambda}\langle f, f'\rangle$ when working with the field of complex numbers. Furthermore, let $l(\mathcal{J})$ be the space of all sequences $\mathbf{c} = \{c_k^j : (j,k) \in \mathcal{J}\}$ such that $\|\mathbf{c}\|_{l(\mathcal{J})}$ is finite where $\|\cdot\|_{l(\mathcal{J})}$ is (quasi-) monotone in the following sense. There exists a positive constant γ such that for all $\mathbf{c} \in l(\mathcal{J})$ and all $\tilde{\mathbf{c}}$ such that for some $\tilde{\mathcal{J}} \subseteq \mathcal{J}$,

$$\tilde{\mathbf{c}}_k^j = \begin{cases} \mathbf{c}_k^j & \text{if} \quad (j,k) \in \tilde{\mathcal{J}}, \\ 0 & \text{if} \quad (j,k) \in \mathcal{J} \setminus \tilde{\mathcal{J}} \end{cases}$$

one has

$$\|\tilde{\mathbf{c}}\|_{l(\mathcal{J})} \leq \gamma \|\mathbf{c}\|_{l(\mathcal{J})}. \tag{3.4}$$

Examples are $\mathcal{F} = L_p(\Omega)$, $l(\mathcal{J}) = l_p(\mathcal{J}), 1 \leq p \leq \infty$.

A collection

$$\Psi = \{\psi_k^j : (j,k) \in \mathcal{J}\}$$

may be called a *Riesz-basis* of $\mathcal{F}$ if every $f \in \mathcal{F}$ admits a *unique* strongly converging expansion

$$f = \sum_{(j,k)\in\mathcal{J}} d_k^j(f)\psi_k^j \tag{3.5}$$

in $\mathcal{F}$ and if

$$\|\mathbf{d}(f)\|_{l(\mathcal{J})} \sim \|f\|_{\mathcal{F}}. \tag{3.6}$$

Since (3.4) implies, in particular, that

$$|d_k^j| \leq \gamma \left\|\{\delta_{(j,k),(l,m)}\}_{(l,m)\in\mathcal{J}}\right\|_{l(\mathcal{J})} \|\mathbf{d}\|_{l(\mathcal{J})}, \tag{3.7}$$

the same reasoning as above yields a collection

$$\tilde{\Psi} = \{\tilde{\psi}_k^j : (j,k) \in \mathcal{J}\}$$

of bounded linear functionals satisfying

$$\langle \psi_k^j, \tilde{\psi}_m^l \rangle = \delta_{j,l}\delta_{k,m}, \quad (j,k),(l,m) \in \mathcal{J}. \tag{3.8}$$

In fact, a straightforward duality argument based on (3.3) combined with (3.6) and (3.7) yields

$$\left\|\tilde{\psi}_k^j\right\|_{\mathcal{F}'} \leq c \left\|\{\delta_{(j,k),(l,m)}\}_{(l,m)\in\mathcal{J}}\right\|_{l(\mathcal{J})} \tag{3.9}$$

for some constant c which does not depend on j,k.

Substituting the expansion (3.5) into $\langle f, f' \rangle$ and using duality shows that every $f' \in \mathcal{F}'$ has a unique expansion

$$f' = \sum_{(j,k)\in\mathcal{J}} \langle \psi_k^j, f' \rangle \tilde{\psi}_k^j. \tag{3.10}$$

To see how to interpret the convergence on the right hand side of (3.10) let for $f \in \mathcal{F}, f' \in \mathcal{F}'$

$$Q_n f := \sum_{(j,k)\in\mathcal{J}_n} \langle f, \tilde{\psi}_k^j \rangle \psi_k^j, \quad Q_n' f' := \sum_{(j,k)\in\mathcal{J}_n} \langle \psi_k^j, f' \rangle \tilde{\psi}_k^j. \tag{3.11}$$

Remark 3.1. *Let Q_n, Q_n' be defined by (3.11).*

(i) *Q_n, Q_n' are linear projectors and are dual to each other, i.e., $\langle Q_n f, f' \rangle = \langle f, Q_n' f' \rangle$.*

(ii) *The* Q_n, Q'_n *are uniformly bounded, and*

$$\|Q_n\|_{\mathcal{F}} = \|Q'_n\|_{\mathcal{F}'}, \tag{3.12}$$

where $\|\cdot\|_{\mathcal{F}}, \|\cdot\|_{\mathcal{F}'}$ *denote also the respective operator norms induced by the norms on* $\mathcal{F}$ *and* $\mathcal{F}'$*, respectively.*

(iii) *The right hand side of (3.10) is* weak**-convergent.*

Proof: (i) is obvious. As for (ii) note that assuming only the strong convergence in (3.5), the uniform boundedness of the Q_n follows from the Uniform Boundedness Principle, while (3.12) is a well-known consequence of duality. Here one doesn't even have to resort to the Uniform Boundedness Principle since the uniform boundedness of the Q_n is also an immediate consequence of (3.4) and (3.6). Since $|\langle f, f' - Q'_n f'\rangle| = |\langle f - Q_n f, f'\rangle| \le \|f'\|_{\mathcal{F}'} \|f - Q_n f\|_{\mathcal{F}}$ (iii) follows as well. ■

Throughout the following we will assume that $\mathcal{F}$ is reflexive. To see that then the convergence in (3.10) is even strong, recall the classical Lebesgue estimate:

Remark 3.2. *For any sequence* $\mathcal{Q}$ *of bounded linear projectors* Q_j *with ranges* $S_j \subset \mathcal{F}$*, one has*

$$\|Q_n f - f\|_{\mathcal{F}} \le (1 + \|Q_n\|_{\mathcal{F}}) \inf_{f_n \in S_n} \|f - f_n\|_{\mathcal{F}} \tag{3.13}$$

Now let $\tilde{\mathcal{S}}$ denote the sequence of ranges $\tilde{S}_j \subset \mathcal{F}'$. The $\tilde{S}_j$ are closed since the Q'_j are bounded projections. Now let

$$V := \overline{\bigcup_{j \in \mathbb{N}_0} \tilde{S}_j},$$

and suppose that $V \ne \mathcal{F}'$. Thus there exists some $f' \in \mathcal{F}' \setminus V$ and some $f \in \mathcal{F} = \mathcal{F}''$ such that

$$\langle f, f'\rangle = 1, \quad \langle f, g'\rangle = 0 \quad \text{for all } g' \in V. \tag{3.14}$$

Since $Q'_n f' \in V$ we have $\langle f, Q'_n f'\rangle = 0$ for all $n \in \mathbb{N}_0$. By Remark 3.1 (iii) this gives $\langle f, f'\rangle = 0$ contradicting (3.14). This shows

$$\overline{\bigcup_{j \in \mathbb{N}_0} \tilde{S}_j} = \mathcal{F}'. \tag{3.15}$$

Thus recalling Remark 3.1 (ii) and applying Remark 3.2 to $\mathcal{Q}'$ yields

Remark 3.3. *The expansion (3.10) converges strongly.*

Let $\langle \mathbf{d}, \mathbf{c}\rangle_{\mathcal{J}} := \sum_{(j,k)\in\mathcal{J}} d_k^j \overline{c_k^j}$ and let $l'(\mathcal{J})$ be the corresponding dual of $l(\mathcal{J})$.

Remark 3.4. *Under the above assumptions one has*

$$\|f'\|_{\mathcal{F}'} \sim \left\|\tilde{\mathbf{d}}\right\|_{l'(\mathcal{J})}, \tag{3.16}$$

where $\tilde{d}^j_k := \langle \psi^j_k, f' \rangle$.

Proof: By (3.5) and (3.6), one has for $d^j_k(f) := \langle f, \tilde{\psi}^j_k \rangle$ and $\langle \psi^j_k, f' \rangle := \tilde{d}^j_k$

$$\begin{aligned} \|f'\|_{\mathcal{F}'} &= \sup_{\|f\|_{\mathcal{F}}=1} |\langle f, f'\rangle| = \sup_{\|f\|_{\mathcal{F}}=1} |\sum_{(j,k)\in\mathcal{J}} d^j_k(f)\langle \psi^j_k, f'\rangle| \\ &= \sup_{\|f\|_{\mathcal{F}}=1} |\langle \mathbf{d}(f), \overline{\tilde{\mathbf{d}}(f')}\rangle_{\mathcal{J}}| \sim \sup_{\|\mathbf{d}\|_{l(\mathcal{J})}=1} |\langle \mathbf{d}, \overline{\tilde{\mathbf{d}}(f')}\rangle|, \end{aligned}$$

which proves the claim. ∎

The above observations show that a Riesz basis leads in a natural way to a *biorthogonal pair* of bases $(\Psi, \tilde{\Psi})$ whose more detailed properties can be desribed in terms of the operators Q_n and their duals. These properties may now easily be complemented by the following simple consequences which parallel the classical framework of biorthogonal wavelets.

Remark 3.5. *Under the above assumptions one has*

(i) *The ranges $S_n, \tilde{S}_n$ of Q_n, Q'_n, respectively, are closed linear subspaces of $\mathcal{F}, \mathcal{F}'$, respectively, satisfying*

$$S_0 \subset S_1 \subset \ldots \subset S_n \subset \ldots \mathcal{F} \tag{3.17}$$

$$\tilde{S}_0 \subset \tilde{S}_1 \subset \cdots \subset \tilde{S}_n \subset \ldots \mathcal{F}'. \tag{3.18}$$

Moreover,

$$\overline{\bigcup_{j\in\mathbb{N}_0} S_j} = \mathcal{F}, \quad \overline{\bigcup_{j\in\mathbb{N}_0} \tilde{S}_j} = \mathcal{F}', \tag{3.19}$$

where the closures are taken with respect to the respective norms.

(ii) *The mappings $Q_n - Q_{n-1}$ and $Q'_n - Q'_{n-1}$ are also projectors with ranges $W_n, \tilde{W}_n$, respectively, and*

$$S_n = S_{n-1} \oplus W_n, \quad \tilde{S}_n = \tilde{S}_{n-1} \oplus \tilde{W}_n.$$

Moreover, the spaces $W_n, \tilde{W}_n$ are biorthogonal relative to $\langle \cdot, \cdot \rangle$, i.e.,

$$\langle (Q_n - Q_{n-1})f, (Q'_j - Q'_{j-1})f' \rangle = 0, \qquad f \in \mathcal{F}, f' \in \mathcal{F}', \ n \neq j, \tag{3.20}$$

or equivalently,

$$(Q_n - Q_{n-1})(Q_j - Q_{j-1}) = \delta_{j,n}(Q_n - Q_{n-1}), \quad j, n \in \mathbb{N}_0. \tag{3.21}$$

However, in practical applications, a Riesz-basis is usually not given. Rather the discussion in the previous section shows that one has to construct one in order to ascertain stability. Instead the starting point is usually an ascending sequences $\mathcal{S}$ of strictly nested closed subspaces S_j of $\mathcal{F}$

$$S_0 \subset S_1 \subset \ldots \subset S_n \subset \ldots \mathcal{F} \tag{3.22}$$

which is *dense*

$$\overline{\bigcup_{j \in \mathbb{N}_0} S_j} = \mathcal{F}. \tag{3.23}$$

One then has to search for suitable complements W_j of S_{j-1} in S_j such that a stable basis Ψ^j of W_j becomes part of a Riesz-basis. Moreover, Remark 3.5 suggests to realize these complements W_j by means of differences of suitable linear projectors Q_j onto S_j, namely $W_j = (Q_j - Q_{j-1})S_j$. Remark 3.5 also shows that these complements should have a special biorthogonality structure. This can be conveniently expressed by a simple property of the Q_j which is, in particular, easily confirmed to hold for the above projectors defined in (3.11). The following well-known facts are relevant in this context.

Remark 3.6. *Let $\mathcal{S}$ satisfy (3.22) and let $\mathcal{Q}$ be an associated sequence of linear projectors Q_j from $\mathcal{F}$ onto S_j. Let $\mathcal{Q}'$ denote the sequence of dual projectors Q_j' with ranges $\tilde{\mathcal{S}}$. The following facts are equivalent:*

(i)

$$Q_j Q_n = Q_j \qquad j \leq n. \tag{3.24}$$

(ii) *The $Q_j - Q_{j-1}$ (and hence the $Q_j' - Q_{j-1}'$) are also projectors and therefore satisfy (3.21) or equivalently*

$$W_j = (Q_j - Q_{j-1})S_j = (Q_j - Q_{j-1})S_n = (Q_j - Q_{j-1})\mathcal{F}, \;\; j \leq n. \tag{3.25}$$

(iii) *$\tilde{S}$ is also nested*

$$\tilde{S}_j \subset \tilde{S}_{j+1}, \quad j \in \mathbb{N}_0. \tag{3.26}$$

(iv) *The corresponding complements $\tilde{W}_j$ satisfy the biorthogonality relation (3.20).*

Thus all the algebraic properties identified above are equivalent to the *commutator relation* (3.24).

We will therefore try to go the other (hopefully more practical) way around starting off with a dense nested sequence $\mathcal{S}$ and an associated sequence $\mathcal{Q}$ of uniformly bounded projectors satisfying (3.24). We wish to explore to what extent and under which circumstances the above facts about Riesz-bases can be redeveloped along this line.

Specifically, defining for convenience $Q_{-1} = 0$, under the assumption (3.24), the telescoping expansion

$$f = \sum_{j \in \mathbb{N}_0} (Q_j - Q_{j-1})f \tag{3.27}$$

readily provides then the components $f_j = (Q_j - Q_{j-1})f$ of $f \in \mathcal{F}$ from W_j, provided the expansion (3.27) converges in a suitable sense. Of course, Remark 3.1 suggests that $\mathcal{Q}$ should be uniformly bounded, i.e., $\|Q_n\|_{\mathcal{F}} = \mathcal{O}(1), n \to \infty$. Then the Lebesgue estimate (3.13) again yields

Remark 3.7. *If $\mathcal{S}$ is dense and $\mathcal{Q}$ is uniformly bounded, then the right hand side of (3.27) converges strongly to f for every $f \in \mathcal{F}$. Moreover, if $\mathcal{S}$ is in addition nested and $\mathcal{Q}$ satisfies (3.24) then $\mathcal{Q}'$ is also strongly convergent. The argument is completely analogous to the one in Remark 3.1 and Remark 3.3.*

(3.27) may be viewed as a *basis free* analog to (3.5). What is missing is the relation of (3.27) to a suitable discrete norm. Of course, in general, this requires identifying suitable bases in the first place. However, for the most important applications we have in mind this can actually be deferred a bit. In fact, take $\mathcal{F} = L_p(\Omega), l(\mathcal{J}) = l_p(\mathcal{J})$ for $1 < p < \infty$. If we had any uniformly l_p-stable bases Ψ^j of W_j at hand, the relation (3.6) would be equivalent to

$$\begin{aligned} \|f\|_{L_p(\Omega)} &\sim \left(\sum_{j\in\mathbb{N}_0} \|(Q_j - Q_{j-1})f\|^p_{L_p(\Omega)} \right)^{1/p} \\ &= \left\| \left\{ \|(Q_j - Q_{j-1})f\|_{L_p(\Omega)} \right\}_{j\in\mathbb{N}_0} \right\|_{l_p(\mathbb{N}_0)}. \end{aligned} \tag{3.28}$$

Thus the issue of stability is ultimately reduced to the following task: Given a dense and nested $\mathcal{S}$ construct an associated uniformly bounded sequence $\mathcal{Q}$ of projectors satisfying (3.24) such that (3.28) holds (for $p = 2$).

We will show next that relations of the type (3.28) and even more general variants do indeed hold if one quantifies the strong convergence of the $\mathcal{Q}$ and $\tilde{\mathcal{Q}}$ a bit and if some mild regularity assumptions are imposed on $\mathcal{S}$ and $\tilde{\mathcal{S}}$. To describe this, let us assume in the following that there exists some one parameter family of nonnegative subadditive functionals $\omega_{\mathcal{F}}(\cdot, t), t \geq 0$, with the following properties: There exists a positive constant c such that

$$\omega_{\mathcal{F}}(f,t) \leq c\ \|f\|_{\mathcal{F}}, \qquad f \in \mathcal{F}. \tag{3.29}$$

$$\lim_{t\to 0+} \omega_{\mathcal{F}}(f,t) = 0 \qquad f \in \mathcal{F}. \tag{3.30}$$

$$\omega_{\mathcal{F}}(f_1 + f_2, t) \leq \omega_{\mathcal{F}}(f_1,t) + \omega_{\mathcal{F}}(f_2,t), \quad f_1, f_2 \in \mathcal{F}. \tag{3.31}$$

Moreover, for each $\lambda > 0$ there exists some $\lambda' > 0$ such that for all $t > 0$, $f \in \mathcal{F}$

$$\omega_{\mathcal{F}}(f, \lambda t) \leq \lambda'\ \omega_{\mathcal{F}}(f,t). \tag{3.32}$$

Thus $\omega_{\mathcal{F}}$ has the properties of a *modulus of smoothness.*

One can quantify now the approximation properties of a given dense sequence $\mathcal{S}$ of closed subspaces of $\mathcal{F}$ by requiring that a *Jackson estimate*

$$\inf_{f_j\in S_j} \|f - f_j\|_{L_p} \leq c\,\omega_{\mathcal{F}}(f, \rho^{-j}) \tag{3.33}$$

holds for some fixed $\rho > 0$.

In addition we will have to impose a mild regularity assumption on $\mathcal{S}$ in terms of the following *inverse* or *Bernstein estimate*

$$\omega_{\mathcal{F}}(f_j, t) \leq c\left(\min\{1, t\rho^j\}\right)^{\gamma} \|f_j\|_{\mathcal{F}} \tag{3.34}$$

which is to hold for some $\gamma > 0$ and all $f_j \in S_j$, uniformly in $j \in \mathbb{N}_0$.

These notions are illustrated by the following example

Example 3.8. *Let Ω be some domain in $\mathbb{R}^s$ and let*

$$\mathcal{F} = L_p(\Omega), \quad \mathcal{F}' = L_{p'}(\Omega) \quad \text{for} \quad 1 < p < \infty, \quad \frac{1}{p} + \frac{1}{p'} = 1. \tag{3.35}$$

Let $|\cdot|$ denote a norm on $\mathbb{R}^s$ and let

$$(\Delta_h^l f)(x) := \sum_{j=0}^{l} \binom{l}{j} (-1)^{l-j} f(x + jh)$$

be the lth order difference operator in the direction h. With

$$\Omega_{h,l} := \{x \in \Omega : x + jh \in \Omega, \; j = 0, \ldots l\},$$

the lth order L_p-modulus of continuity of f is defined as

$$\omega_l(f, t, \Omega)_p := \sup_{|h| \leq t} \left\|\Delta_h^l f\right\|_{L_p(\Omega_{h,l})}.$$

In this case

$$\omega_{L_p(\Omega)}(f, t) := \omega_l(f, t, \Omega)_p \tag{3.36}$$

is known to satisfy (3.30), (3.31), and (3.32), and hence is an admissible choice. For instance, (3.33) is valid for (3.36) with $l \geq 1$ when S_j is spanned by compactly supported functions which sum to one on Ω. In general, (3.33) is known as a Whitney-type estimate. It is typically valid for some $l \in \mathbb{N}$ when S_j contains all polynomials of degree less than l and is spanned by compactly supported functions such that the diameter of the supports is uniformly proportional to ρ^{-j}. For various versions of different moduli see e.g. [10]. Likewise such an inverse estimate for (3.36) is known to hold for essentially all finite element spaces, spline spaces, or multiresolution settings of interest for some $l \in \mathbb{N}$ and some $\gamma > 0$. Usually it is established first for the basis functions and then extended to all elements in S_j by stability.

Furthermore, the above moduli serves to define subspaces of $\mathcal{F}$ for which the above direct and inverse estimates can be made more precise. For fixed $\rho > 1$ and $0 < r < \gamma$, $1 \leq q \leq \infty$ let

$$|f|^q_{B^r_q(\mathcal{F})} := \sum_{j \in \mathbb{N}_0} \rho^{jrq} \omega_{\mathcal{F}}(f, \rho^{-j})^q.$$

Then $B_q^r(\mathcal{F})$ is the space of all functions $f \in \mathcal{F}$ for which

$$\|f\|_{B_q^r(\mathcal{F})} := \|f\|_{\mathcal{F}} + |f|_{B_q^r(\mathcal{F})}$$

is finite. In the special setting of Example 3.8 and $\omega_{\mathcal{F}}$ defined by (3.36), $B_q^r(\mathcal{F})$ is the *Besov space* $B_{p,q}^r(\Omega)$ when $r < \min\{l, \gamma\}$. It is known that under suitable regularity assumptions on Ω one has

$$B_{2,2}^r(\Omega) = H^r(\Omega), \quad r > 0, \quad B_{p,p}^r(\Omega) = W_p^r(\Omega), \quad r > 0, r \notin \mathbb{N},$$

where $H^r(\Omega), W_p^r(\Omega)$ denote the Sobolev-spaces of order r in $L_2(\Omega), L_p(\Omega)$, respectively (see e.g. [31]).

Let us mention next a few consequences of the basic inequalities (3.33) and (3.34).

Remark 3.9. *Let $\mathcal{S}$ satisfy (3.34). Then for each fixed $r < \gamma$ and $1 \le q \le \infty$ one has*

$$S_j \subseteq B_q^r(\mathcal{F}), \quad j \in \mathbb{N}_0,$$

and there exists a constant c independent of $0 < t \le r$ such that

$$\|f_j\|_{B_q^t(\mathcal{F})} \le c\, \rho^{jt}\, \|f_j\|_{\mathcal{F}} \tag{3.37}$$

holds for all $f_j \in S_j$ uniformly in $j \in \mathbb{N}_0$.

Proof: By (3.34), one has for any $f_n \in S_n$

$$\begin{aligned} |f_n|^q_{B_q^r(\mathcal{F})} &= \sum_{j=0}^{\infty} \rho^{jrq} \omega_{\mathcal{F}}(f_n, \rho^{-j})^q \\ &\le c \sum_{j=0}^{n} \rho^{jrq}\, \|f_n\|^q_{\mathcal{F}} + \sum_{j=n+1}^{\infty} \rho^{jrq} \rho^{(n-j)q\gamma}\, \|f_n\|^q_{\mathcal{F}} \\ &\le c\, \rho^{nrq}\, \|f_n\|^q_{\mathcal{F}}, \end{aligned}$$

where c depends only on the constant in (3.34) and on $\gamma - r$. Since one trivially has $\|f_n\|_{B_q^t(\mathcal{F})} \le \|f_n\|_{B_q^r(\mathcal{F})}$ for $t \le r$, the assertion follows. ■

Remark 3.10. *Suppose $\mathcal{S}$ satisfies (3.33) and let $\mathcal{Q}$ be any associated sequence of uniformly bounded linear projectors. Then*

$$\|Q_j f - f\|_{\mathcal{F}} \le c\, \omega_{\mathcal{F}}(f, \rho^{-j}), \tag{3.38}$$

where c is independent of $j \in \mathbb{N}_0$ and $f \in \mathcal{F}$. Moreover, there exists a constant c such that for all $j \in \mathbb{N}_0$, $0 < t \le r < \gamma$ and all $f \in B_q^t(\mathcal{F})$

$$\|Q_j f - f\|_{\mathcal{F}} \le c\, \rho^{-tj}\, \|f\|_{B_q^t(\mathcal{F})}. \tag{3.39}$$

Proof: (3.38) is an immediate consequence of the Lebesgue estimate (3.13) and the uniform boundedness of $\mathcal{Q}$. (3.39) follows from (3.38) and the fact that

$$\omega_{\mathcal{F}}(f,\rho^{-j}) \le \rho^{-jt}\left(\sum_{i\in\mathbb{N}_0}\rho^{itq}\omega_{\mathcal{F}}(f,\rho^{-i})^q\right)^{1/q} = \rho^{-tj}\,\|f\|_{B^t_q(\mathcal{F})} \tag{3.40}$$

holds for all $0 \le t < \gamma$ and $f \in B^t_q(\mathcal{F})$. ∎

Such estimates can be extended by *interpolation* and more can be said about bounding the operators Q_n on the spaces $B^t_q(\mathcal{F})$. Since this will be used later, we pause here to point out a convenient tool for deriving such estimates. To this end, we will need the following result which was established in [11] for the case (3.35), (3.36). The same arguments work to prove the present somewhat more general version.

Theorem 3.11. *Let $\mathcal{S}$ be a nested and dense sequence of closed subspaces of $\mathcal{F}$ and $\mathcal{Q}$ an associated sequence of uniformly bounded projectors. Assume that $\mathcal{S}$ satisfies the Jackson estimate (3.33) and the Bernstein estimate (3.34) for some $\gamma > 0$. Then for all $0 < r < \gamma$ the norm $\|\cdot\|_{B^r_q(\mathcal{F})}$ is equivalent to*

$$\|f\|_{\mathcal{F},q,r} := \left(\sum_{j\in\mathbb{N}_0}\rho^{jrq}\,\|(Q_j - Q_{j-1})f\|^q_{\mathcal{F}}\right)^{1/q}, \tag{3.41}$$

where as before $Q_{-1} = 0$.

Remark 3.12. *Let $\mathcal{S},\mathcal{Q}$ satisfy the assumptions of Theorem 3.11.*

(i) *Then for $\tau \le t \le r < \gamma$ one has*

$$\|f_n\|_{B^t_q(\mathcal{F})} \le c\,\rho^{n(t-\tau)}\,\|f_n\|_{B^\tau_q(\mathcal{F})} \qquad f_n \in S_n,\ n \in \mathbb{N}_0,$$

where c is independent of n, t, τ.

(ii) *If in addition to the above assumptions $\mathcal{Q}$ satisfies (3.24) then*

$$\|Q_n\|_{B^t_q(\mathcal{F})} = \mathcal{O}(1), \qquad n \to \infty. \tag{3.42}$$

(iii) *For $\tau \le t < \gamma$ there exists a positive constant c such that for all $f \in B^t_q(\mathcal{F})$*

$$\|Q_n f - f\|_{B^\tau_q(\mathcal{F})} \le c\,\rho^{n(\tau-t)}\,\|f\|_{B^t_q(\mathcal{F})},\quad n \in \mathbb{N}_0. \tag{3.43}$$

Proof: Since the Q_n are projectors onto S_n, Theorem 3.11 yields for $f_n \in S_n$

$$\|f_n\|^q_{B^t_q(\mathcal{F})} \le c\sum_{j=0}^{n}\rho^{jtq}\,\|(Q_j - Q_{j-1})f_n\|^q_{\mathcal{F}}$$

which, on account of (3.39), gives (i).

Since by (3.24)

$$(Q_j - Q_{j-1})Q_n = \begin{cases} 0 & \text{for} \quad j > n, \\ Q_j - Q_{j-1} & \text{for} \quad j \le n, \end{cases}$$

and similarly

$$(Q_j - Q_{j-1})(Q_n f - f) = \begin{cases} 0 & \text{for} \quad j \le n, \\ (Q_{j-1} - Q_j)f & \text{for} \quad j > n, \end{cases}$$

the remainder of the assertions is easily derived from Theorem 3.11. ■

Returning now to the issue of establishing an equivalence of the type (3.28), one should note that, in general, one cannot pass r to zero to recover $\|\cdot\|_{\mathcal{F}}$ by the right hand side expression in (3.41). Nevertheless, the following main result of this section tells us under which circumstances this is possible.

Theorem 3.13. *Let $\mathcal{S}$ be nested and dense in $\mathcal{F}$. Let $\mathcal{Q}$ be an associated uniformly bounded sequence of projectors satisfying the biorthogonality relation (3.24) and let $\tilde{\mathcal{S}}$ consist of the ranges of the dual sequence $\mathcal{Q}'$. Assume that $\mathcal{S}$ and $\tilde{\mathcal{S}}$ both satisfy Jackson and Bernstein estimates (3.33) and (3.34) (with respect to $\omega_{\mathcal{F}}, \omega_{\mathcal{F}'}$, respectively) for some $\gamma > 0$. Then for $1 \le q \le \infty$ one has*

$$\|f\|_{\mathcal{F}} \sim \left(\sum_{j \in \mathbb{N}_0} \|(Q_j - Q_{j-1})f\|_{\mathcal{F}}^q \right)^{1/q} \tag{3.44}$$

with the usual interpretation for $q = \infty$.

Proof: Let $0 < r < \gamma$. On account of (3.24), (3.38), and the definition of $\|\cdot\|_{B_q^t(\mathcal{F})}$, we obtain for every $f \in \mathcal{F}$ and $0 \le t \le r$

$$\begin{aligned} \sum_{j=0}^{n} \rho^{qtj} \|(Q_j - Q_{j-1})f\|_{\mathcal{F}}^q &= \sum_{j=0}^{n} \rho^{qtj} \|(Q_j - Q_{j-1})Q_n f\|_{\mathcal{F}}^q \\ &\le c \sum_{j=0}^{n} \rho^{qtj} \omega_{\mathcal{F}}(Q_n f, \rho^{-j})^q \\ &\le c \, \|Q_n f\|_{B_q^t(\mathcal{F})}^q, \end{aligned} \tag{3.45}$$

where the constant c in (3.45) does not depend on $0 < t \le r$. Invoking Remark 3.9 (3.37) and using the uniform boundedness of $\mathcal{Q}$, we have

$$\begin{aligned} \left(\sum_{j=0}^{n} \rho^{qtj} \|(Q_j - Q_{j-1})f\|_{\mathcal{F}}^q \right)^{1/q} &\le c \, \rho^{nt} \|Q_n f\|_{\mathcal{F}} \\ &\le c \, \rho^{nt} \|f\|_{\mathcal{F}}, \end{aligned} \tag{3.46}$$

where the constant c in (3.46) does not depend on $0 < t \le r$. Therefore we may let t tend to zero in (3.46) which yields

$$\left(\sum_{j=0}^{n} \|(Q_j - Q_{j-1})f\|_{\mathcal{F}}^q \right)^{1/q} \le c \; \|f\|_{\mathcal{F}}$$

for all $n \in \mathbb{N}_0$. Hence

$$\left(\sum_{j=0}^{\infty} \|(Q_j - Q_{j-1})f\|_{\mathcal{F}}^q \right)^{1/q} \le c \; \|f\|_{\mathcal{F}}, \tag{3.47}$$

which proves the first part of the assertion.

To establish the corresponding lower estimate, recall from Remark 3.7 that for $f \in \mathcal{F}, f' \in \mathcal{F}'$ the expansions

$$f = \sum_{j \in \mathbb{N}_0} (Q_j - Q_{j-1})f, \quad f' = \sum_{j \in \mathbb{N}_0} (Q'_j - Q'_{j-1})f'$$

converge strongly in $\mathcal{F}, \mathcal{F}'$, respectively, since $Q_n f = \sum_{j=0}^{n} (Q_j - Q_{j-1})f$. Thus by Remark 3.6 (ii), (3.3), and Hölder's inequality for $l_p(\mathbb{N}_0)$, we obtain

$$\begin{aligned} \langle f, f' \rangle &= \sum_{j=0}^{\infty} \langle (Q_j - Q_{j-1})f, (Q'_j - Q'_{j-1})f' \rangle \\ &\le \sum_{j=0}^{\infty} \|(Q_j - Q_{j-1})f\|_{\mathcal{F}} \, \|(Q'_j - Q'_{j-1})f'\|_{\mathcal{F}'} \\ &\le \left(\sum_{j=0}^{\infty} \|(Q_j - Q_{j-1})f\|_{\mathcal{F}}^q \right)^{1/q} \left(\sum_{j=0}^{\infty} \|(Q'_j - Q'_{j-1})f'\|_{\mathcal{F}'}^{q'} \right)^{1/q'}, \end{aligned} \tag{3.48}$$

where $\frac{1}{q} + \frac{1}{q'} = 1$. In view of the symmetry of our assumptions, now we can apply (3.47) to the second factor on the right hand side of (3.48) to conclude

$$|\langle f, f' \rangle| \le c \; \|f'\|_{\mathcal{F}'} \left(\sum_{j=0}^{\infty} \|(Q_j - Q_{j-1})f\|_{\mathcal{F}}^q \right)^{1/q}.$$

Thus

$$\begin{aligned} \|f\|_{\mathcal{F}} &= \sup_{f' \in \mathcal{F}' \setminus \{0\}} \frac{|\langle f, f' \rangle|}{\|f'\|_{\mathcal{F}'}} \\ &\le c \left(\sum_{j=0}^{\infty} \|(Q_j - Q_{j-1})f\|_{\mathcal{F}}^q \right)^{1/q}, \end{aligned}$$

which completes the proof of Theorem 3.13. ∎

Relations of the form (3.44) are employed in [26] for establishing the equivalence of Besov-norms of positive and negative order with certain discrete norms of the form (3.41) also for negative r. This in turn is used there in the derivation of multilevel preconditioners for saddle point problems which arise when appending boundary conditions in elliptic problems by Lagrange multipliers.

A few comments on the assumptions in Theorem 3.13 are in order. In view of Remark 3.9, the spaces S_j and $\tilde{S}_j$ both must possess a little more regularity than just belonging to $\mathcal{F}$, $\mathcal{F}'$, respectively, so that by Remark 3.12,

$$\|Q_n\|_{B^t_q(\mathcal{F})} = \mathcal{O}(1), \quad \|Q'_n\|_{B^t_{q'}(\mathcal{F}')} = \mathcal{O}(1), \quad n \to \infty, \tag{3.49}$$

holds for $t < \gamma$. By duality this implies that the Q_n, Q'_n possess uniformly bounded extensions to somewhat larger spaces, namely that

$$\|Q'_n\|_{(B^t_q(\mathcal{F}))'} = \mathcal{O}(1), \quad \|Q_n\|_{(B^t_{q'}(\mathcal{F}'))'} = \mathcal{O}(1), \quad n \to \infty \tag{3.50}$$

is also valid for $t < \gamma$. Playing the duality game again also yields

Remark 3.14. *Under the assumptions in Theorem 3.13, one has*

$$\|Q_n f - f\|_{(B^t_{q'}(\mathcal{F}'))'} \le c\, \rho^{-nt} \|f\|_{\mathcal{F}}, \quad f \in \mathcal{F},\ n \in \mathbb{N}_0, \tag{3.51}$$

if and only if

$$\|Q'_n f' - f'\|_{\mathcal{F}'} \le c\, \rho^{-nt} \|f'\|_{B^t_{q'}(\mathcal{F}')}, \quad f' \in B^t_{q'}(\mathcal{F}'), \quad n \in \mathbb{N}_0. \tag{3.52}$$

So far all the assumptions are symmetric with respect to $\mathcal{S}$ and $\tilde{\mathcal{S}}$. Of course, some properties on the dual side such as uniform boundedness of $\mathcal{Q}'$ are already implied. Therefore the question arises as to what extent assumptions on $\mathcal{S}$ and $\mathcal{Q}$ alone already suffice since this is usually the part which is better controlled. Remark 3.14 says that convergence of $\mathcal{Q}$ in some 'negative norm' already implies an estimate of the type (3.39) for $\mathcal{Q}'$ Thus, for each $g' \in B^t_{q'}(\mathcal{F}')$ one has

$$\begin{aligned} \inf_{f'_n \in \tilde{S}_n} \|f' - f'_n\|_{\mathcal{F}'} &\le \|f' - g'\|_{\mathcal{F}'} + \|g' - Q'_n g'\|_{\mathcal{F}'} \\ &\le c\,(\|f' - g'\|_{\mathcal{F}'} + \rho^{-nt} \|g'\|_{B^t_{q'}(\mathcal{F}')}). \end{aligned}$$

Since g' was arbitrary, one obtains

$$\inf_{f'_n \in \tilde{S}_n} \|f' - f'_n\|_{\mathcal{F}'} \le c\, K(f', \rho^{-n}, t, \mathcal{F}', B^t_{q'}(\mathcal{F}')), \tag{3.53}$$

where

$$K(f,\delta,t,U,V) := \inf_{g\in V}(\|f-g\|_U + \delta^t \|g\|_V)$$

is called K-functional. In the special case described in Example 3.8 (see (3.36)), it is known (cf. [10]) and the literature quoted there) that, under mild assumptions concerning the regularity of the boundary of Ω, one has for instance

$$K(\cdot,\delta,l,L_2(\Omega),H^l(\Omega)) \sim \omega_l(\cdot,\delta,\Omega)_2.$$

Hence the Jackson estimate (3.33) for $\tilde{\mathcal{S}}$ would actually follow in this case from convergence of $\mathcal{Q}$ in the dual spaces H^{-t}.

Let us conclude this section with some brief remarks on how the above observations blend in with the results on biorthogonal wavelets [7,8]. In this case $\mathcal{S}$ and $\tilde{\mathcal{S}}$ are generated by the dilates and integer translates of a single refinable function $\varphi,\tilde{\varphi}$ as in Example 2.3 which are biorthogonal, i.e., $(\varphi,\varphi(\cdot-k))_{\mathbb{R}} = \delta_{0,k}, k\in\mathbb{Z}$. It is known that the integer translates of such functions must sum to a nonvanishing constant which already implies the Jackson estimate (3.33) for $l=1$ in (3.36). Moreover, they must possess some extra regularity [7,8]. Using stability this can be exploited to confirm also the inverse estimate (3.34) as in [11]. However, in this special situation much more is known with regard to to the actual construction of the $\varphi,\tilde{\varphi}$ and hence of $\mathcal{Q}$. We will briefly address this issue in the following section.

§4. Subdivision

Adhering to the notation in the previous section, let $\mathcal{S}$ be again a sequence of closed nested subspaces of $\mathcal{F}$. Moreover, let Φ_j be a stable basis of S_j relative to $\|\cdot\|_{\mathcal{F}}, \|\cdot\|_{l(I_j)}$, i.e.,

$$\|\mathbf{c}\|_{l(I_j)} \sim \left\|\sum_{k\in I_j} c_k\varphi_k^j\right\|_{\mathcal{F}} \tag{4.1}$$

uniformly in $j\in\mathbb{N}_0$. Any bounded linear projector Q_j on S_j then has the form

$$Q_jf = \sum_{k\in I_j}\langle f,\tilde{\varphi}_k^j\rangle\varphi_k^j, \tag{4.2}$$

where the $\tilde{\varphi}_k^j \in \mathcal{F}'$ satisfy

$$\langle\varphi_k^j,\tilde{\varphi}_m^j\rangle = \delta_{k,m}, \quad k,m\in I_j. \tag{4.3}$$

Remark 4.1. *Suppose that $\mathcal{Q}$ is uniformly bounded in $\mathcal{F}$ and that the Φ_j are uniformly stable. Then $\tilde{\Phi}_j$ is uniformly stable relative to $\|\cdot\|_{\mathcal{F}'}, \|\cdot\|_{l'(I_j)}$.*

In fact, by stability of Φ_j, one obtains

$$\begin{aligned}\left\|\sum_{k\in I_j} c'_k\tilde{\varphi}_k^j\right\|_{\mathcal{F}'} &\geq \sup_{f\in S_j}\frac{|\sum_{k\in I_j}\overline{c'_k}\langle f,\tilde{\varphi}_k^j\rangle|}{\|f\|_{\mathcal{F}}} \geq c \sup_{\mathbf{c}\in l(I_j)}\frac{|\sum_{k\in I_j}\overline{c'_k}c_k|}{\|\mathbf{c}\|_{l(I_j)}}\\ &= c\ \|\mathbf{c}'\|_{l'(I_j)}.\end{aligned}$$

Conversely since for $f'_j = \sum_{k\in I_j} c'_k \tilde{\varphi}^j_k$ one has $\langle f, f'_j\rangle = \langle Q_j f, f'_j\rangle$ the uniform boundedness of $\mathcal{Q}$ yields

$$\|f'_j\|_{\mathcal{F}'} \le c \sup_{f\in\mathcal{F}} \frac{|\langle Q_j f, f'_j\rangle|}{\|Q_j f\|_{\mathcal{F}}} \le c\, \|\mathbf{c}'\|_{l'(I_j)},$$

where we have used as before the stability of the Φ_j in the last step.

Clearly stability and nestedness of $\mathcal{S}$ implies refinability of Φ_j with some $\mathbf{A}_{j,0} \in \mathcal{L}(l(I_j), l(I_{j+1}))$ (cf. (2.16)). Moreover, one easily verifies the following fact.

Remark 4.2. *$\mathcal{Q}$ satisfies the commutator relation (3.24) if and only if $\tilde{\Phi}_j$ are also refinable, i.e.,*

$$\tilde{\varphi}^j_k = \sum_{m\in I_{j+1}} b^j_{m,k} \tilde{\varphi}^{j+1}_m, \tag{4.4}$$

where

$$b^j_{m,k} = \langle \varphi^{j+1}_m, \tilde{\varphi}^j_k\rangle. \tag{4.5}$$

Moreover, with $\mathbf{B}_{j,0} := (b^j_{m,k})_{m\in I_{j+1}, k\in I_j}$ one has

$$\mathbf{B}^*_{j,0}\mathbf{A}_{j,0} = I. \tag{4.6}$$

As mentioned before, in practice usually the Φ_j satisfying (2.16) are given or chosen. One then has to find $\mathbf{B}_{j,0}$ satisfying (4.6), e.g., by completing $\mathbf{A}_{j,0}$ to an invertible $\mathbf{A}_j$ and using (2.23) in Proposition 2.4. It now remains to see whether there exists $\tilde{\Phi}_j \subset \mathcal{F}'$ satisfying (4.4).

To this end, let us be more specific about the setting. In the following let $\mathcal{F} = L_p(\Omega)$, $l(I_j) = l_p(I_j)$ for some $1 < p < \infty$ so that $\mathcal{F}' = L_{p'}(\Omega)$ where $\frac{1}{p} + \frac{1}{p'} = 1$ and Ω is (topologically equivalent to) some (possibly unbounded) domain in $\mathbb{R}^s$. Thus $\langle f, g\rangle = \int_\Omega f(x)\overline{g(x)}dx$. Moreover, we will assume that one can associate with each I_j some partition of Ω into cells C^j_k, $k \in I_j$, such that the *mesh size* $\max\{\mathrm{diam}(C^j_k) : k \in I_j\} = h_j$ tends to zero as j tends to infinity. One typically has $h_j = \rho^{-j}$ for some $\rho > 1$. In addition we require these partitions to be *quasi-uniform.* By this we mean that $h_j \sim \max_{k\in I_j} \mathrm{diam}\,(C^j_k) \sim \min_{k\in I_j} \mathrm{diam}\,(C^j_k)$ uniformly in j. Furthermore, the matrices $\mathbf{A}_{j,0} \in \mathcal{L}(l(I_j), l(I_{j+1}))$ are supposed to have uniformly bounded norms $\|\mathbf{A}_{j,0}\|_{\mathcal{L}(l(I_j),l(I_{j+1}))} := \|\mathbf{A}_{j,0}\|$ and to be uniformly sparse. In particular, the entries are uniformly bounded. In addition, let $\sigma^j_k := \{m \in I_j : a^j_{m,k} \ne 0\}$ denote the support of the jth column of $\mathbf{A}_{j,0}$ and more generally, let $\sigma^{j,l}_k := \cup_{m\in\sigma^{j,l-1}_k} \sigma^{j+l}_m$, $\sigma^{j,0}_k = \sigma^j_k$. Finally, for $\sigma \subseteq I_j$ let

$$\Omega(\sigma) := \bigcup_{k\in\sigma} C^j_k.$$

Then we assume that for each $j \in \mathbb{N}_0$, $k \in I_j$ there exists some domain $\Sigma^j_k \subset \Omega$ with

$$\mathrm{diam}\,(\Sigma^j_k) \le c\, h_j \tag{4.7}$$

such that

$$\Omega(\sigma_k^{j,l}) \subseteq \Sigma_k^j \quad \text{for all} \quad k \in I_j, \quad l \in \mathbb{N}_0. \tag{4.8}$$

These properties are easily verified for the stationary matrices $\mathbf{A}_{j,0} = \mathbf{A}_0$ from Example 2.3. when the mask is finitely supported (see e.g. [12]).

Now, viewing Φ_j as a vector with components φ_k^j, (2.16) may be rewritten as $\Phi_j = \mathbf{A}_{j,0}^T \Phi_{j+1}$, and for $\Phi = \{\Phi_j\}_{j\in\mathbb{N}_0}$ and $\mathcal{S}$ as above it will be convenient to abbreviate

$$\sum_{k\in I_j} c_k \varphi_k^j =: \Phi_j^T \mathbf{c}.$$

Thus refinability of Φ with respect to $(\mathbf{A}) = \{\mathbf{A}_{j,0}\}_{j\in\mathbb{N}_0}$ can be expressed for $m > j$ as

$$\Phi_j^T \mathbf{c} = \Phi_m^T \mathbf{A}_j^{m-1} \mathbf{c}, \tag{4.9}$$

where $\mathbf{A}_j^m := \mathbf{A}_{m,0} \cdots \mathbf{A}_{j,0}$. Obviously, one can view $\mathbf{A}_j^{m-1} : l(I_j) \to l(I_m)$ as $m - j$ steps of a (non-homogeneous) *subdivision scheme*. From the theory of *stationary* subdivision schemes one expects a close connection between refinability and *convergence* of such schemes (cf. e.g. [4]). Since in general Φ_{j+1} is not obtained from Φ_j by dilation there is, in contrast to the situation described in Example 2.3, a certain ambiguity in normalizing Φ_j. It is clear that $\mathbf{A}_{j,0}$ depends on this normalization which therefore has to be treated here with some care when addressing the issue of convergence.

To develop a suitable notion of convergence for the present setting let $\lambda_k^{j,p} := (\text{meas}\,(C_k^j))^{-1/p} \chi_{C_k^j}$ so that $\Lambda_j^p = \{\lambda_k^{j,p} : k \in I_j\} \subset L_{p'}(\Omega) \cap L_p(\Omega)$. Now let $\mathbf{C}_j^p := \text{diag}\,((\text{meas}\,(C_k^j))^{1/p} : k \in I_j)$ and suppose that $\hat{\Phi} = \{\hat{\Phi}_j\}_{j\in\mathbb{N}_0}$, $\hat{\Phi}_j = \{\hat{\varphi}_k^j : k \in I_j\}$ satisfies

$$\begin{aligned} (\mathbf{C}_j^p \hat{\Phi})_j^T \mathbf{e}^j \;&=\; 1, \\ \text{supp}\,\hat{\varphi}_k^j \cap C_k^j \neq \emptyset, \qquad & \text{diam}\,(\text{supp}\,\hat{\varphi}_k^j) \leq c\, h_j, \;\; k \in I_j, \end{aligned} \tag{4.10}$$

where c is independent of j, k and $(\mathbf{e}^j)_k = 1, k \in I_j$. Then one can show that for all $f \in L_p(\Omega)$

$$\lim_{j\to\infty} \left\| f - \hat{\Phi}_j^T \Lambda_j^{p'}(f) \right\|_{L_p(\Omega)} = 0, \tag{4.11}$$

where $\Lambda_j^{p'}(f) := \{\langle f, \lambda_k^{j,p'} \rangle : k \in I_j\}$. Here $\hat{\Phi}$ need not be refinable. In particular, choosing $\hat{\Phi}_j = \Lambda_j^p$ and using the uniform stability of Λ_j^p, it is clear that

$$\lim_{j\to\infty} \left\| \Lambda_j^{p'}(f) \right\|_{l_p(I_j)} = 0, \quad \text{implies} \quad f = 0. \tag{4.12}$$

We say $(\mathbf{A})$ converges *strongly* if for every $\mathbf{c} \in l_p(I_j)$ there exists a unique $f_{\mathbf{c}} \in L_p(\Omega)$ such that

$$\lim_{m\to\infty} \left\| \Lambda_m^{p'}(f_{\mathbf{c}}) - \mathbf{A}_j^{m-1} \mathbf{c} \right\|_{l_p(I_m)} = 0. \tag{4.13}$$

It is useful to keep in mind that convergence means that

$$\lim_{m\to\infty} \left\| f_{\mathbf{c}} - \hat{\Phi}_m^T \mathbf{A}_j^{m-1}\mathbf{c} \right\|_{L_p(\Omega)} = 0, \tag{4.14}$$

for any $\mathbf{c} \in l_p(I_j)$ whenever $\hat{\Phi}$ satisfies (4.10). In fact,

$$\begin{aligned}\left\| f_{\mathbf{c}} - \hat{\Phi}_m^T \mathbf{A}_j^{m-1}\mathbf{c} \right\|_{L_p(\Omega)} &\le \left\| f_{\mathbf{c}} - \hat{\Phi}_m^T \Lambda_m^{p'}(f_{\mathbf{c}}) \right\|_{L_p(\Omega)} \\ &+ \left\| \hat{\Phi}_m^T (\Lambda_m^{p'}(f_{\mathbf{c}}) - \mathbf{A}_j^{m-1}\mathbf{c}) \right\|_{L_p(\Omega)},\end{aligned}$$

whence the assertion follows upon using (4.10), (4.11), and (4.13).

Proposition 4.3. *Suppose that under the above assumptions* (**A**) *converges strongly. Then for each* $k \in I_j$, $j \in \mathbb{N}_0$, *there exists a unique* $\varphi_k^j \in \mathcal{F}$ *supported in* Σ_k^j *which is refinable, i.e.,*

$$\Phi_j^T = \Phi_{j+1}^T \mathbf{A}_{j,0}.$$

Moreover, the limit of the sequence $\mathbf{A}_j^m \mathbf{c}$ *in* $L_p(\Omega)$ *is given by*

$$f_{\mathbf{c}} = \Phi_j^T \mathbf{c}. \tag{4.15}$$

Proof: Let $(\mathbf{e}^{j,k})_r = \delta_{k,r}, k, r \in I_j$, and define

$$\varphi_k^j := \lim_{m\to\infty} \hat{\Phi}_m^T \mathbf{A}_j^{m-1}\mathbf{e}^{j,k}, \tag{4.16}$$

where the limit is to be understood in the above sense. The fact that φ_k^j is supported in Σ_k^j can be derived from (4.9). Moreover, since with $\hat{\Phi}_j = \Lambda_j^p$

$$\begin{aligned} & \|\Phi_{j+1}^T \mathbf{A}_{j,0}\mathbf{e}^{j,k} - \varphi_k^j\|_{L_p(\Omega)} \\ \le\ & \|\Phi_{j+1}^T \mathbf{A}_{j,0}\mathbf{e}^{j,k} - (\Lambda_{m+1}^p)^T \mathbf{A}_{j+1}^m \mathbf{A}_{j,0}\mathbf{e}^{j,k}\|_{L_p(\Omega)} \\ +\ & \|(\Lambda_{m+1}^p)^T \mathbf{A}_j^m \mathbf{e}^{j,k} - \varphi_k^j\|_{L_p(\Omega)} \to 0, \quad m \to \infty,\end{aligned}$$

refinability of Φ follows. Now for any neighborhood Γ in Ω let $\Gamma_j := \{k \in I_j : \operatorname{supp}\varphi_k^j \cap \Gamma \ne \emptyset\}$. For Γ sufficiently large relative to the supports Σ_k^j, again (4.8) can be used to conclude that for a suitable subdomain $\tilde{\Gamma} \subset \Gamma$ one has that $f_{\mathbf{c}}$ agrees on $\tilde{\Gamma}$ with $f_{\tilde{\mathbf{c}}}$, where $\tilde{\mathbf{c}} := \mathbf{c}\,|_{\Gamma_j}$. But clearly $f_{\tilde{\mathbf{c}}} = \sum_{k\in\Gamma_j} c_k\varphi_k^j$ since the sum is finite. This completes the proof. ■

As in the stationary case, there is a partial converse (cf. [4]). To this end, it is reasonable to assume that at least constants can be represented by the elements of $\Phi_j, j \in \mathbb{N}_0$, to guarantee a minimum of approximation power. In fact, in the stationary case described in Example 2.3, refinability of φ is known to *imply* that its integer translates sum to some nonvanishing constant. Specifically, it will be convenient to assume that Φ_j satisfies (4.10). Moreover, if Φ and $\tilde{\Phi}$ are both to satisfy (3.4), they must clearly both be *injective*. Here Φ is called injective if for each $j \in \mathbb{N}_0$ the mapping $\mathbf{c} \to \Phi_j^T\mathbf{c}$ is injective on the space of *all* sequences $\mathbf{c}$ on I_j. Also when Φ and $\tilde{\Phi}$ are biorthogonal with all functions satisfying the second part of (4.10) one readily checks that Φ and $\tilde{\Phi}$ are both uniformly stable. These remarks indicate that the assumptions below are in some sense unavoidable in connection with biorthogonality.

Proposition 4.4. *Suppose that Φ is injective, uniformly stable, refinable relative to* ($\mathbf{A}$) *as above, and that Φ_j satisfies (4.10). Then the subdivision scheme induced by* ($\mathbf{A}$) *converges strongly. Moreover one has*

$$(\mathbf{C}_{j+1}^p)^{-1}\mathbf{A}_{j,0}\mathbf{C}_j^p\mathbf{e}^j = \mathbf{e}^{j+1}. \tag{4.17}$$

Conversely if ($\mathbf{A}$) *converges strongly and (4.17) holds, then the corresponding Φ whose existence is asserted by Proposition 4.3 satisfies (4.10).*

Proof: By assumption (4.9) and (4.10), one has

$$1 = (\mathbf{C}_j^p\Phi_j)^T\mathbf{e}^j = \Phi_{j+1}^T\mathbf{A}_{j,0}\mathbf{C}_j^p\mathbf{e}^j, \quad 1 = (\mathbf{C}_{j+1}^p\Phi_{j+1})^T\mathbf{e}^{j+1}.$$

By injectivity, the relation (4.17) follows. Setting for any $j \in \mathbb{N}_0$, $\mathbf{c} \in l_p(I_j)$, $f_\mathbf{c} := \Phi_j^T\mathbf{c} = \Phi_m^T\mathbf{A}_j^{m-1}\mathbf{c}$, stability combined with refinability yields, in view of Proposition 4.3,

$$\begin{aligned}\left\|\Lambda_m^{p'}(f_\mathbf{c}) - \mathbf{A}_j^{m-1}\mathbf{c}\right\|_{l_p(I_m)} &\leq c\,\left\|\Phi_m^T(\Lambda_m^{p'}(f_\mathbf{c}) - \mathbf{A}_j^{m-1}\mathbf{c})\right\|_{L_p(\Omega)}\\ &= c\,\left\|\Phi_m^T\Lambda_m^{p'}(f_\mathbf{c}) - f_\mathbf{c}\right\|_{L_p(\Omega)}.\end{aligned}$$

By (4.11), the right hand side tends to zero as $m \to \infty$. Since by (4.17),

$$\Phi_j^T\mathbf{C}_j^p\mathbf{e}^j = \lim_{m\to\infty}(\Lambda_{m+1}^p)^T\mathbf{A}_j^m\mathbf{C}_j^p\mathbf{e}^j = \lim_{m\to\infty}(\Lambda_{m+1}^p)^T\mathbf{C}_{m+1}^p\mathbf{e}^{m+1} = 1,$$

the rest of the assertion easily follows. ■

Proposition 4.5. *Let Φ satisfy the assumptions of Proposition 4.4. Assume that there exists a uniformly sparse* ($\mathbf{B}$) *which converges strongly in $L_{p'}(\Omega)$ and satisfies (4.6) for all $j \in \mathbb{N}_0$. Then there exists $\tilde{\Phi}_j \subset L_{p'}(\Omega), j \in \mathbb{N}_0$, satisfying (4.3) and (4.4).*

Proof: If ($\mathbf{B}$) converges strongly Proposition 4.3 implies that

$$\tilde{\varphi}_k^j := \lim_{m\to\infty}(\Lambda_{m+1}^{p'})^T\mathbf{B}_j^m\mathbf{e}^{j,k}$$

is refinable. Now write briefly $\langle\Phi_j, \tilde{\Phi}_j^T\rangle := (\langle\varphi_k^j, \tilde{\varphi}_m^j\rangle)_{k,m\in I_j}$ so that

$$\begin{aligned}\left\langle\Phi_j, \tilde{\Phi}_j^T\right\rangle &= \left\langle\Phi_j, \left(\tilde{\Phi}_j^T - \left(\Lambda_{m+1}^{p'}\right)^T\mathbf{B}_j^m\right)\right\rangle\\ &+ \left\langle\Phi_j - (\mathbf{A}_j^m)^T\Lambda_{m+1}^p, \left(\Lambda_{m+1}^{p'}\right)^T\mathbf{B}_j^m\right\rangle\\ &+ \left\langle(\mathbf{A}_j^m)^T\Lambda_{m+1}^p, \left(\Lambda_{m+1}^{p'}\right)^T\mathbf{B}_j^m\right\rangle.\end{aligned}$$

By convergence of (**A**) and (**B**), the first two summands on the right hand side of the above identity tend to zero as $m \to \infty$, while, by (4.6),

$$\langle (\mathbf{A}_j^m)^T \Lambda_{m+1}^p, (\Lambda_{m+1}^{p'})^T \mathbf{B}_j^m \rangle = (\mathbf{A}_j^m)^T \mathbf{B}_j^m = \overline{(\mathbf{B}_j^m)^* \mathbf{A}_j^m} = I.$$

This confirms that (4.3) holds which completes the proof. ■

It is not hard to check that it would actually be sufficient to require in Proposition 4.5 that (**B**) converges weakly, i.e.,

$$\langle f, f_{\mathbf{c}} - (\Lambda_{m+1}^{p'})^T \mathbf{B}_j^m \mathbf{c} \rangle \to 0, \quad m \to \infty,$$

holds for all $f \in L_p(\Omega)$.

Thus, in general, to assure that the multiscale transformations associated with the cascade schemes (2.5), (2.10) are stable, one would have to prove the convergence of the schemes (**A**) and (**B**) where the $\mathbf{A}_{j,0}, \mathbf{B}_{j,0}$ satisfy (4.6). In general this is expected to be a rather difficult task which may strongly depend on the particular case at hand. These observations so far offer only a somewhat more focused formulation of the original problem of stability which, however, might be helpful in concrete cases. When the $\mathbf{A}_{j,0}, \mathbf{B}_{j,0}$ do not depend on j and $\Omega = \mathbb{R}^s$ one obtains *stationary subdivision schemes* which are extensively studied in [4]. In this case and for $p = 2$ necessary and sufficient conditions on those $\mathbf{A}_{j,0} = \mathbf{A}_0$ and $\mathbf{B}_{j,0} = \mathbf{B}_0$ are derived in [7] such that biorthogonal systems $\Phi, \tilde{\Phi}$ exists which give rise to Riesz-bases. In general, given a biorthogonal pair with associated convergent schemes (**A**), (**B**), the construction of the corresponding stable multiscale basis can be summarized as follows. Given completions of $\mathbf{A}_{j,0}, \mathbf{B}_{j,0}$ to $\mathbf{A}_j, \mathbf{B}_j$ such that (2.9) holds, the functions

$$\psi_k^j = \sum_{m \in I_{j+1}} a_{m,k}^j \varphi_m^{j+1}, \quad \tilde{\psi}_k^j = \sum_{m \in I_{j+1}} b_{m,k}^j \tilde{\varphi}_m^{j+1}, \quad k \in J_j,$$

form biorthogonal Riesz-bases provided that the mappings

$$\mathbf{A}_j \in \mathcal{L}(l(I_{j+1}), l(I_{j+1})), \quad \mathbf{B}_j \in \mathcal{L}(l'(I_{j+1}), l'(I_{j+1}))$$

are uniformly bounded [3]. In fact, as mentioned before, this latter requirement ensures that the corresponding $\Psi^j, \tilde{\Psi}^j$ are uniformly stable and the equivalence (3.44) takes the form (3.6) since

$$(Q_{n+1} - Q_n)f = \sum_{k \in J_n} \langle f, \tilde{\psi}^n \rangle \psi_k^n, \quad (Q'_{n+1} - Q'_n)f' = \sum_{k \in J_n} \langle \psi^n, f' \rangle \tilde{\psi}_k^n$$

for $f \in L_p(\Omega)$, $f' \in L_{p'}(\Omega)$.

References

1. de Boor, C., R. A. DeVore, and A. Ron, On the Construction of Multivariate (pre–) Wavelets, preprint 1992.
2. Beylkin, G., R. Coifman, and V. Rokhlin, Fast Wavelet Transforms and Numerical Algorithms I, Comm. Pure and Appl. Math. **XLIV** (1991), 141-183.
3. Carnicer, J. M., W. Dahmen, and J. M. Peña, Local Decomposition of Refinable Spaces, in preparation.
4. Cavaretta, A. S., W. Dahmen, and C. A. Micchelli, *Stationary Subdivision*, Memoirs of the American Math. Soc. Vol. **93**, No. 453, 1991.
5. Chui, C. K., J. Stöckler, and J. D. Ward, Compactly Supported Box Spline Wavelets, Approximation Theory and its Applications **8** (1992), 77-100.
6. Cohen, A., Rapport Scientifique, Université Paris IX Dauphine, June 1992.
7. Cohen, A. and I. Daubechies, Non–Separable Bidimensional Wavelet Bases, Preprint AT & T Bell Laboratories, New Jersey, 1991.
8. Cohen, A., I. Daubechies and J.-C. Feauveau, Biorthogonal Bases of Compactly Supported Wavelets, Comm. Pure and Appl. Math. **45** (1992), 485-560.
9. Coifman, R., Adapted Multiresolution Analysis, in *Computation, Signal Processing and Operator Theory*, Proceedings of the International Congress of Mathematicians, Kyoto, 1990, 879-887.
10. Dahmen, W., R. A. DeVore, and K. Scherer, Multidimensional Spline Approximation, SIAM J. Numer. Anal. **17** (1980), 380-402.
11. Dahmen, W., and A. Kunoth, Multilevel Preconditioning, Numer. Math., **63** (1992), 315–344.
12. Dahmen, W., and C. A. Micchelli, Biorthogonal Wavelet Expansions, in preparation.
13. Dahmen, W., S. Prössdorf, and R. Schneider, Wavelet Approximation Methods for Pseudodifferential Equations I: Stability and Convergence, Math. Z., to appear.
14. Dahmen, W., S. Prössdorf, and R. Schneider, Wavelet Approximation Methods for Pseudodifferential Equations II: Matrix Compression and Fast Solution, Advances in Computational Mathematics **1** (1993), 259-335.
15. Dahmen, W., S. Prössdorf, and R. Schneider, Multiscale Methods for Pseudo-differential Equations, in *Recent Advances in Wavelet Analysis*, L.L. Schumaker, G. Webb (eds.), Academic Press, New York, 1994, 191-235.
16. Dahmen, W., B. Kleemann, S. Prössdorf, and R. Schneider, A Multiscale Method for the Double Layer Potential Equation on a Polyhedron, in *Advances of Computational Mathematics*, H.P. Dikshit, C.A. Micchelli (eds.), to appear.

17. Daubechies, I., Orthonormal Bases of Compactly Supported Wavelets, Communications on Pure and Applied Mathematics, **41** (1988), 909-996.
18. Daubechies, I., *Ten Lectures on Wavelets*, CBMS-NSF Regional Conference Series in Applied Mathematics, SIAM, 1992.
19. Donoho, D. L., Nonlinear Solution of Linear Inverse Problems by Wavelet-vaguelette Decomposition, Technical Report, Department of Statistics, Stanford University, 1992.
20. Jaffard, S., Wavelet Methods for Fast Resolution of Elliptic Problems, SIAM J. Numer. Anal. **29** (1992), 965-987.
21. Gröchenig, K. H., and W. R. Madych, Haar Bases and Self-affine Tilings of $\mathbb{R}^n$, IEEE Trans. Inform. Th. **38** (2), 1992, 556-568.
22. Jia, R. Q., and C. A. Micchelli, Using the Refinement Equation for the Construction of Pre-wavelets II: Powers of Two, in *Curves and Surfaces*, P.J. Laurent, A. LeMéhauté, L.L. Schumaker (eds.), Academic Press, New York, 1991, 209-246.
23. Harten, A., and I. Yad-Shalom, Fast Multiresolution Algorithms for Matrix-vector Multiplication, ICASE Report No. 92-55, October 1992.
24. Harten, A., Multiresolution Representation of Data, CAM Report 93-13, UCLA, 1993.
25. Harten, A., Multiresolution Algorithms for the Numerical Solution of Hyperbolic Conservation Laws, Courant Math. and Comp. Lab. Report, 1993.
26. Kunoth, A., *Multilevel Preconditioning*, Ph. D. thesis, FU Berlin, 1994.
27. Meyer, Y., *Ondelettes et Opérateurs 1:* , Hermann, Paris 1990.
28. Mallat, S. G., Multiresolution Approximation and Wavelet Orthonormal Basis of $L_2(\mathbb{R})$, Trans. Amer. Math. Soc. **315** (1989), 69-87.
29. Micchelli, C. A., Using the Refinement Equation for the Construction of Pre-wavelets, Numerical Algorithms, **1** (1991), 75-116.
30. Riemenschneider, S., and Z. Shen, Wavelets and Pre-wavelets in Low Dimensions, J. Approx. Theory, **71** (1992), 18-38.
31. Triebel, H., *Interpolation Theory, Function Spaces and Differential Operators*, North-Holland, Amsterdam, 1978.
32. de Vore, R., B. Jawerth, and V. Popov, Compression of Wavelet Decompositions, American Journal of Math. **114** (1992), 737-785.
33. Yserentant, H., On the Multilevel Splitting of Finite Element Spaces, Numer. Math. **49** (1986), 379-412.

Wolfgang Dahmen
Institut für Geometrie und Praktische Mathematik
RWTH Aachen
Templergraben 55
52056 Aachen, GERMANY
Telefon (49) 241 80 3950
Telefax (49) 241 8888 317
dahmen@ igpm.rwth-aachen.de

Fractal Interpolation Functions for a Class of Finite Elements

S. De Marchi and M. Morandi Cecchi

Abstract. Classical ways to describe *shape functions* for finite element methods make use of interpolating or approximating schemes like for example, Taylor and Lagrange [8,9] or Bézier [5,8]. In this paper we outline the possibility of using iterative schemes first applied in the computation of fractal curves and surfaces. Our attempt will be restricted to shape functions defined over 2-simplices. Because of the fractal nature of the functions, we get only continuous or uniformly continuous functions. We see that they can be found as an *attractor* of a suitable *Iterated Function System (IFS)* [2,7].

§1. Introduction

A finite element $\mathcal{F}_E$ on $\mathbb{R}^n$ is a triplet (E, P, Γ) (see [3]) where E is a closed polyhedron in $\mathbb{R}^n$; $P \subset \mathcal{C}^s(E)$, $s \in \mathbb{N}$, is a finite dimensional space of real value functions (we shall assume that its dimension is M); Γ is a set of linear functions γ_i, $1 \le i \le M$, linearly independent and defined over the set P. The set Γ is P-unisolvent (see [3,8]). In particular, there exist M functions $p_i \in P$, $1 \le i \le M$, such that $\gamma_j(p_i) = \delta_{ij}$, $1 \le i, j \le M$.

In the following, we use the symbol E to identify the finite element $\mathcal{F}_E$ and its defining polyhedron, and E is a n-simplex defined as usual, by

$$E = \{x \in \mathbb{R}^n : \ x_i = \sum_{j=1}^{n+1} s_{ij}\alpha_j; \ 0 \le \alpha_j \le 1, \ 1 \le i \le n, \ 1 \le j \le n+1\}, \quad (1)$$

where $\{\alpha_j\}$ are the barycentric coordinates of the point x. The functions γ_i and p_i are known as *degrees of freedom* and *shape functions* of E, respectively.

As is well-known, to compute the interpolating polynomial on E, we may use the Neville-Aitken's interpolation scheme. This scheme was generalized in [9] to a family $\mathcal{H} = \{h_i\}_{i=1,\dots,n}$ of continuous functions forming a Chebyshev system over a set of points $\mathcal{S}$, of cardinality n at least.

Wavelets, Images and Surface Fitting
P. J. Laurent, A. Le Méhauté, and L. L. Schumaker (eds.), pp. 189–196.

ISBN 1-56881-040-7.

Again, the interpolating polynomial based on such a system can be expressed by recurrence formulae. The coefficients $(\eta_l(x))_{l=0,\ldots,m}$ in those formulae are a ratio of determinants whose elements $(\beta_{i,l}(x))_{i,l=0,\ldots,m}$ are the interpolation *errors* obtained substituting the function which has to be interpolated with one of the functions of the family. Two interesting relations were obtained

$$\begin{cases} \sum_{l=0}^{m} \eta_l(x) = 1 \\ \sum_{l=0}^{m} \beta_{i,l}(x)\eta_l(x) = 0, \quad i = 1, \ldots, m. \end{cases} \tag{2}$$

The coefficients $(\beta_{j,i}(x))_{i,j=0,\ldots,m}$ are related to the given Chebychev set $\mathcal{H}$ by the following (see [5])

$$\beta_{j,i}(x) = -h_{0,j}^{(i)}(x) = h_j(x) - h_j(x_i)\frac{h_0(x)}{h_0(x_i)}, \qquad i,j = 1,...,m, \tag{3}$$

assuming $\beta_{0,j}(x) = 1 \;\; {}_{j=0,..,m}$.

Some properties of this interpolating scheme are proved in [6]. This framework also includes the hierarchical finite elements [3]. The construction of the *attractor* at which the sequence of hierarchical elements converges is a key tool for this class of elements. This paper presents a way of computing the attractor of an *IFS* that interpolates at the element.

Section 2 briefly reviews a way to compute an attractor of an IFS. Section 3 gives an example of this technique for a function defined over a triangle. An algorithm to compute a fractal interpolating surface over a triangle is provided in Section 4.

§2. Attractors and IFS

Let $X = (X, d_X)$ be a compact metric space or a closed subset of $\mathbb{R}^s$, $s > 0$ with metric d_X and let $\mathcal{H}(X)$ be the set of all nonempty closed subsets of X with Hausdorff metric function h.

On $\mathcal{H}(X)$ we consider the finite dimensional set of continuous functions

$$G = \{g_i : X \longrightarrow X, \;\; i = 1,...,n\}. \tag{4}$$

The pair $G_X = (G, X)$ is known as an *Iterated Function System* (briefly IFS) on X (see [1,2]). If the functions g_i are contractive, then G_X is said to be a *hyperbolic* IFS.

Definition 1. *We say the compact set $K \subset \mathbb{R}^s$ is invariant if there exists a hyperbolic IFS, G_X, such that $K = \cup_{i=1}^{n} g_i(K)$.*

An invariant set is mainly determined by an iterative converging sequence of sets which can be constructed from G_X starting from a polygonal region in $\mathbb{R}^s$. The maps g_i have to satisfy some properties that guarantee the convergence of the approximating sequence of sets.

In the case of a hyperbolic IFS, the invariant set is known as the *attractor* of an IFS. If the functions g_i are contraction maps, we can prove the existence and uniqueness of a compact set K invariant with respect to G (see [4]).

Let $\mathcal{S} = \{(x_i, y_i),\ i = 0, .., n;\ n \geq 1\} \subset \mathbb{R}^2$ be a set of points and let $f : \mathcal{D} = [x_0, x_n] \longrightarrow \mathbb{R}$ be the function interpolating on $\mathcal{S}$. Let $\mathcal{G}_f$ be its graph. The graph $\mathcal{G}_f$ can be seen as the *attractor* of an IFS, $W_X = \{w_1, .., w_n\}$, defined on the set $X \subset \mathbb{R}^2$ [1]. The functions w_i are taken to be *contractive affine maps*, with contractivity factor $0 \leq \lambda_i < 1$ of the form

$$w_i(x, y) = \begin{pmatrix} a_i & 0 \\ c_i & d_i \end{pmatrix} \begin{pmatrix} x \\ y \end{pmatrix} + \begin{pmatrix} e_i \\ g_i \end{pmatrix} \tag{5.1}$$

and

$$w_i(x_0, y_0) = (x_{i-1}, y_{i-1}) \tag{5.2}$$

$$w_i(x_n, y_n) = (x_i, y_i). \tag{5.3}$$

These conditions allow us to determine the parameters a_i, c_i, e_i and g_i whereas the d_i remain *free.* If we choose $d_i = 0$, we recover *piecewise linear interpolation.*

Proposition 1. *$\mathcal{G}_f$ is an invariant set in $\mathbb{R}^2$, i.e. $\mathcal{G}_f = \bigcup_{i=1}^{n} w_i(\mathcal{G}_f)$.*

Proof: For each i, W_X has an attractor which is the (fractal) curve between (x_{i-1}, y_{i-1}) and (x_i, y_i). By union of these subsets we get the conclusion. ■

A function $f(x)$ whose graph $\mathcal{G}_f$ is the attractor of an IFS is said to be a *fractal interpolation function* corresponding to the data set $\mathcal{S}$. Given the IFS W_X on $\mathcal{H}(X)$, we define the function $W : \mathcal{H}(X) \to \mathcal{H}(X)$ by $W(S) = \bigcup_i w_i(S)$ for any $S \in \mathcal{H}(X)$. Thus, the attractor of W_X is any set $A \in \mathcal{H}(X)$ that is a fixed point of W, i.e. $W(A) = A$. We have the following question:

> *"How can we find an IFS in $\mathbb{R}^2$ such that its attractor is a fractal interpolation function $f : [x_0, x_n] \to \mathbb{R}$, and $f(x_i) = y_i$?"*

A well-known technique to compute the attractor of a given IFS in $\mathbb{R}^2$ is the *Random Iteration Algorithm* (RIA) [2]. The RIA allows us to compute the coordinates of the points on the curve by choosing at random one of the n maps w_i. The kernel of the algorithm is to find out the attractor, making use of the RIA. It can be summarized as follows:

Algorithm 1

1) Define the affine transformations. That is, we have to determine the numbers a_i, c_i, e_i and g_i corresponding to the i-th tranformation w_i to be a contraction on the interval of definition, i.e. $Range(w_i)\big|_{[x_0,x_n]} = [y_{i-1}, y_i]$.

2) Apply the RIA to the transformations $w_i(x, y)$, MAX_ITER times.

§3. Fractal Interpolation Functions on General Sets

To extend the theory to $\mathbb{R}^s$, $s > 2$, we have to consider *general fractal interpolation functions.* To this end, let $(\boldsymbol{Y}, d_{\boldsymbol{Y}})$ denote a complete metric space, and let $f : K \subset \mathbb{R} \to \boldsymbol{Y}$ be a function.

Definition 2. *A set of generalized data is the set*

$$GD = \{(x_i, Y_i) \in \boldsymbol{X} = \mathbb{R} \times \boldsymbol{Y}, \quad x_i \in K \text{ in ascending order for } i \in \mathcal{I}\}.$$

where $\mathcal{I} = \{0, 1, ..., n\}$, $n \geq 2$. The continuous function $f : [x_0, x_n] \longrightarrow \boldsymbol{Y}$ such that $f(x_i) = Y_i$, $\forall\, i$ is an *interpolation function* corresponding to this set.

On the set GD, we define n linear functions $L_i : \mathbb{R} \to \mathbb{R}$ where $L_i(x) = a_i x + e_i$ for $i \in \mathcal{I}_0 = \mathcal{I}\backslash\{0\}$. Let c and σ be real numbers such that $c > 0$ and $0 \leq \sigma < 1$. For each $i \in \mathcal{I}_0$ let $M_i : \boldsymbol{X} \longrightarrow \boldsymbol{Y}$ be a function which is Lipschitz in the first variable

$$d_{\boldsymbol{Y}}(M_i(a, y), M_i(b, y)) \leq c|a - b|, \qquad \forall a, b \in \mathbb{R}, \quad y \in \boldsymbol{Y}, \tag{6}$$

and contractive in the second one

$$d_{\boldsymbol{Y}}(M_i(x, a), M_i(x, b)) \leq \sigma d_{\boldsymbol{Y}}(a, b) \quad \forall x \in \mathbb{R}, \; a, b \in \boldsymbol{Y}. \tag{7}$$

From L_i and M_i we define the transformation

$$\begin{aligned} w_i :\ & \boldsymbol{X} \longrightarrow \boldsymbol{X} \\ & (x, Y) \mapsto w_i(x, Y) = (L_i(x), M_i(x, Y)) \end{aligned} \tag{8}$$

One can prove that the IFS, $W_{\boldsymbol{X}} = \{\boldsymbol{X}, w_i, \; i \in \mathcal{I}_0\}$ is hyperbolic with respect to the metric $d_{\boldsymbol{X}}(X_1, X_2) = |x_1 - x_2| + \lambda|Y_1 - Y_2|$, where $X_1 = (x_1, Y_1)$, $X_2 = (x_1, Y_1)$ and $\lambda > 0$ (see [4]).

3.1. Fractal interpolation surfaces

Here we wish to present a technique to construct a fractal interpolation surface over a 2-simplex. We start from the *canonical* 2-simplex Σ that has vertices $\sigma_0 = (0, 0)$, $\sigma_1 = (1, 0)$ and $\sigma_2 = (0, 1)$. Let $K = [0, 1] \times [0, 1]$. On the set $A = K \times \mathbb{R}$ we define the functions $w_i : A \longrightarrow A$ as

$$w_i(x, y, z) = (L_i(x, y), M_i(x, y, z)) \quad for \; i \in \mathcal{I}_0, \tag{9}.$$

where $L_i : K \to K$ and $M_i : A \to \mathbb{R}$. We require that $w_i(\Sigma) = \Sigma_i$, that is they map the simplex onto its $i - th$ subsimplex Σ_i, $L_i(\Sigma) = \Sigma_i$ is a homeomorphism and M_i is Lipschitz in x, y and contractive in z. We also assume $L_1(\sigma_0) = \sigma_0$, $M_1(0, 0, 0) = 0$ and there exist indices $i_0, i_1 \in \mathcal{I}_0$ such that $L_{i_0}(\sigma_1) = \sigma_1$, $L_{i_1}(\sigma_2) = \sigma_2$ and $M_{i_0}(1, 0, 0) = M_{i_1}(0, 1, 0) = 0$. These conditions are analogous to (5.2) and (5.3).

Let consider the IFS $W_A = \{A, w_i, i \in \mathcal{I}_0\}$. It is straightforward to prove that W_A is hyperbolic on the space A with distance function $d_{A,\alpha}$ given by

$$d_{\Sigma,\alpha}((x_1, y_1, z_1), (x_2, y_2, z_2)) = |x_1 - x_2| + |y_1 - y_2| + \frac{1-\tilde{c}}{\alpha L}|z_1 - z_2|.$$

Here $\tilde{c} = \max\{c_1, c_2\}$ with c_1 is the contraction constant of $L_i(x, \cdot)$ and c_2 the corresponding for $L_i(\cdot, y)$; while $L = \max\{L_1, L_2\}$, where L_1 is the Lipschitz constant of $M_i(x, \cdot, \cdot)$, and L_2 the Lipschitz constant for $M_i(\cdot, y, \cdot)$. $\alpha > 1$ is a fixed real number. Since W_A is hyperbolic, we are interested in its attractor. The following proposition characterizes this set.

Proposition 2. *The attractor of W_A is the graph of a continuous function $f : \Sigma \longrightarrow \mathbb{R}$ that passes through the vertices of the subsimplices Σ_i, $i \in \mathcal{I}_0$.*

Definition 3. *We call the graph of a function $f : \mathbb{R}^2 \longrightarrow \mathbb{R}$ that is a fractal a fractal surface.*

As an example, in generalizing (5.1) to 3D, we consider affine maps of the form

$$w_i(x, y, z) = \begin{pmatrix} a_i & b_i & 0 \\ c_i & d_i & 0 \\ g_i & h_i & l_i \end{pmatrix} \begin{pmatrix} x \\ y \\ z \end{pmatrix} + \begin{pmatrix} e_i \\ f_i \\ m_i \end{pmatrix}.$$

The functions L_i and M_i required in (9) are

$$L_i(x, y) = \begin{pmatrix} a_i & b_i \\ c_i & d_i \end{pmatrix} \begin{pmatrix} x \\ y \end{pmatrix} + \begin{pmatrix} e_i \\ f_i \end{pmatrix}. \tag{10}$$

$$M_i(x, y, z) = g_i x + h_i y + l_i z + m_i$$

The constants $a_i, b_i, c_i, d_i, g_i, h_i, e, f_i$ and g_i are found by the conditions on the functions w_i, that is $w_i(\Sigma) = \Sigma_i$, while the parameters l_i, $|l_i| < 1$ are arbitrarily chosen (see previous the section).

Example. Let $\mathcal{B}$ be the triangle depicted in Fig. 1. Assume that on the nodes, the values of the functions are $0, 0, \frac{1}{2}, 1$, respectively. Let $\{\mathcal{B}; w_1, w_2\}$ be the IFS defined on $\mathcal{B}$, with code given in the next table, assuming that the related probabilities are the same, i.e. $p_1 = p_2 = 0.5$.

w	a	b	c	d	g	h	e	f	m
1	1.0	0.5	1.0	1.0	0.5	0.0	0.0	0.0	0.0
2	0.5	1.0	0.5	1.0	1.0	0.0	0.0	0.0	0.0

Table 1. IFS code for the triangle of Fig. 1.

It is easy to see that its attractor is the set $\mathcal{G}$ that is the fractal surface interpolating at $\mathcal{B}$.

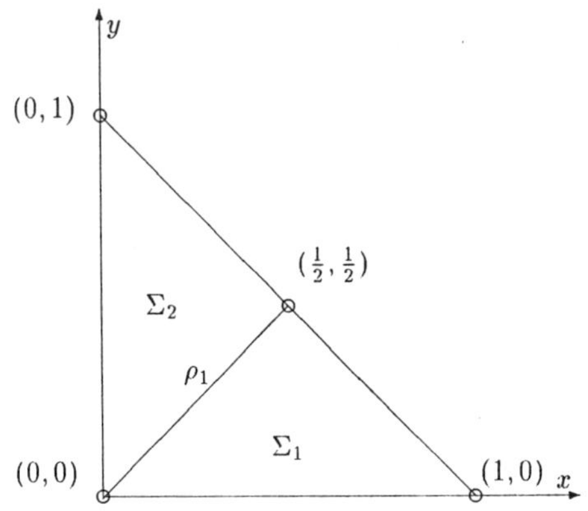

Fig. 1. The *canonical* simplex subdivided in two subtriangles.

§4. An Algorithm for Constructing a Fractal Surface

Let Ω be a closed nondegenerate polygonal region in $\mathbb{R}^2$ containing $n+1$ distinct points $\{(x_i, y_i)\}_{i=0}^n$. We want to find a self-affine fractal interpolating surface of the form $z = \gamma(x, y)$ which interpolates the data $\{(x_i, y_i, z_i)\}_{i=0}^n$. We assume that Ω is triangulable and $\Sigma_\Omega = \{\Sigma_i\}_{i=1}^p$ is such a triangulation with $\mathcal{S} = \{\sigma_1, ..., \sigma_m\}$ as the set of the vertices of Σ_Ω.

Now, let γ be the fractal surface representing the attractor of a given IFS, and let L_i, M_i be the functions defined in the previous section, such that $L_i : \Omega \to \Sigma_i$ and $M_i : \Omega \times \mathbb{R} \to \mathbb{R}$ for $i = 1, ..., p$. By using functions L_i, M_i, we define the mapping $\Psi : \mathcal{C}(\Omega) \to \mathcal{C}(\Omega)$ by

$$(\Psi\gamma)(x) = M_i(L_i^{-1}(x), \gamma(L_i^{-1}(x))) \;\; x \in \Sigma_i \;\; i = 1, ..., p. \tag{11}$$

Ψ is well-defined and contractive in the sup-norm $\|\cdot\|_\infty$ on a suitable subset of $\mathcal{C}(\Omega)$ [7]. Thus, it has a fixed point $\tilde{\gamma}$ that defines the fractal interpolation surface. We show that the $graph(\tilde{\gamma})$ is self-affine and there exists a *deterministic* algorithm to construct it.

Making use of the *recurrent IFS formalism* [2], consider the set $\mathcal{H}$ of all nonempty compact subsets of $\mathbb{R}^3$. We construct its p-cartesian product $\mathcal{H}^p = \underbrace{\mathcal{H} \times \mathcal{H} \times \ldots \times \mathcal{H}}_{p}$. Set $\chi(i) = \{j : \Sigma_j \subset \rho_{\kappa(i)}\}$ where κ is a mapping from $\{1, ..., m\}$ into $\{1, ..., r\}$ such that $L_i(\rho_{\kappa(i)}) = \Sigma_i$. This means that given the triangulation Σ_Ω, we construct the triangulation $R_\Omega = \{\rho_k\}_{k=1}^r$ made by r triangular subregions of Ω obtained as the union of some triangles of Σ_Ω. The vertices of the triangulation Σ_Ω are ordered in such a way that $\{\sigma_1, ..., \sigma_l\} \subset \mathcal{S}$ is the set of the vertices of R_Ω.

On $\mathcal{H}^p$ we define the function $F : \mathcal{H}^p \to \mathcal{H}^p$ whose i-th component is given by

$$F_i(A_1, ..., A_p) = w_i \left(\bigcup_{j \in \chi(i)} A_j \right), \qquad i = 1, ..., p, \tag{12}$$

where the functions $\{w_i\}_{i=1}^p$ are given as in (9). We need the following definitions.

Definition 4. *Let ϑ be a nonvertical plane in $\mathbb{R}^3$. We denote by $\mathcal{C}_\vartheta(\Omega)$ the set*

$$\mathcal{C}_\vartheta(\Omega) = \{f \mid f : \Omega \subset \mathbb{R}^2 \to \mathbb{R} \textit{ with } (x, f(x)) \in \vartheta,\ \forall x \in \partial\Omega\}$$

Definition 5. *On Ω we consider the set of points $\sigma_i = (x_i, y_i)$, and let z_i be the values of a function which interpolates the set $\{(\sigma_i, z_i)\}_{i=1}^n$. We denote by $\mathcal{C}_B(\Omega)$ the set*

$$\mathcal{C}_B(\Omega) = \{f \mid f : \Omega \subset \mathbb{R}^2 \to \mathbb{R} \textit{ with } f(\sigma_j) = z_j,\ \forall \sigma_j \in \partial\Omega\}.$$

We note that Ψ is contractive on both $\mathcal{C}_\vartheta$ and $\mathcal{C}_B$. Finally, we state the main result (proved in [7]) which suggested the algorithm to us:

Proposition 3. *Let $\gamma : \Omega \subset \mathbb{R}^2 \to \mathbb{R}$ be given, and suppose the following properties hold:*

(a) $\gamma \in \mathcal{C}_\vartheta(\Omega)$ *or* $\gamma \in \mathcal{C}_B(\Omega)$;

(b) *given the triangulation Σ_Ω we assume that the graph associated with Σ_Ω has chromatic number equal to 3. Then*

$$graph(\Psi\gamma_{|\Sigma_i}) = F_i(graph(\gamma)_{\Sigma_1}, graph(\gamma)_{\Sigma_2}, ..., graph(\gamma)_{\Sigma_p}) \quad i = 1, ..., p \tag{13}$$

Since $F_i\left(graph(\gamma)_{\Sigma_1}, graph(\gamma)_{\Sigma_2}, ..., graph(\gamma)_{\Sigma_p}\right) = w_i \left(\bigcup_{j \in \chi(i)} A_j \right)$, the previous result leads to the following algorithm:

Algorithm 2

1) Set $A_0 = \left(graph(\gamma)_{\Sigma_1}, graph(\gamma)_{\Sigma_2}, \ldots, graph(\gamma)_{\Sigma_p}\right)$;

2) For $n = 1, 2, \ldots$, set $A_n = F(A_{n-1})$

3) Then, $A_n \to \tilde{\gamma}$, where $\tilde{\gamma}$ is a fractal surface.

To show the convergence in step 3), we have to check whether F is contractive in $\mathcal{H}^p$, or in view of (13), whether $\|\Psi^{\circ\, k}\gamma - \tilde{\gamma}\|_H < \epsilon$ for each $\epsilon > 0$. But Ψ is contractive on $\mathcal{C}_\vartheta$ and $\mathcal{C}_B$, so we can conclude that

$$A_n \to \left(graph(\tilde{\gamma})_{\Sigma_1}, graph(\tilde{\gamma})_{\Sigma_2}, ..., graph(\tilde{\gamma})_{\Sigma_p}\right)$$

and $\bigcup_{k=1}^{n} A_k^{(i)} \to graph(\tilde{\gamma})$.

Acknowledgments. This work was supported by the CNR within the *Progetto Strategico Sistema Lagunare Veneziano* and 40% of MURST on the project *Analisi Numerica e Matematica Computazionale.*

References

1. Barnsley, M., F., *Fractals Everywhere*, Academic Press Inc., 1988.
2. Barnsley, M., F., J. H. Elton, and D. P. Hardin, Recurrent Iterated Functions Systems, Const. Appr. **5** (1989), 3–31.
3. Ciarlet, P., G., and J. L. Lions, *Handbook of Numerical Analysis: Finite Element Methods (Part 1), volume II*, North Holland, 1991.
4. De Marchi, S., A short survey of fractal interpolation curves and surfaces, Report no. 2, Università di Padova, January, 1994.
5. De Marchi, S., and M. Morandi Cecchi, The polynomial approximation in the finite element method, Journal of Computational and Applied Mathematics, to appear.
6. De Marchi, S., and M. Morandi Cecchi, Reference functional characteristic space for Lagrange and Bernstein operators, submitted.
7. Geronimo, J., S., and D. Hardin, Fractal Interpolation Surfaces and a Related 2-D Multiresolution Analysis, J. Math. Anal. Appl. **176** (1993), 561–586.
8. Le Méhauté, A., A Finite Element Approach to Surface Reconstruction, in *Computation of Curves and Surfaces*, W. Dahmen, M. Gasca and C. A. Micchelli (eds.), Kluwer Academic Pubblisher, Dordrecht, 1990, 237–274.
9. Mühlbach, G., The General Neville-Aitken-Algorithm and Some Applications, Numer. Math. **31** (1978), 97–110.

S. De Marchi and M. Morandi Cecchi
Dip. di Matematica Pura ed Applicata
Università di Padova
Via Belzoni, 7 I-35131, Padova, ITALY
{*demarchi*, *mcecchi*}@ pdmat1.math.unipd.it

Adaptive Wavelet Bases for Image Compression

Ronald A. DeVore

Abstract. Wavelets are being suggested as a platform for various tasks in image processing. We shall consider several issues that arise in the application of wavelets to image compression. Our main emphasis will be on the choice of wavelet basis which are best suited for compression algorithms. Some results of numerical experiments will be presented. Finally, we suggest some alternatives to the usual wavelet bases to be used in the design of compression algorithms.

§1. Introduction

The application of wavelet decompositions to image processing brings forward several questions on just how wavelets should be implemented. We shall restrict our attention to the application of wavelets to image compression, and we shall discuss some questions ensuant to this application. Our main point of emphasis will be the question of which wavelet basis is most suitable for the design of compression algorithms. In partial answer to this question, we shall suggest two types of bases for further analysis.

This paper reflects recent research on wavelet decompositions and image processing that the author has undertaken with several collaborators. While their names do not appear as authors of the present paper, their influence is certainly present in this writeup. Foremost to mention is Brad Lucier who has collaborated with the author on most of his work in image processing. During the past year, a working seminar at the University of South Carolina uncovered many of the new ideas presented here.

Motivation for some of the questions on image processing discussed here came from data transmission problems considered by the Partnership in Computational Science (PICS) consortium. PICS is concerned with real time transmission and visualization of large data sets arising from the numerical modeling of groundwater flow. I am very thankful to Dick Ewing and Bob Sharpley for the many interesting discussions about this problem.

Wavelets, Images and Surface Fitting
P. J. Laurent, A. Le Méhauté, and L. L. Schumaker (eds.), pp. 197–219.

ISBN 1-56881-040-7.

§2. Wavelets and Multiresolution Analysis

Wavelet functions give an efficient method for the representation of functions and images. Several properties of wavelet decompositions suggest their superiority to more traditional decompositions such as the Fourier decompositions. These include the fast computation of wavelet coefficients and the fast numerical reconstruction of functions from these coefficients; the localness of the functions in the wavelet basis; the efficient encoding of smoothness and other properties of functions in the wavelet coefficients; and the discretization of many function norms in terms of wavelet coefficients. The latter property for example allows the simple solution of extremal problems, relevant to numerical applications, to be accomplished in terms of wavelet coefficients (see e.g. [10]).

Given a multivariate function g defined on $\mathbb{R}^d$, we use the notation

$$g_{j,k} := g(2^k \cdot -j), \quad \tilde{g}_{j,k} := 2^{kd/2} g_{j,k}, \quad j \in \mathbb{Z}^d, k \in \mathbb{Z},$$

to denote its shifted dilates and its L_2-normalized shifted dilates. For the purposes of this paper, we shall say that a univariate function ψ is a *wavelet* if its L_2-normalized shifted dilates $\tilde{\psi}_{j,k}$, $j,k \in \mathbb{Z}$, are a stable basis for $L_2(\mathbb{R})$. That is, each $f \in L_2(\mathbb{R})$ should have the unique representation

$$f = \sum_{j,k \in \mathbb{Z}} c_{j,k}(f) \tilde{\psi}_{j,k} \tag{2.1}$$

and

$$C_1 \sum_{j,k \in \mathbb{Z}} |c_{j,k}(f)|^2 \le \|f\|^2_{L_2(\mathbb{R})} \le C_2 \sum_{j,k \in \mathbb{Z}} |c_{j,k}(f)|^2$$

with absolute constants C_1, C_2. Here the $c_{j,k}$ are coefficient functionals defined on $L_2(\mathbb{R})$. If the functions $\tilde{\psi}_{j,k}$ form a complete orthonormal system for $L_2(\mathbb{R})$, then we say that ψ is an orthogonal wavelet.

The usual method for constructing wavelets is by multiresolution. In the univariate case, we begin with a function ϕ whose shifts $\phi(\cdot - j)$, $j \in \mathbb{Z}$, are $L_2(\mathbb{R})$-stable and we form the shift-invariant space $\mathcal{S} := \mathcal{S}(\phi)$ which is by definition the closure in $L_2(\mathbb{R})$ of all finite linear combinations $\sum_j c_j \phi(\cdot - j)$. Then each function $s \in \mathcal{S}(\phi)$ has the representation

$$s = \sum_{j \in \mathbb{Z}} c_j(s) \phi(\cdot - j)$$

with

$$C_1 \sum_{j \in \mathbb{Z}} |c_j(s)|^2 \le \|s\|^2_{L_2(\mathbb{R})} \le C_2 \sum_{j \in \mathbb{Z}} |c_j(s)|^2.$$

From the space $\mathcal{S}$, we form by dilation the spaces $\mathcal{S}^k := \{S(2^k \cdot) : S \in \mathcal{S}\}$. We assume that the spaces $\mathcal{S}^k$ are nested: $\mathcal{S}^k \subset \mathcal{S}^{k+1}$, $k \in \mathbb{Z}$. This imposes a severe restriction on ϕ since it says that $\phi \in \mathcal{S}^1$, and therefore

$$\phi(x) = \sum_{j \in \mathbb{Z}} a_j \phi(2x - j), \quad x \in \mathbb{R}, \tag{2.2}$$

for some sequence $(a_j)_{j\in\mathbb{Z}}$ from $\ell_2(\mathbb{Z})$. This is called the *refinement equation* for ϕ. For applications in image processing it is desirable that the mask coefficients a_j, $j \in \mathbb{Z}$, are finite in number.

To construct wavelet functions ψ, we take a projector P which maps $L_2(\mathbb{R})$ onto $\mathcal{S}$ and we let $Q := I - P$ which is also a projector from $L_2(\mathbb{R})$ onto its range. We denote by W the range of $\mathcal{S}^1$ under Q. The space W is called the *wavelet space.* In the case that P is the orthogonal projector onto $\mathcal{S}$ which takes $f \in L_2(\mathbb{R})$ into its best L_2-approximation from $\mathcal{S}$, then W is the orthogonal complement of $\mathcal{S}$ in $\mathcal{S}^1$, i.e. $W = \mathcal{S}^1 \ominus \mathcal{S}$. In general W is the complement of $\mathcal{S}$ in $\mathcal{S}^1$ with respect to the projector P and we have

$$\mathcal{S}^1 = \mathcal{S} + W$$

in the sense that each $f \in \mathcal{S}^1$ has the unique decomposition

$$f = s + w, \quad s = Pf, \quad w = Qf.$$

We obtain wavelet functions by finding generators ψ for W. That is, we should have $W = \mathcal{S}(\psi)$. We shall restrict our discussions in this paper to the following special wavelets.

Haar wavelet. If we take $\phi := \chi_{[0,1]}$, then $\mathcal{S}(\phi)$ is the space of piecewise constant $L_2(\mathbb{R})$-functions with integer breakpoints. With P the orthogonal projector, the Haar function $\psi := \chi_{[0,1/2]} - \chi_{(1/2,1]}$ is a generator of the wavelet space W. The Haar function is an orthogonal wavelet.

Daubechies wavelets. Daubechies has found a class of orthogonal wavelets which have compact support and can have an arbitrarily high smoothness. These wavelets are found by discovering suitable refinement coefficients (2.2). For each r, Daubechies has shown the existence of refinement coefficients $a_j = a_j(r)$, such that the corresponding multiresolution generator $\phi = \phi_r$ has compact support and has orthogonal shifts. The number of nonzero coefficients in the refinement mask is $2r$. Then using the orthogonal projector P onto $\mathcal{S}$, the corresponding orthogonal wavelet $\psi = \mathcal{D}_r$ is also of compact support. The smoothness of the $\mathcal{D}_r$ increase with r (they have roughly $r/2$ order of smoothness in $C(\mathbb{R})$; see [3] for a detailed discussion of smoothness of the Daubechies and the biorthogonal wavelets that follow.

Biorthogonal wavelets. Cohen, Daubechies, and Feauveau [1] (see also Vetterli and Herley [10]) have given a general procedure to construct certain nonorthogonal wavelets with compact support and high smoothness. Their construction begins with two refineable functions ϕ and η which are in duality:

$$\int_{\mathbb{R}} \phi(x)\eta(x-j)\,dx = \delta(j), \quad j \in \mathbb{Z},$$

with δ the Kronecker delta. Under certain conditions, multiresolution generates from this pair a wavelet ψ whose dual functionals $c_{j,k}$ in the wavelet decomposition (2.1) are given by

$$c_{j,k}(f) = \int_{\mathbb{R}} f \mu_{j,k}\, dx$$

for a suitable function μ. The wavelet function ψ and the dual function μ are obtained in a simple way from ϕ, η and the refinement masks (see (2.2)) for ϕ and η (see [3], Chapter 8, but note that our notation is different from that given there).

We shall only be interested in special cases of the biorthogonal wavelets discussed in Chapter 8 of [3]. They are indexed on two integer parameters k, ℓ. The integers k and ℓ relate to the number of coefficients in the refinement equations and indicate that the wavelet function ψ is a spline function of order k (degree $k-1$) with breakpoints at the half integers and that the dual wavelet μ has ℓ vanishing moments. A discussion of these wavelets and graphs of typical examples can be found in Chapter 8 of [3].

§3. Bivariate Wavelets

Images are bivariate. The usual method for representing bivariate functions with wavelets is as follows. From the univariate scaling function ϕ and the corresponding wavelet ψ, we form the set Ψ consisting of the three functions

$$\phi(x)\psi(y) \quad \psi(x)\phi(y), \quad \psi(x)\psi(y). \tag{3.1}$$

The functions

$$\tilde{\psi}_{j,k}, \quad \psi \in \Psi \tag{3.2}$$

then form a stable basis for $L_2(\mathbb{R}^2)$. This can be viewed as multiresolution analysis for the shift invariant spaces $\mathcal{S}(\varphi)$ generated by the tensor product $\varphi(x,y) := \phi(x)\phi(y)$ of the univariate function ϕ with itself.

There is another natural bivariate wavelet basis which can be formed from the univariate wavelet ψ; it consists of the tensor products

$$\tilde{\psi}_{j,k}(x)\tilde{\psi}_{j',k'}(y), \quad j, j', k, k' \in \mathbb{Z}. \tag{3.3}$$

This is sometimes called the *standard basis* in wavelet literature. We shall call it the *hyperbolic basis* to make clear its distinction from the basis in (3.2).

One should note that in the hyperbolic basis, the dilation parameters 2^k and $2^{k'}$ are different in different coordinate directions. The result is that the function $\tilde{\psi}_{j,k}(x)\tilde{\psi}_{j',k'}(y)$ has support associated to the rectangle $[j2^{-k}, (j+1)2^{-k}] \times [j'2^{-k'}, (j'+1)2^{-k'}]$. The linear space H^n spanned by the functions $\tilde{\psi}_{j,k}(x)\tilde{\psi}_{j',k'}(y)$ with $2^k 2^{k'} \le 2^n$ is analogous to the Fourier space spanned by the exponentials $e_k(x)e_{k'}(y)$ with $|kk|' \le 2^n$ and $e_k(x) := e^{ikx}$. This latter region of points (k,k') in $\mathbb{R}^2$ is called a *hyperbolic cross.*

Approximation from the hyperbolic spaces H^n is quite different then approximation from $\mathcal{S}^n$. While the approximation from the $\mathcal{S}^n$ is governed by the usual ways of measuring smoothness (derivatives and moduli of smoothness and Besov spaces), the hyperbolic approximation is governed by certain mixed derivatives and a new modulus of smoothness (see [7]).

§4. Wavelet Representation of Grey Scale Images

We shall discuss here only the case of grey scale images; color images can be handled by using a vector of coefficients but interesting questions arise as to the best choice of representing the three dimensional space of colors. It will also be convenient to restrict our attention to square images with the number of pixel values $N = 2^{2m}$. (Some comments are made later about efficient ways to handle images of different dimensions). Thus, for us, a digitized grey scale image $\mathcal{I}$ can be considered as an array of pixels values p_j, $j = (j_1, j_2)$, with each p_j an integer between 0 (black) and 255 (white) depicting the level of grey scale intensity. It is sometimes useful to view the digitized image as arising from an intensity function F defined on the unit square $\Omega := [0,1]^2$ with the pixel value p_j the average of F over the square with lower left corner $2^{-m}(j_1, j_2)$ and side length 2^{-m}.

In order to utilize wavelet decompositions in image processing, we need a way to go from the image $\mathcal{I}$ (i.e. the set of pixel values (p_j)) to a wavelet decomposition. We shall first discuss this for the usual wavelet bases (3.1). Image representation with hyperbolic wavelet bases is discussed in §6.

We fix a univariate generating function ϕ and its corresponding wavelet ψ. We recall that the usual wavelet basis generated by the shifted dilates of the three functions (3.1) can be obtained by multiresolution with respect to the tensor product φ of ϕ with itself: $\varphi(x,y) := \phi(x)\phi(y)$.

To obtain a wavelet decomposition for the image $\mathcal{I}$, we begin by assigning coefficients c_j, $j \in \mathbb{Z}^2$, and represent $\mathcal{I}$ at the fine dyadic level m as

$$\mathcal{I} \sim \sum_{j\in\mathbb{Z}^2} c_j \varphi_{j,m}(x,y). \tag{4.1}$$

We shall discuss just how to assign the (c_j) in a moment. We next use a change of basis to transform from the representation (4.1) to the wavelet representation

$$I \sim \sum_{j\in\mathbb{Z}^2} c_j \varphi_{j,m}(x,y) = \sum_{j\in\mathbb{Z}^2} d_{j,0}\phi_{j,0} + \sum_{k=0}^{m-1} \sum_{j\in\mathbb{Z}^2} \sum_{\psi\in\Psi} c_{j,k}\psi_{j,k} \tag{4.2}$$

with Ψ the set of the three functions in (3.1).

It is important to note that this change of basis is performed by the Fast Wavelet Transform (see e.g. [3] or [8]) and all computations are made in terms of the refinement coefficients. If, as we assume, these coefficients are finite in number, then the change of basis requires at most $O(N)$ operations with $N = 2^{2m}$ the number of pixels. Similarly, one easily changes from a wavelet representation (4.2) to the representation (4.1) at the fine level using rewrite rules again given by the refinement coefficients. Again these operations require only $O(N)$ computations.

4.1. Assignment of wavelet coefficients

We next discuss the assignment of the coefficients c_j. A naive assignment is to simply take $c_j := p_j$ wherever the p_j are defined and to take $c_j := 0$ otherwise. This works perfectly well for example for the Haar wavelets because the computation of the coefficients $c_{j,k}$ in (4.2) corresponding to functions which do not vanish identically on the unit square never involves any coefficients which have been set to zero.

When employing more general bivariate wavelets, there are functions $\psi_{j,k}$ appearing in (4.2) whose support intersects both Ω and its complement. The coefficient $c_{j,k}$ of such a function, when computed by the Fast Wavelet Transform, will involve pixel values which have been set to zero. The subsequent application of compression strategies such as thresholding and quantization (as discussed in the next section) will generally create artifacts in the compressed image due to the coefficients set to zero.

The appearance of artifacts is illustrated by Figure 1 where the left image of Lena was compressed using Daubechies wavelets $\mathcal{D}_4$. The black lines appearing near the boundary are artifacts caused by defining extended coeficients to be 0.

There is another explanation of the deficiency in simply setting coefficients to be 0 in the representation (4.1). Such an assignment has a debilitating effect on the smoothness of the function (4.1) representing $\mathcal{I}$. This can be seen by the fact that wavelet coefficients (in the representation (4.2)) corresponding to functions $\psi_{j,k}$ supported near the boundary will be artificially large. Subsequent application of compression strategies processes these large coefficients. For example, thresholding will keep some of these coefficients but will discard others in the compressed image.

To avoid the creation of artifacts, we need to assign coefficients c_j so that the function in (4.1) is smooth. This can be accomplished by a smooth extension of the pixel values outside of the unit square. We first need to understand how many coefficients c_j this involves. If $\psi_{j,0}$ is any wavelet function appearing at the coarsest level in the decomposition (4.1), its coefficient (computed by the Fast Wavelet Transform) involves all c_j for which the support of $\phi_{j,m}$ is contained in the support of $\psi_{j,0}$. Thus, this involves c_j corresponding to j that are $O(2^m)$ pixel rows from the boundary. Thus, there are $O(2^{2m})$ coefficients which would need to be assigned new values.

In order to avoid the computational expense of directly defining c_j for all of these j, Brad Lucier has suggested the following multiscale extension strategy. At the finest level, we define coefficient values c_j only for those j such that the support of $\phi_{j,m}$ is contained in the support of some $\psi_{j,m-1}$ which intersects Ω. This guarantees that the wavelet coefficients $c_{j,m-1}$ computed at the dyadic level $m-1$ will not be polluted by a poor definition of the c_j. This is also true for the coefficients $d_{j,m-1}$ of the functions $\phi_{j,m-1}$ which represent the projection (of the image) onto the space $\mathcal{S}^{m-1}$. However, we need to extend the coefficients $d_{j,m-1}$ in order to retain our fidelity at the next dyadic level. That is, we need to extend them to all values j such that

$d_{j,m-1}$ contributes to the computation of one of the coefficients of $\psi_{j,m-2}$ or $\phi_{j,m-2}$. We continue in this way, extending coefficients as needed.

The advantage of the multilevel extension is that at each level we need define only $O(2^k)$ new coefficients, hence $O(2^m)$ coefficients in total, a savings over the $O(2^{2m})$ coefficients which would need to be defined if the extension only occured at the finest level.

We have not discussed what form the extension operator should take. There are several possibilities such as some variant of the Whitney extension operators (see e.g. [8]) which preserve polynomials of a fixed degree. We have found that simply preserving constants performs well in compression algorithms.

The right image in Figure 1 shows Lena after compression when a multi-level extension of coefficients (based on extending constant functions exactly) has been used.

There are other approaches to handling boundary effects in wavelet based image processing. One of these is to modify the wavelet basis to be orthogonal on Ω. We refer the reader to the paper of Cohen, Daubechies, and Vial [2] or the book of Daubechies [3] for a more complete discussion of this subject.

`Naive extension` `Multiscale extension`

Fig. 1. 50-1 Compression with $\mathcal{D}_4$: Naive extension (L); Multiscale extension (R).

§5. Image Compression Algorithms

Some of the main issues confronted in the design of image compression algorithms are:

- choice of wavelet basis;
- choice of metric to measure compression error;
- compression strategy: linear versus nonlinear;
- smoothness of images;

- encoding of compressed wavelet coefficients;
- computational efficiency.

While the choice of wavelet basis is the main topic of the present paper, it will be convenient to postpone its discussion until some of the other issues have been vented.

5.1. Choice of compression metric

The criteria used to evaluate compressed images should depend on the intended application. For example, producing compressed images for video are quite a different matter from producing them for medical imagery. In the former, the human visual system would be the ultimate judge of quality, while retention of diagnostic fidelity is the main issue in the latter. In some applications, image compression is done as a preprocessor to other image processing tasks such as feature classification or object recognition. Compression can then have the effect of reducing the complexity of the postprocessing task and enhancing the features important for the postprocessing application. For example, some compression strategies can be viewed as removing noise from images.

To carry out a sensible mathematical analysis of compression, it is desireable to quantify in mathematical terms the compression criteria. One way of doing this is to attempt to find a metric defined on images to measure the error between the original and the compressed image. The metric chosen should model the goals of the compression algorithm.

Most metrics used for measuring compression error arise by considering images as functions defined on a square or rectangle Ω. Even here there is some ambiguity in how to interpret the image as a function. One possibility is to consider the image as the piecewise constant function given by its pixel representation. Another is to consider the image as the function which is its wavelet representation. Still another possibility is to think of the image as a function defined on a continuum and its pixel values as samples obtained by averaging over small squares (the function F introduced earlier). Fortunately, the decision of how to view the image as a function has no essential effect on the design and analysis of compression algorithms.

For the purposes of the following discussion of compression algorithms it will be convenient to think of an image $\mathcal{I}$ as its wavelet representation, i.e. as the function appearing on the right side of (4.1) (or equivalently (4.2)).

The most frequently used compression metric is the L_2 norm. The reason for this seems to be that it is easiest to describe optimal compression strategies with this choice since we can compute compression error exactly in terms of pixel values or wavelet coefficients (in the orthogonal wavelet case). On the other hand there is much evidence to indicate that this may not always be the best choice given the intended application of compression.

For example, to produce visually pleasing compressed images, it seems more reasonable to try to model the human visual system. Experiments of this type in the psychological community suggest that the L_1-norm models better some aspect of the human visual system (see for example the discussion in [4]).

While other possibilities could be considered, we shall assume that our compression metric comes from one of the L_p-(quasi)norms, $0 < p \leq \infty$.

5.2. Compression strategies: linear and nonlinear

We suppose that we have chosen a compression metric that models our compression criteria; we assume that this is one of the L_p-metrics $0 < p \leq \infty$. We also assume that we have chosen the wavelet basis we shall use to represent our image. We obtain compression by retaining some of these terms in the wavelet decomposition (4.2) of the image and deleting others. But what exactly should be our strategy for the retention of terms.

For notational convenience, we shall assume that all terms corresponding to $\phi_{j,0}$ will be retained in the compressed image; in any case, there are very few of them.

We want to draw the distinction between two classes of compression algorithms: linear and nonlinear. In *linear algorithms*, one chooses a dyadic level $0 \leq K \leq m$ and retains in the compressed image $\tilde{I}$ all terms in (4.2) coresponding to $k < K$. We say that this is a linear algorithm because we can view the compressed image as an approximation to I from the linear space $\mathcal{S}^K = \mathcal{S}^K(\phi)$.

If the L_2-projector was used in the construction of the wavelet ψ, then the linear compression strategy is optimal with respect to compression in the L_2-metric in the sense that once we have decided to approximate the image I by functions from the linear space $\mathcal{S}^K$ then our strategy for choosing the approximation is optimal. Indeed, the portion of the wavelet decomposition (4.2) corresponding to the values of $k < K$ gives the L_2 projection of $\mathcal{I}$ on the space $\mathcal{S}^K$ and is therefore the best L_2 approximation from that space. It is also important to note that in linear methods the compression strategy (the terms that we retain) does not depend on the image.

In *nonlinear compression methods*, we allow the terms that we retain in the compressed image $\tilde{I}$ to depend on the image I. Thus, typically some but not necessarily all terms are chosen from each dyadic level. If n is the total number of terms to be retained, this can be thought of as approximation from the nonlinear manifold consisting of all functions g which are a linear combination of at most n of the functions appearing on the right side of (4.2).

There is one case where it is particularly simple to describe the optimal nonlinear compression strategy, namely, if ϕ and ψ have orthogonal shifts and the compression is done in the metric L_2. If we fix the number n of terms we wish to retain in the compressed image (which obviously is related to the rate of compression), then the optimal strategy is to retain the n terms in (4.2) for which

$$\|c_{j,k}\psi_{j,k}\|_{L_2} = |2^{-k}c_{j,k}|\|\psi\|_{L_2}$$

is largest (with any criteria used to break ties).

In numerical implementation, to avoid the time expensive sorting, the following modification of this strategy is implemented. We fix a positive number $\epsilon > 0$ which will measure the quality of the compressed image. We retain in the compressed image $\tilde{I}$ all terms of (4.2) for which

$$|2^{-k} c_{j,k}| > \epsilon. \tag{5.2.1}$$

The selection (5.2.1) is called *thresholding*. The smaller the value of ϵ, the more terms that are retained in the compressed image and the smaller the compression error.

A disadvantage of thresholding is that one does not know in advance the relationship between the choice of ϵ and the compression ratio. One possible way to circumvent this difficulty is to adaptively choose ϵ. One begins with a large value of ϵ so that no coefficients satisfy the threshold criteria (5.2.1). Given in hand any value of ϵ, we can replace ϵ by $\epsilon/2$ and add to our compressed image the additional terms that satisfy (5.2.1). We proceed until we have the desired quality in the compressed image or the desired compression ratio. The buildup of the compressed image in this way is progressive since we only add to the previous compressed image and do not have to redo any previous calculations. A discussion of progressive transmission for nonlinear based wavelet compression and their encoding can be found in Shapiro [12].

There is another important variant in the nonlinear strategy. Rather than using the entire coefficient $c_{j,k}$ in the terms that are retained, we replace it by an approximation $\bar{c}_{j,k}$ which satisfies

$$2^{-k} |c_{j,k} - \bar{c}_{j,k}| \le \epsilon. \tag{5.2.2}$$

This is called *quantization* and is implemented numerically by retaining only the leading bits of the coefficient $c_{j,k}$. The effect of the L_2-quantization strategy (5.2.2) is to retain one less bit in the coefficients as we move from one dyadic level to a finer dyadic level.

If in place of the L_2-metric we use the L_p-metric, $p \neq 2$, or if we use non-orthogonal wavelets, then it is not so clear what should be the strategy for thresholding or quantization. The following strategy was derived in [5] and shown to be optimal in a sense that will be described in the next subsection. If we consider the compression problem as retaining the best n terms of the wavelet decomposition of the image so as to minimize L_p compression error then the strategy put forth in [5] is to choose the n terms for which

$$\|c_{j,k} \psi_{j,k}\|_p = |2^{-2k/p} c_{j,k}| \|\psi\|_{L_2} \tag{5.2.3}$$

are largest. Again this leads to corresponding thresholding and quantization strategies obtained from (5.2.1) and (5.2.2) by replacing 2^{-k} by $2^{-2k/p}$. For example, the quantization strategy when using the L_1 error metric is to choose two less bits as we move from one dyadic level to the next finer dyadic level.

5.3. Smoothness of images

A common way of making comparisons between compression algorithms is to test their performance on certain standardized images. This obviously has its limitations. Another possibility for comparing performance of compression algorithms is to try to classify images according to their compressibility. Compression is a form of approximation and we know from such questions in approximation theory that the rate of approximation of a function is connected with its smoothness. This leads us to consider the classification of images by their membership in smoothness spaces.

While there are many smoothness spaces, we shall limit our discussion to the Besov spaces. They are robust enough to measure all orders of smoothness in all L_p spaces. As we shall see, there is room for discussion as to whether these are appropriate spaces for measuring the smoothness of images.

For any $0 < s \leq \infty$ and $\alpha > 0$, we let $B_q^\alpha(L_s)$ denote the *Besov space* of smoothness order α in L_s with secondary parameter q. We do not give the definition of these spaces here (see e.g. [5] for a definition). One can heuristically view the space $B_q^\alpha(L_s)$ as consisting of the functions f from L_s whose α-th derivative is in L_s (the parameter q gives a more subtle delineation of smoothness). However, note that we allow α to be noninteger and s to be less than one.

We can classify the smoothness of an image by its membership in the Besov spaces. Since Besov space norms are applied to functions, we need to view the image as a function. One possibility is to consider the image as the underlying function F (introduced earlier) whose cell averages gave the pixel values. It is also reasonable to consider either the pixel representation (piecewise constant function) or the wavelet representation as being the function representing the image. We adopt the latter viewpoint in this section. One could argue that the wavelet function representation is always smooth (because it consists of a finite sum of smooth functions $\psi_{j,k}$) and that the image would be in all Besov spaces within the smoothness of the wavelet. But correctly viewed, the wavelet representation is just a partial sum of some larger series and this larger series has some intrinsic smoothness. For the purposes of the following discussion, one should adopt this latter viewpoint.

The value of measuring smoothness of images is that it gives us a way of evaluating the performance of compression algorithms. For example, if the dual functionals for the wavelet basis have r vanishing moments, then standard results in approximation theory (see e.g. [8]) say that the linear method of compression will approximate a bivariate function f to an order $O(n^{-\alpha/2})$ (with n the number of terms retained and $\alpha < r$) in the L_p-metric if and only if f is in the Besov space $B_\infty^\alpha(L_p)$. Hence, we can expect this performance by the compression algorithm only if the image has this smoothness.

A similar result holds for the nonlinear approximation which gave rise to the nonlinear compression algorithm (see [5]). In this case, there is a characterization of the bivariate functions f which have approximation order like $O(n^{-\alpha/2})$ for $\alpha < r$. The function f should be in the Besov space $B_\tau^\alpha(L_\tau)$

with $\tau = (\alpha/2 + 1/p)^{-1}$.

It is interesting to compare the linear and the nonlinear algorithms from this vantage point. To obtain an L_p-error $O(n^{-\alpha/2})$ (with n the number of terms retained), both require smoothness of order α but the linear algorithm measures this smoothness in the space L_p (the same space as the compression error is measured) while the nonlinear algorithm measures the smoothness in L_τ with $\tau = (\alpha/2 + 1/p)^{-1}$. Note that $\tau < p$ and it is more likely for a function to be in the Besov space for nonlinear approximation than in the one for linear approximation. Thus the nonlinear algorithm will compress more images to order $O(n^{-\alpha/2})$, i.e., require less smoothness for the image to be compressed with this accuracy.

For example, consider compression in the L_2 error metric. For compression error of order $O(n^{-\alpha/2})$, the image should have smoothness of order α as measured in L_2. Since images have edges and other discontinuities they cannot be in these Besov spaces for large values of α. For example, the characteristic function of a square is in the Besov space $B^\alpha_\infty(L_2)$ only if $\alpha \le 1/2$, and hence the linear compression algorithm will not give compression error $O(n^{-\alpha/2})$ for any $\alpha > 1/2$. However, in the case of the nonlinear algorithm, the smoothness is measured in L_τ. For example, this same characteristic function of a square is in the space $B^\alpha_\tau(L_\tau)$, $\tau = (\alpha/2 + 1/2)^{-1}$ provided $\alpha < 1/\tau$, or equivalently if $\alpha < 1$. In other words, images can have higher smoothness orders α when measured in the L_τ spaces and therefore can be compressed more efficiently by the nonlinear algorithms. A similar discussion applies to approximation in the metric L_p.

Choosing the coefficients for which (5.2.3) is largest provides a near optimal strategy for nonlinear approximation in the following sense. For functions in the unit ball of the Besov spaces $B^\alpha_\tau(L_\tau)$, this strategy provides an approximation error $O(n^{-\alpha/2})$. No other compression strategy based on approximation by a sum $\sum_{j,k,\psi\in\Lambda} b_{j,k}\psi_{j,k}$, with $|\Lambda| \le n$, can improve on this error. There are however the constants in the $O(n^{-\alpha/2})$ term about which we are not able to say anything. These constants are of importance in numerical comparisons and it may be that some other strategy could improve on (5.2.3) in these constants.

There is another sense in which the nonlinear strategy based on (5.2.3) is optimal. Given any stable nonlinear approximation scheme, not necessarily based on wavelets, it cannot improve on the error estimate $O(n^{-\alpha/2})$ of (5.2.3) (see [6]). Therefore, if we agree in advance that we want to compress all images from unit ball of the Besov space $B^\alpha_\tau(L_\tau)$) then no stable compression strategy can do better whether it is based on wavelets or some other ideas such as fractals or Fourier expansions.

The above strategy also gives a method for numerically computing the smoothness of an image. One computes the errors from the nonlinear approximation strategy (5.2.3), and graphs them on a log-log scale (error versus number of coefficients). This graph will look linear for a certain range of n (typically for $n \le 30,000$ in the case of 512×512 images). The slope of

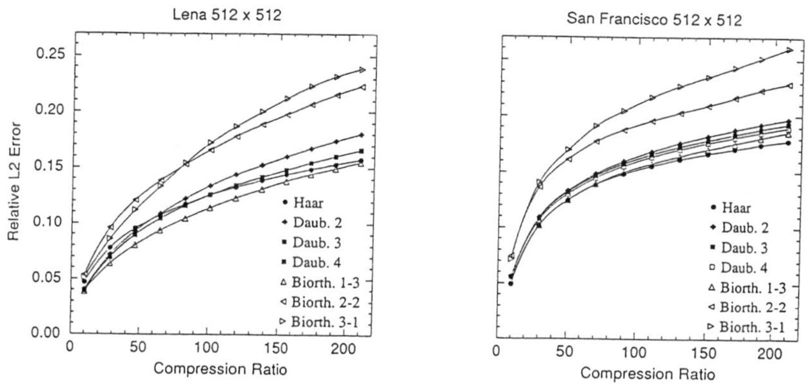

Fig. 2. L_2-error versus compression.

this line gives the largest α for which the image is in $B^\alpha_\tau(L_\tau)$. We refer the reader to [4] for a more detailed discussion of the numerical determination of smoothness of images.

Further discussion of linear and nonlinear compression algorithms can be found in [4]. They show the overall superiority of the nonlinear algorithms. For the remainder of this paper, we shall limit our discussion to nonlinear compression.

5.4. Choice of wavelet: fixed basis

There is the interesting question of whether any particular wavelet is preferable over another in compression algorithms. We have analyzed several wavelets for this purpose. In order to isolate the question of choice of wavelet basis, we have made two restrictions in the analysis that follows. First, in order to not have the compression strategy play a role in this analysis, we have chosen to measure the compression error in the L_2-metric. In this case, when using orthogonal wavelets, we know the optimal compression strategy for thresholding or quantization (we use only thresholding in the analysis that follows). Secondly, our analysis considers compression error (in the L_2-metric) versus the number of wavelet coefficients. A more accurate analysis of the performance of a compression algorithm would study the compression error versus the number of bits needed to encode the compressed image. However the latter would depend on the encoding of the wavelet coefficients (discussed in the last section). The best encoding could very well depend on the wavelet that is utilized in the compression algorithm. To eliminate this dependence on encoding we have simply counted coefficients.

We have tested several images with quite similar results. Figure 2 gives a graphical presentation of our results for two particular images (Lena and San Francisco) that were compressed using Haar, Daubechies ($\mathcal{D}_2$, $\mathcal{D}_3$, $\mathcal{D}_4$) and biorthogonal ((1,3), (2,2), and (3,1)) bivariate bases (3.1). Here are some conclusions that can be drawn from these results.

There seems to be little evidence that smoother wavelets perform better in compression. For example, at high rates of compression, Haar performed better than the smoother Daubechies wavelets. At lower rates of compression, the smoother Daubechies wavelets $\mathcal{D}_r$, $r = 3, 4$, perform slightly worse than $\mathcal{D}_2$. The biorthogonal $(1, 3)$ wavelets which are piecewise constant perform better in all ranges than any of the other wavelet bases.

It seems that a more important issue is the playoff between the support of the coefficients functionals and the number of its vanishing moments. If the coefficient functional has large support then the dominant discontinuities in the image effect many wavelet coefficients. In low compression, the large coefficients from the dominant discontinuities are small in number compared to the number of terms retained and the deciding factor in the compression rate are the coefficients corresponding to smoother parts of the image. These coefficients are smaller because the coefficient functionals have many vanishing moments. For high compression however, the coefficients corresponding to the dominant discontinuities are the deciding factor.

Overall, the biorthogonal spline based wavelets performed better than the Daubechies wavelets. An explanation put forward for this by others is the symmetry in the biorthogonal wavelets. We should also reiterate the advantage of certain of the biorthogonal wavelets in that the computation of the wavelet decomposition can be done in integer arithmetic with shifts.

§6. Hyperbolic Bases

The use of hyperbolic bases in image compression brings forward new questions as to how compression algorithms should be implemented with these wavelets. As with the standard wavelet basis, one can divide the compression question into two parts corresponding to linear methods (approximation from linear spaces) and nonlinear methods. There is a theory (see [7]) for linear methods of compression based on hyperbolic wavelets that is similar in spirit to that given in §5.2. We do not want to discuss in detail these results but mention only a couple highlights. This theory introduces new moduli of smoothness based on certain mixed differences and defines new smoothness spaces in a similar way to the Besov spaces but using the new moduli of smoothness in place of the usual moduli of smoothness. It is shown then that these new smoothness spaces replace the roles of the Besov spaces when characterizing rates of approximation by the linear spaces H^n.

Unfortunately, there is not yet a corresponding theory for the nonlinear compression problem using hyperbolic wavelets except for special case when the error is measured in L_2. One of the reasons for this is that there is no characterization of the approximation spaces for nonlinear approximation using hyperbolic wavelets. This however does not prevent the experimentation

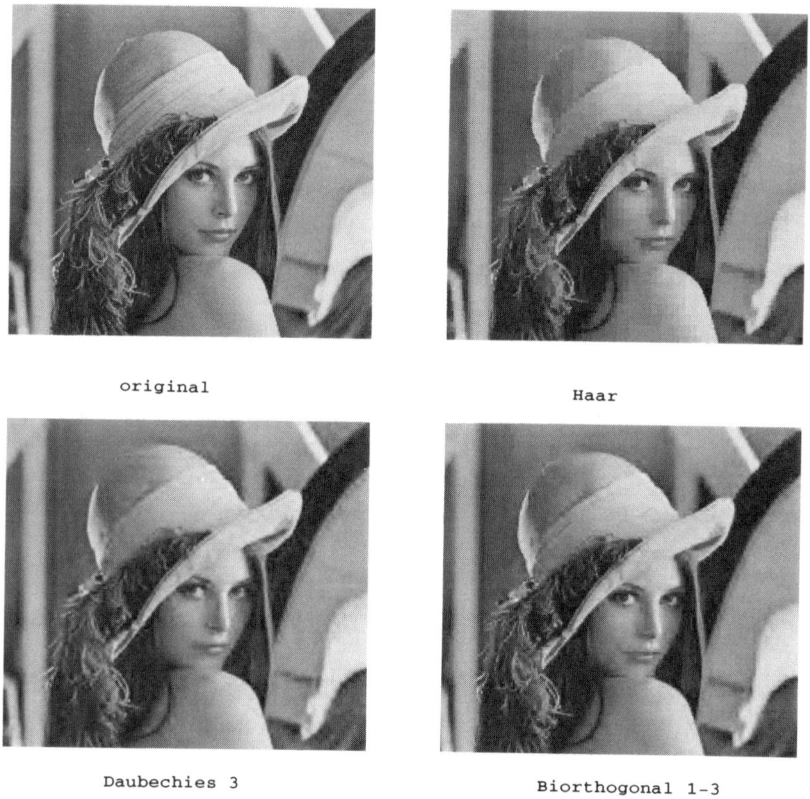

Fig. 3. 60-1 Compression.

with hyperbolic wavelet in image compression. The following is an initial report on some of our experience with hyperbolic wavelets.

We first discuss the representation of digitized images by hyperbolic wavelet bases. In order to simplify the following discussion by not having to deal with extensions near the boundary of the unit square (as discussed in §4), we shall limit our discussion to the case of the hyperbolic Haar basis. Using ideas similar to that presented in §4, we can derive expansions of images using other hyperbolic basis.

For this discussion, we let $\phi = \chi_{[0,1]}$ and let ψ be the univariate Haar function. We start with the representation (4.1) of the image $\mathcal{I}$ as

$$\mathcal{I} \sim \sum_{j \in \mathbb{Z}^2} c_j \varphi_{j,m}(x, y). \tag{6.1}$$

with $\varphi(x, y) := \phi(x)\phi(y)$ and c_j defined to be the given pixel values p_j in case the support of $\varphi_{j,m}$ is contained in Ω and zero otherwise. The univariate functions $\phi_{s,m}$, $0 \le s < 2^m$, are a linear combination of $\phi_{0,0}$ and the univariate Haar functions ψ_{j_1,k_1}, $0 \le j_1 < 2^{k_1}$. Making this change of basis gives the hyperbolic Haar representation

$$\mathcal{I} \sim \sum_{j,k \in \mathbb{Z}^2} c_{j,k} \eta_{j,k}(x,y) \tag{6.2}$$

with $j = (j_1, j_2)$ and $k = (k_1, k_2)$ and

$$\eta_{j,k}(x,y) := \psi_{j_1,k_1}(x)\psi_{j_2,k_2}(y)$$

with the convention that $\psi_{0,0}$ is replaced by $\phi_{0,0}$ and $c_{j,k} = 0$ unless the support of $\eta_{j,k}$ is contained in Ω. As is discussed in the following section, the coefficients in the representation (6.2) can be found from the univariate Fast Wavelet Transform.

Given the hyperbolic wavelet representation (6.2) (or its analogue in the non-Haar case) of an image, we can implement compression strategies to threshold or quantize the coefficients $c_{j,k}$. We have implemented the analogue of the nonlinear L_2 compression strategy (5.2.1) for various hyperbolic wavelets. Figure 4 gives a graphical representation of the results of our analysis of two hyperbolic bases (Haar and Daubechies $\mathcal{D}_3$) versus the usual wavelet bases. We again display L_2-compression error versus the number of coefficients using only thresholding. The results are representative of our usual experience that for some images (such as San Francisco), the hyperbolic wavelet bases performed slightly better than the usual wavelet basis while on other (for example Lena) the opposite is true. It seems therefore that the choice whether to use the standard wavelet basis or the hyperbolic basis in compression depends on the image to be compressed.

§7. Adaptive Bases for Image Compression

The advantages of hyperbolic bases in compression led us to consider the possibility of adaptively chosing a wavelet basis to depend on the image being compressed. To discuss this approach, we suppose that ϕ is a univariate function satisfying the refinement equation. Suppose further that ϕ has orthogonal shifts and that ψ is a univariate orthogonal wavelet derived from ϕ by multiresolution. We have seen two bivariate orthogonal bases which can be derived from ϕ and ψ; namely the usual wavelet bivariate basis given by the shifted dilates of the three functions in (3.1) and secondly the hyperbolic basis given in (3.3). We can consider these as special cases of more general orthogonal bases constructed from ϕ and ψ.

To describe these general orthogonal bases, it is convenient to introduce the following notation. If $I = [j2^{-k}, (j+1)2^{-k}]$ is a dyadic interval, we let $\phi_I(x) := |I|^{-1/2}\phi(2^k x - j)$ denote the L_2 normalized shifted dilate of ϕ corresponding to the interval I. We use a similar notation for ψ. The interval I is associated to the support of ϕ_I and ψ_I. In fact, in the special case that

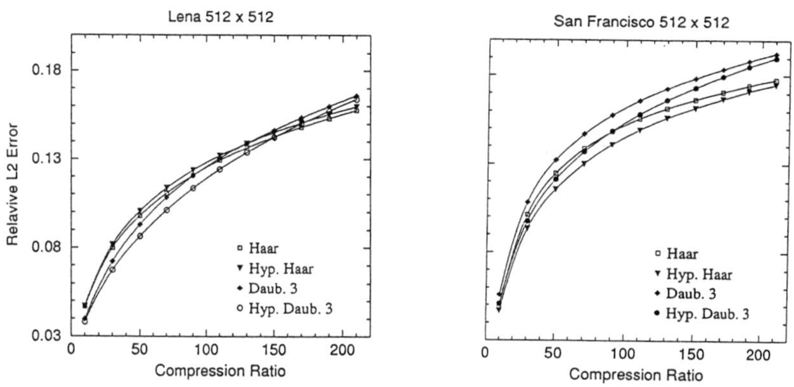

Fig. 4. L_2 error versus compression for hyperbolic and regular bases.

$\phi = \chi_{[0,1]}$ and ψ is the Haar function these are exactly the supports of ϕ_I and ψ_I.

We consider the totality of function

$$\phi_I(x)\phi_J(y), \quad \phi_I(x)\psi_J(y), \quad \psi_I(x)\phi_J(y), \quad \psi_I(x)\psi_J(y), \tag{7.1}$$

where I and J are dyadic intervals. The two wavelet bases already mentioned are only special cases of orthogonal bases for $L_2(\Omega)$, $\Omega := [0,1]^2$, that can be selected from the totality of functions in (7.1). We mention, in the special case of the Haar wavelets, some other orthogonal basis useful in partitioning (segmenting) the image. Starting with any finite partition $\mathcal{R}$ of the unit square into dyadic rectangles $R = I \times J$, then the functions $\phi_I(x)\phi_J(y)$ can be completed into an orthogonal bases for $L_2(\Omega)$ by adjoining a regular or hyperbolic Haar bases for each of these rectangles.

Given an image $\mathcal{I}$, we would like to select an orthogonal basis from the functions (7.1) which is particularly good for compression or some other image processing application. To keep the discussion simple, we shall only discuss here the case where $\phi = \chi_{[0,1]}$ and ψ is the Haar wavelet. Similar ideas can be applied to other pairs ϕ, ψ.

Suppose that $\mathcal{I}$ is a given image of dimension $2^m \times 2^m$. Let $\mathcal{D}_k$ denote the collection of dyadic subintervals of $[0,1]$ of sidelength 2^{-k} and let $\mathcal{D} = \cup_{k=0}^{m} \mathcal{D}_k$. We begin with the representation

$$\mathcal{I} \sim \sum_{I,J \in \mathcal{D}_m} p_{I,J}\phi_I(x)\phi_J(y) := f_{\mathcal{I}}, \tag{7.2}$$

where $p_{I,J}$ is the pixel value associated with the square $I \times J$. This is our usual representation (4.1) of the image. The function $f_{\mathcal{I}}$ is the piecewise constant function which takes the corresponding pixel values on the squares $I \times J$.

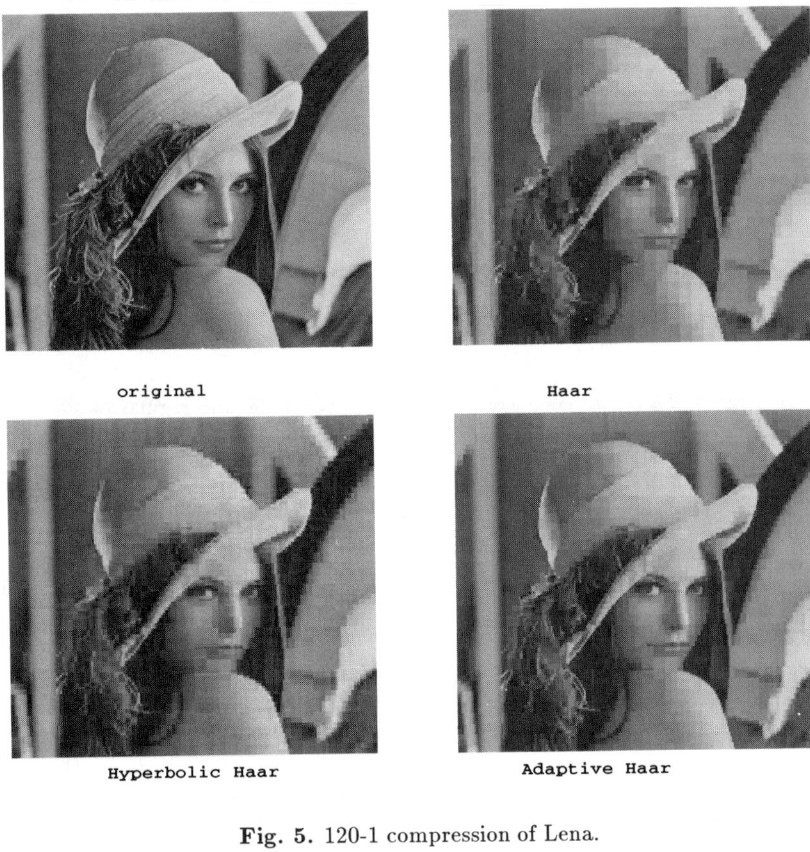

Fig. 5. 120-1 compression of Lena.

We first want to observe that we can efficiently compute the inner products of $f_{\mathcal{I}}$ with respect to all of the functions of (7.1) when $I, J \in \mathcal{D}$. Let r_j denote the univariate piecewise constant function

$$r_j(x) = \sum_{I \in \mathcal{D}_m} p_{I,J} \phi_I(x),$$

where $J = [j2^{-m}, (j+1)2^{-m}]$. Then r_j corresponds to the j-th row of pixel values. We can use the fast one dimensional Haar Transform to calculate the inner products

$$\int_{\mathbb{R}} r_j \phi_I \, dx, \quad \int_{\mathbb{R}} r_j \psi_I \, dx, \quad I \in \mathcal{D}. \tag{7.3}$$

Note that for each fixed j, the fast one dimensional Haar Transform while devised to calculate the integrals in (7.3) with respect to the functions ψ_I, $I \in \mathcal{D}$, computes in the process the integrals with respect to the ϕ_I as well. We replace the j-th row of pixel values by the integrals (7.3). We then repeat the application of the fast one dimensional Haar transform to the column vectors of the resulting matrix. This results in the computation of the inner products

$$\begin{aligned} &\int_{\mathbb{R}^2} f_{\mathcal{I}} \phi_I(x) \phi_J(y)\, dx\, dy, \quad \int_{\mathbb{R}^2} f_{\mathcal{I}} \phi_I(x) \psi_J(y)\, dx\, dy, \\ &\int_{\mathbb{R}^2} f_{\mathcal{I}} \psi_I(x) \phi_J(y)\, dx\, dy, \quad \int_{\mathbb{R}^2} f_{\mathcal{I}} \psi_I(x) \psi_J(y)\, dx\, dy. \end{aligned} \tag{7.4}$$

The number of computations needed to compute the inner products in (7.4) is thus $O(N)$ with N the number of pixel values and all computations can be done in integer arithmetic with shifts.

Suppose now that we wish to find a best orthogonal basis for the compression of a given image from among the functions (7.1). When the compression error is measured in the L_2 metric this is particularly simple to describe. We fix the number of coefficients n that we shall retain in the compressed image. The best orthogonal basis (which depends on the image) is the set of n functions taken from (7.1) which are mutually orthogonal and which have the largest sum of squares for its coefficients (7.4).

While it is a simple matter to describe the desired best basis, it is by no means simple to compute it. There are two difficulties. First of all it is not easy to decide which collections of n functions taken from (7.1) produce mutually orthogonal functions. Secondly the number of such bases is enormous and would challenge even the largest of computers when n is large. Thus, we desire numerically efficient algorithms which can select a "good" (in place of best) basis (for the purposes of compression).

A naive algorithm is the following. We order the coefficients (7.4) according to size (with ties broken in an arbitrary way). We select the function (7.1) corresponding to the largest coefficient and place it in our basis. We then consider the next largest coefficient. We check if its corresponding function is orthogonal to the function already in our basis. If so, we place it in our basis. Otherwise, we proceed to the next largest coefficient and continue in this manner. We stop when we have chosen n functions.

This naive algorithm does not give the best basis. In fact, it is not clear that one can prove anything quantitative about this basis selection. On the other hand, this basis performs well in image compression. For example, Figures 5 and 6 compares the compression of two standard images with the adaptive Haar basis with the fixed regular and hyperbolic Haar bases. A computation of compression error versus the number of coefficients retained given in Figure 7 shows about a 2-1 gain in compression with the adaptive basis selection versus fixed bases. That is, a given L_2-error can be achieved in the adaptive bases selection with about one half the coefficients needed with a fixed basis.

Fig. 6. 120-1 compression of San Francisco.

The naive algorithm is still not numerically efficient. Without modification, the checking of orthogonality which takes place at each step would require $O(j)$ operations with j the number of elements already in the basis. Hence, the total number of operations to construct this basis would be of order $O(N+n^2)$. Fortunately, there is a lot of structure in the basis functions (7.1) which allow for faster checks of orthogonality. For example, if an element $\phi_I\psi_J$ is a member of an orthogonal basis, then no function $\phi_{I'}\psi_J$, with $I \cap I' \neq \emptyset$ can also be a member. Similar ideas exclude many other functions. It is possible to utilize this structure to eliminate many of the remaining functions (7.1) whenever a basis element has been selected. A convenient data structure and fast algorithms for this purpose will be reported on elsewhere. These algorithms construct the same "good" basis with just $O(N + n \log N)$ operations.

We also note that to avoid sorting of coefficients, we could modify the algorithms to be based on thresholding (or quantization).

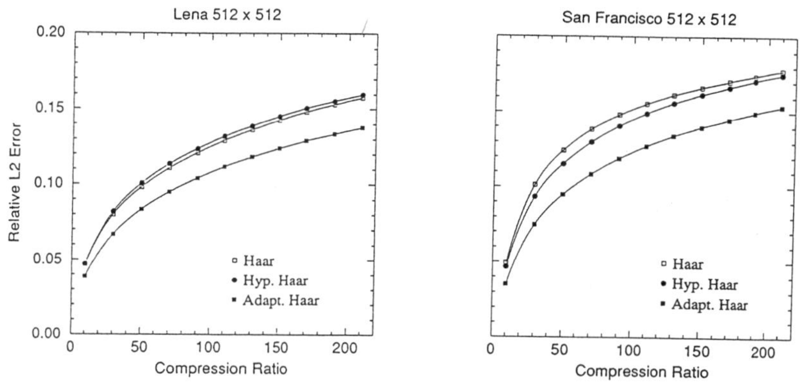

Fig. 7. L_2 error versus compression.

§8. Entropy Encoding of Coefficients

Thus far, we have based our analysis of the performance of a given compression algorithm on the number of coefficients that need to be retained to achieve a given error tolerance. The number of coefficients is not a completely fair assessment. What really counts is the number of bits that must be transmitted in order to reconstruct the compressed image.

If the computation of wavelet coefficients can be done in integer arithmetic with shifts, then each coefficient will have a maximum number of bits (for example, in the case of regular Haar, 10 bits per coefficient suffice for 512×512 images; see the discussion in [4]). If computations are done in real arithmetic, the coefficients will have to be truncated with a truncation error commensurate with the desired compression error (as was already discussed for quantization).

In the case of linear compression algorithms with a fixed wavelet basis, we need only transmit the coefficients in some fixed order, for example, from coarsest level to finest level. An entropy encoder is usually applied to the file of compressed coefficients before transmission. The entropy encoding will add to the compression by a small factor (less than 2 to 1).

In the case of nonlinear compression algorithms, it is not enough to send the coefficients since we must also identify their position. We wish to minimize the number of bits needed to identify this position. Our current algorithms for nonlinear compression methods accomplish this in the following way. We organize all coefficients of the compressed image (even those that are zero) in a file going from the finest to the coarses dyadic levels. Most entries in this file are zero. We then apply an adaptive arithmetic encoder (in our case a modification of Q-coder developed at IBM) to further compress the file. Arithmetic encoders visualize the string of 0's and 1's to be transmitted as a real number and utilize the probability of occurence of the transmitted symbols (in our case 0 and 1) to devise a new arithmetic representation of this number. When the arithmetic encoder is applied at the back end of a Haar based nonlinear compression algorithm utilizing quantization, the result is roughly six bits per nonzero coefficient.

To be fair then, we should return to our comparison of linear versus nonlinear algorithms and analyze their performance based on the number of bits transmitted. This has been done for example with the Haar wavelets with the result that the nonlinear algorithm is still the clear winner (see the discussion in [4]).

However, all of this analysis is dependent on the entropy encoder that is utilized in conjunction with the wavelet compression algorithm. This adds another level of variability into the decision of which wavelets perform the best for compression. This is why in the present paper, we have chosen to simply count wavelet coefficients. A more correct treatment would analyze pairs consisting of a wavelet compression algorithms followed by an entropy encoder. A good encoder to be used in such a pair should take advantage of the data structure of the wavelet coefficients (multiscale and spatial correlation). There are some encoders of this type. For example, Shapiro [12] uses zero quadtrees to design an encoder for nonlinear wavelet based compression. We should also mention that run length encoding on the zeros followed by a Huffman encoder performed well with biorthogonal wavelets in the recent FBI competition (see the recent article in SIAM news).

New problems arise when applying the adaptive wavelet basis algorithms. In this case the total number of potential coefficients (corresponding to (7.4)) is $4N$ rather than N. Thus, if we again organize all compressed coefficients (zero and otherwise) corresponding to (7.4) into a file than the size of this file will be 4 times larger than in the fixed basis case. Even though the number of nonzero entries will be less than in the fixed basis case (reflecting the better performance of the adaptive basis selection), the encoded file may be larger.

Fortunately, the adaptive basis selection has more structure than is being utilized. As we have remarked earlier, when a certain basis element is chosen in the basis selection, it eliminates many other basis functions from being candidates for future addition to the basis. This means that if a certain coefficient in the matrix of compressed coefficients is nonzero, then many other coefficients are forced to be zero. We are currently investigating how to utilize this information to customize encoders for the adaptive basis compression algorithms.

Acknowledgements. This research was supported by an ONR contract N0014-91-51343 and a DOE contract DE-AC05-84OR21400 Martin Marietta subcontract 19X-SK965C

References

1. Cohen, A., I. Daubechies, and J.C. Feauveau, Biorthogonal bases of compactly supported wavelets,Comm. Pure and Applied Math., **XLV** (1992), 485–560.
2. Cohen, A., I. Daubechies, and P. Vial, Wavelets and fast transform on the interval, preprint, Bell Labs., 1992.
3. Daubechies, I., *Ten Lectures on Wavelets*, CBMS-NSF Regional Copnference Series in Applied Mathematics, vol. 61, SIAM, 1992.
4. DeVore R., B. Jawerth, and B. Lucier, Image compression through wavelet transform coding, IEEE Transaction on Information Theory **38** (1992), 719–747.
5. DeVore, R., B. Jawerth, and V. Popov, Compression of wavelet decompositions, Amer. Jour. Math., **114** (1992), 737–785.
6. DeVore, R., G. Kyriazis, D. Leviatan, and V.M. Tikhomirov, Wavelet compression and nonlinear n-widths, Advances Comp. Math., **1** (1993), 197–214.
7. DeVore, R., S. Konjagin, P. Petrushev, V. Temlyakov, Approximation with hyperbolic wavelets, preprint.
8. DeVore, R., and B. Lucier, Wavelets, Acta Numerica **1** (1992), 1–61.
9. DeVore, R., abd G.G. Lorentz, *Constructive Approximation*, Springer Grundlehren, vol. 303, New York, 1993.
10. DeVore, R., and B. Lucier, Fast wavelet techniques for near-optimal image processing, in *1992 IEEE Military Communications Conference*, IEEE, New York, N.Y., 1129–1135.
11. Vetterli, M. and C. Herley, Wavelets and filter banks: Theory and design, IEEE Trans. ASSP **40** (1992), 2207–2232.
12. J. M. Shapiro, An embedded hierarchial image coder using zerotrees of wavelet coefficients, in *Data Compression Conference*, J.A. Storer and M. Cohn (eds.), IEEE Computer Society Press, Los Alamitos, CA, 1993, 214–223.

Ronald A. DeVore
Department of Mathematics
University of South Carolina
Columbia, SC 29208
devore @math.scarolina.edu

A Quasi-Interpolant Box-Spline Formulation for Image Compression and Reconstruction

Harvey Diamond, Louise Arakelian Raphael
and Daniel A. Williams

Abstract. An alternative approach to data compression and reconstruction using multivariate-box splines and "approximate" orthogonal projections is presented. An intensity function of the image is represented by a finite sequence of quasi-interpolant approximations using multivariate box-splines at successively halved resolutions. This is a local method and contrasts with spline and wavelet methods which employ exact orthogonal projections.

§1. Introduction

The use of cubic splines in image processing is well established [6, 16]. Recently applications of wavelets and spline-wavelets to data compression and image reconstruction [7, 8, 9, 13, 14] are receiving much attention. We present an alternative multivariate box-spline and quasi-interpolation approach to this problem. Our algorithm is computationally recursive and is as simple as comparable wavelet decomposition/reconstruction schemes. It is a local method which is well suited for approximating smooth functions.

An elegant spline algorithm for continuous image representation and interpolation is developed in [16]. Using direct and indirect B-spline transforms, their method efficiently transforms signals into a form which can be scaled and then reconstructed with a smoothing effect which results in negligible image resolution loss. Their results were derived initially for one-dimensional signals, but extended to higher dimensions via tensor product polynomial splines.

Our approach differs from [16] in that we apply multivariate box-splines [1] in a quasi-interpolation setting. Our method generalizes the one dimensional curve smoothing box-spline algorithm of Diamond [10]. An advantage of this formulation is that it is computationally simpler. Ultimately we hope our method will serve as a comparison base for image compression and reconstruction as implemented by the wavelets community.

Wavelets, Images and Surface Fitting
P. J. Laurent, A. Le Méhauté, and L. L. Schumaker (eds.), pp. 221–228.

ISBN 1-56881-040-7.

Our algorithm begins by first calculating a sequence of approximations to the intensity function of the images using multivariate box-splines at successively halved resolutions. The key is that these approximations are quasi-interpolants, which are nearly orthogonal projections. This fact ensures the local dependence of the image function approximation. This contrasts with alternative spline and wavelet methods which employ exact orthogonal projections.

As is customary in multiresolution decomposition schemes [2, 3, 7, 8, 9, 13, 16], the original image is decomposed as the sum of the approximation at the coarsest resolution level plus corrections at a succession of finer resolution levels. This sum, up to the finest approximation, yields the exact image. At each level, the image is represented as a sum of multivariate box-splines dilated by a factor of two from the preceding level. Each level is processed so as to reduce the number of non-zero coefficients. The processed data lends itself to further processing using standard lossless compression techniques.

In Section 2 we review the definitions and some properties of multivariate box-splines. In Section 3 we present the algorithm, and in Section 4 we show reconstructed pictures at different threshold levels.

§2. Definitions and Properties

We briefly summarize only those definitions and properties of box-splines which are relevant to our algorithm in the following section. There is an extensive literature on box-splines; see for example the review paper by Höllig [12].

We start with an arbitrary set $\mathcal{V} = \{\mathbf{v}_1, \ldots, \mathbf{v}_k \ : \mathbf{v}_i \in \mathbb{Z}^s\}$ of k vectors in $\mathbb{R}^s$ with integer components. Elements of $\mathbb{R}^s$ or $\mathbb{Z}^s$ are denoted in boldface. In the specific case of $\mathbb{R}^2$ or $\mathbb{Z}^2$ we also employ ordered pair notation.

Definition 1. *The multivariate box-spline $B_{\mathcal{V}}$ associated with the set $\mathcal{V}$ is the probability density of the random linear combination*

$$\sum_{i=1}^{k} t_i \mathbf{v}_i, \qquad \{\mathbf{v}_i\}_{i=1}^{k} \subset \mathcal{V}$$

where the $\{t_i\}_{i=1}^{k}$ are independent random variables each uniformly distributed on $[0,1]$.

Since the probability density of a sum of independent random variables is the convolution of the density function of the summands, we have the following well known property.

$$B_{\mathcal{V}\cup\mathcal{W}}(\mathbf{x}) = \int_{\mathbf{R}^s} B_{\mathcal{V}}(\mathbf{y}) B_{\mathcal{W}}(\mathbf{x}-\mathbf{y}) d\mathbf{y} \tag{2.1}$$

In addition, there is a two scale relation which these splines satisfy. There exists a finite sequence $\{p_{\mathbf{j}}\}$ such that

$$B(\mathbf{x}) = \sum p_{\mathbf{j}} B(2\mathbf{x} - \mathbf{j}). \tag{2.2}$$

§3. Decomposition

We work with the spline space

$$\mathcal{S} = \{\sum_{\mathbf{j}\in\Gamma} c_{\mathbf{j}} B_{\mathcal{V}}(\cdot - \mathbf{j}) : \Gamma \subset \mathbb{Z}^s \ \text{ finite}\}$$

and the scaled spline spaces

$$\mathcal{S}_n = \{\sum_{\mathbf{j}\in\Gamma} c_{\mathbf{j}} B_{\mathcal{V}}(2^{-n} \cdot - \mathbf{j}) : \Gamma \subset \mathbb{Z}^s \ \text{ finite}\}, \quad n = 0, 1, \ldots.$$

Note that $\mathcal{S}_0 = \mathcal{S}$. These spaces satisfy

$$\mathcal{S}_0 \supset \mathcal{S}_1 \supset \cdots \supset \mathcal{S}_N$$

which follows from the two-scale relation (2.2). Given a function $f \in \mathcal{S}$, a sequence of approximations $\{s_n\}_1^N$ to f is constructed such that $s_n \in \mathcal{S}_n$. Then

$$f \equiv s_0 = s_N + (s_{N-1} - s_N) + \cdots + (s_0 - s_1),$$

where it is to be noted that $s_j - s_{j+1} \in \mathcal{S}_j$. Thus, defining

$$u_n = s_n - s_{n+1} = \sum a_{\mathbf{j}}^{(n)} B(2^{-n} \cdot - \mathbf{j}), \qquad n = 0, \ldots, N-1,$$

and $u_N = s_N$, we obtain

$$f = \sum_0^N u_n.$$

This provides a multiresolution representation of f since $u_n \in \mathcal{S}_n$. All information about f is contained in the coefficients $\{a_{\mathbf{j}}^{(n)}\}$. Compression is achieved by thresholding the values of the $\{a_{\mathbf{j}}^{(n)}\}$. The novelty of our approach lies in the technique for the construction of the approximants s_n. Rather than compute an exact orthogonal projection, as is the standard procedure in wavelet techniques, we compute only an *approximate* orthogonal projection s_n. Given a function g and a spline space $\mathcal{S}$, the orthogonal projection of g onto $\mathcal{S}$ satisfies

$$\langle g(\cdot), B(\cdot - \mathbf{k})\rangle = \langle \sum c_{\mathbf{j}} B(\cdot - \mathbf{j}), B(\cdot - \mathbf{k})\rangle.$$

With quasi-interpolation formulas, we are able to define a local relationship between the coefficients $c_{\mathbf{j}}$ and the inner products $\langle g(\cdot), B(\cdot - \mathbf{j})\rangle$ such that the spline sum $\sum c_{\mathbf{j}} B(\cdot - \mathbf{j})$ approximately solves the equation above and provides an approximate orthogonal projection. This local relationship will be used in our image compression application.

We examine details of the decomposition/compression algorithm in $\mathbb{R}^2$. For the general case, see [11]. We start with a photo represented as a raw data file of 256×256 pixel intensities where each pixel's value spans one byte. We associate with the array of pixel intensities $\{c_{i,j}^{(0)}\}$ a spline function f

$$f(x,y) = \sum_{(j,k)\in \mathbb{Z}^2} c_{j,k}^{(0)} B_{\mathcal{V}}(x-j, y-k),$$

where in this case $B_{\mathcal{V}}$ is chosen to be the specific cubic bivariate box-spline B_{4400}, a box-spline where $\mathcal{V}$ consists of four copies of the vector $(1,0)$ and four copies of the vector $(0,1)$.

As discussed above, typically in wavelet decompositions of this type one decomposes the f with respect to orthogonal projections onto a sequence of subspaces spanned by an orthogonal basis. As bivariate box-spline bases are not orthogonal, we approximately determine the coefficients for the orthogonal projection of f at each resolution level.

An approximate projection of the intensity function f onto the space $\mathcal{S}_n$ is

$$s_n(x,y) = \sum_{(j,k)\in \mathbb{Z}^2} c_{j,k}^{(n)} B_{\mathcal{V}}(2^{-n}x - j, 2^{-n}y - k), \tag{2.3}$$

where the coefficient $c_{j,k}^{(n)}$'s are approximated as follows. First we compute the scaled inner products $\gamma_{j,k}^{(n)}$:

$$\gamma_{j,k}^{(n)} = 4^{-n}\langle f(x,y), B_{\mathcal{V}}(2^{-n}x - i, 2^{-n}y - j)\rangle, \tag{2.4}$$

where the inner product of two functions f and g on $\mathbb{R}^2$ is defined by $\langle f, g\rangle = \int_{\mathbb{R}^2} f(x)g(x)dx$. The $\gamma_{j,k}^{(n)}$'s can be recursively calculated starting from $\gamma_{i,j}^{(0)}$. The equations which define the exact orthogonal projection are then

$$\gamma_{j,k}^{(n)} = \sum_{(l,m)\in \mathbb{Z}^2} c_{l,m}^{(n)} B_{2\mathcal{V}}(j-l, k-m). \tag{2.5}$$

Employing two iterations of the Jacobi method, the above linear system yields the approximate solution

$$c_{j,k}^{(n)} = 2\gamma_{j,k}^{(n)} - \sum_{(l,m)\in \mathbb{Z}^2} \gamma_{j-l,k-m}^{(n)} B_{2\mathcal{V}}(l,m). \tag{2.6}$$

Using these $c_{j,k}^{(n)}$ defines our quasi-interpolant s_n in (2.3). Using our two-scale relation for bivariate box-splines (2.2), we obtain

$$u_n(x,y) = s_n(x,y) - s_{n+1}(x,y) = \sum_{j,k} a_{j,k}^{(n)} B(2^{-n}x - j, 2^{-n}y - k),$$

where

$$a_{j,k}^{(n)} = c_{j,k}^{(n)} - \sum c_{l,m}^{(n+1)} p_{2l+j,2m+k}, \qquad n = 0 \ldots N-1, \tag{2.7}$$

is the sequence of arrays which represents the detail lost in representing the photo on a coarser level. Thus our decomposition of the original data $\{c_{j,k}^{(0)}\}$ is the sequence of arrays

$$\{\{a_{j,k}^{(0)}\},\ldots,\{a_{j,k}^{(N)}\}\} \qquad \text{where} \quad a_{i,j}^{(N)} = c_{i,j}^{(N)}.$$

Compression of the data consists of discarding all coefficients $a_{i,j}^{(n)}$ whose absolute values are less than some threshold value epsilon.

§4. Reconstruction

Reconstructing the image is tantamount to successively reconstructing the arrays

$$\{c_{i,j}^{(n)}\} \qquad n = N-1, N-2, \ldots, 0.$$

This is recursively done with the following sum:

$$c_{j,k}^{(n)} = a_{j,k}^{(n)} + \sum_{(l,m)\in\mathbb{Z}^2} c_{l,m}^{(n+1)} p_{2l+j,2m+k},$$

where the $p_{j,k}$'s are the coefficients in the two scale formula (2.2) which depend only on the choice of the original B-spline.

The test image used for this paper is of a magnetic resonance image (MRI) of a human head. MRI is a modern biomedical technology which is used to detect anomalies such as the growth of a tumor. These are reconstructions associated with different compression ratios. The raw data file, and our subsequent processed versions of the image, were rendered into Sun Raster format and displayed on a Sun workstation. The images shown are photographs taken of the screen.

The pictures are arranged as follows. Each corresponds to an intensity threshold value of epsilon used to discard coefficients, $a_{j,k}^{(n)}$. For comparison purposes we show the original in the upper left, corresponding to no coefficients discarded. Three reconstructions are shown, corresponding to different threshold values. The number of coefficients discarded for each image is indicated below the picture. In the bottom row, we show the image generated using only the finest level of coefficients, as well as the image which results when the finest level is subtracted from the original.

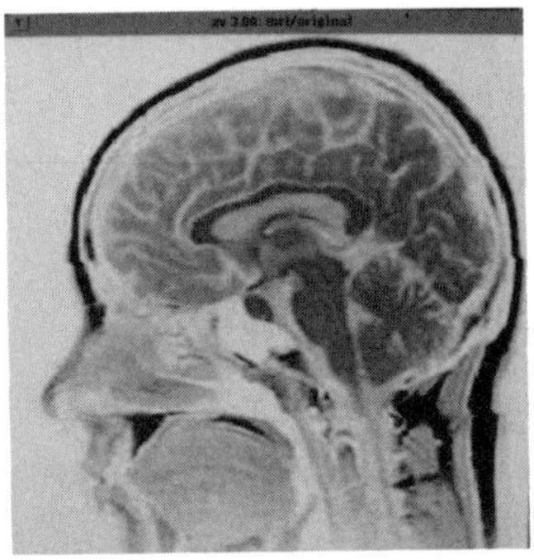

A. Original Image

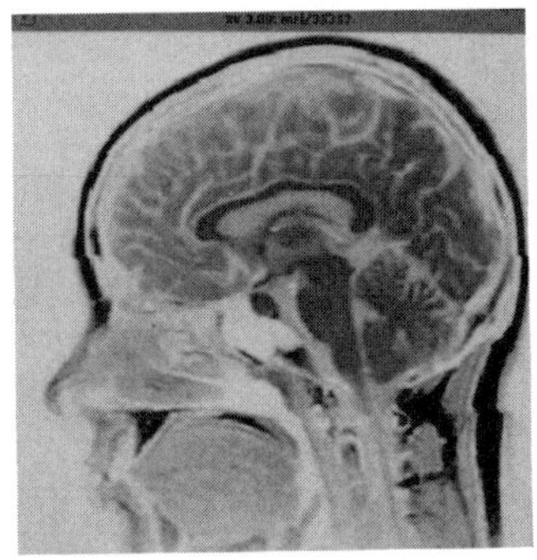

B. 38% coefficients deleted

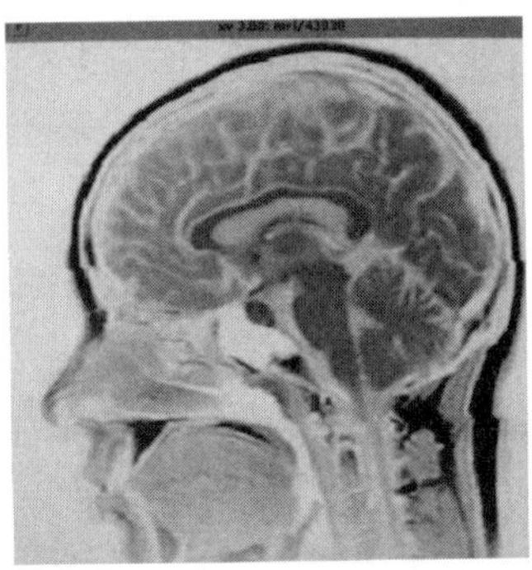

C. 47% coefficients deleted

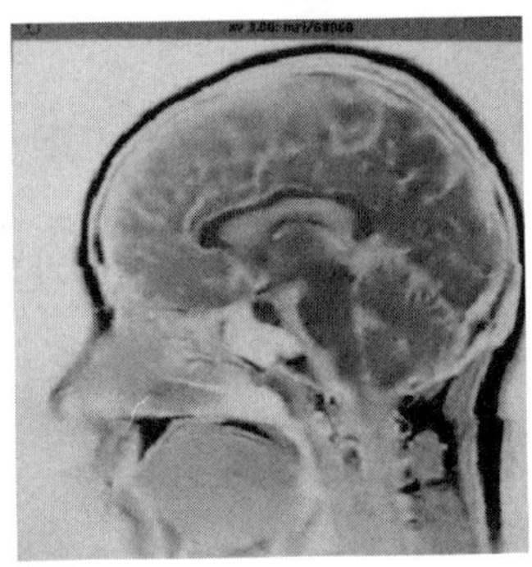

D. 73% coefficients deleted

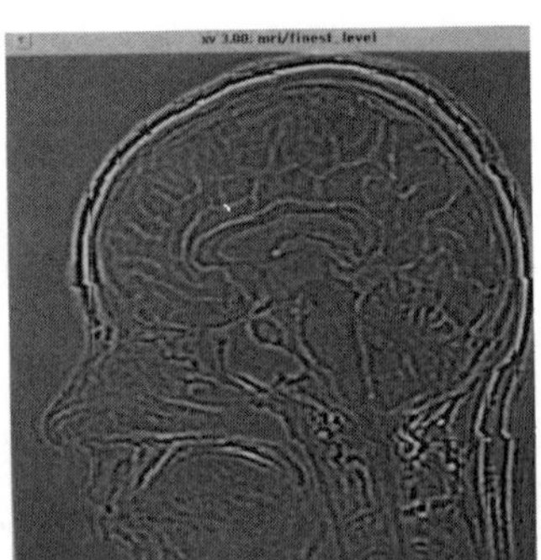

E. Finest Level (Amplified)

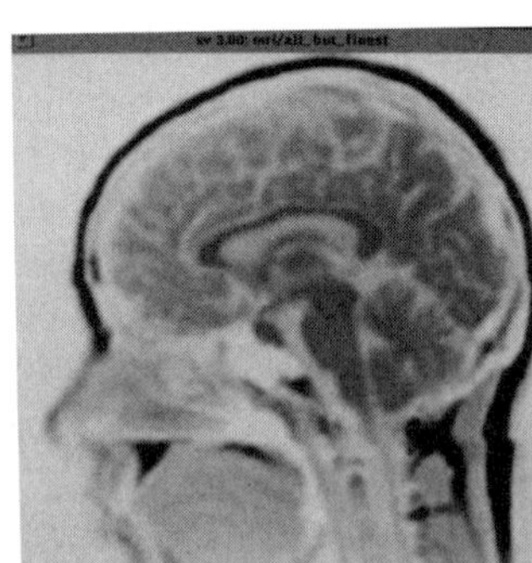

F. Image Minus Finest Level

§5. Conclusion

Our multivariate box-spline quasi-interpolation approximation is especially effective in regions where the pixel values vary smoothly. In regions where there are edges, our algorithm retains more coefficients and delineates the edges. This results in a good approximation of the original picture, and enables us to detect two-dimensional boundaries. Our algorithm efficiently compresses smoothly varying images. We are continuing our work to achieve better compression ratios.

Acknowledgements. This research was partially supported by Air Force Office of Scientific Research Contract 91NM062.

References

1. de Boor, C., and R. De Vore, Approximation by Smooth Multivariate Splines, Trans. Amer. Math. Soc. **289** (1983), 775–788.
2. Burt, P., and E. Adelson, The Laplacian Image Pyramid as a Compact Image Code, IEEE Trans. **31** (1983), 482–540.
3. Burt, P., and E. Adelson, A Multiresolution spline with Applications to Image Mosaics, ACM Trans. in Graphics **2** (1983), 217–236.
4. Chui, C. K., and H. Diamond, A Natural Formulation of Quasi-interpolation by Multivariate Splines, Proc. Amer. Math. Soc. **51** (1987), 643–646.
5. Chui, C. K., and H. Diamond, A Characterization of Multivariate Formulas and Its Applications, Numer. Math. **57** (1990), 105–121.
6. Daehlen, M., and T. Lyche, Decomposition of Splines in *Mathematical Methods in Computer Aided Geometric Design II*, T. Lyche and L. Schumaker (eds.), Acad. Press, Boston, 1992, 135–160.
7. De Vore, R., B. Jawarth, and B. Lucier, Image Compression Through Wavelet Transform Coding, IEEE Trans. on Information Theory **38** no. 2 (1993), 719–746.
8. De Vore, R., B. Jawarth, and B. Lucier, Surface Compression, Computer-Aided Geom. Design **9** no. 3 (1990), 219–239.
9. De Vore, R., B. Jawarth, and B. Lucier, Data Compression Using Wavelets, Error, Smoothness, and Quantization, Proceedings Data Compression **9**, no. 3, 1990.
10. Diamond, H., Multiresolution Representations in CAGD: Observations and Experiments, abstract, Jan '92 7th International Symposium on Approximation Theory.
11. Diamond, H., L. Raphael, and D. Williams, Compressed Representations of Curves and Images Using a Multiresolution Box-Spline Framework, Proceedings of International Conference on Advances of Computational Mathematics 1993, accepted.
12. Hollig, K., Box splines, in *Approximation Theory V*, C. K. Chui, L. L. Schumaker, and J. Ward, Eds., Academic Press, New York, 1986, 71–69.

13. Mallat, S., A Theory for Multiresolution Signal Decomposition: The Wavelet Representation, Trans. Pattern Analysis and Machine Intelligence **11** (1989), 674–692.
14. Mallat, S., Multiresolution Approximation and Wavelets, Trans. Amer. Math. Soc. **315** (1989), 69–88.
15. Strang, G., Wavelet Transforms versus Fourier Transforms, Bull. Amer. Math. Soc. **28** (1993), 288–305.
16. Unser, M., A. Aldroubi and M. Eden, Fast B-spline Transforms for Continuous Image Representation and Interpolation, IEEE Trans. Pattern Analysis and Machine Intelligence **13** (1991), 277–285.

Harvey Diamond
Mathematics Department
West Virginia University
Morgantown, WV 26506
diamond@ math.wvu.edu

Louise Arakelian Raphael
Mathematics Department
Howard University
Washington, D.C. 20059
alama38@ humain.howard.edu

Daniel A. Williams
Mathematics Department
Howard University
Washington, D.C. 20059
daw@ scs.howard.edu

The Subdivision Experience

N. Dyn and D. Levin

Dedicated to the memory of John A. Gregory

Abstract. Subdivision started as a tool for efficient computation of spline functions, and is now an independent subject with many applications. It is used for developing new methods for curve and surface design, for approximation, for generating wavelets and multiresolution analysis, and also for solving some classes of functional equations. This paper reviews recent new directions and new developments in subdivision analysis. Extensions of the uniform stationary binary subdivision process and their analysis are presented. These include non-stationary subdivision, non-uniform subdivision, integral subdivision, distributional subdivision, and the convergence of the above in the strong and in the weak sense.

§1. Introduction

The *stationary uniform binary subdivision scheme* is a process which recursively defines a sequence of sets of control points $\{P_\alpha^k : \alpha \in \mathbb{Z}^s\}_{k\in\mathbb{Z}_+} \subset \mathbb{R}^d$ by a rule of the form

$$P_\alpha^{k+1} = \sum_{\beta\in\mathbb{Z}^s} a_{\alpha-2\beta} P_\beta^k = \sum_{\beta\in\alpha+2\mathbb{Z}^s} a_\beta P_{(\alpha-\beta)/2}^k \ , \tag{1.1}$$

which is denoted formally by $P^{k+1} = S_a P^k$. Here we assume that the set $\mathrm{supp}(a) = \{\alpha : a_\alpha \neq 0\}$ is finite.

The geometric interpretation to this sequence of sets of control points is given by considering a corresponding sequence of geometrical entities

$$F_k(t) = \sum_{\alpha\in\mathbb{Z}^s} P_\alpha^k \psi(2^k t - \alpha) \ , \tag{1.2}$$

where $\psi(t) \in C(\mathbb{R}^s)$ is of compact support and satisfies

$$\sum_{\alpha\in\mathbb{Z}^s} \psi(\cdot - \alpha) = 1 \ , \tag{1.3}$$

Wavelets, Images and Surface Fitting
P. J. Laurent, A. Le Méhauté, and L. L. Schumaker (eds.), pp. 229–244.

ISBN 1-56881-040-7.

$$m\|c\|_\infty \le \|\sum_{\alpha\in\mathbb{Z}^s} c_\alpha \psi(\cdot-\alpha)\|_\infty \le M\|c\|_\infty, \quad c \in \ell_\infty(\mathbb{Z}^s) . \tag{1.4}$$

For $s=1$, $d=2$ or 3, $F_k(t)$ is a curve, while for $s=2$, $d=3$ it is a surface. For $s=1$ and the choice $\psi(t)=1-|t|$, $t\in(-1,1)$ and zero otherwise, $F_k(t)$ is the control polygon through the control points. Stationary uniform binary subdivision schemes received a lot of attention in the literature in recent years. A general analysis of such schemes can be found in [3,8,9,12,13,15,19,20,21]. The subdivision scheme S_a is termed *uniformly convergent* if for the initial set of control points in $\mathbb{R}$, $f^0=\delta=\{\delta_{\alpha,0}\}_{\alpha\in\mathbb{Z}^s}$, $\{F_k(t)\}$ converges uniformly to a non-zero function ϕ. This property of the scheme is independent of the choice of ψ in (1.2) [3] (see also [12]), and it implies that

$$\lim_{k\to\infty} \sup_{2^{-k}\alpha\in\Omega} \|P_\alpha^k - F_\infty(2^{-k}\alpha)\|_\infty = 0 , \tag{1.5}$$

for any compact $\Omega\subset\mathbb{R}^s$, where $F_\infty(t)=\sum_{\alpha\in\mathbb{Z}^s} P_\alpha^0\phi(t-\alpha)$.

A necessary condition on the mask a for the corresponding subdivision process $\{S_a\}$ to be uniformly convergent is [3]:

$$\sum_{\beta\in\mathbb{Z}^s} a_{\alpha-2\beta} = 1, \quad \alpha_i\in\{0,1\}, \quad i=1,\ldots,s,$$

which implies that $\sum_{\alpha\in\mathbb{Z}^s} a_\alpha = 2^s$.

We are interested in the smoothness properties of ϕ which determine the smoothness of the components of $F_\infty(t)$. A scheme is termed C^m if $\phi\in C^m$, and it is termed H^γ with $\gamma\in(0,1)$, if $\phi\in H^\gamma$, namely if

$$|\phi(t)-\phi(\tau)| \le A|t-\tau|^\gamma . \tag{1.6}$$

Thus, the analysis of the convergence of the scheme and the properties of ϕ requires the study of the scalar case $d=1$.

A uniformly convergent $\{S_a\}$ defines its basic function ϕ, which satisfies the functional (refinement) equation

$$\phi(x) = \sum_{\alpha\in\mathbb{Z}^s} a_\alpha \phi(2x-\alpha) . \tag{1.7}$$

This is the case, since $(S_a\delta)_\alpha = a_\alpha$, $\alpha\in\mathbb{Z}^s$, and $\phi(2\cdot)$ is the limit function for the initial data δ in the second level of the subdivision process. Also it is easy to verify that the support of ϕ is contained in the convex hull of the points in $\mathbb{Z}^s$ constituting the support of the mask a [3]. Equation (1.7) indicates that the spaces $V_k=\text{span}\{\phi(2^k\cdot-\alpha):\alpha\in\mathbb{Z}^s\}$ are nested, namely

$$V_0\subset V_1\subset V_2\subset\cdots, \tag{1.8}$$

which is necessary for multiresolution analysis.

Example: *B*-**splines.** A typical example of a stationary uniform binary subdivision scheme is provided by schemes generating B-splines with equally-spaced knots. The B-spline with integer knots of order m (degree $m-1$ and smoothness C^{m-2}) is the basic function of a scheme defined by the mask [5]

$$a_\alpha = 2^{-m+1}\binom{m}{\alpha}, \quad \alpha = 0, 1, \ldots, m \ .$$

In this paper we review several different extensions and generalizations of stationary uniform binary subdivision schemes, and the notion of their convergence. Some of these extensions have geometric advantage and others are of analytical interest. First we discuss briefly the extensions considered in the paper, and list several more, some of which have not yet been investigated.

1. Non-stationary schemes. The mask applied at level k of the subdivision depends on the level:

$$P_a^{k+1} = \sum_{\beta\in\mathbb{Z}^s} a_{\alpha-2\beta}^{(k)} P_\beta^k, \quad \alpha \in \mathbb{Z}^s \ . \tag{1.9}$$

Here a necessary condition for convergence is $\lim_{k\to\infty} \sum_{\beta\in\mathbb{Z}^s} a_{\alpha-2\beta}^{(k)} = 1$ for $\alpha_i \in \{0, 1\}$, $i = 1, \ldots, s$ [16, 17].

2. Non-uniform schemes. The mask depends on the location of the defined point. In the simplest case $\mathbb{R}^s$ is divided into disjoint domains, with different rules in these domains:

$$P_\alpha^{k+1} = \sum_{\beta\in\mathbb{Z}^s} a_{\alpha-2\beta}^{(\ell)} P_\beta^k, \quad 2^{-k-1}\alpha \in \Omega_\ell \ , \tag{1.10}$$

where $\Omega_\ell \cap \Omega_j = \emptyset$ if $j \neq \ell$ and $\cup_\ell \Omega_\ell = \mathbb{R}^s$. The supports of the masks $\{a^{(\ell)}\}$ are uniformly bounded [14].

A more general non-uniform scheme is [2]

$$P_\alpha^{k+1} = \sum_{\beta\in 2\mathbb{Z}^s+\alpha} a_{\alpha,\beta} P_\beta^k, \quad \alpha \in \mathbb{Z}^s \ . \tag{1.11}$$

Here $a_{\alpha,\beta}$ has the property that $a_{\alpha,\beta} = 0$ for $\alpha - 2\beta \notin J$, where J is a finite subset of $\mathbb{Z}^s$, which is termed the "*support*" of the masks. The matrix $A = \{a_{\alpha\beta}\}$ is termed *two-slanted.*

3. Non-uniform and non-stationary schemes. These schemes are obtained from (1.10) and (1.11) by adding dependence of the masks on the subdivision level [2].

4. Integral schemes. All schemes discussed up to now have masks defined on discrete sets of points. Consider a mask which is a function of compact support in $L_1(\mathbb{R}^s)$ and sets of control points depending on a continuous parameter. The process becomes

$$P^{k+1}(t) = \int_{\mathbb{R}^s} a(t-2\tau)P^k(\tau)d\tau \ , \tag{1.12}$$

and defines a sequence of vectors with components in $C(\mathbb{R}^s)$, starting with such a vector $P^0(t)$. The geometrical entities are given by

$$F_k(t) = P^k(2^k t) \ , \tag{1.13}$$

and one can define the process directly in terms of the geometrical entities as

$$F_{k+1} = 2^{ks}a(2^{k+1}\cdot) * F_k = 2^{-s}\int_{\mathbb{R}^s} F_k(\cdot - 2^{-k-1}\tau)a(\tau)d\tau \ , \tag{1.14}$$

with $F_0(t) = P^0(t)$. Equation (1.14) indicates that the process is equivalent to repeated averaging over exponentially decreasing neighborhoods if $\int_{\mathbb{R}^s} a(\tau)d\tau = 2^s$, which is indeed a necessary condition for the convergence of the process [10].

There is a way to unify the representation of discrete and integral subdivision schemes. It is achieved by first extending the notion of convergence.

5. Weak convergence. The subdivision scheme

$$f_\alpha^{k+1} = \sum_{\beta\in\mathbb{Z}^s} a_{\alpha-2\beta} f_\beta^k, \quad \alpha\in\mathbb{Z}^s, \quad k\in\mathbb{Z}_+ \ , \tag{1.16}$$

is *weakly convergent* if for every $g \in C_0^\infty$ and $f^0 = \delta$, the following limit exists [10]:

$$\lim_{k\to\infty} 2^{-ks}\sum_{\alpha\in\mathbb{Z}^s} f_\alpha^k g(2^{-k}\alpha) \ . \tag{1.17}$$

The weak limit of the subdivision scheme is the distribution F_∞ defined by

$$\langle F_\infty, g\rangle = \lim_{k\to\infty} 2^{-ks}\sum_{\alpha\in\mathbb{Z}^s} f_\alpha^k g(2^{-k}\alpha), \quad g\in C_0^\infty \ . \tag{1.18}$$

6. Distributional schemes. The distributional schemes include the discrete and the integral schemes as special cases. A distributional scheme has the form as in (1.14), $F_{k+1} = 2^{ks}a(2^{k+1}\cdot)*F_k$, with $\{F_k\}$ a sequence of distributions and $a(\cdot)$ a distribution of compact support satisfying $\langle a, 1\rangle = 2^s$, where 1 denotes any C_0^∞ function which is 1 on supp(a). For $a \in L_1(\mathbb{R}^s)$ and $F_0 \in C(\mathbb{R}^s)$ it is an integral scheme, while for $F_0 = \delta(\cdot)$, and $a = \sum_{\alpha\in\mathbb{Z}^s} a_\alpha\delta(\cdot-\alpha)$, the distribution F_k has the form

$$F_k = \sum_{\alpha\in\mathbb{Z}^s} f_\alpha^k \delta(2^k\cdot - \alpha) \ , \tag{1.19}$$

with f^k obtained by (1.16). The limit distribution of $\{F_k\}$ is the weak limit of the discrete scheme (1.16) given by (1.18)[10].

Note that in the distributional case, $\delta(\cdot)$ denotes the Dirac distribution $\langle\delta(\cdot), g\rangle = g(0)$.

7. Non-binary distributional schemes. These schemes have a similar form to (1.14)

$$F_{k+1} = p^{ks} a(p^{k+1}\cdot) * F_k \ , \tag{1.20}$$

with $p \in I\!R$, $p > 1$, and $a(\cdot)$ a distribution of compact support satisfying $\langle a, 1\rangle = p^s$. For $p \in Z\!\!Z_+\backslash\{0,1\}$, $a = \sum_{\alpha\in Z\!\!Z^s} a_\alpha \delta(\cdot - \alpha)$ and $F_0 = \delta(\cdot)$, we get the discrete p-nary scheme

$$F_k = \sum_{\alpha\in Z\!\!Z^s} f_\alpha^k \delta(p^k \cdot - \alpha) \ , \tag{1.21}$$

where

$$f_\alpha^{k+1} = \sum_{\beta\in Z\!\!Z^s} a_{\alpha-p\beta} f_\beta^k \ . \tag{1.22}$$

Another case of interest is when p is rational, namely $p = r/q$, $r, q \in Z\!\!Z_+\backslash\{0\}$, $r > q$, and $a(\cdot)$ is the distribution

$$a = \sum_{\alpha\in Z\!\!Z^s} a_\alpha \delta(\cdot - \alpha/q), \quad \sum_{\alpha\in Z\!\!Z^s} a_\alpha = p^s \ . \tag{1.23}$$

Then for $F_0 = \delta(\cdot)$, we get the discrete scheme

$$F_k = \sum_{\alpha\in Z\!\!Z^s} f_a^k \delta\left(p^k \cdot - \frac{\alpha}{q^k}\right) \ , \tag{1.24}$$

$$f_\alpha^{k+1} = \sum_{\beta\in I_\alpha^k} a_{\frac{\alpha-r\beta}{q^k}} f_\beta^k \ , \tag{1.25}$$

where $I_\alpha^k = \{\beta : \alpha - r\beta \text{ is divisible by } q^k\}$.

Note that the set $\mathcal{F}^k = \{f_\alpha^k : \alpha \in q^k Z\!\!Z^s\}$ defines the set $\mathcal{F}^{k+1}$ by the rule

$$f_{\alpha q^{k+1}}^{k+1} = \sum_{\beta\in Z\!\!Z^s} a_{\alpha q - r\beta} f_{\beta q^k}^k, \alpha \in Z\!\!Z^s \ , \tag{1.26}$$

and that $f_{\alpha q^k}^k$ is attached to the parameter value αp^{-k}. Thus (1.26) gives rise to the rational schemes introduced in [1].

A more general non-binary scheme can be defined in terms of a sequence of monotonically increasing reals $\{p_k\}_{k\in Z\!\!Z_+}$ with $p_0 = 1$. This scheme has the form

$$F_{k+1} = p_{k+1}^s a(p_{k+1}\cdot) * F_k \ , \tag{1.27}$$

with $a(\cdot)$, a compactly supported distribution satisfying $\langle a, 1\rangle = 1$.

The scheme (1.27) can be further generalized into a non-stationary scheme of the form

$$F_{k+1} = a_k * F_k, \quad k \in \mathbb{Z}_+ \ , \tag{1.28}$$

where $a_k(\cdot)$ is a distribution supported in $p_{k+1}^{-1}\Omega$, with $\Omega \subset \mathbb{R}^s$ compact, and $\lim_{k\to\infty}\langle a_k, 1\rangle = 1$.

In the following sections we review results about convergence and smoothness of some of the types of schemes presented above. We also consider several interesting examples, and discuss properties of spaces spanned by the shifts of the corresponding basic functions.

§2. Non-Stationary Discrete Schemes

A converging non-stationary scheme $\{S_{a^{(k)}}\}_{k\in\mathbb{Z}_+}$ defines a system of refinement equations [17]:

$$\phi_\ell = \sum_{\alpha\in\mathbb{Z}^s} a_\alpha^{(k)} \phi_{\ell+1}(2\cdot - \alpha), \quad \ell \in \mathbb{Z}_+ \ , \tag{2.1}$$

where ϕ_ℓ is the basic function of the non-stationary scheme $\{S_{a^{(k+\ell)}}\}_{k\in\mathbb{Z}_+}$.

Thus the spaces $V_\ell = \text{span}\{\phi_\ell(2^\ell\cdot - \alpha) : \alpha \in \mathbb{Z}^s\}$, $\ell \in \mathbb{Z}_+$ are nested (satisfy (1.8)), and can be used for multiresolution analysis. The advantage of this set-up over the nested spaces generated by stationary schemes, is the possibility to get $\{\phi_\ell\}$ of compact support and infinitely smooth [10].

Three classes of non-stationary schemes together with typical examples are reviewed in this section. A mask is termed *stationary* if it determines a stationary C^0 scheme. The first class of schemes consists of schemes defined by stationary masks only.

2.1. Schemes defined by stationary masks

The simplest convergence result for non-stationary schemes is a direct consequence of a convergence result for stationary schemes. In what follows $f^k = S_{a^{(k-1)}} \cdots S_{a^{(0)}} f^0$, $k \in \mathbb{Z}_+$ and $e^{(i)} = \delta_{i,j}$, $i,j = 1,\ldots,s$.

Theorem 2.1 (Convergence [17]). *Let $\{S_{a^{(k)}}\}_{k\in\mathbb{Z}_+}$ be a non-stationary scheme such that each $a^{(k)}$ satisfies the conditions*

$$\sum_{\beta\in\mathbb{Z}^s} a_{\alpha-2\beta}^{(k)} = 1, \quad \alpha_i \in \{0,1\}, \quad i = 1,\ldots,s \ . \tag{2.2}$$

Let $S_{a^{(k)}}^{\Delta}$ be the subdivision rule with a matrix mask mapping the set of vectors

$$\Delta f^k = \{\Delta f_j^k = (f_{j+e^{(i)}}^k - f_j^k, \quad i = 1,\ldots,s) \ : \ j \in \mathbb{Z}^s\} \ , \tag{2.3}$$

into the set of vectors Δf^{k+1}. (The existence of $S_{a^{(k)}}^{\Delta}$ is guaranteed by (2.2).) If there exists $\mu \in (0,1)$ such that for all $k \in \mathbb{Z}_+$

$$||S_{a^{(k)}}^{\Delta} \Delta f||_\infty \le \mu ||\Delta f||_\infty, \quad f \in \ell^\infty(\mathbb{Z}^s) \ , \tag{2.4}$$

where $||\Delta f||_\infty = \sup_{j\in\mathbb{Z}^s} ||\Delta f_j||_\infty$, *then* $\{S_{a^{(k)}}\}_{k\in\mathbb{Z}_+}$ *is* H^ν, $\nu = \min\{-\log_2 \mu, 1\}$.

Note that a necessary and sufficient condition for a mask a to be stationary is the contractivity of the stationary scheme $\{S_a^\Delta\}$ as an operator on the space $\{\Delta f : f \in \ell^\infty(\mathbb{Z}^s)\}$(see e.g. [12]). Hence by (2.4) $a^{(k)}$ is stationary.

Theorem 2.2 (Smoothness [17]). *Let* $\{S_{a^{(k)}}\}_{k\in\mathbb{Z}_+}$ *be such that for some* $\lambda \in \mathbb{Z}^s$ *and* $k_0 \in \mathbb{Z}_+$

$$\sum_{\alpha\in\mathbb{Z}^s} a_\alpha^{(k)} z^\alpha = (1+z^\lambda) \sum_{\alpha\in\mathbb{Z}^s} b_\alpha^{(k)} z^\alpha, \quad k \geq k_0 \ . \tag{2.5}$$

If $\{S_{b^{(k)}}\}_{k\in\mathbb{Z}_+}$ *is convergent then* $\{S_{a^{(k)}}\}_{k\in\mathbb{Z}_+}$ *is convergent and the corresponding basic functions denoted by* ϕ_b *and* ϕ_a, *respectively, are related by*

$$\partial^\lambda \phi_a = \sum_{i=1}^s \lambda_i \frac{\partial}{\partial x_i} \phi_a = \phi_b - \phi_b(\cdot - \lambda) \in C(\mathbb{R}^s) \ . \tag{2.6}$$

Note that $\phi_a \in C^1(\mathbb{R}^s)$ if (2.6) holds for s linearly independent vectors $\lambda^1, \ldots, \lambda^s \in \mathbb{Z}^s$.

The first example we present is typical to nonstationary schemes with $\text{supp}(a^{(k)})$ growing linearly with k.

2.2. The up-function

Let $s = 1$ and let $a^{(k)}$ be the stationary mask generating the B-spline of order k,

$$a_i^{(k)} = \frac{1}{2^{k-1}} \binom{k}{i}, \quad i = 0, \ldots, k, \quad k \in \mathbb{Z}_+ \ . \tag{2.7}$$

By Theorems 2.1 and 2.2 the functions $\{\phi_\ell\}$, with ϕ_ℓ the basic function of $\{S_{a^{(k+\ell)}}\}_{k\in\mathbb{Z}_+}$, are in $C_0^\infty(\mathbb{R})$ with $\text{supp}\,\phi_\ell = (0, \ell+2)$. As shown in [10] the function ϕ_0 is the Rvachev's up-function [22]. Also, by (2.1) the functions $\{\phi_\ell\}$ generate a nested sequence of shift invariant spaces $\{V_k\}_{k\in\mathbb{Z}_+}$.

The above example is interesting enough to deserve detailed analysis. Here we cite the major properties of the sequence of nested spaces generated by it.

Theorem 2.3 (Stability [18]). *The function* $\phi_j(2^j\cdot)$ *is a stable generator of* $V_j = \text{span}\{\phi_j(2^j \cdot -\alpha) : \alpha \in \mathbb{Z}\}$, *namely, there exist constants* $0 < R_j < T_j$ *such that for each* $d \in \ell^2(\mathbb{Z})$

$$R_j ||d||_{\ell^2(\mathbb{Z})} \leq || \sum_{\alpha\in\mathbb{Z}} d_\alpha \phi_j(2^j \cdot -\alpha)||_{L_2(\mathbb{R})} \leq T_j ||d||_{\ell^2(\mathbb{Z})} \ , \tag{2.8}$$

with $T_k/R_k = O\left(\left(\frac{\pi}{2}\right)^k\right)$. *Moreover* $\{\phi_j(\cdot - \alpha) : \alpha \in \mathbb{Z}\}$ *are linearly independent.*

Theorem 2.4 (Spectral approximation order [18]). *Let*

$$E_k(f) = \inf\{\|f - v\|_{L_2(\mathbb{R})} : v \in V_k\} .$$

Then for any $r \in \mathbb{Z}_+$ *and* $f \in W_2^r(\mathbb{R})$ *(the Sobolev space of order* r*)*

$$\lim_{k\to\infty} 2^{kr} E_k(f) = 0 \quad \text{as } k \to \infty . \tag{2.9}$$

For the sequence of nested spaces $\{V_k\}_{k\in\mathbb{Z}_+}$ one can construct a sequence of prewavelets $\{\psi_k\}_{k\in\mathbb{Z}_+}$ such that

$$V_k \oplus \text{span}\{\psi_k(\cdot - \alpha) : \alpha \in 2^{-k}\mathbb{Z}\} = V_{k+1} . \tag{2.10}$$

These prewavelets are given in terms of their Fourier transforms:

Theorem 2.5 (Properties of prewavelets [18]). *Let* $\rho_k = \phi_k(2^k\cdot)$ *and let* $\tau_k(\omega) = \sum_{\alpha\in 2^{k+1}\mathbb{Z}} |\hat{\rho}_k(\omega + 2\pi\alpha)|^2$. *Then*

$$\widehat{\psi}_k(\omega) = \frac{1}{2} e^{-i\omega/2^{k+1}} a_k(e^{-i\omega/2^{k+1}})\tau_k(\omega)\widehat{\rho}_{k+1}(\omega) , \tag{2.11}$$

and $|\operatorname{supp}\psi_k| = 2|\operatorname{supp}\rho_k| = \frac{k+2}{2^{k-1}}$. *Also,* ψ_k *generates a stable basis of the orthogonal complement of* V_j *in* V_{j+1}.

2.3. Asymptotically stationary schemes

An interesting class of non-stationary schemes is that of "asymptotically stationary schemes" [16, 17]. The masks defining such a scheme $a^{(k)} = \{a_\alpha^{(k)} : \alpha \in \mathbb{Z}^s\}$, $k \in \mathbb{Z}_+$ satisfy the two conditions

(i) $\text{supp}(a^{(k)}) \subset [0, n]^s$, $k \in \mathbb{Z}_+$ for some $n \in \mathbb{Z}_+^s$.

(ii) There exists a limiting mask a^* supported in $[0, n]^s$ such that

$$\lim_{k\to\infty} \max_{\alpha\in[0,n]^s} |a_\alpha^{(k)} - a_\alpha^*| = 0 . \tag{2.12}$$

Example: The exponential splines. Univariate exponential B-splines with integer knots are basic functions of asymptotically stationary schemes, with masks $a^{(k)}$, $k \in \mathbb{Z}_+$ of the form

$$\frac{1}{2}\sum_{\alpha\in\mathbb{Z}} a_\alpha^{(k)} z^\alpha = \prod_{i=0}^{\ell}(1 + e^{2^{-k-1}\gamma_i} z)/2, \quad k \in \mathbb{Z}_+ . \tag{2.13}$$

For $\ell \geq 1$, the scheme generates a $C^{\ell-1}$ piecewise exponential function in the span of $\{e^{\gamma_i x} : i = 0, \ldots, \ell\}$ with integer knots and support $(0, \ell+1)$ (see [2, 16, 17]). The scheme is asymptotically stationary since

$$a_j^{(k)} = \frac{1}{2^\ell} \sum_{0\leq i_1<\cdots<i_j\leq \ell} \exp\{2^{-k-1}\sum_{r=1}^{j}\gamma_{i_r}\} = \frac{1}{2^\ell}\binom{\ell+1}{j} + 0(2^{-k}) . \tag{2.14}$$

The limiting mask is that of a B-spline of order $\ell+1$.

Similarly in the multivariate case, the sequence of masks

$$\frac{1}{2^s}\sum_{\alpha\in\mathbb{Z}^s} a_\alpha^{(k)} z^\alpha = \prod_{i=0}^{\ell}(1+\exp\{2^{-k-1}\gamma_i\}z^{\lambda^i})/2 ,$$

with directions $\Lambda = \{\lambda^0,\ldots,\lambda^\ell\} \subset \mathbb{Z}^s$, $\ell \geq s$, determines a C^0 scheme with an exponential box-spline as its basic function, if any ℓ vectors from Λ span $\mathbb{R}^s$ and there is a subset of s vectors in Λ with determinant ± 1 [6, 17]. The limiting mask is stationary generating the polynomial box-spline with directions Λ [3].

Theorem 2.6 (Convergence and smoothness [17]). *Let $\{a^{(k)}\}$ satisfy conditions (i), (ii) with a^* a stationary mask, and let*

$$\theta_k = \max\{|a_\alpha^{(k)} - a_\alpha^*| : \alpha\in\mathbb{Z}^s\}, \quad k\in\mathbb{Z}_+ .$$

If for some $m\geq 0$, $\sum_{k\in\mathbb{Z}_+}\theta_k 2^{km} < \infty$ and $\{S_{a^*}\}$ is C^m, then $\{S_{a^{(k)}}\}_{k\in\mathbb{Z}_+}$ is C^m. If $\{S_{a^*}\}$ is H^ν, $\nu\in(0,1)$, and $\theta_k \leq A2^{-k\mu}$, $k\in\mathbb{Z}_+$ for some $\mu\in(0,1)$, then the scheme $\{S_{a^{(k)}}\}_{k\in Z_+}$ is H^γ with $\gamma=\min(\nu,\mu)$.

For asymptotically stationary schemes there is an analogous result to Theorem 2.2.

Theorem 2.7 (Smoothness [17]). *Let $\{a^{(k)}\}_{k\in\mathbb{Z}_+}$ be a sequence of masks satisfying $\mathrm{supp}(a^{(k)})\subset[0,n]^s$, and*

$$a_k(z) = \sum_{\alpha\in\mathbb{Z}^s} a_\alpha^{(k)} z^\alpha = \prod_{i=1}^{s}(1+r_{k,i}z^{\lambda^i})b_k(z), \quad k\geq k_0 , \tag{2.15}$$

with $\lambda^1,\ldots,\lambda^s\in\mathbb{Z}^s$ a basis of $\mathbb{R}^s$. If , moreover, for each $i=1,\ldots,s$, the polynomials $\{a_k(z)/(1+r_{k,i}z^{\lambda^i})\}_{k\in\mathbb{Z}_+}$ determine a C^m scheme and

$$\lim_{k\to\infty}(r_{k,i}^2-1)2^k = c_i < \infty, \quad \sum_{k=k_0}^{\infty} 2^k|r_{k+1,i}-r_{k,i}| < \infty , \tag{2.16}$$

then $\{S_{a^{(k)}}\}_{k\in\mathbb{Z}_+}$ is a C^{m+1} scheme.

Example: Exponential B-splines revisited. For the exponential B-spline scheme (2.13) with $\ell\geq 1$, S_{a^*} generates B-splines of order $\ell+1$ hence S_{a^*} is $C^{\ell-1}$. By (2.14), $\{a^{(k)}\}_{k\in\mathbb{Z}_+}$ satisfies the conditions of Theorem 2.6 with $m=0$. Hence the scheme $\{S_{a^{(k)}}\}_{k\in\mathbb{Z}_+}$ is converging uniformly to a C^0 function. The stronger result in Theorem 2.7 allows us to conclude that $\{S_{a^{(k)}}\}_{k\in\mathbb{Z}_+}$ is $C^{\ell-1}$. Indeed, a repeated application of Theorem 2.7 with $r_k=\exp(2^{-k-1}\gamma_j)$, $k\in\mathbb{Z}_+$ for $j=0,\ldots,\ell-2$, yields the desired property of the scheme, since $r_{k+1}^2=r_k$, $1-r_k^2=\mathcal{O}(2^{-k})$.

2.4. Schemes with C_0^∞ basic functions

A general class of univariate ($s = 1$) non-stationary schemes generating compactly supported C^∞ basic functions is studied in [4]. The interesting examples are the C_0^∞ cardinal interpolatory basic functions ($\phi|_{\mathbb{Z}} = \delta$) and the C_0^∞ orthonormal wavelets. Typical to the masks of these examples is the linear growth of their support with the subdivision level. In this sense it is a generalization of the process leading to the up-function studied in Subsection 2.2, where the Fourier transforms $a_k(\omega) = (1/2)\sum_{\alpha\in\mathbb{Z}} a_\alpha^{(k)} e^{-i\alpha\omega}$, $k \in \mathbb{Z}_+$ of the masks $\{a^{(k)}\}_{k\in\mathbb{Z}_+}$, given by (2.7) are of the form

$$a_k(\omega) = \exp(-i(k+1)(\omega/2))\cos^{k+1}(\omega/2), \quad k \in \mathbb{Z}_+ \,. \tag{2.17}$$

Theorem 2.8 (Convergence and smoothness [4]). *Let $a(\omega)$ be a 2π-periodic function satisfying*

$$a(0) = 1, \quad a(\omega) = \cos^r(\omega/2)q(\omega), \qquad \sup_{\omega\in[0,2\pi]} \prod_{j=1}^{m} |q(2^{j-1}\omega)| < 2^{rm} \,, \tag{2.18}$$

for some $r > 0$ and $m \in \mathbb{Z}_+ \setminus \{0\}$, with q bounded and Hölder continuous at the origin. Let $\{a^{(k)}\}_{k\in\mathbb{Z}_+}$ be a sequence of masks of finite support with Fourier transforms $a_k(\omega)$ satisfying the four requirements

(i) $||a_k||_\infty \le C$, $k \in \mathbb{Z}_+$.

(ii) $|a_k(\omega)| \le (1+\mu_k)|a(\omega)|^k$, $k \in \mathbb{Z}_+$, $\sum_{k\in\mathbb{Z}_+} |\mu_k| < \infty$.

(iii) $\sum_{k\in\mathbb{Z}_+} \deg(a_k)2^{-k} < \infty$.

(iv) $\sum_{k\in\mathbb{Z}_+} |a_k(0) - 1| < \infty$.

Then the subdivision scheme $\{S_{a^{(k)}}\}_{k\in\mathbb{Z}_+}$ is uniformly convergent and its basic function ϕ_0 is a compactly supported C^∞ function. Moreover the sequence of functions

$$F_k = \sum_{\alpha\in\mathbb{Z}} f_\alpha^k \psi(2^k \cdot -\alpha), \quad f^k = S_{a^{(k-1)}} \cdots S_{a^{(0)}}\delta, \quad k \in \mathbb{Z}_+ \,, \tag{2.19}$$

with $\psi = \sin(2\pi\cdot)/(2\pi\cdot)$, and the sequences of their derivatives, converge uniformly to ϕ_0 and its derivatives:

$$\lim_{k\to\infty} ||F_k^{(j)} - \phi_0^{(j)}||_\infty = 0, \quad j \in \mathbb{Z}_+ \,. \tag{2.20}$$

Note that requirement (iii) guarantees the finite support of the basic function ϕ_0 of the scheme $\{S_{a^{(k)}}\}_{k\in\mathbb{Z}_+}$, with

$$\deg(a_k) = \max\{\alpha : a_\alpha^{(k)} \ne 0\} - \min\{\alpha : a_\alpha^{(k)} \ne 0\} \,.$$

This is the case since

$$|\operatorname{supp}\phi_0| \le \sum_{k\in\mathbb{Z}_+} 2^{-k-1}\deg(a_k) \,.$$

Approximation orders of the shift invariant spaces of the multiresolution analysis determined by $\{a^{(k)}\}_{k\in\mathbb{Z}_+}$ are studied in [4], and similar results to those in Theorem 2.4 for approximation in the L_2 – norm are obtained for approximation in Sobolev norms..

Theorem 2.9 (Spectral approximation in Sobolev norms [4]). *Let $\{a^{(k)}\}_{k\in\mathbb{Z}_+}$ satisfy the assumptions of Theorem 2.8 and the additional condition*

$$|a_k(\omega)| \geq (1-\nu_k)|A(\omega)|^k ,$$

where $|\nu_k| < 1$, $\sum_{k\in\mathbb{Z}_+} |\nu_k| < \infty$, $A(0) = 1$ and $A(\cdot)$ is Hölder continuous at the origin. Then the spaces $\{V_j\}_{j\in\mathbb{Z}_+}$ generated by this subdivision have density order r in H^s for all $r \geq s$, namely for $r \geq s$ and $f \in H^r = \{g : \|g\|_r = \int_{\mathbb{R}} |\widehat{g}(\omega)|^2(1+|\omega|^{2r})d\omega < \infty\}$

$$\inf_{v\in V_j} \|f-v\|_s \leq C2^{j(s-r)}\|f\|_r\epsilon(f,j) ,$$

with $0 \leq \epsilon(f,j) \leq 1$ and $\lim_{j\to\infty} \epsilon(f,j) = 0$. Moreover, there exists $t \in (0,\pi]$ such that the sequence of the L_2-projections of S_jf into V_j, $j \in \mathbb{Z}_+$, with $(S_jf)^\wedge = \chi_{[-2^jt,2^jt]}\widehat{f}$, constitutes an approximation scheme which achieves density order r in H^s for all $r \geq s$.

Example: C_0^∞ cardinal interpolatory basic functions. The construction is based on a family of stationary univariate interpolatory subdivision schemes with masks $\{a^{(n)}\}_{n\in\mathbb{Z}_+}$, introduced in [11]. The stationary scheme with the mask $a^{(n)}$ reproduces polynomials of degree $\leq 2n+1$. The non-zero elements of $a^{(n)}$ are

$$a_0^{(n)} = 1, \quad a_{2\alpha+1}^{(n)} = a_{-2\alpha-1}^{(n)} = \ell_\alpha(0), \quad \alpha = 0,1,\ldots,n , \tag{2.21}$$

where $\ell_\alpha(x) \in \pi_{2n+1}$ satisfies $\ell_\alpha\big((2\beta+1)/2\big) = \delta_{\alpha,\beta}$, $\beta = -n-1,-n,\ldots,n$.

The Fourier transform of the mask (2.21) is [11]

$$a_n(\omega) = \cos^{2n+2}(\omega/2)\sum_{j=0}^{n}\binom{n-1-j}{j}y^j, \quad y = \sin^2(\omega/2) . \tag{2.22}$$

Thus $a_n(0) = 1$, $\deg(a_n) = 4n+2$, and requirements (iii) and (iv) of Theorem 2.8 hold.

It is shown in [4] that $|a_n(\omega)| \leq \big(a(\omega)\big)^n$ where

$$a(\omega) = \begin{cases} 1, & |\omega| \leq \frac{\pi}{2} , \\ \sin^2\omega, & \frac{\pi}{2} \leq |\omega| < \pi . \end{cases} \tag{2.23}$$

Hence $||a_n||_\infty \leq 1$, and requirements (i) and (ii) of Theorem 2.8 are valid. Defining $a(\omega) = \cos^2(\omega/2)q(\omega)$ we get from (2.23) that $q(\omega) < q(\pi) = 4$ for $|\omega| < \pi$, and hence (2.18) holds with $r = 2$ and $m = 2$. Thus the non-stationary scheme $\{S_{a^{(k)}}\}_{k\in\mathbb{Z}_+}$ with $a^{(k)}$ given by (2.21) or (2.22) generates a C_0^∞ interpolatory basic function ϕ_0 satisfying $\phi_0|_{\mathbb{Z}} = \delta$. In fact, for any $j \in \mathbb{Z}_+$, the basic function ϕ_j corresponding to the scheme $\{S_{a^{(k+j)}}\}_{k\in\mathbb{Z}_+}$ is C_0^∞ and interpolatory. Also, $\{\phi_j\}$ generates multiresolution analysis, with the property of spectral approximation in Sobolev norms.

It should be noted that although $a^{(n)}$ in this example is stationary for $n \in \mathbb{Z}_+$ [11], Theorem 2.1 does not apply, since

$$||S_{a^{(n)}}^\Delta||_\infty > 1, \quad n \neq 0 .$$

§3. Non-Uniform Discrete Schemes

First we consider the case $s = 1$ and non-uniformity according to domain as in (1.10). Let $\Omega_j = [x_j, x_{j+1})$, $j \in J \subset \mathbb{Z}$, with $\{x_j\}_{j\in J} \subset \mathbb{Z}$ satisfying $x_{j+1} - x_j \geq L > 0$, $j \in J$, and $\bigcup_{j\in J} \Omega_j = \mathbb{R}$. We are concerned with non-uniform schemes of the form

$$f_\alpha^{k+1} = \sum_\beta a_{\alpha-2\beta}^{(\ell)} f_\beta^k, \quad \alpha 2^{-k-1} \in \Omega_\ell \,, \tag{3.1}$$

where $a^{(\ell)}$ is the mask applicable in Ω_ℓ.

To each point x_j there corresponds an index set I_j and a matrix A_j of the form

$$(A_j)_{\alpha\beta} = \begin{cases} a_{\alpha-2\beta}^{(j-1)}, & \alpha < 0 \\ a_{\alpha-2\beta}^{(j)}, & \alpha \geq 0 \end{cases} \qquad (\alpha, \beta) \in I_j \,. \tag{3.2}$$

The index set I_j is given by $I_j = \{\alpha : \alpha + 1 \in \widetilde{I}_j \quad \text{or} \quad \alpha - 1 \in \widetilde{I}_j\}$, where $\widetilde{I}_j$ is the smallest set of integers including zero with the property

$$(-1)^\ell \alpha \geq \ell, \quad \alpha \in \widetilde{I}_j, \quad a_{\alpha\beta}^{(j-\ell)} \neq 0 \Rightarrow \beta \in \widetilde{I}_j \quad \text{for} \quad \ell = 0, 1 \,. \tag{3.3}$$

The cardinality of I_j is denoted by $|I_j|$.

Theorem 3.1 (Convergence and smoothness [14]). *Let $m \in \mathbb{Z}_+$, and let $\{S_{a^{(j)}}\}$ be a C^m stationary scheme for $j \in J$. If for all $j \in J$ the eigenvalues of A_j, $\{\lambda_i^{(j)}\}_{i=1}^{|I_j|}$ have the form*

$$\lambda_i^{(j)} = 2^{-i}, \quad i = 0, \ldots, m, \quad |\lambda_i| < 2^{-m}, \quad i = m+1, \ldots, |I_j| \,, \tag{3.4}$$

then the scheme (3.1) is C^m.

Results on the convergence of non-uniform schemes of the type (1.11) with non-negative masks and of several generalizations are obtained in [2]. Here we cite one result which applies to non-uniform and non-stationary schemes.

Theorem 3.2 (Convergence [2]). *Let $\gamma, \eta \in \mathbb{Z}_+^s$ such that $\gamma_i < \eta_i$, $i = 1, \ldots, s$, and let $\{A^{(k)}\}_{k\in\mathbb{Z}_+}$ be a sequence of stochastic matrices with the property*

$$A_{\alpha\beta}^{(k)} \neq 0 \quad \textit{only if} \quad \gamma_i \leq \alpha_i - 2\beta_i \leq \eta_i, \quad i = 1, \ldots, s \,. \tag{3.5}$$

If moreover, $A_{\alpha\beta}^{(k)} \geq \varepsilon > 0$ for $\alpha_i - 2\beta_i \in \{\gamma_i, \gamma_i + 1, \eta_i - 1, \eta_i\}$, $i = 1, \ldots, s$, then the scheme

$$P^{k+1} = A^{(k)} P^k, \quad k \in \mathbb{Z}_+ \,,$$

is uniformly convergent.

§4. Integral Subdivision and Weak Convergence

The discrete binary subdivision process generates points $\{P_\alpha^{k+1}\}$ at level $k+1$ by a linear operation on $\{P_\alpha^k\}$, where the operator is a two slanted matrix [2]. Viewing $\{P_\alpha^k\}$ as values of a function F_k at the points $\{2^{-k}\alpha\}$ respectively, the slanted matrix operator can be viewed as a discrete convolution operator. In the following we discuss the natural extension of the discrete binary subdivision to integral binary subdivision where the entities are functions instead of sequences of points, and the operators are convolutions. Here the mask is a function $a \in L^1(\mathbb{R}^s)$ of compact support, and starting with an initial 'control' function $F_0 : \mathbb{R}^s \to \mathbb{R}^d$ we recursively define a sequence of functions $\{F_k\}$ by the rule

$$F_{k+1} = 2^{ks}a(2^{k+1}\cdot) * F_k = 2^{ks}\int_{\mathbb{R}^s} F_k(\tau)a\big(2^{k+1}(t-\tau)\big)dt, \quad k \in \mathbb{Z}_+ \ . \tag{4.1}$$

As shown in the introduction, the rule (4.1) is derived from (1.12) by the substitution $F_k(t) = P^k(2^k t)$. It is interesting to note that the two slanted nature of the discrete subdivision, which is still apparent as the two slanted convolution (1.12), is absent when we consider the function $F_k(t) = P^k(2^k t)$. In (4.1) the operator is a simple convolution operator on F_k defined by a weight function whose support is exponentially decreasing with k. A necessary condition for the convergence of the integral subdivision process is that $\int_{\mathbb{R}^s} a(\tau)d\tau = 2^s$, which simply means that $2^{ks}a(2^{k+1}\cdot) \to \delta(\cdot)$ as $k \to \infty$ [10].

Example: The up-function. The up-function derived in Section 2 by a non-stationary subdivision scheme is also the natural example for integral subdivision. We consider $s = d = 1$ and the mask

$$a(x) = \chi_{[-1,1]}(x) = \begin{cases} 1, & -1 \le x \le 1 \ , \\ 0, & \text{otherwise} \ . \end{cases} \tag{4.2}$$

Starting the integral subdivision process with $F_1 = a(2\cdot)$ (corresponding to the initial distribution $F_0 = \delta(\cdot)$), the sequence $\{F_k\}$ is uniformly convergent to the up-function, up $(\cdot) \in C_0^\infty$ [10].

As in the discrete case, the integral subdivision process is related to a dilation equation which is now an integral functional equation

$$F(x) = \int_{\mathbb{R}^s} a(t)F(2x-t)dt = (a * F)(2x) \ . \tag{4.3}$$

For example, the up-function satisfies the functional equation

$$\text{up}(x) = \int_{-1}^{1} \text{up}(2x-t)dt \ . \tag{4.4}$$

The convergence of integral subdivision processes, the nature of the limit functions and the relation to the integral functional equation, are discussed in [10] for $s = 1$ and in [7] for $s \ge 1$. The following theorems and example are straightforward extensions of some of the results in [10] to the case $s \ge 1$. The analysis is based on Fourier methods and the results depend strongly on the decay rate of the Fourier transform $\widehat{a}(\lambda)$ of a as $|\lambda| \to \infty$. Here $|\lambda| = \sum_{i=1}^{s} |\lambda_i|$.

Theorem 4.1 (Convergence [10]) . *If the infinite product $\prod_{j=1}^{\infty}(2^{-s}\widehat{a}(\lambda 2^{-j}))$ is convergent in $L^1(\mathbb{R}^s)$ to $\widehat{F}$, and if we apply the integral subdivision process (1.5) starting with $F_1 = a(2\cdot)$, then $\{F_k\}$ converges uniformly on $\mathbb{R}^s$ to F which is a solution of the integral functional equation (4.3).*

The condition $\widehat{a}(0) = 2^s$ is evidently a necessary condition for the convergence of the above infinite product, and for the existence of an L^1 solution of (4.3) with $\widehat{F}(0) \neq 0$. The limit function F in Theorem 4.1 is termed "the basic function" of the integral subdivision (4.1).

Theorem 4.2 (Uniform convergence to C^∞ functions [10]). *Let a be of finite support and satisfy for some $r > 0$,*

$$\int_{\mathbb{R}^s} a(t)dt = 2^s, \quad \int_{\mathbb{R}^s} |a(t)|dt \le 2^{s+1}, \quad |\widehat{a}(\lambda)| = o(|\lambda|^{-r}) \quad \text{as } |\lambda| \to \infty \,. \tag{4.5}$$

Then the integral subdivision process starting with $F_1 = a(2\cdot)$ is a sequence which converges uniformly on $\mathbb{R}^s$ to a C_0^∞ function F. Moreover any derivative of $\{F_k\}$ is a sequence which converges uniformly on $\mathbb{R}^s$ to the corresponding derivative of F.

The asymptotic behavior of $\widehat{a}(\lambda)$ as $\lambda \to 0$ and as $|\lambda| \to \infty$ gives us indication on the continuity of the limit function F:

Theorem 4.3 (Smoothness of limit function [10]). *Let $\widehat{a}$ satisfy the two conditions*

$$|\widehat{a}(\lambda) - 2^s| = 0(|\lambda|^\gamma) \quad \text{as} \quad |\lambda| \to 0, \quad \gamma > 0 \,, \tag{4.6}$$
$$|a(\lambda)| < q \le 2^{-n-s+1-\varepsilon} \quad \text{for} \quad |\lambda| > R, \quad \varepsilon > 0 \,. \tag{4.7}$$

Then the integral subdivision process converges weakly to $F \in C^n(\mathbb{R}^s)$.

Corollary 4.4. *If $\widehat{a}$ satisfies (4.6) and also*

$$|\widehat{a}(\lambda)| = o(1) \quad \text{as} \quad |\lambda| \to \infty \,, \tag{4.8}$$

then the integral subdivision converges weakly to $F \in C^\infty(\mathbb{R}^s)$.

In the up-function example $\widehat{a}(\lambda) = 2\frac{\sin\lambda}{\lambda}$ and a satisfies the conditions of Theorem 4.2. Thus the uniform convergence and the C^∞ property follow.

The relation between the integral subdivision and the non-stationary subdivision which appears in the case of the up-function is not a coincidence. The following theorem reveals that if the mask $a(\cdot)$ of the integral subdivision process (or the integral functional equation) is itself a basic function of a discrete uniform binary subdivision (stationary or not), then the basic function of the integral subdivision process can be obtained as the limit of a non-stationary discrete subdivision process.

Theorem 4.5 (Relation between integral and non-stationary subdivisions). *Let $a(2\cdot) \in L^1(\mathbb{R}^s)$ be the basic function of a non-stationary subdivision process with coefficients $\{c_\beta^{(k)}\}$ satisfying $\sum_{\beta\in\mathbb{Z}^s} c_\beta^{(k)} = 2^s$. We define another* *non-stationary subdivision process with coefficients $\{b_\alpha^{(k)}\}$ determined by the trigonometric equality*

$$b_k(\omega) = 2^{-s} \sum_{\alpha \in \mathbb{Z}^s} b_\alpha^{(k)} e^{-i\alpha\omega} = \prod_{\ell=0}^{k} c_\ell(\omega), \tag{4.9}$$

where $c_k(\omega) = 2^{-s} \sum_{\beta \in \mathbb{Z}^s} c_\beta^{(k)} e^{-i\beta\omega}$ If the non-stationary process $\{S_{b^{(k)}}\}_{k \in \mathbb{Z}_+}$ converges weakly to an L^1 limit function, then its basic function is the basic function of the integral subdivision process with the mask a, and it is a solution of the integral functional equation with the mask a.

Example: Box-up functions. Consider an integral subdivision process with a mask $a = B(\cdot/2)$, where $B(x)$ is a box-spline function in $\mathbb{R}^s$ centered at the origin and normalized such that $\int_{\mathbb{R}^s} B(x)dx = 1$. Clearly this a satisfies condition (4.5) of Theorem 4.2, and the integral subdivision process with the initial function $F_1 = a$ is uniformly convergent to $F_B \in C^\infty(\mathbb{R}^s)$. Recalling the up-function example, we term F_B a *box-up function.* It follows from the infinite convolution process and the non-negativity of B that $\operatorname{supp} F_B = \operatorname{supp} B(\cdot/2)$. Also, as in the case of the up-function, the integer shifts of F_B, $\{F_B(\cdot - \alpha) : \alpha \in \mathbb{Z}^s\}$ are linearly independent if $\{B(\cdot - \alpha) : \alpha \in \mathbb{Z}^s\}$ are, and both spaces reproduce the same subspace of polynomials. Furthermore, since box-splines based on a set of directions which includes a subset with determinant ± 1, are basic functions of a discrete stationary subdivision process [3], we get by Theorem 4.5 that the box-up function F_B can also be computed by a discrete non-stationary subdivision with easily computable masks.

As is shown in the Introduction, by allowing the mask and the initial F_0 in (4.1) to be distributions of δ-type, the integral subdivision scheme (4.1) gives rise to the discrete subdivision scheme (1.1), and the notion of weak (distributional) convergence becomes natural. The following is a strong result on weak convergence of discrete schemes:

Theorem 4.6 (Weak convergence [10]). *The discrete uniform stationary binary subdivision process (1.1) with a mask a satisfying $\sum_{\alpha \in \mathbb{Z}^s} a_\alpha = 2^s$, is always weakly convergent. Its basic function is the weak solution of the functional equation (1.7).*

References

1. Blu, T., and O. Rioul, Wavelet regularity of iterated filter banks with rational sampling changes, Proceedings ICASSP, 1993.
2. Buhmann M. D., and C. A. Micchelli, Using two slanted matrices for subdivision, DAMTP report NA4, 1992.
3. Cavaretta, A. S., W. Dahmen, and C. A. Micchelli, *Stationary Subdivision*, Memoirs of AMS, No. 453, 1991.
4. Cohen, A., and N. Dyn, Nonstationary subdivision schemes and multiresolution analysis, preprint.

5. Cohen, E., T. Lyche, and R. Riesenfeld, Discrete B-splines and subdivision techniques in Computer-Aided Geometric Design and Computer Graphics, Computer Graphics and Image Processing **14** (1980), 87–111.
6. Dahmen, W., and C. A. Micchelli, On multivariate E-splines, Adv. Math. **76** (1989), 33–93.
7. Dahmen, W., and C. A. Micchelli, Continuous refinement equations and subdivision, Adv. Comp. Math. **1** (1993), 1–37.
8. Daubechies, I., and J. Lagarias, Two scale difference equations I. Existence and global regularity of solutions, SIAM J. Math. Anal. **22** (1991), 1338–1410.
9. Daubechies, I., and J. Lagarias, Two scale difference equations II. Local regularity, infinite products of matrices and fractals, SIAM J. Math. Anal. **23** (1992), 1031–1079.
10. Derfel, G., N. Dyn, and D. Levin, Generalized functional equations and subdivision schemes, J. Approximation Theory, to appear.
11. Deslauriers, G., and S. Dubuc, Symmetric iterative interpolation, Constr. Approx. **5** (1989), 49–68.
12. Dyn, N., Subdivision schemes in computer aided geometric design, in *Advances in Numerical Analysis II, Subdivision algorithms and radial functions*, W.A. Light (ed.), Oxford University Press, 1992, 36–104.
13. Dyn, N., J. A. Gregory, and D. Levin, Analysis of linear binary subdivision schemes for curve design, Constructive Approximation **7** (1991), 127–147.
14. Dyn, N., J. A. Gregory, and D. Levin, Non-uniform binary subdivision schemes, in preparation.
15. Dyn, N., and D. Levin, Interpolating subdivision schemes for the generation of curves and surfaces, in *Multivariate Interpolation and Approximation*, W. Haussmann and K. Jetter (eds.), Birkhauser Verlag, Basel, 1990, 91–106.
16. Dyn, N., and D. Levin, Stationary and non-stationary binary subdivision schemes, in *Mathematical Methods in Computer Aided Geometric Design*, T. Lyche and L. Schumaker (eds.), Academic Press, 1992, 209–216.
17. Dyn, N., and D. Levin, Non-stationary binary subdivision schemes converging to a limit scheme, TAU, Appl. Math. Tech. Rpt. 10-93, 1993.
18. Dyn, N., and A. Ron, Multiresolution Analysis by infinitely differentiable compactly supported functions, CMS TSR #93-4, University of Wisconsin-Madison, 1992.
19. Micchelli, C. A., and H. Prautzsch, Computing curves invariant under halving, Computer Aided Geometric Design **4** (1987), 113–140.
20. Micchelli, C. A., and H. Prautzsch, Uniform refinement of curves, Linear Algebra and Applications **114/115** (1989), 841–870.
21. Rioul, O., Simple regularity criteria for subdivision schemes, SIAM J. Math. Anal. **23** (1992), 1544–1576.
22. Rvachev, V. A., Compactly supported solutions of functional-differential equations and their applications, Russ. Math. Surveys **45** (1990), 87–120.

Banded Matrices with Banded Inverses III: p-Slanted Matrices

M. Gasca, C. A. Micchelli, and J. M. Peña

Abstract. In this paper we study an equation for biinfinite matrices which is important for nonorthogonal wavelet decomposition and give an application of our results to the construction of local interpolation schemes by splines with a given accuracy which improves upon those previously available.

§1. Introduction

The purpose of this paper is to extend some of the results of Dahmen and Micchelli [5] to a larger class of matrices than that considered in [5] and give an application of our results to interpolation and nonorthogonal wavelet decomposition by spline functions on arbitrary knots. The paper is organized as follows. In the second section we provide motivation for the problem studied here in the context of linear spaces of functions on the real line which are invariant under translation by integers. In this setting the matrices that arise extend the classical Hurwitz matrices which are important in determining if a polynomial has all its zeros in the left half plane, cf. Chapter XV of Gantmacher [6].

In Sect. 3 we extend the results of the second section to a class of matrices which we call p-slanted. These matrices have a similar pattern of zero entries as the matrices that occur in Sect. 2 but otherwise are arbitrary. The last section contains two applications of our results from Sect. 3. The first concerns spline interpolation and the second deals with locally finite decompositions of spline spaces.

Wavelets, Images and Surface Fitting
P. J. Laurent, A. Le Méhauté, and L. L. Schumaker (eds.), pp. 245–268.

ISBN 1-56881-040-7.

§2. Translation Invariant Spaces

Let us begin with a continuous function h defined on $\mathbb{R}$ which has finite support. We assume that h satisfies a *refinement equation*, that is, for some constants a_j, $j \in \mathbb{Z}$, with $a_j = 0$ if $j \notin (l, m)$, and a positive integer $p > 1$

$$h(x) = \sum_{j\in\mathbb{Z}} a_j h(px - j), \quad x \in \mathbb{R}. \tag{2.1}$$

This equation, especially in the case $p = 2$, is central to wavelet analysis, cf. Micchelli [10]. It also arises in the study of stationary subdivision schemes for geometric modeling, cf. Cavaretta, Dahmen and Micchelli [3].

Associated with the function h is the linear space V_0 defined to be the algebraic span of the integer translates of h, viz.

$$V_0 = \{\sum_{j\in\mathbb{Z}} c_j h(\cdot - j) : c_i \in \mathbb{R}, \forall i \in \mathbb{Z}\} := \mathrm{span}\{h(\cdot - j) : j \in \mathbb{Z}\}. \tag{2.2}$$

To insure that a typical function in V_0 is faithfully represented by the coefficient sequence $c = (c_j : j \in \mathbb{Z})$ we assume h has integer translates which are algebraically independent, that is, whenever

$$\sum_{j\in\mathbb{Z}} c_j h(x - j) = 0, \quad x \in I, \tag{2.3}$$

where I is some interval, it follows that $c_j = 0$ for all j such that

$$\{x : h(x - j) \neq 0\} \cap I \neq \emptyset.$$

This property implies that whenever a function f in V_0 is of finite support it must be a finite linear combination of integer translates of h, and conversely. An important example of such a function h is the cardinal B-spline of degree n. It is defined as the $(n+1)$st convolution of the characteristic function of the interval [0,1]. Call this function M. Then for any p, M satisfies a refinement equation with coefficients $a_j, j \in \mathbb{Z}$, defined by

$$\sum_{j\in\mathbb{Z}} a_j z^j = p^{-n} \left(\frac{z^p - 1}{z - 1}\right)^{n+1}, \quad z \in \mathbb{C}. \tag{2.4}$$

This can be proved by induction on n, for instance. Also, it is well-known that the integer translates of M are algebraically independent, cf. Schumaker [12]. Moreover, for this example the space V_0 consists of all functions with $n - 1$ continuous derivatives such that on any interval between consecutive integers it is a polynomial of degree at most n. We call this space S_0.

Next, we need the space V_1, defined to be the algebraic span of the functions $h(p \cdot - i)$, $i \in \mathbb{Z}$. For the B-spline example above V_1 consists of all functions f with $n - 1$ continuous derivatives such that on any interval of the form $[\frac{i}{p}, \frac{i+1}{p}]$, $i \in \mathbb{Z}$, it is a polynomial of degree at most n. We call this space S_1.

According to (2.1) it easily follows that

$$V_0 \subseteq V_1 := \text{span}\{h(p \cdot -i) : i \in \mathbb{Z}\}. \tag{2.5}$$

Specifically, for any sequence $c = (c_j : j \in \mathbb{Z})$ we have

$$\sum_{j \in \mathbb{Z}} c_j h(x-j) = \sum_{j \in \mathbb{Z}} d_j h(px-j), \quad x \in \mathbb{R}, \tag{2.6}$$

where

$$d_i = \sum_{j \in \mathbb{Z}} a_{i-pj} c_j, \quad i \in \mathbb{Z}. \tag{2.7}$$

The biinfinite matrix

$$(a_{i-pj})_{i,j \in \mathbb{Z}} \tag{2.8}$$

is called a *subdivision matrix* and is studied, especially for $p = 2$, in Cavaretta, Dahmen and Micchelli [3].

Here we are interested in the following question. We wish to decompose V_1 as a direct sum of V_0 and spaces $W_1, \ldots, W_{p-1}$,

$$V_1 = V_0 + W_1 + \cdots + W_{p-1}. \tag{2.9}$$

We demand two properties of this decomposition. First, each W_l, $l = 1, \ldots, p-1$, is the algebraic span of the integer translates of some function in V_1 of finite support,

$$W_l = \text{span}\{g_l(\cdot - i) : i \in \mathbb{Z}\}, \tag{2.10}$$

that is, $g_l \in V_1$ and the set $\{x : g_l(x) \neq 0\}$ is bounded. Secondly, we require that whenever $f \in V_1$ is of finite support it is a sum of a finite number of integer translates of h and $g_1, \ldots, g_{p-1}$. Under these circumstances we say (2.9) is a *locally finite decomposition* of V_1 through V_0.

Theorem 2.1. *Let V_0 and V_1 be as above. Then there exists a locally finite decomposition of V_1 through V_0.*

This theorem immediately provides

Corollary 2.1. *For any integer p there exists a locally finite decomposition of S_1 through S_0.*

We prove Theorem 2.1 by means of several ancillary facts of some independent interest.

Lemma 2.1. *Let h be a continuous function on $\mathbb{R}$ of finite support which satisfies the refinement equation (2.1) for some integer $p > 1$ and scalars a_j, $j \in \mathbb{Z}$, such that $\#\{j : a_j \neq 0\} < \infty$. Then the Laurent polynomials*

$$a_l(z) := \sum_{j \in \mathbb{Z}} a_{l+jp} z^j, \quad z \in \mathbb{C}\backslash\{0\}, \quad l = 0, 1, \ldots, p-1, \tag{2.11}$$

have no common zeros on $\mathbb{C}\backslash\{0\}$.

Proof: Suppose there is some $\tau \in \mathbb{C}\backslash\{0\}$ such that $a_l(\tau) = 0$, $l = 0, 1, \ldots, p-1$. Specializing formula (2.6) to the sequence $c_j = \tau^{-j}$, $j \in \mathbb{Z}$, we get

$$\sum_{j\in\mathbb{Z}} \tau^{-j} h(x-j) = \sum_{j\in\mathbb{Z}} d_j h(px-j), \quad x \in \mathbb{R},$$

where

$$d_i = \sum_{j\in\mathbb{Z}} a_{i-pj}\tau^{-j}. \tag{2.12}$$

We claim that $d = 0$, which would contradict the algebraic linear independence of the integer translates of h and prove the lemma. To see that $d = 0$ we pick an $i \in \mathbb{Z}$ and write it in the form $i = pr + l$ for some integers $r \in \mathbb{Z}$ and $l \in \{0, 1, \ldots, p-1\}$. Then we get from (2.12) that $d_i = \tau^{-r} a_l(\tau) = 0$. ■

Lemma 2.2. *Let $a_0(z), \ldots, a_{p-1}(z)$ be any Laurent polynomials with no common zeros on $\mathbb{C}\backslash\{0\}$. Then there exists a $p \times p$ matrix*

$$A(z) = (a_{ij}(z))_{i,j=0,1,\ldots,p-1} \quad z \in \mathbb{C}\backslash\{0\},$$

such that each $a_{ij}(z)$ is a Laurent polynomial with $a_{0j} = a_j$, $j = 0, 1, \ldots, p-1$, and $A(z)$ has determinant one on $\mathbb{C}\backslash\{0\}$.

This is a well-known algebraic fact. For the benefit of the reader we comment on its proof.

Proof: The case $p = 2$ means that whenever $a_0(z)$ and $a_1(z)$ have no common zeros on $\mathbb{C}\backslash\{0\}$ there are Laurent polynomials $b_0(z)$, $b_1(z)$ such that

$$a_0(z)b_1(z) - a_1(z)b_0(z) = 1, \quad z \in \mathbb{C}\backslash\{0\}.$$

This fact can be proved by using the Euclidean algorithm, cf. Walker [13]. The general case follows by induction on p. For the details of the induction step we follow Proposition 9.1 of Cavaretta, Dahmen and Micchelli [3]. Suppose the result is true for any $p-1$ Laurent polynomials. There are two cases to consider. If $a_0(z)$,...,$a_{p-2}(z)$ have no common zeros then by the induction hypothesis there is a $(p-1) \times (p-1)$ matrix $A(z)$ of Laurent polynomials whose first row is $(a_0(z), \ldots, a_{p-2}(z))$ and has determinant one on $\mathbb{C}\backslash\{0\}$. Then the $p \times p$ matrix

$$B(z) = \begin{pmatrix} & & & a_{p-1}(z) \\ & & & 0 \\ & A(z) & & \vdots \\ & & & 0 \\ 0 & \cdots & 0 & 1 \end{pmatrix}$$

has all the desired properties and thus advances the induction step. If the Laurent polynomials $a_0(z)$,...,$a_{p-2}(z)$ have common zeros on $\mathbb{C}\backslash\{0\}$, we write them in the form $a_l(z) = q(z)b_l(z)$, $l = 0, 1, \ldots, p-2$, where $b_0(z)$,...,$b_{p-2}(z)$

have no common zeros on $\mathbb{C}\backslash\{0\}$. Our assumptions imply that $a_{p-1}(z)$ and $q(z)$ also do not have common zeros on $\mathbb{C}\backslash\{0\}$. Thus there are Laurent polynomials $r(z)$ and $s(z)$ such that

$$r(z)q(z) - s(z)a_{p-1}(z) = 1, \quad z \in \mathbb{C}\backslash\{0\},$$

and a $(p-1)\times(p-1)$ matrix $C(z)$ whose first row is $(b_0(z),\ldots,b_{p-2}(z))$ and has determinant one. Let $D(z)$ be the $(p-2)\times(p-1)$ submatrix of $C(z)$ consisting of its last $p-2$ rows and consider the $p \times p$ matrix

$$E(z) = \begin{pmatrix} a_0(z) & \ldots & a_{p-2}(z) & a_{p-1}(z) \\ & & & 0 \\ & D(z) & & \vdots \\ & & & 0 \\ s(z)b_0(z) & \ldots & s(z)b_{p-2}(z) & r(z) \end{pmatrix}.$$

Then $\det E(z) = r(z)q(z) - s(z)a_{p-1}(z) = 1$. ■

For the next result we use the following notation. Given any sequence $a = (a_i : i \in \mathbb{Z})$ of finite support we associate a biinfinite matrix A defined by

$$A = (A_{ij})_{i,j\in\mathbb{Z}}, \quad A_{ij} = a_{j-pi}. \tag{2.13}$$

Note that A is the transpose of the subdivision matrix (2.8) and that for some integers $l < m$ $A_{ij} = 0$ if $j < pi + l$ and $j > pi + m$.

Proposition 2.1. *Given a sequence $a^0 = (a_j^0 : j \in \mathbb{Z})$, there are sequences $a^1,\ldots,a^{p-1}$, $b^0,\ldots,b^{p-1}$ of finite support such that the corresponding matrices A^0, $A^1,\ldots,A^{p-1}$, $B^0,\ldots,B^{p-1}$ satisfy the equations*

$$(B^0)^T A^0 + (B^1)^T A^1 + \cdots + (B^{p-1})^T A^{p-1} = I \tag{2.14}$$

and

$$A^i (B^j)^T = I\delta_{ij}, \quad i,j = 0,1,\ldots,p-1 \tag{2.15}$$

if and only if the Laurent polynomials

$$a_l^0(z) := \sum_{j\in\mathbb{Z}} a_{l+pj}^0 z^j, \quad l = 0,1,\ldots,p-1,$$

have no common zeros on $\mathbb{C}\backslash\{0\}$.

Proof: Let us first examine (2.14). A typical term on the left hand side of (2.14) is

$$((B^l)^T A^l)_{ij} = \sum_{k\in\mathbb{Z}} b_{i-pk}^l a_{j-pk}^l.$$

We multiply both sides of this equation by z^i and sum the resulting formula over $i \in \mathbb{Z}$ to get

$$\sum_{i\in\mathbb{Z}} ((B^l)^T A^l)_{ij} z^i = b^l(z) a_j^l(z^{-p}), \tag{2.16}$$

where

$$b^l(z) = \sum_{j\in\mathbb{Z}} b^l_j z^j$$

and

$$a^l_j(z) = \sum_{r\in\mathbb{Z}} a^l_{j+pr} z^r.$$

Thus (2.14) is equivalent to the identity

$$\sum_{l=0}^{p-1} b^l(z) a^l_j(z^{-p}) = z^j,\ j \in \mathbb{Z},\ z \in \mathbb{C}\backslash\{0\}. \tag{2.17}$$

To express this equation in another way, we write each $b^l(z)$ in the form

$$b^l(z) = \sum_{r=0}^{p-1} z^r b^l_r(z^p), \tag{2.18}$$

where

$$b^l_r(z) = \sum_{k\in\mathbb{Z}} b^l_{r+pk} z^k. \tag{2.19}$$

Then equation (2.14) takes the equivalent form

$$B(z)A(z) = I, \quad z \in \mathbb{C}\backslash\{0\}, \tag{2.20}$$

where

$$A(z) = (a_{ij}(z))_{i,j=0,1,\ldots,p}, \quad a_{ij}(z) = a^i_j(z^{-1}), \tag{2.21}$$

and

$$B(z) = (b_{ij}(z))_{i,j=0,1,\ldots,p}, \quad b_{ij}(z) = b^j_i(z). \tag{2.22}$$

Since the elements of both $B(z)$ and $A(z)$ are Laurent polynomials it follows that

$$\det A(z) = \text{constant} \neq 0, \quad z \in \mathbb{C}\backslash\{0\}, \tag{2.23}$$

and hence the polynomials $a^0_0(z^{-1})$,...,$a^0_{p-1}(z^{-1})$, which make up the first row of $A(z)$, have no common zeros.

Conversely, if these polynomials have no common zeros then by Lemma 2.2 there is a $p \times p$ matrix of Laurent polynomials $A(z)$ whose first row is $(a^0_0(z^{-1}), \ldots, a^0_{p-1}(z^{-1}))$ and the remaining rows are chosen so that $\det A(z) = 1$, $z \in \mathbb{C}\backslash\{0\}$. Thus the inverse of $A(z)$, $B(z) := A^{-1}(z)$, satisfies (2.20) and its elements consist of Laurent polynomials. By the equations (2.21) and (2.22) the coefficients of these Laurent polynomials provide matrices which solve (2.14).

As for equation (2.15), we express it in the form

$$(A^i(B^j)^T)_{rs} = \sum_{k\in\mathbb{Z}} a^i_{k-rp} b^j_{k-sp} = \delta_{rs}\delta_{ij}, \quad r, s \in \mathbb{Z}$$

which is also equivalent to

$$\sum_{q=0}^{p-1}\sum_{m\in\mathbb{Z}} a^i_{q+pm}b^j_{q+pm+lp} = \delta_{0l}\delta_{ij} \quad l\in\mathbb{Z}.$$

Multiply both sides of this equation by z^l and sum over $l\in\mathbb{Z}$. This gives us the formula

$$\sum_{q=0}^{p-1} a^i_q(z^{-1})b^j_q(z) = \delta_{ij}, \quad z\in\mathbb{C}\backslash\{0\}.$$

Combining this with definitions (2.21) and (2.22) we see that (2.15) is equivalent to the equation

$$A(z)B(z) = I, \tag{2.24}$$

which, in turn, is equivalent to (2.20). Thus we have shown, in fact, that (2.14) and (2.15) are equivalent conditions. This proves the result. ■

Proof of Theorem 2.1. According to Lemma 2.1 and Proposition 2.1 there are sequences a^0,..., a^{p-1}, b^0,...,b^{p-1} with $a^0 = a$ such that the corresponding matrices A^0,..., A^{p-1}, B^0,..., B^{p-1} satisfy (2.14) (or equivalently (2.15)). Let

$$g_l(x) = \sum_{j\in\mathbb{Z}} a^l_j h(px-j), \quad l = 0,1,\ldots,p-1 \tag{2.25}$$

and $W_l = \operatorname{span}\{g_l(\cdot - i) : i\in\mathbb{Z}\}$. Since $a^0 = a$ we see by the refinement equation (2.1) that $g_0 = h$ and so $W_0 = V_0$. From (2.25) we conclude that

$$g_l(x-i) = \sum_{j\in\mathbb{Z}} a^l_{j-pi}h(px-j), \quad i\in\mathbb{Z}.$$

Equivalently, in matrix form $G_l(\cdot) = A^l H(p\cdot)$, where G_l, H are the sequences of functions $(g_l(x-i) : i\in\mathbb{Z})$, $(h(x-i) : i\in\mathbb{Z})$, respectively. Thus, by (2.14) we get

$$\sum_{l=0}^{p-1}(B^l)^T G_l(\cdot) = H(p\cdot).$$

In other words, this means that

$$h(px-i) = \sum_{l=0}^{p-1}\sum_{j\in\mathbb{Z}} b^l_{i-pj}g_l(x-j),\ x\in\mathbb{R}, \quad i\in\mathbb{Z}. \tag{2.26}$$

To show that this formula provides a decomposition of V_1 as a direct sum of $V_0, W_1, \ldots, W_{p-1}$, we suppose there are sequences $C_0 = ((C_0)_i : i\in\mathbb{Z})$,..., $C_{p-1} = ((C_{p-1})_i : i\in\mathbb{Z})$ such that

$$\sum_{l=0}^{p-1}\sum_{i\in\mathbb{Z}}(C_l)_i g_l(x-i) = 0, \quad x\in\mathbb{R}.$$

Then from the linear independence of the integer translates of h and the formula $G_l = A^l H(p\cdot)$ we get

$$\sum_{l=0}^{p-1}(A^l)^T C_l = 0.$$

For each $r = 0, 1, \ldots, p-1$, apply the matrix B^r to both sides of this equation and use (2.15) to conclude $C_r = 0$. Finally, since $b^0, \ldots, b^{p-1}$ are finitely supported this direct sum decomposition is also locally finite. ■

§3. p-slanted Matrices.

The goal in this section is to provide the necessary matrix theoretic results to extend Corollary 2.1 to spline functions with arbitrary knots. Motivated by the material in Section 2, we say that a biinfinite matrix $A = (a_{ij})_{i,j\in\mathbb{Z}}$ is p-slanted with band (l, m) if there exist integers $l < m$ such that $a_{ij} = 0$ if $j < pi + l$ or $j > pi + m$. Our aim is to extend [5, Theorem 2.1] to p-slanted matrices. Following Dahmen and Micchelli [5] we approach this problem by first factoring the matrix A. The idea is to reduce the band width of A by successively multiplying A by suitably chosen one banded lower triangular matrices. These matrices have all one's on their diagonal and on their secondary diagonal nonzero elements are separated by $p-1$ zero elements.

To explain this procedure we require some notation. Specifically, we let $A[i_1, i_2, \ldots, i_r | j_1, j_2, \ldots, j_s]$ be the submatrix of A formed by the rows $i_1, i_2, \ldots, i_r$ and the columns $j_1, j_2, \ldots, j_s$ and denote its determinant by

$$A\begin{pmatrix} i_1 & i_2 & \ldots & i_r \\ j_1 & j_2 & \ldots & j_r \end{pmatrix}.$$

The following lemma will be useful in the sequel.

Lemma 3.1. *Let A be any biinfinite matrix and assume for some $i, j, k \in \mathbb{Z}$ that the submatrix $A[i+1, i+2, \ldots, i+k | j+1, j+2, \ldots, j+k]$ is nonsingular. Then there exists a biinfinite matrix P such that $\tilde{A} := AP$ has the property that for any $r \in \mathbb{Z}$*

$$\tilde{a}_{rj} = \frac{A\begin{pmatrix} r & i+1 & i+2 & \ldots & i+k \\ j & j+1 & j+2 & \ldots & j+k \end{pmatrix}}{A\begin{pmatrix} i+1 & i+2 & \ldots & i+k \\ j+1 & j+2 & \ldots & j+k \end{pmatrix}}. \tag{3.1}$$

In particular $\tilde{a}_{i+1,j} = \cdots = \tilde{a}_{i+k,j} = 0$.

Proof: Let $S = A[i+1, i+2, \ldots, i+k | j]$, $L = A[i+1, i+2, \ldots, i+k | j+1, j+2, \ldots, j+k]$ and P the biinfinite matrix defined by

$$P[j+1, j+2, \ldots, j+k | j] = -L^{-1}S,$$

$$p_{st} = \delta_{st} \quad (s,t) \notin \{j+1, j+2, \ldots, j+k\} \times \{j\}.$$

Taking into account the above definitions of $\tilde{A}$ and P and restricting ourselves to the matrix $\tilde{A}[r, i+1, i+2, \ldots, i+k|j, j+1, j+2, \ldots, j+k]$ for any $r \in \mathbb{Z}$, we have

$$\begin{pmatrix} a_{rj} & a_{r,j+1} & \cdots & a_{r,j+k} \\ a_{i+1,j} & a_{i+1,j+1} & \cdots & a_{i+1,j+k} \\ a_{i+2,j} & a_{i+2,j+1} & \cdots & a_{i+2,j+k} \\ \cdots & \cdots & \cdots & \cdots \\ a_{i+k,j} & a_{i+k,j+1} & \cdots & a_{i+k,j+k} \end{pmatrix} \begin{pmatrix} 1 & 0 & \cdots & 0 \\ p_{j+1,j} & 1 & \cdots & 0 \\ p_{j+2,j} & 0 & \cdots & 0 \\ \cdots & \cdots & \cdots & \cdots \\ p_{j+k,j} & 0 & \cdots & 1 \end{pmatrix} =$$

$$\begin{pmatrix} \tilde{a}_{rj} & a_{r,j+1} & \cdots & a_{r,j+k} \\ 0 & a_{i+1,j+1} & \cdots & a_{i+1,j+k} \\ 0 & a_{i+2,j+1} & \cdots & a_{i+2,j+k} \\ \cdots & \cdots & \cdots & \cdots \\ 0 & a_{i+k,j+1} & \cdots & a_{i+k,j+k} \end{pmatrix}.$$

Therefore, taking determinants of both sides of this equation gives (3.1). ∎

Remark 3.1. *The element $\tilde{a}_{rj}$ is called the Schur complement of $A[i+1, i+2, \ldots, i+k|j+1, j+2, \ldots, j+k]$ with respect to $A[r, i+1, i+2, \ldots, i+k|j, j+1, j+2, \ldots, j+k]$. From (3.1) it is straightforward that its value does not change if we add to the column $j+t$ $(1 \le t \le k)$ of A a linear combination of the columns $j+h$, $1 \le h \le k, h \ne t$.*

This fact will be often used in the proof of the next theorem, which will prepare us for the desired factorization of A. We assume from now on that the band of the p-slanted matrix A is $(0, m)$, and leave it to the reader to formulate the general case by shifting the column index of A by l.

The elimination process which appears in condition (ii) of the theorem below is closely related to *Neville elimination* for finite matrices studied by Gasca and Peña in [8].

Theorem 3.1. *Suppose $A = (a_{ij})_{i,j\in\mathbb{Z}}$ is a p-slanted matrix with band $(0, m)$. For any integer n, $1 \le n < m$, the following properties are equivalent.*

(i) Let r, q be integers such that $n = q(p-1) + r$, $0 \le r < p-1$, (that is $n \equiv r \bmod (p-1)$). Then for all integers i, s, t such that $0 \le s \le q$ and $1 \le t \le p-1$ if $s < q$ while $1 \le t \le r$ if $s = q$ and $r > 0$, one has, for $j = (p-1)s + t$,

$$A\begin{pmatrix} i & i+1 & \cdots & i+s \\ pi+j & pi+j+1 & \cdots & pi+j+s \end{pmatrix} \neq 0. \tag{3.2}$$

(ii) There exist $G^{(j)} = (g_{hk}^{(j)})_{h,k\in\mathbb{Z}}$, $j = 1, 2, \ldots, n+1$, defined recursively as

$$g_{hk}^{(j)} = \begin{cases} 1 & \text{if} \quad h = k, \\ -a_{i,pi+j-1}^{(j-1)}/a_{i,pi+j}^{(j-1)} & \text{if} \quad (h,k) = (pi+j, pi+j-1), \\ 0 & \text{otherwise}, \end{cases} \tag{3.3}$$

where $A := (a_{iu}^{(0)})_{i,u\in\mathbb{Z}}$ and

$$(a_{iu}^{(j)})_{i,u\in\mathbb{Z}} := A^{(j)} := AG^{(1)}\cdots G^{(j)}, \tag{3.4}$$

such that the matrices $A^{(j)}, j = 0, 1, \ldots n+1$, *satisfy*

$$a_{iu}^{(j)} = 0, \quad \text{if} \quad u < pi + j \quad \text{or} \quad u > pi + m. \tag{3.5}$$

In particular, they are p*-slanted with band-width* $m + j$.

Proof: Since $a_{i,pi+1}^{(0)} = a_{i,pi+1} = A\begin{pmatrix} i \\ pi+1 \end{pmatrix}$, the existence of the matrix $G^{(1)}$ is equivalent to (3.2) for $j = 1$. The matrix $A^{(1)} = AG^{(1)}$ is obtained by adding to the column pi of A, for each integer i, the column $pi+1$ multiplied by $-a_{i,pi}/a_{i,pi+1}$. $A^{(1)}$ is p-slanted, with band width $m-1$ because $a_{i,pi}^{(1)} = 0$.

Now $a_{i,pi+2}^{(1)} = a_{i,pi+2}^{(0)} = A\begin{pmatrix} i \\ pi+2 \end{pmatrix}$ and the existence of $G^{(2)}$ is equivalent to (3.2) for $j = 2$. Analogous reasoning proves that, for any $j = 1, \ldots, p-1$, the existence of $G^{(j)}$ is equivalent to (3.2). In particular $G^{(p-1)}$ exists if and only if $a_{i,pi+p-1}^{(p-2)} = a_{i,pi+p-1}^{(0)} = A\begin{pmatrix} i \\ pi+p-1 \end{pmatrix} \neq 0$.

For $j = p$ we observe that $a_{i+1,pi+p}^{(p-1)} = 0$ because $a_{i+1,pi+p}^{(1)} = 0$. Hence, $a_{i,pi+p}^{(p-1)}$ can be seen to be the Schur complement of $A[i+1|pi+p+1]$ (which is nonzero from (3.2) by choosing $j = 1$ and i replaced by $i+1$) with respect to $A[i, i+1|pi+p, pi+p+1]$, and by Lemma 3.1 we have

$$a_{i,pi+p}^{(p-1)} = \frac{A\begin{pmatrix} i & i+1 \\ pi+p & pi+p+1 \end{pmatrix}}{A\begin{pmatrix} i+1 \\ pi+p+1 \end{pmatrix}}.$$

Therefore, the existence of $G^{(p)}$ is equivalent to (3.2) for $j = p$ (that is $s = 1$, $t = 1$). The matrix $G^{(p)}$ is p-slanted with band width $m-p$. Similar arguments prove that, for $j = p + w \quad (1 \leq w \leq p-2)$,

$$a_{i,pi+j}^{(j-1)} = \frac{A\begin{pmatrix} i & i+1 \\ pi+j & pi+j+1 \end{pmatrix}}{A\begin{pmatrix} i+1 \\ pi+j+1 \end{pmatrix}}$$

and that, for $j = 2p-1 = 2(p-1)+1$, by Lemma 3.1 and Remark 3.1, we obtain

$$a_{i,pi+2p-1}^{(2p-2)} = \frac{A\begin{pmatrix} i & i+1 & i+2 \\ pi+j & pi+j+1 & pi+j+2 \end{pmatrix}}{A\begin{pmatrix} i+1 & i+2 \\ pi+j+1 & pi+j+2 \end{pmatrix}}.$$

In general, for $j = (p-1)s + t$, $1 \le t \le p-1$, $(j \le n)$, the existence of $A^{(j+1)} = A^{(j)}G^{(j+1)}$ is equivalent to the condition that $a^{(j)}_{i,pi+j+1}$ be nonzero. This particular element of $A^{(j)}$ can be seen to be the Schur complement of $A[i+1,\ldots,i+s|pi+j+2,\ldots,pi+j+s+1]$ with respect to $A[i,\ldots,i+s|pi+j+1,\ldots,pi+j+s+1]$ because the columns $pi+j+2,\ldots,pi+j+s+1$ have been used to make zeros in column $pi+j+1$, rows $i+1, i+2, \ldots$ of A. Hence, by Lemma 3.1 and Remark 3.1, we conclude that

$$a^{(j)}_{i,pi+j+1} = \frac{A\begin{pmatrix} i & i+1 & \ldots & i+s \\ pi+j+1 & pi+j+2 & \ldots & pi+j+s+1 \end{pmatrix}}{A\begin{pmatrix} i+1 & i+2 & \ldots & i+s \\ pi+j+2 & pi+j+3 & \ldots & pi+j+s+1 \end{pmatrix}},$$

which is nonzero if and only if (3.2) holds for $j+1$. ■

Remark 3.2. *It is possible to produce a factorization of the form (3.4) even under conditions weaker than the requirement (3.2). Specifically, suppose that whenever*

$$A\begin{pmatrix} i & \ldots & i+s \\ pi+j & \ldots & pi+j+s \end{pmatrix} = 0$$

it follows that

$$a_{i,pi+j-1} = \cdots = a_{i+s,pi+j-1} = 0.$$

In this case (3.4) remains valid for the matrices $G^{(j)}$ whose elements are defined to be

$$g^{(j)}_{hk} = \begin{cases} 1 & \text{if} \quad h = k, \\ -\frac{a^{(j-1)}_{i,pi+j-1}}{a^{(j-1)}_{i,pi+j}} & \text{if} \quad (h,k) = (pi+j, pi+j-1) \text{ and } a_{i,pi+j-1} \neq 0, \\ 0 & \text{otherwise.} \end{cases}$$

To prove this fact we merely follow the same steps as in the proof of Theorem 3.1 taking account of the fact that the determinant

$$A\begin{pmatrix} i & \ldots & i+s \\ pi+j & \ldots & pi+j+s \end{pmatrix}$$

is allowed to be zero when the elements $a_{i+h,pi+j-1}$, $h = 0,\ldots,s$ are zero. In fact, with the matrices $G^{(j)}$ defined above these elements will be always zero.

Corollary 3.1. *If (3.2) holds for $n = m-1$, then*

$$AG^{(1)}\cdots G^{(m)} = DI^{(m)},$$

where the matrices $G^{(j)}$, $j = 1,\ldots,m$, are given by (3.3), D is the diagonal matrix with diagonal entries $d_{ii} = a^{(m)}_{i,pi+m}$, $i \in \mathbb{Z}$ and $I^{(m)}$ is a p-slanted matrix with band (m,m) given explicitly as

$$I^{(m)} = (\delta_{k,pi+m})_{i,k\in\mathbb{Z}}. \tag{3.6}$$

Furthermore, for any $i \in \mathbb{Z}$

$$a^{(m)}_{i,pi+m} = a^{(m-1)}_{i,pi+m} \neq 0,$$

and if the minors of A in (3.2) are positive for $n = m-1$, A is totally positive.

Proof: This corollary follows from Theorem 3.1 by choosing $n = m-1$. As for the claim that A is totally positive we note that when the minors of A in (3.1) are positive, the off-diagonal elements of each $G^{(j)}$ are nonpositive. Hence its lower triangular inverse is a bidiagonal matrix with nonnegative off-diagonal elements . It is a well-known fact that such matrices are totally positive. ■

Corollary 3.2. *If A is totally positive, condition (i) of Theorem 3.1 can be replaced by*

(i) *Let r, q be integers such that $n = q(p-1) + r$, $0 \le r < p-1$ (that is, $n \equiv r \bmod (p-1)$). Then*

$$A\begin{pmatrix} i & \cdots & i+q \\ pi+j & \cdots & pi+j+q \end{pmatrix} \neq 0, \quad j = n, n-1, \ldots, n-r+1 \quad (3.7)$$

and

$$A\begin{pmatrix} i & \cdots & i+q-1 \\ pi+j & \cdots & pi+j+q-1 \end{pmatrix} \neq 0, \quad j = n-r, n-r-1, \ldots, n-p+2. \quad (3.8)$$

Proof: The differences between the indices of column and row on the main diagonal of the determinants of (3.7) are, respectively, $(p-1)i+n, (p-1)i+n-1, \ldots, (p-1)i+n-r+1$, and those of the determinants of (3.8) are $(p-1)i+n-r, (p-1)i+n-r-1, \ldots, (p-1)i+n-p+2$. In other words, if for each $h = 0, 1, \ldots, p-2$ we take the determinant of (3.2) *with maximum order* such that the difference between the indices of row and column on its main diagonal is congruent with h, mod $(p-1)$, we obtain the determinants in (3.7)-(3.8). Therefore, the remaining determinants in (3.2) are principal minors of those in (3.7)-(3.8). For example, the difference between indices of column and row on the main diagonal of $A\begin{pmatrix} i & i+1 \\ pi+p & pi+p+1 \end{pmatrix}$ (which corresponds to $j = p$ in (3.2)) is $(p-1)i + p \equiv 1 \bmod (p-1)$). On the other hand, for $j = n-r+1$, the difference between the indices on the main diagonal of (3.7) is $(p-1)i + n - r + 1 \equiv 1 \bmod (p-1)$. Consequently, $A\begin{pmatrix} i & i+1 \\ pi+p & pi+p+1 \end{pmatrix}$ is a principal minor of

$$A\begin{pmatrix} i-q+1 & \cdots & i+1 \\ p(i-q+1)+n-r+1 & \cdots & p(i-q+1)+n-r+1+q \end{pmatrix}$$

because $p(i-q+1)+n-r+1+q = pi+p+1$. Recall that in a nonsingular totally positive matrix all the principal minors are positive (see Corollary 3.8

of Ando [1], for example). Hence, for a totally positive matrix A, (3.2) is a consequence of (3.7)(3.8) and we deduce the equivalence of (i) of Theorem 3.1 and the condition of this corollary. ■

Remark 3.3. *According to Corollary 3.2, if A is totally positive, condition (3.2) in Corollary 3.1 can be replaced by (3.7)-(3.8), which involve only $p-1$ minors of A. Both Theorem 3.1 and Corollary 3.1 parallel the corresponding results in Dahmen and Micchelli [5] for $p=2$. In that case (3.7) is empty, because r is always zero, and (3.8) involves only one minor.*

As in [5], the main application of Theorem 3.1 yields the following extension of Proposition 2.1.

Theorem 3.2. *Suppose that A is a p-slanted matrix with band $(0,m)$ which satisfies (3.2) with $n=m-1$. Then there exist p-slanted matrices $A_0,\ldots,$ A_{p-1}, $B_0,\ldots,$ B_{p-1} such that $A_0=A$,*

$$A_iB_j^T=\delta_{ij}I, \quad i,j=0,1,\ldots,p-1 \tag{3.9}$$

and

$$\sum_{j=0}^{p-1}B_j^TA_j=I. \tag{3.10}$$

Proof: To construct these matrices we let $G=G^{(1)}\cdots G^{(m)}$ as in Theorem 3.1, so that the lower triangular inverse of G is given by

$$G^{-1}=\left(G^{(m)}\right)^{-1}\cdots\left(G^{(1)}\right)^{-1}.$$

Referring back to (3.6) we observe that

$$\left(\left(I^{(m)}\right)^TI^{(m)}\right)_{ij}=\begin{cases}1 & \text{if } i=j\equiv m \mod p,\\ 0 & \text{otherwise.}\end{cases}$$

Thus it follows that

$$\sum_{h=0}^{p-1}\left(I^{(m+h)}\right)^TI^{(m+h)}=I. \tag{3.11}$$

This suggests that we define

$$B_0{}^T=G\left(I^{(m)}\right)^TD^{-1}, \tag{3.12}$$

where D is the diagonal matrix defined in Corollary 3.1 and, for $j=1,\ldots,$ $p-1$,

$$A_j=I^{(m+j)}G^{-1}, \tag{3.13}$$

$$B_j{}^T=G\left(I^{(m+j)}\right)^T. \tag{3.14}$$

Finally, according to Corollary 3.1 we define

$$A_0 := A = DI^{(m)}G^{-1}.$$

Thus, from (3.11), it follows that

$$\sum_{j=0}^{p-1} B_j{}^T A_j = G(\sum_{j=0}^{p-1}(I^{(m+j)})^T I^{(m+j)})G^{-1} = GG^{-1} = I.$$

As for (3.9) we use the easily verified fact that

$$I^{(m+i)}(I^{(m+j)})^T = \delta_{ij} I, \quad i,j = 0,1,\dots,p-1. \quad \blacksquare$$

Remark 3.4. *Observe that this theorem also holds under the weaker condition described in Remark 3.2.*

Next, we make some remarks about the bands of the matrices A_i, B_i, $i = 0,1,\dots,p-1$ constructed in Theorem 3.2. To this end, we write m in the form

$$m = s(p-1) + r, \quad 0 < r \leq p-1.$$

Then it can be proved inductively on m that,

$$G_{hk} = 0, \quad h < k \quad \text{or} \quad h > k+s+1. \tag{3.15}$$

Moreover, if

$$k = pi + m + t, \quad 0 \leq t \leq p-1 \tag{3.16}$$

then for $0 \leq t \leq p-1-r$

$$G_{k+s+1,k} = 0. \tag{3.17}$$

As an illustration, we choose $p = 3$ and $m = 7$. Then, it follows that $r = 1$ and $s = 3$ and so the zero pattern of the submatrix of G formed from the rows and columns $0, 1, \dots, 10$ is given by

$$\begin{matrix}
\ddots & \vdots & \vdots & \vdots & \vdots & \vdots & \vdots & \vdots & \vdots & \vdots & \vdots & \vdots & \\
\cdots & 1 & 0 & 0 & 0 & 0 & 0 & 0 & 0 & 0 & 0 & 0 & \cdots \\
\cdots & * & 1 & 0 & 0 & 0 & 0 & 0 & 0 & 0 & 0 & 0 & \cdots \\
\cdots & * & * & 1 & 0 & 0 & 0 & 0 & 0 & 0 & 0 & 0 & \cdots \\
\cdots & * & * & * & 1 & 0 & 0 & 0 & 0 & 0 & 0 & 0 & \cdots \\
\cdots & * & * & * & * & 1 & 0 & 0 & 0 & 0 & 0 & 0 & \cdots \\
\cdots & 0 & 0 & * & * & * & 1 & 0 & 0 & 0 & 0 & 0 & \cdots \\
\cdots & 0 & 0 & 0 & * & * & * & 1 & 0 & 0 & 0 & 0 & \cdots \\
\cdots & 0 & 0 & 0 & * & * & * & * & 1 & 0 & 0 & 0 & \cdots \\
\cdots & 0 & 0 & 0 & 0 & 0 & * & * & * & 1 & 0 & 0 & \cdots \\
\cdots & 0 & 0 & 0 & 0 & 0 & 0 & * & * & * & 1 & 0 & \cdots \\
\cdots & 0 & 0 & 0 & 0 & 0 & 0 & * & * & * & * & 1 & \cdots \\
\ddots & \vdots & \vdots & \vdots & \vdots & \vdots & \vdots & \vdots & \vdots & \vdots & \vdots & \vdots &
\end{matrix}.$$

Let us describe carefully the proof of equations (3.15) and (3.17). First, we recall from (3.3) that

$$g_{hk}^{(j)} = \begin{cases} 1 & \text{if} \quad l = k, \\ 0 & \text{if} \quad k < l - 1 \text{ or } k > l \\ 0 & l \not\equiv j \bmod p. \end{cases} \tag{3.18}$$

Hence, when $m = 1$ we have $s = 0$ and $r = 1$. Moreover, for $0 \le t \le p-2$ it follows that $k + 1 \not\equiv 1(\text{mod}p)$ and so $G_{k+1,k} = 0$. This proves (3.15) and (3.17) in this case (it is clear G is lower triangular since each $G^{(j)}$ is lower triangular).

Now, suppose (3.15) and (3.17) are valid for $m - 1$. Set $H := G^{(1)} \cdots$ $\cdots G^{(m-1)}$ and observe that for any $h, k \in \mathbb{Z}$

$$G_{hk} = H_{hk} + H_{h,k+1} g_{k+1,k}^{(m)}. \tag{3.19}$$

To use the induction hypothesis we need to note that

$$m - 1 = \begin{cases} (s-1)(p-1) + p - 1, & \text{if } r = 1 \\ s(p-1) + r - 1, & \text{if } 2 \le r \le p - 1. \end{cases} \tag{3.20}$$

First, we prove (3.15). To this end, suppose $h > k + s + 1$. We divide the argument into two cases. If $2 \le r \le p - 1$ then the induction hypothesis implies that

$$H_{hk} = 0, \quad \text{if} \quad h > k + s + 1 \tag{3.21}$$

(since the value of s doesn't change when m is replaced by $m-1$) and therefore (3.19) implies that

$$G_{hk} = H_{h,k+1} g_{k+1,k}^{(m)}. \tag{3.22}$$

The right hand side of this equation is clearly zero by (3.18) and (3.21) unless

$$h = k + s + 2$$

and

$$k + 1 \equiv m (\text{mod}\, p).$$

The last equation above and (3.16) imply that $t = p - 1$. Hence

$$k + 1 = p(i + 1) + (m - 1) + 1$$

and therefore by the induction hypothesis (use equations (3.16) with k replaced by $k + 1, i$ by $i + 1, m$ by $m - 1, t$ by 1 and (3.17) with G replaced by H and r by $r - 1$; s remains the same)

$$H_{k+s+2,k+1} = 0,$$

which again implies by equation (3.22) that $G_{hk} = 0$.

In the second case, $r = 1$, we use (3.20) and the induction hypothesis to conclude that

$$H_{hk} = 0, \quad h > k + s,$$

(use equation (3.15) with s replaced by $s - 1$ and G by H). This equation and (3.19) clearly imply that

$$G_{hk} = 0, \quad \text{if} \quad h > k + s + 1.$$

Thus, in all cases, we have verified (3.15).

Now, let us confirm (3.17). In this case

$$k = pi + m + t,$$

where $0 \le t \le p - 1 - r \le p - 2$. This implies that

$$k + 1 \not\equiv m (\bmod p),$$

and so (3.18) and (3.19) give

$$G_{hk} = H_{hk}.$$

When $h = k + s + 1$ we claim that $H_{k+s+1,k} = 0$. Again, we consider two cases. When $r = 1$ (3.15) (with G replaced by H, s by $s - 1$) we conclude $H_{hk} = 0$ if $h > k + s$, in particular $H_{k+s+1,k} = 0$, while for $2 \le r \le p - 1$ we have

$$k = pi + m - 1 + \hat{t}$$

where $0 \le \hat{t} := t + 1 \le p - 1 - (r - 1)$ and so (3.17) implies that $H_{k+s+1,k} = 0$. These remarks confirm our claim about G given in equations (3.15) and (3.17).

Next we observe that similarly, for G^{-1} we have, when $t = 0$, that

$$(G^{-1})_{kj} = 0 \quad \text{if} \quad j < k - m \quad \text{or} \quad j > k, \tag{3.23}$$

while for $t = 1, 2 \ldots, p - 1$

$$(G^{-1})_{kj} = 0 \quad \text{if} \quad j < k - m - t + p \quad \text{or} \quad j > k. \tag{3.24}$$

The proof of equations (3.23) and (3.24) also follows by induction on m. As these equations are not used in the sequel we omit their proof. Specializing (3.23) and (3.24) to the case $p = 3$ and $m = 7$ gives, for the submatrix of G^{-1} consisting of the rows and columns $0, 1, \ldots, 10$ the zero pattern

$$\begin{array}{ccccccccccccc}
\ddots & \vdots & \vdots & \vdots & \vdots & \vdots & \vdots & \vdots & \vdots & \vdots & \vdots & \vdots & \\
\cdots & 1 & 0 & 0 & 0 & 0 & 0 & 0 & 0 & 0 & 0 & 0 & \cdots \\
\cdots & * & 1 & 0 & 0 & 0 & 0 & 0 & 0 & 0 & 0 & 0 & \cdots \\
\cdots & * & * & 1 & 0 & 0 & 0 & 0 & 0 & 0 & 0 & 0 & \cdots \\
\cdots & * & * & * & 1 & 0 & 0 & 0 & 0 & 0 & 0 & 0 & \cdots \\
\cdots & * & * & * & * & 1 & 0 & 0 & 0 & 0 & 0 & 0 & \cdots \\
\cdots & * & * & * & * & * & 1 & 0 & 0 & 0 & 0 & 0 & \cdots \\
\cdots & * & * & * & * & * & * & 1 & 0 & 0 & 0 & 0 & \cdots \\
\cdots & * & * & * & * & * & * & * & 1 & 0 & 0 & 0 & \cdots \\
\cdots & 0 & 0 & 0 & * & * & * & * & * & 1 & 0 & 0 & \cdots \\
\cdots & 0 & 0 & 0 & * & * & * & * & * & * & 1 & 0 & \cdots \\
\cdots & 0 & 0 & 0 & * & * & * & * & * & * & * & 1 & \cdots \\
\ddots & \vdots & \vdots & \vdots & \vdots & \vdots & \vdots & \vdots & \vdots & \vdots & \vdots & \vdots &
\end{array}$$

Returning to the general case, we conclude from these facts the following result.

Proposition 3.1. *Let A_i, B_i, $i = 0, 1, \ldots, p-1$ be the p-slanted matrices in Theorem 3.2 and suppose $m = s(p-1) + r$, $\quad 0 < r \leq p-1$. Then for $j = 1, 2 \ldots, p-1$, A_j has band $(p, m+j)$. Moreover, for $j = 0, 1, \ldots, p-1-r$, B_j has band $(m+j, m+j+s)$ while for $j = p-r, \ldots, p-1$, it has band $(m+j, m+j+s+1)$.*

Proof: Indeed, A_j is the matrix formed with the rows $pi + m + j$ $(i \in \mathbb{Z})$ of G^{-1} and B_j^T the matrix formed with the columns $pi + m + j$ $(i \in \mathbb{Z})$ of G. Thus the proposition follows from equations (3.15), (3.17), (3.23)–(3.24). ■

We remark that the two matrices S and T whose elements are defined by

$$S_{pi+h,j} = (A_h)_{ij}, \quad h = 0, 1 \ldots, p-1, \quad i, j \in \mathbb{Z},$$

and

$$T_{i,pj+h} = (B_h)_{ij}, \quad h = 0, 1 \ldots, p-1, \quad i, j \in \mathbb{Z},$$

respectively, are inverses of each other. In fact equation (3.10) is equivalent to the equation $TS = I$ while (3.9) means $ST = I$ and moreover both T and S are banded matrices.

The following result gives a family of solutions (not necessarily banded) of equations (3.9) and (3.10).

Proposition 3.2. *Let $A_i, B_i, i = 0, 1, \ldots, p-1$ be the matrices in Theorem 3.2. For any banded matrices $K_1, \ldots, K_{p-1}$ with banded inverses and any matrices $M_1, \ldots, M_{p-1}$ we define*

$$\tilde{B}_0^T = B_0^T + \sum_{j=1}^{p-1} B_j^T M_j \tag{3.25}$$

and for $j = 1, \ldots, p-1$

$$\tilde{B}_j^T = B_j^T K_j, \quad \tilde{A}_j = K_j^{-1}(A_j - M_j A). \tag{3.26}$$

Then the matrices $A_0, \tilde{A}_1, \ldots, \tilde{A}_{p-1}, \tilde{B}_0, \ldots, \tilde{B}_{p-1}$ satisfy equations (3.9) and (3.10).

Proof: The proof is by direct computation using that A_i, B_j satisfy (3.9), (3.10). ■

§4. Applications

In this section we consider two applications of the results of Section 3. The first concerns spline interpolation and the situation is the following. We begin with a biinfinite sequence of points $\{x_i\}_{i\in\mathbb{Z}}$ such that $x_i < x_{i+1}, i \in \mathbb{Z}$. We wish to interpolate data y_i at x_i for all $i \in \mathbb{Z}$ by a spline function of degree $k-1$, where $k \geq 2$. We choose the knots of the spline on a sequence $\{t_i\}_{i\in\mathbb{Z}}$ such that $t_i < t_{i+1}$ and $x_i = t_{pi}, i \in \mathbb{Z}$, where p is some integer greater than one. Hence in each of the intervals (x_i, x_{i+1}) there are $p-1$ knots of the spline.

Let $\{L_i : i \in \mathbb{Z}\}$ be the Lagrange splines for this interpolation problem and $\{F_i : i \in \mathbb{Z}\}$ the B-splines associated with the knot sequence $\{t_i : i \in \mathbb{Z}\}$. To be precise, F_i is the B-spline with support on the interval $[t_{i+1-k}, t_{i+1}]$. The Lagrange functions satisfy the conditions

$$L_i(x_j) = \delta_{ij}, \quad i, j \in \mathbb{Z} \tag{4.1}$$

and for some biinfinite matrix $B = (b_{ij})_{i,j\in\mathbb{Z}}$ can be expressed in the form

$$L_i(x) = \sum_{j\in\mathbb{Z}} b_{ij} F_j(x). \tag{4.2}$$

We wish to find L_i of compact support which satisfies (4.1). To this end, we suppose B is p-slanted with band (α, β) where naturally $\alpha \leq \beta$. Then according to (4.2)

$$\operatorname{supp} L_i \subseteq [t_{pi+\alpha-k+1}, t_{pi+\beta+1}]. \tag{4.3}$$

Equation (4.1) demands that $x_i = t_{pi} \in (t_{pi+\alpha-k+1}, t_{pi+\beta+1})$ and hence we must require that

$$\alpha - k + 1 < 0, \quad \beta + 1 > 0 \tag{4.4}$$

Next, we let $A = (F_j(x_i))_{i,j\in\mathbb{Z}}$ and observe that (4.1) means that

$$I = BA^T. \tag{4.5}$$

The matrix A is also p-slanted with band $(0, k-2)$ and moreover is almost strictly totally positive in the sense of [7]. Thus every minor of A is nonnegative and is strictly positive if and only if its diagonal elements are positive (see de Boor [2]).

Following Dahmen, Goodman and Micchelli [4] we also require that the Lagrange functions provide an interpolation operator with a prescribed degree of accuracy. By this we mean that for all $x \in \mathbb{R}$

$$x^i = \sum_{j\in\mathbb{Z}} x_j^i L_j(x), \quad i \in J := \{0, 1, \ldots, l\}. \tag{4.6}$$

Here l is a prescribed integer with $-1 \le l \le k-1$ ($l = -1$ means that no condition is imposed by (4.6)). Since there exists a matrix $U = (u_{ij})_{i\in J, j\in \mathbb{Z}}$ such that

$$x^i = \sum_{j\in\mathbb{Z}} u_{ij} F_j(x), \quad x \in \mathbb{R}, \quad i \in J, \tag{4.7}$$

we may use equations (4.2),(4.6) and (4.7) to obtain a matrix equation equivalent to (4.6), namely

$$U = VB, \tag{4.8}$$

where $V = (x_j^i)_{i\in J, j\in\mathbb{Z}}$.

For $p = 2$, Dahmen, Goodman and Micchelli [4] showed that for any $k \ge 3$, $-1 \le l \le k-1$ and integer ν with $1 \le \nu \le k-2+l$ there is an L_i satisfying (4.1) and (4.6) which vanishes off the interval $[x_{i-\nu}, x_{i-\nu+k+l-1}]$. Consequently, this means that B is 2-slanted with band $(k-1-2\nu, 2k+2l-3-2\nu)$. Encouraged by this result we examine equations (4.5) and (4.8) for p-slanted solutions. We begin with

Proposition 4.1. *Let A be a p-slanted matrix which satisfies the hypotheses of Theorem 3.2 and A_j, B_j, $j = 0, 1, \ldots, p-1$, be the matrices constructed there. Suppose U, V are two biinfinite matrices with $l+1$ rows ($l = \infty$ is allowed) such that*

$$UA^T = V. \tag{4.9}$$

Then a biinfinite matrix B satisfies equations (4.5) and (4.8) if and only if there are biinfinite matrices F_j such that

$$UA_i^T = VF_i, \quad i = 1, \ldots, p-1, \tag{4.10}$$

$$B = B_0 + \sum_{j=1}^{p-1} F_j B_j. \tag{4.11}$$

Proof: First we show that B given by (4.11), where F_j satisfy (4.10), solves (4.5) and (4.8). For this purpose, we recall from (3.9) of Theorem 3.2 that

$$B_j A_0^T = \delta_{0j}, \quad j = 0, 1, \ldots, p-1. \tag{4.12}$$

Thus

$$BA^T = BA_0^T = (B_0 + \sum_{j=1}^{p-1} F_j B_j) A_0^T = B_0 A_0^T = I,$$

which confirms equation (4.5). As for (4.8) we have by (4.10)

$$VB = VB_0 + \sum_{j=1}^{p-1} VF_j B_j = VB_0 + \sum_{j=1}^{p-1} UA_j^T B_j,$$

which simplifies, by using (3.10) and (4.9),

$$VB = VB_0 + U(I - A_0^T B_0) = VB_0 + U - VB_0 = U$$

and confirms (4.8).

Conversely, suppose B satisfies (4.5) and (4.8). As was pointed out after Proposition 3.1, the two banded matrices

$$S = (S_{ij})_{i,j\in\mathbb{Z}}, \quad T = (T_{ij})_{i,j\in\mathbb{Z}}$$

given by

$$S_{pi+h,j} = (A_h)_{ij}, \quad h = 0, 1, \ldots, p-1, \quad i, j \in \mathbb{Z},$$

$$T_{j,pi+h} = (B_h)_{ij}, \quad h = 0, 1, \ldots, p-1, \quad i, j \in \mathbb{Z},$$

are inverses of each other. We set $F = BS^T$. Then equation (4.5) can be interpreted as saying that

$$F_{i,pj} = \delta_{ij}, \quad i, j \in \mathbb{Z}.$$

Thus, if we define matrices F_1,..., F_{p-1} by

$$(F_h)_{i,j} = F_{i,pj+h}, \quad h = 0, \ldots, p-1, \quad i, j \in \mathbb{Z},$$

we see that $F_0 = I$ and,

$$B_{ij} = \sum_{r\in\mathbb{Z}} F_{ir}T_{jr} = \sum_{h=0}^{p-1}\sum_{t\in\mathbb{Z}} F_{i,pt+h}T_{j,pt+h} = \sum_{h=0}^{p-1}\sum_{t\in\mathbb{Z}}(F_h)_{it}(B_h)_{tj}$$

or

$$B = B_0 + \sum_{j=1}^{p-1} F_j B_j,$$

which proves (4.11).

There remains the task to verify (4.10). We substitute (4.11) into (4.8) to get

$$U = VB_0 + \sum_{j=1}^{p-1} VF_jB_j.$$

For $i = 1, \ldots, p-1$ we multiply both sides of this equation on the right by A_i^T and use (3.9) to obtain (4.10). ■

We give two possible solutions of equations (4.10) and, hence, B given in (4.11) will solve (4.5) and (4.8). For our first solution we let W be any right inverse of V, that is

$$VW = I. \tag{4.13}$$

Here I is the $(\ell+1)\times(\ell+1)$ identity matrix and W has $l+1$ columns. In fact, we can make all of the rows of W zero except l. For instance choose any $i_0, \ldots, i_l$ and set $W_{ij} = 0$ for $i \notin \{i_0, \ldots, i_l\}$, $j \in J$. Then (4.13) becomes

$$\sum_{t\in J} x_{i_t}^r W_{i_t,j} = \delta_{rj}, \quad r, j \in J. \tag{4.14}$$

Since the Vandermonde matrix $(x_{i_t}^r)_{r,t\in J}$ is invertible (4.14) has a solution. With this choice of W we define

$$B = B_0 + \sum_{j=1}^{p-1} WUA_j^T B_j, \tag{4.15}$$

in other words, we have chosen

$$F_j = WUA_j^T, \quad j = 1, \ldots, p-1,$$

in (4.11) and, in view of (4.13), F_j is clearly a solution of (4.10). Thus (4.15) provides a solution to our problem (which is *not* p-slanted).

Some comments on the form of the fundamental functions are in order. Let

$$Q = \sum_{j=1}^{p-1} WUA_j{}^T B_j$$

so that $B = B_0 + Q$. Since W has all its rows zero except rows $i_0, \ldots, i_l$, Q has the same property. Thus the Lagrange functions

$$L_i(x) = \sum_{j\in\mathbb{Z}} (B_0)_{ij} F_j(x) + \sum_{j\in\mathbb{Z}} Q_{ij} F_j(x)$$

are all of compact support for $i \notin \{i_0, \ldots, i_l\}$. The band width of A is $k-2$ and according to Proposition 3.1, the band of B_0 is $(k-2, k-2+s)$, where s is defined by $k-2 = s(p-1)+r$, $0 < r \le p-1$. Therefore, according to our remark following (4.2), we see that the support of L_i, $i \notin \{i_0, \ldots, i_l\}$, is $[t_{pi-1}, t_{pi+k-1+s}]$. In the case $p=2$ we have $r=1$ and $s=k-3$ and so, in particular, L_i, $i \notin \{i_0, \ldots, i_l\}$, has support in $[t_{2i-1}, t_{2i+2k-4}]$, *independent* of l. In contrast, the fundamental function L_i constructed by Dahmen, Goodman and Micchelli mentioned above has support on the interval $[x_{i-\nu}, x_{i-\nu+k+l-1}]$ where ν is any positive integer not greater than $k+l-2$. Except in the case $l=-1$, we have reduced the support of the fundamental functions $L_i, i \notin \{i_0, \ldots, i_l\}$ at the sake of allowing the remaining ones $L_i, i \in \{i_0, \ldots, i_l\}$ have *infinite* support.

Next, when $l \ge 0$ we present a p-slanted solution to equations (4.5) and (4.7) which will insure that *all* fundamental functions are of compact support. This solution is based on the fact that if F is a banded matrix and B is p-slanted then FB is p-slanted. Specifically, if there are integers l_F, m_F such that whenever $F_{ij} \ne 0$ it follows that $l_F \le j-i \le m_F$ and integers l_B, m_B such that $B_{ij} \ne 0$ implies that $l_B \le j-pi \le m_B$, then FB is p-slanted with band $(l_B + pl_F, m_B + pm_F)$. Thus we should construct matrices F_j, $j = 1, \ldots, p-1$, satisfying (4.10), which are banded. For this purpose, we choose $r_0 \in \mathbb{Z}$ and observe that for any $v \in \mathbb{Z}$ and $j = 1, \ldots, p-1$ there exist numbers $(F_j)_{v-r_0-l+t,v}$, $t = 0, 1, \ldots, l$, which satisfy the equation

$$(UA_j^T)_{uv} = \sum_{t=0}^{l} x_{v-r_0+t-l}^u (F_j)_{v-r_0+t-l,v}$$

for $u = 0, 1, \ldots, l$. We now define $(F_j)_{iv} = 0$ for $i \notin \{v - r_0 - l, \ldots, v - r_0\}$. Keeping in mind that U only has $l + 1$ rows we have that F_j satisfies (4.10) and is banded, in fact, $(F_j)_{iv} \neq 0$ implies $r_0 \leq v - i \leq r_0 + l$. Next, we appeal to Proposition 3.1 and recall that, by our above remark, $F_j B_j$ is p-slanted with band

$$\begin{cases} (k-2+j+pr_0, k-2+j+s+pr_0+pl), & j = 1, \ldots, p-1-r, \\ (k-2+j+pr_0, k-2+j+s+pr_0+pl+1), & j = p-r, \ldots, p-1, \end{cases}$$

where $k - 2 = s(p-1) + r$, $0 < r \leq p - 1$ and in addition B_0 is p-slanted with band $(k-2, k-2+s)$.

The matrix B defined in (4.11) is p-slanted with band given by

$$\begin{cases} (k-2, k-2+s+p+pr_0+p_l), & \text{if } r_0 \geq 0 \\ (k+pr_0-1, k-2+s+p+pr_0+p_l), & \text{if } -l-1 \leq r_0 \leq -1 \\ (k+pr_0-1, k-2+s), & \text{if } r_0 \leq -l-2. \end{cases}$$

To insure that inequalities (4.4) hold whenever $-l - 1 \leq r_0 \leq -1$ we must also demand that $r_0 \geq -l + \frac{1}{p}(1 - k - s)$. With this additional constraint we are led to fundamental functions L_i which have support.

$$\begin{cases} [t_{pi-1}, t_{pi+k-1+p+pr_0+p_l+s}], & \text{if } r_0 \geq 0 \\ [t_{pi+pr_0}, t_{pi+k-1+p+pr_0+p_l+s}], & \text{if } -l-1 \leq r_0 \leq -1 \\ [t_{pi+pr_0}, t_{pi+k-1+s}], & \text{if } r_0 \leq -l-2. \end{cases}$$

Specifically, if $p = 2$ then $r = 1$, $s = k - 3$ and we get, when $-l - 1 \leq r_0 \leq -1$,

$$\text{supp} L_i \subseteq [x_{i+r_0}, x_{i+r_0+l+k-1}],$$

which follows from the result of [4] mentioned earlier.

The case $p = k - 1$ and $l = k - 1$ is also studied in [4], see Theorem 4.2.1. In this case, $s = 0$, $r = k - 2$ and so we get, when $-k \leq r_0 \leq -1$,

$$\text{supp} L_i \subseteq [x_{i+r_0}, x_{i+r_0+k+1}],$$

which agrees with (4.2.2) of [4].

The second application we have in mind is an extension of the spline decomposition problem studied in Dahmen and Micchelli [5]. Let $\{x_i\}_{i \in \mathbb{Z}}$ and $\{t_i\}_{i \in \mathbb{Z}}$ be two knot sets related as before, that is, $x_i = t_{pi}$, $i \in \mathbb{Z}$. Suppose $\{C_i : i \in \mathbb{Z}\}$ and $\{F_i : i \in \mathbb{Z}\}$ are the corresponding B-splines of support $[x_i, x_{i+k}] = [t_{pi}, t_{pi+pk}]$, $[t_i, t_{i+k}]$ respectively. Then there is a matrix $A = (A_{ij})_{i,j \in \mathbb{Z}}$ such that

$$C_i(x) = \sum_{j \in \mathbb{Z}} A_{ij} F_j(x).$$

The transpose of A is known as the discrete B-spline collocation matrix (see [9],[11]). The sum above goes from $j = pi$ to $j = pi + pk - k$ and so again A is p-slanted but in this case with band $(0, (p-1)k)$.

A special case of a result in Jia [9] insures that this matrix is totally positive and that

$$A\begin{pmatrix} i_1 & i_2 & \dots & i_s \\ j_1 & j_2 & \dots & j_s \end{pmatrix} > 0$$

whenever $A_{i_h j_h} > 0$ for $h = 1, \dots, s$ since in our case both sequences of knots are strictly increasing. Moreover, Jia shows for the case we consider that for each i, j $A_{ij} > 0$ if and only if $pi \le j \le pi + (p-1)k$. For a recent commentary on the results in [9], see Morken [11].

Let us now observe that A satisfies condition (i) of Corollary 3.2 for $n = m - 1$, with $m = (p-1)k$ (that is $r = p - 2$, $q = k - 1$). We merely have to check that the diagonal entries of the determinants

$$A\begin{pmatrix} i & \dots & i+k-1 \\ pi+j & \dots & pi+j+k-1 \end{pmatrix}, \quad j = (p-1)k - 1, \dots, (p-1)k - p + 2,$$

$$A\begin{pmatrix} i & \dots & i+k-2 \\ pi+j & \dots & pi+j+k-2 \end{pmatrix}, \quad j = (p-1)k - p + 1,$$

are positive. But this is true because the diagonal entries of these determinants are always in the band of positive elements of the matrix A given above. That is, $A_{i+h, pi+j+h} \neq 0$, for all integers h and j such that $0 \le h \le k - 1$, $(p-1)k - p + 1 \le j \le (p-1)k - 1$. Hence we can apply Theorem 3.2 to obtain matrices $A_1, \dots, A_{p-1}$ and $B_0, \dots, B_{p-1}$ which satisfy (3.10). The corresponding functions

$$\Psi_{ih}(x) = \sum_{j \in \mathbb{Z}} (A_h)_{ij} F_j(x)$$

are of compact support and have the property that for all $i \in \mathbb{Z}$ and $x \in \mathbb{R}$

$$F_i(x) = \sum_{h=0}^{p-1} \sum_{j \in \mathbb{Z}} (b_h)_{ij} \Psi_{ih}(x).$$

Thus, as in (2.25), these functions provide a locally finite decomposition of the spline space with knots $\{t_i\}_{i \in \mathbb{Z}}$ through the spline space with knots $\{x_i\}_{i \in \mathbb{Z}}$, since $\Psi_{i0}(x) = C_i(x), \quad i \in \mathbb{Z}$.

Acknowledgements. M. Gasca and J.M. Peña were partially supported by DGICYT Spain Res. Grant Ps 90-0121. Charles A. Micchelli was partially supported by DGICYT Spain Res. Grant SAB 92-0047.

References

1. Ando, T., Totally positive matrices, Linear Algebra Appl. **90** (1987), 165-219.
2. de Boor, C., Total positivity of the spline collocation matrix, Indiana Univ. J. Math. **25** (1976), 541-551.
3. Cavaretta, A.S., Dahmen, W., and C.A. Micchelli, Stationary subdivision, Memoirs AMS vol.**93**, n.453 (1991).
4. Dahmen, W., Goodman, T.N.T., and C.A. Micchelli, Compactly supported fundamental functions for spline interpolation, Numer. Math. **52** (1988), 629-654.
5. Dahmen, W., and C.A. Micchelli, Banded matrices with banded inverses II: Locally finite decomposition of spline spaces, Constructive Approximation **9** (1993), 263-282.
6. Gantmacher, F.R., *The Theory of Matrices, Vol II.* Chelsea Publ. Co., New York, 1989.
7. Gasca, M., Micchelli, C.A. and J.M. Peña, Almost strictly totally positive matrices, Numerical Algorithms **2** (1992), 225-236.
8. Gasca, M., and Peña, J.M., Total positivity and Neville elimination, Linear Algebra Appl. **165** (1992), 25-44.
9. Jia, R.Q., Total positivity of the discrete spline collocation matrix, J. Approx. Theory, **39** (1983), 11-23.
10. Micchelli, C.A., Using the refinement equation for the construction of pre-wavelets, Numerical Algorithms **1** (1991), 75-116.
11. Morken, K.M., Total positivity of the discrete spline collocation matrix II. Preprint, University of Oslo (1992).
12. Schumaker, L.L., *Spline functions: Basic Theory*, John Wiley and Sons, New York, 1981.
13. Walker, R.J., *Algebraic Curves*, Princeton Univ. Press, Princeton, 1950.

M. Gasca and J.M. Peña
Departamento de Matemática Aplicada.
Universidad de Zaragoza.
50009 Zaragoza, SPAIN
gasca@cc.unizar.es

Charles A. Micchelli
Department of Mathematical Sciences
IBM T.J. Watson Research Center
P.O. Box 218
Yorktown Heights, New York 10598
cam@yktvmx.bitnet

Invariant Approximation of Star-Shaped Form for Medical Applications

F. Ghorbel and V. Burdin

Abstract. In many applications, discrete data need approximations to obtain a continuous representation. Therefore, it is suitable for the used algorithm, called curve algorithm, to be independent of the points location. For medical applications which manipulate 2D star-shaped contours, we propose a radial parameterization which makes any curve algorithm invariant.

§1. Introduction

In many applications, such as medical imaging, data are acquired in a discrete form and approximations are needed to obtain a continuous model. Therefore, it is suitable for the algorithm, called here *curve algorithm*, to be independent of the points location. Euclidean invariance is widely recognized as an important property of curve and surface shape description algorithms for computer-aided design, pattern recognition and image analysis. The Euclidean transformations are defined by combining rotations and translations, and the Euclidean invariance is equivalent to axis-independence.

Some authors [2] have investigated the general functional form of curve algorithms that are invariant under certain transformation groups. This simply means that applying the curve algorithm to the transformed data set is equivalent to apply the algorithm to the original data set and transforming the resultant curve or surface.

In this paper, we use the radial curve representation with arc length parameterization in order to obtain axis-independence for any planar curve algorithm. In pattern recognition or image analysis, this parameterization can be justified by its fitting to a wide range of shapes (for example in medical and biological imaging). However, we restrict ourselves to a particular class of planar shapes namely star-shaped contours. From this particular parameterization, we extract invariant features with respect to Euclidean transformations that are useful in pattern recognition.

Wavelets, Images and Surface Fitting
P. J. Laurent, A. Le Méhauté, and L. L. Schumaker (eds.), pp. 269–276.

ISBN 1-56881-040-7.

§2. Curve Algorithms and Transformation Invariance

Following the notation of [2], we denotes the n-dimensional real space by $\mathbb{R}^n$ and the set of all the finite subsets of $\mathbb{R}^n$ by $\mathcal{D}^n$. An element D of $\mathcal{D}^n$ has the following form:

$$D = \{r_i \mid r_i \in \mathbb{R}^n,\ 0 \le i \le N\},$$

where N is a positive integer.

Definition 1. *A curve algorithm F produces a parametric curve $F(D)$ with the general form:*

$$F(D) = (t, c_F(D,t))\,, \quad t \in I(D),$$

where $I(D) \subset \mathbb{R}$ is an interval and $c_F(D,t) \in \mathbb{R}^n$.

As an example, the cubic spline is a usual curve algorithm.

Definition 2. *A curve algorithm F is said to be invariant under a transformation group G of $\mathbb{R}^n$ if*

$$gF(D) = F(gD), \qquad \forall g \in G \quad \text{and} \quad \forall D \in \mathcal{D}^n.$$

In practice, this means that applying a curve algorithm to a transformed data set is equivalent to apply the curve algorithm to the data and transform the resultant curve.

Instead of looking for some invariant curve algorithms [2], we propose a particular parameterization which makes any curve algorithm invariant. In the next section, we introduce the radial parameterization which fulfils this condition for the class of closed star-shaped contours.

§3. Radial Parameterization

We recall that a parameterization of a closed planar contour is given by an interval, $I(D)$, and an immersion γ of $I(D)$ in $\mathbb{R}^2$. In shape analysis, a normalized arc length of γ, denoted γ^*, is usually employed:

$$\begin{aligned} \gamma\colon [0,T] &\longrightarrow \mathbb{R}^2 \\ t &\longmapsto (x(t), y(t)) \end{aligned} \qquad\qquad \begin{aligned} \gamma^*\colon [0,1] &\longrightarrow \mathbb{R}^2 \\ s &\longmapsto \gamma(t(s)), \end{aligned}$$

where

$$s(t) = \frac{1}{L}\int_0^t \|\gamma'(\ell)\|_2\, d\ell \quad \text{and} \quad L = s(T).$$

We shall first study the radial parameterization in the continuous case. The discrete case will be considered later.

3.1. Continuous case

Let $\mathcal{R} = (O, \vec{\imath}, \vec{\jmath})$ be a cartesian coordinates system of the Euclidean plane and s a normalized arc length.

Theorem 1. *For the determination of a unique closed planar contour up to a rotation with center at the origin O, it is necessary and sufficient to have both the radial function ρ and the sign function σ of the derivative of the polar function θ according to s, defined by*

$$\begin{aligned} \rho\colon [0,1] &\longrightarrow \mathbb{R}_+^* \\ s &\longmapsto \rho(s) \end{aligned} \qquad\qquad \begin{aligned} \sigma\colon [0,1] &\longrightarrow \{-1,1\} \\ s &\longmapsto \operatorname{sign}(\theta'(s)), \end{aligned}$$

in $\mathcal{R}$.

Proof: We will express the angular function θ from both the above functions, ρ and σ. Let $\gamma(s)$ be a normalized arc length parameterization of a curve C. In polar coordinates, this becomes

$$\gamma(s) = (\rho(s)\cos(\theta(s)), \rho(s)\sin(\theta(s)))\,.$$

After derivation under the variable s, we consider the norm of both the members of the obtained equation. Since $\|\gamma'(s)\|_2 = 1$, this gives

$$\theta'^2(s) = \frac{1-\rho'^2(s)}{\rho^2(s)} \quad \text{with } s \in [0,1].$$

The solution of this differential equation is given by

$$\theta(s) = \theta(0) + \int_0^s \sigma(\ell)\,\frac{\sqrt{1-\rho'^2(\ell)}}{\rho(\ell)}\,d\ell.$$

With the assumption of a normalized arc length parameterization, we can change the angular function θ into the sign function of its derivative without losing any information. ■

Remark. Both the functions $\rho(s)$ and $\sigma(s)$ are axis-invariant.

3.2. Discrete case

Theorem 2. *Given a finite sequence $\rho(j)$ of positive numbers, a finite sequence $\sigma(j)$ of elements belonging to $\{-1,1\}$, and any point O in the Euclidean plane, the polygonal line with equal segments made from these given data, is unique up to a rotation with center at the origin O.*

Proof: The construction of the polygonal line will be done recursively. This supposes that the polygonal line corresponding to $\gamma(j)$ has been already drawn, with $\gamma(j) = (\rho(j)\cos(\theta(j)), \rho(j)\sin(\theta(j)))$.

The problem can be summarized by the following assertion: if the knowledge of $\rho(j+1)$ and $\sigma(j+1)$ is sufficient to find the point $\gamma(j+1)$, then $\gamma(j)$ will be joint to $\gamma(j+1)$ by a line with a length equal to $1/N$. The value of $\rho(j+1)$ indicates that $\gamma(j+1)$ belongs to the circle which has the origin O as a center and $\rho(j+1)$ as a radius.

The equation $\|\gamma(j)-\gamma(j+1)\|_2 = 1/N$, which is only verified in the case of regular polygons, implies that $\gamma(j+1)$ belongs to the circle having $\gamma(j)$ as a center. Thus, $\gamma(j+1)$ is the cross-section of the last two circles.

The inequality

$$|\rho(j) - \rho(j+1)| \leq \|\gamma(j) - \gamma(j+1)\|_2,$$

means that the arc length parameterization "speed" is greater than that of the radius and implies that the intersection of the two circles exists and contains either one point or two points.

In the latter case, the two points are located in a symmetrical way on the half-axis $]0, \gamma(j))$. The choice between the two points depends on the sign $\sigma(j)$ of the derivative of the angular function $\theta(j)$.

When the intersection is reduced to one point, the inequality becomes an equality. In this case, we consider that $\sigma(j+1) = \sigma(j)$. ■

3.3. Application to star-shaped contours

In the following proposition, we give a characterization for star-shaped contours that can be met in natural scenes.

Proposition 1. *Let C be a closed planar contour, U its interior, and O a point in U. The following assertions are equivalent:*

(i) *U is star-shaped with respect to the origin O,*
(ii) *the sign function $\sigma(s)$ of the contour C, computed with respect to the origin O, is constant.*

Taking Theorem 1 into account, the only function $\rho(s)$ gives a complete representation of the contour if the function $\sigma(s)$ is constant.

Remark. If U is star-shaped with respect to its centroid G, we choose $O = G$.

Proposition 2. *Let $D_1 \in \mathcal{D}^2$ describing a discrete curve star-shaped with respect to its centroid. A radial parameterization of D_1 can make any curve algorithm invariant.*

Proof: Let D_1 be an element of $\mathcal{D}^2$ having the following form

$$D_1 = \{r_i(x_i, y_i) \mid r_i \in \mathbb{R}^2,\ 0 \leq i \leq N\}.$$

Let $r_G = (x_G, y_G)$ be the centroid of D_1. Using Theorem 2 of Section 3, the radial representation of the star-shaped curve D_1 can be reduced to the following finite sequence:

$$D_1^* = \{\rho_0, \rho_1, \cdots, \rho_N \mid \rho_i \in \mathbb{R}_+\},$$

where $\rho_i = \|r_i - r_G\|_2$ is the Euclidean norm of $r_i - r_G$. We denote the polar angle of r_0 by $\theta_0 = \arctan(\frac{y_0 - y_G}{x_0 - x_G})$.
Let g be a Euclidean transformation,

$$D_2 = gD_1 = \{gr_i(x_i, y_i) \mid r_i \in \mathbb{R}^2,\ 0 \le i \le N\}.$$

Since $\|gr_i - gr_G\|_2 = \|g(r_i - r_G)\|_2 = \|r_i - r_G\|_2$, we have

$$D_2^* = (gD_1)^* = \{\rho_{i_0}, \rho_{i_0+1}, \cdots, \rho_N, \rho_0, \cdots, \rho_{i_0-1} \mid \rho_i \in \mathbb{R}_+\}.$$

Let $\theta = \arctan(\frac{y}{x})$ where (x, y) corresponds to $g(r_0 - r_G)$. After applying the transformation g, we recover the same sequence of ρ, but with a circular permutation because the first chosen point is the one closest to the positive horizontal axis drawn from r_G.

If F is a curve algorithm, then

$$[F(D_2)^*](t_2) = [F(gD_1)^*](t_2) = R_{\theta-\theta_0}[F(D_1)^*](t_1) = g[F(D_1)^*](t_1),$$

where $R_{\theta-\theta_0}$ is the $(\theta - \theta_0)$-angular rotation. The last equality comes from the fact that a translation has no effect on polar coordinates. ■

§4. Fourier Descriptors

In the field of pattern recognition and shape identification, several authors [1,3] have defined invariant features as follow:

Definition 3. *Let $\{I_n\}$ be a sequence of scalars, they are invariant features with respect to a group G of transformations acting on a set of parameterizations if and only if for two curves, C_1 and C_2, having the same shape, we have*

$$I_n(\gamma_1) = I_n(\gamma_2), \qquad \forall n \in \mathbb{N},$$

where γ_1 and γ_2 are the respective parameterizations of C_1 and C_2.

Different sets of invariants have been introduced from moment methods, Fourier coefficients, methods of prediction, etc... The first and practical criterion for choosing the best invariant was the computing time. Crimmins [3] introduced a theoretical criterion for setting smart invariant features which he called *completeness* property. This expresses that two objects have the same shape if and only if they have the same invariant features. An other important property is to obtain invariant features which are *robust* under small distorsions on the curve. We call this property *stability* [1].

In the case of radial parameterization, we can define complete and stable features, which are invariant under rotation, translation, and starting point on a curve.

Theorem 3. *Let p be a positive real number, n_0 a positive integer, $\rho(s)$ the radial parameterization with s the normalized arc length on the continuous curve C. Then the features*

$$\begin{cases} I_{n_0}(\rho) & = & |a_{n_0}(\rho)|^{n_0+p} \\ I_n(\rho) & = & \dfrac{a_n^{n_0}(\rho)}{a_{n_0}^n(\rho)}|a_{n_0}(\rho)|^{n+p}, \end{cases}$$

where $a_n(\rho) = \int_0^1 \rho(s)\mathrm{e}^{-2i\pi ns}\, ds$, are *complete*, *stable*, and invariant under rotation, translation and starting point on the curve.

Proof: Let us consider the radial normalized arc length parameterizations of the curves C_1 and C_2, corresponding to FD_1^* and FD_2^* of Proposition 2, which we denote by

$$\rho_1\left[s_1(t_1)\right] = \|FD_1^*(t_1)\|_2 \qquad \text{and} \qquad \rho_2\left[s_2(t_2)\right] = \|FD_2^*(t_2)\|_2,$$

where $t_i \in [0, T]$ and s_i is a normalized arc length of the respective curve C_i.

Taking account of Proposition 2, the arc length of the first point on the curve is shifted after transformation, *i.e.* $s_2(t_2) = s_1(t_1) + s_0$, $s_0 \in [0, 1]$ and

$$\rho_2\left[s_2(t_2)\right] = \rho_1\left[s_1(t_1)\right] = \rho_1\left[s_2(t_2) - s_0\right].$$

For the Fourier coefficients, we obtain $a_n(\rho_2) = a_n(\rho_1)\, \mathrm{e}^{ins_0}$, and if we consider the invariant features of Theorem 3, then

$$I_n(\rho_2) \;=\; \frac{a_n^{n_0}(\rho_2)}{a_{n_0}^n(\rho_2)}|a_{n_0}(\rho_2)|^{n+p} \;=\; I_n(\rho_1). \quad \blacksquare$$

Remark. The entire proof of the completeness and stability properties of these features can be found in [1].

In the following section, we show the importance of the invariance property when an approximation is needed (for example in medical imaging). The invariant features alone are not sufficient to obtain an entire invariance when a continuous step is required. The curve algorithm used must satisfy the invariancy. Here, we have solved this problem by finding a parameterization which makes invariant any curve algorithm.

§5. Applications

The radial parameterization is well-fitted to natural shapes which do not contain sharp edges. We propose to apply this method to anatomic 3D objects like long bone structures, that are viewed as stacks of cross-sectional slices. As an example, the ulna of Figure 2 contains 423 slices with the first contour, located at the right side, corresponding to the wrist (the ulna is one of the two bones of the forearm). A slice is a gray-level image and the closed contour of the bone region must be extracted [4]. The set of slice contours gives a stack of discrete closed curves (the two thick lines in Figure 1 correspond to the slices number 280, on the left and 355, on the right).

On each contour, a radial parameterization with a normalized arc length is performed. In Figure 1, dark points represent the D_1 set, which will be the input for the curve algorithm, corresponding to the above two slices. The points are regularly spaced on the original contours. The discrete radial function corresponding to the dark points is expanded in Fourier series. The series is truncated in order to keep only the first coefficients. The radial function reconstruction (therefore the curve reconstruction) gives us an approximation of the original contour [5]. As seen in Figure 1, the reconstructed curves (the two thin lines) have lost the high spatial frequencies of the original contours. The smoothing of all the contours leads to a more or less smoothed 3D structure as seen in Figure 2. The surface is drawn by a heuristic method to using triangles between two successive contours (see part 4 of [4]).

As shown in [4], the invariant features are efficient to define a distance between 2D contours, allowing us to quantify the shape differences between the smoothed 3D structure and the original one.

Fig. 1. Original and reconstructed contours of two ulnar slices.

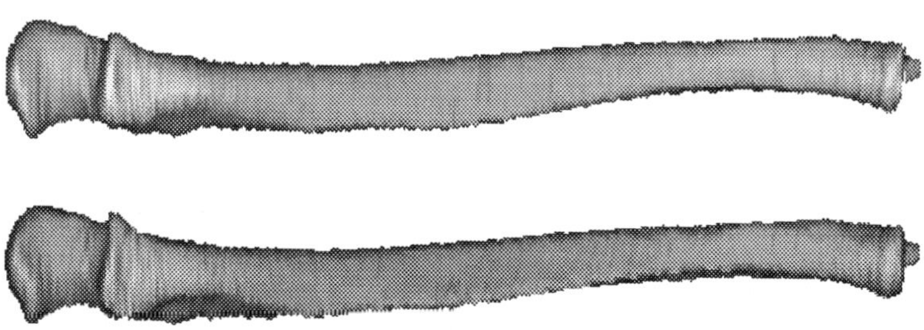

Fig. 2. Original (on the bottom) and reconstructed ulna.

§6. Conclusion

In this paper, we have demonstrated that the invariance property is very important in two ways. In many applications, data are discrete and need a continuous representation which can be provided using approximation or interpolation. The resultant continuous curve must be independent from data locations. Instead of seeking an invariant curve algorithm, we propose a radial parameterization which makes any curve algorithm invariant. We have also developed invariant features which are efficient in pattern recognition to quantify, as an example, a reconstruction error.

In the future, a curve algorithm suitable to the radial parameterization should be developed. At the moment, the approximation is obtained by smoothing the closed contour. While a distance between both the smoothed and original curves is defined, the location differences between the two curves are not yet clearly established.

The robustness of the curve algorithm is also worth studying. It is the same notion than the stability property that the descriptors possess. In addition to its invariance property, a robust curve algorithm must generate two similar continuous curves when the corresponding discrete sets of data are the same except for small distortions.

Acknowledgements. The authors thank Professor A. Le Méhauté for helpful discussions on this work.

References

1. Ghorbel F., "*Vers une approche mathématique unifiée des aspects géométriques et statistiques de la reconnaissance de formes planes*", Doctoral thesis, Rennes I, December 90.
2. Bez H. E., "On analysis of symmetry for plane curves and plane curve algorithms", *Computer-Aided Geom. Design* **9** (1992), 125–142.
3. Crimmins T. R., "A complete set of Fourier descriptors for two dimensional shapes", *I.E.E.E. Trans. S.M.C.* **12** n°6 (1982), 848–855.
4. Burdin-Fracasso V., "*Modélisation et analyse de structures tridimensionnelles – Application à la morphométrie des os longs*", Doctoral thesis, Rennes I, November 92.
5. Burdin V., F. Ghorbel, J. L. de Bougrenet de la Tocnaye et C. Roux, "A geometrical analysis for data compression of 3D anatomical structures", *Curves and Surfaces*, Academic Press, New York, Editeurs P. J. Laurent, A. Le Méhauté et L.L. Schumaker (1991), 57–63.

F. Ghorbel
Groupe Image INT-ENIC
ENIC – Cité scientifique
59658 Villeneuve d'Ascq cedex
FRANCE
Fax: (33) 20.33.55.99

V. Burdin
Groupe Traitement d'Images
Telecom Bretagne – BP 832
29285 Brest cedex
FRANCE
e-mail: burdin@gti.enst-bretagne.fr

Surface Construction Based on Variational Principles

Günther Greiner

Abstract. Variational principles have become quite popular in the design of surfaces. The idea is to consider a class of surfaces having more degrees of freedom than are necessary to fulfill the constraints (e. g. interpolation or boundary conditions). The remaining degrees of freedom are set by minimizing a *fairness functional*. The choice of the 'right' fairness functional is a crucial step. In this paper we describe a systematic approach for choosing an appropriate fairness functional that produces high quality surfaces and which allows the solution of the optimization problem in a reasonable amount of time.

§1. Introduction

The problem of constructing smooth curves or surfaces satisfying certain constraints arises in different areas in CAGD. Typical examples are (scattered data) interpolation and the construction of blending surfaces. In scattered data interpolation one has to construct a surface which does not oscillate too much and goes through a sampled set of points in the space. Finding a blending surface (i.e. a smooth transition surface between primary surfaces) amounts to the construction of a reasonable surface that satisfies certain boundary conditions determined by the primary surfaces.

The traditional approach to these problems is the following: choose an appropriate set of functions that has as many degrees of freedom as are necessary to fulfill the constraints. Determination of the degrees of freedom usually is done by solving a system of equations. Recently, a variational approach for solving these problems has received more and more attention. The idea is to start with a class of surfaces having more degrees of freedom than are strictly necessary to fulfill the constraints. The remaining degrees of freedom are used to smooth the surface so that it assumes a pleasing shape. This is achieved by optimizing a functional which somehow measures the smoothness or fairness.

The paper is organized as follows. In Sect. 2 we describe the general concept. In the main part, in Sect. 3, we will describe how a fairness functional

Wavelets, Images and Surface Fitting
P. J. Laurent, A. Le Méhauté, and L. L. Schumaker (eds.), pp. 277–286.

ISBN 1-56881-040-7.

can be chosen that gives good looking surfaces and can be solved quite easily (compared to existing methods). Finally, in Sect. 4, we describe in detail how the procedure can be applied to construct blending surfaces.

§2. The General Approach

The task is to construct "optimal" curves or surfaces that satisfy certain constraints. This is rather open ended, so we will begin by explaining what "optimal" means and what kinds of constraints we are considering. First of all, we will describe what type of surfaces we plan on working with. In this section we try to be as general as possible. In the following sections more concrete issues are discussed.

We consider some class $\mathcal{S}$ of parametrized surfaces. Each $F \in \mathcal{S}$ is a mapping from a fixed parameter space $\Omega \subset \mathbb{R}^2$ into $\mathbb{R}^3$. $\mathcal{S}$ may consist of tensor product splines (over a rectangular domain), or of piecewise Bézier patches (over a triangulation), or of triangular splines (over a triangulation and an associated cloud of knots, see [4]) or a scheme of split surfaces (e.g., Clough–Tocher or Powell–Sabin). It is very important that the class has (sufficiently) many degrees of freedom (dofs). We require more dofs than are needed to satisfy the constraints which we describe next. The constraints may take the following forms:

Interpolation. $F(u_k) = y_k$ for finitely many parameter values u_k.

Tangent planes. $\left\langle \frac{\partial F}{\partial u_i}(v_k) \middle| p_k \right\rangle = 0$ $(i = 1, 2)$ for finitely many parameter values v_k.

Blending. Along the boundary of the parameter space, F has prescribed values and prescribed cross boundary derivatives.

Fairing. An already existing surface may be changed only in a specified region (in such a way that its shape is improved).

In addition, one could prescribe curvature (e.g. Dupin's indicatrix) at certain points, or require that a given curve be interpolated

"Optimal", has a variety of possible interpretations. It can mean 'of minimal length or area', 'of small variation', or simply 'nice looking'. While the first quantities can be measured exactly, it is not quite clear how to measure 'nice looking'. Usually this is achieved by considering a functional J measuring the fairness (= 'nice looking') of a surface by a scalar value.

Having chosen a class of surfaces $\mathcal{S}$ and a *fairness functional* J (defined for elements of $\mathcal{S}$), one has to solve the variational problem (or optimization problem)

> Find $F \in \mathcal{S}$ such that
> 1) F satisfies the constraints and
> 2) $J(F) \le J(G)$ for any $G \in \mathcal{S}$ satisfying the constraints.

In case $\mathcal{S}$ is a linear space and the constraints are linear as well, the set of all $F \in \mathcal{S}$ satisfying the constraints form an affine subspace $\mathcal{S}_0$. If one chooses an arbitrary surface $F_{bc} \in \mathcal{S}_0$, then the problem can be reduced to an optimization problem without any constraints in the associated linear subspace $\mathcal{S}_{00} = \mathcal{S}_0 - \{F_{bc}\}$. In any case, the consideration of finitely many constraints always can be incorporated in the optimization process using Lagrangian multipliers.

The crucial step of the variational method is the selection of the 'right' fairness functional. There is no canonical choice. Different authors use different functionals and we will only recall a few of them here.

In the *minimum norm network* approach for scattered data interpolation (see [12,13]) one constructs (in a first step) a network of curves $\{F_e\}$ by minimizing

$$\sum_e \int F_e''(t)^2 + \alpha F_e'(t)^2 \, dt \, .$$

Halstead, Kass, DeRose (see [7]) and others consider for surfaces F

$$\int \left(\frac{\partial^2 F}{(\partial u_1)^2}\right)^2 + 2\left(\frac{\partial^2 F}{\partial u_1 \partial u_2}\right)^2 + \left(\frac{\partial^2 F}{(\partial u_2)^2}\right)^2 .$$

This functional is often called *thin plate energy.* Kallay [8] also takes this expression and adds a term describing the deviation of a given surface F_0 . In other papers (e.g. [1]), one tries to minimize curvature by considering

$$\int \kappa_1^2 + \kappa_2^2$$

where κ_1 , κ_2 denote the principle curvatures. Welch and Witkin's starting point (see [14]) is the functional

$$\int \|I_F\|_\alpha^2 + \|I\!I_F\|_\beta^2$$

where I_F and $I\!I_F$ denote the first and second fundamental form resp. They approximate it somehow by a quadratic functional

$$\int \sum_{i,j=1}^{2} \alpha_{ij} \left\langle \frac{\partial F}{\partial u_i} \middle| \frac{\partial F}{\partial u_j} \right\rangle + \beta_{ij} \left(\frac{\partial^2 F}{\partial u_i \partial u_j}\right)^2 .$$

The same functional (only physical motivated) is used in [2]. Moreton and Séquin (see [11]) consider *variation of curvature.* For curves and surfaces this is given by

$$\int \kappa'(t)^2 \, dt \qquad \text{and} \qquad \int \left(\tfrac{\partial \kappa_1}{\partial e_1}\right)^2 + \left(\tfrac{\partial \kappa_2}{\partial e_2}\right)^2 d\omega \qquad \text{resp.}$$

κ_1 , κ_2 as above and e_1 , e_2 the corresponding directions of principle curvature.

When selecting one of these functionals (or any other) two questions have to be adressed: 1) does the functional actually yield 'nice looking' curves/surfaces?, and 2) can these curves/surfaces be determined in a reasonable amount of time? In [10], these aspects are discussed and one conclusion is that the (highly nonlinear) functionals involving curvature yield the best surfaces, but an enormous amount of time is necessary to arrive at the solution. In the next section we describe how fairness functionals can be chosen that yield surfaces of good quality and are quadratic, hence are easy treated numerically.

§3. Choice of the Fairness Functional

Formally there are two different approaches to choosing a functional. Either one selects the functional by physical analogy (e.g. minimizing energy of thin plates) or, by geometric reasoning (e.g. minimzing area, curvature or the variation of curvature). However, both concepts are closely related. Actually, every reasonable physical quantity has to be a generic property of the surface. That is, it may not depend on the special parametrization which is used to describe the surface. This is, by definition, a *geometric property* of the surface. Thus every physical quantity is a geometric one. Normal vectors or tangent planes, area of the surface, principal curvatures and directions of principle curvature are the most important geometric properties. In contrast, partial derivatives (generating the tangent plane) and the first and second fundamental forms depend on the parametrization. They are not intrinsically geometric properties.

We illustrate the relation between a physical and a geometric property by an example. The energy of a thin plate is given as follows (see [3]):

$$J_{ab}(F) = \int_\Omega a(\kappa_1^2 + \kappa_2^2) + 2(1-b)\kappa_1\kappa_2\, d\omega \tag{1}$$

where κ_1 and κ_2 are the principle curvatures of the surface F. a and b are constants describing properties of the material of the thin plate (resistance to bending and sheering). Obviously, (1) defines a geometric quantity, it only depends on the principal curvatures!

Using the notion of *Gaussian* and *mean curvature* $K_F := \kappa_1\kappa_2$ and $H_F := \frac{1}{2}(\kappa_1 + \kappa_2)$ respectively, (1) can be restated as

$$J_{ab}(F) = \int_\Omega 4aH_F(u)^2 + 2(1-a-b)K_F(u)\, d\omega(u). \tag{2}$$

This functional, though describing the exact energy, is rather complicated, and optimizating it is difficult and time consuming. Therefore one often considers the following simpler (linearized) version of the thin plate energy.

$$\int_\Omega a\langle \Delta F | \Delta F\rangle + 2(1-a-b)\left\{\left\langle \frac{\partial^2 F}{(\partial u_1)^2} \middle| \frac{\partial^2 F}{(\partial u_2)^2}\right\rangle - \left\langle \frac{\partial^2 F}{\partial u_1 \partial u_2} \middle| \frac{\partial^2 F}{\partial u_1 \partial u_2}\right\rangle\right\}. \tag{3}$$

Here $\Delta F = \frac{\partial^2 F}{(\partial u_1)^2} + \frac{\partial^2 F}{(\partial u_2)^2}$ denotes the Laplacian. In case $a = b = 1$ it reduces to

$$\int_\Omega \left(\frac{\partial^2 F}{(\partial u_1)^2}\right)^2 + 2\left(\frac{\partial^2 F}{\partial u_1 \partial u_2}\right)^2 + \left(\frac{\partial^2 F}{(\partial u_2)^2}\right)^2, \tag{4}$$

while for $a = 1$ and $b = 0$ we obtain

$$\int_\Omega \langle \Delta F | \Delta F\rangle\ . \tag{5}$$

(3) is good approximation for (1) provided the parametrization is nearly isometric, i.e., the first fundamental form is nearly the unit matrix. The following proposition shows the reason for this.

Proposition 1. *In the case that the parametrization F is isometric, the functionals given by (1) and (3) coincide.*

Proof: Mean curvature H_F and Gaussian curvature K_F of F are given by

$$H_F = \frac{g_{22}h_{11}-2g_{12}h_{12}+g_{11}h_{22}}{2(g_{11}g_{22}-g_{12}^2)} \qquad \text{and} \qquad K_F = \frac{h_{11}h_{22}-h_{12}^2}{g_{11}g_{22}-g_{12}^2} \qquad \text{resp.}$$

Here (g_{ij}) and (h_{ij}) denote the first and second fundamental forms, respectively, of the parametrized surface F. They are given by

$$g_{ij} = \left\langle \frac{\partial F}{\partial u_i} \middle| \frac{\partial F}{\partial u_j} \right\rangle \qquad \text{and} \qquad h_{ij} = \left\langle \frac{\partial^2 F}{\partial u_i \partial u_j} \middle| n_F \right\rangle,$$

where n_F denotes the normal vector to the surface.

Differentiating the constant functions $g_{ij} = \left\langle \frac{\partial F}{\partial u_i} \middle| \frac{\partial F}{\partial u_j} \right\rangle$ yields

$$\left\langle \frac{\partial^2 F}{(\partial u_1)^2} \middle| \frac{\partial F}{\partial u_1} \right\rangle = 0, \ \left\langle \frac{\partial^2 F}{\partial u_1 \partial u_2} \middle| \frac{\partial F}{\partial u_1} \right\rangle = 0, \ \left\langle \frac{\partial^2 F}{\partial u_1 \partial u_2} \middle| \frac{\partial F}{\partial u_2} \right\rangle = 0, \ \left\langle \frac{\partial^2 F}{(\partial u_2)^2} \middle| \frac{\partial F}{\partial u_2} \right\rangle = 0\,,$$

$$\left\langle \frac{\partial^2 F}{(\partial u_1)^2} \middle| \frac{\partial F}{\partial u_2} \right\rangle + \left\langle \frac{\partial F}{\partial u_1} \middle| \frac{\partial^2 F}{\partial u_1 \partial u_2} \right\rangle = 0\,, \quad \left\langle \frac{\partial^2 F}{\partial u_1 \partial u_2} \middle| \frac{\partial F}{\partial u_2} \right\rangle + \left\langle \frac{\partial F}{\partial u_1} \middle| \frac{\partial^2 F}{(\partial u_2)^2} \right\rangle = 0\,.$$

It follows that $\frac{\partial^2 F}{(\partial u_1)^2}$ is orthogonal to $\frac{\partial F}{\partial u_1}$ and to $\frac{\partial F}{\partial u_2}$ as well. Thus it has the same direction as the normal vector n_F. The same holds true for $\frac{\partial^2 F}{(\partial u_2)^2}$ and $\frac{\partial^2 F}{\partial u_1 \partial u_2}$. Thus, we have $\frac{\partial^2 F}{\partial u_i \partial u_j} = \left\langle \frac{\partial^2 F}{\partial u_i \partial u_j} \middle| n_F \right\rangle \cdot n_F = h_{ij} n_F$. We conclude that Gaussian and mean curvature satisfy

$$K_F(u) = \frac{h_{11}h_{22}-h_{12}^2}{1 \cdot 1 - 0} = \left\langle \frac{\partial^2 F}{(\partial u_1)^2} \middle| \frac{\partial^2 F}{(\partial u_2)^2} \right\rangle - \left\langle \frac{\partial^2 F}{\partial u_1 \partial u_2} \middle| \frac{\partial^2 F}{\partial u_1 \partial u_2} \right\rangle$$

$$H_F(u) \cdot n_F(u) = \frac{1 \cdot h_{11} - 2 \cdot 0 \cdot h_{12} + 1 \cdot h_{22}}{2(1 \cdot 1 - 0)} \cdot n_F(u) = \tfrac{1}{2}(h_{11} + h_{22}) \cdot n_F(u) = \tfrac{1}{2}\Delta F(u)$$

From this one easily deduces that (1) and (3) coincide. ■

From the next proposition it follows that in certain cases, the second term in (2) as well as in (3) has no influence on the local minima of these functionals.

Proposition 2. *(a) In the case that one considers parametrizations having at the boundary of the parameter domain Ω prescribed values and prescribed (non-tangential) derivatives, the (local or global) minima of the functionals J_{ab} do not depend on $a > 0$ and b respectively. More precisely, $J_{ab}(F) = aJ_{10}(F) + C_{ab}$ where C_{ab} is a constant depending only on the boundary values of F.*
(b) The same statement is true for the linearized version of this functional as defined in (3).

Proof: (b) The second term in (3) is a divergence term.

$$\operatorname{div}\left(\frac{\partial f}{\partial u_1} \cdot \frac{\partial^2 f}{(\partial u_2)^2}\,, -\frac{\partial f}{\partial u_1} \cdot \frac{\partial^2 f}{\partial u_1 \partial u_2} \right) = \frac{\partial^2 f}{(\partial u_1)^2} \cdot \frac{\partial^2 f}{(\partial u_2)^2} - \frac{\partial^2 f}{\partial u_1 \partial u_2} \cdot \frac{\partial^2 f}{\partial u_1 \partial u_2}.$$

By the Gaussian divergence theorem, the integral of such a term over a domain can be reduced to an integral along the boundary $\partial\Omega$ of Ω . The integrand of this boundary integral has the form $\frac{\partial f}{\partial u_1} \cdot \frac{\partial}{\partial \tau}\left(\frac{\partial f}{\partial u_2}\right)$, where $\frac{\partial}{\partial \tau}$ denotes the derivative along the boundary of $\partial\Omega$. It only depends on the partial derivatives of f along the boundary. These are uniquely determined provided that the values of f and a non-tangential derivative of f are prescribed along $\partial\Omega$.
(a) The proof (as well as the statement) is the non-linear version of (b). It runs along the same line, but is more complicated. Actually, it follows from the famous Gauss–Bonnet Theorem, which states how the integral over Gaussian curvature can be expressed in terms of boundary data (cf. [9]). ■

Two special cases which are considered very often will be stated separately.

Corollary. *(a) Provided that boundary data along $\partial\Omega$ are fixed, the functionals*

$$J_{11}(F) = \int_\Omega \kappa_1^2(u) + \kappa_2^2(u)\, d\omega(u) \qquad \text{and} \qquad J_{10}(F) = \int_\Omega H_F(u)^2\, d\omega(u)$$

have the same local (and global) minima. (b) Provided that boundary data along $\partial\Omega$ are fixed, then (3), (4) and (5) have the same local (and global) minima. Note that (a) and (b) are true, when closed surfaces are considered!

From the discussion above we know simple functionals (namely (4) or (5)) which can replace the realistic functional (1) or (2) provided that parametrization we are looking for is nearly isometric. This is true e.g., if the surface is parametrized in explicit form $F : (u_1, u_2) \mapsto (u_1, u_2, f(u_1, u_2))$ and the partial derivatives of f are small. In fact, then $g_{11} = 1 + \frac{\partial f}{\partial u_1} \approx 1$, $g_{22} = 1 + \frac{\partial f}{\partial u_2} \approx 1$ and $g_{12} = g_{21} = \frac{\partial f}{\partial u_1} \cdot \frac{\partial f}{\partial u_2} \approx 0$.

This assumption will not be fulfilled, in general, unless we choose the parameter space carefully. Ideally, taking the surface itself as parameter space, would give rise to an isometric parametrization. This, however, is not possible, unless we already know what we are looking for. Nevertheless, one can try, a priori, to estimate the final result and choose the parameter space accordingly. E.g., to find a blend surface for the configuration given in the upper left part of Fig. 1, the plane parameter space shown in the lower left part of this figure seems to be a reasonable choice. An even better choice would be a non-plane parameter space as shown in the lower right part of the figure.

Now the question arises: How do we deal with non-plane parameter spaces? This is actually not very difficult. We will see that the discussion can be reduced to the (plane) parameter space used for parametrizing the (non-plane) parameter space. In the example shown in the figure, the rectangle on the upper right can serve as parameter space for parametrizing the non-plane parameter space shown lower right. Then, if we consider the lower right as a parameter space for the blend surface, all manipulations (including

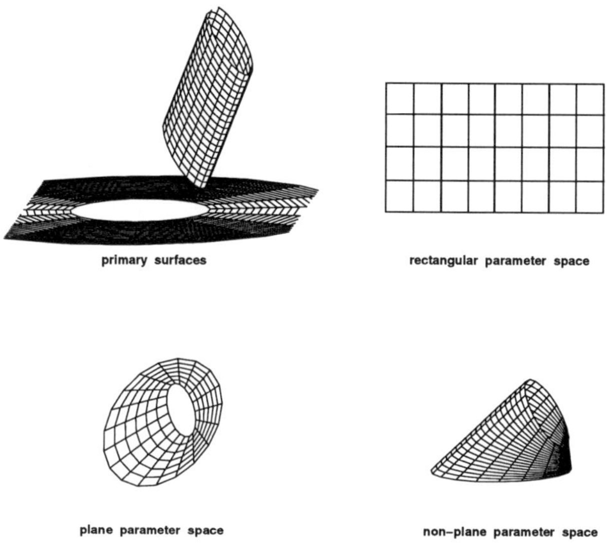

Figure 1. Different parameter spaces.

the parametrization of the blend surface) can be performed in the rectangle. We only have to deal with functions defined on the rectangle!

The concepts needed are well-known in differential geometry. One must consider *gradient, divergence* and *Laplacian* with respect to the Riemannian metric induced by the parametrization. Let us recall some basic definitions (for more details see e.g. [9]). Given a surface $F_0 : \Omega \to \mathbb{R}^3$ (which later will serve as a parameter space), the gradient and divergence with respect to F_0 are defined as follows:

$$\mathrm{grad}_{F_0} f = \left(\frac{\partial f}{\partial u_1}\,,\,\frac{\partial f}{\partial u_2}\right)\begin{pmatrix} g_{011} & g_{012} \\ g_{021} & g_{022} \end{pmatrix}^{-1}$$

$$\mathrm{div}_{F_0}(f_1, f_2) = \frac{1}{\sqrt{g_0}}\left(\frac{\partial}{\partial u_1}(\sqrt{g_0} f_1) + \frac{\partial}{\partial u_2}(\sqrt{g_0} f_2)\right) .$$

f , f_1 and f_2 are scalar functions defined on Ω , (g_{0ij}) denotes the first fundamental form of F_0 and $g_0 = \det(g_{0ij}) = g_{011}g_{022} - {g_{012}}^2$. The Laplacian (with respect to F_0) then is defined as usual. The operator

$$\Delta_{F_0} f = \mathrm{div}_{F_0}(\mathrm{grad}_{F_0} f)$$

is often called the *Laplace–Beltrami* operator corresponding to F_0, and acts on functions defined on the parameter space of F_0. It is a second order elliptic differential operator (with non-constant coefficients). It has the same meaning as the usual Laplacian in the plane, e.g., it governs the heat flow on the surface. Important for our considerations is the fact, that it transforms nicely under parameter transformations. That is, if $\tilde{F_0} = F_0 \circ \Phi$ is another parametrization of the surface, then $\Delta_{\tilde{F_0}} \tilde{f} = \widetilde{\Delta_{F_0} f}$ for any function f.

The Laplacian typically will be applied to scalar functions. When applying it to a vector valued function, $F = (f_1, f_2, f_3)$, we apply it to each component individually $\big($i.e, $\Delta_{F_0} F = (\Delta_{F_0} f_1, \Delta_{F_0} f_2, \Delta_{F_0} f_3)\big)$.

The following proposition shows that Prop. 1 is true in the more general setting. It also indicates how a quadratic functional can be chosen that will be a good approximation for (1) or (2). Namely, choose F_0 such that its shape is close to the solution we are looking for. Then define the functional J_{F_0} as follows

$$J_{F_0}(F) = \int_\Omega \langle \Delta_{F_0} F | \Delta_{F_0} F \rangle \, d\omega. \tag{6}$$

Note that this functional is a geometric quantity of F_0, in the sense that it does not depend on the parametrization used to represent F_0!

Proposition 3. *If the first fundamental forms of F and F_0 coincide, then*

$$J_{F_0}(F) = 4 \int_\Omega H_F(u)^2 \, d\omega(u)$$

where H_F denotes the mean curvature of F.

Proof: By definition, the Laplace–Beltrami operator only depends on (the coefficients of) the first fundamental form. Thus, in the situation of the proposition, we have $\Delta_{F_0} f = \Delta_F f$ for any function f. The statement then follows from the following fact, true for any parametrized surface (see [9]),

$$\Delta_F F(u) = 2 H_F(u) n_F(u). \quad \blacksquare$$

We summarize the consequences of our discussion, explaining how a functional can be chosen, which is sufficiently simple and is a reasonable approximation to the minimal energy functional (1).

1) Choose a surface $F_0 : \Omega \to \mathbb{R}^3$ that has roughly the shape of the surface we are looking for.

2) Consider the Laplace–Beltrami operator Δ_{F_0} and define the quadratic functional J_{F_0} as follows:

$$J_{F_0}(F) = \int_\Omega \langle \Delta_{F_0} F | \Delta_{F_0} F \rangle \, d\omega$$

3) Find the solution to the (quadratic) variational problem
$J_{F_0}(F) = \min$ and F satisfies the constraints.

In the next section, we describe these steps in more detail for the problem of constructing a blend surface.

Obviously, this procedure can be iterated. Starting with F_0 we can solve the variational problem as described above thus obtaining a surface, say F_1.

This now may serve as (non-plane) parameter space allowing an (even better) nearly isometric parametrization, i.e., in the second step we look for the minimum of the quadratic functional $J_{F_1}(F) = \int \langle \Delta_{F_1} F | \Delta_{F_1} F \rangle$. This can be repeated ad infinitum. However, the implementation shows, that after two steps there is hardly any visible difference between successive iterations

§4. Application to Blending Surfaces

We describe the method within a setting of tensor product spline surfaces. As a model problem one may think of the situation described in the upper left part of Figure 1. We assume that the primary surfaces are tensor spline surfaces. The boundary curves γ_0 and γ_1 (which have to be interpolated) are assumed to be closed spline curves. Moreover, along γ_0 and γ_1 the tangential vector fields Γ_0 und Γ_1 are given such that $\Gamma_i(t)$ and $\gamma_i'(t)$ span the tangent space of the primary surface at $\gamma_i(t)$. We also assume that Γ_i are closed spline curves. All four curves are defined over the parameter interval $[0, 2\pi]$. One has to perform the following steps:

1) Consider a space of tensor product spline surfaces $S(\mathbf{T})$ defined over the square $[0, 2\pi] \times [0, 1]$, being periodic with respect to the first component. The boundaries γ_i are parametrized by $[0, 2\pi] \times \{i\}$.

The knots on the given boundary curves γ_i and Γ_i will determine the knot set $\{t_i\}_{0 \leq i \leq N} \subset [0, 2\pi]$. The degree n of the tensor product splines with respect to the first variable is determined by the degree of the spline curves γ_i and Γ_i . There are no restrictions choosing the knots $\{s_j\}_{0 \leq j \leq M} \subset [0, 1]$ (any M , $3 < M$, will be OK). To keep things simple, choose these knots equidistantly. As degree with respect to the second component choose 3 (cubic).

2) The surface F_0 used as non-plane parameter space and determining the quadratic fairing functional will be chosen by linearly interpolating the boundary curves (as shown in the lower right part of Figure 1). Here $F_0(u_1, u_2) := \gamma_0(u_1)(1 - u_2) + \gamma_1(u_1)u_2$.

3) Choose any surface $F_{bc} \in S(\mathbf{T})$ that satisfies the boundary conditions. E.g., one may choose a Hermite interpolant: $F_{bc}(u_1, u_2) := (\gamma_0(u_1)(1 - 2u_2) + \Gamma_0(u_1)u_2)(1 - u_2)^2 + (\Gamma_1(u_1)(1 - u_2) + \gamma_1(u_1)(3 - 2u_2))u_2^2$.

Since the fairness functional J_{F_0} is quadratic and positive definite, minimizing it reduces to solving a linear system. It is the standard procedure for solving least square problcms.

4) Determine the control points of the blending surface F as follows:

a) Consider the periodic B–splines $\{b_1, ..., b_N\}$ over $[0, 2\pi]$ subordinated to $\{t_i\}$, and the cubic B–splines $\{d_0, ..., d_{M+2}\}$ over $[0, 1]$ subordinated to $\{s_j\}$ satisfying $d_k(0) = d_{M+2-k}(1) = \delta_{0k}$ and $d_k'(0) = d_{M+2-k}'(1) = \delta_{1k}$ for $k = 0, 1, 2, ..., M + 2$.

b) Determine the coefficients $B_{jkst} := B_{F_0}(b_j \otimes d_k, b_s \otimes d_t)$ where B_{F_0} is the symmetric bilinear functional associated with J_{F_0}.

c) Determine the control points $a_{jk} = (a_{1jk}, a_{2jk}, a_{3jk})$ of F_{bc} with respect to the tensor product B–splines $b_j \otimes d_k$. (knot insertion)

d) Solve for $i = 1, 2, 3$ the $N \times (M-1)$ linear system

$$\sum_{\substack{1 \le j \le N \\ 2 \le k \le M}} B_{jkst} x_{ijk} = -\sum_{\substack{1 \le j \le N \\ 0 \le k \le M+2}} a_{ijk} B_{jkst} \quad \begin{matrix} 1 \le s \le N \\ 2 \le t \le M \end{matrix}$$

5) With the solutions $x_{jk} = (x_{1jk}, x_{2jk}, x_{3jk})$ one obtains the control points A_{jk} of the surface F as follows:

$$A_{jk} := \begin{cases} a_{jk} + x_{jk} & \text{if} \quad 2 \le k \le M\,, \\ a_{jk} & \text{if} \quad k = 0, 1, M+1, M+2. \end{cases}$$

References

1. Bonneau, G.-P., Hagen, H. and St. Hahmann, Variational Surface Design and Surface Interrogation, Computer Graphics Forum **12**(3), 447–459.
2. Celniker, G. and D. Gossard, Deformable Curve and Surface Finite–Element for Free–Form Shape Design, ACM Computer Graphics, **25** (1991), 257–266.
3. Courant, R. and D. Hilbert, *Methods of Mathematical Physics, Vol. 1*, Wiley, New York, N.Y., 1953.
4. Dahmen, W., Micchelli, C. A. and H.-P. Seidel, Blossoming bedgets B–Spline bases built better by B–Patches, Math. Comp. **59** (1992), 97–115.
5. Greiner, G., Blending techniques based on variational principles, in *Curves and Surfaces* , J. D. Warren (ed.), Proc. SPIE 1830, 1992, 174–183.
6. Greiner G. and H.-P. Seidel, Curvature continuous blend surfaces, in *Modeling in Computer Graphics*, B. Falcidieno, T. L. Kunii (eds.), Springer–Verlag, Berlin 1993, 309–318.
7. Halstead, M., Kass M., and T. DeRose, Efficient, Fair Interpolation using Catmull-Clark surfaces, ACM Computer Graphics **27** (1993), 35–44.
8. Kallay, M., Constrained optimization in surface design, in *Modeling in Computer Graphics*, B. Falcidieno, T. L. Kunii (eds.), Springer–Verlag, Berlin, 1993, 85–93.
9. Klingenberg, W., *A Course in Differential Geometry*, Springer–Verlag, Berlin Heidelberg, 1978.
10. Lounsberry, M., Mann, St, and T. DeRose, Parametric surface interpolation, IEEE Comp. Graphics Appl. **12** (1992), 45–52.
11. Moreton, H. P. and C. Séquin, Functional Optimization for Fair Surface Design, ACM Computer Graphics **26** (1992), 167–176.
12. Nielson, G. M., A method for interpolating scattered data based upon a minimum norm network. Math. Comp. **40** (1983), 253–271.
13. Pottmann, H., Scattered data interpolation based upon generalized minimum norm network, Constr. Approx. **7** (1991) 247–256.
14. Welch, W. and A. Witkin, Variational surface modeling, ACM Computer Graphics, **26** (1992), 157–166.

Direct Methods for Constructing Positive Spline Interpolants

Walter Heß and Jochen W. Schmidt

Abstract. For interpolating data sets under positivity constraints, we use particular rational splines of the continuity classes C^1, C^2, and C^3. The resulting procedures, which are elementary, can be extended to $2D$ and $3D$ problems for gridded data.

§1. Introduction

In this paper, we are concerned with new direct methods for positivity preserving interpolation of gridded data. The idea is to consider particular rational splines defined on the grids introduced by the data. It is of importance that the smoothness conditions do not depend on the occurring rationality parameters. Therefore, these quantities can be used exclusively for meeting the positivity constraints. The resulting procedures are elementary. In the univariate case they are even local. Of course, the interpolants obtained in this way may be improved by applying more expensive methods, e.g., by curvature minimizing methods.

In one dimension, we will give procedures for constructing positive interpolants of the continuity classes C^1, C^2, and C^3. Further, C^1 interpolation of gridded data is considered in $3D$ including, of course, the $2D$ case.

Recently positive interpolation has been studied in several papers. Both direct and variational methods have been developed. A larger list of references can be found, e.g., in [1].

§2. Positive Rational Quadratic C^1 Splines in 1D

We assume that the data points (x_i, h_i), $i = 0(1)n$ with $x_0 < x_1 < \cdots < x_n$ are given, and that

$$h_i > 0, \qquad i = 0(1)n. \tag{1}$$

Wavelets, Images and Surface Fitting
P. J. Laurent, A. Le Méhauté, and L. L. Schumaker (eds.), pp. 287–294.

ISBN 1-56881-040-7.

The step sizes are denoted by $\xi_i = x_i - x_{i-1}$. The aim is to construct C^1 functions s which satisfy the interpolation condition

$$s(x_i) = h_i, \qquad i = 0(1)n \tag{2}$$

and which preserve the nonnegativity on the interval $I = [x_0, x_n]$,

$$s(x) \geq 0 \qquad \text{for } x \in I. \tag{3}$$

The present rational quadratic splines s are defined by

$$s(x) = S_i a_i(u) = (a_i(u)^T S_i) \text{ for } x \in I_i = [x_{i-1}, x_i], \tag{4}$$

with $S_i = S_i^1$ and $a_i(u) = f_1(u) + \varrho_1(u, \alpha_i) g_1(\xi_i)$ where

$$S_i^1 = \begin{pmatrix} h_{i-1} \\ h_i \\ p_{i-1} \end{pmatrix}, \quad f_1(u) = \begin{pmatrix} 1-u \\ u \\ 0 \end{pmatrix}, \quad g_1(\xi) = \begin{pmatrix} 1 \\ -1 \\ \xi \end{pmatrix}, \tag{5}$$

$$\varrho_\kappa(u, \alpha) = \frac{u(1-u)}{1 + \alpha u^\kappa (1-u)^\kappa}. \tag{6}$$

Here u is a local variable on I_i introduced by $u = (x - x_{i-1})/\xi_i$, $i = 1(1)n$. The quantities $\alpha_1 \geq 0, \ldots, \alpha_n \geq 0$ are called rationality parameters. Obviously, the interpolation condition (2) holds. Further, the above splines are from C^1 if and only if

$$p_{i-1} + p_i = \frac{2}{\xi_i}(h_i - h_{i-1}), \qquad i = 1(1)n, \tag{7}$$

and in this case

$$p_i = s'(x_i), \qquad i = 0(1)n. \tag{8}$$

In our context it is of importance that (7) is independent of the rationality parameters. Next, in view of

$$\frac{u(1-u)}{1 + \alpha_i u(1-u)} \leq \frac{1}{4 + \alpha_i} \quad \text{for} \quad 0 \leq u \leq 1, \tag{9}$$

we obtain the lower bound

$$s(x) \geq J(\alpha_i) = A_i + B_i/(4 + \alpha_i) \quad \text{for} \quad x \in I_i \tag{10}$$

with

$$A_i = \min\{ h_{i-1}, h_i \} > 0, \quad B_i = \min\{ 0, h_{i-1} - h_i + \xi_i p_{i-1} \} \leq 0. \tag{11}$$

Hence (3) is satisfied if $J(\alpha_i) \geq 0$ for $i = 1(1)n$, i.e. nonnegativity of s is assured if

$$\alpha_i = \max\{ 0, -4 - B_i/A_i \}, \qquad i = 1(1)n. \tag{12}$$

Thus for constructing nonnegative interpolants (4), (5), (6) we can proceed as follows.

(a) Compute an approximation to the derivative p_0, e.g., by means of the data points,
(b) determine $p_1, \ldots, p_n$ via the C^1 condition (7),
(c) choose the rationality parameters according to (12).

This produces a nonnegative C^1 spline which interpolates the data.

We mention that, in contrast to the positivity, the properties of monotonicity and convexity are not always preserved by the present splines; see [2].

§3. Positive Rational Cubic C² Splines in 1D

In order to get positivity preserving interpolants of C^2 continuity we use certain rational cubic splines. In (4) set $S_i = S_i^2$ and $a_i(u) = f_2(u) + \varrho_2(u, \alpha_i) g_2(u, \xi_i)$ with

$$S_i^2 = \begin{pmatrix} h_{i-1} \\ h_i \\ p_{i-1} \\ p_i \end{pmatrix}, \quad f_2(u) = \begin{pmatrix} 1-u \\ u \\ 0 \\ 0 \end{pmatrix}, \quad g_2(u,\xi) = \begin{pmatrix} 1-2u \\ -1+2u \\ (1-u)\xi \\ -u\xi \end{pmatrix}; \tag{13}$$

for ϱ_2 see (6). Again the interpolation condition (2) is satisfied, and (8) holds. Further the C^2 condition reads

$$\frac{p_{i-1}}{\xi_i} + \left(\frac{2}{\xi_i} + \frac{2}{\xi_{i+1}}\right) p_i + \frac{p_{i+1}}{\xi_{i+1}} = 3\left(\frac{h_{i+1} - h_i}{\xi_{i+1}^2} + \frac{h_i - h_{i-1}}{\xi_i^2}\right), \; i = 1(1)n-1. \tag{14}$$

To derive a sufficient positivity condition, we begin with

$$\varrho_2(u,\alpha) \le \varrho_2^*(\alpha) = \begin{cases} 4/(16+\alpha) & \text{for } \alpha \le 16 \\ 1/(2\sqrt{\alpha}) & \text{for } \alpha \ge 16, \end{cases} \qquad 0 \le u \le 1. \tag{15}$$

Hence we get the estimate

$$s(x) \ge J(\alpha_i) = A_i + \varrho_2^*(\alpha_i)\, B_i \;\; \text{for} \;\; x \in I_i \tag{16}$$

with A_i from (11) and

$$B_i = \min\{\, 0, h_{i-1} - h_i + \xi_i p_{i-1}, h_i - h_{i-1} - \xi_i p_i \,\} \le 0. \tag{17}$$

Thus we obtain $J(\alpha_i) \ge 0$ for $i = 1(1)n$, provided that

$$\alpha_i = \begin{cases} (-B_i)^2/(2A_i)^2 & \text{for } \; -B_i/A_i \ge 8 \\ \max\{\, 0, -16 - 4B_i/A_i \,\} & \text{for } \; -B_i/A_i \le 8, \end{cases} \qquad i = 1(1)n. \tag{18}$$

Consequently, the procedure (a)-(c) from § 2 for computing nonnegative spline interpolants extends immediately to the C^2 case if (14), (18) are substituted for (7), (12).

§4. Positive Rational Quintic C³ Splines in 1D

We are now in the position to offer positive C^3 spline interpolants. They are of the same simplicity as the preceding C^1 and C^2 ones. In (4) we choose $S_i = S_i^3$ and $a_i(u) = f_3(u) + \varrho_3(u, \alpha_i) g_3(u, \xi_i)$ with

$$S_i^3 = \begin{pmatrix} h_{i-1} \\ h_i \\ p_{i-1} \\ p_i \\ P_{i-1} \\ P_i \end{pmatrix}, \quad f_3(u) = \begin{pmatrix} 1-u \\ u \\ 0 \\ 0 \\ 0 \\ 0 \end{pmatrix}, \quad g_3(u,\xi) = \begin{pmatrix} 1+u-9u^2+6u^3 \\ -1-u+9u^2-6u^3 \\ (1-u)^2(3u+1)\xi \\ u^2(3u-4)\xi \\ u(1-u)^2\xi^2/2 \\ u^2(1-u)\xi^2/2 \end{pmatrix}; \tag{19}$$

for ϱ_3 see (6). Conditions (2) and (8) are verified straightforwardly as well as

$$P_i = s''(x_i), \qquad i = 0(1)n, \tag{20}$$

and the following C^3 condition is obtained:

$$\begin{aligned} -\frac{P_{i-1}}{\xi_i} + \left(\frac{3}{\xi_i} + \frac{3}{\xi_{i+1}}\right) P_i - \frac{P_{i+1}}{\xi_{i+1}} &= \frac{8p_{i-1}}{\xi_i^2} + \left(\frac{12}{\xi_i^2} - \frac{12}{\xi_{i+1}^2}\right) p_i - \frac{8p_{i+1}}{\xi_{i+1}^2} \\ &+ 20\left(\frac{h_{i+1}-h_i}{\xi_{i+1}^3} - \frac{h_i - h_{i-1}}{\xi_i^3}\right), \; i = 1(1)n-1. \end{aligned} \tag{21}$$

Further, because of

$$\varrho_3(u,\alpha) \le \varrho_3^*(\alpha) = \begin{cases} 16/(64+\alpha) & \text{for } \alpha \le 32 \\ 2/(3\sqrt[3]{2\alpha}) & \text{for } \alpha \ge 32, \end{cases} \qquad 0 \le u \le 1, \tag{22}$$

we get on I_i

$$s(x) \ge J(\alpha_i) = A_i + \varrho_3^*(\alpha_i) B_i \tag{23}$$

with A_i from (11) and

$$B_i = \min\{\,0, \min\{\, S_i g_3(u,\xi_i) \; : \; 0 \le u \le 1\,\}\} \le 0. \tag{24}$$

Thus, the sufficient positivity condition $J(\alpha_i) \ge 0$, for $i = 1(1)n$ is satisfied if

$$\alpha_i = \begin{cases} (-2B_i)^3/(2(3A_i)^3) & \text{for } -B_i/A_i \ge 6 \\ \max\{\,0, -64 - 16B_i/A_i\,\} & \text{for } -B_i/A_i \le 6, \; i = 1(1)n. \end{cases} \tag{25}$$

To extend the procedure (a)-(c) from §2, in (a) we have to approximate the derivatives $p_0, \ldots, p_n$, P_0, and P_n. Then, in (b) we can compute $P_1, \ldots, P_{n-1}$ by means of the C^3 condition (21) while in (c) the rationality parameters are determined by (25).

§5. Positive Rational Quadratic C^1 Splines in 3D

The above approaches extend to the positive interpolation of gridded data in $2D$ as well as in $3D$. Here the details are described for the C^1 case in $3D$. The given data points now are $(x_i, y_j, z_k, h_{i,j,k})$, $i = 0(1)n, j = 0(1)m, k = 0(1)l$ with $x_0 < x_1 < \ldots < x_n$, $y_0 < y_1 < \ldots < y_m$, $z_0 < z_1 < \ldots < z_l$, and we assume that

$$h_{i,j,k} > 0,\ i = 0(1)n,\ j = 0(1)m,\ k = 0(1)l. \tag{26}$$

On $I_{i,j,k} = [x_{i-1}, x_i] \times [y_{j-1}, y_j] \times [z_{k-1}, z_k]$ we define particular rational triquadratic splines by

$$s(x,y,z) = S_{i,j,k}a_i(u)b_j(v)c_k(w),\ (x,y,z) \in I_{i,j,k} \tag{27}$$

with $a_i(u) = f_1(u) + \varrho_1(u, \alpha_i)g_1(\xi_i)$, $b_j(v) = f_1(v) + \varrho_1(v, \beta_j)g_1(\eta_j)$, $c_k(w) = f_1(w) + \varrho_1(w, \gamma_k)g_1(\zeta_k)$, see (5), (6), and

$$S_{i,j,k}(.)(.)e_1 = \begin{pmatrix} h_{i-1,j-1,k-1} & h_{i-1,j,k-1} & q_{i-1,j-1,k-1} \\ h_{i,j-1,k-1} & h_{i,j,k-1} & q_{i,j-1,k-1} \\ p_{i-1,j-1,k-1} & p_{i-1,j,k-1} & P_{i-1,j-1,k-1} \end{pmatrix},$$

$$S_{i,j,k}(.)(.)e_2 = \begin{pmatrix} h_{i-1,j-1,k} & h_{i-1,j,k} & q_{i-1,j-1,k} \\ h_{i,j-1,k} & h_{i,j,k} & q_{i,j-1,k} \\ p_{i-1,j-1,k} & p_{i-1,j,k} & P_{i-1,j-1,k} \end{pmatrix}, \tag{28}$$

$$S_{i,j,k}(.)(.)e_3 = \begin{pmatrix} r_{i-1,j-1,k-1} & r_{i-1,j,k-1} & R_{i-1,j-1,k-1} \\ r_{i,j-1,k-1} & r_{i,j,k-1} & R_{i,j-1,k-1} \\ Q_{i-1,j-1,k-1} & Q_{i-1,j,k-1} & T_{i-1,j-1,k-1} \end{pmatrix}.$$

Here u, v, w are local variables on $I_{i,j,k}$ introduced by $u = (x - x_{i-1})/\xi_i$, $v = (y - y_{j-1})/\eta_j, w = (z - z_{k-1})/\zeta_k$. The step sizes are denoted by $\xi_i = x_i - x_{i-1}, \eta_j = y_j - y_{j-1}, \zeta_k = z_k - z_{k-1}$ while $\alpha_i \geq 0, \beta_j \geq 0, \gamma_k \geq 0$ are the rationality parameters, $i = 1(1)n, j = 1(1)m, k = 1(1)l$. The interpolation property

$$s(x_i, y_j, z_k) = h_{i,j,k},\ i = 0(1)n, j = 0(1)m, k = 0(1)l \tag{29}$$

is incorporated in this representation. Further, these splines are C^1 continuous on $I = [x_0, x_n] \times [y_0, y_m] \times [z_0, z_\ell]$ if and only if the following conditions hold true,

$$\begin{aligned} p_{i,j,k} + p_{i-1,j,k} &= \frac{2}{\xi_i}(h_{i,j,k} - h_{i-1,j,k}), \\ Q_{i,j,k} + Q_{i-1,j,k} &= \frac{2}{\xi_i}(r_{i,j,k} - r_{i-1,j,k}), \\ P_{i,j,k} + P_{i-1,j,k} &= \frac{2}{\xi_i}(q_{i,j,k} - q_{i-1,j,k}), \end{aligned}$$

$$
\begin{aligned}
T_{i,j,k} + T_{i-1,j,k} &= \frac{2}{\xi_i}(R_{i,j,k} - R_{i-1,j,k}), \; i = 1(1)n, j = 0(1)m, k = 0(1)l, \\
q_{i,j,k} + q_{i,j-1,k} &= \frac{2}{\eta_j}(h_{i,j,k} - h_{i,j-1,k}), \\
P_{i,j,k} + P_{i,j-1,k} &= \frac{2}{\eta_j}(p_{i,j,k} - p_{i,j-1,k}), \\
R_{i,j,k} + R_{i,j-1,k} &= \frac{2}{\eta_j}(r_{i,j,k} - r_{i,j-1,k}), \\
T_{i,j,k} + T_{i,j-1,k} &= \frac{2}{\eta_j}(Q_{i,j,k} - Q_{i,j-1,k}), \; j = 1(1)m, i = 0(1)n, k = 0(1)l, \\
r_{i,j,k} + r_{i,j,k-1} &= \frac{2}{\zeta_k}(h_{i,j,k} - h_{i,j,k-1}), \\
R_{i,j,k} + R_{i,j,k-1} &= \frac{2}{\zeta_k}(q_{i,j,k} - q_{i,j,k-1}), \\
Q_{i,j,k} + Q_{i,j,k-1} &= \frac{2}{\zeta_k}(p_{i,j,k} - p_{i,j,k-1}), \\
T_{i,j,k} + T_{i,j,k-1} &= \frac{2}{\zeta_k}(P_{i,j,k} - P_{i,j,k-1}), \; k = 1(1)l, i = 0(1)n, j = 0(1)m.
\end{aligned}
\tag{30}
$$

In addition, if the equalities (30) are satisfied the mixed partial derivatives $\partial_1\partial_2 s$, $\partial_1\partial_3 s$, $\partial_2\partial_3 s$, and $\partial_1\partial_2\partial_3 s$ are continuous on I, and we have

$$
\begin{aligned}
&p_{i,j,k} = \partial_1 s(x_i, y_j, z_k), \; q_{i,j,k} = \partial_2 s(x_i, y_j, z_k), \; r_{i,j,k} = \partial_3 s(x_i, y_j, z_k), \\
&P_{i,j,k} = \partial_1\partial_2 s(x_i, y_j, z_k), \; Q_{i,j,k} = \partial_1\partial_3 s(x_i, y_j, z_k), \; R_{i,j,k} = \partial_2\partial_3 s(x_i, y_j, z_k), \\
&T_{i,j,k} = \partial_1\partial_2\partial_3 s(x_i, y_j, z_k), \; i = 0(1)n, \; j = 0(1)m, \; k = 0(1)l;
\end{aligned}
\tag{31}
$$

compare with [3], where the consistency of system (30) is proved, namely, the system corresponding to

$$
\begin{aligned}
&p_{0,j,k}, \, q_{i,0,k}, \, r_{i,j,0}, \, P_{0,0,k}, Q_{0,j,0}, \, R_{i,0,0}, \, T_{0,0,0}, \\
&i = 0(1)n, \; j = 0(1)m, \; k = 0(1)l
\end{aligned}
\tag{32}
$$

is uniquely solvable.

Further, because of (9) we get the lower bound

$$
\begin{aligned}
s(x,y,z) \ge J_{i,j,k}(\alpha_i, \beta_j, \gamma_k) &= A_{i,j,k} + \frac{B_{i,j,k}}{4+\alpha_i} + \frac{C_{i,j,k}}{4+\beta_j} + \frac{D_{i,j,k}}{4+\gamma_k} \\
&+ \frac{E_{i,j,k}}{(4+\alpha_i)(4+\beta_j)} + \frac{F_{i,j,k}}{(4+\alpha_i)(4+\gamma_k)} + \frac{G_{i,j,k}}{(4+\beta_j)(4+\gamma_k)} \\
&+ \frac{H_{i,j,k}}{(4+\alpha_i)(4+\beta_j)(4+\gamma_k)}
\end{aligned}
\tag{33}
$$

for $(x, y, z) \in I_{i,j,k}$. Here we use the abbreviations

$$
\begin{aligned}
A_{i,j,k} &= \min\{\, h_{\lambda,\mu,\nu} : \lambda \in \{i-1, i\}, \mu \in \{j-1, j\}, \nu \in \{k-1, k\}\} > 0, \\
B_{i,j,k} &= \min\{0, \min\{S_{i,j,k} g_1(\xi_i) e_\lambda e_\mu : \lambda, \mu \in \{1,2\}\}\} \le 0, \\
C_{i,j,k} &= \min\{0, \min\{S_{i,j,k} e_\lambda g_1(\eta_j) e_\mu : \lambda, \mu \in \{1,2\}\}\} \le 0, \\
D_{i,j,k} &= \min\{0, \min\{S_{i,j,k} e_\lambda e_\mu g_1(\zeta_k) : \lambda, \mu \in \{1,2\}\}\} \le 0, \\
E_{i,j,k} &= \min\{0, \min\{S_{i,j,k} g_1(\xi_i) g_1(\eta_j) e_\lambda : \lambda \in \{1,2\}\}\} \le 0, \\
F_{i,j,k} &= \min\{0, \min\{S_{i,j,k} g_1(\xi_i) e_\lambda g_1(\zeta_k) : \lambda \in \{1,2\}\}\} \le 0, \\
G_{i,j,k} &= \min\{0, \min\{S_{i,j,k} e_\lambda g_1(\eta_j) g_1(\zeta_k) : \lambda \in \{1,2\}\}\} \le 0, \\
H_{i,j,k} &= \min\{0, S_{i,j,k} g_1(\xi_i) g_1(\eta_j) g_1(\zeta_k)\} \le 0.
\end{aligned}
\tag{34}
$$

Hence, we have the sufficient positivity condition

$$J_{i,j,k}(\alpha_i, \beta_j, \gamma_k) \ge 0,\ i = 1(1)n,\ j = 1(1)m, k = 1(1)l \tag{35}$$

which, in the case $\delta = \alpha_i = \beta_j = \gamma_k$, $i = 1(1)n$, $j = 1(1)m$, $k = 1(1)l$, is satisfied by

$$\delta = \max\{\, 0, -4 + \epsilon\,\} \tag{36}$$

where $\epsilon \ge 0$ solves

$$A\epsilon^3 + (B + C + D)\epsilon^2 + (E + F + G)\epsilon + H = 0. \tag{37}$$

Here we have set

$$A = \min A_{i,j,k},\ B = \min B_{i,j,k}, \ldots,\ H = \min H_{i,j,k}. \tag{38}$$

Again the procedure (a)-(c) from §2 extends. In (a) now the initial values (32) are to be computed approximatively while in (b) and (c) we have to substitute (30) and (36) for (7) and (12), respectively.

§ 6. Numerical Example

Figure 1 shows two splines which interpolate the marked $1D$ data set. The dotted one is a quadratic spline obtained by setting $\alpha_i = 0, i = 1(1)10$ in the representation (4)-(6). The derivative p_0 is computed by minimizing the Holladay functional subject to the C^1 condition (7). This spline fails to be everywhere nonnegative. The solid spline (4)-(6), determined by method (a)-(c), is rational quadratic. This spline has the desired property of nonnegativity. In addition to the lower bound zero, here we have also prescribed upper bounds

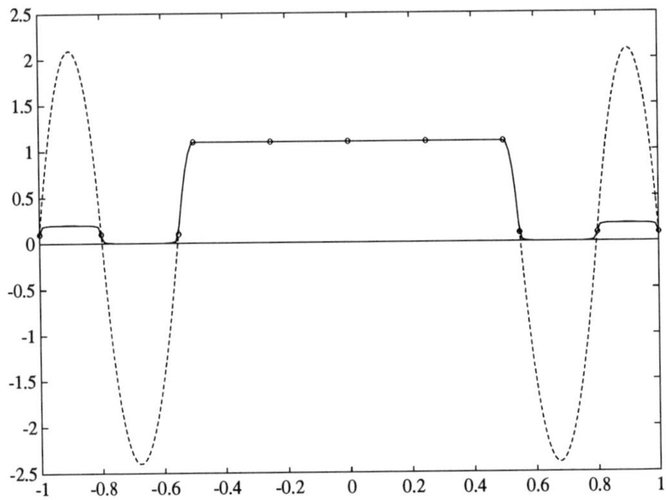

Figure 1. (- - -) quadratic spline not everywhere nonnegative,
(—-) nonnegative rational quadratic spline (4)-(6), (39).

K_i on the subintervals I_i, $i = 1(1)n$. This requirement is easily met in method (a)-(c) by substituting

$$\alpha_i = \max\{\, 0, -4 - B_i/A_i, -4 + M_i/(K_i - L_i)\,\} \tag{39}$$

for (12), where L_i and M_i are defined by

$$L_i = \max\{\, h_{i-1}, h_i \,\} < K_i, \;\; M_i = \max\{\, 0, h_{i-1} - h_i + \xi_i p_{i-1} \,\}. \tag{40}$$

References

1. Mulansky, B., and Schmidt, J. W., Nonnegative C^1-interpolation of scattered data using Powell-Sabin splines, Preprint TU Dresden MATH–NM–02–1993.
2. Schmidt, J. W., Positive, monotone, and S-convex C^1-interpolation on rectangular grids. Computing **48** (1992),363-371.
3. Schmidt, J. W., Positive and S-convex C^1-interpolation of gridded $3D$ data, Preprint TU Dresden, MATH–NM–03–1993.

Walter Heß and Jochen W. Schmidt
Institute of Numerical Mathematics
Technical University of Dresden
D – 01062 Dresden, GERMANY
numath@urzdfn.mathematik.tu-dresden.dbp.de

Conditionally Lower Riesz Bounds for Scattered Data Interpolation

K. Jetter

Abstract. A general approach for deriving lower Riesz bounds for interpolation with conditionally positive definite radial basis functions is presented. The method refers to the representation of the basis function in terms of a Laplace - Stieltjes integral. Therefore, one can apply the well-known fact that (scaled) exponentials are positive definite functions. The method is appropriate for getting upper Riesz bounds as well (and hence condition numbers for the collocation matrix) only for basis functions which are (order 0 conditionally) positive definite. In other cases one is forced to use a different method; e.g., the estimates of [4] can be applied to preconditioned basis functions.

§1. Introduction

A function $\phi \in C(\mathbb{R}^d)$ is said to be *conditionally positive definite of order m*, $m \in \mathbb{N}$, if

$$\sum_{i=1}^{n}\sum_{j=1}^{n} c_i\,\bar{c}_j\,\phi(x_i - x_j) \quad \geq \quad 0 \tag{1.a}$$

for any (pairwise distinct) points $x_1, \ldots, x_n \in \mathbb{R}^d$ and any complex numbers $c_1, \ldots, c_n$ satisfying

$$\sum_{i=1}^{n} \bar{c}_i\, p(x_i) = 0, \qquad \text{for all } p \in \Pi_{m-1}^d. \tag{1.b}$$

Here, Π_{m-1}^d denotes the set of d-variate polynomials of degree at most $m-1$. Functions of this type were studied in detail for interpolation, viz. (1) is related to the following interpolation problem: With the points $x_1, \ldots, x_n$ we consider the evaluation functionals $\delta_{x_j} : f \to f(x_i)$, $i = 1, \ldots, n$, and the interpolation space

$$S := \Pi_{m-1}^d + \text{span}\,\{\phi(\cdot - x_j)\}_{j=1}^n. \tag{2}$$

Wavelets, Images and Surface Fitting
P. J. Laurent, A. Le Méhauté, and L. L. Schumaker (eds.), pp. 295–302.

ISBN 1-56881-040-7.

Whence, a function $s \in S$ has the form

$$s = p + \phi * \nu_c$$

where $p \in \Pi^d_{m-1}$, $*$ denotes convolution and $\nu_c = \sum_{j=1}^n \bar{c}_j \delta_{x_j}$ is a point measure related to the weights $c = (c_1, \ldots, c_n)$. The interpolation problem, then, is to find p and c, for given data $y = (y_1, \ldots, y_n)$, such that

$$s(x_i) = y_i\,, \quad i = 1, \ldots, n, \tag{3}$$

subject to the restriction (1.b). It turns out that the solvability of this problem refers to the condition of (strict) positive definiteness of order m, which can be put as follows:

$$c \neq 0 \quad \text{and} \quad \nu_c \perp \Pi^d_{m-1} \quad \text{implies} \quad \overline{\nu}_c(\phi * \nu_c) > 0. \tag{4}$$

For the special case where ϕ is the semi-Hilbertian kernel of some Sobolev-type space, this interpolation problem was studied in great detail (starting with J. Duchon in 1976) and in a rather abstract setting in France (cf., e.g., Arcangéli [1]). However, for numerical reasons, it is important to have a strong form of (1) such as

$$\sum_{i=1}^n \sum_{j=1}^n c_i \bar{c}_j \phi(x_i - x_j) \quad \geq \quad \text{const} \sum_{j=1}^n |c_j|^2 \tag{5.a}$$

subject to

$$\sum_{i=1}^n \bar{c}_i\, p(x_i) = 0\,, \qquad \text{for all } p \in \Pi^d_{m-1}. \tag{5.b}$$

Here, the constant should not depend on the number n and the particular position of the knots $x_1, \ldots, x_n$, but it may (and indeed will) depend on some measure of spacing of the knots. In what follows, (5) will be called a *conditionally lower Riesz bound* for the following reason: In case $\phi = \varphi * \varphi^*$ (with $\varphi \in L_2(\mathbb{R}^d)$ and $\varphi^*(x) = \overline{\varphi}(-x)$) we have

$$\sum_{i=1}^n \sum_{j=1}^n c_i \bar{c}_j \phi(x_i - x_j) \quad = \quad \Big\| \sum_{j=1}^n c_j \varphi(\cdot - x_j) \Big\|^2_{L_2(\mathbb{R}^d)},$$

and thus the problem studied here is closely related to the question of characterizing Riesz bases of type $\{\varphi(\cdot - x_j)\,;\, j = 1, 2, \ldots\}$.

Bounds of this type have recently turned out to be of particular interest in radial basis function interpolation (Buhmann [2]). They were studied by Narcowich and Ward [7, 8] who have derived several strong results under various restrictions on the basis function ϕ. Using a different approach, Schaback [9] succeeded in getting lower bounds as well.

Let us recall that $\phi \in C(\mathbb{R}^d)$ is called *radial* (or better *radially symmetric*) in case

$$\phi(x) = g(|x|^2) \tag{6}$$

for some $g \in C[0,\infty)$, where $|x|$ denotes the Euclidian norm of the vector x. It has been known for some time that *m-th order completely monotone* functions g, i.e., functions $g \in C[0,\infty) \cap C^\infty(0,\infty)$ satisfying

$$(-1)^j g^{(j)}(t) \geq 0\,, \quad t > 0\,, \quad \text{for all} \quad j \geq m\,, \tag{7}$$

give rise to conditionally positive definite functions ϕ. The inverse is also true due to the following remarkable result:

Theorem GHS [3]. *Given $g \in C[0,\infty)$, the following are equivalent:*

(a) $\phi = g(|\cdot|^2) \in C(\mathbb{R}^d)$ *is conditionally positive definite of order m, for any space dimension d.*

(b) $g \in C^\infty(0,\infty)$ *and g is m-th order completely monotone.*

The great advantage of this result lies in the possibility of representing the quadratic form (1) in terms of the Laplace-Stieltjes integral

$$(-1)^m g^{(m)}(t) = \int_0^\infty e^{-ty} d\mu(y)\,, \quad t > 0\,, \tag{8}$$

where $d\mu$ is a certain positive Borel measure on $(0,\infty)$ satisfying the estimates

$$\int_0^1 d\mu(y) < \infty \quad \text{and} \quad \int_1^\infty \frac{d\mu(y)}{y^m} < \infty\,.$$

We shall elaborate on this in the following sections, and in this way we get the lower Riesz bounds essentially by using the fact that, for any $y > 0$, the exponential function

$$e_y(t) = e^{-ty} \tag{9}$$

is positive definite.

It should be noted that this method does not apply to deriving upper Riesz bounds as well except for the very special case of (0-th order) completely monotone functions. In the higher order case one has to use preconditioning, so that, e.g., the estimates of [4] can be applied.

§2. Involving the Laplace-Stieltjes Representation

Since for $t > 0$,

$$g(t) - \sum_{j=0}^{m-1} \frac{g^{(j)}(1)}{j!}(t-1)^j \quad = \quad \int_1^t \frac{(t-\tau)^m}{m!} g^{(m)}(\tau)\, d\tau\,,$$

we get through (8) and an application of Fubini's theorem that

$$g(t) - \sum_{j=0}^{m-1} \frac{g^{(j)}(1)}{j!}(t-1)^j = \int_0^\infty \left[e^{-ty} - \sum_{j=0}^{m-1} \frac{(-y)^j e^{-y}}{j!}(t-1)^j \right] \frac{d\mu(y)}{y^m}.$$

This in turn yields a representation for $\phi(x) = g(|x|^2)$. Now it is not hard to see that for $f \in \Pi_l^d$ and $\nu_c = \sum_{i=1}^n c_i\delta_{x_i} \perp \Pi_{k-1}^d$ we have $f * \nu_c \in \Pi_{\ell-k}^d$ (in case $\ell \geq k$). In this way we arrive at

Lemma. *For an m-th order completely monotone function $g \in C[0,\infty)$ and the associated radial function $\phi = g(|\cdot|^2)$ the quadratic form (1.a) can be written as*

$$\sum_{i=1}^n \sum_{j=1}^n c_i\bar{c}_j\phi(x_i - x_j) \quad = \quad \int_0^\infty \sum_{i=1}^n \sum_{j=1}^n c_i\bar{c}_j\, e^{-y|x_i-x_j|^2} \frac{d\mu(y)}{y^m} \quad , \tag{10}$$

subject to (1.b); here $d\mu$ is the representing measure in (8).

Now, for $y > 0$ and $c \neq 0$,

$$\sum_{i=1}^n \sum_{j=1}^n c_i\,\bar{c}_j\, e^{-y|x_i-x_j|^2} \quad =$$

$$\frac{1}{(2\pi)^d}\left(\frac{\pi}{y}\right)^{d/2} \int_{\mathbb{R}^d} e^{-|\xi|^2/4y}\,|\sum_{k=1}^n c_k e^{-ix_k\cdot\xi}|^2 d\xi \quad > \quad 0\,,$$

which shows that the integrand in (10) is a continuous and strictly positive function of the variable y. Hence under rather mild assumptions on the measure $d\mu$, we have strict inequality in (1.a).

Let us note that these arguments can be considered as a variation of Micchelli's [5] ideas, but we find it convenient to repeat these in order to keep our paper self-contained.

§3. Estimating the Integrand

We are going to employ the ℓ_∞-separation distance of the point set $X = \{x_1, \ldots, x_n\}$,

$$q \quad := \quad q(X) \quad := \quad \min\{|x_i - x_j|_\infty\,,\, i \neq j\}. \tag{11}$$

Then, putting

$$c(d,q,y) \quad := \quad 1 + 2^d \sum_{k=2}^\infty \{k^d - (k-1)^d\}\, e^{-((k-1)q)^2 y} \tag{12}$$

for $y > 0$, we have

Theorem. *With the assumptions of the lemma we have*

$$\sum_{i=1}^n \sum_{j=1}^n c_i\bar{c}_j\phi(x_i - x_j) \quad \geq \quad \sum_{i=1}^n |c_i|^2 \int_0^\infty [2 - c(d,q,y)]_+ \frac{d\mu(y)}{y^m}\,, \tag{13}$$

where $r_+ := \max\{r, 0\}$ *for* $r \in \mathbb{R}$.

Proof: From the above considerations it is sufficient to find an appropriate (uniform) lower bound for the spectrum of the positive definite matrix

$$E_y = (e^{-y|x_i - x_j|^2})_{i,j=1}^n, \quad y > 0. \tag{14}$$

From Gerschgorin's circle theorem, an obvious lower bound for the smallest eigenvalue $\lambda_{\min}(E_y)$ is given by

$$\lambda_{\min}(E_y) \quad \geq \quad 2 - \max\{\Sigma_i(y),\ i = 1, \ldots, n\},$$

where

$$\Sigma_i(y) \quad := \quad \sum_{j=1}^n e^{-y|x_i - x_j|^2} \quad \leq \quad \sum_{j=1}^n e^{-y|x_i - x_j|_\infty^2} \tag{15}$$

for $i = 1, \ldots, n$. Hence the theorem is an immediate consequence of the lemma provided that the latter sum is bounded by the constant given in (12).

To show this, we apply a counting argument to the sets

$$X_i \quad := \quad \{x_i - x_j; j = 1, \ldots, n\}$$

as follows. Let

$$Q_k := (-kq, +kq)^d, \quad k = 1, 2, \ldots$$

be the open cubes of side-length $2kq$, centered at the origin. Then

$$Q_1 \cap X_i \quad = \quad \{0\},$$

while for $k > 1$ the set difference $Q_k \setminus Q_{k-1}$ contains at most

$$q^{-d}\left[\operatorname{vol} Q_k - \operatorname{vol} Q_{k-1}\right] = (2k)^d - (2k-2)^d$$

points from X_i, and each of these points satisfies the lower bound

$$|x_i - x_j|_\infty \quad \geq \quad (k-1)q.$$

From this the required estimate is obvious. ■

We note that the same method of proof yields the upper bound

$$\sum_{i=1}^n \sum_{j=1}^n c_i \bar{c}_j \phi(x_i - x_j) \quad \leq \quad \sum_{i=1}^n |c_i|^2 \int_0^\infty c(d, q, y) \, \frac{d\mu(y)}{y^m}. \tag{16}$$

This can be seen from the fact that the largest eigenvalue $\lambda_{\max}(E_y)$ of the matrix E_y as defined in (14), is bounded by

$$\lambda_{\max}(E_y) \quad \leq \quad \max\{\Sigma_i(y),\ i = 1, \ldots, n\}.$$

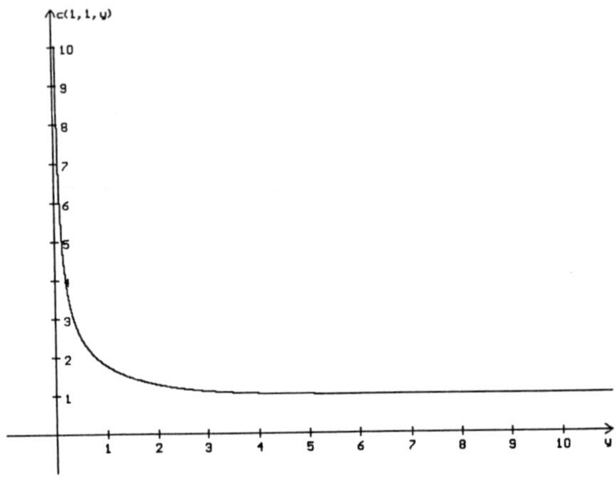

Fig. 1. The constant $c(1,1,y)$.

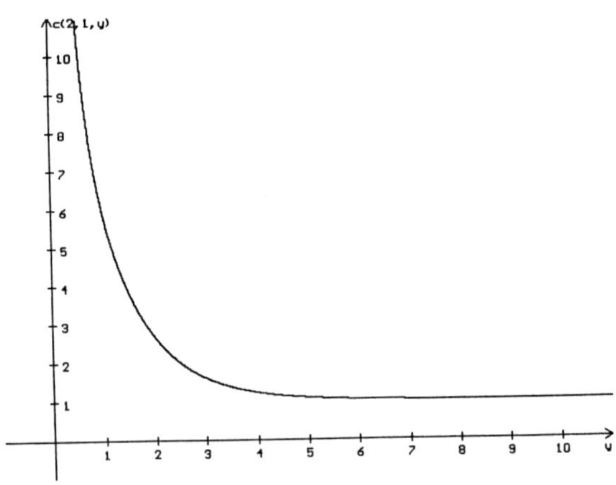

Fig. 2. The constant $c(2,1,y)$.

However, the upper bound (16) is finite only in special cases, e.g., if $m = 0$, or in case the support of the measure $d\mu$ is bounded away from the origin.

It should be noted that the constant (12), for fixed d and q, as a function of the variable y is monotonely decreasing and

$$\lim_{y \to \infty} c(d,q,y) = 1 .$$

Figures 1 and 2 show the graph of $c(1,1,\cdot)$ and $c(2,1,\cdot)$, respectively. Hence it is clear that, for any $\varepsilon > 0$ there is a $y_0 = y_0(d, q, \varepsilon)$ such that the bound (13) can be replaced by the lower bound

$$\sum_{i=1}^{n}\sum_{j=1}^{n} c_i\bar{c}_j\phi(x_i - x_j) \quad \geq \quad (1-\varepsilon)\sum_{i=1}^{n}|c_i|^2 \int_{y_0}^{\infty} \frac{d\mu(y)}{y^m}. \tag{17}$$

In special cases where the measure is explicitly known (as, e.g., in case of thin plate splines, or the multiquadric) this latter integral is easily evaluated.

§4. Additional Remarks

The recent work on quantitative analysis of radial basis interpolation is very well reviewed in Buhmann [2, section 4]. Narcowich and Ward were the first to derive deep results in this direction (except for the case where the points $x_1, \ldots, x_n$ are taken from a regular grid, say, the lattice $\mathbb{Z}^n$). For example, our lemma appears explicitly in [8, Corollary 2.2], but the authors make a different use of this since they base their estimates on the comparison of the Fourier transforms of different basis functions. We think that our method provides at least an alternative (and sometimes simpler) approach.

There is another slight change in the counting argument given above as compared with the Narcowich and Ward papers [7, 8] and the follow-up paper [6]: They use the ℓ_2-separation distance while we prefer to use the (of course, theoretically equivalent) ℓ_∞-separation. This results in an apparently easier counting procedure (since the plane can be trivially covered with cubes); it seems to be also more appropriate in order to keep track of the fact that regular grids give rise to optimal Riesz constants.

Acknowledgements. This research was partially supported by NATO Collaborative Research Grant # 900158.

References

1. Arcangéli, R., Splines d'ordre (m, s), Report # 90/12, Laboratoire de Mathématiques Appliquées, Université de Pau (1990), 49 p.
2. Buhmann, M., New developments in the theory of radial basis function interpolation, in *Multivariate Approximation: From CAGD to Wavelets*, K. Jetter and F. I. Utreras (eds.), World Scientific, Singapore, 1993, 35–75.
3. Guo, K., Hu, S., and X. Sun, Conditionally positive definite functions and Laplace-Stieltjes integrals, J. Approx. Th. **74** (1993), 249–265.
4. Jetter, K., Riesz bounds in scattered data interpolation and L_2-approximation, in *Multivariate Approximation: From CAGD to Wavelets*, K. Jetter and F. I. Utreras (eds.), World Scientific, Singapore, 1993, 167–177.
5. Micchelli, C. A., Interpolation of scattered data: distance matrices and conditionally positive definite functions, Constr. Approx. **2** (1986), 11–22.

6. Narcowich, F. J., Sivakumar, N., and J. D. Ward, On condition numbers associated with radial-function interpolation, CAT Report # 287, Texas A&M University (January 1993).
7. Narcowich, F. J. and J. D. Ward, Norms of inverses and condition numbers for matrices associated with scattered data, J. Approx. Th. **64**, (1991), 69–94.
8. Narcowich, F. J. and J. D. Ward, Norm estimates for the inverses of a general class of scattered-data radial-basis interpolation matrices, J. Approx. Th. **69** (1992), 84–109.
9. Schaback, R., Lower bounds for norms of inverses of interpolation matrices for radial basis functions, preprint (February 1992).

Kurt Jetter
FB Mathematik
Universität Duisburg
D–47048 Duisburg, GERMANY
hn277je@unidui.uni-duisburg.de

The 'Face Lift' Algorithm

Johannes Kaasa and Geir Westgaard

Abstract. A problem often met in modelling of geometry is to modify a master surface for a specific use. For this problem we have developed a new surface editing scheme based on decomposition. We call it the 'face lift' scheme. From the basic surface and scattered points on the modified surface this scheme generates the difference between the existing surface and the desired result. This 'face lift' surface should also satisfy certain smoothness constraints. The 'face lift' surface is modelled by use of scattered data techniques and tensor-product B-spline approximation.

§1. Introduction

This paper introduces a concept we have called *decomposed surface editing*. A surface is changed by adding difference surfaces as revision layers on top of the original surface. The principle is implemented in a new *'face lift'* scheme, which we will illustrate by an application on a turbine blade.

Our paper will concentrate on tensor-product B-spline surfaces. It is then natural to view the difference surfaces as offsets to the original surface, possibly with refined knot vectors to handle local details. This approach is described in [1-3], in these papers the difference surfaces are called 'overlays'. What is new in our scheme is the use of scattered data as input to the revision.

The term *decomposed surface editing* is motivated by the decomposition of the edit into a hierarchy of offsets defined in spline spaces of various resolution.

§2. Decomposed Surface Editing

The need to make edits on a surface turns up in many situations, two of them are:

1) When irregularities have been detected on a surface through some kind of interrogation, and the severity of the flaws demands a mending. For more details about some of the interrogation methods and how to use them in the modification process we refer to [4,6-7].

Wavelets, Images and Surface Fitting
P. J. Laurent, A. Le Méhauté, and L. L. Schumaker (eds.), pp. 303–310.

ISBN 1-56881-040-7.

2) It is profitable for the user to modify a general template surface for use in a specific application. This is the setting behind the 'face lift' scheme. The gain is that major parts of the template can be reused, and the modelling work restricted to minor parts. This is important if the modelling requires time-consuming measuring and simulation methods.

Decomposed surface editing can be regarded as a difference surface d being added to the original surface f_O, giving the revised surface f_R, i.e.,

$$f_R = f_O + d. \tag{1}$$

It is possible to do this repeatedly, using a sequence $d_1, \ldots, d_N$ of difference surfaces. The addition is performed by offsetting, either normally to the original surface or with an angular variation.

All editing operations are directed towards manipulation on the difference surface, the original surface is left unchanged. This separation of the model into a global basis shape with a refined edit on top, offers a number of advantages:

1) The revision process itself is made more secure and stable. The underlying shape is protected from the edit, in the 'face lift' case for instance, this is desirable. It is also trivial to assure that the edit is local and restricted to certain areas, since the difference surface will be different from zero only in these areas.
2) The user gets a better feeling of the quality of the editing along with an improved intuitive understanding of the revision process. This is because the difference surface displays only the changes taking place independently of superfluous information from the complete surface. For instance, it is possible to scale up the difference surface to reveal irregularities which must be removed. A scaling of the revised surface will not in general be equally revealing.

Input to a decomposed surface edit will be the original surface and a set of revision data. The revision data can be represented in a number of ways, dependent on the critical factors motivating the editing: positional deviation, angular inconsistencies, irregular shape etc. It is not uncommon that more than one factor is present in the same process, resulting in multi-format input and multi-step edit. The following list surveys some (more or less common) revision data types:

- Scattered interpolation or approximation points/tangents on the revised surface. This is either the result of the user conducting manual manipulation of points on the original surface, or the result of calculations done on a model of the revised surface.
- Constraints on some parameters on the surface, for instance restrictions on the cross angle along some of the surface edges, or demands on the quality of reflection lines. The revision must ensure that these constraints are met.

- Procedural rules. This is a particularly interesting input format, suggesting a combination of simulation and geometric modelling. Based on the procedural rules, the edit module could access a simulation module (typically a differential equation package) to get information for an optimal revision. It is far more complicated to build a model from the results of a simulation process than to run a simulation on an existing model. In this field there is much left to do.

The revision data can be restricted to a subpart of the total surface, which we call the *revision area.*

In addition to the explicitly stated revision data, there will often be additional implicit data based on various demands on the surface, for instance with regard to smoothness and shape. This must also be taken into consideration by the editing algorithm.

§3. The 'face lift' scheme for Surface Editing

In this section we describe a new 'face lift' scheme we have developed for decomposed surface editing. The term 'face lift' is motivated by the similarity with its surgical analogy, minor changes are made to the face but most of it is saved. Our scheme is motivated by a need to modify template surfaces, where the revision data influence only limited parts of the surface, but can be used in general situations as well. The scheme is illustrated by an example of 'face lift' of a Pelton turbine blade. The background for this example shows the usefulness of such algorithms.

3.1. Background for the Pelton example

A power station Pelton turbine consists of a wheel with a number of double blades attached to its circumference. The wheel and a connected generator is made to rotate by water coming down from a basin on a higher level and ejected onto the blades. It is vital for the efficiency of the turbines to transfer as much kinetic energy as possible from the water to the blades during impact. This efficiency is a function of the shape of the blades. The speed and volume of the incoming water will vary from power station to power station; this means that the optimal shape is not constant. However, the various optimal shapes are equal in large areas; the difference is located to a limited area. It is time-consuming and expensive to make each new model, due to complicated calculations on the shape. If an existing blade is taken as a template and exposed to a 'face lift', these calculations can be restricted to the modification area, saving both time and money.

Input from the turbine manufacturer (Kværner Energy) consisted of an already existing tensor-product B-spline representation of a Pelton blade, used in an earlier power station construction. This surface had been generated from a set of 3D parametric scattered data (measured on a wood model), using tensor-product B-spline approximation. The 3D scattered data and the B-spline approximation are show in Figure 1. More details about the algorithm we developed for this problem are given in [5]. The input revision data was taken from 2D drawings of calculated cross sections of the revised surface, and contained scattered points along parts of the new surface edges and along interior slices (see Figure 2). Subsets of both the edges and the interior of the original surface had to be modified.

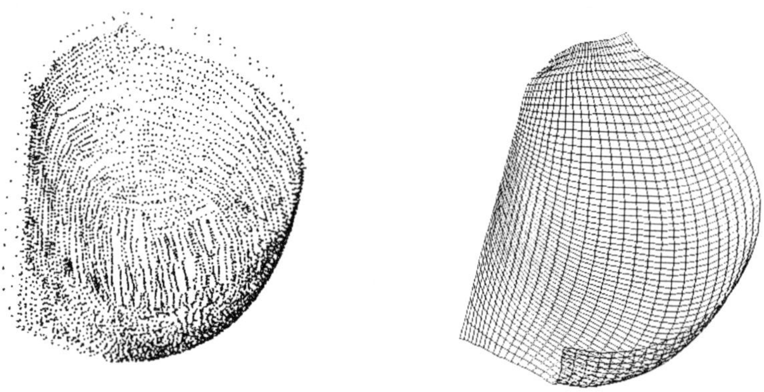

Fig. 1. Existing B-spline represented Pelton blade f_O, generated from the 3D parametric scattered data on the left, by using tensor-product B-spline approximation.

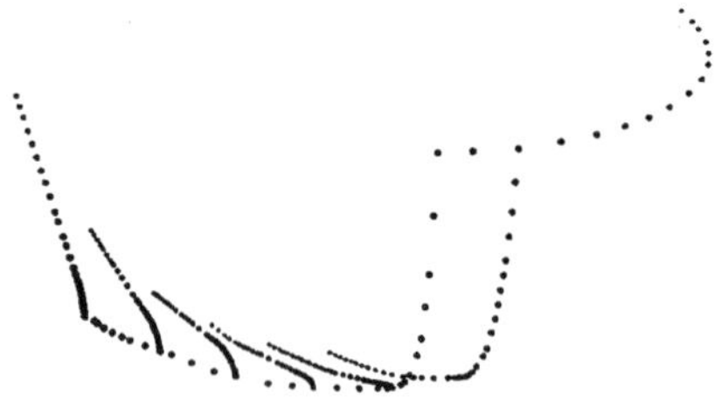

Fig. 2. The 'face lift' data in the front of the blade.

3.2. The algorithm for general 'face lift'

This subsection gives a general description of the 'face lift' algorithm, the figures are taken from the Pelton example.

The 'face lift' algorithm is a surface editor based on offset decomposition, but it also benefits from the principles of multiresolution decomposition. The starting point is a basic tensor-product B-spline surface:

$$f_O(u,v) = \sum_{i=1}^{m_0}\sum_{j=1}^{n_0} c_{i,j} B_{i,\mathbf{s}_0}(u) B_{j,\mathbf{t}_0}(v), \; c_{i,j} \in \mathbb{R}^3, \tag{2}$$

in the linear spline space $S_{\mathbf{s}_0,\mathbf{t}_0}$, $(u,v) \in U$ (parameter area). We approximate the deviation between f_O and the revision points with a difference surface, on the same parameter area U with refined knot vectors $\mathbf{s}_1 \supset \mathbf{s}_0$ and $\mathbf{t}_1 \supset \mathbf{t}_0$:

$$d(u,v) = \sum_{i=1}^{m_1}\sum_{j=1}^{n_1} \hat{c}_{i,j} B_{i,\mathbf{s}_1}(u) B_{j,\mathbf{t}_1}(v), \; \hat{c}_{i,j} \in \mathbb{R}, \tag{3}$$

where $n_1 \geq n_0$ and $m_1 \geq m_0$. The intention of the refinement is to capture details in the revision data. The difference surface d will have constant value 0 except on a parameter subarea $\hat{U} \subset U$, which is the parameter area of what we call the 'face lift' area. Notice that $\hat{U}$ is an arbitrary subarea, generally it is not rectangular.

The difference surface d is added as a normal offset to f_O, to get the new surface f_R, this is illustrated in Figure 6. The revised surface f_R is also defined on U with knot vectors $\mathbf{s}_1$ and $\mathbf{t}_1$.

Generally we assume that the revision points are interior to the basic surface, such that the edges of the basic surface are not to be changed. If this is not the case, a preprocessing is necessary, which we will describe later.

Basically the algorithm is divided into 4 steps:

1. **Calculate the projection distance from the revision points to the basic surface**. The projections are done by a local, iterative closest point procedure. The result of this step is a set of scattered data in $\hat{U}$, where each data point contains the parameter values of the closest point on the basic surface, and the offset distance (see Figure 3 for the Pelton example, the z value is scaled for visual effect).
2. **Approximate the scattered data with a tensor-product B-spline.** The result is a difference surface $d : \mathbb{R}^2 \rightarrow \mathbb{R}$, with constraints that $d = 0$ outside the 'face lift' parameter area $\hat{U}$. New knots may be inserted in the knot vectors of d, to capture details in the scattered data. (Figure 4 shows the difference surface in the Pelton example, with z value scaled.)
3. **Blend the revised and non-revised areas on the difference surface**. An irregular edge may arise in the difference surface, between the areas with and without revision data. To smooth d in the transition area, a blending of the surface across the edge is made. The width of the blending area is set by the user.
4. **Add basic surface and difference surface**. The difference surface d is added to the basic surface f_O as a normal offset. If the knot vectors of$d$ are refined compared to those of f_O, f_O is exposed to a similar knot refinement. The new surface f_R inherits these knot vectors (see Figure 5).

In step 2 we have to use approximation and not interpolation, in fact this is an advantage since it smooths out errors in the revision data. To maintain

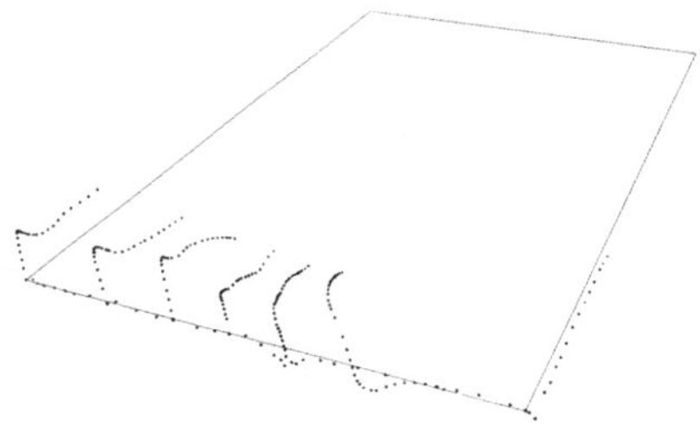

Fig. 3. The 'face-lift' data mapped to the parameter area of the basic surface. Notice that the data exists only in a part of the area. The z-value is scaled for visual effect.

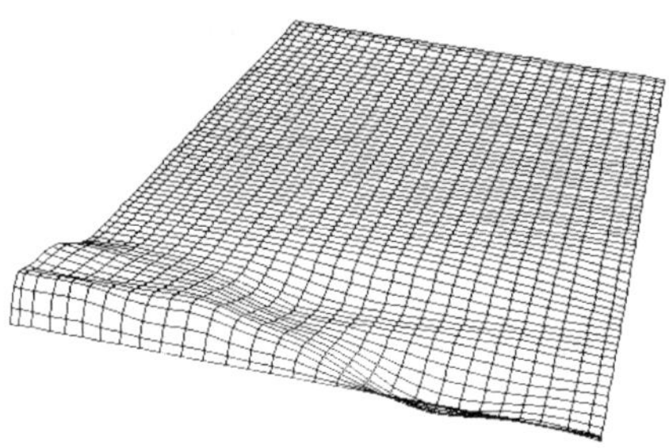

Fig. 4. The difference surface d. This is a tensor-product B-spline approximation of the data in the previous figure, with constraint $d = 0$ outside the 'face lift' parameter area $\hat{U}$.

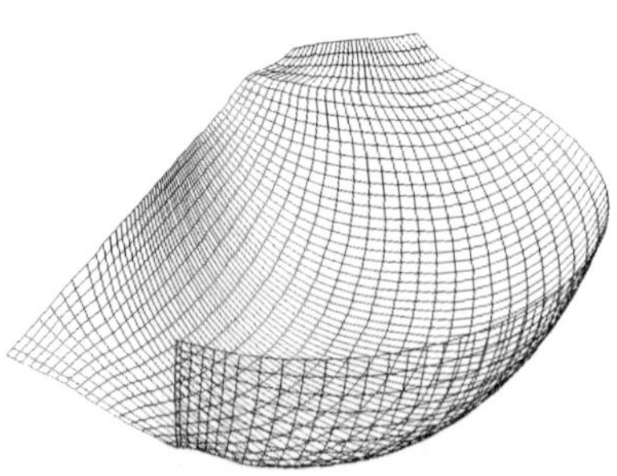

Fig. 5. Final result f_R of the 'face-lift'. The blade is revised in an area around its front.

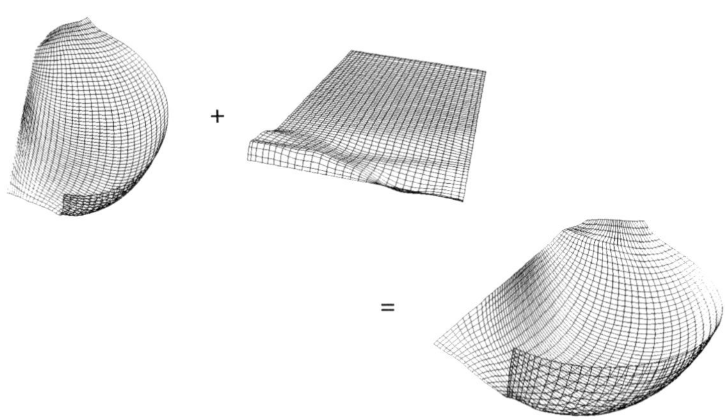

Fig. 6. The different phases of 'face-lift'. The difference surface d is an approximation of the 'face lift' data, and is added to the existing surface f_O to get the revised surface f_R..

the smoothness of the basic surface, various smoothing techniques may be applied to the difference surface. Examples are least square methods, variation diminishing splines and local filter methods (filters in the parameter directions or cross filters).

3.3. Experience with the Pelton example

In the general algorithm it is assumed that the revision points are interior to the basic surface. This was not the case in the Pelton example. It is of course possible to do the offset not parallel to the basic surface normal, but this is less stable than a normal offset scheme. Instead we preprocess to stretch out the basic surface to the new edge points. The stretching is executed by a translation of the coefficients in a zone along the new edge points, and a corresponding stretching of the knot vector. To get an even distribution of the coefficients, new knots and coefficients are inserted in the stretch zone.

Additional knots were inserted in areas of crucial details in the revision data, to get an optimal representation of these. Figure 5 shows the Pelton blade from Figure 1 after it has been 'face lifted'. Figure 6 sums up the phases of the 'face lift' process.

References

1. Bartels, R. H. and D. R. Forsey, Spline overlay surfaces, University of Waterloo, Canada, 1991.
2. Forsey, D. R. and R. H. Bartels, Local refinement editing of B-spline surfaces, in *Proceedings of Graphics Interface '88 (Edmonton, Canada, 6-10 June 1988)*, Canadian Inf. Process Soc., 1988, 125–126.

3. Forsey, D. R. and R. H. Bartels, Hierarchical B-Spline refinement, Computer Graphics **22**, 4 (1988), 205–212.
4. Hagen, H., S. Hahmann, T. Schreiber, Y. Nakajima, B. Wordenweber and P. Hollemann-Grundstedt, Surface interrogation algorithms, IEEE Computer Graphics and Applications **5 12** (1992), 53–60.
5. Kaasa, J. and G. Westgaard, Modelling of CAD models from measured data, in *Proceedings of the Twelfth International Conference on the CAD-CAM, Computer Graphics and Computer Aided Technologies (Paris, France, 9-12 February 1993)*, Hermes, 1993, 187–203.
6. Kaufmann, E. and R. Klass, Smoothing surfaces using reflection lines for families of splines, Computer-Aided Design **6 20** (1988), 312–316.
7. Klass, R., Correction of local surface irregularities using reflection lines, Computer-Aided Design **2**, 12 (1980), 73–77.

Johannes Kaasa
Dept. of Industrial Mathematics
SINTEF
P.O. Box 124 Blindern
N-0314 Oslo, NORWAY
Johannes.Kasa @ si.sintef.no

Geir Westgaard
Dept. of Industrial Mathematics
SINTEF
P.O. Box 124 Blindern
N-0314 Oslo, NORWAY
Geir.Westgaard @ si.sintef.no

Some Generic Properties of the Set of Cross-Sections and the Set of Orthogonal Projections of a Smooth Surface

Y. L. Kergosien

Abstract. Much geometric reasoning in diagnostic medical imaging is based on generic properties of the sections and projections of smooth surfaces. We summarize the tools used for their mathematical study and the results obtained to motivate similar questions in surface approximation.

§1. Motivation: Generic Sign Systems in Medical Imaging

Genericity is fundamental to geometric reasoning in diagnostic medical imaging, especially in such an old technique as conventional radiography [5, 6]. In that method a projection from $\mathbb{R}^3$ (the 3D space) to $\mathbb{R}^2$ (the film) by means of a cone of X-rays is used to produce a density function (optical density of the film) defined on $\mathbb{R}^2$ from a density function (X-ray attenuation power of the body) defined on $\mathbb{R}^3$. The human eye further reduces the relevant information when extracting contrast curves from the images where the optical contrast is high, i.e. approximately at points where the hitting ray is tangent to the surfaces which bound anatomical volumes of different absorbing powers.

Radiologists need to analyze the shapes of anatomical surfaces embedded in $\mathbb{R}^3$ given the images of their sets of singular values under the radiographic projection. They usually rely on a set of correspondence rules such as "the law of tangential incidence" (one sees a line of contrast where the X-ray has been tangent to the surface) or the "silhouette sign" and variants of it (e.g., if two contrast curves cross on the radiograph, then the pieces of surfaces which produced them do not stand at the same distance from the film). Thus using local rules to build a germ of surface embedding over the contrast curves, and given some *a priori* knowledge on the anatomical surface, interpreters can deduce numerous geometrical and topological facts about the surface. It is easy to find configurations where such simple rule systems are confused

Wavelets, Images and Surface Fitting
P. J. Laurent, A. Le Méhauté, and L. L. Schumaker (eds.), pp. 311–318.

ISBN 1-56881-040-7.

(Fig. 1) or configurations which would call for new rules. However, such cases are not practically relevant because they are exceptional. The sign systems in use are found to work for most images in a simple and stable way. The generic properties of surface projections, as introduced by Whitney, are the key to the existence of such sign systems. Similar sign systems can be used in sectional imaging for the analysis and reconstruction of surfaces from their cross-sections, as a consequence of Morse theory.

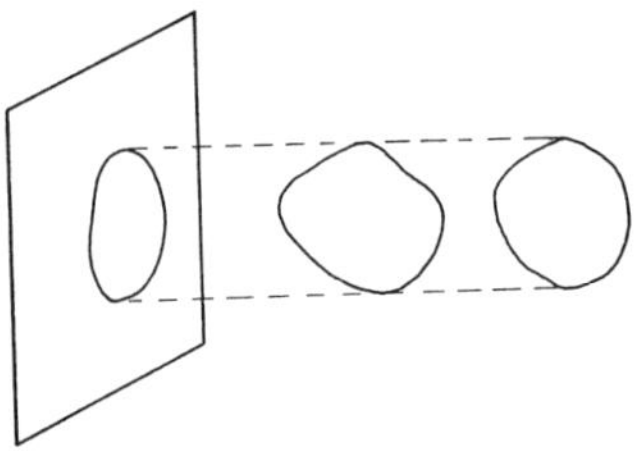

Fig. 1. An exceptional and unstable misleading projection.

Besides the classical results of Morse [7] and Whitney [9] about the generic properties of, respectively, single families of parallel cross-sections and single projections of a surface, we shall summarize some results directly related to medical imaging about the generic angular controls of stacks of cross-sections and projections of a surface. Again generic conditions reduce the number of transitional types to a small number, and one can extend Morse theory to the coding of shapes by recording transitions in projections or cross-sections under higher dimensional controls.

When surface approximation is used for the numerical processing and the display of medical images, it may be of interest to view some generic properties of surfaces as a set of requirements to be met by the approximations if one wants to either allow the radiologist to safely keep his sign systems and geometric reasoning to qualitatively analyze the new images or to use coding techniques as in differential topology.

§2. Definitions and Sketch of the Main Tools

Given C^p manifolds M, N and f, g in $C^p(M,N)$, f, g are called C^p *equivalent* if there exist a C^p diffeomorphism h of M and a C^pdiffeomorphism k of N such that $k \circ f = g \circ h$. (one can also put restrictions on h and k to forbid the use of symmetries). This is an equivalence relation in $C^p(M,N)$, the classes of which are called C^p *types*. Thus, a mapping can be transformed into any equivalent one by changes of coordinates in the source and in the target.

A mapping in $C^p(M,N)$ is defined to be *stable* if it has a neighbourhood of equivalent mappings in the Whitney topology [1,2], i.e., if any perturbation, if small enough, does not change its C^p type. For instance, the function $f(x) = x^3$ is not stable since adding to it a perturbation $g(x) = ax$ will change its type whatever the smallness of a. In fact, the perturbed function $x^3 + ax$

takes a different type according to the sign of a. On the other hand, $f(x) = x^2$ resists small perturbations, taking into account the crucial fact that in order to be small in the topology of $C^2(\mathbb{R}, \mathbb{R})$, a perturbation function should also have small derivatives. The study of stability focuses on singularities.

A property is called *generic* [8] if, given a topology on the set of objects on which this property makes sense, the set of objects that have this property is the countable intersection of open dense sets. Thus, in a way, exceptions to a generic property are rare. In fact, with a probability measure on the set of objects, generic properties can often be shown to be almost sure.

With these definitions, the results of Whitney that are relevant to radiologic interpretation can be stated [9]: Given M, N two C^∞ 2-manifolds, for a mapping f in $C^\infty(M, N)$ it is a generic property to be stable. Moreover, stable mappings are those which are locally either non singular or of one of the following types: the fold ($z = x^2$), the cusp ($z = x^3 - xy$), and the transversal superimposition of two cusps.

Morse theory gives a similar kind of result [7]: Given M a C^∞ n-manifold, for a function f in $C^\infty(M, \mathbb{R})$ it is a generic property to be stable. Moreover, functions are stable iff their singularities can be written $z = -x_1^2 - \cdots - x_k^2 + x_{k+1}^2 + \cdots. + x_n^2$ after a change of coordinates (k is called the *index* of the singularity) and if the images of the critical values are distinct. Parallel cross sections of a surface relate to counter-images of a height function defined on it. Changes in the cross-sections occur for critical values of the height function, and Morse theory uses generic families of cross-sections to study the global topology of compact manifolds from the sequence of indices at critical heights. The generic cross-sections are made of disjoint embedded circles. Transitions occur when the sectional plane is tangent to the surface, generically leading to either: birth of a connected component (plane tangent to a peak), or disappearance of a component (tangent to a well), or else transverse crossings (tangent to a saddle).

Fig. 2. Stable local signs. Left: projections. Right: cross-sections.

We now sketch the main tools [1, 2] used to prove such results. Let S be a k-codimensional submanifold of N. A mapping f in $C^p(M, N)$ is called *transversal* to S if for any $y = f(x)$ in S, $T_{f(x)}N = T_{f(x)}S + T_x f(T_x M)$. One shows that if f is transversal to S, then $f^{-1}(S)$ is a k-codimensional submanifold of M. One also defines A, B, two submanifolds of M to be transversal if either their intersection is void, or at any point of their intersection, $T_x M = T_x A + T_x B$. For instance, two circles in a plane are transversal if they are not tangent (either they cross transversely or they do not meet); two circles in 3D space are transversal if they do not meet (because if they meet, whatever their position, their 1-dimensional tangent spaces cannot generate together the 3-dimensional tangent space of the ambiant space). Non-transversality has an exceptional character. In the given examples one can easily put probability laws on finite dimensional spaces of geometric objects such that transversality is almost sure. Assuming transversality is thus a kind of Bayesian inference. In spaces of smooth mappings, however, constructing probability measures is difficult and one usually knows only generic results.

Two lemmas are basic to transversality theorems.

1. Parametric Transversality Theorem. *Consider a family of smooth mappings f_u from M to N, where the parameter u describes a manifold U, which we note introducing $F : U \times M \rightarrow N$, with $F(u,x) = f_u(x)$. Let S be a submanifold of N. If F is smooth and transversal to S, then the set of parameters u for which f_u is transversal to S is dense in U.*

To show that this is a consequence of Sard's theorem, one first notices that $F^{-1}(S)$ is a submanifold I of $U \times M$, on which one can define a smooth mapping p from I to U by mere restriction to I of the projection from $U \times M$ to U. One then proves that the values of u for which f_u is not transversal to S are precisely the images of the singular values of p, the set of which has null measure in U by Sard's theorem.

Given M, N two manifolds, the k-th order jet space $J^k(M,N)$ is the space where to plot the graph of mappings from M to N together with their partial derivatives up to order k. For f in $C^p(M,N)$, $J^k f : M \rightarrow J^k(M,N)$ associates to each x in M a record of $f(x)$ and all the partial derivatives of f at x up to order k (precise definitions use intrinsic constructions).

2. Jet Transversality Theorem (Thom's Lemma). *Let M, N be smooth manifolds and S be a submanifold of $J^k(M,N)$. In the space of smooth mappings from M to N (with Whitney's topology), the property that $j^k f$ is transversal to S is generic.*

The proof [8] uses the former theorem by introducing a family of local polynomial perturbations of the mappings such that the corresponding family of perturbed $j^k f$ is transversal to S.

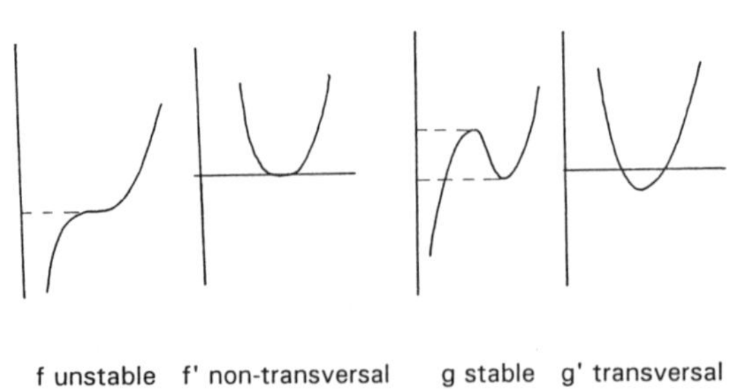

Fig. 3. Relating stability of a singularity to transversality in jet space.

The usual way to prove the genericity of some property about the singularities of smooth mappings is to express it as an incidence relation (transversality) of k-jets with some submanifold of $J^k(M, N)$. The generic presence or absence of associated geometric features follows from dimensional considerations in jet space (transversality can hold vacuously if the intersection is void). Once this geometrical study has been carried out, one often wants to prove (e.g., to deduce stability) that each type of incidence relation for a singularity (together with some inequalities) implies local C^p equivalence with some associated algebraic model. Constructing the equivalence (i.e. the source and target C^p diffeomorphisms) between any such mapping requires using *Malgrange's* or *Mather's Preparation Theorem for Smooth Mappings.*

§3. Controlled Mappings

We now study the generic properties of cross-sections and projections under some kind of control. A mapping control [4-5] is a mapping $F : A\dot{\times}C \to B\dot{\times}C$ where the dot products refer to fiber bundles over the control space C. The notions of C^p equivalence and stability extend to that setting (requiring the diffeomorphisms of $A\dot{\times}C$ and $B\dot{\times}C$ to commute with projections on C). We shall be looking for the structure of the bifurcation set in C of c's for which $f_c = F(\cdot, c)$ is not stable (we call that set the *obturation set*, in reference to the case of controlled projections). C is usually stratified according to the degree of unstability (the codimension) of the singularities of the associated f_c. One should distinguish local unstabillity (to be studied at a single germ of mapping) from multi-local ones (due to some coincidences in the images of different germs). It is often useful to consider the following sets: The union of the sets of critical values of the f_c's (for different values of c) constitute the *singular set* in A. The union of their images under the corresponding f_c's constitute *the diagram of contours* in $B\dot{\times}C$.

§4. Controlled Cross-Sections

In the case of cross-sections the position of a cross-sectional plane depends on 3 parameters. We shall, however, consider stacks of parallel cross-sections, each stack depending on 2 parameters. Given an embedded compact 2-manifold M in $\mathbb{R}^3$ and a line L in $\mathbb{R}^3$, let $p : M \to L$ be the orthogonal projection of M onto L. The surface's cross-sections by the family of (parallel) planes perpendicular to L are just the sets $p^{-1}(y)$ for different y's in L. As stated in the former section, the generic properties of such an object (viewed either as a stack of cross-sections or as a height function defined on M) are a direct application of Morse theory. To account for the family of all possible cross-sections, one makes this object depend on the direction parameter, to be taken as a point on the control sphere S^2 (we neglect antipodal identification and consider oriented directions to avoid introducing 2-dimensional projective space). The mapping control is thus $F : M \times S^2 \to \mathbb{R} \times S^2$.

Given a value of the control, for Morse's theory to apply, it is sufficient that tangencies of the sectional planes occur at elliptic or hyperbolic points.

The generic structure of the set of parabolic points of M follows from Thom's lemma. Representing the surface as $z = f(x, y)$ in local coordinates, the equation $rt - s^2 = 0$ defines a 1-codimensional submanifold of $J^2(M, \mathbb{R})$ after removing the generically non-reached set $rt = s^2 = 0$. The jet of f is generically transversal to that manifold so that the set of parabolic points is generically a 1-codimensional submanifold of M, i.e., a curve on M. The obturation set in S^2 is the set of directions which are normal to one of the tangent planes at the parabolic points, i.e. it is the Gauss image of the parabolic curves (the parabolic points being also the critical points of the Gauss mapping). For these directions, one of the cross-sections in the stack has a cusp, which accounts for the simultaneous disappearance of a crossing and a point in the stack when moving from close directions. By similar transversality arguments the Gauss mapping can generically have only cusps as 2-codimensional local singularities (from "godrons" on the surface, with an associated sectional pattern canceling a crossing and two points). Multi-local strata include curves which can cross transversely on S^2 or end at godrons.

Fig. 4. Generic 1-codimensional local types of cross-sections and projections.

The bifurcation diagram can be computed from the equations of the surface: First compute the curves of parabolic points on the surface; then compute their Gauss images in S^2. Within each of the regions of S^2 bounded by such curves, one is sure that the qualitative structure of the stack of cross-sections is constant. The shape of M can be coded by labeling the corresponding decomposition of S^2 with the corresponding types of stacks.

§5. Controlled Projections

The case of projections is more complex and we only summarize the results [3, 4]. Given a single orthogonal projection, its generic properties are a consequence of the classical work of Whitney on singularities of mappings. As in the case of sections, to get the whole family, one makes the projection depend on the oriented direction, i.e. a point on the sphere S^2, thus getting a mapping control $M \times S^2 \to \mathbb{R}^2 \times S^2$. Due to the possibility of four concurrent curves in some projections with continuously varying C^p type (due to the cross-ratio invariant), the notion of C^p equivalence has to be modified to keep a discrete

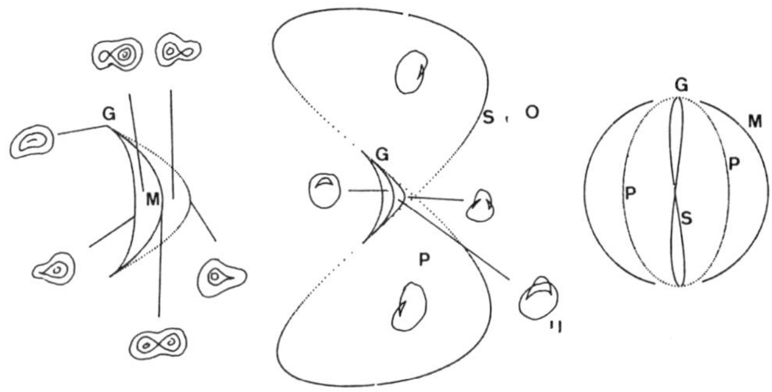

Fig. 5. Cross-sections (left), projections (center), and singular set (right).

classification. As the direction is moved, the observable qualitative changes in the projection are now due to two kinds of local surface configurations: parabolic points (occuring along curves) and swallow-tail points. Swallowtail points generically occur along curves in the hyperbolic parts of the surface at points where the section of the surface by its tangent plane (also the projection of the asymptotic curve on that plane) has an inflection on at least one of its branches. Projecting along the related asymptotic direction gives an unstable singularity with local model $z = x^4 - xy$. In the control space, one describes curves associated either to beek-to-beek or lip projection types (parabolic class) or to swallow-tail projection type; these curves are in correspondence with parabolic and swallowtail curves on the surface, parts of the singular set. Codimension 2 patterns include the "godron" (tangentially merging the parabolic and swallow-tail curves, together with a multilocal curve), the "gouttière" (transition between beek-to-beek and lip points, giving a cusp on the obturation set), and the butterfly (on swallowtail curves, with a cusp on the swallowtail obturation curve). There is also a classification for the multilocal types of each codimension. To compute the bifurcation diagram from the equations of the surface, one first computes the parabolic and swallowtail curves, then one computes the unstable projection direction along them, i.e. the common asymptotic direction for parabolic points and the inflection-related asymptotic direction for the swallowtail points. Multilocal strata involve computing some hypersurface intersections (the self-intersections of the dual surface) with some numerical difficulties near the godron points.

§6. Problems in Surface Approximation

Given a finite dimensional approximating space A, it is easy to build non-singular probability measures on it, and one can ask whether genericity arguments then translate into almost sure geometric properties. As a start, one can ask whether the elements of A generically display the same singularity types as those we singled out in the case of smooth functions. Direct proofs of zero-measure of the set of exceptions can apply to the subset of elements of A which are not transversal to the same associated submanifold of the jet space as used for genericity. If needed, some parameters such as the position of the nodes can be taken randomly to produce results similar to those of the parametric transversality lemma. To better justify the transfer of genericity arguments from the initial setting, it is also natural, given $p : D \to A$ the approximation mapping from the space D of smooth functions to the approximating space A, to ask whether it is a generic property for the elements f of D to be approximated by elements $p(f)$ with singularities from the generic list only. However, if one wants to use the approximation to build a mapping control from it (e.g. to retrieve all the projections or all the cross-sections from a modelled object), the generic properties of the control should be looked for, a much stronger requirement which can be based on the comparative study of the singular set and the associated obturation set.

References

1. Golubitsky, M., Guillemin, V., *Stable Mappings and their Singularities.* Springer, 1973.
2. Hirsch, M. W., *Differential Topology*. Springer, 1976.
3. Kergosien, Y. L. , Thom, R., Sur les points paraboliques des surfaces. Comptes Rendus Acad. Sc. Paris, t.290 (1980), Série A, 705–710.
4. Kergosien, Y. L., La famille des projections orthogonales d'une surface et ses singularités. Comptes Rendus Acad. Sc. Paris, t.292 (1981), Série I, 929–932.
5. Kergosien, Y. L., Medical exploration of some rythmic phenomena, in *Rythms in Biology*, Cosnard, M., Demongeot, J., Lebreton, M. (eds), Lecture Notes in Biomathematics Vol 49, Springer Verlag, 1983.
6. Kergosien, Y. L., Generic sign systems in medical imaging. IEEE CG& A **11** (1991), No 5, 46–65.
7. Morse, M., Cairns, S. S., *Critical Point Theory in Global Analysis and Differential Topology*. Academic Press, 1969.
8. Thom R., Un lemme sur les applications différentiables, Bol. Soc. Math. Mexicana, 1956, 59–71.
9. Whitney H., On Singularities of Mappings of Euclidean Spaces. I : Mappings of the Plane into the Plane. Annals of Math. **62** (1955), 374–410.

Y. L. Kergosien
Dépt. de Mathématiques, Bât 425
Université de Paris-Sud
91405 Orsay, FRANCE
kergos@ psisun.u-psud.fr

Splines Constructed by Pieces of Polyharmonic Functions

O. I. Kounchev

Abstract. In this paper we work with multivariate splines which are piecewise polyharmonic functions. In general, they are defined for scattered data, but here we study the case when the data are located on parallel hyperplanes. This makes possible a comparison with the usual polynomial splines and the polyharmonic splines when the data are on a regular grid.

§1. Introduction

In recent years there have been many attempts to find a proper multivariate paradigma to occupy the position of what has to be called multivariate spline. Recent progress in polyharmonic splines and radial basis functions shows that the search has not finished yet.

The present paper works with the notion of polyspline introduced and studied in [3,4,5,6]. The polysplines of order $2q$ are constructed by pieces of polyharmonic functions of order $2q$, which join up to some order. The very essence of polysplines is that the data are assumed to be given on hypersurfaces lying in a domain D in $\mathbb{R}^n$.

We suppose that $D = \bigcup_{j=1}^{N} D_j$, where the domains $D_j \subseteq D$, $D_j \bigcap D_k = \emptyset$ for $j \neq k$. We assume that $S = \bigcup_{j=1}^{N} \partial D_j$ is a union of smooth manifolds of dimension $n-1$, which do not intersect. The data $f(x)$ are supposed to be given on the set S.

There are two simple nontrivial examples: The first is when D is a strip in $\mathbb{R}^2$ and D_j are strips parallel to it. The second is when D is a circle in $\mathbb{R}^2$ and D_j are annuli concentric with it.

The main purpose of the present paper is to study polysplines defined in the case when D_j are parallel strips in $\mathbb{R}^2$, and more generally, parallel layers in $\mathbb{R}^n$, and to show the analogy to many properties of the univariate splines.

In Section 5 we prove existence of polysplines when D_j are parallel layers, thus making *the present paper completely independent of* [3,4,5]. We also

Wavelets, Images and Surface Fitting
P. J. Laurent, A. Le Méhauté, and L. L. Schumaker (eds.), pp. 319–326.

ISBN 1-56881-040-7.

prove that the approximating power of the polysplines (when the width of the layers goes to zero) is exactly $4q$, i.e. completely similar to that of the univariate splines. The proofs are based on the reduction of the polysplines to Tchebycheffian splines in one dimension (cf. [11, Ch. 9]).

The main technical device is Lemma 3 in Section 4, which provides uniform boundedness of a family of L-splines depending on a parameter.

Let us remark that the polysplines are genetically related to the theory of polyharmonic splines [9,10] and radial basis functions, (cf. [2,8]), since in view of the formula of Green [1, p. 10] they are "integrals over surfaces" of fundamental solutions to the polyharmonic equations up to order $2q$. Due to this fact, the polysplines enjoy increased smoothness as compared to the radial basis functions.

§2. Definition of Polysplines

Let $q \geq 1$ be an integer. In [3,4,5] for bounded domains D, the *interpolation polyspline of order* $2q$, is defined as a solution to the following extremal problem:

$$\int_D \{\Delta^q u(x)\}^2 dx \rightarrow \inf, \tag{2.1}$$

where the infimum is taken over the family of functions $u(x)$ such that

$$u(x) = f(x), \quad x \in S, \tag{2.2}$$

$$\Delta^k u(x) = 0, \quad x \in \partial D, \quad k = 1, \ldots, q-1, \tag{2.3a}$$

$$(\partial/\partial n)\Delta^k u(x) = 0, \quad x \in \partial D, \quad k = 0, \ldots, q-1. \tag{2.3b}$$

In [3,4,5] the uniqueness and the existence is proved in the Sobolev spaces. The space of all such polysplines will be denoted by $PS_{2q}(D)$ or, simply, by PS.

Let us put $T_{ij} = \partial D_i \bigcap \partial D_j$. We shall denote by $u_j(x)$ the restriction of u to the subdomain D_j. There it is proved that in a bounded domain D the solution u to problem (2.1)–(2.3) satisfies the following properties:

$$\Delta^{2q} u_j(x) = 0, \quad x \in D_j, \quad j = 1, 2, \ldots, N; \tag{2.4}$$

$$\Delta^k u(x) = 0, \quad x \in \partial D, \quad k = 1, \ldots, q-1; \tag{2.5a}$$

$$(\partial/\partial n)\Delta^k u(x) = 0, \quad x \in \partial D, \quad k = 0, \ldots, q-1; \tag{2.5b}$$

$$\Delta^p u_i(x) = \Delta^p u_j(x), \quad x \in T_{ij}, \tag{2.6}$$

for $i, j = 1, 2, \ldots, N$, $i \neq j$, and $p = 0, 1, \ldots, 2q-1$;

$$(\partial/\partial n_i)\Delta^p u_i(x) = (\partial/\partial n_i)\Delta^p u_j(x), \quad x \in T_{ij}, \tag{2.7}$$

for $i, j = 1, 2, \ldots, N$, $i \neq j$, and $p = 0, 1, \ldots, 2q-2$.

Equalities (2.4)–(2.7) should be considered like equalities between boundary values of functions in a Sobolev space (traces) (cf. [7]).

It is not difficult to modify the techniques of [5] based on apriori estimates for elliptic boundary value problems (cf. [7]) and to prove the equivalence of (2.1)–(2.3) and (2.2), (2.4)–(2.7) when the domain D is unbounded, and the data f decay at infinity, e.g. $f \in L_2$.

In the next section we consider the particular case when D_j are parallel layers, i.e. when S is a union of parallel hyperplanes and D itself is a layer between two parallel hyperplanes. By a Fourier transform we reduce problem (2.2), (2.4)–(2.7) to Tchebycheffian splines [11].

§3. Polysplines on Parallel Layers

We use the notation $x = (t, y)$, $t \in \mathbb{R}$, $y \in \mathbb{R}^{n-1}$. Consider the numbers $t_1 < \ldots < t_N$. Put $D_j = \{x \in \mathbb{R}^n : t_j < t < t_{j+1}\}$, and consider the domain $D = \bigcup_{j=1}^{N-1} D_j$. We assume that the data f are a function $f(t, y)$ given at least for $t = t_j, j = 1, \ldots, N$.

Proposition 1. *Let u be a solution to problem (2.2), (2.4)–(2.7) and $u_j \in H^{4q}(D_j)$. Then it satisfies*

$$(\partial^k/\partial t^k)u_j(t_{j+1}, y) = (\partial^k/\partial t^k)u_{j+1}(t_{j+1}, y), \quad y \in \mathbb{R}^{n-1}, \tag{3.1}$$

for $k = 0, 1, \ldots, 4q-2$, $j = 1, \ldots, N-2$;

$$(\partial^k/\partial t^k)u_1(t_1, y) = 0, \quad y \in \mathbb{R}^{n-1}, \tag{3.2a}$$

$$(\partial^k/\partial t^k)u_{N-1}(t_N, y) = 0, \quad y \in \mathbb{R}^{n-1}, \tag{3.2b}$$

for $k = 1, ..., 2q-1$, and

$$u_j(t_j, y) = f(t_j, y), \quad y \in \mathbb{R}^{n-1}, \quad j = 1, \ldots, N-1; \tag{3.3a}$$

$$u_{N-1}(t_N, y) = f(t_N, y), \quad y \in \mathbb{R}^{n-1}. \tag{3.3b}$$

Proof: We will show how (3.1) follows. For $p = 0$, from (2.6)–(2.7) we obtain

$$u_j(t_{j+1}, y) = u_{j+1}(t_{j+1}, y), \quad y \in \mathbb{R}^{n-1},$$

and

$$(\partial/\partial t)u_j(t_{j+1}, y) = (\partial/\partial t)u_{j+1}(t_{j+1}, y), \quad y \in \mathbb{R}^{n-1}.$$

Consequently, for every multi-index α for which the derivatives exist, we have

$$D_y^\alpha u_j(t_{j+1}, y) = D_y^\alpha u_{j+1}(t_{j+1}, y), \quad y \in \mathbb{R}^{n-1}, \tag{3.4}$$

and

$$(\partial/\partial t)D_y^\alpha u_j(t_{j+1}, y) = (\partial/\partial t)D_y^\alpha u_{j+1}(t_{j+1}, y), \quad y \in \mathbb{R}^{n-1}.$$

For $p = 1$, (2.6) gives

$$\Delta_y u_j(t_{j+1}, y) + (\partial^2/\partial t^2) u_j(t_{j+1}, y) = \Delta_y u_{j+1}(t_{j+1}, y) + (\partial^2/\partial t^2) u_{j+1}(t_{j+1}, y)$$

for $y \in \mathbb{R}^{n-1}$. Since (3.4) implies $\Delta_y u_j(t_{j+1}, y) = \Delta_y u_{j+1}(t_{j+1}, y)$, we obtain

$$(\partial^2/\partial t^2) u_j(t_{j+1}, y) = (\partial^2/\partial t^2) u_{j+1}(t_{j+1}, y), \quad y \in \mathbb{R}^{n-1}.$$

Thus we have established (2.1) for $k = 0, 1, 2$. Now we can proceed by induction on p. The rest of the equalities are proved in a similar way. ■

Remark 1. Proposition 1 implies that on the hyperplanes $t = t_j$ where the data are located, the polyspline u has more smoothness than the corresponding fundamental solution of Δ^{2q}, resp. the radial basis generated by it (cf. [2,8,9,10]) . Outside these hyperplanes it is a real analytic function [1]. In particular, for $q = 1, u_j$ satisfies $\Delta^2 u_j = 0$ in D_j, and $D_y^\alpha u_j(t_j, y)$ exists if $D_y^\alpha f(t, y)$ exists. Also, $(\partial^2/\partial t^2) u(t, y)$ is continuous throughout the whole D. For a C^2 smoothness only the mixed derivatives $(\partial^2/\partial t \partial y_k) u_j(t_j, y)$ are missing.

We make the Fourier transform $F_{y \to \xi}$ on the y-variable which reduces the above problem to a one-dimensional.

Proposition 2. *For every $\xi \in \mathbb{R}^{n-1}$ the function $Fu(t, \xi) = F_{y \to \xi}[u(t, y)]$ is a L-spline with knots t_j and data $Ff(t_j, \xi) = F_{y \to \xi}[f(t_j, y)]$, where the operator L is given by $L_\xi = (\partial^2/\partial t^2 - \xi^2)^{2q}$. In fact, $Fu(t, \xi)$ is a Tchebycheffian spline. (For the definition of Tchebycheffian and L-splines see [11, Ch. 9 and 10]).*

Proof: The proof easily follows from Proposition 1. To all equalities (3.1)–(3.3), we apply the Fourier transform $F_{y \to \xi}$ and simply replace $u_j(t, y)$ through $Fu_j(t, \xi)$. In fact, property (3.1) implies that $Fu_j(t, \xi) \in C_t^{4q-2}[t_1, t_N]$.

On the other hand, since the Fourier transform maps the Laplace operator Δ_x into $\partial^2/\partial t^2 - \xi^2$, it follows that Δ^{2q} is transformed into L_ξ. Thus the equation $\Delta^{2q} u_j = 0$ in D_j is transformed into the ordinary differential equation $L_\xi F u_j(t, \xi) = 0$, $t_j < t < t_{j+1}$. The last, together with the C^{4q-2} smoothness means exactly that $Fu(t, \xi)$ is a L_ξ-spline. The equality

$$\begin{aligned} L_\xi g(t) &= (\partial/\partial t - \mid \xi \mid)^{2q} (\partial/\partial t + \mid \xi \mid)^{2q} g(t) \\ &= e^{-2|\xi|t} (\partial^{2q}/\partial t^{2q}) e^{2|\xi|t} (\partial^{2q}/\partial t^{2q}) e^{-|\xi|t} g(t) \end{aligned}$$

follows from the easy-to-check

$$(\partial/\partial t - \lambda) g(t) = e^{\lambda t} (\partial/\partial t) e^{-\lambda t} g(t), \quad \lambda \in \mathbb{R}.$$

Consequently, for the Tchebycheffian spline properties [11, §9.1], we have the functions w_j given by $w_1(t) = e^{|\xi|t}$, $w_{2q+1}(t) = e^{-2|\xi|t}$, and $w_j = 1$ for $j = 2, \ldots, 2q, 2q+2, \ldots, 4q$. ■

§4. Some Estimates of L-splines

A very important problem is to study how the L-splines behave when the coefficients of the operator L vary. We provide the following fundamental:

Lemma 3. *Let $v(t,\xi)$ be a L_ξ-spline with data $v(t_j,\xi)=\gamma_j, j=1,...,N$, and boundary conditions (coming from (3.2)):*

$$(\partial^k/\partial t^k)v(t_1,\xi)=(\partial^k/\partial t^k)v(t_N,y)=0, \quad \xi\in\mathbb{R}^{n-1},$$

$k=1,\ldots,2q-1$. *Then*

$$\| v \|_C = \max_{t\in[t_1,t_N]} | v(t,\xi) | \le C \| \gamma \| = C \max_{1\le j\le N} | \gamma_i |, \quad \xi\in\mathbb{R}^{n-1},$$

where the constant C depends on the knots t_i but not on ξ.

The proof is technical and is based on the observation that the system of equations (3.1)–(3.3) is a block diagonal system up to exponentially small terms with respect to the parameter $| \xi |$.

§5. Existence and Approximating Power of Polysplines

Let us denote by $F^{-1}=F_{\xi\to y}$ the inverse Fourier transform. Throughout the rest of the paper we assume that $N\ge 2$q.

Theorem 4. *Let the data $f(t_j,y)\in L_2[\mathbf{R}_y^{n-1}], j=1,\ldots,N$. Then there exists a unique solution u to problem (2.2), (2.4)–(2.7), i.e. interpolation polyspline, such that $u_j\in H^\infty(D_j)$.*

Proof: Let us denote by $\chi_j(t,\xi), j=1,\ldots,N$, the L_ξ-spline such that $\chi_j(t_j,\xi)=1$ and $\chi_j(t_i,\xi)=0$ for $i\ne$ j. Put

$$Fu(t,\xi)=\sum_{j=1}^{N} Ff(t_j,\xi)\chi_j(t,\xi).$$

We have

$$u(t,y)=F_{\xi\to y}[\sum_{j=1}^{N} Ff(t_j,\xi)\chi_j(t,\xi)].$$

The last expression makes sense since $\chi_j(t,\xi)$ is bounded by Lemma 3. The rest follows from Proposition 1 and Proposition 2. ∎

We remark that the existence Theorem 4 shows that Lemma 3 is implicit in the method of apriori estimates for elliptic boundary value problems.

By $L_\infty[t_1,t_N]$ we denote, as usually, the space of bounded functions in $[t_1,t_N]$(cf. [11, p. 14]). Now we can make use of the results on Tchebycheffian splines proved in [11, Ch. 9]. In particular, we use the operator Q introduced in §9.7 there, which maps the space of data $L_\infty[t_1, t_N]$ into the space of L_ξ-splines with knots at $\delta = \{t_1, \ldots t_N\}$. Put $\overline{\delta} = \max_{1 \leq j \leq N-1}(t_{j+1} - t_j)$. We extend the operator Q in a natural way to functions f defined in D by putting

$$Fu(t,\xi) = Q_\xi(t) = Q_\xi[Ff(t_j,\xi), j = 1, \ldots, N], \quad \xi \in \mathbb{R}^{n-1},$$

where $Q_\xi(t)$ is a L_ξ-spline on $[t_1, t_N]$, and we choose the same points for the extended grid for every $\xi \in \mathbb{R}^{n-1}$. The resulting extension is now given by

$$u(t,y) = \tilde{Q}[f] = F^{-1} Q_\xi[Ff(t_j,\xi), j = 1, \ldots, N].$$

In order that the last expression make sense it is sufficient to have

$$\| Q_\xi \| \| Ff(t,\xi) \|_{L_\infty[t_1,t_N]} \in L_1(\mathbb{R}_\xi^{n-1}).$$

Indeed, we have

$$\| Q_\xi[Ff(t_j,\xi), j = 1, \ldots, N](t) \|_{C[t_1,t_N]} \leq \| Q_\xi \| \max_{1 \leq j \leq N} | Ff(t_j,\xi) |,$$

for every $\xi \in \mathbb{R}^{n-1}$, which implies that $Q_\xi(t) \in L_1(\mathbb{R}_\xi^{n-1})$.

In [11, §9.7] the approximating power of Tchebycheffian splines is studied using the operator Q. On the right-hand side there appears a constant C_1, cf. Theorems 9.37, 9.38. In our case, for the operators Q_ξ we have the constant $C_1 = C_1(\xi)$.

In the present paper we will avoid a thorough study of the asymptotics of the constants $C(\xi) = \| Q_\xi \|$ and $C_1(\xi)$, by implementing them in the definition of the necessary space of functions .

For any two integers $p \geq 0$ and $q \geq 1$, we introduce the space $FL_{p,q}(D)$ of measurable functions $g(t,y)$ such that the following integrals are bounded:

$$N(g) = \sum_{j=0}^{2q} \int_{\mathbb{R}^{n-1}} (1 + | \xi |)^{p+2q-j} \| D_t^j Fg(t,\xi) \|_{L_\infty[t_1,t_N]} C_1(\xi) d\xi < \infty,$$

and

$$\int_{\mathbb{R}^{n-1}} \| D_t^j Fg(t,\xi) \|_{L_\infty[t_1,t_N]} C(\xi) d\xi < \infty, \quad j = 0, \ldots, 4q-1.$$

The following result on the approximating power of the polysplines holds:

Theorem 5. *Let $f \in FL_{p,q}(D)$. Then for every $j = 0, 1, ..., 4q-1$, and multi-index $\alpha \in \mathbb{Z}_+^{n-1}, | \alpha | \leq p$, the following inequality holds:*

$$\| D_y^\alpha D_t^j (f - \tilde{Q}f)(t,y) \|_{L_\infty[t_1,t_N]} \leq C \overline{\delta}^{4q-j} N(f).$$

Proof: Since $f \in FL_{p,q}(D)$, according to the definition of the operator $\tilde{Q}$ and the properties of the Fourier transform (cf. [12]) we have that $D_y^\alpha D_t^j \tilde{Q} f(t,y)$ makes sense and

$$I = D_y^\alpha D_t^j f(t,y) - D_y^\alpha D_t^j \tilde{Q} f(t,y)$$

$$= \int_{\mathbb{R}^{n-1}} (-i\xi)^{\alpha} e^{i\xi y} (D_t^j Ff(t,\xi) - D_t^j Q_\xi Ff(t,\xi)) d\xi < \infty.$$

The estimate of Theorem 9.38 in [11] gives

$$\| I \|_{L_\infty[t_1,t_N]}$$

$$\leq C_1 \overline{\delta}^{4q-j} \int_{\mathbb{R}^{n-1}} (1+ | \xi |)^{|\alpha|} \| L_\xi Ff(t,\xi) \|_{L_\infty[t_1,t_N]} C_1(\xi) d\xi$$

$$\leq C \overline{\delta}^{4q-j} \sum_{j=0}^{2q} \int_{\mathbb{R}^{n-1}} (1+ | \xi |)^{p+2q-j} \| D_t^j Ff(t,\xi) \|_{L_\infty[t_1,t_N]} C_1(\xi) d\xi$$

$$= C \overline{\delta}^{4q-j} N(f) < \infty. \quad \blacksquare$$

Acknowledgements. The present research was supported by the Alexander von Humboldt Foundation. Thanks are due to L. Schumaker for the up to date references on *L*-splines. The discussions with K. Jetter on the subject of the paper were very stimulating for the author.

References

1. Aronszajn, N., T. M. Creese, and L. J. Lipkin, *Polyharmonic Functions*, Clarendon Press, Oxford, 1983.
2. Dyn, N., Interpolation and approximation by radial and related functions, in *Approximation Theory VI*, C. K. Chui, L. L. Schumaker and J. D. Ward (eds.), Academic Press, New York, 1989, 211-234.
3. Kounchev, O. I., Definition and basic properties of polysplines, C. R. Acad. bulg. sci. **44** (1991), No. 7, 9–11.
4. Kounchev, O. I., Basic properties of polysplines, C. R. Acad. bulg. sci. **44** (1991), No. 8, 13–16.
5. Kounchev, O. I., Minimizing the integral of the Laplacian of a function squared with prescribed values on interior boundaries - theory of polysplines, I, preprint, University of Duisburg, 1993.
6. Kounchev, O. I., Theory of polysplines - minimizing the Laplacian of a function squared with prescribed values on interior boundaries with singularities, II, preprint, University of Duisburg, 1993.
7. Lions, J. L., and E. Magenes, *Non-Homogeneous Boundary Value Problems and Applications, vol. I*, Springer-Verlag, Berlin, 1972.
8. Light, W., Some aspects of radial basis function approximation, in *Approximation Theory, Spline Functions and Applications*, S. P. Singh (ed.), Kluwer Academic Publishers, Amsterdam, 1992, 163–190.
9. Madych, W. R. and S. A. Nelson, Polyharmonic cardinal splines, J. Approx. Theory **60** (1990), 141–156.

10. Rabut, Ch., Elementary m-harmonic cardinal B-splines, Numerical Algorithms **2** (1992), 39–62.
11. Schumaker, L. L., *Spline Functions: Basic Theory*, Wiley, New York, 1981.
12. Stein, E., and G. Weiss, *Introduction to Fourier Analysis on Euclidean Spaces*, Princeton University Press, Princeton, 1975.

Ognyan I. Kounchev
Current Address:
Dept. of Mathematics
University of Duisburg
Lotharstr. 65
Duisburg, 47058 GERMANY
hn314ko@unidui.uni-duisburg.de

On the Fast Evaluation of Integrals of Refinable Functions

Angela Kunoth

Abstract. Using the refinement equation, the computation of integrals of derivatives of multivariate refinable functions can be reduced to the solution of an eigenvector problem. Based on asymptotic expansions of the corresponding stationary subdivision schemes, and on determining eigenvectors for compressed subdivision schemes, it is shown in [5] how uniqueness can be guaranteed. This paper generalizes the latter part of the theory in [5] to integrals involving directional derivatives.

§1. Introduction

The problem of quickly computing integrals (derivatives) of functions arises in various fields. For example, one might want to generate linear systems derived from Petrov-Galerkin schemes for finding the approximate solution of boundary value problems in Numerical Analysis. In this case, one usually has to compute integrals involving partial and normal derivatives induced by the underlying partial differential operator, and L_2 inner products for the right hand side, see e.g. [3,4] and the references therein. The construction of wavelet-type functions often requires the computation of L_2 inner products as well. Or, since (derivatives) integrals of box splines are again box splines, one can use the calculation of the integrals as a method for evaluating box splines rather than using a recurrence formula. Since the main ingredient of the method proposed here is a refinement equation, all kinds of integrals of products of (derivatives of) wavelet-type functions which are needed for various applications can simply be reduced to the terms treated in the sequel. The method is based on function evaluations of a refinable multivariate function at integers, and can therefore also be used for the fast evaluation of (derivatives of) multivariate functions, see [7] for examples including Daubechies' refinable functions and orthonormal wavelets. The need for evaluating functions typically arises in Computer Aided Geometric Design.

Wavelets, Images and Surface Fitting
P. J. Laurent, A. Le Méhauté, and L. L. Schumaker (eds.), pp. 327–334.

ISBN 1-56881-040-7.

All the desired quantities can be computed quickly in a general shift-invariant framework as follows. Let φ^i, $i = 0, \ldots, m$, be compactly supported continuous functions on $\mathbb{R}^q$ satisfying a *refinement equation*

$$\varphi^i(x) = \sum_{\alpha \in \mathbb{Z}^q} a^i_\alpha \varphi^i(2x - \alpha), \quad x \in \mathbb{R}^q, \tag{1.1}$$

with some *mask coefficients* a^i_α, $\alpha \in \mathbb{Z}^q$. We will always assume that the *masks* $\mathbf{a}^i := \{a^i_\alpha, \alpha \in \mathbb{Z}^q\}$ are *finitely supported* meaning that except for a finite number all coefficients in the refinement equation vanish. In addition to (1.1) let us assume that φ^i is *stable* for each $i = 0, \ldots, m$, i.e.,

$$\left\| \sum_{\alpha \in \mathbb{Z}^q} \lambda_\alpha \varphi^i(\bullet - \alpha) \right\|_{L_p(\mathbb{R}^q)} \sim \left(\sum_{\alpha \in \mathbb{Z}^q} |\lambda_\alpha|^p \right)^{1/p}, \tag{1.2}$$

for any $\lambda \in \ell_p(\mathbb{Z}^q)$ and some $1 \leq p \leq \infty$. As shown in [6], stability for one such p carries over to the whole range $1 \leq p \leq \infty$.

Denote by $D_y := y \cdot \nabla$ the directional derivative in direction y, and let $D_Y := D_{y^1} \cdots D_{y^r}$ for $Y = (y^1, \ldots, y^r)$. We are interested in computing terms of the form

$$\int_{\mathbb{R}^q} \varphi^0(2^{k_0}x - \alpha^0) \prod_{j=1}^m (D_{Y^j}\varphi^j)(2^{k_j}x - \alpha^j)\, dx\,, \tag{1.3}$$

where $\alpha^j \in \mathbb{Z}^q$, $j = 0, \ldots, m$, and $Y^j = (y^{j,1}, \ldots, y^{j,r_j})$, $j = 1, \ldots, m$. This is a generalization of the work in [5] involving only partial derivatives. The special case $m = 1$ is treated in [1] for univariate functions by a similar technique. For the remainder of this note, we follow the lines in [5] with the necessary alterations.

First note that the computation of (1.3) can be reduced to evaluation of the terms

$$H(\alpha) = H(\alpha^1, \ldots, \alpha^m) := \int_{\mathbb{R}^q} \varphi^0(x) \prod_{j=1}^m (D_{Y^j}\varphi^j)(x - \alpha^j)\, dx\,. \tag{1.4}$$

Since each φ^j satisfies a refinement equation, the function

$$F(\alpha) := \int_{\mathbb{R}^q} \varphi^0(x) \prod_{j=1}^m \varphi^j(x - \alpha^j)\, dx\,, \tag{1.5}$$

defined on $\mathbb{R}^s$, $s = mq$, does too:

$$F(x) = \sum_{\mu \in \mathbb{Z}^s} c_\mu F(2x - \mu), \tag{1.6}$$

with mask coefficients

$$c_\mu := 2^{-q} \sum_{\nu \in \mathbb{Z}^q} a_\nu^0 \prod_{j=1}^m a_{\nu-\mu^j}^j \,. \tag{1.7}$$

In view of the definition of F and H, the computation of terms (1.3) reduces to evaluate H at lattice points, followed by employing the refinement equation for H. Note that the latter can be rewritten as

$$2^{-r} H(\alpha) = \sum_{\mu \in \mathbb{Z}^s} c_{2\alpha-\mu} H(\mu) \,, \quad \alpha \in \mathbb{Z}^s \,, \tag{1.8}$$

where $r = r_1 + \cdots + r_m$ is the total number of derivatives in (1.3). Since all masks involved have finite support, this equation can be interpreted as the problem of computing an eigenvector. The remainder of this note will consist of deriving conditions that determine the eigenvector uniquely.

Numerical examples and remarks with respect to the implementation of the method can be found in [7].

§2. Uniqueness Conditions for Eigenvectors

Based on asymptotic expansions of the stationary subdivision operator corresponding to the mask (1.7), it is shown in [5] how additional moment conditions guarantee uniqueness of the eigenvector in the multivariate setting. A second possibility involving appropriate factorizations of the symbols $a^i(z) := \sum_{\alpha \in \mathbb{Z}^q} a_\alpha^i z^\alpha$ corresponding to the masks $\mathbf{a}^i$ will be generalized here to the case of directional derivatives. As in [5], the main ingredient is the notion of compressed subdivision schemes.

For a mask $\mathbf{a}$ on $\mathbb{Z}^s$ of finite support, the *stationary subdivision scheme*

$$(S_{\mathbf{a}}\lambda)_\alpha := \sum_{\beta \in \mathbb{Z}^s} a_{\alpha-2\beta} \lambda_\beta \,, \quad \alpha \in \mathbb{Z}^q \,, \tag{2.1}$$

is said to *converge* if for any $\lambda \in \ell_\infty(\mathbb{Z}^s)$ there is some $f_\lambda \in C(\mathbb{R}^s)$ such that

$$\lim_{k\to\infty} \left(\sup_{\alpha \in \mathbb{Z}^s} \left| (S_{\mathbf{a}}^k \lambda)_\alpha - f_\lambda\left(\frac{\alpha}{2^k}\right) \right| \right) = 0 \,. \tag{2.2}$$

It was shown in [2] Theorem 2.1, p. 14, that a convergent $S_{\mathbf{a}}$ induces the existence of a unique $\varphi \in C_0(\mathbb{R}^s)$ which is refinable with respect to $\mathbf{a}$. Conversely, $\mathbf{a}$-refinability and ℓ_∞-stability of $\varphi \in C_0(\mathbb{R}^s)$ imply convergence in the sense of (2.2) ([2] Proposition 2.3, p. 23).

The general concept of *compressed stationary subdivision schemes* can now be described as follows. For any $X \in \mathbb{Z}^{s\times n}$ with rank $s \le n$, and any $\varphi \in C_0(\mathbb{R}^n)$, refinable with mask $\mathbf{a}$, a distribution $\varphi(\bullet|X)$ on $\mathbb{R}^s$ can be defined by requiring

$$\int_{\mathbb{R}^n} f(Xu)\,\varphi(u)\,du = \int_{\mathbb{R}^s} f(x)\,\varphi(x|X)\,dx \tag{2.3}$$

for any $f \in C_0(\mathbb{R}^s)$. It was then shown in [2] Proposition 4.1, p. 68, that $\varphi(\bullet|X)$ is a function satisfying a.e. the refinement equation

$$\varphi(\bullet|X) = \sum_{\alpha\in\mathbb{Z}^s} a_\alpha^X \varphi(2\bullet - \alpha|X), \quad x \in \mathbb{R}^s, \tag{2.4}$$

with

$$a_\alpha^X := 2^{s-n} \sum_{X\beta=\alpha} a_\beta, \quad \alpha \in \mathbb{Z}^s. \tag{2.5}$$

Our first point of interest is the question of obtaining eigenvectors for compressed schemes. Let us assume that X satisfies in addition

$$\{X\alpha : \alpha \in \mathbb{Z}^n\} = \mathbb{Z}^s. \tag{2.6}$$

Let $Y \in \mathbb{Z}^{n\times r}$ be another matrix with columns $\{y^1, \ldots, y^r\}$. Denote by $(\Delta_y f)(x) := f(x+y) - f(y)$ the forward difference operator, and let $\Delta_Y = \Delta_{y^1}\cdots\Delta_{y^r}$. The operator Δ_{XY} is to be understood as operator Δ_M where $M = XY \in \mathbb{Z}^{s\times r}$ is the usual matrix product. We are now ready to generalize [5] Proposition 3.1 relating stationary subdivision schemes to compressed ones.

Proposition 1. *Suppose* **a** *is a finitely supported mask on* $\mathbb{Z}^n$ *inducing the subdivision operator* $S_{\mathbf{a}}$*, and suppose there exists a finitely supported mask* **b** *such that*

$$\Delta_Y S_{\mathbf{a}} = 2^{-r} S_{\mathbf{b}} \Delta_Y, \tag{2.7}$$

where $S_{\mathbf{b}}$ *is a convergent subdivision scheme. Then the corresponding compressed scheme* $S_{\mathbf{b}}^X$ *converges and satisfies*

$$\Delta_{XY} S_{\mathbf{a}}^X = 2^{-r} S_{\mathbf{b}}^X \Delta_{XY}. \tag{2.8}$$

Proof: Let us recall from [2] Lemma 3.1, p. 57/58, for any finitely supported masks **a**, **b**, that

$$\Delta_y S_{\mathbf{a}} = S_{\mathbf{b}} \Delta_y \qquad \text{if and only if} \qquad a(z) = (1 + z^{-y})\, b(z),$$

where $a(z) = \sum_{\alpha\in\mathbb{Z}^q} a_\alpha z^\alpha$ and $b(z) = \sum_{\alpha\in\mathbb{Z}^q} b_\alpha z^\alpha$. Thus, a repeated application implies that (2.7) is equivalent to

$$a(z) = 2^{-r} \prod_{j=1}^{r} (1 + z^{-y^j})\, b(z). \tag{2.9}$$

Defining $M = XY \in \mathbb{Z}^{s\times r}$ with columns $(m^1, \ldots, m^r) = (Xy^1, \ldots, Xy^r)$, the substitution $z = e^{iX^T\xi}$, $\xi \in \mathbb{Z}^s$, yields

$$a(e^{iX^T\xi}) = 2^{-r} \prod_{j=1}^{r} (1 + e^{-iX^T\xi\cdot y^j})\, b(e^{iX^T\xi}).$$

Thus, since $X^T\xi \cdot y^j = m^j \cdot \xi$, it follows

$$a(e^{iX^T\xi}) = 2^{-r} \prod_{j=1}^{r} (1 + e^{-im^j\cdot\xi})\, b(e^{iX^T\xi})\,. \tag{2.10}$$

According to [2] (4.7), p. 68,

$$a(e^{iX^T\xi}) = 2^{n-s} a^X(e^{i\xi})$$

implying that (2.8) is equivalent to (2.10). In view of the assumption (2.6) on X, by [2] Theorem 4.1, p. 74, the convergence of $S_{\mathbf{b}}^X$ is now an immediate consequence of the convergence of $S_{\mathbf{b}}$. ■

Introducing the mapping on $\ell_\infty(\mathbb{Z}^n)$ with values in $\ell_\infty(\mathbb{Z}^s)$,

$$(I^X\lambda)_\beta := \sum_{X\alpha=\beta} \lambda_\alpha\,, \tag{2.11}$$

the following generalization of [5] Corollary 3.2 connects derivative operators of compressed refinable functions with difference operators.

Proposition 2. *Suppose (2.7) holds, and define* $\mathbf{c} := 2^{s-n} I^X \mathbf{b}$. *If* $S_{\mathbf{b}}$ *is a convergent stationary subdivision scheme, then there exists a unique sequence* $W = (W_\beta)_{\beta\in\mathbb{Z}^s}$ *of finite support satisfying*

$$W_\alpha = \sum_{\beta\in\mathbb{Z}^s} c_{2\alpha-\beta} W_\beta\,, \quad \alpha \in \mathbb{Z}^s, \tag{2.12}$$

$$1 = \sum_{\beta\in\mathbb{Z}^s} W_\beta$$

and

$$(D_{XY}\varphi)(\alpha|X) = (\Delta_{XY}W)_\alpha. \tag{2.13}$$

The proof follows as in [5] by employing Proposition 1 instead of [5] Proposition 3.1.

§3. The Main Theorem

We are now ready to apply these results for proving the main theorem. To this end, define $\Phi(x) := \varphi^0(x^0)\cdots\varphi^m(x^m)$, $x = (x^0, \ldots, x^m) \in \mathbb{R}^n$, where $n = q + s$, $s = mq$. Since each φ^i, $i = 0, \ldots, m$, is refinable (1.1), Φ satisfies the refinement equation

$$\Phi(x) = \sum_{\alpha\in\mathbb{Z}^n} a_\alpha \Phi(2x - \alpha)\,, \quad x \in \mathbb{R}^n\,, \tag{3.1}$$

where

$$a_\alpha = \prod_{i=0}^{m} a^i_{\alpha^i}, \quad \alpha = (\alpha^0, \ldots, \alpha^m). \tag{3.2}$$

Let X now be of the specific form

$$X = \begin{pmatrix} I & -I & \ldots & \ldots & 0 \\ I & 0 & -I & & \vdots \\ \vdots & \vdots & \ddots & \ddots & 0 \\ I & 0 & \ldots & 0 & -I \end{pmatrix}, \tag{3.3}$$

where $I \in \mathbb{Z}^{q\times q}$ is the identity matrix. Obviously X satisfies (2.6). Furthermore, the mask $\mathbf{a}^X$ given in (2.5) coincides with the mask $\mathbf{c}$ for F given by (1.7), and we can identify

$$F(x^1, \ldots, x^m) = \Phi(x^1, \ldots, x^m | X), \quad x^1, \ldots, x^m \in \mathbb{R}^q. \tag{3.4}$$

Consequently, the computation of F respectively H can be performed by evaluating a compressed refinable function. One should mention that the assumptions of the following main result are e.g. satisfied for univariate refinable functions φ^j, $j = 0, \ldots, m$, that are stable in the sense of (1.2) and sufficiently smooth. Examples involving bivariate three-directional box splines can be found in [7].

Theorem 3. *Let φ^j be refinable functions with masks $\mathbf{a}^j$, $j = 0, \ldots, m$. Suppose $S_{\mathbf{a}^0}$ is a convergent subdivision scheme, and assume that, for $j = 1, \ldots, m$, the symbol $a^j(z)$ can be factorized as*

$$a^j(z) = 2^{-r_j} \prod_{\ell=1}^{r_j} (1 + z^{-y^{j,\ell}})\, b^j(z), \quad z \in \mathbb{R}^q, \tag{3.5}$$

where each $\mathbf{b}^j$, $j = 1, \ldots, m$, is a finitely supported mask of a convergent subdivision scheme. If W is the finitely supported solution of (2.12) with $\mathbf{c}$ defined by

$$c_\alpha = 2^{-q} \sum_{\mu \in \mathbb{Z}^q} a^0_\mu \prod_{j=1}^{m} b^j_{\mu - \alpha^j}, \quad \alpha = (\alpha^1, \ldots, \alpha^m), \tag{3.6}$$

we have

$$\int_{\mathbb{R}^q} \varphi^0(x) \prod_{j=1}^{m} (D_{Y^j} \varphi^j)(x - \alpha^j)\, dx = (-1)^r (\Delta_{XY} W)_\alpha, \tag{3.7}$$

where $Y^j = (y^{j,1}, \ldots, y^{j,r_j}) \in \mathbb{Z}^{q\times r_j}$, $j = 1, \ldots, m$, and $Y \in \mathbb{Z}^{n\times r}$, $r = r_1 + \cdots + r_m$, is composed as

$$Y := \begin{pmatrix} 0 & \ldots & & 0 \\ Y^1 & & & \\ & Y^2 & & \\ & & \ddots & \\ 0 & \ldots & & Y^m \end{pmatrix}. \tag{3.8}$$

Proof: Since X has the form (3.3), $\mathbf{c}$ defined by (3.6) is the compressed mask of the tensor product mask

$$b_\alpha := a^0_{\alpha^0} \prod_{j=1}^{m} b^j_{\alpha^j}\,, \qquad \alpha = (\alpha^0, \ldots, \alpha^m) \in \mathbb{Z}^n\,. \tag{3.9}$$

Recall that the refinable function $\Phi(x|X)$, $x = (x^1, \ldots, x^m) \in \mathbb{R}^s$, corresponding to the compressed mask of the tensor product mask (3.2) is $F(x^1, \ldots, x^m)$ as in (3.4). Furthermore, the symbol $a(z)$ of the uncompressed Φ can by (3.5) and (3.9) be written as

$$\begin{aligned} a(\mathbf{z}) &= a^0(z) \prod_{j=1}^{m} a^j(z) \\ &= 2^{-(r_1+\cdots+r_m)} \prod_{j=1}^{m} \prod_{\ell=1}^{r_j} (1 + z^{-y^{j,\ell}})\, b^j(z)\, a^0(z) \\ &= 2^{-r} \prod_{i=1}^{r} (1 + \mathbf{z}^{-y^i})\, b(\mathbf{z})\,, \end{aligned} \tag{3.10}$$

where $\mathbf{z} = (z, \ldots, z)$, $r = r_1 + \cdots + r_m$, $b(\mathbf{z})$ is the symbol corresponding to (3.9), and y^i, $i = 1, \ldots, r$, are the columns of the $n \times r$-matrix Y defined by (3.8). Applying again [2] Lemma 3.1, p. 57/58, we find that (3.10) is equivalent to

$$\Delta_Y S_{\mathbf{a}} = 2^{-r} S_{\mathbf{b}} \Delta_Y\,. \tag{3.11}$$

Thus, the assumptions of Proposition 1 are satisfied so that Proposition 2 can be employed. By definition of X,

$$\begin{aligned} XY &= \begin{pmatrix} I & -I & & 0 \\ \vdots & & \ddots & \\ I & 0 & & -I \end{pmatrix} \begin{pmatrix} 0 & \cdots & 0 \\ Y^1 & & \\ & \ddots & \\ & & Y^m \end{pmatrix} \\ &= \begin{pmatrix} -Y^1 & & 0 \\ & \ddots & \\ 0 & & -Y^m \end{pmatrix} \in \mathbb{Z}^{s \times r} \end{aligned} \tag{3.12}$$

and

$$\begin{aligned} (D_{XY}\Phi)(\alpha^1, \ldots, \alpha^m | X) &= (D_{XY}F)(\alpha^1, \ldots, \alpha^m) \\ &= \int_{\mathbb{R}^q} \varphi^0(x) \prod_{j=1}^{m} (D_{-Y^j}\varphi^j)(x - \alpha^j)\, dx\,. \end{aligned} \tag{3.13}$$

Thus, by (2.13),

$$\int_{\mathbb{R}^q} \varphi^0(x) \prod_{j=1}^{m} (D_{Y^j}\varphi^j)(x - \alpha^j)\, dx = (-1)^r (\Delta_{XY} W)_\alpha$$

for W given by (2.12) with $\mathbf{c}$ as in (3.6). This proves the theorem. ∎

Acknowledgements. I would like to thank Wolfgang Dahmen very much for stimulating discussions. The work of the author is supported by the Deutsche Forschungsgemeinschaft in the Graduiertenkolleg "Algorithmische Diskrete Mathematik" at the Freie Universität Berlin.

References

1. Beylkin, G., On the representation of operators in bases of compactly supported wavelets, Siam J. Numer. Anal. **6** (1992), 1716–1740.
2. Cavaretta, A. S., W. Dahmen, and C. A. Micchelli, *Stationary Subdivision*, Memoirs of Amer. Math. Soc., Vol. 93, # 453, 1991.
3. Dahlke, S., and A. Kunoth, Biorthogonal wavelets and multigrid, in *Proceedings of the 9th GAMM Seminar, Kiel, 1993*, W. Hackbusch, and G. Wittum (eds.), NNFM series, Vieweg, to appear.
4. Dahmen, W., and A. Kunoth, Multilevel preconditioning, Numer. Math. **63** (1992), 315–344.
5. Dahmen, W., and C. A. Micchelli, Using the refinement equation for evaluating integrals of wavelets, Siam J. Numer. Anal. **30** (1993), 507–537.
6. Jia, R. Q., and C. A. Micchelli, Using the refinement equation for the construction of pre-wavelets II: Powers of two, in: *Curves and Surfaces*, P.J. Laurent, A. Le Méhauté and L.L. Schumaker (eds.), Academic Press, New York, 1991, 209–246.
7. Kunoth, A., *Multilevel Preconditioning*, Ph. D. Thesis, Freie Universität Berlin, January, 1994.

Angela Kunoth
Institut für Informatik
Fachbereich Mathematik und Informatik
Freie Universität Berlin
Takustr. 9
D-14195 Berlin, GERMANY
kunoth@inf.fu-berlin.de

Rational C^{k+1} Finite Elements

A. Le Méhauté and K. N. Agbévé

Abstract. Rational $\mathcal{C}^{k+1}$-finite elements are obtained from polynomial $\mathcal{C}^k$-finite elements by adding some suitable rational functions. Using the polynomial interpolant $\mathcal{C}$ associated to the polynomial finite element, the rational global interpolant $\mathcal{P}$ is written in the form $\mathcal{P} = \mathcal{C} + \mathcal{R}$, where $\mathcal{R}$ is a rational functional such that $\mathcal{CR} = 0$, $\mathcal{PC} = \mathcal{C}$ and such that, whenever $\mathcal{C}$ is $\mathcal{C}^k$ continuous then $\mathcal{P}$ is $\mathcal{C}^{k+1}$ continuous.

§1. Interpolation Schemes and Finite Elements

1. Partial Finite Elements

Let V be a linear space, of finite dimension m, and V^* its algebraic dual space. Let $L = \{l_1, \ldots, l_p\}$ a subset of p elements in V^* , and $\Phi = \{\phi_1, \ldots, \phi_p\}$ a set of p real numbers.

In order *to solve the interpolation problem* $Lu = \Phi$, we seek $u \in V$ such that $l_i(u) = \phi_i, \forall i = 1, 2, \ldots, p$. We say that the interpolating scheme L is *V–unisolvent* if the interpolation problem has one and only one solution.

Proposition 1. *The scheme L is V–unisolvent if and only if the mapping* $u \mapsto Lu$ *is injective, or equivalently, if and only if ker* $L = \{0\}$.

Let $\{p_1, p_2, \ldots, p_m\}$ be a basis of V, and let us write $u \in V$ as $u = \sum_{i=1}^m a_i p_i$. Then, *solving the problem* $Lu = F$ is equivalent to solving a linear system

$$\begin{cases} \sum_{i=1}^m a_i\, l_j(p_i) = \phi_j, \\ \forall j = 1, 2, \ldots, p. \end{cases}$$

Obviously, a solution exists if and only if $span(l_i) = V^*$. Thus, hereafter, we will assume that the $l_i, i = 1, 2, \ldots, m$ *are linearly independent in* V^*.

As usual, we will define *a finite element* in the sense of Ciarlet & Raviart [2], *i.e.*, as a triple (K, P, Σ), where K is a closed polyhedron in $\mathbb{R}^n$ (in this paper, K is a triangle in $\mathbb{R}^2$), where $P \subset \mathcal{C}^s(K)$, $s \in \mathbb{N}$, is a space of real valued

Wavelets, Images and Surface Fitting
P. J. Laurent, A. Le Méhauté, and L. L. Schumaker (eds.), pp. 335–342.

ISBN 1-56881-040-7.

functions defined on K, and Σ is a set of linear forms, linearly independent, defined on $\mathcal{C}^s(K)$, in such a way that Σ would be P–unisolvent.

A finite element is said to be *of class* $\mathcal{C}^k$ if, whenever it is used to calculate piecewise a function f over a triangulation of a domain $\Omega \subset \mathbb{R}^n$, then the function f is of class $\mathcal{C}^k$ on Ω .

Definition 1. *An element* (K, P', Σ') *is a partial finite element of* (K, P, Σ) *if* P' *is a subspace of* P, Σ' *a subset of* Σ *and if* Σ' *is* P'*-unisolvent.*

Examples

Let (K, P, Σ) *be the usual* $(\mathcal{C}^0, \mathbb{P}_3)$ *Hermite triangular polynomial finite element in* $\mathbb{R}^2$, *where* $K = [a_1, a_2, a_3]$ *is a non-degenerate triangle and* a_0 *a point in the interior of* K; $\Sigma = \{\partial^\alpha u(a_i), u(a_0); \quad |\alpha| \leq 1; \quad i = 1,2,3\}$ and $P = \mathbb{P}_3(K) = \{\lambda_1^{\alpha_1}\lambda_2^{\alpha_2}\lambda_3^{\alpha_3}, \quad \alpha_1 + \alpha_2 + \alpha_3 = 3; \quad 0 \leq \alpha_i \leq 3; \quad i = 1,2,3\}$

Example 1.

$$\Sigma' = \{u(a_0); \quad i = 1,2,3\}$$
$$P' = \{\lambda_i^3, \quad i = 1,2,3\}$$

Example 2.

$$\Sigma' = \{\partial^\alpha u(a_i), u(a_0); \quad |\alpha| = 1; \quad i = 1,2,3\}$$
$$P' = \{\lambda_1^{\alpha_1}\lambda_2^{\alpha_2}\lambda_3^{\alpha_3}, |\alpha| = 3; 0 \leq \alpha_i < 3; i = 1,2,3\}$$

Example 3.

$$\Sigma' = \{\partial^\alpha u(a_i); \quad |\alpha| = 1; i = 1,2,3\}$$
$$P' = \{\lambda_1^{\alpha_1}\lambda_2^{\alpha_2}\lambda_3^{\alpha_3}, \quad |\alpha| = 3; 0 \leq \alpha_i \leq 3; \alpha_1 \neq \alpha_2 \neq \alpha_3 \neq \alpha_1, i = 1,2,3\}.$$

1.2. Linearly Independent Schemes.

Let $L_i = \{l_j^i; j = 1, \ldots, m_i\}$, $i = 1,2$, be two interpolation schemes defined on the same space V.

Definition 2. *The schemes* L_1 *and* L_2 *are linearly independent if the linear functionals* $l_1^1, \ldots, l_{m_1}^1, l_1^2, \ldots, l_{m_2}^2$ *are linearly independent in* V^*, *where* V^* *is the algebric dual space of the space* V.

Let (K, P', Σ') be a given finite element. We want to construct a finite element (K, P, Σ) which admits (K, P', Σ') as a partial finite element.

Given two linearly independent schemes L_1 and L_2 defined on a space $V = V_1 \oplus V_2$, we know [3] that if L_1 is V_1-unisolvent and if $L = L_1 \cup L_2$ is V-unisolvent then L_2 is V_2-unisolvent.

The next result provides us a condition on L_1 and L_2 which will make L V-unisolvent, if L_i is V_i-unisolvent, $i = 1,2$.

Theorem 1. *Let* $V_i, i = 1,2$, *be two subspaces of* V *and* L_i, $i = 1,2$ *two linearly independent schemes in* $V_i^*, i = 1,2$, *respectively. If* L_i *is* V_i*-unisolvent and if for all* $v \in V_2$ *and for all* $l \in V_1^*$, *we have* $l(v) = 0$, *then* $L = L_1 \cup L_2$ *is* V*-unisolvent, with* $V = V_1 \oplus V_2$.

Proof: We have to prove that for $u \in V$ and for all $l \in V^*$ then $l(u) = 0 \Rightarrow u = 0$, *i.e.*, for a given interpolation problem there exists a unique solution $u \in V$. Let $L_1 = \{l_i^1 \ , i = 1, \ldots, n\}$, $\quad L_2 = \{l_j^2 \ , j = 1, \ldots, m\}$ and $L = L_1 \cup L_2$; let $l_s = l_s^1$ for $s = 1, \ldots, n$ and $l_s = l_{s-n}^2$ for $s = n+1, \ldots, m+n$.

Let $u \in V$. We can write $u = u_1 + u_2$, $u_i \in V_i$ and for any $l_s \in L$,we have $l_s(u) = l_s(u_1) + l_s(u_2)$. From the hypothesis and for $s \leq n$,$l_s(u_2) = 0$, thus $l_s(u) = l_s(u_1)$; that means that for $s \leq n, l_s(u) = 0 \Leftrightarrow l_s^1(u_1) = 0 \Leftrightarrow u_1 = 0$. And for $s \geq n$ we have: $l_s(u) = l_{s-n}^2(u) = l_{s-n}^2(u_1) + l_{s-n}^2(u_2) = l_{s-n}^2(u_2)$ thus $l_s(u) = 0 \Leftrightarrow l_{s-n}^2(u_2) = 0 \Leftrightarrow u_2 = 0$. Finally, we obtain that $l_s(u) = 0 \Leftrightarrow u = u_1 + u_2 = 0$. ■

We define the linear functional $\mathcal{C}_i$ associated to an interpolation scheme $L_i = \{l_1^i, \ldots, l_{m_i}^i\}$ defined on V_i, by

$$\mathcal{C}_i[u] = \sum_{j=1}^{m_i} w_j^i l_j^i[u],$$

where $w_j^i \in V_i$ satisfies $l_r^i(w_s^i) = \delta_{rs}($ Kronecker conditions). The V_i-unisolvence of the interpolation scheme L_i is equivalent to the existence of the functions w_j^i, $j = 1, \ldots, m_i$.

The condition $l(v) = 0$ for all $v \in V_2$ and $l \in V_1^*$ in Theorem 1, is then equivalent to $\mathcal{C}_1\mathcal{C}_2 = 0$

Let (K, P, Σ) be a finite element. Assume that $\Sigma = \{l_i; i = 1, \ldots, m\}$ is a family of m linearly independent linear functionals. Because of the unisolvency, one can find m functions $p_i \in P$, $i = 1, \ldots, m$, such that $l_i(p_j) = \delta_{i,j}$ and $P = \text{Span}\{p_j; j = 1, \ldots, m\}$.They are called *basis functions* in the finite element method used for solving PDE's.

Definition 3. *The interpolant $\mathcal{C}$ associated to the finite element (K, P, Σ) is defined by*

$$\mathcal{C}[u] = \sum_{j=1}^{m} p_j l_j[u].$$

Remark: It is easy to see that the P-unisolvency of the set Σ is equivalent to say that $\mathcal{C}[P] = \mathcal{C}P = P$.

§2. Rational $\mathcal{C}^{k+1}$-Finite Element in $\mathbb{R}^2$.

2.1. Polynomial Finite Element of Type (k, N, χ, p)

Definition 4. *Let $K = [a_1, a_2, a_3]$ be a non-degenerate triangle, σ_s the edge opposite to a_s, with normal ν_s. Let $\partial_{\nu_i}^m u = (\frac{\partial^m u}{\partial \nu_i^m})$. A $\mathcal{C}^k$-polynomial finite element (K, P, Σ) of type (k, N, χ, p) is defined by:*

$$\Sigma = \{\partial^\alpha u(a_i), \partial_{\nu_i}^m u(Q_{is}^m); i = 1, 2, 3; m = 0, \ldots, k; s = 1, \ldots, m+p\} \cup \mathcal{L}$$
$$P = \mathbb{P}_N(K) \text{ with } N = 2(k+\chi) + p + 1 \text{ and } \chi \geq k,$$

where Q^i_{rs} are points on σ_s verifying $Q^i_{rs} \neq A_j, j = 1,2,3$ and $Q^i_{rs} \neq Q^i_{ts}$ if $t \neq r$, and $\mathcal{L}$ is a set of additional degrees of freedom which makes the set Σ, P-unisolvent.

The degrees of freedom in the set $\mathcal{L}$ are called *degrees of freedom of the second type* [3].

Remarks.
1) One can find in [3] a complete study of this family of finite elements (and some others) and an explicit formulation to constructed them.
2) It is easy to see that no degree of freedom of the second type occurs with the $\mathcal{C}^k$-continuity of the finite element.

We obtain obviously:

Lemma 1. *Let (K, P, Σ) be a $\mathcal{C}^k$-polynomial finite element of type (k, N, χ, p). The partial finite element (K, P', Σ') obtained by removing all degrees of freedom of the second type in (K, P, Σ) is still $\mathcal{C}^k$.*

2.2. Rational Finite Element of Type (k, N, χ, p): Construction

In order to increase the smoothness, we will replace the set $\mathcal{L}$ and add some rational functions to P', built in such a way that, restricted to any edge of the triangle, the interpolant will coincide with the underlying polynomial.

We call such an element a *rational finite element of type* (k, N, χ, p), and will make clear the definition by an explicit construction. Let $\mathcal{P}$ be the interpolation functional associated to the rational finite element we are looking for. We'll construct $\mathcal{P}$ as $\mathcal{P} = \mathcal{C} + \mathcal{R}$ where $\mathcal{C}$ is a polynomial functional and $\mathcal{R}$ an irreducible rational functional.

First Step: *Construction of the polynomial partial finite element of type (k, N, χ, p) by removing all second type degree of freedom (i.e., $\mathcal{L}$).* Let $\mathcal{C}$ be the corresponding interpolation functional.

Second Step: *Construction of $\mathcal{P}$ in the form $\mathcal{P} = \mathcal{C} + \mathcal{R}$ satisfying*

$$(a) \quad \mathcal{CR} = 0 \qquad (b) \quad \mathcal{PC} = \mathcal{C} \qquad (c) \quad \mathcal{P} \text{ is } \mathcal{C}^{k+1}.$$

Definition 6. *Let us recall that a function f is k-flat at a point M if and only if $\partial^\alpha f(M) = 0$ for all $\alpha = (\alpha_1, \alpha_2)$, $\alpha_i \in \mathbb{N}$ with $\alpha_1 + \alpha_2 \leq k$ and there exists at least one $\alpha = (\alpha_1, \alpha_2)$, $\alpha_1 + \alpha_2 = k + 1$ with $\partial^\alpha f(M) \neq 0$.*

Lemma 2 [3]. *Let $K = [a_1, \ldots, a_{n+1}]$ be a non-degenerate simplex in $\mathbb{R}^n$. A polynomial p of degre N, in n variables $x_1, x_2, \ldots, x_n$, is k-flat at each vertex of K, if and only if p can be written as:*

$$p(M) = p(x_1, \ldots, x_n) = \tilde{p}(\lambda_1, \ldots, \lambda_{n+1}) = \sum_{\substack{|\alpha|=N \\ \alpha_i < N-k}} c_\alpha \prod_{i=1}^{n+1} \lambda_i^{\alpha_i}$$

where $c_\alpha \in \mathbb{R}$, $\alpha = (\alpha_1, \ldots, \alpha_{n+1}) \in \mathbb{N}^{n+1}$ and $\lambda_i(M) = \lambda_i \in \mathbb{R}$ is the $i-th$ barycentric coordinate of $M = (x_1, \ldots, x_n)$ related to the vertex a_i of K.

Lemma 3. *Let $K = [a_1, a_2, a_3]$ be a non-degenerate triangle; let λ_i be the barycentric coordinates relatively to the vertex a_i, σ_i the edge opposite to a_i,and ν_i a vector normal to the edge σ_i; let p be a polynomial of any degree in two variables and k a positive integer. If for all $m \leq k$ we have $\partial_{\nu_i}^m p\big|_{\sigma_j} = 0$, then p can be divided by Λ^{k+1} where $\Lambda = \lambda_1\lambda_2\lambda_3$.*

Corollary. *Let p and q be two polynomials of any degree in two variables such that $\dfrac{p}{q}$ is irreducible. If for all $m \leq k$ we have $\partial_{\nu_i}^m[\frac{p}{q}]|\sigma_j = 0$ then p can be divided by Λ^{k+1} .*

Theorem 2. *With the notations already defined, let us consider*

$$\Sigma = \{\partial^\alpha u(a_i), \partial_{\nu_i}^m u(Q_{r,i}), |\alpha| \leq k + \chi \text{ }, m = 0, \ldots, k \text{ }, r = 1, \ldots, p + m\}$$

or equivalently, with $I = \frac{3}{2}((k + \chi + 2)(k + \chi + 1) + (k + 1)(k + 2p))$, let us write $\Sigma = \{l_i \text{ }, i = 1, \ldots, I\}$. Let V_k be the subspace of $\mathbb{P}_N(\mathbb{R}^2)$ generated by the functions $\{v_i, i = 1, \ldots, I, \quad with \quad l_i(v_j) = \delta_{ij}\}$, and let $\mathcal{C}$ be the functional defined by $\mathcal{C}[u] = \sum_{i=1}^I v_i l_i[u]$ and $\mathcal{R}$ a rational functional which is W-unisolvent, W being a space of functions to be defined later. Then the functional $\mathcal{P} = \mathcal{C} + \mathcal{R}$ is $V_k \oplus W$-unisolvent if and only if $\chi = k$ or $\chi = k + 1$.

Proof: We want to find $\mathcal{P} = \mathcal{C} + \mathcal{R}$ such that

$$\text{(a)} \quad \mathcal{CR} = 0; \quad \text{(b)} \quad \mathcal{PC} = \mathcal{C}; \quad \text{(c)} \quad \mathcal{P} \text{ is } \mathcal{C}^{k+1}.$$

The second condition implies that $\mathcal{P}$ can be written as: $\mathcal{P} = \mathcal{C} - \mathcal{PC} + \mathcal{P} = \mathcal{C} + \mathcal{P}(\mathcal{I} - \mathcal{C})$ where $\mathcal{I}$ denotes the identity. Using the form $\mathcal{P} = \mathcal{C} + \mathcal{R}$ we have $\mathcal{P}(\mathcal{I} - \mathcal{C}) = \mathcal{R}(\mathcal{I} - \mathcal{C})$; then $\mathcal{P} = \mathcal{C} + \mathcal{R}(\mathcal{I} - \mathcal{C})$. Hereafter, we will write $\mathcal{R}$ instead of $\mathcal{R}(\mathcal{I} - \mathcal{C})$. The condition (a) is equivalent to:

(a-1) $\mathcal{R}$ is $(k + \chi)-$flat at the vertices of $K = [a_1, a_2, a_3]$,

(a-2) $\partial_{\nu_i}^m \mathcal{R}\big|_{\sigma_j} = 0$ for $m \leq k$ and $i, j = 1, 2, 3$.

By the previous lemmas we find that $\mathcal{R}$ can be divided by Λ^{k+1} and then we can write $\mathcal{P}$ as $\mathcal{P} = \mathcal{C} + \Lambda^{k+1}\tilde{\mathcal{R}}(\mathcal{I} - \mathcal{C})$ (In order to simplify the notation we'll still use $\mathcal{R}$ instead of $\tilde{\mathcal{R}}$). ∎

Lemma 4. *The functional $\mathcal{P}$ defined by $\mathcal{P} = \mathcal{C} + \Lambda^{k+1}\mathcal{R}(\mathcal{I} - \mathcal{C})$ is at least $\mathcal{C}^k$.*

Since $\mathcal{C}$ is polynomial of degree N, then $\partial_{\nu_i}^m \mathcal{C}\big|_{\partial K}$ is also polynomial, with degree $N - m$. We ask for $\partial_{\nu_i}^{k+1}\mathcal{R}\big|_{\partial K}$ to be a polynomial of degree $N - k - 1$.

From the data given at the vertices, we are able to evaluate at the end of the edge σ_i the directional derivative of order $m \leq \chi - 1$ of $\partial_{\nu_i}^{k+1}\mathcal{P}$ along the edge. Thus we know 2χ conditions, for a polynomial of degree $N-k-1 = 2\chi+k+p$, therefore, in order to achieve $\mathcal{C}^{k+1}$ continuity, we need another set of $k+p+1$ independant conditions. We choose them as normal cross derivatives evaluated at $k+p+1$ new distinct points R_{is} along σ_i. This defines the new set W.

More precisely, let us define the functional $\mathcal{N}_{is}$ by $\mathcal{N}_{is}[u] = \partial_{\nu_i}^{k+1}u(R_{is})$, where $R_{is}, s = 1,\ldots,k+p+1$, are $k+p+1$ different points, other than the vertices, on σ_i. We can write $\mathcal{R}$ in the form

$$\mathcal{R} = \sum_{i=1}^{3}\sum_{s=1}^{k+p+1}\frac{x_{is}}{g_{is}}\mathcal{N}_{is}(\mathcal{I}-\mathcal{C})$$

where x_{is} and g_{is} are polynomial functions. For simplicity, we choose here $g_{is} = g$.

From the localness properties of finite elements, we need that $\partial_{\nu_i}^{k+1}\mathcal{P}\big|_{\sigma_i}$ depends only upon values given on σ_i. By a direct calculation we have

$$\partial_{\nu_i}^{k+1}\mathcal{P}\big|_{\sigma_i} = \partial_{\nu_i}^{k+1}\mathcal{C}\big|_{\sigma_i} + A\Big[\big(\lambda_j\lambda_l\big)^{k+1}\sum_{i=1}^{3}\sum_{s=1}^{k+p+1}\frac{x_{is}}{g}\widetilde{\mathcal{N}_{is}}\Big]\Big|_{\sigma_i}$$

where $\widetilde{\mathcal{N}_{is}} = \mathcal{N}_{is}(\mathcal{I}-\mathcal{C})$ and A is a constant. Thus $\partial_{\nu_i}^{k+1}\mathcal{P}\big|_{\sigma_i}$ depends on data given along σ_i only if for all i, x_{is} can be divided by $\lambda_j\lambda_l, i\neq l\neq j\neq i$; $i,j,l = 1,2,3$. At this point, we can write $\mathcal{P}$ as

$$\mathcal{P} = \mathcal{C} + \Lambda^{k+1}\sum_{i=1}^{3}\lambda_j\lambda_l\sum_{s=1}^{k+p+1}\frac{\tilde{x}_{is}}{g}\widetilde{\mathcal{N}}_{is}.$$

In order to have $k+p+1$ independent conditions on σ_i, we need $k+p+1$ functions $\tilde{x}_{is}$, $s = 1,\ldots,k+p+1$ linearly independent in the space of polynomial functions. We choose them as Lagrange basis functions, in one variable, at the points R_{is}.

Finally, we have to find a polynomial g if any exists, such that the functions r_{is} defined by

$$r_{is}(M) = \frac{\lambda_i^{k+1}\lambda_j^{k+2}\lambda_l^{k+2}\tilde{x}_{is}}{g}, \quad i\neq l\neq j\neq i, \quad i,j,l = 1,2,3$$

satisfy the following conditions:

1) $(\partial_{\nu_i}^{k+1}r_{is})_{\sigma_i}$ is a polynomial of degree $N-k-1$,

2) $\dfrac{\partial^{k+1}r_{is}}{\partial\nu_i^{k+1}} = B\times\dfrac{\lambda_j^{k+2}\lambda_l^{k+2}\tilde{x}_{is}}{g} + \mathcal{O}(\lambda_i).$

We can choose the function g in different ways, for example as:

1) $\chi = k, \quad g(M) = (\lambda_1\lambda_2 + \lambda_2\lambda_3 + \lambda_3\lambda_1)^2.$

2) $\chi = k+1, \quad g(M) = \lambda_1\lambda_2 + \lambda_2\lambda_3 + \lambda_3\lambda_1.$

§3. Examples

We now consider some examples. We obtain, as particular cases the $\mathcal{C}^1$-finite elements introduced by Zienkiewicz[4] and Herron (see [1] for more details and a presentation of some general families of rational triangular finite elements)

Example 1. A $\mathcal{C}^1$ element with $p = 0$ (Zienkiewicz element)
$\chi = k+1$ **and** $p = 0$
$k = 0$:

$$\Sigma_K = \{\partial^\alpha u(a_i), \quad (\partial_{\nu_i} u)(b_i), \quad i = 1,2,3, \quad |\alpha| \le 1\}$$
$$P_K = V_k \oplus \text{Span}\{r_{i,1}, \quad i = 1,2,3\}$$
with
$$r_{i,1} = \frac{\lambda_i \lambda_{i+1}^2 \lambda_{i+2}^2}{\lambda_1\lambda_2 + \lambda_2\lambda_3 + \lambda_3\lambda_1} \qquad i = 1,2,3$$
$$\dim P_K = 12$$

(b_i can be any point on the edge σ_i, and is not necessarily the midpoint).

Example 2. A $\mathcal{C}^2$ element with $p = 0$.
$\chi = k$ **and** $p = 0$
$k = 1$:

$$\Sigma_K = \{\partial^\alpha u(a_i), \; (\partial^m_{\nu_i} u)(\mathcal{Q}_{i,s}), \; i = 1,2,3, \; s = 1,\dots,2 \; |\alpha| \le 2\}$$
$$P_K = \mathbb{P}_5(K) \oplus \text{Span}\{r_{i,s}, \quad i = 1,2,3; \quad s = 1,2\}$$
with
$$r_{i,1} = \frac{\lambda_i^2 \lambda_{i+1}^3 \lambda_{i+2}^3 (2\lambda_{i+1} - \lambda_{i+2})}{(\lambda_1\lambda_2 + \lambda_2\lambda_3 + \lambda_3\lambda_1)^2}$$
$$r_{i,2} = \frac{\lambda_i^2 \lambda_{i+1}^3 \lambda_{i+2}^3 (2\lambda_{i+2} - \lambda_{i+1})}{(\lambda_1\lambda_2 + \lambda_2\lambda_3 + \lambda_3\lambda_1)^2}$$
$$i = 1,2,3$$
$$\dim P_K = 27$$

Example 3. Another $\mathcal{C}^2$ element but with $p \neq 0$.

$$\Sigma_K = \{\partial^\alpha u(a_i),\ u(b_i),\ (\partial^m_{\nu_i} u)(\mathcal{Q}_{i,s}),\ \ i = 1,2,3,$$
$$s = 0,1,\ldots,2\ \ |\alpha| \leq 2\}$$
$$P_K = \mathbb{P}_6(K) \oplus \text{Span}\{r_{i,s},\quad i = 1,2,3; \quad s = 1,2\}$$

with

$$r_{i,1} = \frac{\lambda_i^2 \lambda_{i+1}^3 \lambda_{i+2}^3 (2\lambda_{i+1} - \lambda_{i+2})}{(\lambda_1\lambda_2 + \lambda_2\lambda_3 + \lambda_3\lambda_1)^2}$$
$$r_{i,2} = \frac{\lambda_i^2 \lambda_{i+1}^3 \lambda_{i+2}^3 (2\lambda_{i+2} - \lambda_{i+1})}{(\lambda_1\lambda_2 + \lambda_2\lambda_3 + \lambda_3\lambda_1)^2}$$
$$i = 1,2,3$$
$$\text{dim} P_K = 36$$

References

1. Agbévé, K., N., Elements finis triangulaires rationnels de classe $\mathcal{C}^k$, Thèse, Université des sciences et techniques de Nantes (1993).
2. Ciarlet, P.G., *The Finite element Method for Elliptic Problems*,Norht-Holland, Amsterdam, 1981.
3. Le Méhauté, A., Interpolation et approximation par des fonctions polynomiales par morceaux dans $\mathbb{R}^n$. Thèse d'Etat, Rennes (1984).
4. Zienkiewicz,O.C., *The finite element method in structural continuum mechanics*, Mc Graw Hill, London, 1967.

K. N. Agbévé
3-135, place de Venise
59650 Villeneuve D'Ascq, FRANCE

A. Le Méhauté
Département de Mathématiques et Systèmes de Communication,
ENST–Bretagne
B.P 832,
29285 Brest Cedex, FRANCE
Alain.LeMehaute@enst-bretagne.fr

Convolution Kernels for Approximation by Radial Basis Functions

J. Levesley

Abstract. Convolution operators can be used to construct approximations to continuous functions on $\mathbb{R}^n$. Using radial basis functions, we construct a wide class of potential kernels for convolution operators, and show that only a very restricted class of such kernels are actually suitable for convolution.

§1. Introduction

One method of approximating a function $f \in \mathbf{C}(\mathbb{R}^n)$ is to convolve it with an *approximation to the identity.* This requires a *convolution kernel*, G, which is both integrable on $\mathbb{R}^n$ and for which $\int_{\mathbb{R}^n} G = 1$. If both of these conditions hold we say that G is suitable for convolution. Then, if we set $G_h(x) := h^{-n}G(x/h)$, $h \in \mathbb{R}$, $G_h * f \to f$ uniformly on compact subsets of $\mathbb{R}^n$, as $h \to 0$; see Stein and Weiss [9, page 11] for further details.

In recent years *radial basis functions* have come to the fore as tools for multivariable approximation. Given a function $\phi : \mathbb{R}^+ \to \mathbb{R}$, a radial basis function is of the form $\phi \circ \|\ \|$, where $\|\ \|$ is the Euclidean norm. Typical choices of the function ϕ are

$$\phi(r) = \begin{cases} r, & \text{norm,} \\ r^2 \log r & \text{thin-plate spline,} \\ (r^2+c^2)^{\frac{1}{2}}, & \text{multiquadric,} \\ e^{-r^2}, & \text{Gaussian.} \end{cases}$$

For general information on radial approximation, see Powell [7].

In this paper we are interested in the construction of convolution kernels using radial basis functions. The first stage in our construction is to average a shifted radial function over the unit sphere S^{n-1} in $\mathbb{R}^n$:

$$g(x,t) = \frac{1}{\omega_{n-1}} \int_{S^{n-1}} \phi(\|x - tu\|)\, dS^{n-1}(u), \quad x \in \mathbb{R}^n, \quad t \in \mathbb{R}, \tag{1.1}$$

Wavelets, Images and Surface Fitting
P. J. Laurent, A. Le Méhauté, and L. L. Schumaker (eds.), pp. 343–350.

ISBN 1-56881-040-7.

where ω_{n-1} is the surface area of S^{n-1}; we will assume here that $n \geq 2$. This approach was introduced by Jackson [2], who considered convolution kernels of the form $\sum_{l\in L} \alpha_l g(x,t_l)$, where $\alpha_l, t_l \in \mathbb{R}$, $l \in L$, where L is a finite index set. In this paper we consider kernel functions G of the form

$$G(x) = \int_{\mathbb{R}} g(x,t)d\mu(t), \qquad x \in \mathbb{R}^n, \tag{1.2}$$

where μ is a finite, regular, signed Borel measure of finite total variation and compact support.

For different choices of the function ϕ in (1.1) we will give necessary conditions on the measure μ which ensure that the associated kernel G in (1.2) is suitable for convolution. These conditions will be in terms of the moments of μ, m_j, defined by

$$m_j = \int_{\mathbb{R}} t^j d\mu(t). \tag{1.3}$$

We shall consider functions ϕ of the form

$$\phi(r) = \begin{cases} \sum_{k=0}^{[n+\alpha]} a_k r^{\alpha-k} + \phi_1(r), & r > A > 0, \qquad (1.4a) \\ \log r \sum_{k=0}^{[n+\alpha]} a_k r^{\alpha-k} + \phi_1(r), & r > A > 0, \qquad (1.4b) \end{cases}$$

for some $\alpha > -n$, where $\phi_1 \circ \|\ \| \in L_1(\mathbb{R}^n)$, and $[\,]$ denotes the integer part. Functions of the form (1.4a) include all powers of the norm and the multiquadric family, whilst functions of the form (1.4b) include the polyharmonic splines in even space dimension.

In what follows, ψ will always denote $\phi \circ \|\ \|$, and we shall always assume that ϕ is such that ψ is locally integrable. G will always be a function of the form (1.2), and C will be a constant which is not necessarily the same at each occurrence.

§2. Results

In this section we shall examine the Fourier transform of G in order to determine whether or not, for a given function ϕ, G is suitable for convolution. The Fourier transform, $\hat{G}$, is given by

$$\hat{G}(\xi) = \int_{\mathbb{R}^n} G(x)e^{-ix\xi}dx\,,$$

where $x\xi$ denotes the usual inner product, and if G is not integrable, we interpret this as the distributional Fourier transform; see Friedlander [1] for details. In particular, we shall often use the simple result that if $\langle f_\nu, \lambda\rangle \to \langle f, \lambda\rangle$ for all *rapidly decaying* functions λ, then $\hat{f}_\nu \to \hat{f}$ distributionally. This is a consequence of the definition of the Fourier transform of a distribution, $\langle \hat{f}, l\rangle = \langle f, \hat{l}\rangle$, for all $l \in \mathbf{C}_0^\infty(\mathbb{R}^n)$. Here, λ is rapidly decaying if

$\|x^\beta[D^\gamma\lambda](x)\|_\infty < \infty$ for all multiindices β and γ. Whenever we say $f_\nu \to f$ we shall assume that this is a distributional limit with rapidly decaying test functions.

If G is integrable it is clear that $\hat{G}$ is continuous. As we shall see, $\hat{G}$ is continuous only if a number of the moments m_{2k}, $k \in \mathbb{Z}_+$, are zero. It is shown in Levesley [5] and Levesley et. al. [6] that it is also sufficient that these moments be zero. It will be shown that, in many cases, if the relevant moments of μ are zero, then $\hat{G}(0) = 0$, i.e., the integral of G is zero, and such a kernel is not suitable for convolution. Thus, the balance between producing an integrable kernel and one with nonzero integral is delicate and it is only possible to obtain both in a restricted set of cases.

The pivotal result of this section is

Theorem 1. *Let G be given by (1.2). Then,*

$$\hat{G}(\xi) = \frac{\omega_{n-2}}{\omega_{n-1}}\hat{\psi}(\xi)\sum_{k=0}^{\infty}(-1)^k\frac{b_{2k}m_{2k}}{(2k)!}\|\xi\|^{2k},$$

where

$$b_k := \int_{-1}^{1} t^k(1-t^2)^{\frac{n-3}{2}}dt.$$

Proof: Let

$$\phi_R(r) = \begin{cases} \phi(r), & r < R, \\ 0, & r \geq R, \end{cases}$$

$$\text{and } G_R(x) = \frac{1}{\omega_{n-1}}\int_{\mathbb{R}}\int_{S^{n-1}}\phi_R(\|x-tu\|)dS^{n-1}(u)d\mu(t).$$

Further, let λ be a rapidly decaying function. Then, with $B_M = \{x : \|x\| < M\}$, we have,

$$\begin{aligned}
&|\langle G,\lambda\rangle - \langle G_R,\lambda\rangle| \\
&= \left|\int_{\mathbb{R}^n}\frac{1}{\omega_{n-1}}\int_{\mathbb{R}}\int_{S^{n-1}}(\phi(\|x-tu\|)-\phi_R(\|x-tu\|))dS^{n-1}(u)d\mu(t)dx\right| \\
&\leq \int_{B_{\frac{R}{2}}}\frac{1}{\omega_{n-1}}\int_{\mathbb{R}}\int_{S^{n-1}}|\phi(\|x-tu\|)-\phi_R(\|x-tu\|)|dS^{n-1}(u)d\mu(t)dx \\
&\quad + \int_{\mathbb{R}^n\setminus B_{\frac{R}{2}}}\frac{1}{\omega_{n-1}}\int_{\mathbb{R}}\int_{S^{n-1}}|\phi(\|x-tu\|)-\phi_R(\|x-tu\|)|dS^{n-1}(u)d\mu(t)dx.
\end{aligned}$$

For sufficiently large R, $\phi_R(\|x-tu\|) = \phi(\|x-tu\|)$ on $B_{\frac{R}{2}}$, because $|t| < a$ for some a as μ is compactly supported, and $\|u\| = 1$. Hence, for sufficiently large R,

$$\begin{aligned}
|\langle G,\lambda\rangle - \langle G_R,\lambda\rangle| &\leq \int_{\mathbb{R}^n\setminus B_{\frac{R}{2}}}\frac{2}{\omega_{n-1}}\int_{\mathbb{R}}\int_{S^{n-1}}|\phi(\|x-tu\|)|dS^{n-1}(u)d\mu(t)dx \\
&\leq C\int_{\mathbb{R}^n\setminus B_{\frac{R}{2}}}\|x\|^\alpha\|x\|^{-\gamma}dx,
\end{aligned}$$

for some $\gamma > n+\alpha+2$. The last equation follows as, on $B_{\frac{R}{2}}$ for sufficiently large R, $\|x-tu\| \le 2\|x\|$ and $|\phi(r)| \le Cr^\alpha$, and, because λ is rapidly decaying, $\|\lambda(x)\| \le C\|x\|^{-\gamma}$. Hence,

$$|\langle G,\lambda\rangle - \langle G_R,\lambda\rangle| \le C\int_R^\infty \frac{dr}{r^2} \to 0 \text{ as } R\to\infty.$$

Therefore, $\hat{G} = \lim_{R\to\infty}\hat{G}_R$.

Now, using Fubini's theorem we can write

$$\begin{aligned}\hat{G}_R(\xi) &= \frac{1}{\omega_{n-1}}\int_{\mathbf{R}}\int_{S^{n-1}}\int_{\mathbf{R}^n}\phi_R(\|x-tu\|)dS^{n-1}(u)d\mu(t)e^{-ix\xi}dx\\ &= \frac{1}{\omega_{n-1}}\int_{\mathbf{R}}\int_{S^{n-1}}\int_{\mathbf{R}^n}\phi_R(\|x-tu\|)e^{-ix\xi}dxdS^{n-1}(u)d\mu(t).\end{aligned}$$

Then, making the change of variable $y = x - tu$, we obtain

$$\begin{aligned}\hat{G}_R(\xi) &= \frac{1}{\omega_{n-1}}\int_{\mathbf{R}}\int_{S^{n-1}}\int_{\mathbf{R}^n}\phi_R(\|y\|)e^{-i(y+tu)\xi}dxdS^{n-1}(u)d\mu(t)\\ &= \frac{1}{\omega_{n-1}}\int_{\mathbf{R}}\int_{S^{n-1}}\int_{\mathbf{R}^n}\phi_R(\|y\|)e^{-iy\xi}dxe^{-itu\xi}dS^{n-1}(u)d\mu(t)\\ &= \frac{1}{\omega_{n-1}}\hat{\psi}_R(\xi)\int_{\mathbf{R}}\int_{S^{n-1}}e^{-itu\xi}dS^{n-1}(u)d\mu(t),\end{aligned}$$

where $\psi_R := \phi_R \circ \|\ \|$. Then, again changing the order of integration,

$$\begin{aligned}\hat{G}_R(\xi) &= \frac{1}{\omega_{n-1}}\hat{\psi}_R(\xi)\int_{S^{n-1}}\int_{\mathbf{R}}e^{-itu\xi}d\mu(t)dS^{n-1}(u)\\ &= \frac{1}{\omega_{n-1}}\hat{\psi}_R(\xi)\int_{S^{n-1}}\int_{\mathbf{R}}\sum_{k=0}^{\infty}\frac{(-itu\xi)^k}{k!}d\mu(t)dS^{n-1}(u)\\ &= \frac{1}{\omega_{n-1}}\hat{\psi}_R(\xi)\sum_{k=0}^{\infty}\int_{S^{n-1}}\frac{(-iu\xi)^k}{k!}\int_{\mathbf{R}}t^k d\mu(t)dS^{n-1}(u), \qquad (2.1)\end{aligned}$$

where the change of order of integration and summation is justified by the dominated convergence theorem; see Rudin [8]. Now, from John [3, 1.2],

$$\begin{aligned}\int_{S^{n-1}}(u\xi)^k dS^{n-1}(u) &= \omega_{n-2}\int_{-1}^{1}t^k(1-t^2)^{\frac{n-3}{2}}dt\|\xi\|^k\\ &= \omega_{n-2}b_k\|\xi\|^k.\end{aligned}$$

However, if k is odd then $b_k = 0$. Combining the last equation with (2.1) gives

$$\hat{G}_R(\xi) = \frac{\omega_{n-2}}{\omega_{n-1}}\hat{\psi}_R(\xi)\sum_{k=0}^{\infty}(-1)^k\frac{b_{2k}m_{2k}}{(2k)!}\|\xi\|^{2k}.$$

The result follows because $\psi_R \to \psi$. ■

Corollary 2. *If $\psi \in L_1(\mathbb{R}^n)$ and $m_0 = 0$, then G has zero integral.*

Proof: If $\psi \in L_1(\mathbb{R}^n)$, then $\hat{\psi}$ is continuous on $\mathbb{R}^n$ and, in particular, at the origin. Thus, if $m_0 = 0$, $\hat{G}(\xi) = \mathcal{O}(\|\xi\|^2)$ as $\xi \to 0$. ■

An immediate consequence of Corollary 2 is

Corollary 3. *If ϕ has compact support and $m_0 = 0$ then G has zero integral.*

Before we prove Theorem 1, we need two lemmas and corollaries. We prove only the first lemma and its corollary; the proofs of the second pair are essentially the same.

Lemma 4. *Suppose $\phi(r) = r^{-n}$, $r > A > 0$. Then,*

$$\hat{\psi}(\xi) = \mathcal{O}(\log \|\xi\|) \text{ as } \xi \to 0.$$

Proof: Let $\phi = \phi_1 + \phi_2$, where

$$\phi_1(r) = \begin{cases} r^{-n}, & r \geq 1, \\ 0, & 0 < r < 1, \end{cases}$$

and $\phi_2 = \phi - \phi_1$. Then, $\psi_2 = \phi_2 \circ \| \ \|$ is compactly supported and locally integrable, and therefore has a continuous Fourier transform.

Let

$$\phi_1^R(r) = \begin{cases} \phi_1(r), & r < R, \\ 0, & r \geq R, \end{cases}$$

and $\psi_1^R = \phi_1^R \circ \| \ \|$. Then, as $\psi_1^R \to \psi$ as $R \to \infty$, $\hat{\psi}_1^R \to \hat{\psi}_1$ as $R \to \infty$.

Now, because ψ_1^R is locally integrable and compactly supported, we have, from Stein and Weiss [8, Theorem 3.10],

$$\begin{aligned} \hat{\psi}_1^R(\xi) &= (2\pi)^{n/2} \|\xi\|^{1-\frac{n}{2}} \int_0^\infty \phi_1^R(r) J_{\frac{n}{2}-1}(\|\xi\| r) r^{\frac{n}{2}} \, dr \\ &= (2\pi)^{n/2} \|\xi\|^{1-\frac{n}{2}} \int_1^R r^{-\frac{n}{2}} J_{\frac{n}{2}-1}(\|\xi\| r) dr \\ &= (2\pi)^{n/2} \int_{\|\xi\|}^{R\|\xi\|} s^{-\frac{n}{2}} J_{\frac{n}{2}-1}(s) ds, \end{aligned}$$

where J_ν is the Bessel function of order ν. We estimate this integral in two parts, assuming that $\|\xi\| < 1$. Because $J_\nu(t) = \mathcal{O}(t^\nu)$ as $t \to 0$ (Tranter [10], eq.(1.5)),

$$\left| \int_{\|\xi\|}^1 s^{-\frac{n}{2}} J_{\frac{n}{2}-1}(s) ds \right| \leq C \int_{\|\xi\|}^1 \frac{ds}{s} = -C \log \|\xi\|, \tag{2.2}$$

where C is a constant independent of ξ. Now, as $t \to \infty$, $J_\nu(t) = \mathcal{O}(t^{-\frac{1}{2}})$ ([10, 3.28]), so that

$$\left| \int_1^{R\|\xi\|} s^{-\frac{n}{2}} J_{\frac{n}{2}-1}(s) ds \right| \leq C \int_1^\infty s^{-\frac{n+1}{2}} ds$$
$$< \infty, \tag{2.3}$$

as $n \geq 2$. From (2.2) and (2.3) we see that

$$|\hat{\psi}_1^R(\xi)| \leq -C \log \|\xi\| \text{ as } \xi \to 0,$$

where C is independent of ξ and R. Thus, $\hat{\psi}_1(\xi) = \mathcal{O}(\log \|\xi\|)$ as $\xi \to 0$. The result follows by the linearity of the Fourier transform and the fact that $\hat{\psi}_2$ is continuous. ■

Corollary 5. *Suppose $\phi(r) = r^{-n}$, $r > A$. G is integrable only if $m_0 = 0$, in which case G has zero integral.*

Proof: From Theorem 1 and Lemma 4, if $m_0 \neq 0$, $\hat{G}(\xi) = \mathcal{O}(\log \|\xi\|)$ as $\xi \to 0$ and so G is not integrable. If $m_0 = 0$, $\hat{G}(\xi) = \mathcal{O}(\|\xi\|^2 \log \|\xi\|)$ as $\xi \to 0$ and so $\hat{G}(0) = 0$. ■

Lemma 6. *Suppose $\phi(r) = r^{-n} \log r$, $r > A > 0$. Then*

$$\hat{\psi}(\xi) = O((\log \|\xi\|)^2) \text{ as } \xi \to 0.$$

Corollary 7. *Suppose $\phi(r) = r^{-n} \log r$, $r > A$. G is integrable only if $m_0 = 0$, in which case G has zero integral.*

Corollaries 5 and 7 are examples of the problem described at the beginning of this section, that is, conditions necessary for G to be integrable force the integral of G to be zero.

It is easy to show that if ψ is integrable then G is integrable. Because of this and Corollary 2, if $m_0 = 0$, the presence of ϕ_1 in the representations (1.4) effects neither the integrability or the integral of G. As we commented at the beginning of this section, the conditions which are necessary for the integrability of G are also sufficient. Thus, from Corollaries 5 and 7, if $m_0 = 0$, we can omit the terms r^{-n} and $r^{-n} \log r$ from the expansion for ϕ without altering the integrability or integral of G. Finally, if $\phi(r) = \mathcal{O}(r^{-n+\epsilon})$ as $r \to 0$, for some $\epsilon > 0$, and is also compactly supported, then ψ is integrable. Hence, from Theorem 1, if $m_0 = 0$, the integrability and integral of G are unaffected if we consider, functions ϕ of the form

$$\phi(r) = \begin{cases} \sum_{k=0}^{K(n,\alpha)} a_k r^{\alpha-k}, & r > 0, \quad (2.4a) \\ \log r \sum_{k=0}^{K(n,\alpha)} a_k r^{\alpha-k}, & r > 0, \quad (2.4b) \end{cases}$$

where $K(n,\alpha) = \max\{k \in \mathbb{N} : k < n + \alpha\}$.

Thus, to use Theorem 1, we need the Fourier transforms of $\psi = \phi \circ \| \, \|$, for ϕ with the above two forms. The results in the following lemma are from [4], Theorems 7.31, 7.32 and 7.34.

Lemma 8. *Let $\beta > -n$. Further, let $f_\beta(x) = \|x\|^\beta$ and $g_\beta(x) = \|x\|^\beta \log \|x\|$. Then,*

$$\hat{f}_\beta(\xi) = \begin{cases} C\|\xi\|^{-n-\beta}, & \beta \notin 2\mathbb{Z}_+, \\ C\triangle^{\frac{\beta}{2}}\delta(\xi), & \beta \in 2\mathbb{Z}_+ \end{cases}$$

where $\triangle$ is the Laplace operator and δ is the Dirac distribution,

$$\hat{g}_\beta(\xi) = \begin{cases} C\|\xi\|^{-n-\beta}(D - \log\|\xi\|), & \beta \notin 2\mathbb{Z}_+, \\ C\|\xi\|^{-n-\beta}, & \beta \in 2\mathbb{Z}_+, \end{cases}$$

where C and D are constants which depend on n and β but are not necessarily the same at each occurrence. ■

Now, using Theorem 1, we can read off necessary conditions for the continuity of the Fourier transform at the origin. We can also see straightaway, that we can get a nonzero integral only if the singularity in the Fourier transform is the inverse of an even power of $\|\xi\|$.

Theorem 9.

(a) *Let ϕ have representation (2.4a). Then, G is integrable only if $m_{2k} = 0$, $k = 0, 1, \ldots, Q$, where $Q = \max\{p_k : 0 \le k \le n + \alpha$ and $a_k \ne 0\}$,*

$$p_k = \begin{cases} \frac{\alpha}{2}, & \alpha \in 2\mathbb{Z}_+, \\ \frac{n+\alpha}{2} - 1, & n+\alpha \in 2\mathbb{Z}_+ \text{ and } \alpha \notin 2\mathbb{Z}_+, \\ \left[\frac{n+\alpha}{2}\right], & \text{otherwise.} \end{cases}$$

Furthermore, if $m_{2k} = 0$, $k = 0, 1, \ldots, Q$, then G has nonzero integral if and only if either

(i) *$n, \alpha > -n$, are odd,*

(ii) *n and $0 > \alpha > -n$, are even, and $m_{n+\alpha} \ne 0$.*

(b) Let ϕ have representation (2.4b). Then, G is integrable only if $m_{2k} = 0$, $k = 0, 1, \ldots, R$, where

$$R = \begin{cases} \frac{n+\alpha}{2} - 1, & n, \alpha \in 2\mathbb{Z}_+, \\ \left[\frac{n+\alpha}{2}\right], & \text{otherwise.} \end{cases}$$

Furthermore, if $m_{2k} = 0$, $k = 0, 1, \ldots, R$, then G has nonzero integral if and only if $n, \alpha \in 2\mathbb{Z}_+$ and $m_{n+\alpha} \ne 0$.

Proof: These results are clear since $\|\xi\|^{2k}\triangle^j\delta(\xi) = 0$ if $k > j$. ■

References

1. Friedlander, F. G., *Introduction to the Theory of Distributions*, Cambridge University Press, 1982.
2. Jackson, I. R. H., Convergence properties of radial basis functions, Constr. Approx. **4** (1988), 243–264.
3. John, F., *Plane Waves and Spherical Means Applied to Partial Differential Equations*, Interscience, New York, 1955.
4. Jones, D. S., *Generalised Functions*, McGraw and Hill, London, 1966.
5. Levesley, J., Convolution kernels based on thin-plate splines, Technical Report 1993/8, Department of Mathematics and Computer Science, Univesity of Leicester.
6. Levesley, J., Xu, Y., Light, W. A., and Cheney E. W., Convolution Kernels for Approximation with Radial Basis Functions, Report 268, Center for Numerical Analysis, University of Austin, July 1, 1993.
7. Powell, M. J. D., The theory of radial basis functions in 1990, in *Advances in Numerical Analysis, vol. II*, W. A. Light (ed.), Oxford University Press, 1992, 105–210.
8. Rudin, W., *Real and Complex Analysis*, McGraw and Hill, 1966.
9. Stein, E. M., and Weiss, G., *Introduction to Fourier Analysis on Euclidean Spaces*, Princeton University Press, 1971.
10. Tranter, C. J., *Bessel Functions with some Physical Applications*, The English Universities Press, London, 1968.

J. Levesley
Department of Mathematics and Computer Science
University of Leicester
University Road
Leicester LE1 7RH
UK
jl1@mathematics-and-computer-science.leicester.ac.uk

Using Radial Functions on Compact Domains

Will Light

Abstract. It is already well-understood that approximation by radial functions is a useful practical tool, especially when the data is high-dimensional, or scattered. The theoretical understanding of the approximating power of these functions has improved dramatically over the last 10 years, particularly with the use of Fourier transform methods for data which is regularly distributed throughout $\mathbb{R}^n$. Good progress has now been made towards understanding a variety of aspects of approximating on compact domains, and using scattered data sites. We shall discuss a selection of these advances with particular references to our own interests in this area.

§1. Introduction

Our main purpose in this section is to introduce the problems, and to review very briefly some of the work by other authors, which is tangential to our later discussions. We then devote the remaining sections to a description of work which pertains to our main interest of approximating on compact domains using quasi-interpolants.

A radial function on $\mathbb{R}^n$ is any function of the form $x \mapsto \phi(\|x\|)$, where $\|\cdot\|$ is the Euclidean norm and ϕ is any real-valued function defined on $[0, \infty)$. In the 1970's Hardy [9] discovered that such functions were useful in the interpolation of scattered data. The usual situation is that if the data is defined at $x_1, x_2, \ldots, x_m$ in $\mathbb{R}^n$, then an interpolant of the form

$$x \mapsto \sum_{i=1}^{m} a_i \phi(\|x - x_i\|)$$

is used, where the $a_1, a_2, \ldots, a_m$ are chosen to satisfy the interpolation conditions

$$\sum_{i=1}^{m} a_i \phi(\|x_j - x_i\|) = d_j, \quad j = 1, 2, \ldots, m,$$

Wavelets, Images and Surface Fitting
P. J. Laurent, A. Le Méhauté, and L. L. Schumaker (eds.), pp. 351–370.

ISBN 1-56881-040-7.

and the $d_1, d_2, \ldots, d_m$ are real data values. It is sometimes beneficial to augment this radial approximating subspace of $C(\mathbb{R}^n)$ by low degree polynomials. The interpolation problem then usually takes the form

$$\sum_{i=1}^{m} a_i \phi(\|x_j - x_i\|) + \sum_{i=1}^{Q} b_i p_i(x_j) = d_j, \quad j = 1, 2, \ldots, m,$$
$$\sum_{i=1}^{m} a_i p_j(x_i) = 0, \quad j = 1, 2, \ldots, Q,$$

where Q is the dimension of the polynomial subspace which is augmenting the radial function subspace. Henceforward, we will assume that $p_1, \ldots, p_Q$ are a basis for π_q, the subspace of polynomials of total degree at most q. This system can be written as

$$\begin{pmatrix} A & P \\ P^T & 0 \end{pmatrix} \begin{pmatrix} a \\ b \end{pmatrix} = \begin{pmatrix} d \\ 0 \end{pmatrix}. \tag{$\dagger$}$$

In many cases it is a remarkable fact that this system of equations has a unique solution *independent of the location of the data points* – see Micchelli [13] and Schoenberg [16]. When one attempts to solve this system of equations there are conditioning problems, rather similar to those experienced with finite element methods for solving partial differential equations. The interested reader can look at Dyn, Levin and Rippa [7], Ward and Narcowich [14], and Baxter [4] to get started in this topic.

Our main interest is the approximating power of such interpolants. To set the problem in a manageable context, we let D be a compact set in $\mathbb{R}^n$. Let $\mathcal{N} = \{x_1, x_2, \ldots, x_m\}$ be a set of 'centres' in D. Let $B_r(x)$ denote the Euclidean ball of radius r around the point $x \in \mathbb{R}^n$. Then define the 'density' function

$$h_r(x) = \max_{y \in B_r(x)} \min_{z \in \mathcal{N}} \|y - z\|.$$

Somehow h measures the density of points in $\mathcal{N}$ around the point x in $\mathbb{R}^n$. Suppose $f \in C(\mathbb{R}^n)$ is a function to be approximated. Then our goal is to find an approximation s in the subspace

$$\mathcal{S} = \operatorname{span}\{\phi(\|\cdot - x_1\|), \ldots, \phi(\|\cdot - x_m\|)\}$$

such that

$$|f(x) - s(x)| \leq C h_r^k(x), \quad x \in D.$$

Here k will usually be a natural number depending to some extent on the differentiability of f, and C is a parameter which should be independent of x, k and $h_r(x)$. From such a pointwise estimate, there follows immediately the (weaker) estimate $\|f - s\|_{D,\infty} \leq C\|h_r\|_{D,\infty}^k$, where $\|\cdot\|_{D,\infty}$ is the usual supremum norm on D.

It is possible to consider a slightly different definition of density, which is given in terms of the set D alone. We take

$$h(D,\mathcal{N}) = \sup_{y\in D}\inf_{x\in\mathcal{N}} \|y-x\|.$$

Before using these density functions, we want to consider briefly a particularly elegant result in this area is due to Brown [5]. This work does not obtain convergence orders, but deals very nicely with density results. Let D be a compact subset of $\mathbb{R}^n$. Let $\phi \in C(\mathbb{R})$. Let H be the subspace of $C(D)$ given by

$$\operatorname{span}\{x \mapsto \phi(\|x-y\|) : y \in D\}.$$

For many of the common choices of ϕ, Brown proves that H is dense in $C(D)$. The argument is via the standard technique from functional analysis. If H is not dense in $C(D)$, then there must exist a non-trivial linear functional in $[C(D)]^*$ which annihilates all the elements in H. The proof then proceeds by showing that every such functional which annihilates H is necessarily trivial. The proof is quite closely linked to Micchelli's ideas about invertibility of the interpolation matrices, and one can see the connection illuminated in Light [11].

Two more approaches, which are closely related, and which do yield convergence orders, are due to Madych and Nelson [12] and Schaback and Wu [15]. Madych and Nelson created a variational theory for interpolation using translates of a prescribed function g. Let $\mathcal{D}(\mathbb{R}^n)$ be the space of infinitely differentiable, compactly supported functions on $\mathbb{R}^n$. Let

$$\mathcal{D}_\ell = \{\theta \in \mathcal{D} : \textstyle\int x^\alpha \theta(x)\,dx = 0 \text{ for all } |\alpha| < \ell.\}$$

It is assumed that $g \in C(\mathbb{R}^n)$ is conditionally positive definite of order ℓ. One now considers the set of all $f \in C(\mathbb{R}^n)$ such that there is a constant $c(f)$ for which

$$\int f(x)\theta(x)\,dx \le c(f)\left\{\int\int g(x-y)\theta(x)\theta(y)\,dxdy\right\}^{\frac{1}{2}}, \quad \text{for all } \theta \in \mathcal{D}_\ell.$$

The smallest constant $c^*(f)$ for which the above inequality holds defines a semi-norm on the set of admissible functions, with kernel $\pi_{\ell-1}$, the space of polynomials of total degree at most $\ell-1$. This is the setting for the variational theory. We will denote the set of admissible functions by $C_{g,\ell}$. There are several ways of defining conditional positive definiteness, but the essential property needed is a result from Gelfand and Vilenkin [8]. If g is conditionally positive of order ℓ, then there exists a positive Borel measure μ on $\mathbb{R}^n \setminus\{0\}$, constants a_γ, $\gamma \in \mathbb{Z}^n_+$, $|\gamma| \le 2\ell$ and a function χ in $\mathcal{D}(\mathbb{R}^n)$ such that:

1) $1-\widehat{\chi}(\nu)$ has a zero of order $2\ell+1$ at $\nu = 0$.
2) both of the integrals $\int_{0<\|\nu\|<1} \|\nu\|^{2\ell}\,d\mu(\nu)$ and $\int_{\|\nu\|\ge 1} d\mu(\nu)$ are finite.

3) for all $\theta \in \mathcal{D}(\mathbb{R}^n)$,

$$\int g(x)\theta(x)\,dx = \int \Big[\widehat{\theta}(\nu) - \widehat{\chi}(\nu) \sum_{|\gamma|<2\ell} D^\gamma \widehat{\theta}(0)\frac{\nu^\gamma}{\gamma!}\Big]\, d\mu(\nu) + \sum_{|\gamma|\le 2\ell} D^\gamma \widehat{\theta}(0)\frac{a_\gamma}{\gamma!}.$$

In [12] it is now assumed that there is a natural number $k > \ell$ such that

$$\int \|\nu\|^{2k}\, d\mu(\nu) < \infty.$$

Furthermore, one needs some mild assumptions on D. For brevity, we will assume that D satisfies a cone condition, although this is unnecessarily restrictive.

Theorem 1.1 (Madych and Nelson). *Let $\{\mathcal{N}_t\}_{t>0}$ be a sequence of sets of points in a compact subset D of $\mathbb{R}^n$, such that D satisfies a cone condition and $h(D, \mathcal{N}_t) \to 0$ as $t \to 0$. Let $f \in C_{g,\ell}$. Let $s_t \in \operatorname{span}\{h(\cdot - y) : y \in \mathcal{N}_t\}$ be such that s_t interpolates f on $\mathcal{N}_t$. Then $\|f - s_t\|_{D,\infty} = \mathcal{O}(t^k)$ as $t \to 0$.*

Example. If $g(x) = -\sqrt{1 + \|x\|^2}$, $x \in \mathbb{R}^n$, then g is conditionally positive definite of order 1. In this case, the parameter k can be arbitrarily large, so that arbitrarily high approximation orders are attainable.

This example is somewhat at variance with the results of Jackson [10] and Buhmann [6] for gridded data uniformly distributed on $\mathbb{R}^n$. There the maximum order achievable is roughly $\mathcal{O}(t^{n+1})$ for functions with sufficient smoothness. However, the key to this seeming discrepancy is that the space $C_{g,\ell}$ is a quite 'small' space consisting of functions which are intrinsically well-behaved from the point of view of interpolation.

The approach by Schaback and Wu yields similar results, but is rather different in character. We adopt the notation used in the setting up of (†). Let

$$R(x) = \big(\phi(\|x - x_1\|), \ldots, \phi(\|x - x_m\|)\big)$$

and

$$S(x) = \big(p_1(x), \ldots, p_Q(x)\big).$$

Then look for a pair of vectors

$$U(x) = \big(u_1(x), \ldots, u_m(x)\big), \qquad V(x) = \big(v_1(x), \ldots, v_Q(x)\big), \quad x \in D,$$

such that

$$\begin{pmatrix} A & P \\ P^T & 0 \end{pmatrix} \begin{pmatrix} U(x) \\ V(x) \end{pmatrix} = \begin{pmatrix} R(x) \\ S(x) \end{pmatrix}. \tag{‡}$$

Suppose this search is successful. Then, since

$$\begin{pmatrix} U(x) \\ V(x) \end{pmatrix} = \begin{pmatrix} A & P \\ P^T & 0 \end{pmatrix}^{-1} \begin{pmatrix} R(x) \\ S(x) \end{pmatrix},$$

each u_i is a linear combination of the functions $\phi(\|x - x_1\|), \ldots, \phi(\|x - x_m\|)$ and $p_1, \ldots, p_Q$. Furthermore, denoting the jth unit vector by e_j, $1 \le j \le m$, it is easily seen that

$$\begin{aligned} \begin{pmatrix} U(x_j) \\ V(x_j) \end{pmatrix} &= \begin{pmatrix} A & P \\ P^T & 0 \end{pmatrix}^{-1} \begin{pmatrix} R(x_j) \\ S(x_j) \end{pmatrix} \\ &= \begin{pmatrix} A & P \\ P^T & 0 \end{pmatrix}^{-1} \begin{pmatrix} A & P \\ P^T & 0 \end{pmatrix} e_j \\ &= e_j. \end{aligned}$$

It follows that $u_i(x_j) = \delta_{ij}$, $1 \le i, j \le m$. Now let f be a continuous, real-valued function on D. Consider the function

$$s(x) = \sum_{i=1}^{m} f(x_i) u_i(x), \quad x \in D.$$

It can be easily seen that s is the required interpolant to f, that is, $f(x_j) = s(x_j)$, $1 \le j \le m$.

The main idea now is to view the solution of (†) as defining the solution of the minimisation problem

$$\min\{U^T A U - 2U^T R(x) + \phi(0) : U \in \mathbb{R}^m,\, P^T U = S(x)\}.$$

Thus $V(x)$ is regarded as a Lagrange multiplier. With enough differentiability on ϕ one can differentiate (‡) and see that $U^{(\mu)}(x)$ satisfies

$$\begin{pmatrix} A & P \\ P^T & 0 \end{pmatrix} \begin{pmatrix} U^{(\mu)}(x) \\ V^{(\mu)}(x) \end{pmatrix} = \begin{pmatrix} R^{(\mu)} x) \\ S^{(\mu)}(x) \end{pmatrix}.$$

It is also true that $U^{(\mu)}(x)$ is the solution of the problem

$$\min\{{U^{(\mu)}}^T A U^{(\mu)} - 2{U^{(\mu)}}^T R(x) + \phi^{(2\mu)}(0) : U^{(\mu)} \in \mathbb{R}^m,\, P^T U^{(\mu)} = S^{(\mu)}(x)\}.$$

One now introduces the 'Kriging function'. Define

$$K_{\mu,Q}(x) = \left\{ u \in \mathbb{R}^m : \sum_{i=1}^{m} u_i p(x_i) = p^{(\mu)}(x), \text{ for all } p \in \pi_q \right\}.$$

The Kriging function κ satisfies

$$\kappa_Q^{(\mu)}(x) = \min\{{U^{(\mu)}}^T(x) A U^{(\mu)}(x) - 2{U^{(\mu)}}^T R(x) + \phi^{(2\mu)}(0) : U^{(\mu)} \in K_{\mu,Q}(x)\}.$$

One cannot do better now than direct the reader with a quote from [15]: it is there proved that

1) the interpolation error can be bounded by the Kriging function,
2) the Kriging function can be expressed as an integral,
3) this integral can be bounded by a power of $h_r(x)$.

Again, the net effect of [15] is that Hardy multiquadrics, inverse multiquadrics and Gaussian radial basis functions give arbitrarily high approximation order. Once again, there is a very limiting hypothesis on the admissible f for which such a convergence order is obtainable (essentially this is the same as the condition in [12]).

§2. One-dimensional Edge Effects

In this section, most of the work described is based on research by Beatson and Powell. These authors have two main papers [2], [3] on univariate multiquadric approximation, and both dwell at some length on a number of important aspects of this problem. Our purpose here is to extract virtually the simplest case of multiquadric quasi-interpolation on a finite, regular grid. This enables us to illustrate the fundamental principles at work in such a quasi-interpolant, without these same principles being obscured by the technicalities which are present in the two-dimensional case.

We begin with the univariate multiquadrics defined by $\theta : \mathbb{R} \to \mathbb{R}$, where $\theta(x) = \sqrt{x^2 + c^2}$, and $c \geq 0$. From these functions, we want to be able to construct a function $\psi : \mathbb{R} \to \mathbb{R}$ which is suitable for quasi-interpolation on the whole of $\mathbb{R}$. We take

$$\psi(x) = \tfrac{1}{2}\{\theta(x+1) - 2\theta(x) + \theta(x-1)\}, \quad x \in \mathbb{R},$$

which may be thought of as a finite-difference approximation to the second derivative of θ. With this observation in mind, it is easy to check that $|\psi(x)| = \mathcal{O}\left(|x|^{-3}\right)$ as $|x| \to \infty$. The aforementioned quasi-interpolant is

$$(L_h f)(x) = \sum_{\nu=-\infty}^{\infty} f(\nu h)\psi\left(\frac{x}{h} - \nu\right), \quad x \in \mathbb{R}.$$

Some restrictions must be placed on f for the image $L_h f$ to be well-defined. For simplicity, let us assume that $f \in C(\mathbb{R})$ and has at most linear growth at infinity, that is, there exists a constant $A > 0$ such that $|f(x)| \leq A(1 + |x|)$, $x \in \mathbb{R}$. Then the sum defining $L_h f$ is absolutely convergent for each value of x.

The following lemma, and particularly its simple proof, will turn out to be most instructive.

Lemma 2.1. *We have*

$$\sum_{\nu \in \mathbb{Z}} \psi(x + \nu) = 1, \quad \text{for all } x \in \mathbb{R}.$$

Proof: Fix $x \in \mathbb{R}$ and write

$$2\sum_{\nu\in\mathbb{Z}} \psi(x+\nu) = \lim_{N\to\infty} \sum_{\nu=-N+1}^{N-1} \{\theta(x+\nu+1) - 2\theta(x+\nu) + \theta(x+\nu-1)\}$$
$$= \lim_{N\to\infty} \{\theta(x+N) - \theta(x+N-1) + \theta(x-N) - \theta(x-N+1)\}.$$

Since $\theta(x) = \sqrt{x^2+c}$, we have that $\theta(x) \sim |x|$ as $x \to \infty$. Hence,

$$2\sum_{\nu\in\mathbb{Z}} \psi(x+\nu) = \lim_{N\to\infty} \{|x+N| - |x+N-1| + |x-N| - |x-N+1|\}$$
$$= \lim_{N\to\infty} \{(x+N) - (x+N-1) + (N-x) - (N-x-1)\}$$
$$= 2. \quad \blacksquare$$

Theorem 2.2. *There is a constant $A > 0$ such that for all $f \in C^{(1)}(\mathbb{R}) \cap W^1_\infty(\mathbb{R})$, $\|f - L_n f\|_\infty \le A\|f'\|_\infty$.*

Proof: We do not furnish the elementary and well-known details here. The interested reader should wait until later to see the simple Taylor series argument which establishes this result. $\blacksquare$

Although no details of the proof of Theorem 2.2 were given, it must be observed that one of the vital ingredients is Lemma 2.1, since from that result we deduce that $(L_h 1)(x) = 1$ for all $x \in \mathbb{R}$, where $1 : \mathbb{R} \to \mathbb{R}$ is the function defined by $1(x) = 1$ for all $x \in \mathbb{R}$.

We now shift our attention to the interval $[-1,1]$, and suppose that $f \in C[-1,1]$. We want to construct a quasi-interpolant Q_h which uses only the given information about f and produces a good approximation to f on $[-1,1]$. We want to avoid in particular extending f either to a finite interval containing $[-1,1]$ or to the whole of $\mathbb{R}$. From the proof of Lemma 2.1 we obtain immediately the following result.

Lemma 2.3. *For all $x \in \mathbb{R}$,*

$$\sum_{|\nu|\le N-1} \psi(x-\nu) + \tfrac{1}{2}\{\theta(x+N-1) - \theta(x+N) - \theta(x-N) + \theta(x-N+1)\} = 0.$$

We now define our quasi-interpolant by

$$(Q_h f)(x) = \sum_{\substack{\nu\in\mathbb{Z}\\ |\nu h|\le 1-h}} f(\nu h)\psi\left(\frac{x}{h}-\nu\right) + \tfrac{1}{2}f(1)\left\{1 + \theta\Big(\frac{x}{h}-\frac{1}{h}+1\Big) - \theta\Big(\frac{x}{h}-\frac{1}{h}\Big)\right\}$$
$$+ \tfrac{1}{2}f(-1)\left\{1 + \theta\Big(\frac{x}{h}+\frac{1}{h}-1\Big) - \theta\Big(\frac{x}{h}+\frac{1}{h}\Big)\right\}.$$

Notice that the first term on the right-hand side is the truncation of the "full" quasi-interpolant on $\mathbb{R}$, so that information about f is only drawn from $[-1+h, 1-h]$. The remaining terms are the two "edge effects".

Lemma 2.4. *Suppose $h^{-1} \in \mathbb{N}$. Then $(Q_h 1)(x) = 1$ for all $x \in \mathbb{R}$.*

Proof: Set $h^{-1} = M \in \mathbb{N}$. Then, for $x \in \mathbb{R}$,

$$\begin{aligned}(Q_h 1)(x) &= 1 + \sum_{\substack{\nu\in\mathbb{Z}\\ |\nu h|\le 1-h}} \psi\Big(\frac{x}{h}-\nu\Big)\\ &\quad + \frac{1}{2}\Big\{\theta\Big(\frac{x}{h}-\frac{1}{h}+1\Big)-\theta\Big(\frac{x}{h}-\frac{1}{h}\Big)+\theta\Big(\frac{x}{h}+\frac{1}{h}-1\Big)-\theta\Big(\frac{x}{h}+\frac{1}{h}\Big)\Big\}\\ &= 1 + \sum_{\nu=-M+1}^{M-1} \psi\Big(\frac{x}{h}-\nu\Big)\\ &\quad + \frac{1}{2}\Big\{\theta\Big(\frac{x}{h}-M+1\Big)-\theta\Big(\frac{x}{h}-M\Big)+\theta\Big(\frac{x}{h}+M-1\Big)-\theta\Big(\frac{x}{h}+M\Big)\Big\}\\ &= 1,\end{aligned}$$

by Lemma 2.3. ■

This result is supposed to illuminate the method by which the edge effects are constructed. Firstly, one adds to the truncated quasi-interpolant the appropriate terms

$$\theta\Big(\frac{x}{h}-\frac{1}{h}+1\Big),\ -\theta\Big(\frac{x}{h}-\frac{1}{h}\Big),\ \theta\Big(\frac{x}{h}+\frac{1}{h}-1\Big),\ -\theta\Big(\frac{x}{h}+\frac{1}{h}\Big).$$

Here one has the objective that for the function 1, these added terms should cancel completely the effect of the truncated sum, giving zero. Secondly, the two appearances of the number 1, one in each curly parenthesis comes from the fact that the terms

$$\theta\Big(\frac{x}{h}-\frac{1}{h}+1\Big)-\theta\Big(\frac{x}{h}-\frac{1}{h}\Big) \quad\text{and}\quad \theta\Big(\frac{x}{h}+\frac{1}{h}-1\Big)-\theta\Big(\frac{x}{h}+\frac{1}{h}\Big)$$

both "look like" -1 as x gets large. The motivation for this is perhaps not yet so clear, until we look in detail at the Taylor series argument.

Suppose $f \in C^{(1)}[-1,1]$. Define f_h by $f_h(x) = f(hx)$, $x \in [-h^{-1}, h^{-1}]$. Fix $x \in [-1,1]$ and define p_h to be the Taylor polynomial of degree 0 for f_h at x/h. Set $r_h = f_h - p_h$. Define $\widetilde{Q}$ on $C[-h^{-1}, h^{-1}]$ by

$$\begin{aligned}(\widetilde{Q}_h f)(x) = &\sum_{\substack{\nu\in\mathbb{Z}\\ |\nu h|\le 1-h}} f(\nu)\psi(x-\nu)\\ &+ \tfrac{1}{2} f(h^{-1})\Big\{1+\theta\Big(x-\frac{1}{h}+1\Big)-\theta\Big(x-\frac{1}{h}\Big)\Big\}\\ &+ \tfrac{1}{2} f(-h^{-1})\Big\{1+\theta\Big(x+\frac{1}{h}-1\Big)-\theta\Big(x+\frac{1}{h}\Big)\Big\}.\end{aligned}$$

Note that $(\widetilde{Q}_h 1)(x) = 1$ for all $x \in \mathbb{R}$ as before. Then

$$\begin{aligned}
|f(x) - (Q_h f)(x)| &= \left| f_h\left(\frac{x}{h}\right) - \left(\widetilde{Q}_h f_h\right)\left(\frac{x}{h}\right)\right| = \left|\left(\widetilde{Q}_h r_h\right)\left(\frac{x}{h}\right)\right| \\
&\le \sum_{\substack{\nu\in\mathbb{Z}\\ |\nu h|\le 1-h}} \left| r_h(\nu)\psi\left(\frac{x}{h}-\nu\right)\right| + \tfrac{1}{2}\left|r_h(h^{-1})\right|\left|1+\theta\Big(\frac{x}{h}-\frac{1}{h}+1\Big)-\theta\Big(\frac{x}{h}-\frac{1}{h}\Big)\right| \\
&\qquad + \tfrac{1}{2}\left|r_h\left(-h^{-1}\right)\right|\left|1+\theta\Big(\frac{x}{h}+\frac{1}{h}-1\Big)-\theta\Big(\frac{x}{h}+\frac{1}{h}\Big)\right| \\
&= \sum_{\substack{\nu\in\mathbb{Z}\\ |\nu h|\le 1-h}} \left|\left(\nu-\frac{x}{h}\right) f_h'(\xi_\nu)\psi\left(\frac{x}{h}-\nu\right)\right| \\
&\qquad + \tfrac{1}{2}\left|\left(h^{-1}-\frac{x}{h}\right) f_h'\left(\xi_{h^{-1}}\right)\left\{1+\theta\left(\frac{x-1}{h}+1\right)-\theta\left(\frac{x-1}{h}\right)\right\}\right| \\
&\qquad + \tfrac{1}{2}\left|\left(-h^{-1}-\frac{x}{h}\right) f_h'\left(\xi_{-h^{-1}}\right)\left\{1+\theta\left(\frac{x+1}{h}-1\right)-\theta\left(\frac{x+1}{h}\right)\right\}\right|,
\end{aligned}$$

where ξ_ν lies between ν and x/h, with similar assumptions about $\xi_{h^{-1}}$ and $\xi_{-h^{-1}}$. Now,

$$\begin{aligned}
|f(x) - (Q_h f)(x)| &\le \|f_h'\|_\infty \sum_{\nu\in\mathbb{Z}}\left|\left(\nu-\frac{x}{h}\right)\psi\left(\frac{x}{h}-\nu\right)\right| \\
&\quad + \tfrac{1}{2}\|f_h'\|_\infty\left|\frac{1-x}{h}\left\{1+\theta\left(\frac{x-1}{h}+1\right)-\theta\left(\frac{x-1}{h}\right)\right\}\right| \\
&\quad + \tfrac{1}{2}\|f_h'\|_\infty\left|\frac{1+x}{h}\left\{1+\theta\left(\frac{x+1}{h}-1\right)-\theta\left(\frac{x+1}{h}\right)\right\}\right|.
\end{aligned}$$

Also,

$$\sum_{\nu\in\mathbb{Z}}\left|\left(\frac{x}{h}-\nu\right)\psi\left(\frac{x}{h}-\nu\right)\right| \le \max_{y\in[0,1]}\sum_{\nu\in\mathbb{Z}}|(y-\nu)\psi(y-\nu)|$$

by periodicity. Hence there is a constant $A > 0$ such that

$$\begin{aligned}
&|f(x) - (Q_h f)(x)| \\
&\quad \le Ah\|f'\|_\infty\left\{1+\frac{1}{2}\left|\frac{1-x}{h}\left(1+\theta\Big(\frac{x-1}{h}+1\Big)-\theta\Big(\frac{x-1}{h}\Big)\right)\right|\right. \\
&\qquad\qquad \left.+\frac{1}{2}\left|\frac{1+x}{h}\left(1+\theta\Big(\frac{x+1}{h}-1\Big)-\theta\Big(\frac{x+1}{h}\Big)\right)\right|\right\}.
\end{aligned}$$

We would like to obtain $|f(x) - (Q_h f)(x)| \le Bh\|f'\|_\infty$ for some suitable constant $B > 0$. This entails showing

$$\left|\frac{1-x}{h}\left\{1+\theta\Big(\frac{x-1}{h}+1\Big)-\theta\Big(\frac{x-1}{h}\Big)\right\}\right|$$

and

$$\left|\frac{1+x}{h}\left\{1+\theta\left(\frac{x+1}{h}-1\right)-\theta\left(\frac{x+1}{h}\right)\right\}\right|$$

are uniformly bounded as $h \to 0$. At this point, we see why it is important that for example

$$\theta\left(\frac{x-1}{h}+1\right)-\theta\left(\frac{x-1}{h}\right)$$

closely resembles -1 as $h \to 0$. In fact, for a fixed x in $[-1,1]$ it is essential that this "resemblance" be $\mathcal{O}(h)$ as $h \to 0$. The next two lemmas examine this question.

Lemma 2.5. *Suppose $0 < h < 1$ and $x \in [-1, 1-h^*]$ where $h^* = (c+1)h$. Then*

$$\left|\left(\frac{1-x}{h}\right)\left(1+\theta\left(\frac{x-1}{h}+1\right)-\theta\left(\frac{x-1}{h}\right)\right)\right| \le \frac{1}{2}(1+c).$$

Proof: We can write

$$\left|\left(\frac{1-x}{h}\right)\left(1+\theta\left(\frac{x-1}{h}+1\right)-\theta\left(\frac{x-1}{h}\right)\right)\right| = \frac{1-x}{h}\left|1+\theta'(\delta)\right|$$

where

$$\frac{x-1}{h} < \delta < \frac{x-1}{h}+1.$$

Notice that

$$\delta < \frac{x-1}{h}+1 \le -c-1+1 = -c < 0.$$

Hence

$$\begin{aligned}\frac{1-x}{h}\left|1+\theta'(\delta)\right| &= \frac{1-x}{h}\left(1+\frac{\delta}{\sqrt{\delta^2+c^2}}\right)\\ &= \frac{1-x}{h}\left(1-\frac{|\delta|}{\sqrt{\delta^2+c^2}}\right)\\ &= \frac{1-x}{h}\left(1-\frac{1}{(1+c^2/\delta^2)^{1/2}}\right).\end{aligned}$$

We now utilise the inequality $(1+u^2)^{-1/2} \ge 1-\frac{1}{2}u^2$, $0 \le u < 1$. Observing that $|\delta| > c$, we can write

$$\begin{aligned}\frac{1-x}{h}\left|1+\theta'(\delta)\right| &\le \frac{1-x}{h}\left(1-\left\{1-\frac{c^2}{2\delta^2}\right\}\right)\\ &= \frac{1-x}{h}\frac{c^2}{2\delta^2}\\ &\le \left(1+\left(\frac{1-x}{h}-1\right)\right)\frac{c^2}{2\left(\frac{1-x}{h}-1\right)^2}\\ &= \frac{c^2}{2}\left(\left(\frac{1-x}{h}-1\right)^{-2}+\left(\frac{1-x}{h}-1\right)^{-1}\right)\\ &\le \frac{c^2}{2}\left(c^{-2}+c^{-1}\right) = \frac{1}{2}(1+c). \quad \blacksquare\end{aligned}$$

Lemma 2.6. *Suppose* $0 < h < 1$ *and* $x \in (1-h^*, 1]$ *where* $h^* = (c+1)h$. *Then*

$$\left|\left(\frac{1-x}{h}\right)\left(1+\theta\left(\frac{x-1}{h}+1\right)-\theta\left(\frac{x-1}{h}\right)\right)\right| \leq 2(c+1).$$

Proof: We can write

$$\left|\left(\frac{1-x}{h}\right)\left(1+\theta\left(\frac{x-1}{h}+1\right)-\theta\left(\frac{x-1}{h}\right)\right)\right| \leq \frac{h^*}{h}|1+\theta'(\delta)|$$

where $\frac{x-1}{h} < \delta < \frac{x-1}{h}+1$. Now, since $h^* = (c+1)h$,

$$\begin{aligned}\left|\left(\frac{1-x}{h}\right)\left(1+\theta\left(\frac{x-1}{h}+1\right)-\theta\left(\frac{x-1}{h}\right)\right)\right| &\leq (c+1)(1+|\theta'(\delta)|)\\ &\leq (c+1)\left(1+\sup_{\delta\in\mathbb{R}}\frac{|\delta|}{\sqrt{\delta^2+c^2}}\right) \leq 2(c+1).\end{aligned}$$

∎

Notice how the inequalities in Lemma 2.6 are much cruder than those in Lemma 2.5. In Lemma 2.5 we require $\theta\left(\frac{x-1}{h}\right)-\theta\left(\frac{x-1}{h}+1\right)$ to model 1 closely, in fact so that the difference between this term and 1 is $\mathcal{O}(h)$. This may be expressed as saying that the "edge effect" terms must be small far away from the edge to which they correspond (in this case the edge $x = 1$).

§3. Thin-plate Splines in $\mathbb{R}^2$

Having seen the underlying principles of edge effects in Section 2, we now go on to consider the more difficult case of thin-plate splines in $\mathbb{R}^2$. We are going to follow Beatson and Light [1], but give a rather different flavour from that presentation. We hope that this new approach will simplify the analytical difficulties of [1]. If $\alpha = (\alpha_1, \ldots, \alpha_n)$ is a multi-index in $\mathbb{Z}^2$, then $|\alpha| = \alpha_1 + \ldots + \alpha_n$. We will make some attempt to reserve the characters α, β, μ, ν for multi-integers. Define $(T_y f)(x) = f(x+y)$ for $x, y \in \mathbb{R}^n$. For any $\alpha \in \mathbb{Z}^n$, $\alpha \geq 0$, $\alpha = (\alpha_1, \ldots, \alpha_n)$ and for $x = (x_1, \ldots, x_n) \in \mathbb{R}^n$, V_α will denote the normalised monomial

$$V_\alpha(x) = \frac{x^\alpha}{\alpha!} = \frac{x_1^{\alpha_1}\cdots x_n^{\alpha_n}}{\alpha_1!\cdots\alpha_n!}.$$

The thin-plate spline radial basis theory begins with the function $\theta(x) = \|x\|^2\log(\|x\|^2)$, $x \in \mathbb{R}^2$. From this one forms a rapidly decreasing function $\psi \in C(\mathbb{R}^2)$ as follows. We take first the finite difference approximation δ to the Laplacian, where for $f \in C(\mathbb{R}^2)$,

$$(\delta^2 f)(x) = f(x-e_1) + f(x-e_2) - 4f(x) + f(x+e_1) + f(x+e_2), \quad x \in \mathbb{R}^2,$$

and $e_1 = (1,0)$ and $e_2 = (0,1)$. We let $\phi = \delta^2\theta$ and $\psi = \delta^2\phi$. Of course, ψ and ϕ are linear combinations of translates of θ. The precise nature of these combinations is best illustrated by the *stencils* for δ^2 and δ^4.

$$\begin{array}{ccc} & 1 & \\ 1 & -4 & 1 \\ & 1 & \end{array}$$

Figure 1. Stencil for δ^2.

$$\begin{array}{ccccc} & & 1 & & \\ & 2 & -8 & 2 & \\ 1 & -8 & 20 & -8 & 1 \\ & 2 & -8 & 2 & \\ & & 1 & & \end{array}$$

Figure 2. Stencil for δ^4.

One can easily verify that there is a constant $A > 0$ such that

$$|\psi(x)| \leq A(1 + \|x\|)^{-4}, \qquad x \in \mathbb{R}^2,$$

and so ψ is a decaying function at infinity. If we now define a quasi-interpolant

$$(Q_h f)(x) = \sum_{\nu \in \mathbb{Z}^2} f(\nu h)\psi\left(\frac{x}{h} - \nu\right), \qquad x \in \mathbb{R}^2,$$

then $Q_h f$ is certainly well-defined for functions $f \in C(\mathbb{R}^n)$ whose rate of growth at infinity is linear. In fact, if $f \in \pi_1$, then $Q_h f = f$. Again an elementary Taylor series argument similar to that in Section 2 shows that

$$\|f - Q_h f\|_\infty = \mathcal{O}(h^2 |\log h|) \quad \text{as} \quad h \to 0,$$

for all $f \in C^{(2)}(\mathbb{R}^2)$ whose derivatives of order not exceeding 2 are bounded.

The situation now feels very similar to that of Section 2. We have a function θ with polynomial growth at infinity. From θ we manufacture ψ by taking a second difference approximation to a suitable differential operator of order 2. In Section 2 the operator is the second derivative. Here it is the Laplacian. The function ψ has sufficiently rapid decay at ∞ for us to be able to construct a quasi-interpolant whose action on low degree polynomials is well-defined. The fact that ψ is constructed from θ via approximations to second-order differential operators allows us to deduce polynomial reproduction properties for the quasi-interpolant.

The next stage is to try to construct suitable "edge functions". Our aim will be modest, we want only to produce a quasi-interpolant on $C\left([-1,1]^2\right)$ which will reproduce constants. Again drawing inspiration from Section 2, we expect to be able to build the quasi-interpolant by "truncating" the operator Q_h so that function values outside $I = [-1,1]^2$ are not used, and then building in a few special functions which are mostly linear combinations of θ, and give the correct behaviour at each edge and corner.

However, we now introduce a twist on the exposition of Section 2. We ignore for the moment the fact that $\phi = \delta^2\theta$, and concentrate on the relationship $\psi = \delta^2\phi$. More or less the first step in Section 2 was to observe that

$$\sum_{\nu=-N+1}^{N-1} \{\theta(x + \nu + 1) - 2\theta(x + \nu) + \theta(x + \nu - 1)\}$$
$$= \theta(x + N) - \theta(x + N - 1) + \theta(x - N) - \theta(x - N + 1).$$

This occured in Lemma 2.1, and is effectively an evaluation of the truncation of $Q_h 1$. The following result was shown to us by Dr. M. Marletta. Similar ideas may be found in [7].

Lemma 3.1. *We have the following summation by parts formula:*

$$\begin{aligned}\sum_{|\nu|\le M-1} f(\nu)(\delta^2 g)(\nu) = -\sum_{|\nu|\le M-1} \Big[&\{f(\nu)-f(\nu-e_1)\}\{g(\nu)-g(\nu-e_1)\}\\ &+\{f(\nu)-f(\nu-e_2)\}\{g(\nu)-g(\nu-e_2)\}\Big]\\ +\sum_{\nu_2=-M+1}^{M-1}\Big[&f(M-1,\nu_2)\{g(M,\nu_2)-g(M-1,\nu_2)\}\\ &-f(-M,\nu_2)\{g(-M+1,\nu_2)-g(-M,\nu_2)\}\Big]\\ +\sum_{\nu_1=-M+1}^{M-1}\Big[&f(\nu_1,M-1)\{g(\nu_1,M)-(\nu_1,M-1)\}\\ &+f(\nu_1,-M)\{g(\nu_1,-M+1)-g(\nu_1,-M)\}\Big].\end{aligned}$$

The proof of this result is a straightforward and tedious calculation. We are interested in the case where $f(x) = 1$, $x \in \mathbb{R}$. The following result is immediate.

Lemma 3.2. *We have*

$$\begin{aligned}&\sum_{|\nu|\le M-1} (\delta^2 g)(\nu) =\\ &\sum_{\nu_2=-M+1}^{M-1} [\{g(M,\nu_2)-g(M-1,\nu_2)\}+\{g(-M,\nu_2)-g(-M+1,\nu_2)\}]\\ &+\sum_{\nu_1=-M+1}^{M-1} [\{g(\nu_1,M)-g(\nu_1,M-1)\}+\{g(\nu_1,-M)-g(\nu_1,-M+1)\}].\end{aligned}$$

It is sometimes helpful to think of Lemma 3.2 in a pictorial manner. If we write

$$\sum_{|\nu|\le M-1} \left(\delta^2 g\right)(\nu) = \sum_{|\nu|\le M} a_\nu g(\nu)$$

then the array of coefficients $\{a_\nu : |\nu| \le M\}$ has the following form

$$\begin{bmatrix} 0 & 1 & 1 & \dots & 1 & 1 & 0\\ 1 & -2 & -1 & \dots & -1 & -2 & 1\\ 1 & -1 & 0 & \dots & 0 & -1 & 1\\ \cdot & \cdot & \cdot & & \cdot & \cdot & \cdot\\ \cdot & \cdot & \cdot & & \cdot & \cdot & \cdot\\ \cdot & \cdot & \cdot & & \cdot & \cdot & \cdot\\ 1 & -1 & 0 & \cdots & 0 & -1 & 1\\ 1 & -2 & -1 & \cdots & -1 & -2 & 1\\ 0 & 1 & 1 & \cdots & 1 & 1 & 0 \end{bmatrix}.$$

The next step in an exploratory treatment of our question would be to try and assess the "edge effects". This was not done in Section 2. There a *fait accompli* was announced in Lemmas 2.5 and 2.6, where the edge effect

$$\theta\left(\frac{x}{h}-\frac{1}{h}\right)-\theta\left(\frac{x}{h}-\frac{1}{h}+1\right)$$

was asserted to be essentially like 1 as $h \to 0$. If we assume $h^{-1} = M \in \mathbb{N}$, then the assertion is that $\theta(y-M)-\theta(y-M+1)$ looks like 1 as $M \to \infty$.

We have to try and resolve this same question in 2 space dimensions. It would help greatly if we could assume that $\phi = \nabla^2\theta$ and $\psi = \delta^2\phi$, so that ϕ is the exact Laplacian of θ, while ψ is a discrete Laplacian of ϕ. The drawback is that

$$(\nabla^2\theta)(r) = 4(\log r^2 + 2), \quad r^2 = x^2+y^2,\ (x,y) \in \mathbb{R}^2 \setminus \{0\},$$

and so ϕ would have a singularity at 0. This is an irksome technicality. To overcome it, we will consider the function ϕ defined by $\phi(r) = \log r^2$, $r > 1/2$, $\phi(r) = \log(1/4)$, $0 \le r \le 1/2$.

It is elementary to check that there exists a constant $A > 0$ such that $|\psi(x)| \le A(1+\|x\|)^{-4}$, $x \in \mathbb{R}^2$, $\|x\|_\infty > 1/2$, so that ψ can be used to construct a quasi-interpolant. We consider

$$(L_h f)(x) = \sum_{\nu\in\mathbb{Z}^2} f(h\nu)\psi\left(\frac{x}{h}-\nu\right), \quad x \in \mathbb{R}^2.$$

Let us look initially at

$$(L_h 1)(x) = \sum_{\nu\in\mathbb{Z}^2}\psi\left(\frac{x}{h}-\nu\right) = \lim_{M\to\infty}\sum_{\substack{\nu\in\mathbb{Z}^2\\ \|\nu\|\le M}}\psi(y-\nu) = \lim_{M\to\infty}\sum_{\substack{\nu\in\mathbb{Z}^2\\ |\nu\|\le M}}(T_y\psi)(\nu)$$

where $y = x/h$. From Lemma 3.2 we have to evaluate a term of the form

$$A(M) = \sum_{\nu_2=-M+1}^{M-1}\{(T_y\phi)(M,\nu_2) - (T_y\phi)(M-1,\nu_2)\}.$$

Let us assume that the mean-value theorem can be invoked in the following manner

$$\begin{aligned} A(M) &\simeq \sum_{\nu_2=-M+1}^{M-1}\phi_u(y_1-M, y_2-\nu_2) \\ &\simeq \int_{-M}^{M}\frac{2(y_1-M)}{(y_1-M)^2+(y_2-v)^2}\,dv \\ &= 2\tan^{-1}\left(\frac{M+y_2}{M-y_1}\right) + 2\tan^{-1}\left(\frac{M-y_2}{M-y_1}\right). \end{aligned}$$

This can be regarded as an estimate of the right-hand edge effect. The other edge effects can be estimated as follows:

$$\begin{aligned}
\text{Left-hand}: &\quad 2\tan^{-1}\left(\frac{M+y_2}{M+y_1}\right)+2\tan^{-1}\left(\frac{M-y_2}{M+y_1}\right)\\
\text{Top}: &\quad 2\tan^{-1}\left(\frac{M+y_1}{M-y_2}\right)+2\tan^{-1}\left(\frac{M-y_1}{M-y_2}\right)\\
\text{Bottom}: &\quad 2\tan^{-1}\left(\frac{M+y_1}{M+y_2}\right)+2\tan^{-1}\left(\frac{M-y_1}{M+y_2}\right)
\end{aligned}$$

From this one sees that, for $y \notin \mathbb{Z}^2$, it is possible that

$$\lim_{M\to\infty} \sum_{\substack{\nu\in\mathbb{Z}^2\\ \|\nu\|\le M}} \psi(y-\nu) = 4\pi.$$

Accordingly, we should redefine L_h as

$$(L_h f)(x) = \frac{1}{4\pi}\sum_{\nu\in\mathbb{Z}^2} f(h\nu)\psi\left(\frac{x}{h}-\nu\right), \quad x\notin h\mathbb{Z}^2.$$

At the same time, we see that our sketchy analysis seems to indicate that for large M,

$$\begin{aligned}
&\sum_{\nu_2=-M+1}^{M-1} \{\phi(M-y_1,\nu_2-y_2)-\phi(M-1-y_1,\ \nu_2-y_2)\}\\
&\qquad\approx 2\tan^{-1}\left(\frac{M+y_2}{M-y_1}\right)+2\tan^{-1}\left(\frac{M-y_2}{M-y_1}\right).
\end{aligned}$$

This suggests the following definition for a quasi-interpolant.

Definition 3.3. *Let $x\in[-1,1]^2$, and suppose $x\notin h\mathbb{Z}^2$. Define*

$$\begin{aligned}
\phi_1^h(x_1,x_2) &= -\sum_{\nu_2=-M+1}^{M-1}\left\{\phi\left(\frac{x_1}{h}-M,\frac{x_2}{h}-\nu_2\right)-\phi\left(\frac{x_1}{h}-M+1,\frac{x_2}{h}-\nu_2\right)\right\}\\
&\qquad + 2\tan^{-1}\left(\frac{1+x_2}{1-x_1}\right)+2\tan^{-1}\left(\frac{1-x_2}{1-x_1}\right),\\
\phi_2^h(x_1,x_2) &= -\sum_{\nu_2=-M+1}^{M-1}\left\{\phi\left(\frac{x_1}{h}+M,\frac{x_2}{h}-\nu_2\right)-\phi\left(\frac{x_1}{h}+M-1,\frac{x_2}{h}-\nu_2\right)\right\}\\
&\qquad + 2\tan^{-1}\left(\frac{1+x_2}{1+x_1}\right)+2\tan^{-1}\left(\frac{1-x_2}{1+x_1}\right),\\
\phi_3^h(x_1,x_2) &= -\sum_{\nu_1=-M+1}^{M-1}\left\{\phi\left(\frac{x_1}{h}-\nu_1,\frac{x_2}{h}-M\right)-\phi\left(\frac{x_1}{h}-\nu_1,\frac{x_2}{h}-M+1\right)\right\}\\
&\qquad + 2\tan^{-1}\left(\frac{1+x_1}{1-x_2}\right)+2\tan^{-1}\left(\frac{1-x_1}{1-x_2}\right),
\end{aligned}$$

and

$$\phi_4^h(x_1,x_2) = -\sum_{\nu_1=-M+1}^{M-1}\left\{\phi\left(\frac{x_1}{h}-\nu_1,\frac{x_2}{h}+M\right)-\phi\left(\frac{x_1}{h}-\nu_1,\frac{x_2}{h}+M-1\right)\right\}$$
$$+2\tan^{-1}\left(\frac{1+x_1}{1+x_2}\right)+2\tan^{-1}\left(\frac{1-x_1}{1+x_2}\right).$$

Let $y_1=(1,x_2)$, $y_2=(-1,x_2)$, $y_3=(x_1,1)$ *and* $y_4=(x_1,-1)$. *For* $x\notin h\mathbb{Z}^2$, *set*

$$4\pi\left(Q_hf\right)(x) = \sum_{\substack{\nu\in\mathbb{Z}^2\\ \|\nu\|\le M-1}} f(h\nu)\psi\left(\frac{x}{h}-\nu\right)+\sum_{i=1}^4 f(y_i)\phi_i^h(x).$$

Theorem 3.4. *For all* $x\in[-1,1]^2$ *with* $x\notin h\mathbb{Z}^2$, $(Q_h1)(x)=1$.

Proof: Using Lemma 3.2, we see that, for $x\notin h\mathbb{Z}^2$,

$$\begin{aligned}4\pi\left(Q_h1\right)(x) &= 2\tan^{-1}\left(\frac{1+x_2}{1-x_1}\right)+2\tan^{-1}\left(\frac{1-x_2}{1-x_1}\right)+2\tan^{-1}\left(\frac{1+x_2}{1+x_1}\right)\\ &\quad+2\tan^{-1}\left(\frac{1-x_2}{1+x_1}\right)+2\tan^{-1}\left(\frac{1+x_1}{1-x_2}\right)+2\tan^{-1}\left(\frac{1-x_1}{1-x_2}\right)\\ &\quad+2\tan^{-1}\left(\frac{1+x_1}{1+x_2}\right)+2\tan^{-1}\left(\frac{1-x_1}{1+x_2}\right)\\ &= 4\pi. \quad\blacksquare\end{aligned}$$

Another Taylor series argument similar to that in Section 2 establishes the following result.

Theorem 3.5. *Adopt the notation of Definition 3.3 and set* $I=[-1,1]$. *Suppose that there exists a constant* $\Lambda>0$ *such that*

$$\frac{|y_i-x|^j}{h}|\phi_i(x)|\le\Lambda,\quad x\in I^2,\ 0<h<1,\ j=0,1,\ i=1,\dots,4.$$

Then for $f\in C^{(1)}\left(I^2\right)$, $\|f-Q_hf\|=\mathcal{O}(h)$.

We will examine the possibility of such a bound in the following lemmas.

Lemma 3.6. *Define* $g:[0,1]\times[-1,1]\to\mathbb{R}$ *by*

$$g(t,v)=\begin{cases}0, & \text{if } t^2+v^2=0\\ \dfrac{2t}{t^2+v^2}, & \text{if } t^2+v^2>0\end{cases},\quad t\in[0,1],\ v\in[-1,1].$$

Let $x_1,x_2\in[-1,1]$, $0<h<1$ *and* $\nu_2\in\{-M+1,\dots,M-1\}$. *If* $1-x_1>3h/2$, *then*

$$\phi\left(\frac{x_1}{h}-M,\frac{x_2}{h}-\nu_2\right)-\phi\left(\frac{x_1}{h}-M+1,\frac{x_2}{h}-\nu_2\right)$$
$$=-h\int_0^1 g(x_1-1+ht,x_2-\nu_2h)\,dt.$$

If $1 - x_1 \le 3h/2$ but $\nu_2 h \notin [x_2 - h/2, x_2 + h/2]$ then the same identity continued to hold.

Proof: Using the fundamental theorem of calculus gives

$$\begin{aligned}
&\phi\left(\frac{x_1}{h} - M + 1, \frac{x_2}{h} - \nu_2\right) \\
&\quad = \phi\left(\frac{x_1}{h} - M, \frac{x_2}{h} - \nu_2\right) + \int_{x_1/h-M}^{x_1/h-M+1} \phi_u\left(t, \frac{x_2}{h} - \nu_2\right) dt \\
&\qquad = \phi\left(\frac{x_1}{h} - M, \frac{x_2}{h} - \nu_2\right) + \int_0^1 \frac{2\,(x_1/h - M + t)}{(x_1/h - M + t)^2 + (x_2/h - \nu_2)^2}\, dt \\
&\qquad\quad = \phi\left(\frac{x_1}{h} - M, \frac{x_2}{h} - \nu_2\right) + \int_0^1 hg\,(x_1 - 1 + ht, x_2 - \nu_2 h)\, dt.
\end{aligned}$$

This process is legal provided

$$\left(\frac{x_1}{h} - M + 1\right)^2 + \left(\frac{x_2}{h} - \nu_2\right)^2 > \frac{1}{4} \quad \text{and} \quad \left(\frac{x_1}{h} - M\right)^2 + \left(\frac{x_2}{h} - \nu_2\right)^2 > \frac{1}{4}.$$

If $1 - x_1 > 3h/2$ then

$$\left|\frac{x_1}{h} - M + 1\right| \ge \frac{1 - x_1}{h} - 1 > \frac{3}{2} - 1 = \frac{1}{2},$$

while

$$\left|\frac{x_1}{h} - M\right| = \frac{1 - x_1}{h} > \frac{3}{2}.$$

If $\nu_2 h \notin [x_2 - h/2, x_2 + h/2]$, then $|x_2 - \nu_2 h| > h/2$. ■

Lemma 3.7. *Let $h > 0$, $\alpha \in \mathbb{R}$ and $N \in \mathbb{N}$. Let g be a non-negative, continuous function with a finite number m (possibly zero) of strong monotonicity changes on $[\alpha, \alpha + Nh]$. Then*

$$\left|\int_\alpha^{\alpha+Nh} g(y)dy - \sum_{\nu=0}^{N} hg(\alpha + \nu h)\right| \le (2m + 1)h\|g\|_\infty.$$

Proof: See Beatson and Light [1]. ■

Lemma 3.8. *Suppose $0 < h < 1$, $3h/2 < 1 - x_1 \le 2$ and $-1 \le x_2 \le 1$. Then*

$$\left|\int_{-1+h}^{1-h} g(x_1-1+ht, x_2-y)dy - \sum_{\nu_2=-M+1}^{M-1} hg(x_1-1+ht, x_2-\nu_2 h)\right| \le \frac{6h}{1 - x_1 - ht},$$

for $0 \le t \le 1$.

Proof: Apply Lemma 3.7. ■

Lemma 3.9. *Suppose* $0 < h < 1$, $3h/2 < 1 - x_1 \leq 2$, *and* $-1 \leq x_2 \leq 1$. *Then*

$$\frac{1-x_1}{h}\left|\tan^{-1}\left(\frac{1-h-x_2}{1-ht-x_1}\right) - \tan^{-1}\left(\frac{1-x_2}{1-x_1}\right)\right| \leq \frac{1}{2},$$

for $0 \leq t \leq 1$.

Proof: One simply applies the Mean Value Theorem in a similar way to that seen in [1]. ∎

Let us now review the progress made by Lemmas 3.6, 3.7, 3.8 and 3.9. Suppose for the moment $0 < h < 1$, $3h/2 < 1 - x_1 \leq 2$ and $-1 \leq x_2 \leq 1$. Then,

$$\begin{aligned}
|\phi_1(x)| &= \Bigg| - \sum_{\nu_2=-M+1}^{M-1} \left\{\phi\left(\frac{x_1}{h} - M, \frac{x_2}{h} - \nu_2\right) - \phi\left(\frac{x_1}{h} - M + 1, \frac{x_2}{h} - \nu_2\right)\right\} \\
&\qquad + 2\tan^{-1}\left(\frac{1+x_2}{1-x_1}\right) + 2\tan^{-1}\left(\frac{1-x_2}{1-x_1}\right)\Bigg| \\
&= \Bigg| \sum_{\nu_2=-M+1}^{M-1} \int_0^1 hg\left(x_1 - 1 + ht, x_2 - \nu_2 h\right) dt \\
&\qquad + 2\tan^{-1}\left(\frac{1+x_2}{1-x_1}\right) + 2\tan^{-1}\left(\frac{1-x_2}{1-x_1}\right)\Bigg| \\
&\leq \int_0^1 \Bigg| \sum_{\nu_2=-M+1}^{M-1} hg\left(x_1 - 1 + ht, x_2 - \nu_2 h\right) \\
&\qquad + 2\tan^{-1}\left(\frac{1+x_2}{1-x_1}\right) + 2\tan^{-1}\left(\frac{1-x_2}{1-x_1}\right)\Bigg| dt \\
&\leq \int_0^1 \Bigg| \sum_{\nu_2=-M+1}^{M-1} hg\left(x_1 - 1 + ht, x_2 - \nu_2 h\right) - \int_{-1+h}^{1-h} g\left(x_1 - 1 + ht, x_2 - y\right) dy\Bigg| dt \\
&\quad + \int_0^1 \Bigg| \int_{-1+h}^{1-h} g\left(x_1 - 1 + ht, x_2 - y\right) dy \\
&\qquad + 2\tan^{-1}\left(\frac{1+x_2}{1-x_1}\right) + 2\tan^{-1}\left(\frac{1-x_2}{1-x_1}\right)\Bigg| dt \\
&=: I_1 + I_2.
\end{aligned}$$

Now by Lemma 3.8,

$$I_1 \leq \int_0^1 (1-t)\frac{6h}{1-x_1-ht}\, dt \leq \frac{6h}{1-x_1}.$$

Furthermore,

$$\int_{-1+h}^{1-h} g\left(x_1 - 1 + ht, x_2 - y\right) dy$$

$$
\begin{aligned}
&= \int_{-1+h}^{1-h} \frac{2\,(x_1+ht)}{(x_1-1+ht)^2+(x_2-y)^2}\,dy \\
&= 2\left[\tan^{-1}\left(\frac{y-x_2}{x_1-1+ht}\right)\right]_{-1+h}^{1-h} \\
&= 2\tan^{-1}\left(\frac{1-h-x_2}{x_1-1+ht}\right) - 2\tan^{-1}\left(\frac{-1+h-x_2}{x_1-1+ht}\right) \\
&= -2\tan^{-1}\left(\frac{1-h-x_2}{1-x_1-ht}\right) - 2\tan^{-1}\left(\frac{1+x_2-h}{1-x_1-ht}\right).
\end{aligned}
$$

Now using Lemma 3.9,

$$
\begin{aligned}
\left(\frac{1-x_1}{h}\right) I_2 = \int_0^1 \frac{1-x_1}{h}\Bigg| &-2\tan^{-1}\left(\frac{1-h-x_2}{1-x_1-ht}\right) - 2\tan^{-1}\left(\frac{1+x_2-h}{1-x_1-ht}\right) \\
&+ 2\tan^{-1}\left(\frac{1+x_2}{1-x_1}\right) + 2\tan^{-1}\left(\frac{1-x_2}{1-x_1}\right)\Bigg|\,dt \\
\le 2.
\end{aligned}
$$

It now follows that for $0 < h < 1$, $3h/2 < 1-x_1 \le 2$ and $-1 \le x_2 \le 1$,

$$
\left(\frac{1-x_1}{h}\right)|\phi_1(x)| \le 8.
$$

There remains the strip $0 < h < 1$, $0 \le 1-x_1 \le 3h/2$ and $-1 \le x_2 \le 1$, and it is in this strip that we encounter the change in ϕ. We leave this computation for the interested reader to figure out!

References

1. Beatson, R.K. and W.A. Light, Quasi-interpolation with thin plate splines on a square, Constr. Approx. (to appear).
2. Beatson, R.K. and M.J.D. Powell, Univariate multiquadric approximation: quasi-interpolation to scattered data, Constructive Approximation **8** (1992), 275–288.
3. Beatson, R.K. and M.J.D. Powell, Univariate interpolation on a regular finite grid by a multiquadric plus a linear polynomial, J. Inst. Math. Applics. **12** (1992), 107-133.
4. Baxter, B.J.C., The interpolation theory of radial basis function, Ph.D. Thesis, University of Cambridge, 1992.
5. Brown, A.L., Uniform approximation by radial basis functions in Chapter 3, Appendix B in *Advances in numerical analysis Volume 2: Wavelets, subdivision algorithms and radial basis function*, ed. W.A. Light, Oxford University Press, 1992, 203-206.
6. Buhmann, M.D., Multivariate cardinal interpolation with radial basis functions, Constr. Approx. **6** (1990), 225-255.

7. Dyn, N., Levin, D. and S. Rippa, Numerical procedures for surface fitting of scattered data by radial functions, SIAM J. Sci. Stat. Comp. **7** (1986), 639-659.
8. Gelfand, I.M. and N. Ya. Vilenkin, Generalized functions, Vol. 4, Academic Press, New York, 1964.
9. Hardy, R.L., Multiquadric equations of topography and other irregular surfcase, J. Geophys. Res., **76** (1971), 1905-1915.
10. Jackson, I.R.H., Convergence properties of radial basis functions, Constr. Approx. **4** (1988), 243-264.
11. Light, W.A., Some aspects of radial basis function approximation, in *Approximation Theory, Spline Functions and Applications,* ed. S.P. Singh, Kluwer (Dordrecht), 1992, 163-212.
12. Madych, W. and S. Nelson, Multivariate interpolation and conditionally positive definite functions II, Math. Comp. **54** (1990), 211–230.
13. Micchelli, C.A., Interpolation of scattered data: distance matrices and conditionally positive definite functions, Constr. Approx. **2**, 11-22.
14. Narcowich, F.J. and J.D. Ward, Norms of inverses and condition numbers of matrices associated with scattered data, J. Approx. Th. **69** (1991), 69-94.
15. Schaback, R. and Z. Wu, Local error estimates for radial basis function interpolation of scattered data, J. Inst. Math. Applics. **13** (1993), 13-27.
16. Schoenberg, I.J., Metric spaces and completely monotone functions, Annals of Math. **39** (1938), 811-841.

Will Light
Department of Mathematics and Computer Science
University of Leicester, Leicester LE1 7RH
ENGLAND
pwl@mcs.le.ac.uk

Smoothing Noisy Data by Kriging with Nugget Effects

Pierre Montès

Abstract. In the problem of smoothing noisy data, a functional containing a term of smoothness of the smoothing function plus a term of fidelity of the smoothing function to the noisy data is to be minimized. In this functional a smoothing parameter which is a compromise between smoothness and fidelity can be obtained by minimizing the well-known Wahba's generalized cross-validation function $GCV(\lambda)$. The minimization of the GCV function is generally done without knowing the variance of the noise. In this paper, it is shown how to obtain a good estimate of the noise variance from the gross variogram of the noisy data before cross-validation.

§1. Introduction

This paper deals with the problem of smoothing noisy data. It is well known that a functional containing a term of smoothness of the smoothing function plus a term of fidelity of the smoothing function to the noisy data is to be minimized. This minimization is made for a given smoothing parameter which is a compromise between smoothness and fidelity. The optimum smoothing parameter can be obtained by minimizing the well-known Wahba's generalized cross-validation function $GCV(\lambda)$. The minimization of the GCV function is generally done without knowing the variance of the noise. But, in some cases, it is reported (e.g. Wahba (1990), Utreras (1993)) that the minimization of the GCV function yielded an unsatisfactory smoothing parameter: 0 *or* ∞. For these pathological cases, according to Wahba (1990), if even the order of magnitude of the error variance were known, it would be possible to find the good optimum parameter. In this paper, it is shown how to obtain a good estimate of the error variance from the gross variogram of the noisy data. In fact, the nugget effect of the variogram (the value at the origin, i.e. at 0^+), is a good estimate of the error variance. With this estimate of the error variance, the optimum smoothing parameter is obtained by minimization of an unbiased estimate of the true mean square error averaged over the data.

Wavelets, Images and Surface Fitting
P. J. Laurent, A. Le Méhauté, and L. L. Schumaker (eds.), pp. 371–378.

ISBN 1-56881-040-7.

§2. Background

2.1. Definition of the Problem

Given N observed data $z(u_i) = P(u_i) + \epsilon(u_i)$, $i = 1, \ldots, N$, where u_i belong to $\Omega \subseteq \mathbb{R}^n$ and $\epsilon(u_i)$ are independent normal random errors with zero mean and known or unknown variance σ^2 ($P(u)$ is only known to be smooth) we want to estimate $P \in H_m(\Omega)$, a reproducing kernel Hilbert space, to minimize the positive functional

$$\Phi(P_\lambda) = \frac{1}{N\lambda} \sum_{i=1}^{N} [z(u_i) - P_\lambda(u_i)]^2 + J_m^n(P_\lambda), \tag{1}$$

where the smoothing term is

$$J_m^n(P_\lambda) = \sum_{|\alpha|=m} \frac{m!}{\alpha_1! \ldots \alpha_n!} \int_{-\infty}^{\infty} \cdots \int_{-\infty}^{\infty} \left(\frac{\partial^m P_\lambda}{\partial x_1^{\alpha_1} \ldots \partial x_n^{\alpha_n}} \right)^2 \prod_j dx_j. \tag{2}$$

Here λ is the smoothing parameter, and

$$u = (x_1, \ldots, x_n) \in \Omega, \qquad |\, \alpha \,| = \alpha_1 + \ldots + \alpha_n.$$

2.2. The Solution

Given a value of the smoothing parameter $\lambda > 0$, the minimizer of Φ is

$$P_\lambda(u_0) = \sum_{i=1}^{N} b_i K(u_0, u_i) + \sum_{l=1}^{D(n,k)} a_l f_l(u_0), \qquad k = m - 1, \tag{3}$$

where

- $D(n,k) = \frac{(n+k)!}{n!k!}$, $D(n,k) \leq N$
- $f_l(u_0)$ are monomials of the coordinates of degree 0 up to k; they are components of the polynomial precision of the approximation
- $K(u_0, u_i) = K(h_{0i})$, $h_{0i} = \|u_0 - u_i\|$,

$$K(h) = \begin{cases} (-1)^{\frac{n}{2}+m+1} \, |h|^{2m-n} \, Log \, |h|^{2m-n}, & \text{if } 2m - n \text{ is even} \\ (-1)^m \, |h|^{2m-n}, & \text{otherwise.} \end{cases}$$

The function P_λ belongs to the space of functions having partial derivatives of total order m in $L_2(\mathbb{R}^n)$. In addition, the space endowed with the seminorm $J_m^n(P)$ must be a reproducing kernel Hilbert space, that is, the evaluation functionals must be bounded in the space. This condition is satisfied iff $2m - n > 0$.

With the solution P_λ given in (3), we have

$$J_m^n(P_\lambda) = \sum_{i=1}^{N}\sum_{j=1}^{N} b_i b_j K(u_i, u_j) \tag{4}$$

under the conditions

$$\sum_{i=1}^{N} b_i f_l(u_i) = 0. \tag{5}$$

The N coefficients b_i and the $D(n,k)$ coefficients a_l are the solution of the linear system of equations

$$\begin{pmatrix} K(u_i, u_j) + (N\lambda)\delta_{ij} & f_l(u_j) \\ f_l^T(u_i) & 0 \end{pmatrix} \begin{pmatrix} b_i \\ a_i \end{pmatrix} = \begin{pmatrix} z_j \\ 0 \end{pmatrix}. \tag{6}$$

In this system, δ_{ij} is the Kronecker delta.

The system (6) has a unique solution because the matrix K is positive definite under the condition (5), and the matrix f_l is of rank $D(n,k)$, the data points being distinct.

2.3. Primal Kriging and Dual Kriging

The solution $P_\lambda(u_0)$ given in equation (3) may be written by using a new inner product as

$$P_\lambda(u_0) = \langle \Psi \mid \Gamma \rangle = \Psi^T(M^{-1}\Gamma) = (\Psi^T M^{-1})\Gamma, \tag{7}$$

where

$$\Psi(u_0) = \{K(u_0, u_1), \ldots, K(u_0, u_N) : \ f_1(u_0), \ldots, f_{D(n,k)}(u_0)\}^T \tag{8}$$

and

$$M = \begin{pmatrix} K + (N\lambda)I & f_l \\ f_l^T & 0 \end{pmatrix} \tag{9}$$

$$\Gamma = \{z_1, \ldots, z_N : \ 0, \ldots, 0\}^T. \tag{10}$$

In the kriging matrix M, the term $(N\lambda)$ added to the diagonal of K is the *nugget effect* adjusted to the kernel $K(.,.)$. Usually, in Geostatistics, the nugget effect σ^2 is used in place of $(N\lambda)$, and the $K(.,.)$ terms are adjusted. Evidently, both adjustments are equivalent from the standpoint of estimation. So, it is evident that

$$P_\lambda(u_0) = (\Psi^T M^{-1})\Gamma = \sum_{i=1}^{N} p_i z(u_i) \tag{11}$$

is the well-known *primal kriging estimate* with coefficients $\{p_i\} = \{M^{-1}\Psi\}$, and

$$P_\lambda(u_0) = \Psi^T(M^{-1}\Gamma) = \sum_{i=1}^{N} b_i K(u_0, u_i) + \sum_{l=1}^{D(n,k)} a_l f_l(u_0) \tag{12}$$

is the well-known *dual kriging estimate* with coefficients $\{b_i, a_l\}^T = \{M^{-1}\Gamma\}$. The adjusted nugget effect $(N\lambda)$ is not used in $\Psi(.)$. We can see that $K(.,.)$ is the reproducing kernel for the Hilbert space $H_m(\Omega)$. The reproducing kernel $K(.,.)$ is k-conditionally positive definite. For more details about the theory summarized in this section, see for example, Matheron (1981), Duchon (1977), Dowd (1985), Wahba (1990), Attéia (1992).

2.4. Choosing the Optimum Smoothing Parameter

The solution P_λ obtained in Subsection 2.2 guarantees a minimum for $\Phi(P_\lambda)$ for all $\lambda > 0$. This minimum yields an estimate of P which is the best compromise between the smoothness of the solution and its fidelity to the data. However, the problem remains of choosing the best value of λ. Among the methods that have been proposed for choosing the optimum smoothing parameter, the most popular one is cross-validation as developed by Grace Wahba. She obtains (Wahba(1975)) that the optimum smoothing parameter minimizes the true mean square error averaged over the data points:

$$R(\lambda) = \frac{1}{N}\sum_{j=1}^{N}[P_\lambda(u_j) - P(u_j)]^2. \tag{13}$$

She defines the *influence matrix* $A(\lambda)$, which from the standpoint of geostatistics, I call the *matrix of the primal kriging coefficients* (discarding the nugget effects in the right-hand side of the primal kriging system $M\{p\} = \{\Psi\}$):

$$[A(\lambda)]\{z(u_i)\} = \{P_\lambda(u_j)\}, \quad \lambda > 0, \tag{14}$$

where

$$[A(\lambda)] = [M^{-1}\Psi(u_j)] = [\Psi^T(u_j)M^{-1}]. \tag{15}$$

Then, we take the mathematical expectation of (13):

$$E\ R(\lambda) = E\ \frac{1}{N}\|[P - A(\lambda)z]\ \|^2 = \frac{1}{N}\ \|\ [I - A(\lambda)P]\ \|^2 + \frac{\sigma^2}{N}Tr[A^2(\lambda)]. \tag{16}$$

Theorem 1. *(Craven and Wahba(1979)). An unbiased estimate of* $ER(\lambda)$ *is given by*

$$\hat{R}(\lambda) = \frac{1}{N}\|[I - A(\lambda)]z\ \|^2 - \frac{\sigma^2}{N}Tr[I - A(\lambda)]^2 + \frac{\sigma^2}{N}Tr[A^2(\lambda)]. \tag{17}$$

that is

$$E\ \hat{R}(\lambda) = E\ R(\lambda). \tag{18}$$

In practice, equation (17) is replaced by the equivalent

$$\hat{R}(\lambda) = \frac{1}{N}\|[I - A(\lambda)]z\|^2 + 2\left(\frac{\sigma^2}{N}\right) Tr[A(\lambda)] - \sigma^2. \tag{19}$$

According to Craven and Wahba(1979), the minimizer of $\hat{R}(\lambda)$ can be a good choice of the smoothing parameter, but it requires knowledge of the variance of the noise (nugget effect). Craven and Wahba (1979) obtain a good estimate of the minimizer of $E\ R(\lambda)$ which does not require the knowledge of σ^2: this is the generalized cross-validation (GCV) estimate of the optimum smoothing parameter taken as the minimizer of

$$GCV(\lambda) = \frac{\frac{1}{N}\|[I - A(\lambda)]z\|^2}{\{\frac{1}{N}Tr[I - A(\lambda)]\}^2}. \tag{20}$$

§3. Nugget Effect Estimation from Gross Variogram

In this section, a method is proposed to estimate the variance of the noise before cross-validation. This estimate can be used in the minimization of the function (19). We assume the following decomposition:

$$z(u) = P(u) + \epsilon(u) = [m(u) + Y(u)] + \epsilon(u),$$

where

- $m(u)$ is the mean of the phenomenon (*polynomial drift of degree k*);
- $Y(u)$ is the fluctuation around the mean;
- $\epsilon(u)$ is the noise.

We assume that

$$EY = E\ \epsilon = E\ Y\epsilon = E\ Ym = E\ m\epsilon = 0$$

and

$$E\epsilon(u+h)\epsilon(u) = 0 \quad \text{if} \quad h < 0 \quad \text{or} \quad h > 0, \quad \text{and} \quad E\epsilon^2 = \sigma^2$$

We define the *semi-variogram* (see, e.g., Journel and Huijbregts (1978), which we will call variogram for short:

$$\gamma(u, u+h) = \frac{1}{2}Var[z(u+h) - z(u)]$$

or

$$\gamma(u, u+h) = \frac{1}{2}E[z(u+h) - z(u)]^2 - \frac{1}{2}[m(u+h) - m(u)]^2 \tag{21}$$

But

$$\begin{aligned} E[z(u+h) - z(u)]^2 = & E[Y(u+h) - Y(u)]^2 + \\ & E[\epsilon(u+h) - \epsilon(u)]^2 + [m(u+h) - m(u)]^2. \end{aligned} \tag{22}$$

Then

$$\gamma(u, u+h) = \frac{1}{2}E[Y(u+h) - Y(u)]^2 + \frac{1}{2}E[\epsilon(u+h) - \epsilon(u)]^2. \tag{23}$$

We let

$$\gamma_R(h) = \frac{1}{2}E[Y(u+h) - Y(u)]^2, \quad \text{and} \quad \sigma^2 = \frac{1}{2}E[\epsilon(u+h) - \epsilon(u)]^2.$$

Then, equation (23) becomes

$$\gamma(u, u+h) = \gamma_R(h) + \sigma^2. \tag{24}$$

Consequently, equation (21) becomes

$$\frac{1}{2}E[z(u+h) - z(u)]^2 = \gamma_R(h) + \sigma^2 + \frac{1}{2}[m(u+h) - m(u)]^2. \tag{25}$$

The term on the left-hand side of equation (25) is often called abusively the *gross variogram*. If we designate it by $\gamma_z(h)$, then we can write

$$\gamma_z(h) = \gamma_R(h) + \sigma^2 + \frac{1}{2}E[m(u+h) - m(u)]^2. \tag{26}$$

Then we have

$$\gamma_z(0^+) = \sigma^2. \tag{27}$$

To estimate the gross variogram from the data, we define the *experimental gross variogram* as

$$\hat{\gamma}_z(h) = \frac{1}{2N(h)} \sum_{j=1}^{N(h)} [z(u+h) - z(u)]^2, \tag{28}$$

where $N(h)$ is the number of pairs of data points separated by the distance h. The first values of the experimental gross variogram in the vicinity of the origin allow us to estimate the variance of the noise if it is present in the data and is not negligible:

$$\hat{\sigma}^2 = \hat{\gamma}(0^+) \geq 0. \tag{29}$$

In general, the function $P(u)$ being smooth and the noise being relatively small, the gross variogram has a parabolic behavior at the origin. Therefore, it may be helpful to plot $\sqrt{\hat{\gamma}_z(h)}$, to draw the straight line passing through these points, and to find the value at the origin which is a good estimation of the square root of the noise variance. Moreover, it must be clearly understood that equations (27) and (29) remain rigorously exact even if the phenomenon is non-stationary (non constant mean), and that the nugget effect is the unique value that is taken from the gross variogram for subsequent use in the smoothing process i.e. the kriging system is built with the kernel (generalized covariance) $K(.,.)$, but never with the variogram.

§4. Minimization of Equation (19)

Once the value of σ^2 is estimated, it is convenient to proceed with the minimization of equation (19). We rewrite (19) as

$$\hat{R}(\lambda) = \frac{1}{N}\|[I - A(\lambda)]z\|^2 + 2\left(\frac{\hat{\sigma}^2}{N}\right) Tr[A(\lambda)] - \hat{\sigma}^2. \tag{30}$$

We can see that

$$\hat{R}(0) = \hat{\sigma}^2. \tag{31}$$

The minimization of equation (30) gives a good estimate of the optimum value of the smoothing parameter: $\lambda_{optimum}$. My numerical computations and theoretical results from Wahba (1975) allow us to write

$$0 \quad < \quad \hat{\lambda}_{optimum} \quad < \quad \lambda_{Reinsch}, \tag{32}$$

where $\lambda_{Reinsch}$ is the value of the smoothing parameter proposed earlier by Reinsch(1971) as the solution, when σ^2 is known, of the following equation:

$$\frac{1}{N}\sum_{i=1}^{N}[P_\lambda(u_i) - z(u_i)]^2 = \sigma^2. \tag{33}$$

To minimize equation (30), one can apply for example the golden section algorithm, like Utreras (1980) did with equation (20). The trace of the influence matrix, i.e. its first invariant, can be found by the sum of the eigenvalues of this matrix. These eigenvalues can be estimated by the method proposed by Utreras (1980, 1983) and Silverman (1984) when it is possible, or, if the noise is normally distributed, $\epsilon \sim N(0,1)$, the trace may be estimated by the method of Girard (Girard and Laurent(1987)) based on the generation of simulated vectors of errors that satisfy the equality

$$E\{\epsilon\}^T[A(\lambda)]\{\epsilon\} = Tr[A(\lambda)]. \tag{34}$$

§5. Applications

The method described above was applied to noisy data concerning curves and surfaces. The gross variogram computed in each case gives a very good estimation of the noise variance and the minimization of equations (20) and (30) gives practically the same optimum smoothing parameter. The computations carried out show that the idea proposed by Hall and Titterington (1987) to estimate the optimum smoothing parameter cannot yield any satisfactory result.

References

1. Atteia, M., *Hilbertian Kernels and Spline Functions*, Elsevier Science Publishers, Amsterdam, 1992.
2. Craven, P. and Wahba, G., Smoothing Noisy Data with Spline Functions: Estimating the Correct Degree of the Smoothing by the Method of Generalized Cross-Validation, Num. Math. **31** (1979), 377–403.
3. Dowd, P., A Review of Geostatistical Techniques for Contouring, in *Fundamental Algorithms for Computer Graphics*, Nato Asi Series, F17, R.A. Earnshaw (ed.), Springer-Verlag, 1985, 483–529.
4. Duchon, J., Splines Minimizing Rotation-Invariant Semi-norms in Sobolev Spaces, in *Constructive Theory of Functions of Several Variables*, Springer-Verlag, Berlin, 1977, 85–100.
5. Girard, D. and Laurent, P.J., Splines and Estimation of Nonlinear Parameters, in *Mathematical Methods in Computer-Aided Geometric Design*, T. Lyche and L. Schumaker, (eds.), Academic Press, New York, 1989, 273–298.
6. Hall, P. and Titterington, D., Common Structure of Techniques for Smoothing Parameters in Regression Problems, J. Roy. Statist. Soc., Ser. B, **49** (1987), 184–198.
7. Journel, A. and Huijbregts, C., *Mining Geostatistics*, Academic Press, London, 1978.
8. Matheron, G., Splines and Kriging: their Formal Equivalence, in *Down to the Earth Statistics*, D.F. Merriam, (ed.), Syracuse University, Geology Contributions 8, 1981, 77–95.
9. Reinsch, C., Smoothing by Spline Functions II, Num. Math. **16** (1971), 451–454.
10. Silverman, B., A Fast and Efficient Cross Validation Method for Smoothing Parameter Choice in Spline Regression, J. Amer. Statist. Assoc. **79** (1984), 584–589.
11. Utreras, F., Sur le choix du paramètre d'ajustement dans le lissage par les fonctions splines, Num. Math. **34** (1980), 15–28.
12. Utreras, F., Natural Spline Functions: their Associated Eigenvalue Problem, Num. Math. **42** (1983), 349–362.
13. Utreras, F., private communication, 1993.
14. Wahba, G., Smoothing Noisy Data by Spline Functions, Num. Math. **24** (1975), 383–393.
15. Wahba, G., *Spline Models for Observational Data*, CBMS-NSF Regional Conference Series in Applied Mathematics, 59, SIAM, Philadelphia, Pa., 1990.

Pierre Montès
Ecole Polytechnique de Montréal
Département de Génie Civil
5182 Snowdon, Montréal
Québec, CANADA H3W 2G1

Nonnegative Interpolation by Biquadratic Splines on Refined Rectangular Grids

B. Mulansky and J. W. Schmidt

Abstract. This paper is concerned with nonnegativity preserving bivariate C^1 interpolation to gridded data by biquadratic tensor product splines on a refined rectangular grid. The presented method requires the suitable specification of values for the first partial derivatives and the mixed second partial derivative at each data site. Sufficient nonnegativity conditions are derived utilizing the tensor product structure of the spline space and corresponding results for univariate quadratic splines with additional knots. Since the resulting system of linear inequalities for the parameters is separated with respect to the data sites, the proposed fit-and-modify approach reduces to the solution of independent local quadratic programs.

§1. Introduction

Several problems of shape preserving *bivariate* interpolation have received considerable attention during the last years, see the survey articles [5,11] and the references cited therein. In the present paper we will only consider the construction of *nonnegativity preserving* C^1 interpolants to data values given on a rectangular grid.

To be precise, given a rectangular grid

$$\Delta_x \times \Delta_y = \{x^1 < \cdots < x^n\} \times \{y^1 < \cdots < y^m\} \tag{1}$$

of data sites and corresponding *nonnegative* data values $z_{ij} \geq 0$, we are interested in a C^1 function s satisfying the interpolation conditions

$$s(x^i, y^j) = z_{ij}, \qquad i = 1, \ldots, n, \quad j = 1, \ldots, m \tag{2}$$

and preserving the nonnegativity on the domain Ω, *i.e.*,

$$s(x, y) \geq 0, \qquad (x, y) \in \Omega = [x^1, x^n] \times [y^1, y^m]. \tag{3}$$

Wavelets, Images and Surface Fitting
P. J. Laurent, A. Le Méhauté, and L. L. Schumaker (eds.), pp. 379–386.

ISBN 1-56881-040-7.

Depending on the practical background of the interpolation problem, the preservation of nonnegativity can be an essential property of the method. For example, if the data represent a material concentration or pressure, a probability density, or the progress of an irreversible process, negative values of an interpolant are physically meaningless.

The stated problem has been investigated recently by Schmidt [8] using rational biquadratic splines. Concerning the nonnegativity preserving interpolation to *scattered* data, we would like to mention the proposal by Mulansky & Schmidt [7] using quadratic Powell-Sabin splines and the variational approach by Utreras [10].

In our approach the interpolant is chosen from a space of biquadratic splines on a refined rectangular grid. Each rectangle of the original grid is divided into four subrectangles. The spline space is formed as the tensor product of univariate quadratic C^1 splines with additional knots. These univariate spline spaces were successfully used for convexity and monotonicity preserving interpolation by Schumaker [9] and for nonnegativity preserving interpolation by Lahtinen [6] and Mulansky & Schmidt [7]. In Asaturyan & Unsworth [1], the same tensor product space of biquadratic splines has been used to develop a bivariate monotonicity preserving interpolation method.

The proposed method requires the suitable specification of certain parameters, namely of the values of the first partial derivatives and the mixed second partial derivative at each data site. Our derivation of sufficient conditions for the nonnegativity of a biquadratic spline on the refined grid utilizes the tensor product structure of the spline space and is based on corresponding results for the univariate case [7]. This approach avoids the technical details concerning the explicit representation of the biquadratic spline. Furthermore, it can be applied to tensor products of other univariate spline spaces as well, which will be discussed in more generality in a forthcoming paper.

The derived sufficient nonnegativity conditions establish a system of linear inequalities for the parameters, which is separated with respect to the data sites. This system is always solvable for given nonnegative data values, which assures the existence of a nonnegative biquadratic spline interpolant. It is remarkable that the refinement of the grid can be fixed in advance independently of the data values.

In general there exists an infinite number of interpolants meeting the constraints. Since the interactive variation of the parameters is inappropriate, the choice of a suitable objective functional is essential. Here, the selection of a visually pleasant solution is based on a fit-and-modify approach. Due to the separation of the inequality system, this proposal reduces to the solution of independent small quadratic programs and results in a *local* nonnegativity preserving interpolation method.

§2. Interpolation by Biquadratic Splines

In order to explain the tensor product structure of our bivariate interpolation method, we start with a short description of univariate interpolation by quadratic C^1 splines with additional knots. Given a grid

$$\Delta_x : \ x^1 < x^2 < \cdots < x^n \tag{4}$$

in the x-direction, denote its step sizes by $h_x^i = x^{i+1} - x^i$, $i = 1, \ldots, n-1$. We define a refined grid

$$\tilde{\Delta}_x : \ x^1 < \xi^1 < x^2 < \cdots < x^{n-1} < \xi^{n-1} < x^n \tag{5}$$

using the additional knots

$$\xi^i = \beta_x^i x^i + \alpha_x^i x^{i+1}, \qquad i = 1, \ldots, n-1, \tag{6}$$

where the ratios $\alpha_x^i, \beta_x^i > 0$, $\alpha_x^i + \beta_x^i = 1$ are arbitrary, but fixed.

The $2n$-dimensional space of quadratic C^1 splines s_x on the refined grid $\tilde{\Delta}_x$ is denoted by $S_2^1(\tilde{\Delta}_x)$. We state the interpolation conditions using the linear functionals λ_x^i, $i = 1, \ldots, 2n$, on $S_2^1(\tilde{\Delta}_x)$, which are defined by

$$\begin{aligned} \lambda_x^i s_x &= s_x(x^i), && i = 1, \ldots, n, \\ \lambda_x^{n+i} s_x &= s_x'(x^i), && i = 1, \ldots, n, \end{aligned} \tag{7}$$

for all $s_x \in S_2^1(\tilde{\Delta}_x)$.

It follows from general results on spline interpolation [3] and is discussed more explicitly in [9] that, for given z_i, p_i, $i = 1, \ldots, n$, the following interpolation problem is uniquely solvable:

$$\begin{aligned} &\text{Find } s_x \in S_2^1(\tilde{\Delta}_x) \text{ such that} \\ &\lambda_x^i s_x = z_i, \qquad i = 1, \ldots, n, \\ &\lambda_x^{n+i} s_x = p_i, \qquad i = 1, \ldots, n. \end{aligned} \tag{8}$$

Analogously, let $\Delta_y : \ y^1 < y^2 < \cdots < y^m$ be a given grid in y-direction, which is refined to $\tilde{\Delta}_y : \ y^1 < \eta^1 < y^2 < \cdots < y^{m-1} < \eta^{m-1} < y^m$ by the additional knots $\eta^j = \beta_y^j y^j + \alpha_y^j y^{j+1}$, $j = 1, \ldots, m-1$, where $\alpha_y^j, \beta_y^j > 0$, $\alpha_y^j + \beta_y^j = 1$. The linear functionals λ_y^j, $j = 1, \ldots, 2m$ on $S_2^1(\tilde{\Delta}_y)$ are defined in the same way as in (7).

For the justification of the following tensor product constructions we refer to [3] and [4]. The $(2n)(2m)$-dimensional tensor product space

$$S(\tilde{\Delta}_x \times \tilde{\Delta}_y) = S_2^1(\tilde{\Delta}_x) \otimes S_2^1(\tilde{\Delta}_y) \tag{9}$$

is built up from all finite linear combinations of functions of the form $s_x \otimes s_y$, which are defined by

$$(s_x \otimes s_y)(x, y) = s_x(x) \cdot s_y(y), \qquad (x, y) \in \Omega = [x^1, x^n] \times [y^1, y^m]. \tag{10}$$

The space $S(\tilde{\Delta}_x \times \tilde{\Delta}_y)$ consists of biquadratic splines on the refined rectangular grid $\tilde{\Delta}_x \times \tilde{\Delta}_y$, which divides each rectangle of the grid $\Delta_x \times \Delta_y$ into four subrectangles. These splines are continuous and possess continuous first partial derivatives and continuous mixed second partial derivatives.

The tensor product functional $\lambda_x \otimes \lambda_y$ on $S(\tilde{\Delta}_x \times \tilde{\Delta}_y)$ of two linear functionals λ_x on $S_2^1(\tilde{\Delta}_x)$ and λ_y on $S_2^1(\tilde{\Delta}_y)$ is defined by linear extension from the basic rule

$$(\lambda_x \otimes \lambda_y)(s_x \otimes s_y) = \lambda_x s_x \cdot \lambda_y s_y, \qquad s_x \in S_2^1(\tilde{\Delta}_x), \quad s_y \in S_2^1(\tilde{\Delta}_y). \tag{11}$$

We consider the tensor product of the interpolation problem (8) with its analogue in y-direction. The tensor products of the linear functionals defined in (7) and its counterparts in y-direction can be easily identified as

$$\begin{aligned}
(\lambda_x^i \otimes \lambda_y^j)s &= s(x^i, y^j), && i = 1,\dots,n,\ j = 1,\dots,m,\\
(\lambda_x^{n+i} \otimes \lambda_y^j)s &= \frac{\partial s}{\partial x}(x^i, y^j), && i = 1,\dots,n,\ j = 1,\dots,m,\\
(\lambda_x^i \otimes \lambda_y^{m+j})s &= \frac{\partial s}{\partial y}(x^i, y^j), && i = 1,\dots,n,\ j = 1,\dots,m,\\
(\lambda_x^{n+i} \otimes \lambda_y^{m+j})s &= \frac{\partial^2 s}{\partial x \partial y}(x^i, y^j), && i = 1,\dots,n,\ j = 1,\dots,m,
\end{aligned} \tag{12}$$

for $s \in S(\tilde{\Delta}_x \times \tilde{\Delta}_y)$. Therefore, the tensor product interpolation problem reads:

$$\begin{aligned}
&\text{Find } s \in S(\tilde{\Delta}_x \times \tilde{\Delta}_y) \text{ such that}\\
&s(x^i, y^j) = z_{ij}, && i = 1,\dots,n,\ j = 1,\dots,m,\\
&\frac{\partial s}{\partial x}(x^i, y^j) = p_{ij}, && i = 1,\dots,n,\ j = 1,\dots,m,\\
&\frac{\partial s}{\partial y}(x^i, y^j) = q_{ij}, && i = 1,\dots,n,\ j = 1,\dots,m,\\
&\frac{\partial^2 s}{\partial x \partial y}(x^i, y^j) = r_{ij}, && i = 1,\dots,n,\ j = 1,\dots,m.
\end{aligned} \tag{13}$$

It follows from a general result in [3], that problem (13) is uniquely solvable for given $z_{ij}, p_{ij}, q_{ij}, r_{ij}$, $i = 1,\dots,n$, $j = 1,\dots,m$.

§3. Sufficient Nonnegativity Conditions

In order to derive our sufficient nonnegativity conditions we define the linear functionals μ_x^i, $i = 1,\dots,2n$ on $S_2^1(\tilde{\Delta}_x)$ by

$$\begin{aligned}
\mu_x^1 s_x &= s_x(x^1), \qquad \mu_x^{2n} s_x = s_x(x^n),\\
\mu_x^{2i} s_x &= 2s_x(x^i) + \alpha_x^i h_x^i s_x'(x^i), && i = 1,\dots,n-1,\\
\mu_x^{2i-1} s_x &= 2s_x(x^i) - \beta_x^{i-1} h_x^{i-1} s_x'(x^i), && i = 2,\dots,n,
\end{aligned} \tag{14}$$

for $s_x \in S_2^1(\tilde{\Delta}_x)$. The cone $C_x \subset S_2^1(\tilde{\Delta}_x)$ is defined by

$$C_x = \left\{ s_x \in S_2^1(\tilde{\Delta}_x) : \ \mu_x^i s_x \geq 0, \ i = 1, \ldots, 2n \right\}. \tag{15}$$

As established in [7], the cone C_x describes sufficient nonnegativity conditions for quadratic C^1 splines with additional knots, *i.e.*,

$$s_x \in C_x \ \Rightarrow \ s_x(x) \geq 0 \ \text{ for } \ x \in [x^1, x^n]. \tag{16}$$

The linear functionals $\mu_y^j, j = 1, \ldots, 2m$ on $S_2^1(\tilde{\Delta}_y)$, and the cone $C_y \subset S_2^1(\tilde{\Delta}_y)$ are introduced analogously.

Now, the projective cone $C = C_x \otimes C_y \subset S(\tilde{\Delta}_x \times \tilde{\Delta}_y)$ is formed by

$$C = \left\{ \sum_{\ell=1}^{r} s_x^\ell \otimes s_y^\ell : \ r \in \mathbb{N}, s_x^\ell \in C_x, s_y^\ell \in C_y \right\}. \tag{17}$$

It is obvious from the definition that the cone C gives sufficient nonnegativity conditions for biquadratic splines, *i.e.*,

$$s \in C \quad \Rightarrow \quad s(x,y) \geq 0 \ \text{ for } \ (x,y) \in \Omega. \tag{18}$$

Moreover, the cone C can be characterized by a finite number of linear inequalities using the tensor products of the functionals μ_x^i and μ_y^j.

Theorem 1. *For $s \in S(\tilde{\Delta}_x \times \tilde{\Delta}_y)$, the property $s \in C$ is equivalent to*

$$(\mu_x^i \otimes \mu_y^j)s \geq 0, \qquad i = 1, \ldots, 2n, \quad j = 1, \ldots, 2m. \tag{19}$$

Proof: Let us assume $s \in C$. Then s has a representation

$$s = \sum_{\ell=1}^{r} s_x^\ell \otimes s_y^\ell, \ s_x^\ell \in C_x, s_y^\ell \in C_y,$$

which implies

$$(\mu_x^i \otimes \mu_y^j)s = \sum_{\ell=1}^{r} \mu_x^i s_x^\ell \cdot \mu_y^\ell s_y^\ell \geq 0$$

for $i = 1, \ldots, 2n, j = 1, \ldots, 2m$.

Conversely, since the functionals $\mu_x^i, \ i = 1, \ldots, 2n$ are linearly independent, there exists a dual basis $\{B_x^k, \ k = 1, \ldots, 2n\}$ of $S_2^1(\tilde{\Delta}_x)$ such that $\mu_x^i B_x^k = \delta_{ik}$, which implies $B_x^k \in C_x$. Using a corresponding basis $\{B_y^\ell, \ \ell = 1, \ldots, 2m\}$ of $S_2^1(\tilde{\Delta}_y)$, the tensor products $B_x^i \otimes B_y^\ell, \ i = 1, \ldots, 2n, j = 1, \ldots, 2m$ form a basis of $S(\tilde{\Delta}_x \times \tilde{\Delta}_y)$. Since any $s \in S(\tilde{\Delta}_x \times \tilde{\Delta}_y)$ has the representation

$$s = \sum_{i=1}^{2n} \sum_{j=1}^{2m} (\mu_x^i \otimes \mu_y^j)s \cdot (B_x^i \otimes B_y^j),$$

we get $s \in C$ whenever (19) holds. ∎

Using Theorem 1 and the particular form of the functionals μ_x^i, μ_y^j, and of their tensor products, a system of linear inequalities is established, which is sufficient for the nonnegativity of the solution of the interpolation problem (13).

Theorem 2. *Suppose the data* $z_{ij}, p_{ij}, q_{ij}, r_{ij},\ i = 1,\ldots,n, j = 1,\ldots,m$ *satisfy the inequalities*

$$z_{ij} \geq 0 \quad \text{for} \quad i = 1, n,\ j = 1, m, \tag{20}$$

at the corners of the domain Ω,

$$\begin{aligned} &2z_{ij} + \alpha_x^i h_x^i p_{ij} \geq 0 \quad \text{for} \quad i = 1,\ldots,n-1,\ j = 1, m, \\ &2z_{ij} - \beta_x^{i-1} h_x^{i-1} p_{ij} \geq 0 \quad \text{for} \quad i = 2,\ldots,n,\ j = 1, m, \\ &2z_{ij} + \alpha_y^j h_y^j q_{ij} \geq 0 \quad \text{for} \quad i = 1, n,\ j = 1,\ldots,m-1, \\ &2z_{ij} - \beta_y^{j-1} h_y^{j-1} q_{ij} \geq 0 \quad \text{for} \quad i = 1, n,\ j = 2,\ldots,m, \end{aligned} \tag{21}$$

on the boundary of Ω, *and*

$$\begin{aligned} &4z_{ij} + \alpha_x^i h_x^i p_{ij} + \alpha_y^j h_y^j q_{ij} + \alpha_x^i h_x^i \alpha_y^j h_y^j r_{ij} \geq 0 \\ &\qquad \text{for } i = 1,\ldots,n-1,\ j = 1,\ldots,m-1, \\ &4z_{ij} - \beta_x^{i-1} h_x^{i-1} p_{ij} + \alpha_y^j h_y^j q_{ij} - \beta_x^{i-1} h_x^{i-1} \alpha_y^j h_y^j r_{ij} \geq 0 \\ &\qquad \text{for } i = 2,\ldots,n,\ j = 1,\ldots,m-1, \\ &4z_{ij} + \alpha_x^i h_x^i p_{ij} - \beta_y^{j-1} h_y^{j-1} q_{ij} - \alpha_x^i h_x^i \beta_y^{j-1} h_y^{j-1} r_{ij} \geq 0 \\ &\qquad \text{for } i = 1,\ldots,n-1,\ j = 2,\ldots,m, \\ &4z_{ij} - \beta_x^{i-1} h_x^{i-1} p_{ij} - \beta_y^{j-1} h_y^{j-1} q_{ij} + \beta_x^{i-1} h_x^{i-1} \beta_y^{j-1} h_y^{j-1} r_{ij} \geq 0 \\ &\qquad \text{for } i = 2,\ldots,n,\ j = 2,\ldots,m, \end{aligned} \tag{22}$$

in the interior of Ω. *Then the solution of the interpolation problem (13) is nonnegative, i.e.,*

$$s(x, y) \geq 0 \quad \text{for} \quad (x, y) \in \Omega. \tag{23}$$

A closer inspection of the system (20)–(22) of linear inequalities shows its separation with respect to the data sites (x^i, y^j). A set of four inequalities involving only the data $z_{ij}, p_{ij}, q_{ij}, r_{ij}$ corresponds to each data site (x^i, y^j), $i = 1,\ldots,n$, $j = 1,\ldots,m$, respectively. These sets are independent of each other.

§4. Existence and Construction of Nonnegative Interpolants

Now we return to the interpolation problem (1), (2) stated in the introduction. The data values $z_{ij} \geq 0$, $i = 1,\ldots,n$, $j = 1,\ldots,m$ are assumed to be given, while we can dispose of the parameters p_{ij}, q_{ij}, r_{ij}, $i = 1,\ldots,n$, $j = 1,\ldots,m$ in order to preserve the nonnegativity. Therefore, we will try to find a suitable solution of the system (21), (22) of sufficient nonnegativity conditions. Explicit C^1 conditions do not occur in the present situation. As pointed out above, this system can be written in the form

$$(p_{ij}, q_{ij}, r_{ij}) \in G_{ij}, \qquad i = 1,\ldots,n, \quad j = 1,\ldots,m. \tag{24}$$

Each of the sets $G_{ij} \subset \mathbb{R}^3$ is described by four of the inequalities (21), (22) (respectively, three inequalities in the cases $i = 1, n$ and $j = 1, m$, *i.e.*, at the four corners of the domain Ω).

The solvability of the system (24), *i.e.*, the *existence* of a nonnegative interpolant, is always assured by the feasible choice

$$(p_{ij}, q_{ij}, r_{ij}) = (0, 0, 0), \qquad i = 1, \ldots, n, \quad j = 1, \ldots, m. \tag{25}$$

To obtain a visually more pleasant solution, we propose a fit-and-modify approach, which utilizes the separation of the nonnegativity conditions, compare [2], [7]. Initial estimates $p_{ij}^0, q_{ij}^0, r_{ij}^0$ for the parameters, which define an initial interpolant s^0, are computed as usual approximations by divided differences, see [1] for details. These estimates have to be modified as little as possible in order to fulfill the sufficient nonnegativity conditions, which leads to the following optimization problem:

$$\begin{aligned} &\text{Minimize} \sum_{i=1}^{n} \sum_{j=1}^{m} [(p_{ij} - p_{ij}^0)^2 + (q_{ij} - q_{ij}^0)^2 + (r_{ij} - r_{ij}^0)^2] \\ &\text{subject to } (p_{ij}, q_{ij}, r_{ij}) \in G_{ij}, \qquad i = 1, \ldots, n, \quad j = 1, \ldots, m. \end{aligned} \tag{26}$$

This problem reduces to $n \cdot m$ independent quadratic programs involving three variables only, which can be efficiently solved by standard methods of quadratic optimization.

Concerning the invariance of the interpolant with respect to scalings of the variables x and y, it seems to be more appropriate to replace the objective functional in (26) by

$$\sum_{i=1}^{2n} \sum_{j=1}^{2m} [(\mu_x^i \otimes \mu_y^j)(s - s^0)]^2. \tag{27}$$

Omitting the details, we only mention that this functional results in the same advantage of the separation into local problems.

Another proposal is to minimize the bivariate Holladay functional

$$\int_\Omega \left(\frac{\partial^4 s}{\partial^2 x \partial^2 y} \right)^2 dxdy, \tag{29}$$

or the thin plate functional

$$\int_\Omega \left[\left(\frac{\partial^2 s}{\partial^2 x} \right)^2 + 2 \left(\frac{\partial^2 s}{\partial x \partial y} \right)^2 + \left(\frac{\partial^2 s}{\partial^2 y} \right)^2 \right] dxdy \tag{28}$$

with respect to the constraints (24). The resulting quadratic optimization problems do not separate using these functionals, and the special structure of the functionals and constraints should be utilized in a solution method. However, these functionals could give some hints on the introduction of additional weights in the fit-and-modify approach.

Finally, we would like to mention that the presented results can be easily extended to range restricted interpolation involving constant lower and upper bounds for the splines on Ω.

Once the parameters have been obtained, the interpolant s can be computed efficiently by tensor product methods as described in [3]. Further applications of Theorem 1 and results of our numerical experiments will be reported elsewhere.

References

1. Asaturyan, S. and K. Unsworth, A C^1 monotonicity preserving surface interpolation scheme, in *Mathematics of Surfaces III*, D. C. Handscomb (ed.), Oxford University Press, Oxford, 1989, 243–266.
2. Beatson, R. K. and H. Wolkowicz, Post-processing piecewise cubics for monotonicity, SIAM J. Numer. Anal. **26** (1989), 480–502.
3. de Boor, C., *A Practical Guide to Splines*, Springer-Verlag, New York, 1978.
4. Ewald, S., B. Mulansky, and H. Mühlig, Bivariate interpolating and smoothing tensor product splines, in *Splines in Numerical Analysis*, J. W. Schmidt and H. Späth (eds.), Akademie-Verlag, Berlin, 1989, 55–68.
5. Fontanella, F., Shape preserving interpolation, in *Computation of Curves and Surfaces*, W. Dahmen, M. Gasca, and C. A. Micchelli (eds.), Kluwer Academic Publ., Dordrecht, 1990, 183–214.
6. Lahtinen, A., Positive Hermite interpolation by quadratic splines, SIAM J. Math. Anal. **24** (1993), 223–233.
7. Mulansky, B. and J. W. Schmidt, Nonnegative C^1 interpolation of scattered data using Powell-Sabin splines, preprint MATH-NM-02-1993, TU Dresden, 1993.
8. Schmidt, J. W., Positive, monotone, and S-convex C^1-interpolation on rectangular grids, Computing **48** (1992), 363–371.
9. Schumaker, L. L., On shape preserving quadratic spline interpolation, SIAM J. Numer. Anal. **20** (1983), 854–864.
10. Utreras, F. I., Positive thin plate splines, J. Approximation Theory Applications **1** (1985), 77–108.
11. Utreras, F. I., Constrained surface construction, in *Topics in Multivariate Approximation*, C. K. Chui, L. L. Schumaker, and F. I. Utreras (eds.), Academic Press, New York, 1987, 233–254.

Bernd Mulansky
Jochen W. Schmidt
Institut für Numerische Mathematik
Technische Universität
D-01062 Dresden, GERMANY
mulansky@ math.tu-dresden.de

Spline Wavelets with Higher Defect

G. Plonka

Abstract. In this paper a generalized multiresolution analysis, generated by cardinal B-splines of degree m and defect r, is considered. Using the cardinal Hermite fundamental splines of degree $2m+1$ and defect r new spline wavelets with defect r are represented. In constrast with other papers dealing with wavelets with higher defect (cf. [3, 4]) the two-scale symbol $\mathbf{Q}_m^r$ of the wavelet vector can explicitly be given.

§1. Introduction

The subject of this paper is a natural generalization of the concept of interpolatory spline wavelets introduced in [2, p. 177]. Let $\{V_j^m\}$ $(j \in \mathbb{Z})$ be the multiresolution analysis of $L_2(\mathbb{R})$ generated by the cardinal B-spline N_m of degree m. Further, with $\{W_j^m\}$ $(j \in \mathbb{Z})$ we denote the sequence of wavelet spaces, in the sense that

$$V_{j+1}^m = V_j^m \oplus W_j^m,$$

where $\oplus$ indicates the orthogonal summation.

Let $\mathcal{T} := \{z \in \mathbb{C}, |z| = 1\}$. With the help of the Euler–Frobenius polynomial of degree $2m+1$

$$\Phi_{2m+1}^1(z) := \sum_{l=\infty}^{\infty} N_{2m+1}(l) z^l \quad (z \in \mathcal{T})$$

we introduce the cardinal fundamental spline L_{2m+1} of degree $2m+1$

$$L_{2m+1}(w) := \frac{1}{2\pi} \int_{-\infty}^{\infty} \frac{\hat{N}_{2m+1}(u)}{\Phi_{2m+1}^1(e^{-iu})} e^{-iuw} \, du$$

satisfying

$$L_{2m+1}(n) = \delta_{0n} \quad (n \in \mathbb{Z})$$

Wavelets, Images and Surface Fitting
P. J. Laurent, A. Le Méhauté, and L. L. Schumaker (eds.), pp. 387–398.

ISBN 1-56881-040-7.

with the Kronecker symbol δ. For $m \in \mathbb{N}$ the *interpolatory wavelet* $\psi_{I,m}$ is defined by

$$\psi_{I,m} := \mathrm{D}^{m+1} L_{2m+1}(2 \cdot -1),$$

where D denotes the differential operator. Then $\psi_{I,m}$ generates the wavelet spaces W_j^m ($j \in \mathbb{Z}$)

$$W_j^m := \operatorname{clos}_{L_2}\left(\operatorname{span}\{\psi_{I,m}(2^j \cdot -l);\ l \in \mathbb{Z}\}\right)$$

(cf. [2], p. 178).

We want to generalize this concept in the following way. Let $m \in \mathbb{N}_0$ and $r \in \mathbb{N}$ be given integers. We consider equidistant knots of multiplicity r

$$x_l^r := \left\lfloor \frac{l}{r} \right\rfloor, \tag{1.1}$$

where $\lfloor x \rfloor$ means the integer part of $x \in \mathbb{R}$.

Let $N_k^{m,r} \in C^{m-r}(\mathbb{R})$ ($r \leq m$, $k \in \mathbb{Z}$) denote the normalized B-splines of degree m and defect r with the knots $x_k, \ldots, x_{k+m+1}$. For $r = m+1$ the B-splines $N_k^{m,m+1}$ ($k = 0, \ldots, m$) coincide with the well-known Bernstein polynomials. According to the distribution theory, let $N_k^{m,r}$ be defined for $r > m+1$ and $k = 0, \ldots, r-m-2$ as follows

$$N_k^{m,r} := \frac{1}{r-1-k}\,\mathrm{D}^{r-m-2-k}\,\delta, \tag{1.2}$$

where δ denotes the Dirac distribution.

Using the ideas in [3, 4] in Section 2, we shall consider the *generalized multiresolution analysis* $\{V_j^{m,r}\}$ ($j \in \mathbb{Z}$) *of multiplicity* r of $L_2(\mathbb{R})$ generated by the linearly independent scaling functions $N_k^{m,r}$ ($k = 0, \ldots, r-1$), that is

$$V_j^{m,r} := \operatorname{clos}_{L_2}\left(\operatorname{span}\{N_k^{m,r}(2^j \cdot -l);\ k = 0, \ldots, r-1\}\right). \tag{1.3}$$

In particular, an explicit formula for the two-scale symbol $\mathbf{P}_m^r$ of the B-spline vector $\mathbf{N}_m^r := (N_k^{m,r})_{k=0}^{r-1}$ can be given (cf. [7]).

Let $\{W_j^{m,r}\}$ ($j \in \mathbb{Z}$) denote the sequence of wavelet spaces determined by

$$V_{j+1}^{m,r} = V_j^{m,r} \oplus W_j^{m,r}.$$

In Section 3 we shall introduce the cardinal Hermite fundamental splines $L_k^{2m+1,r}$ ($k = 0, \ldots, r-1$) satisfying for $n \in \mathbb{Z}$ the interpolation conditions

$$\mathrm{D}^\nu L_k^{2m+1,r}(n) = \delta_{0n}\,\delta_{\nu k} \quad (\nu, k = 0, \ldots, r-1). \tag{1.4}$$

We put

$$\psi_k^{m,r} := \mathrm{D}^{m+1} L_k^{2m+1,r}(2 \cdot -1) \quad (k = 0, \ldots, r-1).$$

In constrast with [3, 4] we can firstly give an explicit formula for the two–scale symbol $\mathbf{Q}_m^r$ of the wavelet vector $\mathbf{\Psi}_m^r := (\psi_k^{m,r})_{k=0}^{r-1}$. This two–scale symbol $\mathbf{Q}_m^r$ can be used for the computation of the *wavelets* $\psi_k^{m,r}$ $(k = 0, \ldots, r-1)$ *with defect* r as well as for deriving the Riesz basis property in the wavelet space $W_0^{m,r}$. For $r \geq m+1$ the wavelets $\psi_k^{m,r}$ $(k = 0, \ldots, r-1)$ are compactly supported, for $r \leq m$ they have exponential decay.

Note that in [3] other wavelets are constructed, which are derived from special compactly supported splines, firstly introduced in [9].

In Section 4 we show the close connection between the wavelet space $W_0^{m,r}$ and the subspace $V_{1,0}^{2m+1,r} \subset V_1^{2m+1,r}$, which contains splines with degree $2m+1$ and defect r satisfying some interpolation conditions. Analogous assertions for the simple case $r = 1$ can be found in [2].

Finally, in Section 5 the obtained formulas are applied to the case of cubic spline wavelets $(m = 3)$ with defect $r = 2$.

§2. Multiresolution Analysis of Multiplicity r

For a summary of basic properties of B-splines with multiple knots we refer to [1, 7]. Here we recall only the following important relations.

Let $\hat{\mathbf{N}}_m^r := (\hat{N}_k^{m,r})_{k=0}^{r-1}$ be the vector of Fourier transformed B-splines

$$\hat{N}_k^{m,r} := \int_{-\infty}^{\infty} N_k^{m,r}(x)\, e^{-i \cdot x}\, \mathrm{d}x.$$

For the Fourier transformed B-spline vector $\hat{\mathbf{N}}_m^r$ of length $r > m+1 \geq 1$ we find by (1.2)

$$\hat{\mathbf{N}}_m^r(u) = \Big(\frac{(iu)^{r-m-2}}{r-1}, \ldots, \frac{(iu)^0}{m+1}, \hat{\mathbf{N}}_m^{m+1}(u)^{\mathrm{T}} \Big)^{\mathrm{T}},$$

where $\mathbf{N}_m^{m+1}$ denotes the vector of the $m+1$ Bernstein polynomials of degree m. Further, we put

$$\hat{\mathbf{N}}_{-1}^r(u) := \Big(\frac{(iu)^{r-1}}{r-1}, \ldots, \frac{(iu)^1}{1}, 1 \Big)^{\mathrm{T}} \quad (u \in \mathbb{R}). \tag{2.1}$$

Then for $m \in \mathbb{N}_0$ and $r \in \mathbb{N}$ the following recursion relation can be found:

$$(iu)\, \hat{\mathbf{N}}_m^r(u) = \mathbf{A}_m^r(e^{-iu})\, \hat{\mathbf{N}}_{m-1}^r(u) \quad (u \in \mathbb{R}). \tag{2.2}$$

The (r, r)-matrices $\mathbf{A}_m^r(z)$ $(z \in \mathcal{T})$ are defined for $m > r-1$ by

$$\mathbf{A}_m^r(z) := m \begin{pmatrix} \frac{1}{x_m^r} & -\frac{1}{x_{m+1}^r} & \cdots & 0 & 0 \\ 0 & \frac{1}{x_{m+1}^r} & \cdots & 0 & 0 \\ \vdots & \vdots & \ddots & \vdots & \vdots \\ 0 & 0 & \cdots & \frac{1}{x_{m+r-2}^r} & -\frac{1}{x_{m+r-1}^r} \\ -\frac{z}{x_m^r} & 0 & \cdots & 0 & \frac{1}{x_{m+r-1}^r} \end{pmatrix}, \tag{2.3}$$

where x^r_{m+k} $(k = 0, \ldots, r-1)$ are given in (1.1). For $m = r-1 > 0$ let

$$\mathbf{A}_m^{m+1}(z) := m \begin{pmatrix} 1 & -1 & \ldots & 0 & 0 \\ 0 & 1 & \ldots & 0 & 0 \\ \vdots & \vdots & \ddots & \vdots & \vdots \\ 0 & 0 & \ldots & 1 & -1 \\ -z & 0 & \ldots & 0 & 1 \end{pmatrix} \tag{2.4}$$

and for $0 \le m < r-1$

$$\mathbf{A}_m^r(z) := \begin{pmatrix} \mathbf{I}_{r-m-1} & \mathbf{0} \\ \mathbf{0} & \mathbf{A}_m^{m+1}(z) \end{pmatrix}, \tag{2.5}$$

where $\mathbf{A}_0^1(z) := 1 - z$. Further, $\mathbf{I}_{r-m-1}$ denotes the $(r-m-1)$-th unit matrix and $\mathbf{0}$ a zero matrix (cf. [7]).

Note that $\det \mathbf{A}_m^r(z) = c_m^r\,(1-z)$. The Fourier transformed *two–scale relation of* $\mathbf{N}_m^r$ is given by

$$\hat{\mathbf{N}}_m^r = \mathbf{P}_m^r(e^{-i\cdot/2})\,\hat{\mathbf{N}}_m^r(\cdot/2) \quad (m \in \mathbb{N}_0,\ r \in \mathbb{N}).$$

The *two–scale symbol (or refinement mask) of* $\mathbf{N}_m^r$

$$\mathbf{P}_m^r(z) := \frac{1}{2} \sum_{n=-\infty}^{\infty} \mathbf{P}_n\, z^n \quad (z \in \mathcal{T}) \tag{2.6}$$

is a finite sum and satisfies for $m \ge 0$ the following recursion formula

$$\begin{aligned} \mathbf{P}_m^r(z) &= \frac{1}{2}\mathbf{A}_m^r(z^2)\,\mathbf{P}_{m-1}^r(z)\,\mathbf{A}_m^r(z)^{-1} \quad (z \in \mathcal{T},\ z \ne 1) \\ \mathbf{P}_m^r(1) &= \frac{1}{2}\lim_{u \to 0} \mathbf{A}_m^r(e^{-2iu})\,\mathbf{P}_{m-1}^r(e^{-iu})\,\mathbf{A}_m^r(e^{-iu})^{-1} \quad (u \in \mathbb{R}) \end{aligned} \tag{2.7}$$

with $\mathbf{A}_m^r(z)$ defined in (2.3) – (2.5) and

$$\mathbf{P}_{-1}^r(z) := \operatorname{diag}\,(2^{r-1}, \ldots, 2^0)^{\mathrm{T}} \tag{2.8}$$

(cf. [7]). In particular, the two–scale symbol $\mathbf{P}_m^r(z)$ is a matrix polynomial in z with

$$\det \mathbf{P}_m^r(z) = 2^{-rm+r(r-3)/2}\,(1+z)^{m+1} \quad (z \in \mathcal{T}).$$

The functions $N_k^{m,r}(\cdot - l)$ form a Riesz basis (or $L_2(\mathbb{R})$-stable basis) of $V_0^{m,r}$ (cf. [1]). The Riesz basis property is equivalent to the assertion that the *autocorrelation symbol* $\mathbf{\Phi}_m^r$, defined by

$$\mathbf{\Phi}_m^r(e^{-iu}) := \sum_{n=-\infty}^{\infty} \hat{\mathbf{N}}_m^r(u + 2\pi n)\,\hat{\mathbf{N}}_m^r(u + 2\pi n)^\star \tag{2.9}$$

with $\hat{\mathbf{N}}_m^r(u)^\star := \overline{\hat{\mathbf{N}}_m^r(u)^{\mathrm{T}}}$ is positive definite (cf. [4, 5, 7]).

Further, we introduce the following *Euler–Frobenius matrix*

$$\mathbf{H}^r_{2m+1} := (H^\nu_k)^{r-1}_{\nu,k=0} \tag{2.10}$$

with

$$H^\nu_k(z) := \sum_{l=-\infty}^{\infty} \mathrm{D}^\nu N^{2m+1,r}_k(l) z^l \quad (k,\, \nu = 0, 1, \ldots, r-1,\, z \in \mathcal{T}). \tag{2.11}$$

For $2m+1-\nu \leq r-1$, the functions $\mathrm{D}^\nu N^{2m+1,r}_k$ are understood according to the distribution theory. For $r = 1$ we obtain the well-known Euler–Frobenius polynomial

$$\mathbf{H}^1_{2m+1}(z) = H^0_{2m+1}(z) = \sum_{l=-\infty}^{\infty} N_{2m+1}(l)\, z^l.$$

By the Poisson summation formula the matrix $\mathbf{H}^r_{2m+1}$ reads for $z = e^{-iu}$ as follows

$$\mathbf{H}^r_{2m+1}(e^{-iu}) = \sum_{l=-\infty}^{\infty} \Big((i(u+2\pi l))^k\Big)^{r-1}_{k=0} \hat{\mathbf{N}}^r_{2m+1}(u+2\pi l)^{\mathrm{T}} \quad (u \in \mathbb{R}). \tag{2.12}$$

For $m \in \mathbb{N}_0$ and $r \in \mathbb{N}$ we have the following relationship:

$$\begin{aligned} \mathbf{\Phi}^r_m(z) &= \mathbf{D}^r_{m,0}(z)\, \mathbf{D}_r\, \overline{\mathbf{H}^r_{2m+1}(z)}\, (\mathbf{D}^r_{m,1}(z)^\star)^{-1} \quad (z \in \mathcal{T},\, z \neq 1), \\ \mathbf{\Phi}^r_m(1) &= \lim_{u \to 0} \mathbf{D}^r_{m,0}(e^{-iu})\, \mathbf{D}_r\, \mathbf{H}^r_{2m+1}(e^{iu})\, (\mathbf{D}^r_{m,1}(e^{-iu})^\star)^{-1} \quad (u \in \mathbb{R}) \end{aligned} \tag{2.13}$$

with

$$\begin{aligned} \mathbf{D}^r_{m,0}(z) &:= \mathbf{A}^r_m(z)\, \mathbf{A}^r_{m-1}(z) \ldots \mathbf{A}^r_0(z), \\ \mathbf{D}^r_{m,1}(z) &:= \mathbf{A}^r_{2m+1}(z)\, \mathbf{A}^r_{2m}(z) \ldots \mathbf{A}^r_{m+1}(z), \end{aligned} \tag{2.14}$$

where the (r,r)-matrices $\mathbf{A}^r_k$ $(k = 0, \ldots, 2m+1)$ are defined in (2.3) – (2.5),

$$\mathbf{D}_r := (-1)^{m+1} \begin{pmatrix} 0 & 0 & \ldots & 0 & \frac{(-1)^{r-1}}{r-1} \\ 0 & 0 & \ldots & \frac{(-1)^{r-2}}{r-2} & 0 \\ \vdots & \vdots & & \vdots & \vdots \\ 0 & (-1)^1 & \ldots & 0 & 0 \\ (-1)^0 & 0 & \ldots & 0 & 0 \end{pmatrix} \quad (r > 1)$$

and $\mathbf{D}_1 := (-1)^{m+1}$. In particular, the invertibility of the autocorrelation matrix $\mathbf{\Phi}^r_m(z)$ for $z \in \mathcal{T}$ causes the invertibility of the Euler–Frobenius matrix $\mathbf{H}^r_{2m+1}(z)$ for $z \in \mathcal{T}$ (cf. [7]).

Since $V_j^{m,r}$ contains the space $V_j^{m,1}$ generated by the cardinal B-spline N_m (cf. [2]), it follows that

$$\operatorname{clos}_{L_2} \bigcup_{j=-\infty}^{\infty} V_j^{m,r} = L_2(\mathbb{R}).$$

The Riesz basis property and the partition of unity property for the B-splines $N_k^{m,r}(\cdot - l)$ $(k = 0, \ldots, r-1,\ l \in \mathbb{Z})$ also lead to

$$\bigcap_{j=-\infty}^{\infty} V_j^{m,r} = \{0\}.$$

Thus, the sequence $\{V_j^{m,r}\}$ $(j \in \mathbb{Z})$ generates a multiresolution analysis of $L_2(\mathbb{R})$ with multiplicity r (cf. [3,4]).

§3. The Wavelet Space $W_j^{m,r}$

In this section we want to find spline wavelets with defect r

$$\psi_k^{m,r} \qquad (k = 0, \ldots, r-1), \tag{3.1}$$

such that the integer translates of (3.1) form a Riesz basis of the wavelet space $W_0^{m,r} := V_1^{m,r} \ominus V_0^{m,r}$.

First we want to introduce cardinal Hermite fundamental splines. Let $\hat{\mathbf{L}}_{2m+1}^r := (\hat{L}_k^{2m+1,r})_{k=0}^{r-1}$ be the Fourier transformed vector of spline functions $L_k^{2m+1,r}$ $(k = 0, \ldots, r-1)$ defined by

$$\hat{\mathbf{L}}_{2m+1}^r(u) := (\mathbf{H}_{2m+1}^r(e^{-iu})^{\mathrm{T}})^{-1}\, \hat{\mathbf{N}}_{2m+1}^r(u), \tag{3.2}$$

and $\mathbf{L}_{2m+1}^r := (L_k^{m,r})_{k=0}^{r-1}$. Then we have:

Theorem 3.1. *The spline functions $L_k^{2m+1,r}$ $(k = 0, \ldots, r-1)$ are cardinal Hermite fundamental splines in $V_0^{2m+1,r}$, i.e., for $n \in \mathbb{Z}$ the interpolation conditions*

$$\mathrm{D}^\nu L_k^{2m+1,r}(n) = \delta_{0n}\, \delta_{\nu k} \quad (\nu, k = 0, \ldots, r-1) \tag{3.3}$$

are satisfied.

Proof: By $\mathcal{W}$ we define the Wiener class. Since $\det \mathbf{H}_{2m+1}^r(z) \in \mathcal{W}$, it follows that there exists a representation

$$[\mathbf{H}_{2m+1}^r(e^{-iu})]^{-1} = \sum_{n=-\infty}^{\infty} \mathbf{H}_n\, e^{-iun},$$

where the elements of the (r,r)-matrices $\mathbf{H}_n$ $(n \in \mathbb{Z})$ lie in l_1. Thus, (3.2) implies that the functions $L_k^{2m+1,r}$ $(k = 0, \ldots, r-1)$ are contained in $V_0^{2m+1,r}$.

It remains to show that the interpolation conditions (3.3) hold. Putting

$$[\mathrm{D}^\nu \mathbf{L}^r_{2m+1}]^\sim := \sum_{n=-\infty}^{\infty} [\mathrm{D}^\nu \mathbf{L}^r_{2m+1}]^\wedge(\cdot + 2\pi n)$$

it follows by the Poisson summation formula that

$$[\mathrm{D}^\nu \mathbf{L}^r_{2m+1}]^\sim = \sum_{l=-\infty}^{\infty} \mathrm{D}^\nu \mathbf{L}^r_{2m+1}(l)\, e^{-i\cdot l}.$$

Therefore we have to show that $[\mathrm{D}^\nu \mathbf{L}^r_{2m+1}]^\sim \equiv \mathbf{e}_\nu$, where $\mathbf{e}_\nu := (\delta_{k\nu})_{k=0}^{r-1}$ are the unit vectors. By

$$[\mathrm{D}^\nu \mathbf{N}^r_{2m+1}]^\sim = \sum_{l=-\infty}^{\infty} \mathrm{D}^\nu \mathbf{N}^r_{2m+1}(l)\, e^{-i\cdot l} = (H^\nu_k)_{k=0}^{r-1}$$

the relation (3.2) leads to

$$\begin{aligned}
&([\mathrm{D}^0 \mathbf{L}^r_{2m+1}]^\sim(u), \ldots, [\mathrm{D}^{r-1} \mathbf{L}^r_{2m+1}]^\sim(u)) \\
&= [\mathbf{H}^r_{2m+1}(e^{-iu})^{\mathrm{T}}]^{-1} \left((H^0_k(e^{-iu}))_{k=0}^{r-1}, \ldots, (H^{r-1}_k(e^{-iu}))_{k=0}^{r-1}\right) \\
&= [\mathbf{H}^r_{2m+1}(e^{-iu})^{\mathrm{T}}]^{-1}\, \mathbf{H}^r_{2m+1}(e^{-iu})^{\mathrm{T}} = \mathbf{I}. \quad \blacksquare
\end{aligned}$$

Now let

$$\psi_k^{m,r} := \mathrm{D}^{m+1} L_k^{2m+1,r}(2\cdot -1) \qquad (k = 0, \ldots, r-1) \tag{3.4}$$

and $\mathbf{\Psi}^r_m := (\psi_k^{m,r})_{k=0}^{r-1}$. We shall show that the spline wavelets $\psi_k^{m,r}$ ($k = 0, \ldots, r-1$) and their integer translates form a Riesz basis of $W_0^{m,r}$.

Using the relations (2.2) and (3.2) we obtain for the vector $\hat{\mathbf{\Psi}}^r_m := (\hat{\psi}_k^{m,r})_{k=0}^{r-1}$ of Fourier transformed wavelets

$$\begin{aligned}
\hat{\mathbf{\Psi}}^r_m(u) &= [\mathrm{D}^{m+1} \mathbf{L}^r_{2m+1}(2\cdot -1)]^\wedge(u) \\
&= \tfrac{1}{2}(iu/2)^{m+1}\, e^{-iu/2}\, \hat{\mathbf{L}}^r_{2m+1}(u/2) \\
&= \tfrac{1}{2} e^{-iu/2}\, [\mathbf{H}^r_{2m+1}(e^{-iu/2})^{\mathrm{T}}]^{-1}\, \mathbf{D}^r_{m,1}(e^{-iu/2})\, \hat{\mathbf{N}}^r_m(u/2)
\end{aligned}$$

with $\mathbf{D}^r_{m,1}$ defined in (2.14). Thus, we have the two–scale relation

$$\hat{\mathbf{\Psi}}^r_m = \mathbf{Q}^r_m(e^{-i\cdot/2})\, \hat{\mathbf{N}}^r_m(\cdot/2) \tag{3.5}$$

with the *two–scale symbol of* $\mathbf{\Psi}^r_m$

$$\mathbf{Q}^r_m(z) := \tfrac{1}{2} z\, [\mathbf{H}^r_{2m+1}(z)^{\mathrm{T}}]^{-1}\, \mathbf{D}^r_{m,1}(z) \quad (z \in \mathcal{T}). \tag{3.6}$$

Observe that the elements of the matrix $\mathbf{Q}_m^r$ belong to the Wiener class. The two–scale relation (3.5) implies that the functions $\psi_k^{m,r}$ $(k = 0, \ldots, r-1)$ lie in $V_1^{m,r}$. The functions $\psi_k^{m,r}$ belong to $W_0^{m,r}$ if and only if for $k, \nu = 0, \ldots, r-1$ and $l \in \mathbb{Z}$,

$$\langle N_k^{m,r}(\cdot - l), \psi_\nu^{m,r} \rangle := \int_{-\infty}^{\infty} N_k^{m,r}(x-l)\, \overline{\psi_\nu^{m,r}(x)}\, \mathrm{d}x = 0,$$

i.e., if and only if the condition

$$\mathbf{P}_m^r(z)\, \mathbf{\Phi}_m^r(z)\, \mathbf{Q}_m^r(z)^\star + \mathbf{P}_m^r(-z)\, \mathbf{\Phi}_m^r(-z)\, \mathbf{Q}_m^r(-z)^\star = \mathbf{0} \quad (z \in \mathcal{T}) \tag{3.7}$$

is satisfied (cf. [4]).

Theorem 3.2. *The functions $\psi_k^{m,r}$ $(k = 0, \ldots, r-1)$ belong to $W_0^{m,r}$.*

Proof: From the recursion relation (2.7) it follows with (2.8) and (2.14)

$$\mathbf{P}_m^r(z) = \frac{1}{2^{m+1}}\, \mathbf{D}_{m,0}^r(z^2)\, \mathbf{P}_{-1}^r\, \mathbf{D}_{m,0}^r(z)^{-1} \quad (z \in \mathcal{T}), \tag{3.8}$$

where for $z = 1$, (3.8) is understood according to (2.7). Using the relations (2.13) and (3.6) we obtain

$$\mathbf{P}_m^r(z)\, \mathbf{\Phi}_m^r(z)\, \mathbf{Q}_m^r(z)^\star = \frac{z}{2^{m+2}}\, \mathbf{D}_{m,0}^r(z^2)\, \mathbf{P}_{-1}^r\, \mathbf{D}_r,$$

where $\mathbf{P}_{-1}^r$ and $\mathbf{D}_r$ do not depend on z. Thus, (3.7) is satisfied. ■

We can even prove:

Theorem 3.3. *The functions $\psi_k^{m,r}$ $(k = 0, \ldots, r-1)$ form a Riesz basis of W_0, i.e., there exist Riesz bounds $0 < \alpha \le \beta < \infty$ such that*

$$\alpha \sum_{l=-\infty}^{\infty} \sum_{k=0}^{r-1} |c_l^k|^2 \le \| \sum_{l=-\infty}^{\infty} \sum_{k=0}^{r-1} c_l^k\, \psi_k^{m,r}(\cdot - l) \|_{L_2}^2 \le \beta \sum_{l=-\infty}^{\infty} \sum_{k=0}^{r-1} |c_l^k|^2$$

for any sequences $(c_l^k)_{l=-\infty}^{\infty} \in l_2$ $(k = 0, \ldots, r-1)$.

Proof: For $(c_l^k)_{l=-\infty}^{\infty} \in l_2$ $(k = 0, \ldots, r-1)$ let C_k denote their 2π-periodic symbols,

$$C_k := \sum_{l=-\infty}^{\infty} c_l^k\, e^{-iu\cdot} \quad (k = 0, \ldots, r-1).$$

Put $\mathbf{C} := (C_0, \ldots, C_{r-1})^{\mathrm{T}}$. Then by the Parseval identity we find

$$\begin{aligned}
\| \sum_{l=-\infty}^{\infty} \sum_{k=0}^{r-1} c_l^k\, \psi_k^{m,r}(\cdot - l) \|_{L_2}^2 &= \frac{1}{2\pi} \int_{-\infty}^{\infty} |\mathbf{C}(u)^{\mathrm{T}}\, \hat{\mathbf{\Psi}}_m^r(u)|^2\, \mathrm{d}u \\
&= \frac{1}{2\pi} \sum_{l=-\infty}^{\infty} \int_0^{2\pi} \mathbf{C}(u)^{\mathrm{T}}\, \hat{\mathbf{\Psi}}_m^r(u + 2\pi l)\, \hat{\mathbf{\Psi}}_m^r(u + 2\pi l)^\star\, \overline{\mathbf{C}(u)}\, \mathrm{d}u.
\end{aligned}$$

Using the two–scale relation it follows

$$
\begin{aligned}
&\| \sum_{l=-\infty}^{\infty} \sum_{k=0}^{r-1} c_l^k \, \psi_k^{m,r}(\cdot - l)\|_{L_2}^2 \\
&= \frac{1}{2\pi} \sum_{l=-\infty}^{\infty} \int_0^{2\pi} \mathbf{C}(u)^{\mathrm{T}} \mathbf{Q}_m^r(e^{-i(u/2+\pi l)})\, \hat{\mathbf{N}}_m^r(u/2+\pi l)\, \hat{\mathbf{N}}_m^r(u/2+\pi l)^\star \\
&\cdot \mathbf{Q}_m^r(e^{-i(u/2+\pi l)})^\star\, \overline{\mathbf{C}(u)}\, \mathrm{d}u \\
&= \frac{1}{2\pi} \int_0^{2\pi} \mathbf{C}(u)^{\mathrm{T}} \left(\mathbf{Q}_m^r(e^{-iu/2})\, \mathbf{\Phi}_m^r(e^{-iu/2})\, \mathbf{Q}_m^r(e^{-iu/2})^\star \right) \overline{\mathbf{C}(u)}\, \mathrm{d}u \\
&+ \frac{1}{2\pi} \int_0^{2\pi} \mathbf{C}(u)^{\mathrm{T}} \left(\mathbf{Q}_m^r(-e^{-iu/2})\, \mathbf{\Phi}_m^r(-e^{-iu/2})\, \mathbf{Q}_m^r(-e^{-iu/2})^\star \right) \overline{\mathbf{C}(u)}\, \mathrm{d}u.
\end{aligned}
\tag{3.9}
$$

Recall that $\mathbf{\Phi}_m^r(z)$ is Hermitian and positive definite for $z \in \mathcal{T}$. Thus, the matrix $\mathbf{Q}_m^r(z)\, \mathbf{\Phi}_m^r(z)\, \mathbf{Q}_m^r(z)^\star$ is Hermitian and positive definite for $z \in \mathcal{T}$, $z \neq 1$ and positive semidefinite for $z = 1$. It follows that for $\|\mathbf{C}\|_{L_2}^2 := \sum_{k=0}^{r-1} \|C_k\|_{L_2}^2 > 0$ the terms in (3.9) are nonnegative for all $z \in \mathcal{T}$ and at least one term in (3.9) is positive. By the definitions of $\mathbf{Q}_m^r$ and $\mathbf{\Phi}_m^r$ the terms in (3.9) are bounded for fixed m and r. ■

§4. The Spline Space $V_{1,0}^{2m+1,r}$

In view of the preceding result let us consider the subspace

$$
V_{1,0}^{2m+1,r} := \{ s \in V_1^{2m+1,r} : \mathrm{D}^\nu s(n) = 0,\ \nu = 0, \ldots, r-1;\ n \in \mathbb{Z} \}
$$

of cardinal splines of degree $2m+1$ and defect r with the knot sequence $2^{-1}\mathbb{Z}$. Then the functions

$$
\Lambda_k^{2m+1,r} := L_k^{2m+1,r}(2 \cdot -1)
$$

belong to $V_{1,0}^{2m+1,r}$ and we have:

Theorem 4.1. *For $m \in \mathbb{N}_0$ and $r \in \mathbb{N}$ the functions*

$$
\Lambda_k^{2m+1,r}(\cdot - l) \quad (k = 0, \ldots, r-1;\ l \in \mathbb{Z}) \tag{4.1}
$$

form a Riesz basis of $V_{1,0}^{2m+1,r}$.

Proof: The cardinal Hermite fundamental splines $L_k^{2m+1,r}(2 \cdot -l)$ $(k = 0, \ldots, r-1;\ l \in \mathbb{Z})$ form a basis of $V_1^{2m+1,r}$, i.e., an arbitrarily chosen element $G \in V_{1,0}^{2m+1,r} \subset V_1^{2m+1,r}$ can be uniquely represented in the form

$$
G = \sum_{k=0}^{r-1} \sum_{l=-\infty}^{\infty} a_l^k \, L_k^{2m+1,r}(2 \cdot -l).
$$

The conditions $\mathrm{D}^\nu G(n) = 0$ ($\nu = 0, \ldots, r-1$; $n \in \mathbb{Z}$) imply that $a_{2l}^k = 0$ ($k = 0, \ldots, r-1$; $l \in \mathbb{Z}$), i.e.,

$$G = \sum_{k=0}^{r-1} \sum_{l=-\infty}^{\infty} b_l^k \, \Lambda_k^{2m+1,r}(\cdot - l)$$

with $b_l^k := a_{2l+1}^k$ ($k = 0, \ldots, r-1$; $l \in \mathbb{Z}$). Thus, the functions in (4.1) form a basis of $V_{1,0}^{2m+1,r}$. To show the Riesz basis property we note that

$$\hat{\mathbf{\Lambda}}_{2m+1}^r := \left(\hat{\Lambda}_k^{2m+1,r}\right)_{k=0}^{r-1} = \mathbf{R}_{2m+1}^r(e^{-i\cdot/2})\,\hat{\mathbf{N}}_{2m+1}^r(\cdot/2)$$

with $\mathbf{R}_{2m+1}^r(z) := z\,(\mathbf{H}_{2m+1}^r(z)^{\mathrm{T}})^{-1}$ ($z \in \mathcal{T}$). Following the ideas in the proof of Theorem 3.3, we only have to consider the matrices

$$\mathbf{R}_{2m+1}^r(z)\,\mathbf{\Phi}_{2m+1}^r(z)\,\mathbf{R}_{2m+1}^r(z)^\star \quad (z \in \mathcal{T}),$$

where $\mathbf{\Phi}_{2m+1}^r(z)$ denotes the autocorrelation symbol for $\hat{\mathbf{N}}_{2m+1}^r$. $\mathbf{H}_{2m+1}^r(z)$ is invertible and $\mathbf{\Phi}_{2m+1}^r(z)$ is positive definite for $z \in \mathcal{T}$, and it follows that $\mathbf{R}_{2m+1}^r(z)\,\mathbf{\Phi}_{2m+1}^r(z)\,\mathbf{R}_{2m+1}^r(z)^\star$ is positive definite for $z \in \mathcal{T}$. Thus, the Riesz basis property is satisfied. ■

As a consequence of Theorems 4.1 and 3.3 we have the following result (cf. [2], p. 190 for $r = 1$).

Theorem 4.2. *For $m \in \mathbb{N}_0$ and $r \in \mathbb{N}$ the $(m+1)$-th order differential operator D^{m+1} maps the spline space $V_{1,0}^{2m+1,r}$ one–to–one onto the wavelet space $W_0^{2m+1,r}$. Moreover, the Riesz basis $\{\Lambda_k^{2m+1,r}(\cdot - l);\ k = 0, \ldots, r-1;\ l \in \mathbb{Z}\}$ of $V_{1,0}^{2m+1,r}$ corresponds to the Riesz basis $\{\psi_k^{m,r}(\cdot - l);\ k = 0, \ldots, r-1;\ l \in \mathbb{Z}\}$ of $W_0^{m,r}$ via the relation $\psi_k^{m,r} = \mathrm{D}^{m+1}\,\Lambda_k^{2m+1,r}$ ($k = 0, \ldots, r-1$).*

§5. An Example

We want to apply the obtained formulas to the case $m = 3$, $r = 2$ of cubic spline wavelets with defect 2.

With $\hat{\mathbf{N}}_{-1}^2 = (iu,\ 1)^{\mathrm{T}}$ and

$$\begin{aligned}\mathbf{D}_{3,0}^2(z) &= \mathbf{A}_3^2(z)\,\mathbf{A}_2^2(z)\,\mathbf{A}_1^2(z)\,\mathbf{A}_0^2(z)\\ &= 6\begin{pmatrix} 1 & -1/2 \\ -z & 1/2 \end{pmatrix}\begin{pmatrix} 1 & -1 \\ -z & 1 \end{pmatrix}\begin{pmatrix} 1 & -1 \\ -z & 1 \end{pmatrix}\begin{pmatrix} 1 & 0 \\ 0 & 1-z \end{pmatrix}\\ &= 3\begin{pmatrix} 2+4z & -5+4z+z^2 \\ -4z-2z^2 & 1+4z-5z^2 \end{pmatrix}\end{aligned}$$

it follows from (2.2) that

$$\hat{\mathbf{N}}_3^2(u) = \frac{3}{(iu)^4}\begin{pmatrix} (2iu-5) + 4(iu+1)e^{-iu} + e^{-2iu} \\ 1 + 4(-iu+1)e^{-iu} + (-2iu-5)e^{-2iu} \end{pmatrix}.$$

For the two–scale symbol satisfying $\hat{\mathbf{N}}_3^2 = \mathbf{P}_3^2(e^{-i\cdot/2})\,\hat{\mathbf{N}}_3^2(\cdot/2)$ we find with (2.8) and (3.8)

$$\begin{aligned}\mathbf{P}_3^2(z) &= \frac{1}{16}\,\mathbf{D}_{3,0}^2(z^2)\,\mathbf{P}_{-1}^2(z)\,\mathbf{D}_{3,0}^2(z)^{-1}\\ &= \frac{1}{16}\begin{pmatrix} 2+6z+z^2 & 5+2z\\ 2z+5z^2 & 1+6z+2z^2\end{pmatrix}.\end{aligned}$$

The autocorrelation symbol reads

$$\mathbf{\Phi}_3^2(z) = \frac{1}{560}\begin{pmatrix} 9z^{-1}+128+9z & 53z^{-1}+80+z\\ z^{-1}+80+53z & 9z^{-1}+128+9z\end{pmatrix}.$$

The Euler–Frobenius matrix $\mathbf{H}_7^2$ is given by

$$\mathbf{H}_7^2(z) = \frac{1}{432}\begin{pmatrix} 37z+176z^2+3z^3 & 3z+176z^2+37z^3\\ 175z-224z^2-21z^3 & 21z+224z^2-175z^3\end{pmatrix},$$

such that (2.13) can simply be verified with

$$\begin{aligned}\mathbf{D}_{3,1}^2(z) &= \mathbf{A}_7^2(z)\,\mathbf{A}_6^2(z)\,\mathbf{A}_5^2(z)\,\mathbf{A}_4^2(z)\\ &= \frac{35}{9}\begin{pmatrix} 6+26z+3z^2 & -17-18z\\ -18z-17z^2 & 3+26z+6z^2\end{pmatrix}.\end{aligned}$$

The matrix $\mathbf{H}_7^2(z)$ is invertible on the unit circle $z \in \mathcal{T}$ and we have

$$\mathbf{H}_7^2(z)^{-1} = \frac{12}{7z\,\Delta_7^2(z)}\begin{pmatrix} 21+224z-175z^2 & -3-176z-37z^2\\ -175+224z+21z^2 & 37+176z+3z^2\end{pmatrix}$$

with

$$\Delta_7^2(z) = 1-72z+262z^2-72z^3+z^4.$$

Thus, the two–scale symbol $\mathbf{Q}_3^2$ of the wavelet vector $\mathbf{\Psi}_3^2$ is given by

$$\begin{aligned}\mathbf{Q}_3^2(z) &= \tfrac{1}{2}z\,(\mathbf{H}_7^2(z)^{\mathrm{T}})^{-1}\,\mathbf{D}_{3,1}^2(z)\\ &= \frac{60}{\Delta_7^2(z)}\begin{pmatrix} 7(1+40z+30z^2 & 7(-7-64z+30z^2\\ -64z^3-7z^4) & +40z^3+z^4)\\ & \\ 1+100z+478z^2 & 9+252z+478z^2\\ +252z^3+9z^4 & +100z^3+z^4\end{pmatrix}.\end{aligned}$$

Acknowledgements

This research was supported by the Deutsche Forschungsgemeinschaft.

References

1. de Boor C., Splines as linear combinations of B-splines. A survey, in *Approximation Theory II*, G. Lorentz, C. K. Chui, L. Schumaker (eds.), Academic Press, New York, 1976, 1–47.
2. Chui, C. K., *An Introduction to Wavelets*, Academic Press, New York, 1992.
3. Goodman, T. N. T., Interpolatory Hermite spline wavelets, J. Approx. Th. , to appear.
4. Goodman, T. N. T., and Lee, S. L., Wavelets of multiplicity r, Applied Analysis Report AA/921, University of Dundee, 1992 (submitted).
5. Goodman, T. N. T., Lee, S. L., and Tang, W. S., Wavelets in wandering subspaces, Trans. Amer. Math. Soc. **338** (1993), 639–654.
6. Lipow, P. R., and Schoenberg, I. J., Cardinal interpolation and spline functions III. Cardinal Hermite interpolation, Linear Algebra Appl. **6** (1973), 273–96
7. Plonka, G., Two–scale symbol and autocorrelation symbol for B-splines with multiple knots, preprint (submitted).
8. Schoenberg, I. J., Cardinal interpolation and spline functions, J. Approx. Th. **2** (1969), 288–297.
9. Schoenberg, I. J., and Sharma, A., Cardinal interpolation and spline functions V. The B-splines for cardinal Hermite interpolation. Linear Algebra Appl. **7** (1973), 1–42.

Gerlind Plonka
FB Mathematik
Universität Rostock
D-18051 Rostock, GERMANY
nfa016@ nfa.uni-rostock.d400.de

Regularity Analysis of Non-uniform Data

Christine Potier and Christine Vercken

Abstract. A particular class of wavelets, derivatives of B-splines, leads to fast and efficient algorithms for contour detection of 1D and 2D signals by a discrete dyadic finite multiscale wavelet analysis. To detect higher order singularities, the multiscale analysis algorithm, with higher order spline wavelets is very simply expressed in terms of B-splines subdivision and differentiation formulas. This formulation allows us to generalize the algorithm for nonuniformly sampled data. This algorithm is used to locate the singularities of a function.

§1. Introduction

The detection of singularities with multiscale transforms has been widely studied not only in mathematics but also in signal processing, computer vision, etc. For contour detection, the signal can be smoothed at different scales and the edges are found for example as local extrema of the gradient modulus of the smoothed signal (Canny) or as zero-crossings of the Laplacian of the smoothed signal (Marr and Hilbreth). Mallat and al. [3] showed the equivalence between the Canny edge detector and the local maxima of a wavelet transform modulus detection. The signal regularity at each point x is measured by the decay across scales of its wavelet coefficients centered at x. For this analysis, wavelet transforms with dyadic sampling, particularly orthonormal bases, may introduce distortion since they are not translation invariant. A particular class of wavelet, derivatives of B-splines, leads to fast and efficient algorithms for edge detection of 1D and 2D signals by a discrete dyadic finite scale analysis (Mallat and Zhong). For this class, the scaling function ϕ is a quadratic spline if the wavelet ψ is the 1st derivative of a cubic spline. More generally, a wavelet which is the pth derivative of a smoothing function is well suited to characterize singularities of a $(p-1)^{th}$ derivative through multiscale analysis using B-spline functions. The two scale difference equation is a subdivision equation and the wavelet coefficients are obtained at each scale from the differentiation formulas. This formulation allows us to generalize the algorithm to nonuniformly sampled data to locate singularities.

Wavelets, Images and Surface Fitting
P. J. Laurent, A. Le Méhauté, and L. L. Schumaker (eds.), pp. 399–406.

ISBN 1-56881-040-7.

§2. Dyadic Multiscale Wavelet Transform and Singularities Analysis

Let ψ be a function $\in L_1 \cap L_2$ so that the function $x\psi$ is integrable. The function ψ can be considered as a wavelet if $\int \psi = 0$. The *dilation* of ψ at the scale s is defined by

$$\psi_s(x) = \frac{1}{s}\psi(\frac{x}{s}). \tag{1}$$

Let f be a function $\in L_1 \cap L_2$. The *wavelet transform* $W_s f(x)$ of f at a scale s and a position x is defined by the convolution product

$$W_s f(x) = f * \psi_s(x) \tag{2}$$

The *dyadic wavelet transform* is the sequence $\{W_{2^j} f(x)\}$ given by $\{f * \psi_{2^j}(x)\}_{j \in \mathbb{Z}}$. Mallat[2] showed that the analysis is complete and redundant and $f(x)$ can be recovered from its dyadic wavelet transform if there exist positive constants A and B such that

$$A \leq \sum_{-\infty}^{+\infty} |\hat{\psi}(2^j \omega)|^2 \leq B, \qquad \forall \omega \in \mathbb{C}. \tag{3}$$

We now introduce a particular set $S(\theta, p)$ of wavelets for singularities analysis:

$$S(\theta, p) = \{\psi \mid \exists\, \theta \in C^{2p}, \text{ compactly supported}, \int \theta dx = 1,\ \psi = \theta^{(p)}\}$$

The function θ is called the *smoothing function.* From the definition of $S(\theta, p)$, $\psi \in C^p(\mathbb{R})$ and is a wavelet with p vanishing moments :

$$\int_{-\infty}^{+\infty} x^i \psi(x) dx = 0, \qquad 0 \leq i < p. \tag{4}$$

The wavelet transform of $f(x)$ at a scale s and a position x, satisfies

$$W_s^p f(x) = f * \psi_s(x) = f * (s^p \frac{d^p \theta(x)}{dx^p})(x) = s^p \frac{d^p}{dx^p}(f * \theta_s)(x), \tag{5}$$

and $W_s^p f(x)$ is proportional to the pth derivative of f smoothed by $\theta_s(x)$. If $f^{(p-1)}$ has an isolated singularity at x_0, $W_s^p f$ has a local maximum at x_0. If ψ^1 is the first derivative of θ, ψ^1 is well suited to localize a discontinuity of f (a sharp edge). The local extrema of $W_s^1 f(x)$ correspond to the zero-crossings of $W_s^2 f(x)$ and to the inflection points of $f * \theta_s(x)$.

More precisely [3], the decay of wavelet moduli across scales depends on the Lipschitz regularity of f. If $f \in L^2$ is uniformly Lipschitz $\alpha < p$ over $]a, b[$, there exists a positive constant K such that, for $x \in]a, b[$, $s > 0$ and $\psi \in S(\theta, p)$,

$$|W_s^p f(x)| \leq K s^\alpha. \tag{6}$$

To analyze singularities of $f^{(p-1)}$ we use a wavelet $\psi \in S(\theta, p)$ and study $|W_s^p f(x)|/s^\alpha$.

§3. Discrete Dyadic Wavelet Transform

In numerical applications, the input signal $\{x_i, y_i\}_{0\leq i \leq N-1}$ is measured at a finite resolution that we call the *scale* $s = 1$. The input signal $S_1 f$ is a smoothed approximation of f at scale $s = 1$, and we can write

$$S_1 f(x) = f * \phi(x), \tag{7}$$

where $\phi(x)$ is a smoothing function satisfying $\int \phi(x) = 1$ that we call the *scaling function.* The dilation of the scaling function is defined by equation (1). We suppose that $\phi(\frac{x}{2})$ and $\psi(\frac{x}{2})$ can be expressed as linear combinations of translated $\phi(x)$:

$$\phi(\frac{x}{2}) = \sum_{n=a}^{n=b} \alpha_n \phi(x-n) \quad \text{and} \quad \psi(\frac{x}{2}) = \sum_{n=c}^{n=d} \beta_n \phi(x-n). \tag{8}$$

The scaling function can be used at scale 2^j to obtain a smoothed signal $S_{2^j} f(x) = f * \phi_{2^j}(x)$. The smoothed signal represents the trend of the signal at the scale 2^j whereas $W_{2^j} f(x) = f * \psi_{2^j}(x)$ represents the details of the signal at the scale 2^j. $S_{2^j} f(x)$ may be viewed as a low-pass filter and $W_{2^j} f(x)$ as a high-pass filter. The discrete dyadic wavelet transform at scale 2^J is the sequence of details $\left\{W_{2^j} f = f * \psi_{2^j}(x)\right\}_{1\leq j<J}$ and the tendency $S_{2^J} f$.
If $\{S_{2^j} f(n) = d_n^j\}_{n\in Z}$ are the coefficients of the wavelet transform at the scale 2^j, $S_{2^{j+1}} f(n)$ and $W_{2^{j+1}} f(n)$ can be computed by using relations (8) as

$$S_{2^{j+1}} f(m) = \sum_{n=a}^{n=b} \alpha_n d^j_{m-2^j*n} \quad \text{and} \quad W_{2^{j+1}} f(m) = \sum_{n=c}^{n=d} \beta_n d^j_{m-2^j*n}. \tag{9}$$

§4. Discrete Dyadic Spline Wavelet Transform

It is well known that the polynomial spline spaces are well suited to multiscale wavelet transform [2,5]. A spline function of order a is a piecewise polynomial function of degree $a-1$ which is C^{a-2} for $a \geq 2$. Its pth derivative may be used to detect singularities of $f^{(p-1)}$ if $2p = a-2$. If the wavelet $\psi(x)$ is the pth derivative of a spline of order $2p+2$, to satisfy (8) ψ must be of the same order $k = p+2$ as the scaling function ϕ. For example, let θ be a spline of order 6, and $\psi = \theta$". Then ψ is a cubic spline which is C^2 and is well suited to detect discontinuities of degree 1.

The splines of order k with minimal support are called B-spline. The cardinal B-spline of order k is $B_k(x) = B_{k-1}(x) * B_1(x)$ where $B_1(x)$ is the B-spline of order 1 defined on $[-1/2, 1/2]$. It is defined on knot sequence $[\frac{-k}{2}, \frac{(-k+2)}{2}, \ldots, \frac{k}{2}]$. Usually the cardinal B-spline $B_k(x)$ is defined on knot sequence $[0, 1, \ldots, k]$. To obtain functions centered at 0, we use translated cardinal B-splines.

The coefficients $\{\alpha_n\}$ and $\{\beta_n\}$ can be obtained by using the dilation properties of polynomial B-splines and differentiation formulas. If B_k denotes the kth order cardinal B-spline defined on knot sequence $[\frac{-k}{2}, \frac{(-k+2)}{2}, \ldots, \frac{k}{2}]$, $B_k(\frac{x}{2})$ is the B-spline of order k defined on knot sequence $[-k, -k+2, \ldots, k]$. Then we can write

$$B_k(\frac{x}{2}) = \sum_{n=-k/2}^{n=k/2} a_j B_k(x-n) = \sum_{n=-k/2}^{n=k/2} \frac{1}{2^{k-1}} \binom{k}{n+k/2} B_k(x-n). \tag{10}$$

Similarly, the pth derivative of a $(k+p)^{th}$ order B-spline can be expressed:

$$B_{k+p}^{(p)}(x) = \sum_{n=-p/2}^{n=p/2} d_j B_k(x-n) = \sum_{n=0}^{n=p} (-1)^n \binom{p}{n} B_k(x-n-\frac{p}{2}). \tag{11}$$

In case of a regular knot sequence $\{t_i = i\}$, the coefficients $\{a_n\}$ and $\{d_n\}$ depend only on the order k of the spline since $p = k-2$. The cardinal B-splines $B_k(x-i)$ currently denoted $B_{i,k}$ satisfy two normalizing relations $\sum_{i=-\infty}^{i=+\infty} B_{i,k}(x) = 1$ and $\int B_{i,k}(x)dx = 1$. Hence, $\phi = B_k(x)$ can be taken as a scaling function.

§5. Irregular Sampling

In case of an irregular knot sequence $\{t_i\}$, the B-spline $B_{i,k}(x)$ defined on knots $[t_{i-k/2}, \ldots, t_{i+k/2}]$ no longer satisfies the relation $\int B_{i,k}(x)dx = 1$. Since we need the scaling function ϕ and the smoothing function θ to satisfy $\int f(x) = 1$, we have to use normalized spline functions with compact support. The spline functions with minimal support satisfying this normalizing equation are called the M-splines and denoted by $M_{i,k}$. The relation between $B_{i,k}$ and $M_{i,k}$ is

$$M_{i,k}(x) = \frac{k}{(t_{i+k/2} - t_{i-k/2})} B_{i,k}(x). \tag{12}$$

The relation corresponding to the relation (8) that expresses the scaling function ϕ_{2s} in terms of ϕ_s, may be obtained by applying the subdivision formulae on $B_{i,k}$ defined on knot sequence $[\ldots, t_{i-2}, t_i, t_{i+2}, \ldots]$ if k is even and on knot sequence $[\ldots, t_{i-1}, t_{i+1}, \ldots]$ if k is odd. The subdivision is made by introducing one knot in each interval [1]. We obtain the coefficients $\{a_{i,n}\}$ and then the coefficients $\{h_{i,n}\}$ are obtained by using the relation (12):

$$B_{i,k}(\frac{x}{2}) = \sum_{n=-k/2}^{n=k/2} a_{i,n} B_{i+n,k}(x) \quad , \quad M_{i,k}(\frac{x}{2}) = \sum_{n=-k/2}^{n=k/2} h_{i,n} M_{i+n,k}(x). \tag{13}$$

Obviously, the coefficients $h_{i,n}$ satisfy $\sum_{j=-k/2}^{j=k/2} h_{i,j} = 1$ because $\int M_{i,k}(x)dx = 1$ and $\int M_{i,k}(\frac{x}{2})dx = 1$ for all i and k.

The subdivision is made at each scale $s = 2^j$ and we obtain coefficients that depend on the position i and on scale $s = 2^j$. For example, if $k = 4$ the relation between coefficients $a^s_{i,n}$ and $h^s_{i,n}$ at scale s is:

$$h^s_{i,n} = a^s_{i,n} \frac{(t_{i+(2+n)*s} - t_{i-(2-n)*s})}{(t_{i+4s} - t_{i-4s})} \tag{14}$$

and the coefficients $h^s_{i,n}$ are

$$\begin{aligned}
h^s_{i,-2} &= \frac{(t_{i-s} - t_{i-4s})(t_{i-3s} - t_{i-4s})}{(t_{i+2s} - t_{i-4s})(t_{i+4s} - t_{i-4s})} \\
h^s_{i,-1} &= \frac{(t_{i-s} - t_{i-4s})(t_{i+s} - t_{i-3s})}{(t_{i+2s} - t_{i-4s})(t_{i+4s} - t_{i-4s})} \\
h^s_{i,0} &= 1 - h^s_{i,-2} - h^s_{i,-1} - h^s_{i,1} - h^s_{i,2} \\
h^s_{i,1} &= \frac{(t_{i+4s} - t_{i+s})(t_{i+3s} - t_{i-s})}{(t_{i+4s} - t_{i-2s})(t_{i+4s} - t_{i-4s})} \\
h^s_{i,2} &= \frac{(t_{i+4s} - t_{i+s})(t_{i+4s} - t_{i+3s})}{(t_{i+4s} - t_{i-2s})(t_{i+4s} - t_{i-4s})}
\end{aligned}$$

The coefficients $g^s_{i,n}$ at scale s are given by differentiation formulae:

$$\begin{aligned}
g^s_{i,-1} &= \frac{30s^2}{(t_{i+3s} - t_{i-3s})(t_{i+2s} - t_{i-3s})} \\
g^s_{i,1} &= \frac{30s^2}{(t_{i+3s} - t_{i-3s})(t_{i+3s} - t_{i-2s})} \\
g^s_{i,0} &= -(g^s_{i,-1} + g^s_{i,1}).
\end{aligned}$$

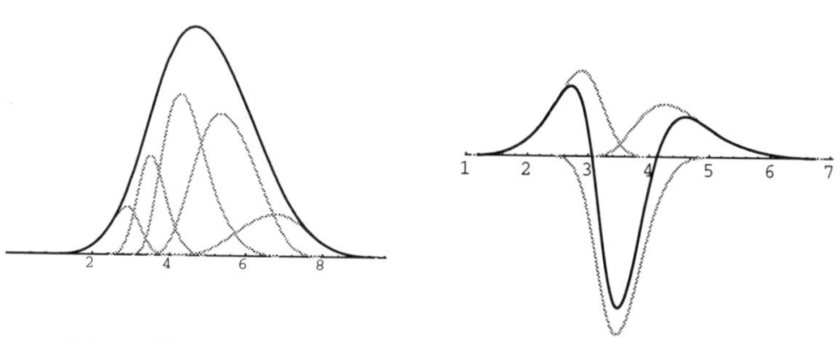

Fig. 1. (a) $M_4(\frac{x}{2})$ and $h_i M_{i,4}(x)$ defined on knots $[1, 2.4, 3, 3.4, 4, 5, 7, 8, 9.5]$, (b) $M_6''(x)$ and $g_i M_{i,4}(x)$ defined on same knots.

Fig. 1(a) represents the cubic M-spline $M_4(\frac{x}{2})$ defined on the knot sequence $[1, 3, 4, 7, 9.5]$. To obtain relation (13) we have introduced knots

$[2.4, 3.4, 5, 8]$. We have drawn $\{h_i M_{i,4}(x)\}_{i=1,5}$, where functions $M_{i,4}(x)$ are defined on five successive knots. Fig. 1(b) represents the function $M_6''(x)$ where M_6 is defined on the knot sequence $[1, 2.4, 3, 3.4, 4, 5, 7]$, and the functions $\{g_i M_{i,4}(x)\}_{i=1,3}$

Practically, if the subdivision schemes are applied from the coarse to the fine grid, the discrete sequences converge

- to a B-spline for an initial sequence $y_i^0 = \delta_{i0}$
- to ψ for an initial sequence $y_i^0 = d_i$

At each step, we compute values y_j^{p+1} by

$$y_{2*i}^{p+1} = \sum_n a_{2(i-n),2n} y_{i-n}^p, \qquad y_{2*i+1}^{p+1} = \sum_n a_{2(i-n),n} y_{i-n}^p,$$

where the coefficients $a_{i,j}$ satisfy

$$\sum_n a_{2(i-n),2n} = 1, \qquad \sum_n a_{2(i-n),n} = 1. \tag{15}$$

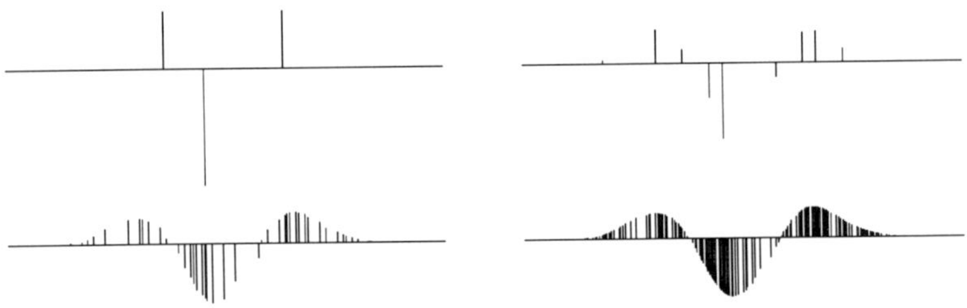

Fig. 2. Subdivision converging to ψ.

§6. Discontinuity Detection

In practice, the original discrete data have a finite number N of values. To solve the boundary problems, the signal is periodized by symmetry at each extremity as in a cosine transform. This periodization often introduces an artificial discontinuity at the boundary and the periodization should depend on the shape of the signal at the boundary.

We have compared the wavelet transform analysis on uniform sampling and on irregular sampling. Fig. 3(a) represents the discrete wavelet transform of the function $f(x) = \sin(2\pi x) + |\sin(8/3\pi x)|$ for $x \in [0.2, 1.7]$ uniformly sampled on $N = 256$ points $\{t_i\}$. To analyze the singularities of the first derivative of these data, we use the wavelet $\psi = M_6''$ and the scaling function $\phi = M_4$. Fig. 3(a) shows the wavelet transform computed at scale $s = 1, 2$ and 4. At each scale the local maxima have the same locations.

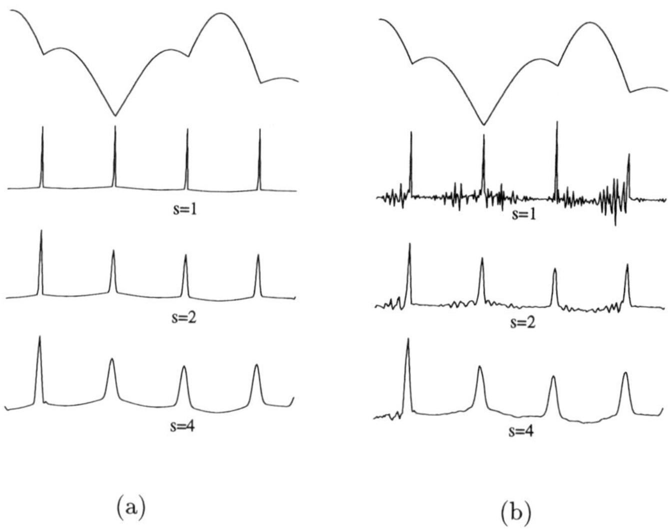

Fig. 3. (a) $|W_s f(x)|/s$ on regular sampling at scale $s = 1, 2$ and 4 and (b) on irregular sampling at same scale.

In Fig. 3(b), the function is sampled on points t_i satisfying $t_{i+1} - t_i = (1+\epsilon_i) * h$, where ϵ_i is a uniform random variable $\in [-0.2, 0.2]$. At each scale, we compute

$$y_m^{2*s} = \sum_{n=-2}^{n=2} h_{i,n}^s y_{m-2^j*n}^s \quad \text{and} \quad v_m^{2*s} = \sum_{n=-1}^{n=1} g_{i,n}^s y_{m-2^j*n}^s.$$

The values $\{v_m^{2*s}\}_{m=1,N}$ are the wavelet coefficients at scale s. Fig. 3(b) represents the wavelet transform computed at scale $s = 1, 2$ and 4. At scale 1, we can see a very oscillating function between two successive peaks corresponding to the maxima of Fig. 3(a). At scale 2, the function $W_s f$ is smoother and at scale 4, the wavelet transform on irregular sampling looks like the wavelet transform obtained on regular sampling. Irregular sampling behaves as if it were noise. The number of maxima due to noise and their modulus decrease when scale increases. To distinguish a local maximum due to a singularity from a local maximum due to the noise, we compute the wavelet coefficients at a few scales. The useful maxima remain stable in location and modulus.

The values of the maxima are shown to be constant at each scale if the wavelet transform is computed exactly and not by discretization. To correct the error due to the discretization process [3,4], we have to normalize the wavelet coefficients at each scale $s = 2^j$ by a factor that corresponds to the maximum modulus of the discrete wavelet transform of $f(x) = x_+^{p-1}$ if the analyzing wavelet is the pth derivative of a B-spline. Table 1 gives the normalizing coefficients for various wavelet analysis computed on regular sampling. On irregular sampling, the coefficients which depend on the knot values, could be computed at each point.

scale	s=1	s=2	s=4	s=8	s=16
$\psi = B_4' \quad \phi = B_3$	$\frac{3}{2}$	$\frac{9}{8}$	$\frac{33}{32}$	$\frac{129}{128}$	$\frac{513}{512}$
$\psi = B_6'' \quad \phi = B_4$	$\frac{20}{11}$	$\frac{25}{22}$	$\frac{725}{704}$	$\frac{11345}{11264}$	$\frac{180545}{180224}$

Table 1. Normalizing coefficients.

§7. Concluding Remarks

This multiscale wavelet analysis allows us to localize the singularities of irregularly sampled data. This algorithm is very fast and very efficient and depends only on the number N of data points. After the singularities localization, one can use appropriate functions to approximate the data. For approximation, the basis functions may be chosen as B-spline with multiple knots at each point where we have detected a singularity and simple knot elsewhere, the knot multiplicity corresponding to the degree of the singularity.

References

1. Goldman R., Blossoming and Knot Insertion Algorithms for B-spline Curves, CAGD **7** (1990), 69–81.
2. Mallat S., A Theory for Multiresolution Signal Decomposition : the Wavelet Representation, IEEE Trans. on PAMI **11** (1989), 674–692.
3. Mallat S. and W. Hwang, Singularity Detection and Processing with Wavelets, IEEE Trans. on Information Theory **38** (1992), 617–642.
4. Unser M., A. Aldroubi and M. Eden, Fast B-spline Transforms for Continuous Image Representation and Interpolation, IEEE Trans. on PAMI, **13** (1991), 277–285.
5. Unser M. and A. Aldroubi, Polynomial splines and Wavelets: A signal Processing Perspective, in *Wavelets : A tutorial*, C.K.Chui (ed.), Academic Press, (1991).

Christine Potier and Christine Vercken
Telecom Paris, 46 rue Barrault
74634, Paris cedex 13, FRANCE
Christine.Potier@ inf.enst.fr
Christine.Vercken@ inf.enst.fr

A Duality Principle for Trigonometric Wavelets

J. Prestin and E. Quak

Abstract. The aim of this paper is to give a direct proof of the duality principle for decomposition and reconstruction algorithms for nested spaces of trigonometric polynomials. The scaling functions of these spaces are defined as fundamental polynomials of Lagrange interpolation. The decomposition matrix for these scaling functions and wavelets is the transpose of the reconstruction matrix for the dual scaling functions and dual wavelets. Analogously, the original reconstruction matrix is the transpose of the decomposition matrix for the dual functions.

§1. Introduction

The trigonometric polynomials used in this paper were introduced by A. A. Privalov for the construction of orthogonal bases of the space $C_{2\pi}$ [6]. His results were very recently generalized by R. A. Lorentz and A. A. Sahakian [4] using a wavelet approach. The investigation of nested spaces of trigonometric polynomials from a wavelet point of view started with a paper by C. K. Chui and H. N. Mhaskar on trigonometric wavelets [2]. The basic tool in [6] is the construction of a suitable interpolation operator of Lagrange type using de la Vallée Poussin kernels. In [5], the exact structure of the decomposition and reconstruction matrices associated with the nested spaces generated by these Lagrange interpolants was investigated. The use of dual scaling functions and wavelets, as first introduced in [3], turned out to be a crucial step. The aim of this paper is to prove a duality principle for these nested spaces of trigonometric polynomials. After defining the scaling functions and wavelets and listing their basic properties in Section 2, the corresponding two-scale relations are recalled from [5] in Section 3. In Section 4, the relationship between the inner-product matrices of scaling functions and wavelets at a level j and the inner-product matrix of scaling functions at the next level $j+1$ is described using a convenient matrix notation. Also, the dual scaling functions and wavelets are introduced. The more complicated decomposition relations are again recalled from [5] in Section 5, so that in Section 6 the duality principle can be proven, saying that the decomposition matrix for the

Wavelets, Images and Surface Fitting
P. J. Laurent, A. Le Méhauté, and L. L. Schumaker (eds.), pp. 407–418.

ISBN 1-56881-040-7.

original scaling functions and wavelets is the transpose of the reconstruction matrix for the dual scaling functions and dual wavelets. Analogously, the original reconstruction matrix is the transpose of the decomposition matrix for the dual functions.

§2. Definitions

For $\ell \in \mathbb{N}$, the Dirichlet kernel $D_\ell \in T_\ell$ is defined as

$$D_\ell(x) = \frac{1}{2} + \sum_{k=1}^{\ell} \cos kx = \begin{cases} \frac{\sin(\ell+\frac{1}{2})x}{2\sin\frac{x}{2}} & \text{for } x \notin 2\pi\mathbb{Z}, \\ \ell + \frac{1}{2} & \text{for } x \in 2\pi\mathbb{Z}, \end{cases}$$

where T_ℓ denotes the linear space of trigonometric polynomials of degree ℓ.

In the following, two different kinds $\phi_{j,0}^D$ and $\phi_{j,0}^F$ of de la Vallée Poussin kernels are used to construct certain interpolatory operators. Therefore, let

$$\begin{aligned}\phi_{j,0}^D(x) &= \frac{1}{3\cdot 2^{2j+1}} \sum_{\ell=2^{j+1}}^{2^{j+2}-1} D_\ell(x) \\ &= \begin{cases} \frac{\sin(3\cdot 2^j x)\sin(2^j x)}{3\cdot 2^{2j+2}\sin^2(\frac{x}{2})} & \text{for } x \notin 2\pi\mathbb{Z}, \\ 1 & \text{for } x \in 2\pi\mathbb{Z} \end{cases}\end{aligned}$$

and

$$\begin{aligned}\phi_{j,0}^F(x) &= \frac{1}{3\cdot 2^{j+1}} \sum_{\ell=3\cdot 2^j-1}^{3\cdot 2^j} D_\ell(x) \\ &= \begin{cases} \frac{\sin(3\cdot 2^j x)\cot\frac{x}{2}}{3\cdot 2^{j+1}} & \text{for } x \notin 2\pi\mathbb{Z}, \\ 1 & \text{for } x \in 2\pi\mathbb{Z}\,. \end{cases}\end{aligned}$$

While the first kernel represents the classical de la Vallée Poussin approach, the second one is only a slight modification of a simple Dirichlet kernel. In the following, the superscripts will be dropped whenever a statement holds for both cases.

The crucial interpolatory property of $\phi_{j,0}$ is

$$\phi_{j,0}\left(\frac{k\pi}{3\cdot 2^j}\right) = \delta_{k,0}, \quad k = 0, 1, \ldots, 3\cdot 2^{j+1} - 1.$$

Definition 2.1. *For $j \in \mathbb{N}_0$, the spaces V_j are defined by $V_j = span\{\phi_{j,n} : n = 0, \ldots, 3 \cdot 2^{j+1} - 1\}$, where $\phi_{j,n}(x) = \phi_{j,0}(x - \frac{n\pi}{3 \cdot 2^j})$.*

For notational convenience, let $\phi_{j,n} = \phi_{j,n \bmod 3 \cdot 2^{j+1}}$ for any $n \in \mathbb{Z}$. Note that V_j^D as generated by $\phi_{j,n}^D$ and V_j^F generated by $\phi_{j,n}^F$ are different spaces. Nevertheless, they have the same dimension $3 \cdot 2^{j+1}$ as can be deduced from the corresponding interpolatory property $\phi_{j,n}(\frac{k\pi}{3 \cdot 2^j}) = \delta_{k,n}$, for all $k, n \in \mathbb{Z}$.

Definition 2.2. *For any $j \in \mathbb{N}_0$, the interpolation operator L_j mapping any real-valued 2π-periodic function f into the space V_j is defined as*

$$L_j f(x) = \sum_{n=0}^{3 \cdot 2^{j+1}-1} f(\frac{n\pi}{3 \cdot 2^j}) \phi_{j,n}(x).$$

The following properties of the operators L_j are well-known ([6,7]):

(i) $$L_j^D f \in T_{2^{j+2}-1}, \quad L_j^F f \in T_{3 \cdot 2^j}$$

(ii) $$L_j f(\frac{k\pi}{3 \cdot 2^j}) = f(\frac{k\pi}{3 \cdot 2^j}), k \in \mathbb{Z}$$

(iii) $$L_j^D f = f \quad \text{for all} \quad f \in T_{2^{j+1}} \cup V_j^D,$$
$$L_j^F f = f \quad \text{for all} \quad f \in T_{3 \cdot 2^j - 1} \cup V_j^F,$$

hence

(iv) $$T_{2^{j+1}} \subset V_j^D \subset T_{2^{j+2}-1},$$
$$T_{3 \cdot 2^j - 1} \subset V_j^F \subset T_{3 \cdot 2^j}.$$

Moreover, property *(iv)* implies that

$$V_j \subset V_{j+1},$$

i.e., the spaces V_j form a sequence of nested subspaces of $L^2[0, 2\pi]$, the space of 2π-periodic square integrable functions. Setting $V_{-1} = \{0\}$, it is also clear that

$$L^2 = L^2 - \text{closure of} \quad \bigcup_{j=-1}^{\infty} V_j \quad \text{and} \quad \bigcap_{j=-1}^{\infty} V_j = \{0\}.$$

As the next step, the orthogonal complement of V_j relative to V_{j+1}, i.e., the so-called wavelet space W_j needs to be described in more detail.

Definition 2.3. *For $j \in \mathbb{N}_0$, the spaces W_j are defined by $W_j = span\{\psi_{j,n} : n = 0, \ldots, 3 \cdot 2^{j+1} - 1\}$, where*

$$\psi_{j,n}(x) = 2\phi_{j+1,2n+1}(x) - \phi_{j,n}(x - \frac{\pi}{3 \cdot 2^{j+1}}) \in V_{j+1}.$$

The functions $\psi_{j,n}$ also show interpolatory properties, namely for all $k \in \mathbb{Z}$

$$\psi_{j,n}(\frac{(2k+1)\pi}{3\cdot 2^{j+1}}) = \delta_{k,n},$$
$$\psi_{j,n}(\frac{k\pi}{3\cdot 2^{j}}) = -\phi_{j,n}(\frac{(2k-1)\pi}{3\cdot 2^{j+1}}) .$$

Therefore, $\dim W_j = 3\cdot 2^{j+1}$. Let $\langle \cdot,\cdot \rangle$ denote the inner-product of two functions f and g in $L^2[0,2\pi]$, i.e.,

$$\langle f,g\rangle = \frac{1}{2\pi}\int_0^{2\pi} f(x)g(x)dx.$$

The following fundamental result was established by A. A. Privalov [6]:

Theorem A. *The spaces V_j and W_j are orthogonal, i.e.,*

$$\langle \phi_{j,n}, \psi_{j,k}\rangle = 0 \quad \text{for all} \quad n,k \in \mathbb{Z}.$$

Therefore,

$$V_{j+1} = V_j \oplus W_j,$$

where $\oplus$ denotes orthogonal summation.

§3. Two-scale Relations

As $V_j \subset V_{j+1}$, there have to exist specific coefficients $p_{j,n,s}$ such that $\phi_{j,n} = \sum_s p_{j,n,s}\phi_{j+1,s}$. The following result establishes their precise values.

Theorem 3.1. [5] *For $j \in \mathbb{N}_0$ and $n = 0,1\ldots,3\cdot 2^{j+1}-1$,*

$$\phi_{j,n}(x) = \phi_{j+1,2n}(x) + \sum_{s=0}^{3\cdot 2^{j+1}-1} \phi_{j,n}(\frac{(2s+1)\pi}{3\cdot 2^{j+1}})\phi_{j+1,2s+1}(x).$$

From $W_j \subset V_{j+1}$, it is clear that there also must be coefficients $q_{j,n,s}$ such that $\psi_{j,n} = \sum_s q_{j,n,s}\phi_{j+1,s}$. These are determined using

Theorem 3.2. [5] *For $j \in \mathbb{N}_0$ and $n = 0,1\ldots,3\cdot 2^{j+1}-1$, the two-scale relation for the wavelet functions $\psi_{j,n}$ is*

$$\psi_{j,n}(x) = \phi_{j+1,2n+1}(x) - \sum_{s=0}^{3\cdot 2^{j+1}-1} \phi_{j,n}(\frac{(2s-1)\pi}{3\cdot 2^{j+1}})\phi_{j+1,2s}(x).$$

Let $\underline{\phi}_j$ denote the vector $(\phi_{j,0},\phi_{j,1},\ldots,\phi_{j,3\cdot 2^{j+1}-1})^T$ and, analogously, $\underline{\psi}_j$ the vector $(\psi_{j,0},\psi_{j,1},\ldots,\psi_{j,3\cdot 2^{j+1}-1})^T$. Furthermore, we define a reordering for the vector of scaling functions by

$$P_j\underline{\phi}_{j+1} = (\phi_{j+1,0}, \phi_{j+1,2}, \ldots, \phi_{j+1,2m}, \ldots, \phi_{j+1,3\cdot 2^{j+2}-2},$$
$$\phi_{j+1,1}, \phi_{j+1,3}, \ldots, \phi_{j+1,2m+1}, \ldots, \phi_{j+1,3\cdot 2^{j+2}-1})^T,$$

i.e., P_j is chosen to be the suitable permutation matrix for this ordering.

Consequently, the two-scale relation or *reconstruction matrix* C_j has the form

$$C_j = \begin{pmatrix} I_j & K_j \\ -K_j^T & I_j \end{pmatrix},$$

where I_j is an identity matrix of dimension $3\cdot 2^{j+1}$ and K_j is a knot evaluation matrix:

$$K_j = \left(\phi_{j,r}\left(\frac{(2s+1)\pi}{3\cdot 2^{j+1}}\right)\right)_{r,s=0}^{3\cdot 2^{j+1}-1}.$$

Due to the definition of the functions $\phi_{j,r}$, C_j is a square matrix of dimension $3\cdot 2^{j+2}$ with circulant blocks. Theorems 3.1 and 3.2 can now be expressed as

$$\begin{pmatrix} \underline{\phi}_j \\ \underline{\psi}_j \end{pmatrix} = C_j P_j \underline{\phi}_{j+1}.$$

As a consequence of Theorem A, both the sets $\{\phi_{j,r}, \psi_{j,r}\}_{r=0}^{3\cdot 2^{j+1}-1}$ and $\{\phi_{j+1,r}\}_{r=0}^{3\cdot 2^{j+2}-1}$ are bases of V_{j+1}. Therefore, the reconstruction matrix C_j is non-singular and its inverse D_j is the decomposition matrix such that

$$P_j\underline{\phi}_{j+1} = D_j \begin{pmatrix} \underline{\phi}_j \\ \underline{\psi}_j \end{pmatrix}.$$

For a detailed analysis of the entries of D_j, it is necessary to investigate inner-products of scaling functions and to introduce dual scaling functions.

§4. Inner Products of Scaling Functions and Wavelets

In order to investigate dual functions and to be able to give a concise formulation of the duality principle for trigonometric wavelets, in this section we study the inner-products of the functions $\phi_{j,k}$ and $\psi_{j,k}$.

For any level $j \in \mathbb{N}_0$, the inner-product matrix G_j for scaling functions is defined as

$$G_j = \left(\langle \phi_{j,k}, \phi_{j,l}\rangle\right)_{k,l=0}^{3\cdot 2^{j+1}-1}.$$

Our next goal is to express the matrix G_j in terms of the matrix G_{j+1}, i.e., inner-products at level j by means of inner-products at level $j+1$.

Lemma 4.1. *For $j \in \mathbb{N}_0$, let K_j be the knot evaluation matrix, I_j the identity matrix and P_j the permutation matrix defined in Section 3. Then*

$$G_j = (I_j \quad K_j) P_j G_{j+1} P_j^T \begin{pmatrix} I_j \\ K_j^T \end{pmatrix}.$$

Proof: Using the two-scale relation for $\phi_{j,k}$ of Theorem 3.1 yields for $k, l = 0, \ldots, 3 \cdot 2^{j+1} - 1$

$$\begin{aligned}\langle \phi_{j,k}, \phi_{j,l} \rangle &= \langle \phi_{j+1,2k}, \phi_{j+1,2l} \rangle \\ &\quad + \sum_{r=0}^{3\cdot 2^{j+1}-1} \phi_{j,k}\left(\frac{(2r+1)\pi}{3 \cdot 2^{j+1}}\right) \langle \phi_{j+1,2r+1}, \phi_{j+1,2l} \rangle \\ &\quad + \sum_{s=0}^{3\cdot 2^{j+1}-1} \phi_{j,l}\left(\frac{(2s+1)\pi}{3 \cdot 2^{j+1}}\right) \langle \phi_{j+1,2k}, \phi_{j+1,2s+1} \rangle \\ + \sum_{r=0}^{3\cdot 2^{j+1}-1} \sum_{s=0}^{3\cdot 2^{j+1}-1} &\phi_{j,k}\left(\frac{(2r+1)\pi}{3 \cdot 2^{j+1}}\right) \phi_{j,l}\left(\frac{(2s+1)\pi}{3 \cdot 2^{j+1}}\right) \langle \phi_{j+1,2r+1}, \phi_{j+1,2s+1} \rangle.\end{aligned}$$

Introducing the following square matrices of dimension $3 \cdot 2^{j+1}$, namely, $\Phi_e = (\langle \phi_{j+1,2k}, \phi_{j+1,2l} \rangle)_{k,l=0}^{3\cdot 2^{j+1}-1}$, $\Phi_o = (\langle \phi_{j+1,2k+1}, \phi_{j+1,2l+1} \rangle)_{k,l=0}^{3\cdot 2^{j+1}-1}$ and $\Phi_m = (\langle \phi_{j+1,2k}, \phi_{j+1,2l+1} \rangle)_{k,l=0}^{3\cdot 2^{j+1}-1}$, the relations above can be written in matrix notation as

$$\begin{aligned}G_j &= \Phi_e + \Phi_m K_j^T + K_j \Phi_m^T + K_j \Phi_o K_j^T \\ &= (I_j \quad K_j) \begin{pmatrix} \Phi_e & \Phi_m \\ \Phi_m^T & \Phi_o \end{pmatrix} \begin{pmatrix} I_j \\ K_j^T \end{pmatrix}.\end{aligned}$$

By careful inspection, it is obtained that indeed

$$P_j G_{j+1} P_j^T = \begin{pmatrix} \Phi_e & \Phi_m \\ \Phi_m^T & \Phi_o \end{pmatrix},$$

which concludes the proof of the lemma. ■

A similar statement holds for the inner-product matrix of the wavelets

$$H_j = (\langle \psi_{j,k}, \psi_{j,l} \rangle)_{k,l=0}^{3\cdot 2^{j+1}-1}.$$

Lemma 4.2. *For $j \in \mathbb{N}_0$,*

$$H_j = (-K_j^T \quad I_j) P_j G_{j+1} P_j^T \begin{pmatrix} -K_j \\ I_j \end{pmatrix}.$$

Proof: This time we use the two-scale relation for the wavelets given by Theorem 3.2, to obtain that

$$\begin{aligned}H_j &= \Phi_o - K_j^T \Phi_m - \Phi_m^T K_j + K_j^T \Phi_e K_j \\ &= (-K_j^T \quad I_j) \begin{pmatrix} \Phi_e & \Phi_m \\ \Phi_m^T & \Phi_o \end{pmatrix} \begin{pmatrix} -K_j \\ I_j \end{pmatrix}.\end{aligned}$$ ■

In a similar way, the orthogonality relations

$$\langle \phi_{j,k}, \psi_{j,l} \rangle = 0, \quad j,k = 0,\ldots,3\cdot 2^{j+1}-1$$

of Theorem A can be written in matrix terms using Theorems 3.1 and 3.2:

$$(I_j \quad K_j)P_jG_{j+1}P_j^T\begin{pmatrix} -K_j \\ I_j \end{pmatrix} = (-K_j^T \quad I_j)P_jG_{j+1}P_j^T\begin{pmatrix} I_j \\ K_j^T \end{pmatrix} = (0).$$

Putting all the above matrix relations together, one obtains

$$\begin{pmatrix} I_j & K_j \\ -K_j^T & I_j \end{pmatrix} P_jG_{j+1}P_j^T \begin{pmatrix} I_j & -K_j \\ K_j^T & I_j \end{pmatrix} = \begin{pmatrix} G_j & 0 \\ 0 & H_j \end{pmatrix}.$$

Thus we have

Proposition 4.1. *For $j \in \mathbb{N}_0$, the two-scale relation for the inner-product matrices G_j, H_j and G_{j+1} is given by*

$$\begin{pmatrix} G_j & 0 \\ 0 & H_j \end{pmatrix} = C_jP_jG_{j+1}P_j^TC_j^T.$$

As a next step, it is possible to describe in detail the entries of the inverse matrix of G_j.

Theorem 4.1. [5] *For any $j \in \mathbb{N}_0$, the inverse matrix for the inner-product matrix G_j is given by*

$$G_j^{-1} = (\alpha_{j,r,s})_{r,s=0}^{3\cdot 2^{j+1}-1},$$

where the coefficients for $r, s = 0,\ldots,3\cdot 2^{j+1}-1$ are given as

$$\alpha_{j,r,s}^F = 3\cdot 2^{j+1}\delta_{r,s} + (-1)^{r-s} \quad \text{and}$$

$$\alpha_{j,r,s}^D = 3\cdot 2^{j+1}\delta_{r,s} + \sum_{\ell=2^{j+1}+1}^{2^{j+2}-1}\left(\frac{2^{2j+1}}{\ell^2 - 3\ell 2^{j+1} + 5\cdot 2^{2j+1}} - 1\right)\cos\frac{\ell(r-s)\pi}{3\cdot 2^j}.$$

In order to find the decomposition sequences for the orthogonal sum $V_{j+1} = V_j \oplus W_j$, it is convenient to make use of the so-called dual scaling functions and wavelets. The concept of duality was introduced in [3] for a multiresolution analysis of $L^2(\mathbb{R})$ and is used here in a form suitably adapted to our purposes.

Definition 4.1. *For any $j \in \mathbb{N}_0$, the functions $\tilde{\phi}_{j,r} \in V_j, r = 0,\ldots,3\cdot 2^{j+1}-1$, uniquely determined by the conditions*

$$\langle \tilde{\phi}_{j,r}, \phi_{j,k} \rangle = \delta_{r,k} \quad \text{for all} \quad r,k = 0,\ldots,3\cdot 2^{j+1}-1,$$

are called dual scaling functions (or dual to the functions $\phi_{j,r}$).

Note that the dual scaling functions $\tilde{\phi}_{j,r}$ lie in the same space V_j as the original scaling functions $\phi_{j,k}$. Consequently, the dual functions can be written as linear combinations of these scaling functions. In fact, for any $j \in \mathbb{N}_0$,

$$\tilde{\phi}_{j,r} = \sum_{s=0}^{3\cdot 2^{j+1}-1} \alpha_{j,r,s}\phi_{j,s}.$$

Thus, the duality conditions lead to a linear system of equations for each dual function $\tilde{\phi}_{j,r}$, $r = 0, \ldots, 3 \cdot 2^{j+1} - 1$, namely,

$$\sum_{s=0}^{3\cdot 2^{j+1}-1} \alpha_{j,r,s}\langle \phi_{j,k}, \phi_{j,s}\rangle = \delta_{r,k}, \quad k = 0, \ldots, 3 \cdot 2^{j+1} - 1,$$

with G_j as its coefficient matrix and different right hand sides for the different dual functions.

Definition 4.2. *For $j \in \mathbb{N}_0$, the functions $\tilde{\psi}_{j,r} \in W_j, r = 0, \ldots, 3 \cdot 2^{j+1} - 1$, uniquely determined by the conditions*

$$\langle \tilde{\psi}_{j,r}, \psi_{j,k}\rangle = \delta_{r,k} \quad \text{for all} \quad r, k = 0, \ldots, 3 \cdot 2^{j+1} - 1,$$

are called dual wavelets (or dual to the functions $\psi_{j,r}$).

As for the scaling functions, the dual wavelets $\tilde{\psi}_{j,r}$ lie in the same space W_j as the original wavelets $\psi_{j,k}$. Therefore, also the dual wavelets can be written as linear combinations of the original ones, i.e., for $r = 0, \ldots, 3 \cdot 2^{j+1} - 1$, $\tilde{\psi}_{j,r} = \sum_{s=0}^{3\cdot 2^{j+1}-1} \beta_{j,r,s}\psi_{j,s}$, where the $\beta_{j,r,s}$ are the entries of the matrix H_j^{-1}.

Remark. The dual scaling functions are again translates of a single function just like the original ones, as the inner-product matrices are circulant, i.e.,

$$\begin{aligned}\tilde{\phi}_{j,r}(x) &= \sum_{s=0}^{3\cdot 2^{j+1}-1} \alpha_{j,r,s}\phi_{j,0}(x - \frac{s\pi}{3\cdot 2^j}) = \sum_{s=0}^{3\cdot 2^{j+1}-1} \alpha_{j,0,s-r}\phi_{j,0}(x - \frac{s\pi}{3\cdot 2^j}) \\ &= \sum_{s=0}^{3\cdot 2^{j+1}-1} \alpha_{j,0,s}\phi_{j,0}(x - \frac{r\pi}{3\cdot 2^j} - \frac{s\pi}{3\cdot 2^j}) = \tilde{\phi}_{j,0}(x - \frac{r\pi}{3\cdot 2^j})\end{aligned}$$

and, analogously, for the dual wavelets

$$\tilde{\psi}_{j,r}(x) = \tilde{\psi}_{j,0}(x - \frac{r\pi}{3\cdot 2^j}), \quad r = 0, \ldots, 3 \cdot 2^{j+1} - 1.$$

§5. Decomposition Sequences

The other pair of important sequences are the so-called decomposition sequences. As $V_{j+1} = V_j \oplus W_j$ for $j \in \mathbb{N}_0$, any $\phi_{j+1,n} \in V_{j+1}$ can be written as a linear combination of the basis functions of V_j and W_j, i.e., $\phi_{j,k}$ and $\psi_{j,k}$. In the following, we will distinguish two cases.

Theorem 5.1. [5] *For any $j \in \mathbb{N}_0$ and $m = 0, \ldots, 3 \cdot 2^{j+1} - 1$,*

$$\phi_{j+1,2m}(x) = \sum_{k=0}^{3\cdot 2^{j+1}-1} \left(a_{j,m,k}\phi_{j,k}(x) + b_{j,m,k}\psi_{j,k}(x) \right),$$

where for $n = 0, \ldots, 3 \cdot 2^{j+1} - 1$, the decomposition coefficients $a_{j,m,n}$ and $b_{j,m,n}$ are given by

$$\begin{aligned} a_{j,m,n} &= \frac{\alpha_{j,m,n}}{3 \cdot 2^{j+2}}, \\ b_{j,m,n} &= -\sum_{s=0}^{3\cdot 2^{j+1}-1} \frac{\alpha_{j,m,s}}{3 \cdot 2^{j+2}} \phi_{j,s}\left(\frac{(2n+1)\pi}{3 \cdot 2^{j+1}} \right) \\ &= -\frac{\tilde{\phi}_{j,m}\left(\frac{(2n+1)\pi}{3\cdot 2^{j+1}} \right)}{3 \cdot 2^{j+2}}, \end{aligned}$$

with the terms $\alpha_{j,m,n}$ being the coefficients of the dual scaling functions as described in Theorem 4.1.

For an even more detailed description of the entries $a_{j,m,n}$ and $b_{j,m,n}$, see Corollary 5.1 in [5].

For the basis functions in V_{j+1} with odd indices the result is

Theorem 5.2. [5] *For any $j \in \mathbb{N}_0$ and $m = 0, \ldots, 3 \cdot 2^{j+1} - 1$,*

$$\phi_{j+1,2m+1}(x) = \sum_{k=0}^{3\cdot 2^{j+1}-1} \left(\tilde{a}_{j,m,k}\phi_{j,k}(x) + \tilde{b}_{j,m,k}\psi_{j,k}(x) \right),$$

where for $n = 0, \ldots, 3 \cdot 2^{j+1} - 1$, the decomposition coefficients $\tilde{a}_{j,m,n}$ and $\tilde{b}_{j,m,n}$ are given as

$$\begin{aligned} \tilde{a}_{j,m,n} &= -b_{j,n,m}, \\ \tilde{b}_{j,m,n} &= a_{j,m,n}, \end{aligned}$$

and the terms $a_{j,m,n}$ and $b_{j,m,n}$ are defined in Theorem 5.1.

By slightly modifying the symmetric and circulant duality matrix of Section 4, i.e.,

$$\tilde{G}_j^{-1} = \frac{1}{3 \cdot 2^{j+2}} G_j^{-1} = \left(\frac{\alpha_{j,r,s}}{3 \cdot 2^{j+2}} \right)_{r,s=0}^{3\cdot 2^{j+1}-1},$$

one obtains by using the notation of Section 3 that the decomposition relations of Theorems 5.1 and 5.2 can be expressed as

$$P_j \underline{\phi}_{j+1} = D_j \begin{pmatrix} \underline{\phi}_j \\ \underline{\psi}_j \end{pmatrix},$$

using the *decomposition matrix* with circulant blocks

$$D_j = C_j^{-1} = \begin{pmatrix} \tilde{G}_j^{-1} & -\tilde{G}_j^{-1} K_j \\ K_j^T \tilde{G}_j^{-1} & \tilde{G}_j^{-1} \end{pmatrix} .$$

As $V_{j+1} = V_j \oplus W_j$, a function $f_{j+1} \in V_{j+1}$ can be written uniquely as

$$f_{j+1} = f_j + g_j, \quad \text{with} \quad f_j \in V_j \quad \text{and} \quad g_j \in W_j \quad .$$

Using the basis functions of these spaces, one obtains

$$f_{j+1}(x) = \sum_{s=0}^{3\cdot 2^{j+2}-1} c_s^{j+1} \phi_{j+1,s}(x),$$

$$f_j(x) = \sum_{s=0}^{3\cdot 2^{j+2}-1} c_s^j \phi_{j,s}(x) \quad \text{and} \quad g_j(x) = \sum_{s=0}^{3\cdot 2^{j+2}-1} d_s^j \psi_{j,s}(x) \quad .$$

Denoting coefficient vectors by $\underline{c}_j^T = (c_0^j, c_1^j, \ldots, c_{3\cdot 2^{j+1}-1}^j)$ and also $\underline{d}_j^T = (d_0^j, d_1^j, \ldots, d_{3\cdot 2^{j+1}-1}^j)$, respectively, one obtains

$$f_{j+1} = \underline{c}_{j+1}^T \underline{\phi}_{j+1}, \quad f_j = \underline{c}_j^T \underline{\phi}_j \quad \text{and} \quad g_j = \underline{d}_j^T \underline{\psi}_j \quad .$$

As $\underline{c}_{j+1}^T \underline{\phi}_{j+1} = (P_j \underline{c}_{j+1})^T P_j \underline{\phi}_{j+1}$, the matrix form of the decomposition relation yields

$$f_{j+1} = (P_j \underline{c}_{j+1})^T P_j \underline{\phi}_{j+1} = (P_j \underline{c}_{j+1})^T D_j \begin{pmatrix} \underline{\phi}_j \\ \underline{\psi}_j \end{pmatrix} .$$

On the other hand,

$$f_j + g_j = \begin{pmatrix} \underline{c}_j^T & \underline{d}_j^T \end{pmatrix} \begin{pmatrix} \underline{\phi}_j \\ \underline{\psi}_j \end{pmatrix} .$$

Comparing coefficients leads to

$$\begin{pmatrix} \underline{c}_j^T & \underline{d}_j^T \end{pmatrix} = (P_j \underline{c}_{j+1})^T D_j,$$

and taking the transpose finally gives the matrix form of one step of the *decomposition algorithm*

$$\begin{pmatrix} \underline{c}_j \\ \underline{d}_j \end{pmatrix} = D_j^T P_j \underline{c}_{j+1} \quad .$$

Multiplying by the inverse $(D_j^T)^{-1} = C_j^T$ yields the matrix representation of one step of the *reconstruction algorithm*

$$P_j \underline{c}_{j+1} = C_j^T \begin{pmatrix} \underline{c}_j \\ \underline{d}_j \end{pmatrix} .$$

§6. The Duality Principle

The duality principle in the sense that will be used here was first introduced by C. K. Chui and J. Z. Wang in [3] for spline wavelets on the real axis and described in a much more general form in the monograph [1]. In short, it says that the decomposition sequences for the original scaling functions and wavelets become the reconstruction sequences for the dual scaling functions and wavelets, as well as the reconstruction sequences for the original functions are the decomposition sequences for the duals.

In this context of trigonometric scaling functions and wavelets, the duality principle can be formulated as follows.

Theorem 6.1. *(The duality principle). For any $j \in \mathbb{N}_0$, the decomposition matrix for the dual functions is C_j^T and the reconstruction matrix for the duals is D_j^T, i.e., for the original scaling functions and wavelets,*

$$\begin{pmatrix} \underline{\phi}_j \\ \underline{\psi}_j \end{pmatrix} = C_j P_j \underline{\phi}_{j+1} \qquad (\textit{reconstruction}),$$

$$P_j \underline{\phi}_{j+1} = D_j \begin{pmatrix} \underline{\phi}_j \\ \underline{\psi}_j \end{pmatrix} \qquad (\textit{decomposition}),$$

while for the dual functions

$$\begin{pmatrix} \underline{\tilde{\phi}}_j \\ \underline{\tilde{\psi}}_j \end{pmatrix} = D_j^T P_j \underline{\tilde{\phi}}_{j+1} \qquad (\textit{reconstruction}),$$

$$P_j \underline{\tilde{\phi}}_{j+1} = C_j^T \begin{pmatrix} \underline{\tilde{\phi}}_j \\ \underline{\tilde{\psi}}_j \end{pmatrix} \qquad (\textit{decomposition}).$$

Proof: Using $D_j = C_j^{-1}$ and $P_j^{-1} = P_j^T$, Proposition 4.1 can be rewritten as

$$G_{j+1} = P_j^T D_j \begin{pmatrix} G_j & 0 \\ 0 & H_j \end{pmatrix} D_j^T P_j.$$

Therefore,

$$G_{j+1}^{-1} = P_j^T C_j^T \begin{pmatrix} G_j^{-1} & 0 \\ 0 & H_j^{-1} \end{pmatrix} C_j P_j.$$

Consequently, using the reconstruction relation for $P_j \underline{\phi}_{j+1}$,

$$\begin{aligned} P_j \underline{\tilde{\phi}}_{j+1} &= P_j G_{j+1}^{-1} \underline{\phi}_{j+1} \\ &= C_j^T \begin{pmatrix} G_j^{-1} & 0 \\ 0 & H_j^{-1} \end{pmatrix} C_j P_j \underline{\phi}_{j+1} \\ &= C_j^T \begin{pmatrix} G_j^{-1} & 0 \\ 0 & H_j^{-1} \end{pmatrix} \begin{pmatrix} \underline{\phi}_j \\ \underline{\psi}_j \end{pmatrix} \\ &= C_j^T \begin{pmatrix} \underline{\tilde{\phi}}_j \\ \underline{\tilde{\psi}}_j \end{pmatrix} \end{aligned}$$

and the reconstruction relation for the duals follows by a simple matrix inversion. ■

References

1. Chui, C. K., *An Introduction to Wavelets*, Academic Press, New York, 1992.
2. Chui, C. K. and H. N. Mhaskar, On trigonometric wavelets, Constr. Approx. **9** (1993), 167-190.
3. Chui, C. K. and J. Z. Wang, On compactly supported spline wavelets and a duality principle, Trans. Amer. Math. Soc. **330** (1992), 903–915.
4. Lorentz, R. A. and A. A. Sahakian, Orthogonal trigonometric Schauder bases of optimal degree for $C(0, 2\pi)$, The Journal of Fourier Analysis and Applications (to appear).
5. Prestin, J. and E. Quak, Trigonometric interpolation and wavelet decompositions, CAT Report 296, Texas A&M University, 1993 (submitted).
6. Privalov, A. A., On an orthogonal trigonometric basis, Mat. Sbornik **182** **(3)** (1991), 384–394.
7. Zygmund, A., *Trigonometric Series*, Cambridge University Press, Cambridge, 1959.

Jürgen Prestin
FB Mathematik
Universität Rostock
Universitätsplatz 1
18051 Rostock, GERMANY
prestin@ mathematik.uni-rostock.d400.de

Ewald Quak
Center for Approximation Theory
Department of Mathematics
Texas A&M University
College Station, Texas 77843-3368
quak@ euler.math.tamu.edu

Decomposition and Reconstruction Algorithms for Bivariate Spline Wavelets on the Unit Square

E. Quak and N. Weyrich

Abstract. The aim of this paper is to describe decomposition and reconstruction algorithms for bivariate spline wavelets on the unit square $[0,1]^2$ using a tensor-product approach based on the univariate results in [2] and [5]. As an application, an example for image compression using bivariate spline wavelets is included.

§1. Introduction and Preliminaries

Univariate spline wavelets for the bounded interval were introduced by Chui and Quak in [2], and the relevant decomposition and recontruction algorithms were described in detail in [5]. The purpose of this paper is to provide a discussion of the corresponding algorithms for a tensor-product approach on $[0,1]^2$. After a brief review of the necessary univariate results, tensor-product wavelets are introduced in Section 2, the bivariate decomposition and reconstruction algorithms are treated in Sections 3 and 4, respectively. In Section 5, a numerical example illustrates the use of these techniques for image compression.

The order $m \in \mathbb{N}$ of the splines will be fixed. If possible, the index m will be suppressed for notational convenience. Let $\mathbf{t}_m^{(j)} := \{t_k^{(j)}\}_{k=-m+1}^{2^j+m-1}$, with $t_{-m+1}^{(j)} = t_{-m+2}^{(j)} = \cdots = t_0^{(j)} = 0$, $t_{2^j}^{(j)} = t_{2^j+1}^{(j)} = \cdots = t_{2^j+m-1}^{(j)} = 1$, and $t_k^{(j)} = k2^{-j} \quad (k = 1, \ldots, 2^j - 1)$, be a knot sequence on $[0,1]$ for any $j \in \mathbb{N}_0$. A nested sequence of subspaces $V_{[0,1]}^j$ is obtained by setting

$$V_{[0,1]}^j := S_m^j = \left\{ s \in C^{m-2}[0,1] \ : \ s_{|(t_k^{(j)}, t_{k+1}^{(j)})} \in \Pi_{m-1} \ (k = 0, \ldots, 2^j - 1) \right\},$$

and the spaces $W_{[0,1]}^j$ are the corresponding orthogonal complements such that

$$V_{[0,1]}^{j+1} = V_{[0,1]}^j \oplus W_{[0,1]}^j \quad (j \in \mathbb{Z}_+). \tag{1.1}$$

Wavelets, Images and Surface Fitting
P. J. Laurent, A. Le Méhauté, and L. L. Schumaker (eds.), pp. 419–428.

ISBN 1-56881-040-7.

B-splines are defined as $B^j_{m,i}(x) := (t^{(j)}_{i+m} - t^{(j)}_i)[t^{(j)}_i, t^{(j)}_{i+1}, \ldots, t^{(j)}_{i+m}]_t (t-x)^{m-1}_+$, where $[\cdot, \ldots, \cdot]_t$ is the m-th divided difference of $(t-x)^{m-1}_+$ with respect to the variable t. Note that $B^j_{m,i}(x) = N_m(2^j x - i) \quad (i = 0, \ldots, 2^j - m)$, where $N_m(x) := m[0,1,\ldots,m]_t(t-x)^{m-1}_+$ is the cardinal B-spline of order m, while for $i = -m+1, \ldots, -1$, $B^j_{m,i}$ contains a multiple knot at 0 and for $i = 2^j - m + 1, \ldots, 2^j - 1$, a multiple knot at 1. To separate the effects of the two endpoints, it is necessary to restrict to $j \geq j_0$, where j_0 is the smallest integer such that $2^{j_0} \geq 2m - 1$. A basis for $V^j_{[0,1]}$, $j \geq j_0$, is then given by

$$\phi^j_k(x) = \begin{cases} B^{j_0}_{m,k}(2^{j-j_0}x) & \text{if } k = -m+1, \ldots, -1, \\ B^{j_0}_{m,2^j-m-k}(1 - 2^{j-j_0}x) & \text{if } k = 2^j - m + 1, \ldots, 2^j - 1, \\ N_m(2^j x - k) & \text{if } k = 0, \ldots, 2^j - m. \end{cases} \tag{1.2}$$

Similarly, a basis for $W^j_{[0,1]}$, $j \geq j_0$, can be found as

$$\psi^j_k(x) = \begin{cases} \psi^{j_0}_k(2^{j-j_0}x) & \text{if } k = -m+1, \ldots, -1, \\ \psi^{j_0}_{2^j-2m+1-k}(1 - 2^{j-j_0}x) & \text{if } k = 2^j - 2m + 2, \ldots, 2^j - m, \\ \psi^{j_0}_0(2^j x - k) & \text{if } k = 0, \ldots, 2^j - 2m + 1, \end{cases} \tag{1.3}$$

where the inner wavelets are again dilated and translated versions of the well-known Chui-Wang spline wavelets. For $k = -m+1, \ldots, -1$ and $k = 2^j - 2m + 2, \ldots, 2^j - m$, special boundary spline wavelets are obtained for the endpoints 0 and 1, respectively. For the details of this construction see [2,5].

The use of dual scaling functions and dual wavelets for splines originated in [3] and was treated in far greater generality in [1].

Definition 1.1. *(i) The functions* $\tilde{\phi}^j_k \in V^j_{[0,1]}$ $(k = -m+1, \ldots, 2^j - 1)$ *are called dual to* $\phi^j_{\tilde{k}} \in V^j_{[0,1]}$ *(or dual scaling functions) if*

$$\left\langle \tilde{\phi}^j_k, \phi^j_{\tilde{k}} \right\rangle = \int_0^1 \tilde{\phi}^j_k(x) \phi^j_{\tilde{k}}(x)\, dx = \delta_{k,\tilde{k}} \quad (k, \tilde{k} = -m+1, \ldots, 2^j - 1).$$

(ii) The functions $\tilde{\psi}^j_k \in W^j_{[0,1]}$ $(k = -m+1, \ldots, 2^j - m)$ *are called dual to* $\psi^j_{\tilde{k}} \in W^j_{[0,1]}$ *(or dual wavelets) if*

$$\left\langle \tilde{\psi}^j_k, \psi^j_{\tilde{k}} \right\rangle = \int_0^1 \tilde{\psi}^j_k(x) \psi^j_{\tilde{k}}(x)\, dx = \delta_{k,\tilde{k}} \quad (k, \tilde{k} = -m+1, \ldots, 2^j - m).$$

Note that on each level the dual scaling functions span the same space as the scaling functions and the dual wavelets span the same space as the wavelets.

As $V^j_{[0,1]} \subset V^{j+1}_{[0,1]}$ and $W^j_{[0,1]} \subset V^{j+1}_{[0,1]}$, we have

$$\phi^j_k(x) = \sum_{l=-m+1}^{2^{j+1}-1} p^j_{l,k}\phi^{j+1}_l(x) \quad (k = -m+1, \ldots, 2^j - 1), \tag{1.4}$$

and

$$\psi^j_k(x) = \sum_{l=-m+1}^{2^{j+1}-1} q^j_{l,k}\phi^{j+1}_l(x) \quad (k = -m+1, \ldots, 2^j - m). \tag{1.5}$$

Taking inner products with $\tilde{\phi}^{j+1}_{\tilde{k}}$ yields

$$p^j_{\tilde{k},k} = \left\langle \tilde{\phi}^{j+1}_{\tilde{k}}, \phi^j_k \right\rangle \quad (\tilde{k} = -m+1, \ldots, 2^{j+1} - 1\ ;\ k = -m+1, \ldots, 2^j - 1),$$

$$q^j_{\tilde{k},k} = \left\langle \tilde{\phi}^{j+1}_{\tilde{k}}, \psi^j_k \right\rangle \quad (\tilde{k} = -m+1, \ldots, 2^{j+1} - 1\ ;\ k = -m+1, \ldots, 2^j - m).$$

For algorithms to compute the values of $p^j_{\tilde{k},k}$ and $q^j_{\tilde{k},k}$, see Lemma 4.1 in [5].

§2. Tensor-Product Wavelets

Starting from the univariate wavelets on the interval $[0,1]$, bivariate wavelets on the unit square $T := [0,1] \times [0,1]$ are obtained by using tensor-products. Let

$$V^j_T := V^j_{[0,1]} \times V^j_{[0,1]},\ i.e.,$$

$$V^j_T = \text{span}\ \left\{\phi^j_k(x)\phi^j_l(y) :\ (k, l \in I_s(j))\right\},$$

where for notational convenience

$$I_s(j) := \{-m+1, \ldots, 2^j - 1\} \text{ and } I_w(j) := \{-m+1, \ldots, 2^j - m\}.$$

This results in a nested sequence of subspaces $V^0_T \subset V^1_T \subset \cdots \subset V^j_T \cdots$ with

$$\text{clos}_{L^2}\left(\bigcup_{j\geq 0} V^j_T\right) = L^2(T).$$

The complementary orthogonal subspaces W^j_T are

$$\begin{aligned} W^j_T = \text{span}\ \{ & \psi^j_k(x)\phi^j_l(y) :\ (k \in I_w(j),\ l \in I_s(j)), \\ & \phi^j_k(x)\psi^j_l(y) :\ (k \in I_s(j),\ l \in I_w(j)), \\ & \psi^j_k(x)\psi^j_l(y) :\ (k \in I_w(j),\ l \in I_w(j))\ \} \quad (j \in \mathbb{N}). \end{aligned}$$

Indeed (1.1) yields that $W^j_T \perp V^j_T$ and $V^j_T \oplus W^j_T = V^{j+1}_T$ $(j \in \mathbb{N})$. Hence

$$L^2(T) = V^0_T \bigoplus_{j\in\mathbb{N}} W^j_T.$$

Note that by definition

$$W^j_T = \left(W^j_{[0,1]} \times V^j_{[0,1]}\right) \oplus \left(V^j_{[0,1]} \times W^j_{[0,1]}\right) \oplus \left(W^j_{[0,1]} \times W^j_{[0,1]}\right).$$

§3. The Decomposition Algorithm

Our aim is to identify different frequency components of a given function $f \in L^2(T)$ (later it will be sufficient to know only discrete values z_k of f at points (x_k, y_k), i.e., $z_k = f(x_k, y_k)$). Using spline interpolation, f is mapped to a function $f^N \in V_T^N$ for a sufficiently large $N \in \mathbb{N}$. Then the multiresolution analysis can be used to obtain the decomposition equation

$$f^{j+1}(x,y) = f^j(x,y) + g^j(x,y) \quad (j = j_0, \ldots, N-1), \tag{3.1}$$

where $j_0 \in \mathbb{N}$ is the smallest number satisfying $2^{j_0} \geq 2m-1$.

In order to manipulate $f^j \in V_T^j$ and $g^j \in W_T^j$ by only working with their coefficient sequences, the basis representations

$$f^j(x,y) = \sum_{k=-m+1}^{2^j-1} \sum_{l=-m+1}^{2^j-1} f_{k,l}^j \phi_k^j(x)\phi_l^j(y) \tag{3.2}$$

and

$$g^j(x,y) = \sum_{k=-m+1}^{2^j-m} \sum_{l=-m+1}^{2^j-1} {}^1g_{k,l}^j \psi_k^j(x)\phi_l^j(y) \tag{3.3}$$

$$+ \sum_{k=-m+1}^{2^j-1} \sum_{l=-m+1}^{2^j-m} {}^2g_{k,l}^j \phi_k^j(x)\psi_l^j(y) + \sum_{k=-m+1}^{2^j-m} \sum_{l=-m+1}^{2^j-m} {}^3g_{k,l}^j \psi_k^j(x)\psi_l^j(y)$$

are used for $(j = j_0, \ldots, N)$, where ϕ_k^j $(k \in I_s(j))$ and ψ_l^j $(l \in I_w(j))$ are given by (1.2) and (1.3). The first step of the algorithm is a transformation of the basis representation (3.2), (3.3) into the dual basis representation. Therefore,

$$\sum_{k=-m+1}^{2^j-1} \sum_{l=-m+1}^{2^j-1} f_{k,l}^j \phi_k^j(x)\phi_l^j(y) + \sum_{k=-m+1}^{2^j-m} \sum_{l=-m+1}^{2^j-1} {}^1g_{k,l}^j \psi_k^j(x)\phi_l^j(y) \tag{3.4}$$

$$+ \sum_{k=-m+1}^{2^j-1} \sum_{l=-m+1}^{2^j-m} {}^2g_{k,l}^j \phi_k^j(x)\psi_l^j(y) + \sum_{k=-m+1}^{2^j-m} \sum_{l=-m+1}^{2^j-m} {}^3g_{k,l}^j \psi_k^j(x)\psi_l^j(y)$$

$$= \sum_{k=-m+1}^{2^j-1} \sum_{l=-m+1}^{2^j-1} \tilde{f}_{k,l}^j \tilde{\phi}_k^j(x)\tilde{\phi}_l^j(y) + \sum_{k=-m+1}^{2^j-m} \sum_{l=-m+1}^{2^j-1} {}^1\tilde{g}_{k,l}^j \tilde{\psi}_k^j(x)\tilde{\phi}_l^j(y)$$

$$+ \sum_{k=-m+1}^{2^j-1} \sum_{l=-m+1}^{2^j-m} {}^2\tilde{g}_{k,l}^j \tilde{\phi}_k^j(x)\tilde{\psi}_l^j(y) + \sum_{k=-m+1}^{2^j-m} \sum_{l=-m+1}^{2^j-m} {}^3\tilde{g}_{k,l}^j \tilde{\psi}_k^j(x)\tilde{\psi}_l^j(y),$$

where the dual functions $\tilde{\phi}_k^l$ and $\tilde{\psi}_k^l$ were introduced in Definition 1.1. Taking inner products with $\phi_{\bar{k}}^j(x)\phi_{\bar{l}}^j(y)$ yields for $j = N$

$$\tilde{f}_{\bar{k},\bar{l}}^N = \sum_{k=-m+1}^{2^N-1} \sum_{l=-m+1}^{2^N-1} f_{k,l}^N \langle \phi_k^N, \phi_{\bar{k}}^N \rangle \langle \phi_l^N, \phi_{\bar{l}}^N \rangle \text{ and } \tilde{F}^N = \left(C^N\right)^T F^N C^N,$$

respectively, with

$$\tilde{F}^j := \left(\tilde{f}^j_{k,l}\right)^{2^j-1}_{k=-m+1,}{}^{2^j-1}_{l=-m+1}, \quad F^j := \left(f^j_{k,l}\right)^{2^j-1}_{k=-m+1,}{}^{2^j-1}_{l=-m+1},$$

and

$$C^j := \left(\left\langle \phi^j_k, \phi^j_l \right\rangle\right)^{2^j-1}_{k,l=-m+1} \qquad (j = j_0, \ldots, N).$$

In the second step of the algorithm, the dual coefficients are decomposed using the two-scale relations (1.4) and (1.5). Indeed, by writing the decomposition equation (3.1) in terms of the dual basis and taking inner products with the appropriate basis functions, we obtain for $j = j_0, \ldots, N-1$

$$\begin{aligned}\sum_{k=-m+1}^{2^{j+1}-1} \sum_{l=-m+1}^{2^{j+1}-1} \tilde{f}^{j+1}_{k,l} \tilde{\phi}^{j+1}_k(x) \tilde{\phi}^{j+1}_l(y) &= \sum_{k=-m+1}^{2^j-1} \sum_{l=-m+1}^{2^j-1} \tilde{f}^j_{k,l} \tilde{\phi}^j_k(x) \tilde{\phi}^j_l(y) \\ &+ \sum_{k=-m+1}^{2^j-m} \sum_{l=-m+1}^{2^j-1} {}^1\tilde{g}^j_{k,l} \tilde{\psi}^j_k(x) \tilde{\phi}^j_l(y) + \sum_{k=-m+1}^{2^j-1} \sum_{l=-m+1}^{2^j-m} {}^2\tilde{g}^j_{k,l} \tilde{\phi}^j_k(x) \tilde{\psi}^j_l(y) \\ &+ \sum_{k=-m+1}^{2^j-m} \sum_{l=-m+1}^{2^j-m} {}^3\tilde{g}^j_{k,l} \tilde{\psi}^j_k(x) \tilde{\psi}^j_l(y)\end{aligned}$$

with

$$\tilde{f}^j_{\tilde{k},\tilde{l}} = \sum_{k=-m+1}^{2^{j+1}-1} \sum_{l=-m+1}^{2^{j+1}-1} \tilde{f}^{j+1}_{k,l} \left\langle \tilde{\phi}^{j+1}_k, \phi^j_{\tilde{k}} \right\rangle \left\langle \tilde{\phi}^{j+1}_l, \phi^j_{\tilde{l}} \right\rangle, \quad (\tilde{k}, \tilde{l} \in I_s(j)),$$

$${}^1\tilde{g}^j_{\tilde{k},\tilde{l}} = \sum_{k=-m+1}^{2^{j+1}-1} \sum_{l=-m+1}^{2^{j+1}-1} f^{j+1}_{k,l} \left\langle \tilde{\phi}^{j+1}_k, \psi^j_{\tilde{k}} \right\rangle \left\langle \tilde{\phi}^{j+1}_l, \phi^j_{\tilde{l}} \right\rangle, \quad (\tilde{k} \in I_w(j)\,; \tilde{l} \in I_s(j)),$$

$${}^2\tilde{g}^j_{\tilde{k},\tilde{l}} = \sum_{k=-m+1}^{2^{j+1}-1} \sum_{l=-m+1}^{2^{j+1}-1} f^{j+1}_{k,l} \left\langle \tilde{\phi}^{j+1}_k, \phi^j_{\tilde{k}} \right\rangle \left\langle \tilde{\phi}^{j+1}_l, \psi^j \right\rangle, \quad (\tilde{k} \in I_s(j)\,; \tilde{l} \in I_w(j)),$$

$${}^3\tilde{g}^j_{\tilde{k},\tilde{l}} = \sum_{k=-m+1}^{2^{j+1}-1} \sum_{l=-m+1}^{2^{j+1}-1} f^{j+1}_{k,l} \left\langle \tilde{\phi}^{j+1}_k, \psi^j_{\tilde{k}} \right\rangle \left\langle \tilde{\phi}^{j+1}_l, \psi^j_{\tilde{l}} \right\rangle, \quad (\tilde{k}, \tilde{l} \in I_w(j)).$$

To state a matrix version of the above equations, the following notations are used

$${}^1\tilde{G}^j := \left({}^1\tilde{g}^j_{k,l}\right)^{2^j-m}_{k=-m+1,}{}^{2^j-1}_{l=-m+1}, \quad {}^2\tilde{G}^j := \left({}^2\tilde{g}^j_{k,l}\right)^{2^j-1}_{k=-m+1,}{}^{2^j-m}_{l=-m+1},$$

$${}^3\tilde{G}^j := \left({}^3\tilde{g}^j_{k,l}\right)^{2^j-m}_{k=-m+1,}{}^{2^j-m}_{l=-m+1},$$

$$P^j := \left(p^j_{\tilde{k},k}\right)^{2^{j+1}-1 \quad 2^j-1}_{\tilde{k}=-m+1\,,\,k=-m+1}, Q^j := \left(q^j_{\tilde{k},k}\right)^{2^{j+1}-1 \quad 2^j-m}_{\tilde{k}=-m+1\,,\,k=-m+1},$$

where the matrices P^j and Q^j can be computed according to Lemma 4.1 in [5]. Hence the decomposition equations are

$$\tilde{F}^j = \left(P^j\right)^T \tilde{F}^{j+1} P^j, \quad {}^1\tilde{G}^j = \left(Q^j\right)^T \tilde{F}^{j+1} P^j,$$
$${}^2\tilde{G}^j = \left(P^j\right)^T \tilde{F}^{j+1} Q^j, \quad {}^3\tilde{G}^j = \left(Q^j\right)^T \tilde{F}^{j+1} Q^j.$$

In order to compute the coefficients in the original basis, it is necessary to perform a back transformation of the dual coefficients. Therefore, equation (3.4) is used one more time. By taking inner products with the relevant basis functions, we obtain for $j = j_0, \ldots, N$

$$\sum_{k=-m+1}^{2^j-m} \sum_{l=-m+1}^{2^j-1} {}^1g^j_{k,l} \left\langle \psi^j_k, \psi^j_{\tilde{k}} \right\rangle \left\langle \phi^j_l, \phi^j_{\tilde{l}} \right\rangle = {}^1\tilde{g}^j_{\tilde{k},\tilde{l}}, \quad (\tilde{k} \in I_w(j)\,; \tilde{l} \in I_s(j)),$$

$$\sum_{k=-m+1}^{2^j-1} \sum_{l=-m+1}^{2^j-m} {}^2g^j_{k,l} \left\langle \phi^j_k, \phi^j_{\tilde{k}} \right\rangle \left\langle \psi^j_l, \psi^j_{\tilde{l}} \right\rangle = {}^2\tilde{g}^j_{\tilde{k},\tilde{l}}, \quad (\tilde{k} \in I_s(j)\,; \tilde{l} \in I_w(j)),$$

$$\sum_{k=-m+1}^{2^j-m} \sum_{l=-m+1}^{2^j-m} {}^3g^j_{k,l} \left\langle \psi^j_k, \psi^j_{\tilde{k}} \right\rangle \left\langle \psi^j_l, \psi^j_{\tilde{l}} \right\rangle = {}^3\tilde{g}^j_{\tilde{k},\tilde{l}}, \quad (\tilde{k}, \tilde{l} \in I_w(j)).$$

Together with equation (3.4), this yields for $j = j_0, \ldots, N$, the matrix equations

$$\left(C^j\right)^T F^j C^j = \tilde{F}^j, \quad \left(D^j\right)^T {}^1G^j C^j = {}^1\tilde{G}^j$$
$$\left(C^j\right)^T {}^2G^j D^j = {}^2\tilde{G}^j, \quad \left(D^j\right)^T {}^3G^j D^j = {}^3\tilde{G}^j,$$

with

$$D^j := \left(\left\langle \psi^j_k, \psi^j_l \right\rangle\right)^{2^j-m}_{k,l=-m+1} = Q^{jT} C^{j+1} Q^j, \quad {}^1G^j := \left({}^1g^j_{k,l}\right)^{2^j-m \quad 2^j-1}_{k=-m+1,\, l=-m+1},$$

$${}^2G^j := \left({}^2g^j_{k,l}\right)^{2^j-1 \quad 2^j-m}_{k=-m+1,\, l=-m+1}, \quad {}^3G^j := \left({}^3g^j_{k,l}\right)^{2^j-m \quad 2^j-m}_{k=-m+1,\, l=-m+1}.$$

Hence the following decomposition algorithm is obtained.

Algorithm A1. Decomposition (bivariate case).

Input: $N \in \mathbb{N}$ starting level (sufficiently large),
$(x_k, y_l, z_{k,l})$ knots and function values
$(k, l = 1, \ldots, 2^N + m - 1)$,
$(x_k, y_l) \in T$, $z_{k,l} = f(x_k, y_l)$.

(i) Interpolate f in V^N_T, i.e., compute F^N such that

$$z_{k,l} = f(x_k, y_l) = \sum_{i=-m+1}^{2^N-1} \sum_{j=-m+1}^{2^N-1} f_{i,j}^N \phi_i^N(x_k)\phi_j^N(y_l) .$$

(ii) Compute $\tilde{F}^N = \left(C^N\right)^T F^N C^N$.

(iii) Compute for each level j $(j = N-1, \ldots, j_0)$

$$M = \left(P^j\right)^T \tilde{F}^{j+1}, \quad \tilde{F}^j = M \cdot P^j, \quad {}^2\tilde{G}^j = M \cdot Q^j,$$

$$M = \left(Q^j\right)^T \tilde{F}^{j+1}, \quad {}^1\tilde{G}^j = M \cdot P^j, \quad {}^3\tilde{G}^j = M \cdot Q^j.$$

(iv) Compute F^{j_0} from the matrix equation

$$\left(C^{j_0}\right)^T F^{j_0} C^{j_0} = \tilde{F}^{j_0} .$$

(v) Compute ${}^iG^j$ $(i = 1, 2, 3)$ for each level j $(j = j_0, \ldots, N-1)$ from the matrix equations

$$\left(D^j\right)^T {}^1G^j C^j = {}^1\tilde{G}^j, \quad \left(C^j\right)^T {}^2G^j D^j = {}^2\tilde{G}^j, \quad \left(D^j\right)^T {}^3G^j D^j = {}^3\tilde{G}^j.$$

Output: coefficient matrices F^{j_0}, ${}^iG^{N-1}, \ldots, {}^iG^{j_0}$ $\quad (i = 1, 2, 3)$

The basic memory requirement of this algorithm is for two matrices of dimension $N \times N$. The total number of multiplications needed to proceed from level $j+1$ to level j is $(28m - 14)2^j 2^j + \mathcal{O}\left(2^j\right)$, where $(2^j + m - 1)^2$ is the number of coefficients on the j-th level. Note that the local support of the scaling functions and wavelets means that C^j and D^j are banded matrices with bandwidth $2m - 1$ and $4m - 3$, respectively. Due to the dilation and translation relations (1.2) and (1.3), it is sufficient to compute the matrices C^j and D^j only for the level $j = j_0$. The same is true for the matrices P^j and Q^j, although they are not square matrices. In order to minimize the computational cost the symmetry of the two-scale relation sequences is used. To solve the matrix equations in step (iv) and (v) of the algorithm, a Cholesky decomposition of the matrices C^j and D^j is necessary. However, this can be done in advance and just once, since this decomposition does not depend on any input data.

§4. The Reconstruction Algorithm

In this section, the goal is to reconstruct f^N from the coefficient matrices obtained by Algorithm A1. Therefore, a tensor-product version of the reconstruction algorithm for spline wavelets in [1] is used, taking into account the different two-scale relation sequences for the boundary splines and boundary wavelets, respectively.

Inserting the basis representations (3.2) and (3.3) into the decomposition equation (3.1) and taking inner products with $\tilde{\phi}_{\tilde{k}}^{j+1}\tilde{\phi}_{\tilde{l}}^{j+1}$ yields

$$\begin{aligned} f_{\tilde{k},\tilde{l}}^{j+1} = & \sum_{k=-m+1}^{2^j-1} \sum_{l=-m+1}^{2^j-1} f_{k,l}^j \left\langle \phi_k^j, \tilde{\phi}_{\tilde{k}}^{j+1} \right\rangle \left\langle \phi_l^j, \tilde{\phi}_{\tilde{l}}^{j+1} \right\rangle \\ & + \sum_{k=-m+1}^{2^j-m} \sum_{l=-m+1}^{2^j-1} {}^1g_{k,l}^j \left\langle \psi_k^j, \tilde{\phi}_{\tilde{k}}^{j+1} \right\rangle \left\langle \phi_l^j, \tilde{\phi}_{\tilde{l}}^{j+1} \right\rangle \\ & + \sum_{k=-m+1}^{2^j-1} \sum_{l=-m+1}^{2^j-m} {}^2g_{k,l}^j \left\langle \phi_k^j, \tilde{\phi}_{\tilde{k}}^{j+1} \right\rangle \left\langle \psi_l^j, \tilde{\phi}_{\tilde{l}}^{j+1} \right\rangle \\ & + \sum_{k=-m+1}^{2^j-m} \sum_{l=-m+1}^{2^j-m} {}^3g_{k,l}^j \left\langle \psi_k^j, \tilde{\phi}_{\tilde{k}}^{j+1} \right\rangle \left\langle \psi_l^j, \tilde{\phi}_{\tilde{l}}^{j+1} \right\rangle . \end{aligned}$$

Hence for $j = j_0, \ldots, N-1$

$$F^{j+1} = P^j F^j \left(P^j\right)^T + Q^j \, {}^1G^j \left(P^j\right)^T + P^j \, {}^2G^j \left(Q^j\right)^T + Q^j \, {}^3G^j \left(Q^j\right)^T ,$$

and the following algorithm is obtained.

Algorithm A2. (Reconstruction.)

Input:	$N \in \mathbb{N}$	highest level as used in Algorithm A1,
	F^{j_0}	coefficients of the spline part on the lowest level - obtained from Algorithm A1,
	${}^iG^{j_0}, \ldots, {}^iG^{N-1}$	(i=1,2,3) coefficients of the wavelet parts - obtained from Algorithm A1.

(i) Compute for each level j ($j = j_0, \ldots, N-1$)

$$F^{j+1} = P^j \left(F^j \left(P^j\right)^T + {}^2G^j \left(Q^j\right)^T\right) + Q^j \left({}^1G^j \left(P^j\right)^T + {}^3G^j \left(Q^j\right)^T\right).$$

Output: F^N the spline coefficients of the highest level.

§5. Numerical Examples

The boundary wavelet approach was tested on the problem of image compression. The 512×512 pixel so-called "Lena" image was decomposed into 3 levels using cubic boundary wavelets. After decomposition the total number of nonzero spline and wavelet coefficients was reduced to 5% of the total number of original coefficients (compression ratio 1 : 20). A threshold criterion was used for this reduction of the number of coefficients (see [4]). Note that in this application there is no need to perform the interpolation (step (i)) in the decomposition algorithm. If the input data are taken to be the coefficients

Fig. 1. Top left (a), top right (b), bottom (c).

for step (ii) of Algorithm A1, the wavelet decomposition works as a subband filter. The number of multiplications per pixel for the decomposition algorithm is 40 in the cubic case and 17 in the linear case. To improve the quality of the compressed image, Algorithm A1 was modified to perform a wavelet packet decomposition using boundary wavelets (see [6]). Figure 1a shows the original image, while Figure 1b shows the compressed image and Figure 1c the compressed image with the same compression ratio of 1:20 using wavelet packets.

Acknowledgements. The second author was supported by Deutscher Akademischer Austauschdienst (DAAD).

References

1. Chui, C. K., *An Introduction to Wavelets*, Academic Press, Boston, 1992.
2. Chui, C. K. and E. Quak, Wavelets on a bounded interval, in *Numerical Methods of Approximation Theory, Vol. 9*, D. Braess and L. L. Schumaker (eds.), Birkhäuser, Basel, 1992, 53–75.
3. Chui, C. K. and J. Z. Wang, On compactly supported spline wavelets and a duality principle, Trans. Amer. Math. Soc. **330** (1992), 903–915.
4. Liu, Q., Image compression using spline wavelets and wavelet packets, Ph.D. thesis, Texas A&M University, 1993.
5. Quak, E. and N. Weyrich, Decomposition and reconstruction algorithms for spline wavelets on a bounded interval, CAT Report 294, 1993 (submitted).
6. Quak, E. and N. Weyrich, Wavelet packets for spline wavelets on a bounded interval, in preparation.

Ewald Quak and Norman Weyrich
Center for Approximation Theory
Department of Mathematics
Texas A&M University
College Station, Texas 77843-3368
quak@ euler.math.tamu.edu weyrich@ math.tamu.edu

Some Criteria to Evaluate the Quality of a Quasi-Interpolant

Christophe Rabut

Abstract. This paper is a reflection on the main properties we demand of a quasi-interpolant; we give and compare various interesting properties of a quasi-interpolant, and explain the importance of the *transfer function* and of the "*interpolation power of B*" to evaluate how far from the data $(jh, y_j)_{j\in\mathbb{Z}^d}$ the function $\sum_{j\in\mathbb{Z}^d} y_j B(\bullet - j)$ is.

§1. Introduction

Given a positive real number h, and a function B from $\mathbb{R}^d$ into $\mathbb{R}$, we call "*cardinal Schoenberg's approximation using B*", or simply "*Schoenberg's approximation*", the operation which, from a vector $y=(y_j)_{j=1,..,n}$ such that the summation in (1.1) is convergent, gives the function σ_y (or simply denoted σ) defined by

$$\sigma := \sigma_y := \sum_{j\in\mathbb{Z}^d} y_j\ B(\bullet/h - j). \tag{1.1}$$

This operation was introduced by I.J. Schoenberg [14] in one dimension ($d=1$) when B is a polynomial B-spline (in this case, there is no problem of convergence since B has compact support); this is the reason why this operator is called *Schoenberg's approximation.* For details on, or around, Schoenberg's approximation, see the survey papers [2,3,4,8,10].

An analog operation can be defined for a finite set of scattered data (see a forthcoming paper of the same author), but, for sake of brevity, this paper is devoted to the cardinal case.

The vector y is called the "*input vector*", or the "*control vector*". The points $(jh, y_j)_{j\in\mathbb{Z}^d}$ are called the "*control points*" of the curve or of the (hyper)-surface associated to σ.

Obviously Schoenberg's approximation may be defined for any function B (for those vectors y such that the sum (1.1) is convergent); however the

Wavelets, Images and Surface Fitting
P. J. Laurent, A. Le Méhauté, and L. L. Schumaker (eds.), pp. 429–436.

ISBN 1-56881-040-7.

most commonly used functions B are B-splines, tensor products of B-splines, or linear combinations (usually a small number) of translates of some radial functions ϕ (*i.e.*, $B = \sum_{j\in\mathbb{Z}^d} \lambda_j \,\phi(\|\bullet - j\|)$ for some function ϕ and some compactly supported vector λ). Three properties are usually required for B: first a sufficient decay of $B(x)$ when $\|x\| \to \infty$ in order to ensure the convergence of the summation in (1.1) for a large class of y and to help for the fast numerical computation of the summation (1.1)); second $\sum_{j\in\mathbb{Z}^d} B(j) = 1$, in order to ensure $\sigma(j) = c$ whenever $\forall j \in \mathbb{Z}^d,\ y_j = c$; and third, unicity of decomposition, *i.e.*, if $\sum_{j\in\mathbb{Z}^d} y_j\, B(\bullet - j) = \sum_{j\in\mathbb{Z}^d} z_j\, B(\bullet - j)$, then $y = z$. Hereafter we always suppose this latter condition satisfied.

In this paper, we want to give tools to evaluate the quality of a Schoenberg's approximation; to do so, we examine which are the important properties for the function B, and we propose some condition for B to be called a "quasi-interpolant": at a first glance, we could say that B (resp. the B_j's) may be called a quasi-interpolant if, for some class of vector y the function σ_y is quite close to the data $(jh, y_j)_{j\in\mathbb{Z}^d}$ (resp. $(x_j, y_j)_{j=1,..,n}$), *i.e.*, if for any j in $\mathbb{Z}^d$, $\sigma(jh) \simeq y_j$ (resp. $\sigma(x_j) \simeq y_j$). Obviously this is not precise enough to be usable in mathematics, and we will give which properties are to be satisfied.

This paper is devoted to the case of cardinal data; some properties are extendable in the scattered data case (however, the idea of transfer function, which is a very important one, is not valid in this case).

Notations

We will use *cardinal data*, *i.e.*, $\forall j \in \mathbb{Z}^d,\ x_j = jh$ for some real number h. Moreover,

$\mathbb{P}_k$ denotes the set of all polynomials of degree at most k on $\mathbb{R}^d$.
$\delta = (\delta_j)_{j\in\mathbb{Z}^d}$ will denote the vector defined by

$$\delta_0 = 1, \qquad \forall j \in \mathbb{Z}^d - \{0\}, \qquad \delta_j = 0. \tag{1.2}$$

$b = (b_j)_{j\in\mathbb{Z}^d}$ will denote the vector defined by

$$\forall j \in \mathbb{Z}^d, \qquad b_j \;=\; B(j). \tag{1.3}$$

$Y = (Y_j)_{j\in\mathbb{Z}^d}$ will denote the vector defined by

$$\forall j \in \mathbb{Z}^d, \qquad Y_j \;=\; \sigma(jh). \tag{1.4}$$

(where $\sigma = \sigma_y$ is defined by (1.2)).

As a direct consequence of (1.4), (1.3) and (1.2), we have the relation

$$Y \;=\; b * y\ , \tag{1.5}$$

where $*$ is the usual convolution product.

Now, for any function f, with variable in $\mathbb{R}^d$, any h in $\mathbb{R}_*^+$, we denote by $\sigma_{f,h}$ the $\mathcal{B}$-Schoenberg's approximation of the data $(jh, f(jh))_{j\in\mathbb{Z}^d}$, i.e., the function defined by

$$\sigma_{f,h} = \sum_{j\in\mathbb{Z}^d} f(jh)\,B(\bullet/h - j). \tag{1.6}$$

We denote by L the (unique) function such that there exists a vector λ satisfying

$$\begin{cases} L = \sum_{j\in\mathbb{Z}^d} \lambda_j B(\bullet - j) \\ \forall j\in\mathbb{Z}^d,\ \ L(j)=\delta_j. \end{cases} \tag{1.7}$$

§2. Transfer Function, Interpolation Weakness and Power

In this section, we take a purely discrete point of view: we examine some tools to evaluate the quality, as a quasi-interpolant of a function B, by looking at the properties of σ on $\mathbb{Z}^d$. To do so, we examine the properties of the vectors $Y = (\sigma(x_j))_{j\in\mathbb{Z}^d}$, for various classes of control vectors y, *i.e.*, the properties of the discrete mapping $y \rightarrow Y = b * y$; we will focus our attention on $Y - y$.

2.1. Transfer function

Notations. Let $r \in \mathbb{R}_+$. We define $\mathcal{Y}_r$ to be the set of vectors y satisfying

$$|y_j| = \mathcal{O}\left(\|j\|_\infty^r\right) \quad \text{when } \|j\|_\infty \rightarrow \infty. \tag{2.1}$$

We use the norm $\|\bullet\|_r$ on $\mathcal{Y}_r$ defined by

$$\forall y \in \mathcal{Y}_r\ ,\ \ \|y\|_r = \max\left(\|y_0\|,\ \ \sup_{j\in\mathbb{Z}^d-\{0\}} \frac{|y_j|}{\|j\|_\infty^r}\right). \tag{2.2}$$

Theorem 2.1. *Let $r \in \mathbb{R}_+$; $r' \in \mathbb{R}$ be such that $r' > r + d$. Let B be a function from $\mathbb{R}^d$ to $\mathbb{R}$ such that $b \in \mathcal{Y}_{-r'}$ (b is defined from B by (1.3)). Let f_B be the mapping from $\mathcal{Y}_r$ to $\mathcal{Y}_r$ defined by*

$$\forall y \in \mathcal{Y}_r\ ,\ \ f_B(y) = b * y. \tag{2.3}$$

Then f_B is a linear function, continuous for the norm $\|\bullet\|_r$, and is shift-invariant (i.e. for any $k \in \mathbb{Z}^d$ and any vector y, $f_B(T_k(y)) = T_k(f_B(y))$, where $T_k(u)$ is defined for any vector u by $(T_k(u))_j = u_{j+k}$ for any $j \in \mathbb{Z}^d$).

For these reasons, f_B is a (digital) filter (see [6]).

Theorem 2.1.a. *Under the hypothesis of Theorem 2.1, for given* $\omega \in \mathbb{R}$, *let* $\mathrm{e}_{h\omega}$ *be the vector defined by*

$$\forall j \in \mathbb{Z}^d \ , \ (\mathrm{e}_{h\omega})_j = \mathrm{e}^{2i\pi jh\omega}. \tag{2.4}$$

Then $\mathrm{e}_{h\omega}$ *is an eigenvector of* f_B, *the associated eigenvalue being*

$$\widetilde{B}(h\omega) = \sum_{j \in \mathbb{Z}^d} B(j)\, \mathrm{e}^{-2i\pi jh\omega}. \tag{2.5}$$

$\widetilde{B}$ *is called the "transfer function" of the filter* f_B, *or the "transfer function associated to the B-approximation", or simply the "transfer function of* B*", or the "symbol function of* B*".*

Theorem 2.1.b. *Let* B *be a function,* $\widetilde{B}$ *the transfer function associated to the B-approximation; let* $n \in \mathbb{N}$. *It is equivalent to say* (D^α *is the derivative of order* α; δ *has been defined in (1.2)) that*

i) B *is* $\mathbb{P}_n$*-interpolating*
(i.e. for any p in $\mathbb{P}_n$ and any k in $\mathbb{Z}^d$, $\sum_{j \in \mathbb{Z}^d} p(j)B(k-j) = p(k)$).

ii) $\sum_{j \in \mathbb{Z}^d} B(j) = 1 \quad ; \quad \forall \alpha \in \mathbb{N}^d \ , \ 0 < |\alpha| \le n \Longrightarrow \sum_{j \in \mathbb{Z}^d} j^\alpha B(j) = 1.$

iii) $\widetilde{B}(0) = 1 \quad ; \quad \forall \alpha \in \mathbb{N}^d \ , \ 0 < |\alpha| \le n \Longrightarrow \left(\mathrm{D}^\alpha \widehat{B}\right)(0) = 0.$

For the proofs of Theorems 2.1.a and 2.1.b, we refer to [13].

Remarks. $|\widetilde{B}(\omega)|$ is the coefficient of the amplification (or of the attenuation) of a (co)sinusoidal signal with frequency ω; $\arg(\widetilde{B}(\omega))$ is the change of phase of the same signal.

Theorem 2.1.b shows that $\mathbb{P}_k$*-interpolation is linked to the shape of the transfer function at the origin.* Since the whole shape of $\widetilde{B}$ is important for the interpolating quality of a function B, we clearly see here that the $\mathbb{P}_k$-interpolation (or $\mathbb{P}_k$-reproduction) criterion **is not** a sufficient one to evaluate the quality of a quasi-interpolant.

2.2. Interpolation power and interpolation weakness

Though it is interesting to look at the whole shape of $\widetilde{B}$, it is also worthwhile having a single number to evaluate if the B-approximation is far from the interpolation or not. So we give the following two definitions:

Definitions 2.2. *Let* δ *and* b *be as defined in (1.2) and (1.3). We define the* interpolation power *and the* interpolation weakness *of a B-approximation as being the quantities*

$$\overline{B} \ = \ 1 - \|\delta - b\|_2 \ = \ 1 - \left\|1 - \widetilde{B}\right\|_2 \tag{2.6}$$

$$\underline{B} = \|\delta - b\|_2 = \left\|1 - \widetilde{B}\right\|_2. \tag{2.7}$$

The equivalence of the two expressions proposed above for $\overline{B}$, as well as for $\underline{B}$, is a consequence of Poisson's formula, (see [13]).

Remarks. The interpolation weakness and power are synthetic numbers measuring the distance from $\widetilde{B}$ to 1, *i.e.*, how far from interpolation a B-approximation is: the smaller the interpolation weakness is (or the bigger the interpolation power is), the closer to interpolation the B-approximation is.

Note that, in the case of scattered data, there is no possibility of defining a transfer function. However, interpolation weakness and interpolation power of a family $\mathcal{B}=(B_j)_{j=1,..N}$ can be directly defined by

$$\underline{\mathcal{B}} = (1/N)^d \sum_{k\in\mathbb{Z}^d} \left(\sum_{j\in\mathbb{Z}^d} \left(\delta_j - B_k(j)\right)^2 \right)^{1/2} \quad \text{and} \quad \overline{\mathcal{B}} = 1 - \underline{\mathcal{B}}. \tag{2.8}$$

§3. Total Positivity

On may think that the positivity of B is an interesting property, as then σ remains in the convex hull of the data $(jh, y_j)_{j\in\mathbb{Z}^d}$. Of course, this may be important for some applications, but we must notice that the B-interpolation of some points **is not** usually in the convex hull of the data $(jh, y_j)_{j\in\mathbb{Z}^d}$. Now, since B-approximation is often used as some fastly computed approximation of B-interpolation, it does not seem necessary to demand that B-approximation has a property that B-interpolation does not have.

However, we would like a related property. For us, the right property seems to be the "*total positivity*":

Definition 3.1. *A function B (from $\mathbb{R}^d$ to $\mathbb{R}$) is said to be "totally positive" if and only if for any nonzero natural number n, any family of n complex numbers $(\xi_j)_{j=1,..,n}$ and any family of n elements of $\mathbb{R}^d$ $(x_j)_{j=1,..,n}$, we have*

$$\sum_{i=1}^{n}\sum_{j=1}^{n} \xi_i \xi_j B(x_i - x_j) \geq 0,$$

which is equivalent to

$$\forall \omega \in \mathbb{R}^d, \qquad \widehat{B}(\omega) \geq 0$$

($\widehat{B}$ is the Fourier transform of B).

Remarks. This definition —and the equivalence using the Fourier transform of B— is quite classical (see for example [5, Section 3]). For sake of brevity, it is not possible to give here in detail the various advantages of using a totally positive function. Let us mention however that if B is a totally positive function, the function H defined by $H(x,y)=B(x-y)$ is a reproducing Kernel of some Hilbert space, that the functions $\left(B(\bullet-j)\right)_{j\in\mathbb{Z}^d}$ form a Riesz basis of the space they span, and finally, that, if B enjoys a refinement equation, it is possible to build a multi-resolution analysis (thus wavelets) using B.

§4. Appraisal

We think that the following properties are important to evaluate the qualities of a quasi-interpolant.

4.1. Essential properties

Proposition 4.1. *The following two properties are essential for a quasi-interpolant B:*

a) *B is $\mathbb{P}_1$-reproducing, i.e., , for any p in $\mathbb{P}_1$, $\sum_{j\in\mathbb{Z}^d} p(j)\,B(\bullet-j)=p$.*

b) $\sum_{j\in\mathbb{Z}^d}\left(\delta_j - B(j)\right)^2 < 1$ *(which is equivalent to $\left\|1-\widetilde{B}\right\|_2 < 1$).*

4.2. Important properties

Proposition 4.2. *Interesting properties for a quasi-interpolant are the following ones, presented in increasing order of importance (of course, depending on the utilisation of the quasi-interpolant, this order may be modified):*

a) $\left\|1-\widetilde{B}\right\|_2$ *is small.*

b) $\forall\omega\in\mathbb{R}^d,\ \widehat{B}(\omega)\geq 0$.

c) *B is $\mathbb{P}_n$-interpolating, for some $n>1$, i.e., there exists some $n>1$ such that, for any p in $\mathbb{P}_n$, $\forall k\in\mathbb{Z}^d$, $\sum_{j\in\mathbb{Z}^d} p(j)\,B(k-j)=p(k)$.*

d) *B is $\mathbb{P}_{n'}$-reproducing, for some n' , $1<n'\leq n$, i.e., there exists some n' $(1<n'\leq n)$ such that, for any p in $\mathbb{P}_{n'}$, $\sum_{j\in\mathbb{Z}^d} p(j)\,B(\bullet-j)=p$.*

4.3. Examples

Univariate case: polynomial B-splines. *Let B_m be the usual polynomial B-spline, of degree m, centered at the origin. B_m is a quasi-interpolant. The value of interpolation weakness for lowest values of m are: $\underline{B_1}=0$, $\underline{B_2}=\sqrt{3/32}\simeq 0.31$, $\underline{B_3}=\sqrt{1/6}\simeq 0.41$, $\underline{B_4}=\sqrt{35270}/384\simeq 0.49$, $\underline{B_5}=\frac{\sqrt{4270}}{120}\simeq 0.54$, $\underline{B_7}=\frac{\sqrt{9751140}}{5040}\simeq 0.62$, $\underline{B_7}\simeq 0.67$.*

B_m is totally positive if and only if m is odd (that is if and only if the associated splines satisfy a minimizing norm property, and B_m is a univariate polyharmonic B-spline). For any k in $\mathbb{N}_*$, there are coefficients $(\alpha_j)_{j=-k,..,k}$ such that $B_{m,k} = \sum_{j=-k}^{k} \alpha_j B_m(\bullet - j)$ is a $\mathbb{P}_{\inf(k,m)}$-reproducing quasi-interpolant, thus it is possible to build better quasi-interpolants using linear combinations of translates of B_m (for more details, see [7] and [10]).

Multivariate case: polyharmonic B-splines. *Let $m > d/2$, and let $B^m = \Delta_1^m v_m$, where*

$$v_m = \begin{cases} \frac{1}{2^{2m}\pi^{d/2}} \frac{(-1)^{m-d/2+1}}{(m-d/2)!\Gamma(m)} \|\bullet\|^{2m-d} \ln \|\bullet\|^2 & \text{if } d \text{ is even} \\ \frac{1}{2^{2m}\pi^{d/2}} \frac{\Gamma d/2-m}{\Gamma(m)} \|\bullet\|^{2m-d} & \text{if } d \text{ is odd} \end{cases}$$

is a solution of $\Delta^m f =$ Dirac (Δ^m is the mth-iterated Laplacean operator), and where Δ_1^m is the mth-iterated discrete Laplacean operator, defined by (e_i is the ith canonical vector of $\mathbb{R}^d$)

$$\begin{cases} \Delta_1 = \sum_{i=1}^{d} f(\bullet + e_i) - 2f + f(\bullet - e_i) \\ \Delta_1^{n+1} f = \Delta_1(\Delta_1^n f). \end{cases}$$

Then B^m is a totally positive quasi-interpolant, whose Fourier transform is

$$\widehat{B^m}(\omega) = \frac{\left(\sum_{i=1}^{d} \sin^2 \pi\omega_i\right)^m}{\|\pi\omega\|^{2m}}.$$

The value of interpolation weakness for lowest values of m are :
$d=2$: $\underline{B^2} \simeq 0.37$; $\underline{B^3} \simeq 0.61$; $\underline{B^4} \simeq 0.63$;
$d=3$: $\underline{B^2} \simeq 0.33$; $\underline{B^3} \simeq 0.58$; $\underline{B^4} \simeq 0.74$;

B^m is called an "*elementary m-harmonic B-spline*". It is also possible to define "*high level m-harmonic B-spline*", which are up to $\mathbb{P}_{2m-1}$-reproducing; more details are given in [11] and [12]. As shown in [7], it is possible to build a multiresolution analysis using B^m, and associated semi-orthogonal wavelets are easily built from their Fourier transform.

References

1. Cohen, Albert, Ondelettes, analyses multi-resolutions et traitement numerique du signal, Thèse, Université de Paris-Dauphine, France, 1990.
2. de Boor, C., Quasi-interpolants and approximation power of multivariate splines, in *Computation of curves and surfaces*, M. Gasca and C.A. Micchelli eds., Dordrecht, Netherlands, Kluwer Academic Publishers, 1990, 313–345.

3. Chui, C.K., "*Multivariate splines*", CBMS-NSF Regional Conference Series in Applied Mathematics, #54, SIAM, (1988) Philadelphia.
4. Dyn, N., Interpolation and Approximation by Radial functions, in *Approximation VI*, C.K. Chui, L.L. Schumaker and J.D. Ward (eds), Academic Press, Boston, 1989, 219–266.
5. Gelfand, I.M., Vilenkin, N.Y. *Les distributions* tome 4, Dunod, Paris, 1967.
6. Hamming, R., *Digital Filters*, Prentice Hall, Englewood cliffs, NJ, 1977.
7. Micchelli, C.A., C. Rabut, and F. Utreras, Using the refinement equation for the construction of pre-wavelets. III: Elliptic splines, Numerical Algorithms **1** (1991), No 4, 331–352.
8. Powell, M. J. D., The theory of radial basis function approximation in 1990, DAMTP report NA 11, University of Cambridge, 1990.
9. Rabut, C., B-splines polyharmoniques cardinales : interpolation, quasi-interpolation, filtrage,Thèse d'Etat, University of Toulouse, 1990.
10. Rabut, C., An Introduction to Schoenberg's Approximation, Computers and Mathematics with Applications **24** 12 (1992), 149–175.
11. Rabut, C., Elementary m-harmonic cardinal B-splines, Numerical Algorithms **2** (1992), 39–62.
12. Rabut, C., High level m-harmonic cardinal B-splines, Numerical Algorithms **2** (1992), 63–84.
13. Rabut, C., Some Tools about Quasi-interpolation on Cardinal Grids, in *Mathematical Methods in Computer Aided Geometric Design*, T. Lyche and L.L. Schumaker (eds) Academic Press, New York, 1992, 481–496.
14. Schoenberg, I. J., Contributions to the problem of approximation of equidistant data by analytic functions, Quarterly Applied Mathematics IV. (1946), 45–99 and 112–137.

Christophe Rabut
Centre de Mathématiques
Institut National des Sciences Appliquées
Complexe Scientifique de Rangueil
F-31077 Toulouse cédex
FRANCE
Rabut @ cict.fr

Hölder Regularity of Subdivision Schemes and Wavelets

O. Rioul

Abstract. We study special functions obtained as limits of dyadic and p/q-adic subdivision schemes in one dimension. While the former are used for designing compactly supported wavelet bases, the latter are a flexible generalization which has already found application in digital signal processing. For $q > 1$, however, we obtain an infinite set of different limit functions instead of shifted copies of a single one, and a direct application of former ideas becomes impossible. However, our "discrete" approach allows us to extend the results of Daubechies and Lagarias (on Hölder regularity estimates based on infinite products of matrices) to p/q-adic schemes. We obtain easily implementable, sharp Hölder regularity estimates.

§1. Introduction

A *subdivision scheme* in one dimension can be defined as follows: consider a finite sequence $\{g_n\}$ ($g_n \neq 0$ for a finite number of values of n), referred to as the *subdivision mask* in the sequel, and the discrete operator acting on finite sequences

$$\mathcal{G} : u_n \to v_n = \sum_k u_k g_{qn-pk}, \tag{1}$$

where p and q are positive integers.

Definition 1. *A p/q-adic subdivision scheme is a collection of sequences g_n^j, computed using the recursion $g_n^{j+1} = \mathcal{G}(g_n^j)$, where $\mathcal{G}$ is given by (1).*

We assume that p and q are coprime, otherwise replace g_n by g_{dn} where $d = \gcd(p, q)$. Note that a subdivision scheme is uniquely defined given a subdivision mask g_n and an initial sequence g_n^1. For $p/q = 2$, we recover the classical binary subdivision schemes, which have long been used in computer-aided geometric design [6] and yield compactly supported wavelets under additional conditions on g_n [4,5].

Wavelets, Images and Surface Fitting
P. J. Laurent, A. Le Méhauté, and L. L. Schumaker (eds.), pp. 437–444.

ISBN 1-56881-040-7.

Being a natural extension to binary subdivision schemes, p/q-adic subdivision schemes may find application in computer-aided geometric design. However, the motivation for this study comes from signal decomposition problems: for $q > 1$, subdivision schemes are strongly related to filter banks with rational sampling changes, as studied by Kovačević and Vetterli [7]. Some of the material presented here and the derivation of "rational" pseudo-wavelets were investigated by Blu and the author in [1,2]. For many signal processing applications, p/q-adic schemes ($1 < p/q < 2$) form a more flexible tool than "pure" dyadic wavelets as they allow signal decomposition on fractions $\log_2 p/q$ of an octave using orthonormal bases [1,2,7].

In this paper we study special functions obtained as limits of p/q-adic subdivision schemes using the "'discrete approach" of [8]. We first define uniform convergence of the $\{g_n^j\}$'s.

Definition 2. *A p/q-adic subdivision scheme $\{g_n^j\}$ converges uniformly to the (compactly supported) function $\phi(x)$ if*

$$\lim_{j\to\infty} \sup_x |\phi(x) - g_{n_j}^j| = 0, \tag{2}$$

where n_j is a sequence of integers such that $|n_j - (p/q)^j x|$ is bounded as j increases.

Compact supports of limit functions were estimated by Blu in [1]. For $p/q = 2$ and initial sequence $g_n^1 = g_n$, we obtain the "father wavelet" $\phi(x)$ as studied in [4,5,8]. In this case every uniform limit is a finite linear combination of translates of ϕ, since replacing g_n^1 by g_{n-k}^1 gives $\phi(x-k)$. However, this shift invariance is never satisfied in the rational case ($q > 1$), as first noticed by Cohen and Daubechies in [3]. Therefore, we consider an infinite set of limit functions instead of one:

$$\phi^s(x) = \text{ uniform limit of } g_n^{j,s}, \tag{3}$$

where $s \in \mathbb{Z}$ and $g_n^{j,s}$ is defined by

$$g_n^{j,s} = \mathcal{G}^j(\delta_{n-s}). \tag{4}$$

Here the initial sequence is an *impulsion at time* $n = s$, i.e., $\delta_{n-s} = 1$ if $n = s$, 0 otherwise. It can be easily shown that any uniform limit is a linear combination of the $\phi^s(x)$, and that we obtain a pseudo-wavelet transform [1] for which the "wavelets" at resolution $(p/q)^{-j}$ are in the form $\psi_k((p/q)^j x)$. It is known that these wavelets can never be put in the exact form $\psi((p/q)^j x - k)$ (hence the name "pseudo-wavelets"), but can be approximated in this manner under conditions on g_n (see below). Our goal is to answer the following questions:

1) Can we find conditions on the subdivision mask $\{g_n\}$ such that uniform limit functions $\phi^s(x)$ exist for all s, yielding a "pseudo-wavelet" transform?

2) What are the conditions on g_n such that the obtained pseudo-wavelets (or the limit functions $\phi^s(x)$) are "smooth"? This amounts to determining e.g. the Hölder regularity of $\phi^s(x)$.
3) For $q > 1$, can we find conditions on g_n such that "pseudo-wavelets" are almost shifted versions of each other, i.e., $\psi_k((p/q)^j x) \approx \psi((p/q)^j x - k)$ within a smal error (to be estimated). This amounts to estimating $\epsilon = \inf_{s,s',x} |\phi^s(x+s) - \phi^{s'}(x+s')|$.

This paper is organized as follows. First, we introduce some useful polynomial notation. Then, we state a necessary and sufficient condition for uniform convergence, thereby answering question 1. We then answer question 2 by deriving sharp Hölder regularity estimates, which are known to be optimal in the case $q = 1$. This is a direct extension of Daubechies and Lagarias estimates [5] to the rational case, and a small example ("pseudo B-splines") is provided. Finally, as an answer to question 3, we show that regularity is also useful for achieving shift invariance within a small error.

§2. Notations and Preliminaries

In the sequel we shall heavily use the Laurent polynomial notation:

$$G(X) = \sum_n g_n X^n \tag{5}$$

for any finite sequence g_n. Polynomial multiplication amounts to performing a discrete convolution by the sequence g_n, which we shall denote by $\mathbf{G}$.

Equation (1) can be seen as resulting from three successive operations: first, insert $p-1$ zeros between every other sample of u_n (discrete dilation $U(X^p) = \mathbf{D}_p(U(X))$). Then convolve the result with g_n (operator $\mathbf{G}$), and finally retain every qth sample of the result (discrete contraction $w_{qn} = \mathbf{C}_q(w_n)$). In other words, $\mathcal{G} = \mathbf{C}_q \mathbf{G} \mathbf{D}_p$.

Using some flow graph algebra it can be easily shown that when p and q are coprime, the iterated operator $\mathcal{G}^j$ can be put in a similar form, namely $\mathcal{G}^j = \mathbf{C}_{q^j} \mathbf{G}^j \mathbf{D}_{p^j}$, where $\mathbf{G}^j$ corresponds to polynomial multiplication by

$$\begin{aligned} G^j(X) &= G(X^{q^{j-1}}) G(X^{pq^{j-1}}) \cdots G(X^{p^{j-1}q}) G(X^{p^{j-1}}) \\ &= \prod_{i=0}^{j-1} G(X^{p^i q^{j-1-i}}). \end{aligned} \tag{6}$$

By definition (4) we obtain the fundamental relation

$$g_n^{j,s} = g^j_{q^j n - p^j s}, \tag{7}$$

which can be used to compute a subdivision scheme $g_n^{j,s}$ from the sequence g_n^j associated to polynomial $G^j(X)$. Using (6), it is easily seen that these polynomials can in turn be computed recursively in two ways:

$$G^{j+1}(X) = G(X^{q^j}) G^j(X^p) = G^j(X^q) G(X^{p^j}). \tag{8}$$

In the following we shall use the notation $U^j(X) = \prod_i U(X^{p^i q^{j-1-i}})$ for any polynomial $U(X)$. The most general recursion formula for $U^j(X)$ is

$$U^{i+l}(X) = U^l(X^{p^i})U^i(X^{q^l}). \tag{8'}$$

§3. Uniform Convergence

Uniform convergence of $\{g_n^{j,s}\}$ (Definition 2) requires an important condition to be fulfilled by g_n, which was first mentioned in [7] and proved in [1].

Proposition 1. *If $g_n^{j,s}$ converges uniformly to $\phi^s(x) \not\equiv 0$, then*

$$\sum_k g_{n-pk} = 1$$

for all $n \in \mathbb{Z}$.

This condition can be rewritten as

$$G(1) = p \text{ and } \frac{1-X^p}{1-X} \text{ divides } G(X). \tag{9}$$

Now consider $F(X) = \frac{1-X^q}{1-X^p}G(X)$. It is easy to show that

$$F^j(X) = \frac{1-X^{q^j}}{1-X^{p^j}}G^j(X). \tag{10}$$

Hence the first-order differences $d_n^{j,s} = g_n^{j,s} - g_{n-1}^{j,s}$ follow a p/q-adic scheme with subdivision mask $F(X)$ (and initial sequence associated to $(1-X)X^s$).

Theorem 2. *The p/q-adic subdivision scheme $g_n^{j,s}$ converges uniformly to continuous limit function $\phi^s(x)$ for all s if and only if (9) holds and*

$$\max_n |g_{n+1}^{j,s} - g_n^{j,s}| \to 0 \text{ as } j \to \infty. \tag{11}$$

Proof: Condition (11) is clearly necessary. Now, using (10) we have $d_n^{j,s} = f^j_{q^j n - p^j s} - f^j_{q^j n - p^j s - p^j}$, hence $f^j_{q^j n - p^j s} = \sum_{k\geq 0} d_n^{j,s+k}$ where the sum is finite. Assuming (10) holds for all s it follows that $f^j_{q^j n - p^j s}$ tends uniformly to 0. On the other hand, using recursion (8') for $F^j(X)$ we end up with

$$\max_n |d_n^{i+l,s}| \leq \max_n \sum_k |f^j_{q^j n - p^j k}| \max_k |d_k^{l,s}|,$$

which by induction gives

$$\max_n |d_n^{j,s}| \leq c \left(\max_n \sum_k |f^i_{n-p^j k}| \right)^{j/i}.$$

Since $f^j_{q^j n - p^j s}$ tends uniformly to 0, there exists $\alpha > 0$ such that

$$\max_n |g^{j,s}_{n+1} - g^{j,s}_n| \le c\,(p/q)^{-j\alpha}, \tag{11'}$$

which is clearly stronger than (11). Now, by adapting a proof in [8], (11') implies

$$\max_{n_j} |g^{j+1,s}_{n_j+1} - g^{j,s}_{n_j}| \le c'(p/q)^{-j\alpha},$$

where n_j is chosen as in Definition 2. From this follows that $g^{j,s}_{n_j}$ is a uniform Cauchy sequence, hence converges to $\phi^s(x)$ which can be easily shown to be continuous. ■

Theorem 2 is powerful for proving uniform convergence for all s: in fact, it is sufficient that $\max_n \sum_k |f^j_{n-q^j k}| = (p/q)^{-i\alpha} < 1$ *for some* i, to ensure that all $\phi^s(x)$ are continuous. Moreover, this gives α ($0 < \alpha < 1$), and as can be shown from (11'), the $\phi^s(x)$ are all Lipschitz of order α.

§4. Hölder Regularity Estimates

We say that $\phi^s(x)$ is *(Hölder) regular of order* $N + \alpha$ if it is C^N and its Nth derivative is Lipschitz of order α. The preceding section has given a condition on g_n such that $\phi^s(x)$ is regular of order α, where $0 < \alpha < 1$. For higher regularity orders we apply this result to Nth order finite differences of the $g^{j,s}_n$. The first-order difference is $\Delta g^{j,s}_n = (g^{j,s}_n - g^{j,s}_{n-1})/(p/q)^{-j}$, and Nth order finite differences are obtained by applying N times the operator Δ. First, we relate finite differences to derivatives of limit functions, using the following theorem proven in [1].

Theorem 3. *If $\Delta^N g^{j,s}_n$ converges uniformly, then so do $\Delta^k g^{j,s}_n$ for $0 \le k \le N$, whose respective limit functions are the kth-order derivatives of $\phi^s(x)$. Moreover, $\phi^s(x)$ is C^N and*

$$\left(\frac{1 - X^p}{1 - X}\right)^{N+1} \quad \textit{divides } G(X) \tag{12}$$

Condition (12) is clearly a generalization of (9). Note that the subdivision mask associated to $\Delta^k g^{j,s}_n$ is

$$\left(\frac{p}{q}\frac{1 - X^q}{1 - X^p}\right)^k G(X),$$

hence, in some sense, multiplying the subdivision mask polynomial by $\frac{p}{q}\frac{1-X^q}{1-X^p}$ amounts to differentiating the limit function [1].

From Theorem 3 and the preceding section it follows that uniform convergence of the Nth-order finite differences $\Delta^N g^{j,s}_n$ can be tackled similarly as that of the $g^{j,s}_n$. That is, we have the following

Corollary 4. *If $G(X)$ satisfies (9), (12) and*

$$\max_n |\Delta^N g_{n+1}^{j,s} - \Delta^N g_n^{j,s}| \leq c\,(p/q)^{-j\alpha} \tag{13}$$

for some $\alpha < 1$, then $\phi^s(x)$ is regular of order $N + \alpha$.

In fact we can show that this result is valid for negative values of α (a proof for the case $p/q = 2$ can be found in [8]), which means that we can obtain positive regularity orders $N + \alpha$ by estimating (13) even though the $\Delta^N g_n^{j,s}$ diverges. In particular we can still apply this criterion for the maximal value of N such that (12) holds. This yields the following Hölder regularity estimate for p/q-adic subdivision schemes:

Theorem 5 (Hölder regularity estimate). *Assume $G(X)$ satisfies (9) and (12) and set*

$$F_N(X) = \left(\frac{p}{q}\right)^N \left(\frac{1 - X^q}{1 - X^p}\right)^{N+1} G(X). \tag{14}$$

Define the Hölder regularity estimate $N + \alpha_N^j$ (for a fixed value of j) by

$$\left(\frac{p}{q}\right)^{-j\alpha_N^j} = \max_{0 \leq n < p^j} \sum_k |(f_N^j)_{n - p^j k}|, \tag{15}$$

and let $\alpha = \sup_j \alpha_N^j$. Then for any j such that $N + \alpha_N^j > 0$, $\phi^s(x)$ is Hölder regular of order $N + \alpha_N^j$, for all s. Moreover α_N^j tends to α as $j \to \infty$, and $\phi^s(x)$ is Hölder regular of order $N + \alpha - \epsilon$ for any $\epsilon > 0$.

For $p/q = 2$ we recover the estimate derived in [8]. Moreover, for any value of p/q, we can rewrite (15) using p square matrices of small order $\mathbf{F}_r$:

$$\left(\frac{p}{q}\right)^{-j\alpha_N^j} = \max_{0 \leq r_i \leq p-1} \|\prod_{i=0}^{j-1} \mathbf{F}_{r_i}\|_1 \tag{15'}$$

where $\|\cdot\|_1$ denotes the ℓ^1 matrix norm and where

$$\mathbf{F}_r = \begin{pmatrix} f_r & f_{r-q} & f_{r-2q} & \cdots \\ f_{r+p} & f_{r+p-q} & f_{r+p-2q} & \cdots \\ \vdots & & & \ddots \end{pmatrix}. \tag{16}$$

This matrix is simply a convolution matrix in which one keeps every pth line and every qth column. For $p = 2$ and $q = 1$ we recover Daubechies and Lagarias estimate [5], which is shown to be optimal under very weak conditions on g_n in [8].

In general, using the matrix form we obtain lower and upper bounds on regularity as

$$N - \log_{p/q} \max_{0 \leq r < p} \|\mathbf{F}_r\| \tag{17}$$

and

$$N - \log_{p/q} \max_{0 \leq r < p} \rho(\mathbf{F}_r), \tag{17'}$$

respectively, where $\rho(A)$ denotes the spectral radius of matrix A.

§5. An Example: "Rational B-Splines"

This example is

$$G(X) = \frac{1}{p^N} \left(\frac{1 - X^p}{1 - X} \right)^{N+1} \tag{18}$$

for which $F_N(X)$ (14) is obtained by replacing p by q in (18). For $p/q = 2$, we recover B-spline functions, whose Hölder regularity is $N - \epsilon$ (see e.g. [8]). Let us derive Hölder regularity orders in the case $p/q = 3/2$ and $N = 1$, 2, and 3.

First, in the case $N = 1$, we obtain three 1×1 matrices $\mathbf{F}_r$, namely (1/2), (1), and (1/2). Hence it follows from Theorem 5 with (15') that all $\phi^s(x)$ are Hölder regular of order 1 (that is, almost continuously differentiable).

In the case $N = 2$, we obtain

$$\mathbf{F}_0 = \begin{pmatrix} 1/4 & 0 \\ 1/4 & 3/4 \end{pmatrix}, \qquad \mathbf{F}_1 = \begin{pmatrix} 3/4 & 0 \\ 0 & 3/4 \end{pmatrix}, \qquad \mathbf{F}_2 = \begin{pmatrix} 3/4 & 1/4 \\ 0 & 1/4 \end{pmatrix}.$$

Here lower and upper bounds (17) are equal to $2 - \log(3/4)/\log(3/2)$ which gives $2.709\ldots$ as the regularity order of the $\phi^s(x)$'s.

For $N = 3$, we obtain similarly three 3×3 matrices

$$\begin{pmatrix} 1/8 & 0 & 0 \\ 1/2 & 1/2 & 0 \\ 0 & 1/8 & 3/4 \end{pmatrix}, \begin{pmatrix} 1/2 & 0 & 0 \\ 1/8 & 3/4 & 1/8 \\ 0 & 0 & 1/2 \end{pmatrix}, \begin{pmatrix} 3/8 & 1/8 & 0 \\ 0 & 1/2 & 1/2 \\ 0 & 0 & 1/8 \end{pmatrix}$$

whose norms and spectral radii are all equal to 3/4, hence the Hölder regularity order is $3 - \log(3/4)/\log(3/2) = 3.709\ldots$.

§6. Achieving Shift Invariance of Regular Pseudo-Wavelets

As pointed out in the introduction, the major difficulty in the rational case ($q > 1$) as opposed to the dyadic case is the lack of shift invariance of $\phi^s(x) \neq \phi(x - s)$. In other words, we obtain only a pseudo-wavelet transform [1] in which wavelet bases are not shifted versions of each other at a given resolution level. However, it can be shown that shift invariance is "almost" achieved within an arbitrary small error by taking N in (12) sufficiently large (thereby increasing the regularity order of the $\phi^s(x)$'s). The following theorem is due to Th. Blu (a proof is outlined in [2]), and explains a numerical observation made in [7].

Theorem 6. *Let $G(X)$ satisfy the conditions of Theorem 5, where N is given such that (12) holds. Then, provided that the $\phi^s(x)$ are regular for all s,*

$$\lim_{N\to\infty} \sup_{x,s,s'} |\phi^s(x+s) - \phi^{s'}(x+s')| = 0. \tag{19}$$

Acknowledgements. This work was done in part with Thierry Blu of CNET Paris B.

References

1. T. Blu, Iterated filter banks with rational rate changes. Connection with discrete wavelet transforms, IEEE Trans. Signal Processing, **41**-12 (1993).
2. T. Blu and O. Rioul, Wavelet regularity of iterated filter banks with rational sampling changes, IEEE Int. Conf. Acoust., Speech, Signal Processing, Minneapolis, MN (1993), Vol. III, 213–216.
3. A. Cohen and I. Daubechies, Orthonormal bases of compactly supported wavelets. III; Better frequency resolution, SIAM J. Math. Anal., **24**-2 (1993), 420–527.
4. I. Daubechies, Orthonormal bases of compactly supported wavelets, Com. Pure Applied Math., **XLI** (1988), 909–996.
5. I. Daubechies and J. Lagarias, Two-scale difference equations. II. Local regularity, infinite products of matrices and fractals, SIAM J. Math. Anal. **23** no. 4 (1992), 1031–1079.
6. N. Dyn, Subdivision schemes in CADG, in *Advances in Numerical Analysis. II. Wavelets, Subdivision Algorithms and Radial Functions*, W. A. Light (ed.), Oxford University Press, 1991, 36–104.
7. J. Kovačević and M. Vetterli, Perfect reconstruction filter banks with rational sampling rate changes, in *Proc. IEEE Int. Conf. Acoust., Speech, Signal Processing*, Toronto, Canada (1991), 1785–1788.
8. O. Rioul, Simple regularity criteria for subdivision schemes, SIAM J. Math. Anal, **23**-6 (1992), 1544-1576.

O. Rioul
CNET Paris B
38-40, rue du Général Leclerc
92 131 Issy-Les-Moulineaux, FRANCE
rioul@balsa.issy.cnet.fr

Hermite and Lagrange Interpolation by Quadratic Splines on Non-uniform Criss-Cross Triangulations

Paul Sablonnière and Fatima Jeeawock-Zedek

Abstract. Let $\Omega \subset \mathbb{R}^2$ be a rectangular domain subdivided into mn macrorectangles whose set of vertices is denoted by $\mathcal{A}$. An additional set of $m+n$ node lines divides each of them into four rectangles in which the diagonals are drawn. Let $S_2^1(\Omega, \Delta)$ be the space of C^1 quadratic splines on the triangulation Δ obtained in this way and let $S_2^1(\Omega, \Delta)$ be the subspace of functions $g \in S_2^1(\Omega, \Delta)$ whose gradient is linear on the edges of macrorectangles. We first construct a Hermite interpolant in this space whose data values are $g(A)$ and $Dg(A)$ at all points $A \in \mathcal{A}$. Then, by replacing partial derivatives by finite differences, we deduce a Lagrange interpolant on $\mathcal{A}$. We give error bounds for C^3 functions and show the influence of node lines on the norm of the Lagrange projector.

§1. Introduction

Let $\Omega = [x_0, x_{2m}] \times [y_0, y_{2n}]$ be a rectangular domain of $\mathbb{R}^2$ subdivided into mn macrorectangles $R_{ij} = [x_{2i}, x_{2i+2}] \times [y_{2j}, y_{2j+1}], 0 \le i \le m-1, 0 \le j \le n-1$, by the partitions $X_0 = \{x_{2i}\}$ and $Y_0 = \{y_{2j}\}$. The set of their vertices $\mathcal{A} = \{A_{ij} = (x_{2i}, y_{2j})\}$ is the set of *interpolation points.* Each R_{ij} is subdivided into four rectangles by the finer partitions $X_1 \times Y_1$ where $X_1 = \{x_i, 0 \le i \le 2m\}$ and $Y_1 = \{y_j, 0 \le j \le 2n\}$. Finally we denote by Δ the triangulation of Ω obtained by drawing diagonals in each rectangle $[x_i, x_{i+1}] \times [y_j, y_{j+1}]$ of the partition $X_1 \times Y_1$. Let $S_2^1(\Omega, \Delta)$ be the space of C^1 quadratic splines on Δ and let $LS_2^1(\Omega, \Delta)$ be the subspace of functions $g \in S_2^1(\Omega, \Delta)$ whose gradient is linear on the edges of macrorectangles.

In Section 2 we prove that for each macrorectangle R, a function $g \in LS_2^1(\Omega, \Delta)$ is uniquely defined by the values of g and Dg at its vertices. This result extends a previous one given by Sibson and Thomson [10] where the partition is uniform. Then in Section 3, we deduce immediately the existence

Wavelets, Images and Surface Fitting
P. J. Laurent, A. Le Méhauté, and L. L. Schumaker (eds.), pp. 445–452.

ISBN 1-56881-040-7.

and unicity of $g \in LS_2^1(\Omega, \Delta)$ interpolating f and Df on $\mathcal{A}$, for some given function $f \in C^1(\Omega)$. We give error bounds for $f \in C^3(\Omega)$. In Section 4, replacing partial derivatives by finite differences, we deduce a Lagrange interpolant on $\mathcal{A}$ (a similar idea was developed by Powell in [9]), for which we derive error bounds when $f \in C^3(\Omega)$. Finally, in Section 5, we shortly discuss the influence of the choice of node lines on the norm of the Lagrange projector (for the max norm on the space $C(\Omega)$). Some of the previous results are proved in detail, while some others are only sketched, due to space limitation (complete proofs of them will be given elsewhere).

Some notations : $h_i = x_{2i+2} - x_{2i}, k_j = y_{2j+2} - y_{2j}; |v| =$ Euclidean norm of $v \in \mathbb{R}^2$; for $f \in C^3(\Omega)$, we use the notations given in [2] and [3] :

$$|D^3 f(X)| = \sup_{|v_i| \leq 1} |D^3 f(X) \cdot (v_1, v_2, v_3)|,$$

$$\|D^3 f\|_R = \max\{|D^3 f(X)|, X \in R\}, \text{ for } R \subset \Omega,$$

$$D^2 f(X) \cdot (v)^2 = D^2 f(X) \cdot (v, v) \qquad \text{(quadratic form)}$$

$$D^3 f(X) \cdot (v)^3 = D^3 f(X) \cdot (v, v, v) \qquad \text{(cubic form)}$$

$$D_x f = \frac{\partial f}{\partial x}, \quad D_y f = \frac{\partial f}{\partial y}.$$

§2. The Local Hermite Problem

We first study the local Hermite problem on some rectangle $R = R_{ij}$.

Theorem 1.
(i) $\dim LS_2^1(R, \Delta) = 12$
(ii) each $g \in LS_2^1(R, \Delta)$ is uniquely defined by the values of g and Dg at the four vertices of R.

Proof: For sake of simplicity, we study the problem on the macrorectangle R with vertices $A_i, i = 1,2,3,4$ (see Figure 1). Each $g \in LS_2^1(R, \Delta)$ is defined by its B (=Bernstein)-coefficients in the 16 triangles determined on R by the triangulation Δ, (see e.g Chui [1], Chapter 5, or Farin [6] for the Bernstein Bézier representation of polynomials on triangles). For example, in the triangle $A_1 M_1 \Omega_1$ with barycentric coordinates (u, v, w), the equation of the restriction of g is given by :

$$g(u, v, w) = a_1 u^2 + m_1 v^2 + \omega_1 w^2 + 2b_1 uv + 2d_1 uw + 2p_1 vw.$$

We will now prove that the 12 parameters $\{a_i, b_i, c_i, 1 \leq i \leq 4\}$ uniquely determine the other B-coefficients of g on R. Since $M_1 = (1 - \alpha')A_1 + \alpha' A_2$, where $\alpha' = h'/h$, we obtain, by C^1-continuity at M_1 :

$$m_1 = (1 - \alpha')b_1 + \alpha' c_2 \text{ (and similarly } m_3 = (1 - \alpha')c_4 + \alpha' b_3).$$

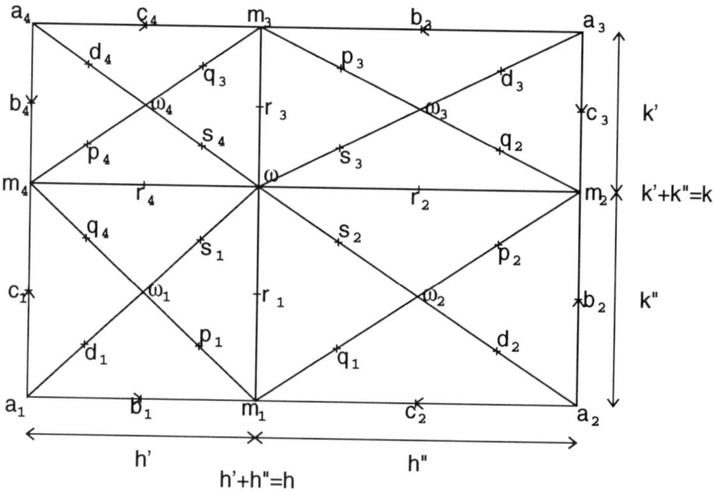

Fig. 1. Macrorectangle R.

Also, since $M_2 = (1-\beta')A_2 + \beta' A_3$, with $\beta' = k'/k$, we obtain :

$$m_2 = (1-\beta')b_2 + \beta' c_3 \quad (\text{and similarly } m_4 = (1-\beta')c_1 + \beta' b_4).$$

Since $k'\frac{\partial g}{\partial y}(A_1) = 2(c_1 - a_1)$, $k'\frac{\partial g}{\partial y}(M_1) = 2(r_1 - m_1)$ and $k'\frac{\partial g}{\partial y}(A_2) = 2(b_2 - a_2)$, the linearity of $\frac{\partial g}{\partial y}$ on A_1A_2 implies :

$$r_1 - m_1 = (1-\alpha')(c_1 - a_1) + \alpha'(b_2 - a_2)$$

which gives the value of r_1. Similar formulae give respectively r_2, r_3 and r_4. By C^1 continuity accross the diagonals of subrectangles $A_iM_i\Omega M_{i-1}, i = 1, 2, 3, 4$ (with the classical convention on indices : $M_0 = M_4$ and $M_5 = M_1$), we obtain successively :

$$d_i = \frac{1}{2}(b_i + c_i), p_i = \frac{1}{2}(b_i + r_i), q_i = \frac{1}{2}(c_{i+1} + r_i), p_i = \frac{1}{2}(b_i + r_i).$$

It remains to compute $\{\omega_i, 1 \le i \le 4\}$ and ω. In fact, we have to prove that $\omega_i = \frac{1}{2}(d_i + s_i) = \frac{1}{2}(q_{i-1} + p_i)$ and $\omega = (1-\alpha')r_4 + \alpha' r_2 = (1-\beta')r_1 + \beta' r_3$.

The first property is verified since

$$4\omega_i = 2(d_i + s_i) = b_i + c_i + r_{i-1} + r_i = 2(p_i + q_{i-1}).$$

As for ω, computing successively

$$r_1 = m_1 + (1-\alpha')(c_1 - \alpha_1) + \alpha'(b_2 - a_2)$$
$$r_3 = m_3 + (1-\alpha')(b_4 - a_4) + \alpha'(c_3 - a_3)$$
$$m_1 = (1-\alpha')b_1 + \alpha' c_2 \text{ and } m_3 = (1-\alpha')c_4 + \alpha' b_3$$

we first obtain

$$(1-\beta')r_1 + \beta' r_3 = (1-\beta')(1-\alpha')(b_1 + c_1 - a_1) + (1-\beta')\alpha'(b_2 + c_2 - a_2) + (1-\alpha')\beta'(b_4 + c_4 - a_4) + \alpha'\beta'(b_3 + c_3 - a_3).$$

The same expression is obtained by computing the quantity $(1-\alpha')r_4 + \alpha' r_2$. Therefore, we have proved that $\dim LS_2^1(R,\Delta) = 12$.

Now it is easy to compute the 12 parameters $\{a_i, b_i, c_i; 1 \le i \le 4\}$ from the values $g(A_i)$ and $Dg(A_i)$. For example, $a_1 = g(A_1), h'\frac{\partial g}{\partial x}(A_1) = 2(b_1 - a_1), k'\frac{\partial g}{\partial y}(A_1) = 2(c_1 - a_1)$, and similar results hold for the three other vertices of R. ■

§3. The Global Hermite Problem

Theorem 2.

(i) $\dim LS_2^1(\Omega,\Delta) = 3(m+1)(n+1)$

(ii) each $g \in LS_2^1(\Omega,\Delta)$ *is uniquely defined by the values of* g *and* Dg *on the set* $\mathcal{A}$.

(iii) $\mathbb{P}_2$ *is included in* $LS_2^1(\Omega,\Delta)$.

Proof: In each rectangle R_{ij}, g is uniquely defined by its values and those of Dg on $\mathcal{A}$. Moreover g is in $C^1(\Omega)$ since the linearity of normal derivatives on edges of macrorectangles obviously implies their continuity across these edges. The dimension of $LS_2^1(\Omega,\Delta)$ is therefore equal to the number of Hermite data, q.e.d. The fact that $\mathbb{P}_2 \subset LS_2^1(\Omega,\Delta)$ is obvious, from the unicity of the Hermite interpolant. ■

Now it is easy to construct compactly supported Hermite functions $\{\phi_A, \chi_A, \psi_A, A \in \mathcal{A}\}$ satisfying the classical conditions :

$$\phi_A(B) = D_x\chi_A(B) = D_y\psi_A(B) = \delta_{AB}$$

and $D\phi_A(B) = D_y\chi_A(B) = D_x\psi_A(B) = \chi_A(B) = \psi_A(B) = 0$ for all $(A,B) \in \mathcal{A}\times\mathcal{A}$. For all $f \in C^1(\Omega)$, we can write the Hermite interpolant of f as follows :

$$Hf = \sum_{A\in\mathcal{A}} (f(A)\phi_A + D_x f(A)\chi_A + D_y f(A)\psi_A).$$

It is clear that $\sum_{A\in\mathcal{A}} \phi_A = 1$. Moreover, it is easy to prove that on each macrorectangle R_{ij} :

$$\omega_0(x,y) = \sum_{A\in\mathcal{A}} (|\chi_A| + |\psi_A|)(x,y) = u\left(1 - \frac{u}{h_i}\right) + v\left(1 - \frac{v}{k_j}\right),$$

where $u = x - x_{2i}, v = y - y_{2j}$ and thus $\max\{\omega_0(x,y), (x,y) \in R_{ij}\} = \frac{1}{4}(h_i + k_i)$.

Theorem 3. *On the macrorectangle R of diameter δ, we obtain the local error bound in uniform norm :*

$$\|f - Hf\|_R \leq \frac{5}{12}\|D^3 f\|_R \delta^3.$$

Proof: By Taylor's formula, we have for $X \in R$ and $A \in V(R)$, the set of vertices of R :

$$f(A) = f(X) + Df(X) \cdot (XA) + \frac{1}{2}D^2 f(X) \cdot (XA)^2 + J_A f(X),$$

where

$$J_A f(X) = \frac{1}{2}\int_0^1 (1-t)^2 D^3 f(X + tXA) \cdot (XA)^3 dt$$

$$D_x f(A) = D_x f(X) + D \cdot D_x f(X) \cdot (XA) + J_A^x f(X)$$
$$D_y f(A) = D_y f(X) + D \cdot D_y f(X) \cdot (XA) + J_A^y f(X)$$

and

$$J_A^x f(X) = \int_0^1 (1-t) D^2 D_x f(X + tXA) \cdot (XA)^2 dt$$

$$J_A^y f(X) = \int_0^1 (1-t) D^2 D_y f(X + tXA) \cdot (XA)^2 dt.$$

Substituting $f(A), D_x f(A)$ and $D_y f(A)$ in the expression of Hf and using the fact that $\mathbb{P}_2 \subset LS_2^1(R, \Delta)$, the error can be written as

$$\begin{aligned} Hf(X) - f(X) &= \sum_{A \in V(R)} (\phi_A(X) J_A f(X) + \chi_A(X) J_A^x f(X) + \psi_A(X) J_A^y f(X)) \\ &\leq \frac{\sqrt{2}}{4}\delta. \end{aligned}$$

Using $\sum \phi_A = 1, \sum(|\chi_A| + |\psi_A|) \leq \frac{1}{4}(h+k) \leq \frac{\sqrt{2}}{4}\delta$ and bounds on remainders : $|J_A f(X)| \leq \frac{1}{6}\|D^3 f\|_R \delta^3, |J_A^x f(X)|$ and $|J_A^y f(X)| \leq \frac{1}{2}\|D^3 f\|_R \delta^2$, we get the result. ∎

Bounds for first and second derivatives can be obtained in a similar way and will be detailed elsewhere : the error constants strongly depend on the position of node lines and on the mesh ratios of the given partitions X_0 and Y_0.

§4. The Lagrange Interpolation Problem

For sake of simplicity, we assume that the function f to be interpolated is known outside Ω at the points needed to define approximate partial derivatives. If this is not the case, we implicity assume that we take values of f obtained by parabolic extrapolation on horizontal and vertical meshlines with even indices. We now replace $D_x f(A)$ (resp. $D_y f(A)$) by the value $d_x f(A)$ of the derivative of the unique parabola interpolating f at the vertex A and at the two left and right (resp. lower and upper) neighbours of A in $\mathcal{A}$. We then define the Lagrange interpolant of A by setting

$$\mathcal{L}f = \sum_{A \in \mathcal{A}} \Big(f(A)\phi_A + d_x f(A)\chi_A + d_y f(A)\psi_A \Big).$$

Let $\Gamma_{ij} = \{(i,j), (i-1,j), (i,j-1), (i-1,j-1)\}$,

$$\hat{R}_{ij} = \bigcup_{\gamma \in \Gamma_{ij}} R_\gamma, \hat{M}_{ij} = \|D^3 f\|_{\hat{R}_{ij}}, \tilde{R}_{ij} = \bigcup \{\hat{R}_\gamma, \gamma \in \Gamma_{i+1,j+1}\}, \tilde{\Omega} = \bigcup \tilde{R}_{ij}.$$

By the Taylor expansion, we readily obtain

$$|D_x f(A_{ij}) - d_x f(A_{ij})| \le \frac{1}{6} \hat{M}_{ij} h_{i-1} h_i$$
$$|D_y f(A_{ij}) - d_y f(A_{ij})| \le \frac{1}{6} \hat{M}_{ij} k_{j-1} k_j.$$

Theorem 4. *For $f \in C^3(\tilde{\Omega})$ and $X \in R_{ij}$, the following error bound holds :*

$$|f(X) - \mathcal{L}f(X)| \le \frac{1}{2} \tilde{M}_{ij} \tilde{\delta}_{ij}^3,$$

where $\tilde{M}_{ij} = \|D^3 f\|_{\tilde{R}_{ij}}$ and $\tilde{\delta}_{ij} = \max\{\text{diam } R, R \subset \tilde{R}_{ij}\}$.

Proof:

$$|f(X) - \mathcal{L}f(X)| \le |f(X) - Hf(X)| + |Hf(X) - \mathcal{L}f(X)|$$

and

$$|Hf(X) - \mathcal{L}f(X)| \le \sum_{A \in V(R_{ij})} \{|D_x f(A) - d_x f(A)| |\chi_A|(X)$$
$$+ |D_y f(A) - d_y f(A)| |\psi_A(X)|\}.$$

Using the error bounds given above for approximate derivatives and for $\omega_0 = \sum(|\chi_A| + |\psi_A|)$, we get the desired result. ■

§5. Lebesgue Function and Norm of the Lagrange Projector

Since $d_x f(A_{ij}) = (1-\sigma_i)(f_{ij}-f_{i-1,j})/h_{i-1}+\sigma_i(f_{i+1,j}-f_{ij})/h_i$ and $d_y f(A_{ij}) = (1-\tau_j)(f_{ij}-f_{i,j-1})/k_{j-1}+\tau_j(f_{i,j+1}-f_{ij})/h_j$, where $\sigma_i = h_{i-1}/(h_{i-1}+h_i)$ and $\tau_j = k_{j-1}/(k_{j-1}+k_j)$, one can write $\mathcal{L}f$ in the form

$$\mathcal{L}f = \sum_{A\in\tilde{\mathcal{A}}} f(A)\cdot L_A,$$

where $\tilde{\mathcal{A}}$ is the set of vertices of macrorectangles in $\tilde{\Omega}$ and L_A is a fundamental function with compact support. More precisely, setting $L_{ij} = L_{A_{ij}}$, we get

$$L_{ij} = \phi_{ij} + \left(\frac{1-\sigma_i}{h_{i-1}} - \frac{\sigma_i}{h_i}\right)\chi_{ij} + \frac{\sigma_{i-1}}{h_{i-1}}\chi_{i-1,j} - \left(\frac{1-\sigma_{i+1}}{h_i}\right)\chi_{i+1,j} + \left(\frac{1-\tau_j}{k_{j-1}} - \frac{\tau_j}{k_j}\right)\psi_{ij} + \frac{\tau_{j-1}}{k_{j-1}}\psi_{i,j-1} - \left(\frac{1-\tau_{j+1}}{k_j}\right)\psi_{i,j+1}.$$

(see also [4] and [5] for related results).

It is well known that the norm of the Lagrange projector $\mathcal{L}$ is equal to the ∞-norm of the associated Lebesgue function :

$$\|\mathcal{L}\| = \max_{X\in\Omega}\wedge(X), \quad \text{with} \quad \wedge(X) = \sum_{A\in\tilde{\mathcal{A}}} L_A(X).$$

It is possible to determine the sign of L_A on its support and to compute exactly the maximum of $\wedge$ on each macrorectangle, and consequently the norm of $\mathcal{L}$. The computation is rather technical and cannot be done here. The norm of $\mathcal{L}$ depends on the mesh spacing in x and y and also on the position of intermediate node lines. We only discuss here an interesting particular case. Let

$$p_i = h_i^2/h_{i-1}(h_{i-1}+h_i), q_i = h_i^2/h_{i+1}(h_{i+1}+h_i),$$
$$r_j = k_i^2/k_{j-1}(k_{j-1}+k_j), s_j = k_j^2/k_{j+1}(k_{j+1}+k_j),$$

then we choose $\alpha_i = q_i/(p_i+q_i)$, i.e., $x_{2i+1} = (1-\alpha_i)x_{2i} + \alpha_i x_{2i+2}$ (resp. $\beta_j = s_j/(r_j+s_j)$ and $y_{2j+1} = (1-\beta_j)x_{2j} + \beta_j x_{2j+2}$), and we obtain

$$\max\{\wedge(X), X\in R_{ij}\} = 1 + \frac{p_i q_i}{p_i+q_i} + \frac{r_j s_j}{r_j+s_j}.$$

(In particular, for uniform partitions X_1 and Y_1, we obtain $\|\wedge\|_\infty = \|\mathcal{L}\| = \frac{3}{2}$).

Since the right member of the above equality is also equal to

$$1 + \left[\frac{h_{i+1}}{h_i}\left(1+\frac{h_{i+1}}{h_i}\right) + \frac{h_{i+1}}{h_i}\left(1+\frac{h_{i+1}}{h_i}\right)\right]^{-1} + \left[\frac{k_{j+1}}{k_j}\left(1+\frac{k_{j+1}}{k_j}\right) + \frac{k_{j+1}}{k_j}\left(1+\frac{k_{j+1}}{k_j}\right)\right]^{-1},$$

we see that the norm depends on the local mesh ratios of partitions X_0 and Y_0. Let $r\geq 1$ be such that $\frac{1}{r} \leq \frac{h_{i+1}}{h_i}, \frac{k_{j+1}}{k_j} \leq r$, then we obtain $\|\mathcal{L}\| \leq r + \frac{1}{r+1}$, which seems reasonable. Another type of Lagrange interpolant in a space of quadratic splines has been described in [7] and [8].

References

1. Chui C.K., *Multivariate Splines*, CBMS-NSF. Reg. Conf. Series in Applied Mathematics **54**, SIAM, Philadelphia, 1988.
2. Ciarlet P.G. and Raviart P.A., General Lagrange and Hermite interpolation in $\mathbb{R}^n$ with applications to finite elements methods, Arch. Rat. Mech. Anal. **46** (1972), 177–199.
3. Ciarlet P.G., Basic Error Estimates for Elliptic Problems, in *Handbook of Numerical Analysis, volume II (part 1)* P.G. Ciarlet and J.J. Lions (eds.), North Holland, Amsterdam, 1991, 17–351.
4. Dahmen W., Goodman T.N.T. and Micchelli C.A., Local Spline Interpolation Schemes in one and several Variables, in *Approximation and Optimization*, A. Gomez, F. Guerra, M.A. Jimenez, G. Lopez (eds.), LNM 1354, Springer, Berlin, 1987, 11–24.
5. Dahmen W., Goodman T.N.T. and Micchelli C.A., Compactly supported fundamental functions for spline interpolation, Numer. Math., (1988), 639-664.
6. Farin G., Triangular Bernstein Bézier patches, Computer Aided Geometric Design **3**, (1986), 83–127.
7. Jeeawock-Zedek F., Lagrange Interpolation by quadratic Splines on a Quadrilateral Domain of $\mathbb{R}^2$, in *Curves and Surfaces*, P.-J. Laurent, A. Le Méhauté, and L. L. Schumaker (eds.), Academic Press, New York, 1991, 511–514.
8. Jeeawock-Zedek F., Interpolation de Lagrange par des Splines quadratiques sur un Quadrilatère de $\mathbb{R}^2$, Math. Modelling and Numer. Analysis (M^2AN) **26** n° 5 (1992), 575–594.
9. Powell M.J.D., Piecewise quadratic surface fitting for contour plotting, in *Software for Numerical Mathematics*, D.J. Evans (ed.), Academic Press, London, 1974, 253-271.
10. Sibson R. and Thomson G.D., A seamed quadratic element for contouring, Computer J. **24** (4) (1981), 378-382.

Paul Sablonnière
INSA 20, Avenue des Buttes de Coësmes
35042 Rennes cédex FRANCE

Jeeawock-Zedek
Université de Lille, Département de Mathématiques
59655 Villeneuve d'Ascq cédex FRANCE

Partial Differential Equation Techniques for Surface Determination

C. Sbert

Abstract. We present two direct methods to detect a surface in a 3D image: 1) a smoothing by anisotropic diffusion according to the model proposed by Alvarez-Lions-Morel, followed by classical thresholding to detect contours, 2) a new model of active contours (snakes) based on a partial differential equation proposed by Caselles-Catte-Coll-Dibos. In this case we obtain the surface of the object as the limit of the evolution of an initial surface. To visualize the detected surface, we construct a wire-frame model from the contours of the object and render it with the classical techniques of computer graphics.

§1. Introduction

The purpose of this work is study some methods for edge detection in three-dimensional images, i.e., we want to detect the surface of the objects that appear in an image. The initial data come from tomography or magnetic resonance, and are given by a series of two-dimensional slices (images). We do not try to determine explicitly the surface of the object, but to detect it and display it. For this we need sophisticated methods to detect the object.

The basic idea of edge detection comes from Marr-Hildreth [15] who proposed to convolve the image with Gaussians of increasing variance and then define the edges at each scale as the set of points where the norm of the gradient of the resulting smoothed signal has a local maximum. This theory has been improved by Witkin, Koenderink [20,12], etc, who noticed that the convolution of the signal with Gaussians at each scale was equivalent to solving the heat equation with the signal as initial datum:

$$\begin{aligned} u_t &= \triangle u \\ u(0) &= u_0 \end{aligned} \tag{1}$$

The solution of this equation for an initial datum with bounded quadratic norm is $u = G_t * u_0$, where G_t is the Gaussian function of variance t. It is well known that the edges at low scales give an inexact account of the boundaries.

Wavelets, Images and Surface Fitting
P. J. Laurent, A. Le Méhauté, and L. L. Schumaker (eds.), pp. 453–458.

ISBN 1-56881-040-7.

An important improvement in this sense has been introduced by Malik and Perona [17]. They proposed to replace the heat equation by a nonlinear equation:

$$u_t = div(g(|Du|)Du)$$
$$u(0) = u_0,$$

where g is a smooth nonincreasing function with $g(0) = 1$, $g \geq 0$ and $g(x)$ tends to zero at infinity. The smoothing with this model is conditional: if Du is large then the diffusion will be low ; if Du is small the model tends to be the heat equation (1). This model has several theoretical and practical difficulties.

In order to avoid the above mentioned difficulties, Catte-Coll-Lions-Morel [5] proposed a new model which is robust in the presence of noise and consistent from the formal point of view. They define a "selective smoothing" of u_0 at scale $t^{\frac{1}{2}}$ as

$$\begin{aligned} u_t &= div(g(|G_\sigma * Du|)Du) \\ u(0) &= u_0. \end{aligned} \tag{2}$$

The term $G_\sigma * Du$ appears to be an estimate of the gradient of u at point x. This new model avoids the inconsistencies of the Malik and Perona model.

§2. The Model of Anisotropic Diffusion

It seems natural to modify the diffusion operator so that it diffuses more in the direction parallel to the edge and less in the perpendicular one. The diffusion equation which does not diffuse at all in the direction of the gradient can be written as

$$u_t = \Delta u - \frac{1}{|Du|^2}(D^2u(Du), Du),$$

and in a quasi divergence form the equation can be also written as

$$u_t = |Du| div\left(\frac{Du}{|Du|}\right).$$

Taking into account the estimate of the gradient proposed in (2), Alvarez-Lions-Morel [1] proposed the following model:

$$\begin{aligned} u_t &= g(|G * Du|)|Du| div\left(\frac{Du}{|Du|}\right) \\ u(0) &= u_0, \end{aligned} \tag{3}$$

where u_0 is the grey level image to be processed, u represents its smoothed version depending on the scale parameter t and G is a smoothing kernel

Given a 3D image we smooth it using equation (3) and then use a classical threshold to detect the contours of the image.

In order to define a numerical scheme associated with the equation,we consider an approximate model whose numerical treatment will be easier.

Equation (3) represents a diffusion in the hyperplane orthogonal to the gradient. In the 3D case this direction is a plane. Let us denote it by P, and the anisotropic diffusion term by $\Delta_P u$. This term can be easily discretized at a point (i, j, l) of the cube only if in the plane P one can find points of the sampled cube near it.

Now, we consider a new formulation of the equation which will take into account the discrete number of diffusion planes. Let $P_1, \ldots, P_n$ be the discrete diffusion planes. Let ξ_i^1 and ξ_i^2 be an orthonormal basis of the plane P_i. Then, the anisotropic diffusion operator associated to this plane is

$$\Delta_{P_i} u = \frac{\partial^2 u}{\partial {\xi_i^1}^2} + \frac{\partial^2 u}{\partial {\xi_i^2}^2}.$$

Now, we may decompose the diffusion in the plane P orthogonal to the gradient into a linear nonnegative combination of the fixed diffusion operators $\Delta_{P_i} u$. That is, we consider

$$Au = \sum_{j=1}^{n} f_j \left(\frac{Du}{|Du|} \right) \Delta_{P_j} u, \tag{4}$$

where $f_j \geq 0$ are designed to be active only if the plane P is close to P_j and $\sum_{j=1}^{n} f_j \longrightarrow 1 \quad as \quad n \to \infty$.

Now we discretize the equation (3) using the semi-implicit form

$$\frac{u^{k+1} - u^k}{\Delta t} = g(|G_\sigma * Du^k|) \sum_{j=1}^{9} f_j \left(\frac{Du^k}{|Du^k|} \right) \Delta_{P_j} u^{k+1}),$$

which gives u^{k+1} as the solution of a linear system.

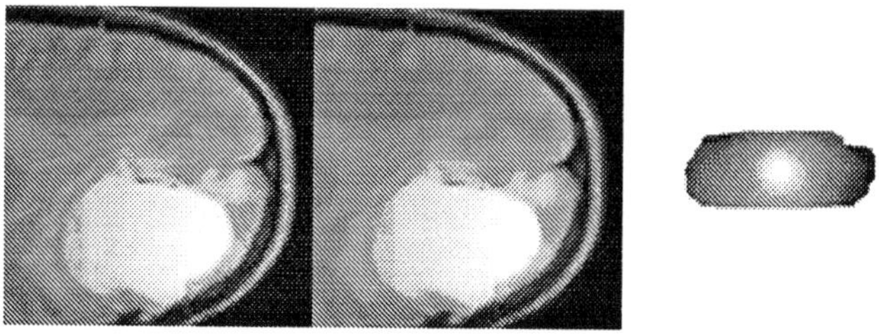

Fig. 1. Left: a slice of a brain tumour. Middle: processed image. Right: reconstructed tumor.

§3. A Geometric Model for Active Contour

The classical active contour method starts with an initial curve called a *snake* near some contour in the image, and looks for admissible deformations of the snake, which let it move toward the desired contour. These deformations are obtained by trying to minimize an energy functional. The original idea is due to Kass et al [11] (see also [7,19], etc.). Now we present a new model for active contour based on a partial differential equation.

Taking into account that the equation

$$u_t = |Du| div \left(\frac{Du}{|Du|} \right) \tag{5}$$

describes the motion of the level sets of the function u, which evolve following the normal direction with speed depending on the mean curvature. Caselles-Catte-Coll-Dibos [4], propose a new model for active contours based on a geometric partial differential equation. They add to (5) a term representing a constant force in the direction of the normal to the level sets and then multiply it by a function $g(x)$ to stop the followed level set there where we want to reproduce the desired contour. Then the model is

$$\begin{aligned} u_t &= g(x)|Du|(div \left(\frac{Du}{|Du|} \right) + \nu) \\ u(0) &= u_0 \\ g(x) &= \frac{1}{1 + |D(G_\sigma * I_0)|^2}, \end{aligned} \tag{6}$$

ν is a real constant, $G_\sigma * I_0$ is the convolution of the image I_0 where we are looking for the contour of an object O with the Gaussian G_σ and u_0 is the initial datum which is taken as a smoothed version of the function $1 - \chi_C$, where χ_C is the characteristic function of the set C containing O.

In [4] Caselles et al prove the existence and uniqueness of solutions of (6) in the viscosity sense, and the geometrical correctness of the method.

In 3D we study the asymptotic behavior of the equation in two cases:

1) (the zero level set of u_0 is a set containing the object we are looking for). In this case we choose $\nu > 0$ in such way that the term $div(\frac{Du}{|Du|}) + \nu$ remains positive, then the grey level at a point of the level set will increase and one expects it to reproduce the boundary of the object as $t \longrightarrow \infty$. ν may be interpreted as a force pushing the level set of u towards O when the curvature of ∂C is null or negative.

2) (the zero level set of u_0 is contained in the object). We choose $\nu < 0$ such that the term $div(\frac{Du}{|Du|}) + \nu$ remains negative. Then the grey level at a point decreases and the level set is pushed towards the contour of the object.

In both cases we suppose that $S = \{x : g(x) = 0\}$ is a compact surface of class C^2 and $\chi(S) = 2$. Let $u(t,x)$ the unique viscosity solution of equation (6). We want to follow the evolution of the sets $G(t) = \{x : u(t,x) = 0\}$ whose boundary $S(t)$ we are interested in. In fact we prove the following result.

Theorem 1. *If the constant $|\nu|$ is sufficiently large, then $S(t)$ converges to S in the Hausdorff distance as $t \longrightarrow \infty$.*

We see in the numerical experiments that this equation can change the topology of the evolving surface, for example, if we have a torus and we take as initial condition a sphere containing the torus, the sphere evolves until it reproduces the contour of the torus (see Fig. 2). The analog of Theorem 1 holds if S is a smooth surface with $\chi(S) = 0$

We adapt the approximated model (4) explained above to discretize the operator $|Du|div(\frac{Du}{|Du|})$, to this equation.

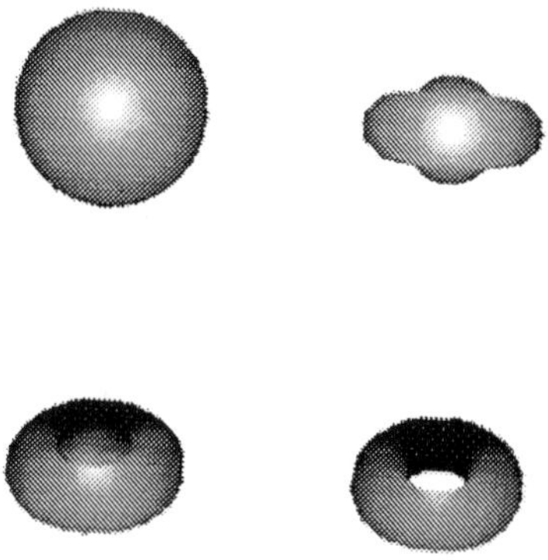

Fig. 2. Four steps of the evolution of a sphere until it reproduces the torus.

§4. Modelling the Surface

We obtain with the above segmentation a set of points on the boundary of the object. The data are organized in a set of planar contours representing the cross-sections through the object. We reconstruct a three-dimensional surface from the cross-sections, contructing a triangular mesh between the points on pairs of contours from adjacent sections. After this we render the triangles with usual techniques of computer graphics.

Acknowledgements. I'm grateful to Vicent Caselles, Tomeu Coll and Jean Michel Morel for their help.

References

1. Alvarez L., P. L. Lions, J. M. Morel, Image selective smoothing and edge detection by nonlinear diffusion (II), SIAM J. Numer. Anal. **29** (1992), 845–866.
2. Ayache N., J. D. Boissonat, J. D. Brunet, L. Cohen, J. P. Chieze, B. Geiger, O. Monga, J. M. Rocchisani, and P. Sander, Building highly

structured volume representations in 3d medical images, Computer Aided Radiology, 1989. Berlin.
3. Canny, J., Finding edges and lines in images, Technical Report 720, MIT, Artificial Intelligence Laboratory, 1983.
4. Caselles, V., F. Catte, B. Coll, and F. Dibos, A geometric model for active contours in image processing, Numer. Math. , to appear.
5. Catte, F., B. Coll, P. L. Lions, and J. M. Morel, Image selective smoothing and edge detection by nonlinear diffusion (I), Cahier du CEREMADE.
6. Cinquin, P., Un modèle pour la représentation d'images medicales 3d, Proceedings Euromedicine, 86.
7. Cohen, L.D., On active contour Models and balloons. CVGIP: Image Understanding **53** (1991), 211–218.
8. Crandall, M.G., H. Ishii, and P. L. Lions P.L., User's guide to viscosity solution of second order partial differential equation, Bull. Amer. Math. Soc. **27** (1992), 1–67.
9. Evans, L.C., and J. Spruck, Motion of level sets by mean curvature Diff Geometry **33** (1991), 635–681.
10. Foley, Van Dam, Feiner, Hughes, *Computer Graphics Principles and Practice*, Addison Wesley, 1991.
11. Kass, M., A. Witkin, and D. Terzopoulos, Snakes: Active contour models, Int Journal Comp. Vision **1** (1988), 321–331.
12. Koenderink, J.J., The structure of images, Biol. Cybern. **50** (1984), 363–370.
13. McLendon, P., Graphics Library Programming Guide. Silicon Graphics Inc. Document Number: 007-1210-040, 1991.
14. Marr, D., *Vision*, Freeman, San Francisco, 1982.
15. Marr, D, and E. Hildreth, A theory of edge detection, Proc Roy. Soc. London **B207** (1980), 187–217.
16. Osher, S., and J. A. Sethian, Fronts propagating with curvature dependent speed: Algorithms based on Hamilton-Jacobi Formulations, Jour. Compt. Phys **79** (1988), 12–49.
17. Perona, P., and J. Malik, Scale space and edge detection using anisotropic diffusion, Proc IEEE Comp.Soc. Workshop on computer vision, 1987.
18. Shirai, Y., *Three-Dimensional Computer Vision*, Springer Verlag, 1987.
19. Terzopoulos, D., A. Witkin, and M. Kass, Constraints on Deformable Models: Recovering 3d shape and nonrigid motion, Artificial Intelligence, (1988), 91-123.
20. Witkin, A. P., Scale-space filtering. Proc of IJCAI, Karlsruhe 1983, 1019–1021.

C. Sbert
Dept. of Mathematics and Computer Science
Univ. de les Illes Balears
Ctra Valldemossa Km. 7,5
07071 Palma de Mallorca, SPAIN
dmicsj0@ ps.uib.es

Approximation of Polynomials by Radial Basis Functions

Robert Schaback

Abstract. For radial basis function interpolation of scattered data in $\mathbb{R}^d$, the approximation of high-degree polynomials is studied. Results include uniform error bounds and convergence orders on compact sets.

§1. Introduction

We consider interpolation of real-valued functions f defined on a set $\Omega \subseteq \mathbb{R}^d$, $d \geq 1$. These functions are interpolated on a set $X := \{x_1, \ldots, x_{N_X}\}$ of $N_X \geq 1$ pairwise distinct points $x_1, \ldots, x_{N_X}$ in Ω. Interpolation is done by linear combinations of translates $\Phi(x - x_j)$ of a single continuous real-valued function Φ defined on $\mathbb{R}^d$. For various reasons it is sometimes necessary to add the space $\mathbb{P}_m^d$ of d–variate polynomials of order not exceeding m to the interpolating functions. Interpolation is uniquely possible under the assumption

$$\text{if } p \in \mathbb{P}_m^d \text{ satisfies } p(x_i) = 0 \text{ for all } x_i \in X, \text{ then } p = 0, \tag{1}$$

and if Φ is conditionally positive definite of order m (see e.g. [8]).

Definition 1. *A function $\Phi : \mathbb{R}^d \to \mathbb{R}$ with $\Phi(-x) = \Phi(x)$ is conditionally positive definite of order m on $\mathbb{R}^d$ if for all sets $X = \{x_1, \ldots, x_{N_X}\} \subset \mathbb{R}^d$ with N_X distinct points and all vectors $\alpha := (\alpha_1, \ldots, \alpha_{N_X}) \in \mathbb{R}^{N_X}$ with*

$$\sum_{j=1}^{N_X} \alpha_j p(x_j) = 0 \quad \text{for all } p \in \mathbb{P}_m^d, \tag{2}$$

the quadratic form $\sum_{j,k=1}^{N_X} \alpha_j \alpha_k \Phi(x_j - x_k)$ is always nonnegative, and vanishes only if $\alpha = 0$.

Wavelets, Images and Surface Fitting
P. J. Laurent, A. Le Méhauté, and L. L. Schumaker (eds.), pp. 459–466.

ISBN 1-56881-040-7.

We list a few examples where $\Phi(x) := \phi(\|x\|_2)$ is truly radial:

- multiquadrics $\phi(r) = (c^2 + r^2)^{\beta/2}$ for $\beta \in \mathbb{R}_{>-d} \setminus 2\mathbb{Z}$ and $2m > \beta$,
- thin-plate splines $\phi(r) = r^\beta$ for $\beta \in \mathbb{R}_{>0} \setminus 2\mathbb{Z}$ and $2m > \beta$,
- thin-plate splines $\phi(r) = (-1)^{\beta/2+1} r^\beta \log r$ for $\beta \in 2\mathbb{N}$, $2m > \beta$,
- Sobolev splines $\phi(r) = \frac{2\pi^{d/2}}{\Gamma(k)} K_{k-d/2}(r)(r/2)^{k-d/2}$ for $k > d/2$ and $m \geq 0$, using the Macdonalds (or modified spherical Bessel) function K_ν,
- Gaussians $\phi(r) = e^{-cr^2}$ for $c > 0$ and $m \geq 0$.

There has been some discussion concerning the relative merits of these radial basis functions. Gaussians and other integrable radial basis functions were said to have “a severe disadvantage”, because they are not exact on constant functions [1, p. 93]. The other choices, especially multiquadrics were said to be “far superior”.

On the other hand, Gaussians have an even better convergence behavior than multiquadrics when used on scattered data (Madych/Nelson [6]) for c fixed. Thus, it is questionable whether the advantage of a method must be necessarily tied to its polynomial reproduction properties. The link between convergence orders and polynomial reproduction properties is clearly visible in theories on grids that make use of Strang–Fix conditions (see the review by W. Light [3]), but it is not at all evident for interpolation of scattered data on irregular domains. Exact reproduction may well be replaced by high-order approximation without any loss in theory and numerical practice.

Numerical results indicate that radial basis function interpolation behaves very well indeed on data from polynomials, provided that the centers are dense enough in and around the domain where the interpolant is evaluated. To add further support to this statement, this paper investigates the polynomial reproduction properties of radial basis function interpolation in the scattered data case.

§2. Spaces for Radial Basis Functions

Each conditionally positive definite function Φ allows two constructions of an inner-product space of functions. Both constructions are based on ideas of Madych and Nelson [4, 5], but we adopt the terminology of [8] here and omit details.

The *algebraic* approach introduces a space F_Φ by direct reference to the conditional positive definiteness of Φ of order m. Functions of the form

$$f_\alpha(x) = \sum_{j=1}^{N} \alpha_j \Phi(x - x_j), \tag{3}$$

with (2) are in F_Φ and have a norm $\|f_\alpha\|_\Phi^2 = \sum_{j,k=1}^N \alpha_j \alpha_k \Phi(x_j - x_k)$. The *Fourier transform* approach makes use of the fact that all of the radial basis functions given in the introduction have a generalized Fourier transform in the sense of tempered distributions, and the latter coincides with a positive and continuous function φ on $\mathbb{R}^d\setminus\{0\}$ in the sense of Jones [2]. In this paper we assume the function φ is continuous and positive on $\mathbb{R}^d\setminus\{0\}$, and satisfies the conditions

$$\begin{aligned}|\varphi(\omega)| &\geq C\cdot\|\omega\|^{-d-s_0} &&,\omega\in U,\\ |\varphi(\omega)| &\geq c\cdot e^{-\gamma\|\omega\|^2} &&,\omega\notin U,\\ |\varphi(\omega)| &\leq c\cdot\|\omega\|^{-d-s_\infty} &&,\omega\notin U.\end{aligned} \tag{4}$$

on a neighborhood $U = K_R(0) = \{x : \|x\|_2 < R,\ x\neq 0\}$ of the origin in $\mathbb{R}^d\setminus\{0\}$. Examples are compiled conveniently in [1] and [8], and we add $\varphi(\omega) = (1+\|\omega\|_2^2)^{-k}$ for Sobolev radial basis functions of order k. This function, for instance, satisfies (4) for $s_0=-d$, $s_\infty = 2k-d$, and an arbitrary $\gamma>0$ related to the radius R of the ball U.

The Fourier transform approach introduces the scalar product

$$(f,g)_\varphi := (2\pi)^{-d}\int_{\mathbb{R}^d}\frac{\hat f(\omega)\overline{\hat g(\omega)}}{\varphi(\omega)}\,d\omega$$

and the space

$$\mathcal{F}_\varphi := \{f : \mathbb{R}^d\to\mathbb{R}\mid \hat f\in L_1(\mathbb{R}^d),\ \|f\|_\varphi<\infty\}, \tag{5}$$

that occurs in [9] and dates back to a similar but different space defined by Madych and Nelson [5], using a nonstandard theory of generalized functions. Functions of the form (3) with (2) for some $m > s_0/2$ are also in $\mathcal{F}_\varphi$ and have a norm

$$\|f_\alpha\|_\varphi^2 = (2\pi)^{-d}\int_{\mathbb{R}_d}\frac{|\hat f_\alpha(\omega)|^2}{\varphi(\omega)}\,d\omega = \|f_\alpha\|_\Phi^2.$$

We finally note that the Fourier transform approach adds $\mathbb{P}_q^d$ to $\mathcal{F}_\varphi$ to perform interpolation, where $q > s_0/2$ holds for the smallest possible s_0 in (4). Reasons for this will become apparent later.

§3. Approximation of Polynomials

We shall now approximate polynomials by functions in the space $\mathcal{F}_\varphi$ of (5). First we represent the delta functional δ by the sequence of good functions $\gamma_k(\omega) = \left(\frac{k}{\pi}\right)^{d/2} e^{-k\|\omega\|^2}$ in the sense of Jones [2]. Then the generalized derivative $\mathcal{D}^\alpha\delta$ for a multi-index α is represented by the sequence $D^\alpha\gamma_k$, and the generalized Fourier transform of this sequence represents the polynomial $\widehat{\mathcal{D}^\alpha\delta}(x) = (+ix)^\alpha$. We now fix z and consider the test functions $f_{k,\alpha,z}(x) := (i(x+z))^\alpha\hat\gamma_k(x+z) = \widehat{D^\alpha\gamma_k}(x+z)$ which satisfy $\delta(f_{k,\alpha,z}) = f_{k,\alpha,z}(0) = (iz)^\alpha\hat\gamma_k(z) = (iz)^\alpha e^{-\|z\|^2/4k}$ and converge to $(iz)^\alpha$ for $k\to\infty$, proving that the δ functional is bounded on these functions when $k\to\infty$. We shall make use of this fact later. The functions $f_{k,\alpha,z}(x)$ converge uniformly to the polynomial $(i(x+z))^\alpha$ on compact sets $\Omega \subset \mathbb{R}^d$, as k tends to infinity. This is due to the fact that

$$\hat{\gamma}_k(u) = e^{-\|u\|^2/4k} = 1 - \frac{\|u\|^2}{4k} + \frac{\|u\|^4}{16k^2} \pm \cdots \geq 1 - \frac{\|u\|^2}{4k}$$

for all $u \in \mathbb{R}^d$. More precisely, for each α and each compact set $\Omega \subset \mathbb{R}^d$, there is a constant c_0, independent of $k \in \mathbb{N}$ and $x \in \Omega$, such that

$$|(ix)^\alpha - f_{k,\alpha,0}(x)| \leq \frac{\|x\|^2}{4k} \tag{6}$$

for all $x \in \Omega$ and all $k \in \mathbb{N}$. In what follows, we shall use the symbols $c, c_0, c_1, \ldots$ to denote generic constants which may differ by certain factors, and we just state the variables that the constants do *not* depend on. Such "constants" may, however, depend on other variables that occur in the context, and these are mainly α, Φ, and Ω.

Now we assert that all the functions $f_{k,\alpha,z}(x)$ are in the spaces $\mathcal{F}_\varphi$ for all radial basis functions that satisfy the basic assumptions (4).

Lemma. *Let the given radial basis function satisfy (4). Then there are positive constants c_1 and c_2, independent of k and z, such that*

$$|f_{k,\alpha,z}|_\varphi^2 = \int \frac{|\widehat{f_{k,\alpha,z}}(\omega)|^2}{\varphi(\omega)}\, d\omega \leq c_1 \cdot k^{|\alpha| - s_0/2} \tag{7}$$

for all $z \in \mathbb{R}^d$ and all $k \geq c_2$.

Proof: By elementary calculations,

$$\begin{aligned}\widehat{f_{k,\alpha,z}}(\omega) &= (2\pi)^d e^{i\langle z,\omega\rangle} D^\alpha \gamma_k(\omega)\\ &= (2\pi)^d e^{i\langle z,\omega\rangle} \left(\frac{k}{\pi}\right)^{d/2} D^\alpha e^{-k\|\omega\|^2},\end{aligned}$$

and we have to check these derivatives for $k \to \infty$. Using Hermite polynomials $H_n(x)$, defined here as $H_n(x) = (-1)^n e^{x^2/2} D^n e^{-x^2/2}$, we first treat the univariate case by

$$\frac{d^n}{dx^n}\, e^{-x^2/2} = (-1)^n H_n(x) e^{-x^2/2} = (\sqrt{2k})^{-n} \frac{d^n}{d\omega^n}\, e^{-k\omega^2}$$

with $x = \omega\sqrt{2k}$, giving

$$D^n e^{-k\omega^2} = (\sqrt{2k})^n (-1)^n H_n(\omega\sqrt{2k}) e^{-k\omega^2}.$$

The multivariate case is a product of univariate cases, because

$$\begin{aligned}D^\alpha e^{-k\|\omega\|^2} &= \prod_{j=1}^d \frac{\partial^{\alpha_j}}{\partial \omega_j^{\alpha_j}}\, e^{-k\omega_j^2}\\ &= \prod_{j=1}^d (\sqrt{2k})^{\alpha_j} (-1)^{\alpha_j} H_{\alpha_j}(\omega_j\sqrt{2k}) e^{-k\omega_j^2}\\ &= (-\sqrt{2k})^{|\alpha|} e^{-k\|\omega\|^2} H_\alpha(\omega\sqrt{2k}),\end{aligned}$$

where we defined a multivariate Hermite polynomial $H_\alpha(z) = \prod_{j=1}^d H_{\alpha_j}(z_j)$ for $z \in I\!R^d$, $\alpha \in \mathbb{Z}^d_{\geq 0}$. Using (4), the integral of (7) on U is bounded by

$$k^{d+|\alpha|} \int_U \|\omega\|^{d+s_0} \cdot e^{-2k\|\omega\|^2} \cdot |H_\alpha(\omega\sqrt{2k})|^2 d\omega,$$

up to constants which are independent of k and z. By transformation $u = \omega\sqrt{2k}$ we get the bound

$$k^{d+|\alpha|}k^{-d-s_0/2} \int_{\|u\|\leq R\sqrt{2k}} \|u\|^{d+s_0} e^{-\|u\|^2} |H_\alpha(u)|^2 du \leq c \cdot k^{|\alpha|-s_0/2}.$$

On the complement of U and for $2k > \gamma$ we can use the bound $|H_n(x)| \leq c(n) \cdot |x|^n$ for Hermite polynomials $H_n(x)$ for large arguments, and we find

$$\begin{aligned} & k^{2|\alpha|}k^d \int_R^\infty r^{d-1} e^{\gamma r^2} r^{2|\alpha|} e^{-2kr^2} dr \\ \leq \quad & c \cdot \frac{k^{2|\alpha|}k^d}{(2k-\gamma)^{(d+2|\alpha|)/2}} \int_{R\sqrt{2k-\gamma}}^\infty s^{d-1+2|\alpha|} e^{-s^2} ds, \end{aligned}$$

which vanishes exponentially for $k \to \infty$. Thus we finally have (7) with constants that depend only on α, Ω, d, φ, and R, but not on z or k. ∎

The above argument proves that for $s_0 > 0$ the delta functional is not continuous on $\mathcal{F}_\varphi$, because one can pick any α with $0 \leq |\alpha| < s_0/2$ to get

$$\lim_{k\to\infty} \frac{|\delta(f_{k,\alpha,z})|}{\|f_{k,\alpha,z}\|_\varphi} = \infty,$$

for all $z \in \mathbb{R}^d \setminus \{0\}$. More generally,

Theorem 1. *A functional of the form*

$$\epsilon_x f = f(x) - \sum_{j=1}^N u_j(x) f(x_j) \tag{8}$$

is bounded on $\mathcal{F}_\varphi$ iff it vanishes on all polynomials of degree less than $s_0/2$.

Proof: The functional is continuous if it vanishes on $\mathbb{P}^d_q$ (see [5,9]). To prove the converse, assume ϵ_x to be bounded on $\mathcal{F}_\varphi$. Then

$$\lim_{k\to\infty} |\epsilon_x f_{k,\alpha,z}| \leq c(x) \cdot \lim_{k\to\infty} \|f_{k,\alpha,z}\|_\varphi = 0$$

for all $|\alpha| < s_0/2$ and all z. On the other hand, by the mollification and (6),

$$\lim_{k\to\infty} \epsilon_x f_{k,\alpha,z} = (i(x+z))^\alpha - \sum_j u_j(x)(i(x_j+z))^\alpha,$$

for all $|\alpha| < s_0/2$ and all z. But this means that ϵ_x must vanish on all polynomials of degree $< s_0/2$ if it is continuous on $\mathcal{F}_\varphi$. ■

Theorem 1 sheds some light on the admissible values of q. It is possible to define the space $\mathcal{F}_\varphi$ of (5) with values of q that do not satisfy $q > s_0/2$, but this inequality is required to permit the usual radial basis function interpolants to be contained in $\mathcal{F}_\varphi$. At the same time $q > s_0/2$ ensures that functionals of the form (8) are continuous on $\mathcal{F}_\varphi$ if they are exact on $\mathbb{P}_q^d$, and Theorem 1 gives a partial converse, making reproduction of low-degree polynomials a *necessary* condition for boundedness of the error functional. The consequence is that one cannot use values of q that do not satisfy $q > s_0/2$, without throwing the interpolants out of $\mathcal{F}_\varphi$ or making the error functional discontinuous on this space.

Due to Theorem 1, there are two situations:

1) For $s_0 \notin 2\mathbb{N}$, continuity of the error functional is equivalent to exactness on $\mathbb{P}_q^d$ for the smallest $q > s_0/2$. This value of q yields the smallest possible error bound.
2) For $s_0 = 2k \in 2\mathbb{N}$, the necessary and sufficient conditions differ. Exactness on $\mathbb{P}_q^d$ for any $q > k$ is sufficient for continuity, and exactness on $\mathbb{P}_q^d$ for all $q \leq k$ is necessary. This gap occurs for thin–plate splines $\phi(r) = r^{2k}\log r$.

Furthermore, the proof of Theorem 1 now implies that interpolants from multiquadrics $\Phi(\|x\|) = \sqrt{c^2 + \|x\|^2}$ without addition of constant polynomials will not lead to a bounded error functional on the space $\mathcal{F}_\varphi$.

On the space $\mathcal{F}_\varphi$ defined for thin-plate splines $\Phi(x) = \|x\|^3$, interpolation by multiquadrics will have an unbounded error, if linear polynomials are not added to the multiquadrics.

However, there can possibly exist larger and more interesting spaces than $\mathcal{F}_\varphi$ that still contain the interpolants, and may lead to error estimates in weaker norms. These spaces could possibly allow smaller values of q, but they are still to be found.

§4. Approximate Polynomial Reproduction

We use results concerning the approximation order of radial basis function interpolants [4,5,8]. For a given radial basis function Φ with a generalized Fourier transform coinciding with φ on $\mathbb{R}^d \setminus \{0\}$ satisfying (4), a given compact set $\Omega \subset \mathbb{R}^d$ and a given constant $\rho > 0$, there are positive constants c_3 and c_4 such that for all finite sets $X = \{x_1, \ldots, x_{N_X}\} \subset \mathbb{R}^d$ of centers and all points $x \in \Omega$ with

$$h_{X,\rho}(x) := \sup_{\|y-x\|_2 \leq \rho} \min_{1 \leq j \leq N_X} \|y - x_j\|_2 \leq c_3, \tag{9}$$

the interpolation error of x is bounded by $c_4 \cdot \|f\|_\varphi h_{X,\rho}^{s_\infty/2}(x)$, where s_∞ is determined by φ at infinity via (4). Here, the constants c_3 and c_4 are independent of x, X, and $h_{X,\rho}(x)$. With this, (7) yields

$$|f_{k,\alpha,z}(x) - \sum_j u_j^*(x) f_{k,\alpha,z}(x_j)|^2 \le c \cdot k^{|\alpha|-s_0/2} \cdot h_{X,\rho}^{s_\infty}(x) \tag{10}$$

for all $z \in I\!R^d$, where $u_j^*(x)$ are optimal for Φ, and where c is independent of k, z, X, and x. This estimate interests us for $|\alpha| \ge s_0/2$, in order to prove something about the behaviour of optimal interpolation processes on high–degree polynomials. Defining $s_{X,k,\alpha}(x) := \sum_{j=1}^{N_X} u_j^*(x) f_{k,\alpha,0}(x_j)$ as a quasi-interpolant that is an interpolant to a mollification of the polynomial $(ix)^\alpha$, we get

$$|f_{k,\alpha,0}(x) - s_{X,k,\alpha}(x)|^2 \le c_5 k^{|\alpha|-s_0/2} \cdot h_{X,\rho}^{s_\infty}(x),$$

and add (6) to yield

$$|x^\alpha - (-i)^{|\alpha|} s_{X,k,\alpha}(x)| \le \frac{c_6}{k} + c_7 k^{|\alpha|/2 - s_0/4} h_{X,\rho}^{s_\infty/2}(x)$$

with constants that are independent of k, X, and $h_{X,\rho}(X)$.

This bounds the approximation error of a polynomial in the space of radial basis function interpolants. Special approximations of the monomial x^α are obtained here via interpolation of the test functions $f_{k,\alpha,0}$ that uniformly converge to x^α on compact sets at the rate $\mathcal{O}(k^{-1})$ for $k \to \infty$. The error depends on the centers X and their local density $h_{X,\rho}(x)$ at x, of course, but also on the free parameter k that controls the mollification.

We now tie k to $h_{X,\rho}(x)$ by making the two terms in the error bound approximately equal, and we are interested in the convergence rate of the error for centers X with densities $h_{X,\rho}(x)$ tending to zero. This is done by choosing $k \approx h_{X,\rho}(x)^{-s_\infty/(2+|\alpha|-s_0/2)}$ which tends to infinity for $h_{X,\rho}(x)$ tending to zero, whenever the inequalities $|\alpha| > s_0/2 - 2$, $s_\infty > 0$ hold. Under these conditions, we finally have

$$|x^\alpha - (-i)^{|\alpha|} s_{X,k,\alpha}(x)| \le c_7 (h_{X,\rho}(x))^{s_\infty/(2+|\alpha|-s_0/2)} \tag{11}$$

for all sets X of centers and all points $x \in \Omega$ satisfying (9), and with a constant c_7 that is independent of x, X, and $h_{X,\rho}(x)$. From (11) we can read off a number of statements:

Theorem 2. *In the space $\mathcal{F}_\varphi$ generated by any of the usual radial basis functions Φ, polynomials are dense with respect to uniform convergence on compact sets. Polynomials of degree less than $s_0/2$ must necessarily be exactly reproduced to make error functionals of the form (8) continuous on $\mathcal{F}_\varphi$. Polynomials of degree $n \ge s_0/2$ can be approximated on compact sets by interpolants of certain mollifications of themselves with a guaranteed uniform convergence rate that depends on the radial basis function and on the local density $h_{X,\rho}$ of the set X of centers. For $h_{X,\rho}(x) \to 0$ the convergence order in terms of $h_{X,\rho}$ is $s_\infty/(2+n-s_0/2)$ with the following special implications:*

1) *For multiquadrics, inverse multiquadrics, and Gaussians with fixed parameters c, the convergence order is arbitrarily large.*
2) *For thin-plate splines with parameter β, the order is at least $\beta/(2+n-\beta/2)$.*
3) *For splines in Sobolev spaces of order k, the approximation order is at least $k/(2+n-k/2)$.*

Proof: For the last three statements we simply use the appropriate values of s_0 and s_∞. ■

Acknowledgements

Help in proofreading was provided by Armin Iske and Marko Weinrich.

References

1. Buhmann, M. D., Multivariable interpolation using radial basis functions, Ph. D. dissertation, Cambridge, 1989.
2. Jones, D.S., *The Theory of Generalised Functions*, Cambridge University Press, 1982.
3. Light, W. , Some aspects of radial basis function approximation, in: *Approximation Theory, Spline Functions and Applications*, S.P. Singh (ed.), Kluwer Academic Publishers, Dordrecht, 1992, 163–190.
4. Madych, W. R., and Nelson, S. A., Multivariate Interpolation and conditionally positive definite functions, Approx. Theory Applic. **4** (1988), 77–89.
5. Madych, W. R., and Nelson, S. A., Multivariate Interpolation and conditionally positive definite functions II, Math. Comp. **54** (1990), 211–230.
6. Madych, W. R., and Nelson, S. A., Bounds on Multivariate Polynomials and Exponential Error Estimates for Multiquadric Interpolation, in *J. Approx. Th.* **70** (1992), 94–113.
7. Micchelli, C. A., Interpolation of scattered data: distance matrices and conditionally positive definite functions, Constr. Approx. **2** (1986), 11–22
8. Schaback, R., Comparison of radial basis function interpolants, in *Multivariate Approximations: From CAGD to Wavelets*, K. Jetter and F. Utreras (eds.), World Scientific Publishing Co., 1993, 293–305.
9. Wu, Zong-min, and R. Schaback, Local Error Estimates for Radial Basis Function Interpolation IMA J. of Numer. Anal. **13** (1993), 13–27.

Robert Schaback
Institut für Numerische und Angewandte Mathematik
Georg-August-Universität Göttingen
Lotzestr. 16–18
D-37083 Göttingen, GERMANY
schaback@namu01.gwdg.de

Modeling of Geological Surfaces Using Finite Elements

J. Springer

Abstract. Geological surfaces are generally known incompletely and have to be reconstructed by use of interpolation methods. Some of these surfaces may be described by a continuous function $z = u(x,y)$, where z is the vertical coordinate (depth). However, tectonic stresses cause discontinuities (faults) and complicated surfaces (folds). In order to interpolate geological surfaces an algorithm is proposed which uses triangular finite elements. Among all functions satisfying the given constraints (u and possibly u_x and u_y at the data points) the sought function shall minimize an integral, the potential energy of a thin elastic plate.

§1. Introduction

Many geological structures are described by surfaces. Since these geological surfaces can frequently only be observed at single points, an interpolation method is necessary. For a long time interpolation has been done manually by geologists; today also computers and numerical methods are used for this purpose. Some specific geological problems make the mathematical treatment of these surfaces interesting.

§2. Bivariate interpolation

In the simplest case, we have the problem of bivariate interpolation of scattered data: Given the depth z_i of a surface at certain points (x_i, y_i), $i = 1, \ldots, n$, find a function $u = u(x,y)$ such that $u(x_i, y_i) = z_i$, $i = 1, \ldots, n$. The surface described by the function $u(x,y)$ should be *as smooth as possible* to show no more features than are inherent in the data.

A variety of algorithms are used to solve this problem: linear interpolation in triangles [8], weighted averaging of data values [9], piecewise polynomial interpolation [1], kriging, splines [5] and others. Many of them are of heuristical nature, they do not really claim to find the smoothest surface in any sense.

An approach which introduces an exact definition of the smoothest surface is the thin plate spline interpolation by Duchon [6]. This method solves the following problem.

Wavelets, Images and Surface Fitting
P. J. Laurent, A. Le Méhauté, and L. L. Schumaker (eds.), pp. 467–474.

ISBN 1-56881-040-7.

Problem 1. *Given n points $(x_i, y_i) \in \mathbb{R}^2$ find a function $u(x,y) \in C^2(\mathbb{R}^2)$ which minimizes the measure of roughness*

$$\|u\| = \iint_{\mathbb{R}^2} (u_{xx}^2 + 2u_{xy}^2 + u_{yy}^2)\, dxdy$$

subject to the constraints $u(x_i, y_i) = z_i, \quad i = 1, \ldots, n$ with given values $z_i \in \mathbb{R}$.

The exact solution has the form

$$u(x,y) = a_0 + a_1 x + a_2 y + \sum_{i=1}^{n} b_i d_i^2 \ln(d_i)$$

$$\text{with } d_i = \sqrt{(x - x_i)^2 + (y - y_i)^2}.$$

The $n+3$ coefficients a_0, a_1, a_2, $b_1, \ldots, b_n$ are found as solution of a system of $n+3$ linear equations:

$$\begin{aligned} u(x_i, y_i) = z_i, &\qquad i = 1, \ldots, n, \\ \sum_{i=1}^{n} b_i = \sum_{i=1}^{n} b_i x_i = \sum_{i=1}^{n} b_i y_i = 0 & \end{aligned} \tag{1}$$

This method usually provides surprisingly smooth surfaces in comparison to other methods.

§3. Including Additional Information

A legion of software exists for the solution of the bivariate interpolation problem stated above. Nevertheless the results are often not satisfactory for the geologist. The reason is simple: a geologist knows more about the surface than he can put into the computer.

Based on heuristic geological methods several attempts have been made to include additional data into the interpolation process. Stenzel [7] suggests the addition of artificial "data" points to reflect known discontinuities or general shapes of expected structures. A procedure of this type frequently does not produce satisfactory results and can hardly be automatized.

In this paper a direct way is proposed to add geological knowledge. Two types of additional information are considered.

1. Slope of the surface at data points.

The slope can be measured by hand at the earth's surface or by dipmeters in boreholes. In geology it is usually given by two angles (dip and strike). For the surface representation $z = u(x,y)$ these angles can be converted to the gradient of u.

2. Lines in the xy-plane along which the interpolation function is allowed to be discontinuous.

Such lines correspond to geological faults which result from tectonic forces. Their location is often detected by remote sensing or geophysical methods. In this context the faults are assumed to be vertical.

Taking these types of additional information into account, the problem stated above can be modified:

Problem 2. *Given $n = n_1 + n_2 + n_3$ points $(x_i, y_i) \in \Omega \subset \mathbb{R}^2$ and a subset $\ell \in \Omega$ consisting of some lines, find a function $u(x,y) \in C^2(\Omega \setminus \ell)$ that minimizes*

$$\iint\limits_{\Omega\setminus\ell} (u_{xx}^2 + 2u_{xy}^2 + u_{yy}^2)\, dxdy \tag{2}$$

subject to the constraints

$$\begin{aligned}
&u(x_i, y_i,) = z_i, \quad i = 1, \ldots, n_1 \\
&\frac{\partial u}{\partial x}(x_i, y_i) = g_i, \quad \frac{\partial u}{\partial y}(x_i, y_i) = h_i, \quad i = n_1{+}1, \ldots, n_1{+}n_2 \\
&u(x_i, y_i,) = z_i, \quad \frac{\partial u}{\partial x}(x_i, y_i) = g_i, \quad \frac{\partial u}{\partial y}(x_i, y_i) = h_i, \quad i = n_1{+}n_2{+}1, \ldots, n
\end{aligned} \tag{3}$$

with given values z_i,g_i,$h_i \in \mathbb{R}$.

Note that the function $u(x,y)$ is allowed to be discontinuous along the lines in ℓ and that at the data points the value, the gradient or both is given.

The next two sections describe an algorithm which provides an approximate solution to Problem 2.

§4. Using Finite Elements

Bivariate interpolation may also be done by use of the finite element method. For this purpose the domain Ω is to be approximated by the union of simple subdomains, the finite elements. For scattered data the triangle is the natural form of these elements.

The interpolation function, usually a polynomial within an element, is described by a finite number of parameters, usually the function values and derivatives up to a certain degree at the nodes (corners of the triangles). If the data points are chosen as nodes some of the parameters are given by the constraints (3) while the others may be used to minimize the integral (2). These considerations lead to a quadratic programming problem and finally to a (sparse) system of linear equations.

By this interpolation method two goals are achieved: Slope information can easily be incorporated, and the interpolation function minimizes a measure of roughness, although only in a finite-dimensional subspace of functions. As will be shown later discontinuities can also be allowed.

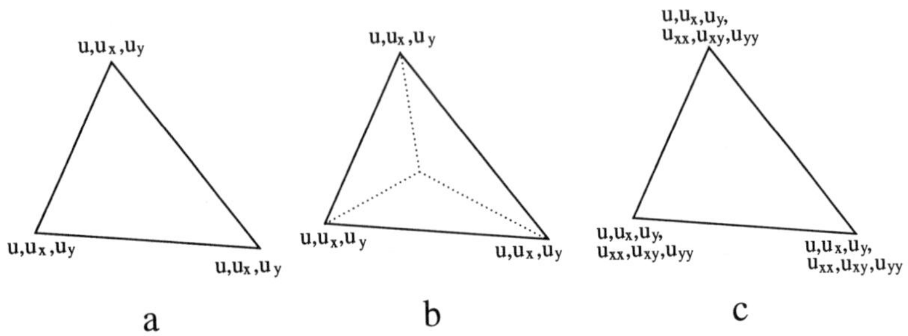

Fig. 1. Three triangular finite elements:
a) Zienkiewicz, b) Clough-Tocher, c) Bell.

Various triangular finite element types are available. To allow prescribed gradients at least three degrees of freedom at each node are necessary: function value and the two components of the gradient. This gives 9 degrees of freedom per triangle. Three finite element types have been tested (Fig. 1):

1) The *Zienkiewicz element* [2] uses a polynomial of third degree in two variables but reduces the number of parameters from 10 to 9 (u, u_x and u_y at each node). This element provides quite good results for most geological purposes, however, the interpolation function is not in C^1: the gradient is continuous at the nodes and within the triangles, but generally not along the triangle edges.

2) The *Clough-Tocher element* [4] uses the same parameters as the Zienkiewicz element, but subdivides the triangle into three subtriangles. In each subtriangle a complete polynomial of third order is used. The $3 \times 10 = 30$ parameters are determined by continuity conditions and the assumption that the normal slope changes linearly along the sides of the original triangle. The latter assumption makes the gradient along an edge uniquely determined by the gradient at the corresponding nodes. Therefore, the gradient is forced to be continuous across the triangles and the interpolation function is in C^1. Another disadvantage (common to the Zienkiewicz element) remains: if all slopes are given at the data points, there is no parameter left for minimizing the measure of roughness — the function is already uniquely determined. As a result the surface often does not look very smooth. The triangle edges are visible as zones of higher curvature.

3) The best results are provided by a triangle with 18 degrees of freedom (u, u_x, u_y, u_{xx}, u_{xy} and u_{yy} at each node) proposed by Bell [3]. This triangle uses a complete polynomial of fifth degree and assumes the normal slope to change cubically along the edges. The three additional parameters per node allow the creation of a very smooth surface which should be close to the smoothest surface at all.

§5. Discontinuities

Using finite elements, it is possible to include discontinuities directly into the interpolation process (Fig. 2):

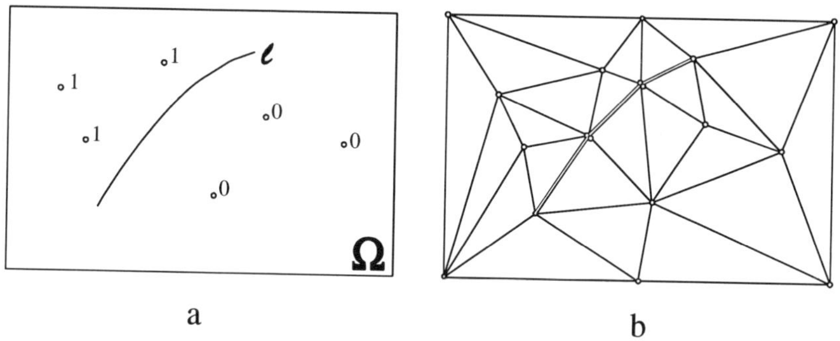

Fig. 2. Interpolation with discontinuities: a) Domain Ω, data points with function values and line of discontinuity ℓ, b) Triangularization.

Each line of discontinuity is approximated by a polyline, a series of connected line segments. The endpoints of these line segments are added to the set of data points and the triangularization is forced to use the line segments themselves as edges of triangles.

To represent the discontinuity the nodes within the polyline (but in general not its endpoints) are doubled. The two nodes of identical coordinates are then properly attached to the triangles at both sides of the discontinuity. *After this modification the whole process of interpolation can be done exactly as before!* No parameters are prescribed at the additional points along a discontinuity, values and slopes of the function are computed, and they usually differ at the two sides of the discontinuity (Fig. 3).

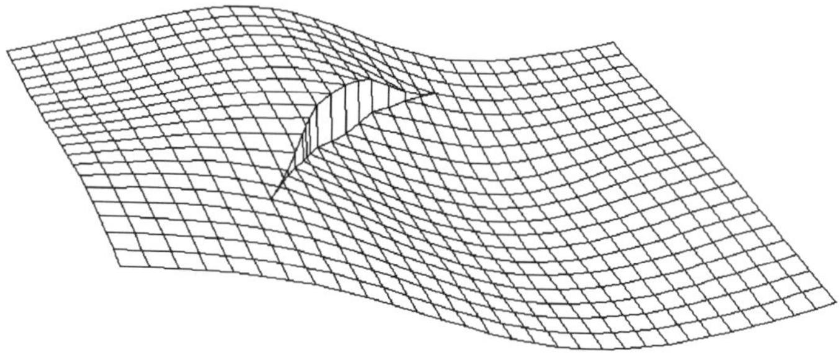

Fig. 3. Interpolated surface.

§6. General Surfaces

Many geological surfaces may not be represented by a single-valued function $z = u(x, y)$. Horizontal tectonic forces compress the sedimentary layers and produce complex overlapping folds. The resulting surfaces can only be represented in parametric form:

$$\mathbf{x} = \mathbf{x}(p, q) \text{ with } \mathbf{x} = (x, y, z)^T.$$

Minimizing the curvature of such a general surface leads to a non-linear problem. This paper does not treat this problem, but restricts itself to the *interpolation between points with given slopes.* The slopes may be defined by the normals:

Problem 3. *Given n points* $\mathbf{x}_i \in \mathbb{R}^3$, *n unit vectors* $\mathbf{n}_i \in \mathbb{R}^3$ *and a triangular mesh connecting the points, find a surface through the points* $\mathbf{x}_i$, *normal to the* $\mathbf{n}_i$ *at these points and topologically equivalent to the triangular mesh.*

"Topologically equivalent" means here the existence of a continuous map from the triangular mesh to the interpolated surface which maps the data points onto themselves.

Using a local coordinate system for each triangle allows the application of the two-dimensional finite element approach. To get a continuous surface the local coordinate systems must be distorted. This problem can be treated by defining "normal" axes $\mathbf{d}_i$ at each data point which are linear interpolated within the triangles (Fig. 4). A good choice for the $\mathbf{d}_i$'s is any kind of mean of the normals of the surrounding triangles.

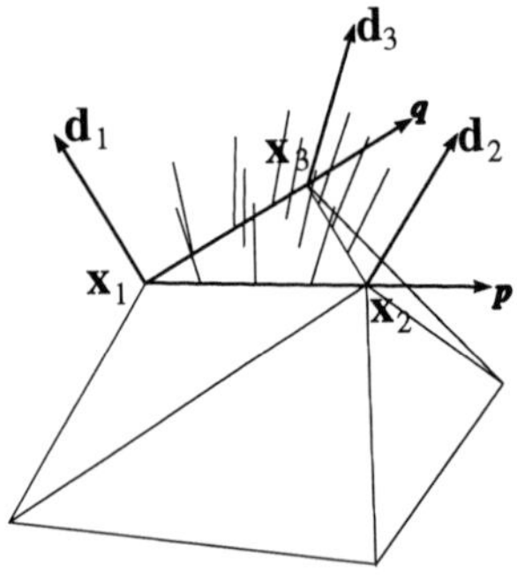

Fig. 4. Local coordinate system for interpolation of a generalized surface.

The part of the surface corresponding to a triangle is described by

$$\mathbf{x}(p, q) = \mathbf{x}_1 + p \cdot (\mathbf{x}_2 - \mathbf{x}_1) + q \cdot (\mathbf{x}_3 - \mathbf{x}_1) + u(p, q) \cdot \mathbf{d}(p, q), \tag{4}$$

$$\mathbf{d}(p, q) = (1 - p - q) \cdot \mathbf{d}_1 + p \cdot \mathbf{d}_2 + q \cdot \mathbf{d}_3, 0 \le p \le 1, \quad 0 \le q \le 1 - p, \tag{5}$$

where $\mathbf{x}_1$, $\mathbf{x}_2$, $\mathbf{x}_3$ and $\mathbf{d}_1$, $\mathbf{d}_2$, $\mathbf{d}_3$ are the triangle corners (data points) and corresponding "normal" axes, respectively. The parameters p and q are the local coordinates in the triangle plane, the corners have the coordinates (p_i, q_i) with $p_1 = q_1 = 0$, $p_2 = 1$, $q_2 = 0$, $p_3 = 0$, $q_3 = 1$. The remaining problem is the appropriate choice of the interpolation function $u(p, q)$.

Theorem 1. *The surface described by (4) passes through the data points* $\mathbf{x}_1, \mathbf{x}_2, \mathbf{x}_3$ *and has there the normals* $\mathbf{n}_1, \mathbf{n}_2, \mathbf{n}_3$, *if* $u(p,q)$ *fulfils for* $i = 1,2,3$ *the following conditions:*

$$u(p_i, q_i) = 0, \tag{6}$$

$$u_p(p_i, q_i) = -\frac{(\mathbf{x}_2 - \mathbf{x}_1)^T \mathbf{n}_i}{\mathbf{d}_i^T \mathbf{n}_i}, \quad u_q(p_i, q_i) = -\frac{(\mathbf{x}_3 - \mathbf{x}_1)^T \mathbf{n}_i}{\mathbf{d}_i^T \mathbf{n}_i}. \tag{7}$$

Proof: From (4) and (6) it follows immediately $\mathbf{x}(p_i, q_i) = \mathbf{x}_i$, thus the surface passes through the data points. The tangential plane at the data points is spanned by the vectors

$$\mathbf{x}_p(p_i, q_i) = \mathbf{x}_2 - \mathbf{x}_1 + u_p(p_i, q_i) \cdot \mathbf{d}(p_i, q_i) + u(p_i, q_i) \cdot \mathbf{d}_p(p_i, q_i),$$
$$\mathbf{x}_q(p_i, q_i) = \mathbf{x}_3 - \mathbf{x}_1 + u_q(p_i, q_i) \cdot \mathbf{d}(p_i, q_i) + u(p_i, q_i) \cdot \mathbf{d}_q(p_i, q_i).$$

Because of (6) the last terms vanish. Equation (5) yields $\mathbf{d}(p_i, q_i) = \mathbf{d}_i$, so we have

$$\mathbf{x}_p(p_i, q_i) = \mathbf{x}_2 - \mathbf{x}_1 + u_p(p_i, q_i) \cdot \mathbf{d}_i, \quad \mathbf{x}_q(p_i, q_i) = \mathbf{x}_3 - \mathbf{x}_1 + u_p(p_i, q_i) \cdot \mathbf{d}_i.$$

The dot products of these vectors with $\mathbf{n}_i$ gives

$$\mathbf{x}_p(p_i, q_i)^T \cdot \mathbf{n}_i = \mathbf{x}_q(p_i, q_i)^T \cdot \mathbf{n}_i = 0$$

because of (7). This means that the tangential plane and so the interpolated surface itself are normal to $\mathbf{n}_i$. ■

Theorem 1 provides a natural way for interpolation: each of the given triangles is replaced by a finite element with 9 degrees of freedom. The interpolation function $u(p,q)$ in (4) is determined by its values and first derivatives at the corners which in turn are given by (6) and (7). By this method the interpolation of the whole surface can be done independently within each triangle (Fig. 5).

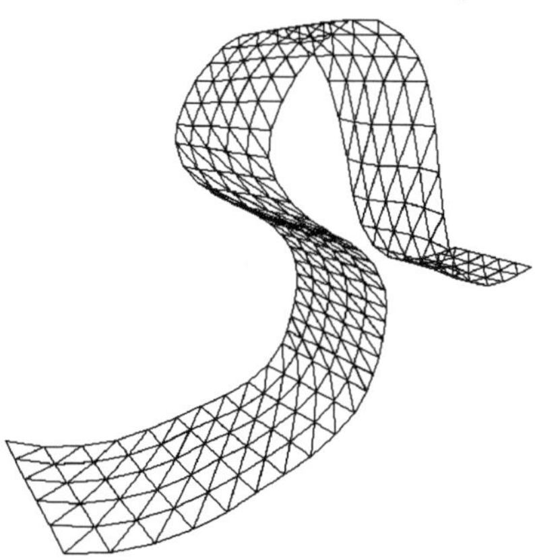

Fig. 5. Simple fold interpolated from 20 points.

§7. Summary

The usage of the finite element method for interpolation of geological surfaces is suggested. In the case of single-valued functions it offers the possibility to include additional information like known slopes or known lines of discontinuity. The resulting surface satisfies the given constraints exactly and minimizes a measure of roughness in a finite-dimensional subspace of functions.

Multi-valued surfaces can be interpolated if a triangularization and slopes at all data points are given.

Acknowledgements. I would like to thank U. Bayer and R. Ondrak, GFZ Potsdam, for reviewing the manuscript and helpful discussions.

References

1. Akima, H., A method of bivariate interpolation and smooth surface fitting for irregularly distributed data points, ACM Trans. Math. Softw. **4** (1978), 148–159.
2. Bazeley, G. P., Y. K. Cheung, B. M. Irons, and O. C. Zienkiewicz, Triangular elements in bending - conforming and non-conforming solutions, in *Proc. 1st Conf. Matrix Methods in Struct. Mech.*, Air Force Inst. of Tech. Wright Patterson A.F. Base, Ohio, 1965, 547–576.
3. Bell, K., A refined triangular plate bending finite element, Int. J. Num. Meth. Eng. **1** (1969), 101–122.
4. Clough, R. W. and J. L. Tocher, 1965, Finite element stiffness matrices for analysis of plates in bending, in *Proc. 1st Conf. Matrix Methods in Struct. Mech.*, Air Force Inst. of Tech. Wright Patterson A.F. Base, Ohio, 1965, 515–545.
5. Dubrule, O., 1984, Comparing splines and kriging, Computers & Geosciences **10** (1984), 327–338.
6. Duchon, J., Splines minimizing semi-norms in Sobolev spaces, in *Constructive Theory of Functions of Several Variables*, Proc. Conf. Oberwolfach 1976, Lecture Notes in Mathematics 571, Springer, Berlin, 1977.
7. Stenzel, P., An attempt of including some geological information into interpolation process, Sci. de la Terre, Sér. Inf. **27** (1988), 69–77.
8. Watson, D. F., ACORD: Automatic contouring of raw data, Computers & Geosciences **8** (1982), 97–101.
9. Watson, D. F. and G. M. Philip, A refinement of inverse distance weighted interpolation, Geoprocessing **2** (1985), 315–327.

Jörn Springer
GeoForschungsZentrum Potsdam
Telegrafenberg A51
D-14473 Potsdam, GERMANY
pr@gfz-potsdam.dbp.de

Orthogonal Periodic Spline Wavelets

Manfred Tasche

Abstract. For many applications in signal processing and numerical analysis, it is necessary to use periodic wavelets. In this paper, a new constructive approach to orthogonal periodic spline wavelets is presented. Using Euler–Frobenius functions and two–scale symbol (2,2)–matrices, we obtain new simple, efficient decomposition and reconstruction algorithms which are mainly based on the Fourier technique.

§0. Introduction

In order to investigate periodic processes in physics, signal processing and engineering by wavelets, a theory of periodic wavelets is necessary. Y. Meyer [5] was the first to study periodic wavelets by periodization of cardinal wavelets. A different approach to orthogonal periodic spline wavelets based on periodization of orthogonal Battle–Lemarié wavelets [1, 4] was described by V. Perrier and C. Basdevant [6]. Semiorthogonal periodic spline wavelets obtained by periodization of Chui–Wang wavelets [2] were recently considered by G. Plonka and M. Tasche [7].

Our aim is a short introduction to orthogonal periodic spline wavelets and to corresponding decomposition and reconstruction algorithms. First we summarize some basic facts on periodic splines, periodic multiresolution and Euler–Frobenius functions. In Sections 2 and 3, we construct orthonormal bases of the sample space V_j and the wavelet space W_j, respectively. Further, we discuss the relations to semiorthogonal periodic spline wavelets [7]. Using two–scale symbol (2,2)–matrices, we obtain in Section 4 new decomposition and reconstruction algorithms based on discrete Fourier transform (DFT). Applying fast Fourier transform (FFT), these algorithms are fast and numerically stable. Both algorithms are free of truncation errors.

Wavelets, Images and Surface Fitting
P. J. Laurent, A. Le Méhauté, and L. L. Schumaker (eds.), pp. 475–484.

ISBN 1-56881-040-7.

§1. Preliminaries

Let $m, M \in \mathbb{N}$ be fixed. Put $M_j := 2^j M$ $(j \in \mathbb{N}_0)$. Consider the Hilbert space L^2_M of all M–periodic, square integrable functions $f, g, \ldots$ with the scalar product

$$\langle f, g\rangle := \int_0^M f(t)\,\overline{g(t)}\,\mathrm{d}t \qquad (n \in \mathbb{Z})$$

and the corresponding norm $\|\cdot\|$. The Fourier coefficients of $f \in L^2_M$ are denoted by

$$c_n(f) := \frac{1}{M}\int_0^M f(t)\,e^{-2\pi int/M}\,\mathrm{d}t \qquad (n \in \mathbb{Z}).$$

Let φ_j $(j \in \mathbb{N}_0)$ be the M–periodization of the m-th order cardinal B-spline N_m. Then the *Fourier–transformed two–scale relation* of φ_j reads as follows

$$c_n(\varphi_j) = 2\,P(w^n_{M_{j+1}})\,c_n(\varphi_{j+1}) \qquad (n \in \mathbb{Z}) \tag{1.1}$$

with

$$P(z) := \left(\frac{1+z}{2}\right)^m \qquad (z \in \mathbb{C}),$$
$$w_{M_j} := e^{-2\pi i/M_j} \qquad (j \in \mathbb{N}_0).$$

We introduce the *sample space* V_j *of level* j $(j \in \mathbb{N}_0)$ as

$$V_j := \operatorname{span}\{\varphi_{j,k};\ k = 0, \ldots, M_j - 1\}$$

with $\varphi_{j,k} := \varphi_j(\cdot - 2^{-j}k)$. Then V_j contains all M–periodic splines of order m with the knot set $2^{-j}\mathbb{Z}$. Clearly, V_j is a 2^{-j}–shift–invariant space with $\dim V_j = M_j$. All sample spaces V_j $(j \in \mathbb{N}_0)$ form an *M–periodic multiresolution* [3], i.e.,

$$\begin{aligned} &V_j \subset V_{j+1} \qquad (j \in \mathbb{N}_0),\\ &\operatorname{clos}\Big(\bigcup_{j=0}^{\infty} V_j\Big) = L^2_M\,. \end{aligned} \tag{1.2}$$

We define the *wavelet space* W_j *of level* j $(j \in \mathbb{N}_0)$ as the orthogonal complement of V_j in V_{j+1}, i.e.

$$W_j := V_{j+1} \ominus V_j\,. \tag{1.3}$$

Note that $\dim W_j = M_j$.

As is well known, the $2m$–th Euler–Frobenius function Φ defined by

$$\Phi(z) := \sum_{k=1-m}^{m-1} N_{2m}(m+k)\,z^k \qquad (z \in \mathbb{C}\setminus\{0\})$$

plays a crucial role in periodic spline interpolation. Later we will apply the following properties of Φ:

$$M_j^2 \sum_{r=-\infty}^{\infty} |c_{k+rM_j}(\varphi_j)|^2 = \Phi(w_{M_j}^k) \quad (k=0,\dots,M_j-1;\ j\in\mathbb{N}_0), \tag{1.4}$$

$$P(z)P(z^{-1})\Phi(z) + P(-z)P(-z^{-1})\Phi(-z) = \Phi(z^2) \quad (z\in\mathbb{C}\setminus\{0\}). \tag{1.5}$$

Further, the 2π–periodic even function $\Phi(e^{i\,\cdot})$ is strictly decreasing on $[0,\pi]$ and we have

$$\Phi(-1) \le \Phi(e^{iu}) \le \Phi(1) = 1 \quad (u\in[0,\pi]) \tag{1.6}$$

with

$$\Phi(-1) = \left(\frac{2}{\pi}\right)^{2m} \sum_{n=-\infty}^{\infty} (2n+1)^{-2m} = \frac{2^{2m}\,(2^{2m}-1)}{(2m)!}\,|B_{2m}|,$$

where B_{2m} denotes the $2m$–th Bernoulli number.

§2. Orthonormal Basis of V_j

Let us first establish the following lemmas.

Lemma 2.1. *Let $j\in\mathbb{N}_0$, $0<\alpha\le\beta<\infty$ and $\chi\in L_M^2$ be given. Then the following statements are equivalent:*
(i) *For all $(c_k)_{k=0}^{M_j-1}\in\mathbb{C}^{M_j}$,*

$$\alpha \sum_{k=0}^{M_j-1} |c_k|^2 \le \|\sum_{k=0}^{M_j-1} c_k\, 2^{j/2}\,\chi(\cdot\,-\,2^{-j}k)\|^2 \le \beta \sum_{k=0}^{M_j-1} |c_k|^2\,. \tag{2.1}$$

(ii) *For all $k=0,\dots,M_j-1$,*

$$\alpha \le M_j^2 \sum_{r=-\infty}^{\infty} |c_{k+rM_j}(\chi)|^2 \le \beta. \tag{2.2}$$

The proof based on Parseval identity is omitted here.

Lemma 2.2. *Let $j\in\mathbb{N}_0$ and $0<\alpha\le\beta<\infty$. Let $\chi\in L_M^2$ with (2.2) be given. Then the following statements are equivalent:*

(i) $f\in\operatorname{span}\{\chi(\cdot\,-\,2^{-j}k);\ k=0,\dots,M_j-1\}$.
(ii) $c_n(f)=\hat a_n\,c_n(\chi)$ *with* $\hat a_n\in\mathbb{C}$, $\hat a_n=\hat a_{n+M_j}$ $(n\in\mathbb{Z})$.

Proof: By Lemma 2.1, $\{\chi(\cdot\,-\,2^{-j}k);\ k=0,\dots,M_j-1\}$ is a basis of its span. Assume that (i) is true. Then f can be expressed by

$$f = \sum_{k=0}^{M_j-1} a_k\,\chi(\cdot\,-\,2^{-j}k) \qquad (a_k\in\mathbb{C}). \tag{2.3}$$

Hence we obtain $c_n(f) = \hat{a}_n \, c_n(\chi)$ $(n \in \mathbb{Z})$ with

$$\hat{a}_n := \sum_{k=0}^{M_j-1} a_k \, w_{M_j}^{kn} \qquad (n \in \mathbb{Z}). \tag{2.4}$$

For arbitrary given $\hat{a}_n \in \mathbb{C}$, $\hat{a}_n = \hat{a}_{n+M_j}$ $(n \in \mathbb{Z})$, there exist unique $a_k \in \mathbb{C}$ $(k = 0, \ldots, M_j - 1)$ with (2.4). If (ii) is fulfilled, then for all $n \in \mathbb{Z}$,

$$c_n(f) = \hat{a}_n \, c_n(\chi) = c_n\Big(\sum_{k=0}^{M_j-1} a_k \, \chi(\cdot - 2^{-j}k)\Big),$$

and (2.3) holds. ■

Lemma 2.3. *Let $j \in \mathbb{N}_0$ and $\chi \in L_m^2$ be given. Then the following statements are equivalent:*
(i) *For all $k = 0, \ldots, M_j - 1$,*

$$2^j \, \langle \chi(\cdot - 2^{-j}k), \, \chi \rangle = \delta_{k,0}.$$

(ii) *For all $k = 0, \ldots, M_j - 1$,*

$$h_k := M_j^2 \sum_{r=-\infty}^{\infty} |c_{k+rM_j}(\chi)|^2 = 1. \tag{2.5}$$

Proof: By Parseval identity, we have for $k = 0, \ldots, M_j - 1$

$$2^j \langle \chi(\cdot - 2^{-j}k), \, \chi \rangle = M_j \sum_{n=-\infty}^{\infty} c_n(\chi(\cdot - 2^{-j}k)) \, c_{-n}(\chi)$$

$$= M_j \sum_{n=-\infty}^{\infty} |c_n(\chi)|^2 \, w_{M_j}^{nk} = M_j^{-1} \sum_{l=0}^{M_j-1} h_l \, w_{M_j}^{kl}.$$

Hence, (i) is fulfilled if and only if $h_l = 1$ $(l = 0, \ldots, M_j - 1)$. ■

Consider $\varphi_j^\star \in L_M^2$ $(j \in \mathbb{N}_0)$ given by its Fourier coefficients

$$c_n(\varphi_j^\star) := \frac{c_n(\varphi_j)}{\Phi(w_{M_j}^n)^{1/2}} \quad (n \in \mathbb{Z}). \tag{2.6}$$

Note that $\Phi(w_{M_j}^n) \geq \Phi(-1) > 0$ (see (1.6)). By (1.4), we have for $k = 0, \ldots, M_j - 1$

$$M_j^2 \sum_{r=-\infty}^{\infty} |c_{k+rM_j}(\varphi_j^\star)|^2 = 1. \tag{2.7}$$

Hence from Lemma 2.3, it follows that $2^{j/2}\,\varphi^\star_{j,k} := 2^{j/2}\varphi^\star_j(\cdot - 2^{-j}k)$ $(k = 0,\dots,M_j-1)$ form an orthonormal system. By (1.1) and (2.6), the *Fourier–transformed two–scale relation* of $\varphi^\star_j$ reads as follows

$$c_n(\varphi^\star_j) = 2\,P^\star(w^n_{M_{j+1}})\,c_n(\varphi^\star_{j+1}) \quad (n \in \mathbb{Z}) \tag{2.8}$$

with

$$P^\star(z) := P(z)\,\Phi(z)^{1/2}\,\Phi(z^2)^{-1/2} \quad (n \in \mathbb{Z}),$$
$$\mathcal{T} := \{z \in \mathbb{C};\ |z| = 1\}.$$

By (1.5), we have

$$|P^\star(z)|^2 + |P^\star(-z)|^2 = 1 \quad (z \in \mathcal{T}). \tag{2.9}$$

From (2.6) and Lemma 2.2 (with $\chi := \varphi_j$ and $f := \varphi^\star_j$), it follows that $\varphi^\star_j \in V_j$ and thus $\varphi^\star_{j,k} \in V_j$ $(k = 0,\dots,M_j-1)$. We summarize:

Theorem 2.4. *Let $j \in \mathbb{N}_0$. Then $\{2^{j/2}\,\varphi^\star_{j,k};\ k = 0,\dots,M_j-1\}$ is an orthonormal basis of V_j.*

§3. Orthonormal Basis of W_j

Now we are able to construct an orthonormal basis of W_j. Define $\psi^\star_j \in L^2_M$ $(j \in \mathbb{N}_0)$ by its Fourier coefficients

$$c_n(\psi^\star_j) = 2\,Q^\star(w^n_{M_{j+1}})\,c_n(\varphi^\star_{j+1}) \quad (n \in \mathbb{Z}) \tag{3.1}$$

with

$$Q^\star(z) := (-z)\,P^\star(-z^{-1}) \quad (z \in \mathcal{T}). \tag{3.2}$$

Note that

$$P^\star(z)\,Q^\star(z^{-1}) + P^\star(-z)\,Q^\star(-z^{-1}) = 0 \quad (z \in \mathcal{T}). \tag{3.3}$$

Using (3.1), (3.2), (2.7) and (2.9), we obtain that for $k = 0,\dots,M_j-1$

$$\begin{aligned} M_j^2 \sum_{r=-\infty}^{\infty} |c_{k+rM_j}(\psi^\star_j)|^2 &= M_{j+1}^2\,|P^\star(-w^{-k}_{M_{j+1}})|^2 \sum_{r=-\infty}^{\infty} |c_{k+rM_{j+1}}(\varphi^\star_{j+1})|^2 \\ &\quad + M_{j+1}^2\,|P^\star(w^{-k}_{M_{j+1}})|^2 \sum_{r=-\infty}^{\infty} |c_{k+M_j+rM_{j+1}}(\varphi^\star_{j+1})|^2 \\ &= |P^\star(-w^{-k}_{M_{j+1}})|^2 + |P^\star(w^{-k}_{M_{j+1}})|^2 = 1. \end{aligned}$$

Then by Lemma 2.3 (with $\chi := \psi^\star_j$), the functions

$$2^{j/2}\,\psi^\star_{j,k} := 2^{j/2}\,\psi^\star_j(\cdot - 2^{-j}k) \quad (k = 0,\dots,M_j-1)$$

form an orthonormal system. By (3.1), Lemma 2.2 (with $\chi := \varphi^\star_{j+1}$ and $f := \psi^\star_j$) and Theorem 2.4, we have $\psi^\star_{j,k} \in V_{j+1}$ $(k = 0,\dots,M_j-1)$.

Theorem 3.1. *Let $j \in \mathbb{N}_0$. Then $\{2^{j/2}\, \psi^\star_{j,k};\ k = 0,\dots,M_j - 1\}$ is an orthonormal basis of W_j.*

Proof: By (1.3) and dim $W_j = M_j$, it remains to show that $\psi^\star_j \in W_j$, i.e.

$$\varphi^\star_{j,k} \perp \psi^\star_j \quad (k = 0,\dots,M_j - 1).$$

Using the Parseval identity, we have for $k = 0,\dots,M_j - 1$

$$\langle \varphi^\star_{j,k}, \psi^\star_j \rangle = M \sum_{n=-\infty}^{\infty} w^{nk}_{M_j}\, c_n(\varphi^\star_j)\, c_{-n}(\psi^\star_j).$$

Applying the Fourier–transformed two–scale relations (2.8) and (3.1), this yields

$$\langle \varphi^\star_{j,k}, \psi^\star_j \rangle = 4M \sum_{n=-\infty}^{\infty} w^{nk}_{M_j}\, P^\star(w^n_{M_{j+1}})\, Q^\star(w^{-n}_{M_{j+1}})\, |c_n(\varphi^\star_{j+1})|^2.$$

Substituting $n := l + sM_{j+1}$ ($l \in \{0,\dots,M_{j+1} - 1\}$, $s \in \mathbb{Z}$), we obtain by (2.7) and (3.3) that

$$\begin{aligned}
\langle \varphi^\star_{j,k}, \psi^\star_j \rangle &= \frac{M}{M_j^2} \sum_{l=0}^{M_{j+1}-1} w^{lk}_{M_j}\, P^\star(w^l_{M_{j+1}})\, Q^\star(w^{-l}_{M_{j+1}}) \\
&= \frac{M}{M_j^2} \sum_{l=0}^{M_j-1} w^{lk}_{M_j} \left(P^\star(w^l_{M_{j+1}})\, Q^\star(w^{-l}_{M_{j+1}}) + P^\star(-w^l_{M_{j+1}})\, Q^\star(-w^{-l}_{M_{j+1}})\right) \\
&= 0.
\end{aligned}$$

This completes the proof. ■

Since

$$V_{j+1} = V_0 \oplus W_0 \oplus \ldots \oplus W_j, \qquad j \in \mathbb{N}_0,$$

it follows from Theorems 2.4 and 3.1 that the system

$$\{\varphi^\star_{0,k};\ k = 0,\dots,M - 1\} \cup \bigcup_{j=0}^{\infty} \{2^{j/2}\, \psi^\star_{j,k};\ k = 0,\dots,M_j - 1\}$$

is an orthonormal spline basis of L^2_M.

§4. Decomposition and Reconstruction Algorithms

Introducing for $z \in \mathcal{T}$ the functions

$$P_0^\star(z^2) := \frac{1}{2}(P^\star(z) + P^\star(-z)), \quad P_1^\star(z^2) := \frac{1}{2z}(P^\star(z) - P^\star(-z)), \tag{4.1}$$

it follows from (3.2) that

$$\frac{1}{2}(Q^\star(z) + Q^\star(-z)) = P_1^\star(z^{-2}), \quad \frac{1}{2z}(Q^\star(z) - Q^\star(-z)) = -P_0^\star(z^{-2}).$$

Then we can write the Fourier–transformed two–scale relations (2.8) and (3.1) in the compressed form

$$\frac{1}{2}\begin{pmatrix} c_n(\varphi_j^\star) \\ c_n(\psi_j^\star) \end{pmatrix} = \mathbf{S}_{j,n} \begin{pmatrix} c_n(\varphi_{j+1}^\star) \\ c_n(\varphi_{j+1,1}^\star) \end{pmatrix} \quad (n \in \mathbb{Z})$$

with the *two–scale matrices of the j-th level*

$$\mathbf{S}_{j,n} := \begin{pmatrix} P_0^\star(w_{M_j}^n) & P_1^\star(w_{M_j}^n) \\ P_1^\star(w_{M_j}^{-n}) & -P_0^\star(w_{M_j}^{-n}) \end{pmatrix}. \tag{4.2}$$

Lemma 4.1. *All two–scale symbol matrices $\mathbf{S}_{j,n}$ are invertible, and for $j \in \mathbb{N}_0$ and $n \in \mathbb{Z}$,*

$$\mathbf{S}_{j,n}^{-1} = 2\begin{pmatrix} P_0^\star(w_{M_j}^{-n}) & P_1^\star(w_{M_j}^n) \\ P_1^\star(w_{M_j}^{-n}) & -P_0^\star(w_{M_j}^n) \end{pmatrix}. \tag{4.3}$$

Proof: For $z \in \mathcal{T}$, we obtain by (4.1) and (2.9) that

$$-P_0^\star(z^2)\,P_0^\star(z^{-2}) - P_1^\star(z^2)\,P_1^\star(z^{-2}) = -\frac{1}{2}(|P^\star(z)|^2 + |P^\star(-z)|^2) = -\frac{1}{2}.$$

Hence we have

$$\det \mathbf{S}_{j,n} = -\frac{1}{2}.$$

Thus we can invert $\mathbf{S}_{j,n}$ and obtain (4.3). ∎

In order to decompose a given function $f_{j+1} \in V_{j+1}$ ($j \in \mathbb{N}_0$) of the form

$$f_{j+1} = \sum_{l=0}^{M_{j+1}-1} c_{j+1,l}\,\varphi_{j+1,l}^\star \quad (c_{j+1,l} \in \mathbb{C}) \tag{4.4}$$

into the sum

$$f_{j+1} = f_j + g_j \tag{4.5}$$

of uniquely determined functions $f_j \in V_j$ and $g_j \in W_j$, we describe an efficient decomposition algorithm based on Fourier technique. Assume that the coefficients $c_{j+1,l} \in \mathbb{C}$ $(l = 0, \ldots, M_{j+1} - 1)$ of f_{j+1} or their DFT

$$\hat{c}_{j+1,k} := \sum_{l=0}^{M_{j+1}-1} c_{j+1,l}\, w_{M_{j+1}}^{kl} \quad (k = 0, \ldots, M_{j+1} - 1) \tag{4.6}$$

are known. The desired functions f_j, g_j can be uniquely represented by

$$f_j = \sum_{n=0}^{M_j-1} c_{j,n}\, \varphi_{j,n}^{\star}, \quad g_j = \sum_{n=0}^{M_j-1} d_{j,n}\, \psi_{j,n}^{\star} \tag{4.7}$$

with unknown coefficients $c_{j,n}$, $d_{j,n} \in \mathbb{C}$.

In order to reconstruct $f_{j+1} \in V_{j+1}$ $(j \in \mathbb{N}_0)$ given in the decomposed form (4.5) with $f_j \in V_j$ and $g_j \in W_j$, we present an efficient reconstruction algorithm based on DFT. Assume that $c_{j,n}$, $d_{j,n} \in \mathbb{C}$ $(n = 0, \ldots, M_j - 1)$ in (4.7) or their DFT's

$$\hat{c}_{j,k} := \sum_{n=0}^{M_j-1} c_{j,n}\, w_{M_j}^{kn}, \quad \hat{d}_{j,k} := \sum_{n=0}^{M_j-1} d_{j,n}\, w_{M_j}^{kn} \quad (k = 0, \ldots, M_j - 1) \tag{4.8}$$

are known. By $V_{j+1} = V_j \oplus W_j$, the function $f_{j+1} \in V_{j+1}$ can be uniquely represented in the form (4.4) with unknown coefficients $c_{j+1,l}$ $(l = 0, \ldots, M_{j+1} - 1)$.

The related decomposition and reconstruction algorithms are based on the following

Theorem 4.2. *For $j \in \mathbb{N}_0$ let $f_{j+1} \in V_{j+1}$, $f_j \in V_j$ and $g_j \in W_j$ with (4.4) – (4.8) be given. Then we have for $n = 0, \ldots, M_{j+1} - 1$*

$$\hat{c}_{j+1,n} = 2\,\hat{c}_{j,n}\, P^{\star}(w_{M_{j+1}}^{n}) + 2\,\hat{d}_{j,n}\, Q^{\star}(w_{M_{j+1}}^{n}) \tag{4.9}$$

and for $k = 0, \ldots, M_j - 1$

$$2\,(\hat{c}_{j,k}\ \hat{d}_{j,k})\,\mathbf{S}_{j,k} = (\hat{c}_{j+1,k}^{(0)}\ \hat{c}_{j+1,k}^{(1)}) \tag{4.10}$$

or

$$(\hat{c}_{j,k}\ \hat{d}_{j,k}) = \frac{1}{2}\,(\hat{c}_{j+1,k}^{(0)}\ \hat{c}_{j+1,k}^{(1)})\,\mathbf{S}_{j,k}^{-1} \tag{4.11}$$

with

$$\begin{aligned} \hat{c}_{j+1,k}^{(0)} &:= \frac{1}{2}\,(\hat{c}_{j+1,k} + \hat{c}_{j+1,M_j+k}), \\ \hat{c}_{j+1,k}^{(1)} &:= \frac{1}{2}\, w_{M_{j+1}}^{-k}\,(\hat{c}_{j+1,k} - \hat{c}_{j+1,M_j+k}). \end{aligned} \tag{4.12}$$

Proof: From

$$c_n(f_{j+1}) = c_n(f_j) + c_n(g_j) \quad (n \in \mathbb{Z}),$$

it follows by (4.6), (4.8), (2.8) and (3.1) that for all $n \in \mathbb{Z}$,

$$\begin{aligned}\hat{c}_{j+1,n}\, c_n(\varphi^\star_{j+1}) &= \hat{c}_{j,n}\, c_n(\varphi^\star_j) + \hat{d}_{j,n}\, c_n(\psi^\star_j)\\ &= \big(2\,\hat{c}_{j,n}\, P^\star(w^n_{M_{j+1}}) + 2\,\hat{d}_{j,n}\, Q^\star(w^n_{M_{j+1}})\big)\, c_n(\varphi^\star_{j+1})\end{aligned}$$

and hence (4.9), since $c_n(\varphi^\star_{j+1}) \neq 0$ $(n = 0, \ldots, M_{j+1} - 1)$. Using (3.2), we obtain for $k = 0, \ldots, M_j - 1$

$$\begin{aligned}\hat{c}_{j+1,k} &= 2\,\hat{c}_{j,k}\, P^\star(w^k_{M_{j+1}}) - 2\,\hat{d}_{j,k}\, w^k_{M_{j+1}}\, P^\star(-w^{-k}_{M_{j+1}}),\\ \hat{c}_{j+1,M_j+k} &= 2\,\hat{c}_{j,k}\, P^\star(-w^k_{M_{j+1}}) + 2\,\hat{d}_{j,k}\, w^k_{M_{j+1}}\, P^\star(w^{-k}_{M_{j+1}}).\end{aligned}$$

By (4.12) and (4.1), this yields

$$\begin{aligned}\hat{c}^{(0)}_{j+1,k} &= 2\,\hat{c}_{j,k}\, P^\star_0(w^k_{M_j}) + 2\,\hat{d}_{j,k}\, P^\star_1(w^{-k}_{M_j}),\\ \hat{c}^{(1)}_{j+1,k} &= 2\,\hat{c}_{j,k}\, P^\star_1(w^k_{M_j}) - 2\,\hat{d}_{j,k}\, P^\star_0(w^{-k}_{M_j}).\end{aligned}$$

But this implies (4.10) and (4.11) by (4.2) and (4.3). ■

Algorithm 4.3. (Decomposition Algorithm)

Input: $m \in \mathbb{N}$; $j \in \mathbb{N}_0$;
M power of 2; $M_j := 2^j M$;
$\mathbf{S}^{-1}_{j,k}$ $(k = 0, \ldots, M_j - 1)$ given by (4.3);
$\hat{c}_{j+1,l} \in \mathbb{C}$ $(l = 0, \ldots, 2M_j - 1)$.

Step 1: Form $\hat{c}^{(0)}_{j+1,k}$, $\hat{c}^{(1)}_{j+1,k}$ $(k = 0, \ldots, M_j - 1)$ by (4.12).

Step 2: Compute $\hat{c}_{j,k}$, $\hat{d}_{j,k}$ $(k = 0, \ldots, M_j - 1)$ by (4.11).

Output: $\hat{c}_{j,k}$, $\hat{d}_{j,k}$ $(k = 0, \ldots, M_j - 1)$.

Algorithm 4.4 (Reconstruction Algorithm)

Input: $m \in \mathbb{N}$; $j \in \mathbb{N}_0$;
M power of 2; $M_j := 2^j M$;
$\mathbf{S}_{j,k}$ $(k = 0, \ldots, M_j - 1)$ given by (4.2);
$\hat{c}_{j,k}$, $\hat{d}_{j,k} \in \mathbb{C}$ $(k = 0, \ldots, M_j - 1)$.

Step 1: Compute $\hat{c}^{(0)}_{j+1,k}$, $\hat{c}^{(1)}_{j+1,k}$ $(k = 0, \ldots, M_j - 1)$ by (4.10).

Step 2: Form for $k = 0, \ldots, M_j - 1$

$$\begin{aligned}\hat{c}_{j+1,k} &:= \hat{c}^{(0)}_{j+1,k} + w^k_{M_{j+1}}\, \hat{c}^{(1)}_{j+1,k},\\ \hat{c}_{j+1,k+M_j} &:= \hat{c}^{(0)}_{j+1,k} - w^k_{M_{j+1}}\, \hat{c}^{(1)}_{j+1,k}.\end{aligned}$$

Output: $\hat{c}_{j+1,k}$ $(k = 0, \ldots, 2M_j - 1)$.

Both algorithms are very similar, require only $\mathcal{O}(M_j)$ complex flops, and are numerically stable by using FFT. Applying (4.9), we get another simple reconstruction algorithm.

Acknowledgements. This research was supported by the Deutsche Forschungsgemeinschaft.

References

1. Battle, G., A block spin construction of ondelettes. Part 1: Lemarié functions, Comm. Math. Phys. **110** (1987), 601–615.
2. Chui, C. K., and Wang, J. Z., On compactly supported spline wavelets and a duality principle, Trans. Amer. Math. Soc. **330** (1992), 903–915.
3. de Boor, C., DeVore, R. A., and Ron, A., On the construction of multivariate (pre)wavelets, Constr. Approx. **9** (1993), 123–166.
4. Lemarié, P. G., Ondelettes à localisation exponentielle, J. Math. Pures Appl. (9) **67** (1988), 227–236.
5. Meyer, Y., Wavelets and operators, Rapport CEREMADE 8704, Univ. Paris–Dauphine, 1987.
6. Perrier, V., and Basdevant, C., Periodic wavelet analysis, a tool for inhomogeneous field investigation. Theory and algorithms, Rech. Aérospat. **3** (1989), 53–67.
7. Plonka, G., and Tasche, M., Periodic spline wavelets, Preprint, Univ. Rostock, 1993.
8. Schoenberg, I. J., Cardinal interpolation and spline functions, J. Approx. Th. **2** (1969), 167–206.

Manfred Tasche
Fachbereich Mathematik
Universität Rostock
D–18051 Rostock, GERMANY
tasche@ mathematik.uni-rostock.d400.de

Approximation of Parametric Surfaces with Discontinuities by Discrete Smoothing D^{m}-splines

Juan José Torrens

Abstract. In this paper we study a smoothing method for fitting discontinuous parametric surfaces when the data points are part of the nodes of a curvilinear grid. This problem may appear, for example, in geophysics, for modelling reverse faults. We proceed as follows. First, a uniform parameterization of the surface is introduced in order to compensate for the absence of a natural parameterization of the data points. An approximating surface is then obtained by solving a minimization problem in a suitable finite element space, whose solution is called *discrete smoothing D^m-spline*. We finally establish the convergence of the method.

§1. Preliminaries

Let $n \in \mathbb{N}^*$. We denote by $\|\cdot\|_{\mathbb{R}^n}$ and $\langle\cdot,\cdot\rangle_{\mathbb{R}^n}$ the Euclidean norm and its associated inner product in $\mathbb{R}^n$. Likewise, the closure, boundary and interior of a given set $E \subset \mathbb{R}^n$ are denoted, respectively, by $\overline{E}$, ∂E and $\mathrm{Int}(E)$. For all $m, n \in \mathbb{N}^*$, we write $\mathbb{R}^{m,n}$ for the space of real matrices with m rows and n columns. For all $\alpha = (\alpha_1, \alpha_2) \in \mathbb{N}^2$, we set $|\alpha| = \alpha_1 + \alpha_2$ and $|\alpha|_\infty = \max\{\alpha_1, \alpha_2\}$.

Let ω be an open subset of $\mathbb{R}^2$ and let τ be a closed subset of $\partial\omega$ without isolated points. We say that ω is an *open set with cut* τ if $\mathrm{Int}(\overline{\omega})$ has a Lipschitz–continuous boundary (in the Nečas[8] sense) and there exists a finite family $\{\omega_1, \ldots, \omega_M\}$ of pairwise disjoint, connected, open subsets of ω with Lipschitz–continuous boundary such that $\overline{\omega} = \cup_{i=1}^{M} \overline{\omega}_i$ (see Figure 1). In this case, we also say that *the family* $\{\omega_1, \ldots, \omega_M\}$ *represents* τ *in* ω. In many practical situations, the cut τ is a collection of (possibly intersecting) polygonal lines.

Let ω be an open set in $\mathbb{R}^2$. For all $m \in \mathbb{N}$, we denote by $H^m(\omega; \mathbb{R}^3)$ the Sobolev space of (classes of) functions φ which belong to $L^2(\omega; \mathbb{R}^3)$, together

Wavelets, Images and Surface Fitting
P. J. Laurent, A. Le Méhauté, and L. L. Schumaker (eds.), pp. 485–492.

ISBN 1-56881-040-7.

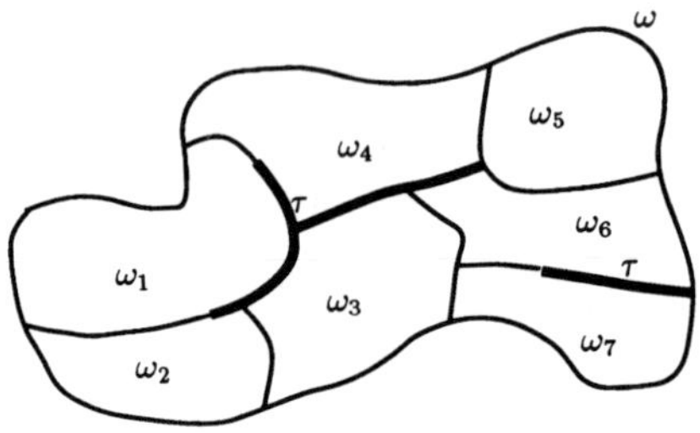

Fig. 1. Open set with cut.

with all their partial derivatives $\partial^\alpha \varphi = \dfrac{\partial^{|\alpha|}\varphi}{\partial x_1^{\alpha_1} \partial x_2^{\alpha_2}}$, in the distribution sense, of order $|\alpha| \le m$. It is well known that $H^m(\omega; \mathbb{R}^3)$ is a Hilbert space for the norm

$$\|\varphi\|_{m,\omega,\mathbb{R}^3} = \left(\sum_{|\alpha|\le m} \int_\omega \|\partial^\alpha \varphi(x)\|^2_{\mathbb{R}^3}\, dx \right)^{1/2} .$$

We shall also use the inner semi–product

$$(\varphi, \psi)_{m,\omega,\mathbb{R}^3} = \sum_{|\alpha|=m} \int_\omega \langle \partial^\alpha \varphi(x), \partial^\alpha \psi(x) \rangle_{\mathbb{R}^3}\, dx ,$$

and its associated semi-norm

$$|\varphi|_{m,\omega,\mathbb{R}^3} = (\varphi, \varphi)^{1/2}_{m,\omega,\mathbb{R}^3} .$$

Let ω be an open set in $\mathbb{R}^2$ with cut τ represented by $\{\omega_1, \ldots, \omega_M\}$. For all $k \in \mathbb{N}$, we define $C^k_\tau(\omega; \mathbb{R}^3)$ as the space of functions φ with values in $\mathbb{R}^3$ which are continuous on ω, together with all their partial derivatives $\partial^\alpha \varphi$ of order $|\alpha| \le k$, and such that, for $i = 1, \ldots, M$ and for all $\alpha \in \mathbb{N}^2$ with $|\alpha| \le k$, $\partial^\alpha(\varphi|_{\omega_i})$ is bounded and uniformly continuous on ω_i. We remark that every function in $C^k_\tau(\omega; \mathbb{R}^3)$ can be considered as a function defined on $\overline{\omega}$ such that all its partial derivatives of degree $\le k$ and the function itself are continuous on $\overline{\omega}$, except on τ, where some of them may have jump discontinuities.

Finally, the following results can be stated :

Theorem 1.1. *$C^k_\tau(\omega; \mathbb{R}^3)$ is a Banach space with norm*

$$\|\varphi\|_{C^k_\tau(\omega;\mathbb{R}^3)} = \max_{1\le i\le M} \max_{0\le|\alpha|\le k} \sup_{x\in\omega} \|\partial^\alpha(\varphi|_{\omega_i})(x)\|_\infty ,$$

where $\|y\|_\infty = \max_{1\le i\le 3} |y_i|$ if $y = (y_1, y_2, y_3) \in \mathbb{R}^3$. Moreover, this space and its norm are independent of the family that represents τ in ω.

Proof: (Cf. [9, Theorem 3.3]). ∎

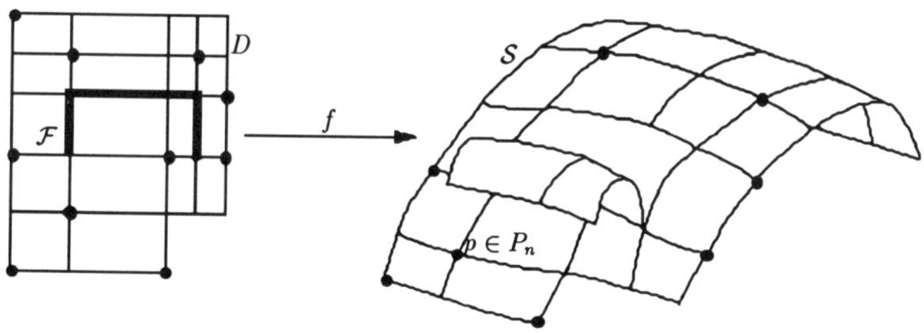

Fig. 2. The parameterization f of the surface S and a set P_n of data points.

Theorem 1.2. *If $m > k+1$, $H^m(\omega;\mathbb{R}^3)$ is a subset of $C^k_\tau(\omega;\mathbb{R}^3)$ with compact injection.*

Proof: (Cf. [9, Theorem 3.9]). ■

§2. Modelling the Problem

Let $\mathcal{S}$ be a surface that presents faults or holes. We assume that $\mathcal{S}$ can be represented by a vector function f (a parameterization of $\mathcal{S}$), defined on a bounded, connected, open set $D \subset \mathbb{R}^2$ with cut $\mathcal{F}$, that is discontinuous on $\mathcal{F}$ (except, at most, at a finite number of points), and that belongs, for some integer $m \geq 2$, to the Sobolev space $H^m(D;\mathbb{R}^3)$ (and hence to $C^{m-2}_{\mathcal{F}}(D;\mathbb{R}^3)$).

For all $n \in \mathbb{N}$, we suppose given a set P_n made up by part (or all) of the vertices of a curve network lying on $\mathcal{S}$ that divides this surface into a finite number of rectangular (eventually triangular) patches. These data points are the values of the function f at the vertices of a lattice, adapted to $\mathcal{F}$, that covers $\overline{D}$ (see Figure 2). In this situation, we consider the problem of constructing, from the set P_n, an approximating surface of $\mathcal{S}$.

In practice, no information, other than the coordinates of the data points, is available. Thus, the first difficulty that must be overcome is to assign parameter values to the data points. Taking into account their distribution on the surface, they can be parameterized by choosing a suitable planar domain $\overline{\Omega}$, tesselated into a grid of equal squares, and attaching, in order, the points in P_n to part of the vertices of these squares. This fact justifies the following hypothesis :

$$\left\{\begin{array}{l}\text{There exist :}\\ \bullet\ \text{a bounded, connected, open set } \Omega \subset \mathbb{R}^2 \text{ with cut } \Phi \text{, such}\\ \quad \text{that } \partial\Omega \text{ is formed by segments parallel to the coordinate}\\ \quad \text{axes,}\\ \bullet\ \text{a sequence } (h_n)_{n\in\mathbb{N}} \text{ of real positive numbers which converges}\\ \quad \text{to } 0,\\ \bullet\ \text{a sequence } (\mathcal{T}_n)_{n\in\mathbb{N}} \text{ of triangulations of } \overline{\Omega} \text{ made with equal}\\ \quad \text{squares with sides of length } h_n,\\ \bullet\ \text{an integer } m \geq 2,\\ \bullet\ \text{a sequence } (g_n)_{n\in\mathbb{N}} \subset H^m(\Omega;\mathbb{R}^3) \text{ of parameterizations of } \mathcal{S},\\ \text{such that :}\\ \bullet\ \text{for all } n \in \mathbb{N},\ \Phi \text{ is union of sides of squares of } \mathcal{T}_n,\\ \bullet\ \text{for all } n \in \mathbb{N},\ P_n \subset g_n(B_n),\ B_n \text{ being the set of vertices of}\\ \quad \mathcal{T}_n,\\ \bullet\ \text{there exists a parameterization } g \text{ of } \mathcal{S} \text{ to which } (g_n)_{n\in\mathbb{N}} \text{ is}\\ \quad \text{weakly convergent in } H^m(\Omega;\mathbb{R}^3).\end{array}\right. \tag{2.1}$$

In this way, the initial fitting problem has been reduced to the smoothing, for all $n \in \mathbb{N}$, of a vector function g_n on part of the vertices of a uniform mesh.

Remark 2.1. It can be shown that, under additional reasonable conditions, (2.1) holds. To see this, it suffices to construct a suitable sequence $(\varphi_n)_{n\in\mathbb{N}}$ of parameter changes from $\overline{\Omega}$ to $\overline{D}$ and then define g_n, for all $n \in \mathbb{N}$, as $f \circ \varphi_n$. The reader is referred to [9], where full details and proofs are given.

§3. The Smoothing Method

For approximating every parameterization g_n, in this section we shall develop a method based on the theory of discrete smoothing D^m-splines (see, for example, [4, 5]), which extends a previous method due to D. Apprato (cf. [1, 2, 3]) to the case of discontinuous parametric surfaces.

3.1. Definition of the Approximation Space

Let $k \in \mathbb{N}^*$ and $n \in \mathbb{N}$. Let $\tilde{\mathcal{T}}_n$ be another triangulation of $\overline{\Omega}$ made with rectangles of diameter $\leq \tilde{h}_n$ and such that Φ is union of sides of elements belonging to $\tilde{\mathcal{T}}_n$.

We first construct on $\tilde{\mathcal{T}}_n$ a finite element space $\tilde{X}_n$ from the Bogner–Fox–Schmit generic finite element of class C^k. Let us remember that such element may be defined as a triple (K, P_K, Σ_K) (see [6]), where K denotes, in this case, any rectangle in $\tilde{\mathcal{T}}_n$, $P_K = Q_{2k+1}(K)$ (space of restrictions to K of polynomials of degree $\leq 2k+1$ with respect to each variable), and Σ_K is the following set of degrees of freedom :

$$\Sigma_K = \{\, v \mapsto \partial^\alpha v(b) : \ b \in K \cap \tilde{B}_n,\ \alpha \in \mathbb{N}^2,\ |\alpha|_\infty \leq k \,\},$$

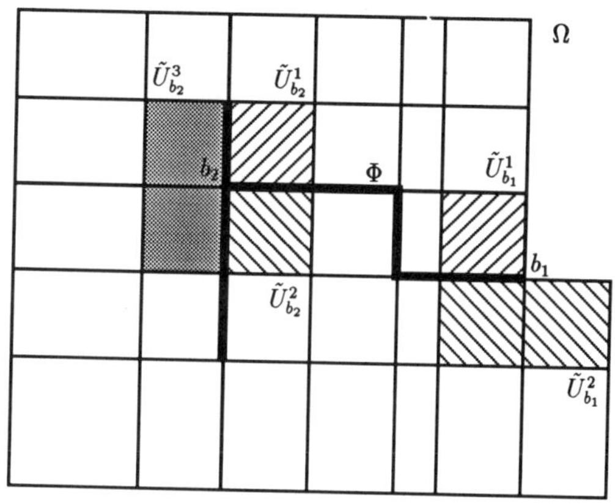

Fig. 3. Connected components $\tilde{u}_b^j$ for nodes b_1 and b_2.

where $\tilde{B}_n$ is the set of vertices of $\tilde{\mathcal{T}}_n$. Likewise, for every $b \in \tilde{B}_n$ and $\alpha \in \mathbb{N}^2$, with $|\alpha|_\infty \leq k$, we denote by w_b^α the basis function of $\tilde{X}_n$ associated with the degree of freedom $v \mapsto \partial^\alpha v(b)$.

Since functions in $\tilde{X}_n$ are C^k-continuous on $\overline{\Omega}$ (cf. [6, 9]), it is necessary to adapt the definition of the finite element space to deal with discontinuous functions on the set Φ. For this, we are going to introduce a new family of basis functions.

For all $b \in \tilde{B}_n$, let $\tilde{\gamma}_b$ be the number of connected components of the set $\mathrm{Int}(\tilde{U}_b) \setminus \Phi$, where $\tilde{U}_b = \cup\{ K \in \tilde{\mathcal{T}}_n : b \in K \}$, and let us denote them by $\tilde{U}_b^j$, for $j = 1, \ldots, \tilde{\gamma}_b$ (see Figure 3). We then define:

$$\forall b \in \tilde{B}_n,\ \forall j = 1, \ldots, \tilde{\gamma}_b,\ \forall \alpha \in \mathbb{N}^2,\ |\alpha|_\infty \leq k, w_b^{\alpha,j} = w_b^\alpha \mathbf{1}_{\overline{\tilde{U}_b^j}},$$

$\mathbf{1}_{\overline{\tilde{U}_b^j}}$ being the characteristic function of $\overline{\tilde{U}_b^j}$. Let us remark that almost every basis function associated to a node belonging to Φ has been "doubled" in two or more new functions whose support is limited to only one "side" of Φ.

Let $\tilde{V}_n$ be the linear space generated by the family $\{ w_b^{\alpha,j} \}$. Assuming that

$$m \leq k + 1, \tag{3.1}$$

where m is the integer introduced in (2.1), it is easily proved (cf. [9]) that

$$(\tilde{V}_n)^3 \subset C_\Phi^k(\Omega; \mathbb{R}^3) \cap H^m(\Omega; \mathbb{R}^3).$$

Therefore, the parametric finite element space $(\tilde{V}_n)^3$ is a suitable approximation space for fitting vector functions having discontinuities on the set Φ, in particular, the parameterization g_n.

3.2. Finding the Approximating Surface

Let B_n^1 be the set of nodes of $\mathcal{T}_n$ associated with at least one data point, and let $N = N(n)$ denote the cardinality of B_n^1. For the sake of simplicity, we suppose that $B_n^1 \cap \Phi = \emptyset$. Thus, since $H^m(\Omega; \mathbb{R}^3)$ is continuously imbedded into $C^0_\Phi(\Omega; \mathbb{R}^3)$, the linear operator

$$\rho^n : v \in H^m(\Omega; \mathbb{R}^3) \mapsto \big(v(b)\big)_{b \in B_n^1} \in \mathbb{R}^{N,3}$$

is well defined. See [9] for the case $B_n^1 \cap \Phi \neq \emptyset$.

We also assume that

$$\forall n \in \mathbb{N},\ \ker \rho^n \cap P_{m-1}(\Omega; \mathbb{R}^3) = \{\mathbf{0}\}\,, \tag{3.2}$$

where $P_{m-1}(\Omega; \mathbb{R}^3)$ is the space of functions with values in $\mathbb{R}^3$ whose component functions are restrictions to Ω of polynomials of degree $\leq m-1$.

Given a real positive number ε_n, we set

$$J_n(v) = \langle \rho^n v - \rho^n g_n \rangle_n^2 + \varepsilon_n |v|^2_{m,\Omega,\mathbb{R}^3}, \qquad \forall v \in H^m(\Omega; \mathbb{R}^3)\,,$$

where $\langle \cdot \rangle_n$ is the Euclidean (or Frobenius) norm in $\mathbb{R}^{N,3}$. Then we consider the following minimization problem: find σ_n such that

$$\begin{cases} \sigma_n \in (\tilde{V}_n)^3, \\ \forall v \in (\tilde{V}_n)^3,\ J_n(\sigma_n) \leq J_n(v)\,. \end{cases} \tag{3.3}$$

Theorem 3.1. *Suppose that (2.1), (3.1) and (3.2) hold. Then, for all $n \in \mathbb{N}$, problem (3.3) has a unique solution, called the $(\tilde{V}_n)^3$-discrete smoothing D^m-spline relative to ρ^n, $\rho^n g_n$ and ε_n, which is also the unique solution of the following variational problem : find σ_n satisfaying*

$$\begin{cases} \sigma_n \in (\tilde{V}_n)^3, \\ \forall v \in (\tilde{V}_n)^3,\ \langle \rho^n \sigma_n, \rho^n v \rangle_n + \varepsilon_n (\sigma_n, v)_{m,\Omega,\mathbb{R}^3} = \langle \rho^n g_n, \rho^n v \rangle_n\,, \end{cases} \tag{3.4}$$

where $\langle \cdot, \cdot \rangle_n$ is the Euclidean inner product in $\mathbb{R}^{N,3}$.

Proof: (Cf. [9, Theorem 6.2]). ■

Remark 3.2. In Theorem 4.1 we shall state that the sequence $(\sigma_n)_{n \in \mathbb{N}}$ converges to a parameterization of the surface $\mathcal{S}$. Hence, every smoothing D^m-spline σ_n defines an approximating surface of $\mathcal{S}$.

Let M denote the dimension of $\tilde{V}_n$ and let $\{w_i\}_{1 \leq i \leq M}$ be a basis of this space. The D^m-spline σ_n can be expressed in the following form :

$$\sigma_n = \sum_{j=1}^{M} \beta_j w_j, \quad \beta_j \in \mathbb{R}^3,\ j = 1, \ldots, M\,.$$

Let $b_1, \ldots, b_N$ be the points of B_n^1 and consider the matrices

$$A = \Big(w_j(b_i)\Big)_{\substack{1\le i\le N\\ 1\le j\le M}} \in \mathbb{R}^{N,M} \quad \text{and} \quad R = \Big((w_j, w_i)_{m,\Omega}\Big)_{1\le i,j\le M} \in \mathbb{R}^{M,M},$$

where $(\,\cdot\,,\,\cdot\,)_{m,\Omega}$ is the inner semi–product of order m in the Sobolev space $H^m(\Omega)$. Then, it is readily proved that (3.4) is equivalent to the problem : find $\beta = (\beta_j)_{1\le j\le M}$ such that

$$\begin{cases} \beta \in \mathbb{R}^{M,3}, \\ (A^T A + \varepsilon_n R)\beta = A^T(\rho^n g_n)\,. \end{cases}$$

Therefore, problem (3.3) has been reduced to solving three linear systems, each one of which has the same symmetric, positive definite matrix $A^T A + \varepsilon_n R$. Notice that $A^T A$ is the least–squares matrix associated to the basis $\{w_i\}_{1\le i\le M}$.

§4. Study of the Convergence

The convergence of the smoothing method is established in the following

Theorem 4.1. *Suppose $m, m', k \in \mathbb{N}$ satisfy (3.1) and that*

$$m' - 2 \ge m \ge 2\,. \tag{4.1}$$

Let $m^* = \min(m', 2k+2)$. Assume that (3.2) holds, as well as (2.1) with m' instead of m. Assume also that

for all $n \in \mathbb{N}$, elements in $\tilde{\mathcal{T}}_n$ are union of elements in $\mathcal{T}_n$,

$$\exists C > 0,\ \forall n \in \mathbb{N},\ \tilde{h}_n \le C h_n\,,$$

$$\lim_{n\to+\infty} \sup_{x\in\Omega} \min_{b\in B_n^1} \|x - b\|_{\mathbb{R}^2} = 0\,,$$

$(\varepsilon_n)_{n\in\mathbb{N}}$ is a bounded sequence,

and that

$$\lim_{n\to+\infty} \frac{h_n^{m^*-m-1}}{\varepsilon_n^{1/2}} = 0\,. \tag{4.2}$$

Finally, suppose that there exist a constant $C > 0$ and, for all $n \in \mathbb{N}$, a function $\zeta_n \in (\tilde{V}_n)^3$ such that

$$\forall n \in \mathbb{N},\ \|g_n - \zeta_n\|_{m,\Omega,\mathbb{R}^3} \le C\,\tilde{h}_n^{m^*-m}\,\|g_n\|_{m^*,\Omega,\mathbb{R}^3}\,. \tag{4.3}$$

Then,

$$\lim_{n\to+\infty} \sigma_n = g \quad \text{in} \quad H^m(\Omega;\mathbb{R}^3)\,,$$

where σ_n is the solution of (3.3) and g is the parameterization of $\mathcal{S}$ introduced in (2.1).

Proof: (Cf. [9, Theorem 7.2]). ■

Remark 4.2. Hypothesis (4.2) imposes a relationship between the sequences $(h_n)_{n\in\mathbb{N}}$ and $(\varepsilon_n)_{n\in\mathbb{N}}$. When the latter is bounded above by a positive real number, (4.2) is a consequence of (2.1), (3.1) and (4.1).

Remark 4.3. If $m' \geq 2k+2$, (4.3) holds choosing as ζ_n the Hermite interpolant of g_n in the space $(\tilde{V}_n)^3$. In general, it would be possible to take ζ_n as an interpolant of "Clément type" (cf. [7]).

References

1. Apprato, D., Approximation de surfaces paramétrées par éléments finis, Thèse d'État, Université de Pau et des Pays de l'Adour, 1987.
2. Apprato, D., and R. Arcangéli, Approximation de surfaces paramétrées régulières par éléments finis. Partie I : Modélisation du problème, Publication UA 1204 CNRS n° 1989/6.
3. Apprato, D., Approximation de surfaces paramétrées régulières par éléments finis. Partie II : Approximation et étude de l'erreur, Publication UA 1204 CNRS n° 1989/7.
4. Arcangéli, R., D^m-splines sur un domaine borné de $\mathbb{R}^n$, Publication UA 1204 CNRS n° 1986/2.
5. Arcangéli, R., Some applications of discrete D^m splines, in *Mathematical Methods in Computer Aided Geometric Design*, T. Lyche and L. L. Schumaker (eds.), Academic Press, New York, 1989, 35–44.
6. Ciarlet, P. G., *The Finite Element Method for Elliptic Problems*, North–Holland, 1978.
7. Clément, P., Approximation by finite element functions using local regularization, *RAIRO R–2* (1975), 77–84.
8. Nečas, J., *Les Méthodes Directes en Théorie des Équations Elliptiques*, Masson, 1967.
9. Torrens, J. J., Approximation de surfaces paramétrées présentant des discontinuités par D^m-splines d'ajustement discrètes, Publication UA 1204 CNRS n° 1993/9.

Juan José Torrens
Departamento de Matemática e Informática
Universidad Pública de Navarra
Campus de Arrosadía s/n
31006 Pamplona, SPAIN
jjtorrens@ upna.es

Wavelet Analysis of Refinable Functions

Lars F. Villemoes

Abstract. We describe a simple relation between the smoothness of a refinable function in the Besov space sense and the $\ell^p(\mathbb{Z})$ spectral radius of an associated subdivision operator. A proof using wavelet analysis is sketched. We then focus on the case $p = 1$ which leads to estimates of the fractal dimension of the graph of a refinable function.

§1. Introduction

Let $\ldots, c_{-1}, c_0, c_1, c_2, \ldots$ be a sequence of real numbers with finitely many non-zero elements such that $\sum_k c_k = 1$. Then up to normalization, there is a unique compactly supported distribution φ solving the refinement equation

$$\varphi(t) = 2 \sum_{k \in \mathbb{Z}} c_k \varphi(2t - k). \tag{1.1}$$

(See [1].) We call this solution a *refinable function*, although it is not necessarily a function.

A square integrable refinable function is almost always the scaling function of a multiresolution analysis in $L^2(\mathbb{R})$. It can only fail if the integer translates of φ are linearly dependent in the sense that for some $\xi \in \mathbb{R}$, $\sum_k e^{i\xi k} \varphi(t - k) = 0$ for almost all $t \in \mathbb{R}$.

If the coefficient set $2c_k$ is the mask of convergent stationary subdivision scheme, then φ is continuous, by definition. The limit curve resulting from this scheme when starting from the control points $\mathbf{p}_k \in \mathbb{R}^2$ can then be parametrized by

$$\mathbf{r}(t) = \sum_{k \in \mathbb{Z}} \mathbf{p}_k \varphi(t - k). \tag{1.2}$$

(See [3].)

Motivated by these two applications, our main goal is to describe how the regularity, or smoothness, of φ depends on the coefficients (c_k).

Wavelets, Images and Surface Fitting
P. J. Laurent, A. Le Méhauté, and L. L. Schumaker (eds.), pp. 493–500.

ISBN 1-56881-040-7.

Explicit solutions to (1.1) are few. All the known continuous ones are built from the basic spline solutions $B_1, B_2, \ldots$, where

$$B_n(t) = \frac{1}{n!}\sum_{k=0}^{n+1}(-1)^k \binom{n+1}{k} (\max\{0, t-k\})^n.$$

The corresponding $c_k \neq 0$ are the binomial coefficients $2^{-(n+1)}\binom{n+1}{k}$ for $k = 0, 1, \ldots, n+1$. B_n has $n-1$ continuous derivatives, is equal to a polynomial of degree n on each of the intervals $[0,1], [1,2], \ldots, [n, n+1]$ and vanishes outside $[0, n+1]$. If maximum regularity for a fixed number of non-zero coefficients is desired, the B-spline is the choice to make. But given further restrictions, excluding this explicit solution, matters become far more complicated.

In §2 we state our main result which connects the precise smoothness in the Besov space sense with the spectral radius of a subdivision operator. Then in §3 we use a result from [2], which relates the critical Besov exponent for $p = 1$ of a real function with the fractal dimension of its graph, to estimate the fractal dimension for graphs of refinable functions.

It should be stressed that we will not give all details of the proofs here, nor will we state the sharpest possible results. More detailed information can be found in [8] and the references therein. In several dimensions, many of the ideas can still be used to estimate regularity, at least when some power of the dilation matrix is proportional to the identity, [9]. Finally, for a recent study of the spectral radius of subdivision operators and more general infinite *slanted* matrices, see [4] and [7].

§2. Besov Regularity and Spectral Radius

Let the *symbol* of the refinement equation (1.1) be the trigonometric polynomial

$$m(\omega) = \sum_k c_k e^{-ik\omega}.$$

Normalizing the Fourier transform by $\hat{f}(\omega) = \int f(t)e^{-i\omega t}\,dt$ for integrable f, the unique compactly supported φ with $\hat{\varphi}(0) = 1$ solving (1.1) is given by

$$\hat{\varphi}(\omega) = \prod_{j=1}^{\infty} m(2^{-j}\omega). \tag{2.1}$$

We can always factor m in two parts,

$$m(\omega) = \left(\frac{1+e^{-i\omega}}{2}\right)^M \widetilde{m}(\omega), \tag{2.2}$$

where M is the order, (possibly zero), of the zero of m at $\omega = \pi$, and $\widetilde{m}(\omega) = \sum_k \widetilde{c}_k e^{-ik\omega}$ is a trigonometric polynomial not vanishing at $\omega = \pi$.

The first factor in (2.2) is the symbol of a refinable function with smoothness of order M, namely the B-spline B_{M-1}. It is therefore reasonable that we only have to study the second factor of (2.2), in order to analyze the smoothness of φ. Indeed, if $\widetilde{\varphi}$ is the refinable function (distribution) with symbol $\widetilde{m}$, we have from (2.1) that $\varphi = B_{M-1} * \widetilde{\varphi}$.

Hence, we will consider the subdivision operator $T = T_{\widetilde{m}}$ associated with the coefficients of the symbol $\widetilde{m}$,

$$(T\lambda)_k = 2\sum_l \widetilde{c}_{k-2l}\lambda_l. \tag{2.3}$$

For each $1 \le p \le \infty$, the operator T maps $\ell^p(\mathbb{Z})$ continuously into itself and has a spectral radius $\rho_p(T)$.

We say that a compactly supported distribution u has *stable integer translates* if $\sum_k e^{ik\xi}u(t-k)$ is different from the zero distribution for all $\xi \in \mathbb{R}$. For refinable functions, stability can be formulated in terms of the zeros of the symbol. For instance, a simple sufficient condition for stability is that the symbol does not vanish on $[-\pi/2, \pi/2]$.

Theorem 2.1. *Assume φ has stable integer translates and define M and $\rho_p(T)$ as above. Then*

$$\sup\{s \mid \varphi \in B_p^{s,q}(\mathbb{R})\} = M + \frac{1}{p} - \log_2 \rho_p(T).$$

The parameters $s \in \mathbb{R}$, $1 \le p, q \le \infty$ of the Besov space $B_p^{s,q}$ should be understood as follows. If $f \in B_p^{s,q}$, then f has "smoothness s in the L^p sense". q is merely a subtle refinement to this idea. For example $B_\infty^{s,\infty}$ is equal to the Hölder class C^s for all non-integer $s > 0$, and $B_2^{s,2}$ is the Sobolev space H^s consisting of tempered distributions u such that $(1+\omega^2)^{s/2}\hat{u} \in L^2(\mathbb{R})$.

Theorem 2.1 therefore says that the critical smoothness in the L^p sense of a refinable function can be found explicitly in terms of the order of the zero of the symbol at $\omega = \pi$ and the ℓ^p spectral radius of the subdivision operator (2.3) corresponding to the residual factor $\widetilde{m}$ in (2.2).

Remark 2.2. Before we sketch a proof of Theorem 2.1, let us be slightly more precise about the Besov spaces. For a function f in L^p, we can form the forward difference $\Delta_h f(t) = f(t+h) - f(t)$ and the L^p modulus of continuity $\omega_p(\tau, f) = \sup_{|h|<\tau} \|\Delta_h f\|_{L^p}$. For $0 < s < 1$, we then have $f \in B_p^{s,\infty}$ if and only if $\omega_p(\tau, f) \le C\tau^s$ for $0 < \tau < 1$. One equivalent norm for $B_p^{s,\infty}$ is then the sum of the infimum of possible constants C and the L^p norm of f. The case $q < \infty$ only differs in the way in which $\omega_p(\tau, f)$ is compared to τ^s. The inclusion $B_p^{s_1,q_1} \subset B_p^{s_2,q_2}$ holds for $s_1 > s_2$ independently of the values of p, q_1, and q_2, see [6]. Therefore the supremum in Theorem 2.1 does not depend on q, and we will only consider the case $q = \infty$.

Proof: We will analyze φ using the Meyer wavelet ψ. This function has a smooth Fourier transform with compact support disjoint from the origin, and the system $\psi_{jk}(t) = 2^{j/2}\psi(2^j t - k)$, $j, k \in \mathbb{Z}$ is an orthonormal basis for the Hilbert space of square integrable functions $L^2(\mathbb{R})$. Moreover, a compactly supported distribution u belongs to $B_p^{s,\infty}$ if and only if

$$\sup_{j\geq 0} 2^{j(\frac{1}{2}-\frac{1}{p}+s)}\|\alpha_j\|_p < \infty, \tag{2.4}$$

where $\alpha_j(k) = u(\psi_{jk})$ and $\|\cdot\|_p$ is the norm in $\ell^p(\mathbb{Z})$. (See [5].)

With $u = \varphi$, repeated use of (1.1) and a change of variables leads to the formula

$$\alpha_j = 2^{-j/2}\alpha_0 * T_m^j \delta_0,$$

where T_m is the subdivision operator associated to the mask $2c_k$, and δ_0 is the discrete Dirac sequence. The factorization (2.2) together with the fact that $\hat{\psi}$ vanishes in a neighborhood of the origin gives that $g(\omega) = (1 - e^{-i\omega})^{-M}\hat{\varphi}(\omega)\overline{\hat{\psi}(\omega)}$ is smooth and compactly supported. The sequence γ defined from $\gamma(k) = h(k)$ where $\hat{h} = g$ is therefore rapidly decreasing for $|k| \to \infty$, in particular $\gamma \in \ell^1$. By construction, α_0 can now be expressed as the M-fold convolution of $\delta_0 - \delta_1$ with γ,

$$\alpha_0 = (\delta_0 - \delta_1)^{*M} * \gamma.$$

Again using the factorization (2.2) we get

$$\alpha_0 * T_m^j \delta_0 = \gamma * (\delta_0 - \delta_{2^j})^{*M} * 2^{-jM} T^j \delta_0. \tag{2.5}$$

(Recall that we put $T = T_{\widetilde{m}}$.) Hence

$$\|\alpha_0 * T_m^j \delta_0\|_p \leq C 2^{-jM}\|T^j \delta_0\|_p, \tag{2.6}$$

with $C = 2^M\|\gamma\|_1$. Inserting this into (2.4) together with the trivial inequality

$$\|T^j \delta_0\|_p \leq \|T^j\|_p, \tag{2.7}$$

where the norm on the right hand side is the operator norm $\ell^p \to \ell^p$, we see that $\varphi \in B_p^{s,\infty}$ if

$$\sup_{j\geq 0}\left\|\left(2^{(s-M-\frac{1}{p})}T\right)^j\right\|_p < \infty.$$

By the spectral radius formula, this will be true if $2^{(s-M-\frac{1}{p})}\rho_p(T) < 1$, or equivalently, if $s < M + \frac{1}{p} - \log_2 \rho_p(T)$.

The same argument will show that $\varphi \notin B_p^{s,\infty}$ for $s > M + \frac{1}{p} - \log_2 \rho_p(T)$ if the inequalities (2.6-7) can be reversed modulo multiplicative constants. Here (2.7) can be reversed due to the fact that $\widetilde{m}$ is a trigonometric polynomial. For the same reason, we have $\|(\delta_0 - \delta_{2^j})^{*M} * T^j \delta_0\|_p \geq C_1\|T^j \delta_0\|_p$. Thus, by (2.5), the theorem will follow if we can choose γ to be invertible in $\ell^1(\mathbb{Z})$ with respect to convolution.

To do this, we need some more flexibility in the choice of ψ. It is not hard to show that (2.4) is still necessary for φ to be in $B_p^{s,\infty}$, when ψ is any function whose Fourier transform is smooth and has compact support disjoint from the origin. A choice of the form $(1-e^{-i\omega})^M\hat{\psi} = \omega^{2M}\eta\hat{\varphi}$ with a suitable smooth compactly supported $\eta \geq 0$ will then lead to a γ with smooth and strictly positive Fourier transform, provided that φ has stable integer translates. ■

Theorem 2.1 reduces the study of the smoothness of φ to the problem of finding $\rho_p(T)$. There are several methods for the estimation of these spectral radii. For $p=\infty$ such estimates leads to known criteria for the Hölder regularity of φ. For $p=2$ the problem can be reduced to finding the spectral radius of a finite matrix. For $p=1$ we will describe some upper and lower bounds in the next section.

Let us just mention one important special case where $\rho_p(T)$ can be computed for all p.

Proposition 2.3. *The subdivision operator defined by* $(T\lambda)_{2k} = (1+\alpha)\lambda_k$ *and* $(T\lambda)_{2k+1} = (1-\alpha)\lambda_k$ *has the spectral radius*

$$\rho_p(T) = \|(1+\alpha, 1-\alpha)\|_p$$

when regarded as acting on $\ell^p(\mathbb{Z})$.

Proof: For $\lambda \in \ell^p(\mathbb{Z})$ with $p<\infty$ we have

$$\|T\lambda\|_p^p = \sum_k |(T\lambda)_{2k}|^p + \sum_k |(T\lambda)_{2k+1}|^p = (|1+\alpha|^p + |1-\alpha|^p)\|\lambda\|_p^p,$$

and by an easier argument for the case $p=\infty$, $\|T\lambda\|_p = \|(1+\alpha,1-\alpha)\|_p\|\lambda\|_p$ for all $1\leq p\leq\infty$. Repeated use of this gives $\|T^j\|_p^{1/j} = \|(1+\alpha,1-\alpha)\|_p$ for all $j\geq 1$ and the result follows from the spectral radius formula. ■

§3. Fractal Dimension of the Graph

The *fractal dimension* of a compact subset of the plane is defined by

$$\mathrm{Fdim}(E) = \limsup_{\epsilon\to 0} \frac{\log N(\epsilon, E)}{\log(1/\epsilon)},$$

where $N(\epsilon, E)$ is the number of balls of radius ϵ needed to cover E. This dimension is bounded from below by the Hausdorff dimension, but it has the advantage of being much easier to compute. For instance, Deliu and Jawerth have connected the fractal dimension of the graph of a real function with its critical smoothness in the L^1 sense, [2]. For a continuous refinable function we get the following easy consequence of their result and Theorem 2.1. Note that it is important here that the coefficients c_k are real.

Theorem 3.1. *With assumptions and notation as in Theorem 2.1, suppose* $\rho_\infty(T) < 2^M$. *Then*

$$\mathrm{Fdim}(\mathrm{graph}\,\varphi) = \max\{\ 1\ \ ,\ \ \log_2 \rho_1(T) + 1 - M\ \}.$$

Proof: The condition $\rho_\infty(T) < 2^M$ implies that the refinable function φ is in $B_\infty^{s,\infty} = C^\epsilon$ for some $\epsilon > 0$, by Theorem 2.1. Thus φ is a continuous compactly supported function and it follows from [2, Theorem 4.2] that $\mathrm{Fdim}(\mathrm{graph}\,\varphi) = 2-\gamma$ if $0 < \gamma < 1$ is such that $\varphi \in B_1^{s,\infty}$ for all $s < \gamma$ but for no $s > \gamma$. Theorem 2.1 shows that $\gamma = M + 1 - \log_2 \rho_1(T)$. Remark 2.2 shows that $\varphi \in B_1^{\epsilon,\infty}$ so $\gamma > 0$, and the theorem follows in the case where $\log_2 \rho_1(T) > M$.

If $\log_2 \rho_1(T) \le M$, Theorem 2.1 gives that $\varphi \in B_1^{s,\infty}$ for all $s < 1$. By the discussion at the end of [2, §4] we then have $\mathrm{Fdim}(\mathrm{graph}\,\varphi) = 1$. ■

Proposition 3.2. *Let T be defined by (2.3). Then for $j \ge 1$,*

$$2\left|\prod_{l=1}^{j} \widetilde{m}(\omega_l)\right|^{1/j} \le \rho_1(T) \le \|T^j\delta_0\|_1^{1/j},$$

where $\omega_1 \mapsto \omega_2 \mapsto \ldots \mapsto \omega_j = \omega_1$ by the map $\omega \mapsto 2\omega \bmod 2\pi$.

Proof: The action of T on the Fourier series side is $T\Lambda(\omega) = 2\widetilde{m}(\omega)\Lambda(2\omega)$. Thus $T^j\Lambda(\omega) = 2^j\left(\prod_{l=0}^{j-1}\widetilde{m}(2^l\omega)\right)\Lambda(2^j\omega)$. Since the scaling $\omega \mapsto 2^j\omega$ does not affect ℓ^1 norms, this shows that $\|T^j\lambda\|_1 \le \|T^j\delta_0\|_1\|\lambda\|_1$, hence $\|T^j\|_1 = \|T^j\delta_0\|_1$. This certainly proves the second inequality of the proposition, and since a Fourier series is dominated by the sum of the absolute values of its coefficients we get that for all $n \ge 1$,

$$\|T^{jn}\|^{1/jn} \ge 2\left|\prod_{l=0}^{jn-1}\widetilde{m}(2^l\omega_1)\right|^{1/jn} = 2\left|\prod_{l=1}^{j}\widetilde{m}(\omega_l)\right|^{1/j}.$$

The first inequality of the proposition now follows by letting $n \to \infty$. ■

Even though the upper bound of Proposition 3.2 converges to $\rho_1(T)$ as $j \to \infty$, estimates relying on the fact that $\log_2 \rho_p(T)$ as a function of $\frac{1}{p}$ is nondecreasing, convex, and of slope less or equal to one often give better results. See [8] for more details on this.

Example 3.3. Let $\varphi_{M,\alpha}$ correspond to the symbol

$$m(\omega) = \left(\frac{1+e^{-i\omega}}{2}\right)^M \frac{1+\alpha+(1-\alpha)e^{-i\omega}}{2}, \qquad \alpha > 0.$$

It is not hard to show that all these $\varphi_{M,\alpha}$ have stable integer translates since $\widetilde{m}(\omega) \ne 0$ for all $\omega \in \mathbb{R}$. Combining Theorems 2.1 and 3.1 with Proposition 3.2, we find that $\varphi_{M,\alpha} \in C^s$ for all $s < s_0 = M - \log_2(1+\alpha)$, provided that $\alpha < 2^M - 1$. (In fact $\varphi_{M,\alpha} \in B_\infty^{s_0,\infty}$ can be seen to hold by inspecting the proofs of Theorem 2.1 and Proposition 3.2.) For the same values of α, also $\mathrm{Fdim}(\mathrm{graph}\,\varphi_{M,\alpha}) = \log_2 \max(1,\alpha) - M + 2$.

In Figure 1, we plot the graph of $\varphi_{M,\alpha}$ for four choices of (M,α), namely $(2, 2\sqrt{2})$, $(4, 3+8\sqrt{2})$, $(4, 8\sqrt{2})$, and $(2, 2\sqrt{2}-3/4)$. The graphs are achieved by first solving an eigenvalue problem for the values of φ at the integers, and then inserting dyadic points down to level 2^{-8} according to (1.1). On each graph we mark the pair (Hölder regularity, Fractal dimension).

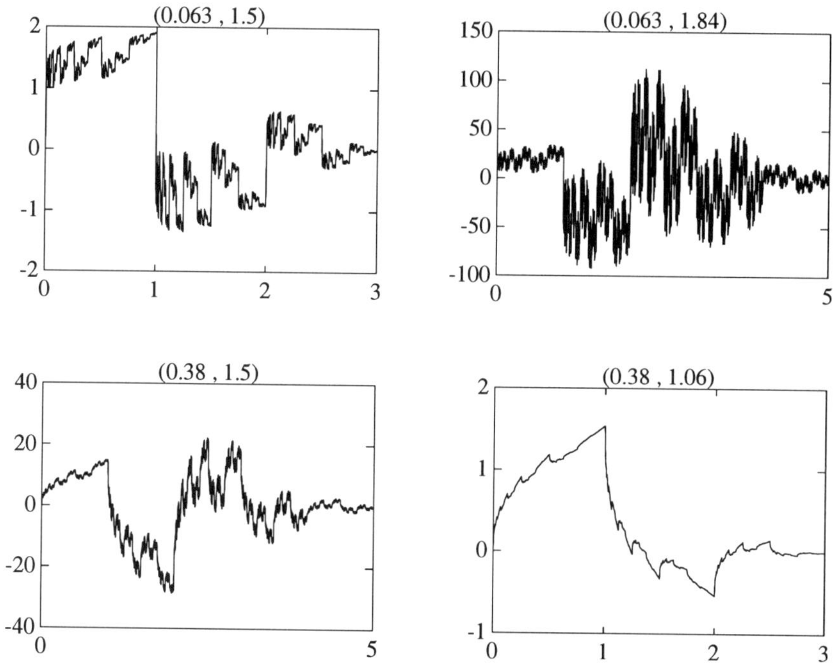

Fig. 1. The graphs of four different $\varphi_{M,\alpha}$.

We observe that fractal dimension seems to give more information about the geometry of the graphs than Hölder regularity. Indeed, each row of graphs is less homogeneous in quality than the first column of graphs.

Example 3.4. For each $-\frac{1}{2} < \beta < \frac{1}{2}$ the symbol $m(\omega) = \cos^2\frac{\omega}{2}(1 + 16\beta \sin^2\frac{\omega}{2}\cos\omega)$ leads to a Hölder continuous φ_β with stable integer translates, (see [8]). This φ_β is interpolating in the sense that $\varphi_\beta(n) = 0$ for all $n \in \mathbb{Z}$ except for $n = 0$ where $\varphi_\beta(0) = 1$. Hence in (1.2) the limit curve interpolates the control points, and this is the reason why this family was studied by Dyn et. al., see [3, Example 3.3].

When β approaches the endpoints $\pm\frac{1}{2}$, the Hölder regularity of φ_β tends to zero. However, using the methods of Proposition 3.2 we will now show that the fractal dimension behaves differently near the two endpoints.

For $\beta = -\frac{1}{2}$, we have $M = 2$ and $\widetilde{m}(\omega) = e^{i\omega}(1 - 8\sin^2\frac{\omega}{2}\cos\omega)$. With $\omega_1 = \frac{2\pi}{3}$ Proposition 3.2 gives that $\rho_1(T) \geq 8$. Hence by Theorem 3.1, $\text{Fdim}(\text{graph}\,\varphi_\beta) \to 2$ as $\beta \to -\frac{1}{2}_+$.

For $\beta = \frac{1}{2}$, we have $M = 2$ again and a brute force numerical computation gives $\|T^{20}\delta_0\|_1 = 2^{20}27055845097$. Thus $\text{Fdim}(\text{graph}\,\varphi_\beta) \to d \leq 1.733$ for $\beta \to \frac{1}{2}_-$.

It is impossible to get $\varphi_\beta \in C^2$. The best possible choice is $\beta = \frac{1}{16}$ where $\varphi_\beta \in C^{s-\epsilon}$ for all $\epsilon > 0$. See [1] where it is also shown that $\varphi_{\frac{1}{16}}$ is almost everywhere twice differentiable. Let us try to prove this with the methods of Proposition 3.2. For $\beta = \frac{1}{16}$ we have $M = 4$ and $\widetilde{m}(\omega) = e^{2i\omega}(2-\cos\omega)$. Then $\|T\delta_0\|_1 = \|(-1, 4-1)\|_1 = 6$ so $\varphi_{\frac{1}{16}} \in B_1^{s,\infty}$ for $s < 5 - \log_2 6 \approx 2.4150$. This implies that $\varphi''_{\frac{1}{16}} \in L^1$ so $\varphi'_{\frac{1}{16}}$ is absolutely continuous, hence differentiable almost everywhere.

References

1. Daubechies, I. and J. Lagarias, Two-scale difference equations II. Existence and global regularity of solutions. SIAM J. Math. Anal. **22** (1991), 1031–1079.
2. Deliu, A. and B. Jawerth, Geometrical dimension versus smoothness, Constr. Approx. **8** (1992), 211–222.
3. Dyn, N., Subdivision schemes in computer aided design, in *Advances in Numerical Analysis, Vol. II*, W. Light (ed.), Clarendon Press, Oxford 1992.
4. Goodman, T. N. T., C. A. Micchelli, and J. D. Ward, Spectral radius formulas for subdivision operators, in *Recent Advances in Wavelet Analysis*, L. L. Schumaker and G. Webb (eds.), Academic Press, Boston, 1994, 335–360.
5. Meyer, I., *Ondelettes et Opérateurs I. Ondelettes*, Hermann, Paris 1990.
6. Peetre, J, *New Thoughts on Besov Spaces*, Duke Univ. Math. Series 1976.
7. Rioul, O., Hölder regularity of subdivision schemes and wavelets, in *Wavelets, Images, and Surface Fitting*, P.-J. Laurent, A. Le Méhauté, and L. L. Schumaker (eds.), A K Peters, Wellesley, 1994, 437–444.
8. Villemoes, L. F., Wavelet analysis of refinement equations, to appear in SIAM J. Math. Anal. .
9. Villemoes, L. F., Continuity of nonseparable quincunx wavelets, submitted to Applied and Computational Harmonic Analysis.

Lars F. Villemoes
Department of Mathematics
Royal Institute of Technology
S-10044 Stockholm, SWEDEN
larsv@ math.kth.se

Interpolation of an Arbitrary Rectangular Mesh with Local Control and Prescribed Continuity

Marshall Walker

Abstract. A theorem in general topology is used to formulate an explicit solution to the problem of interpolating an arbitrary curved rectangular mesh using a surface with prescribed continuity and local control.

§1. Introduction

The task of interpolating a curved mesh by a suitable surface is of fundamental importance in many areas of computer-aided geometric modeling. The general problem in which both the topology and the constituent curves of the mesh are arbitrary has proven difficult. The difficulty depends of course on what one means by a suitable surface. For a surface constructed as a conglomerate of surface patches, one would ideally like to be able to specify both the degree of continuity of each patch and the degree of geometric continuity at the patch boundaries. Suitable surfaces are also required to be "affine invariant" in the sense that given an affine transformation T, a mesh M, and a corresponding interpolatory surface $S(M)$, then if $S(T(M))$ is the interpolating surface corresponding to $T(M)$, it follows that $S(T(M)) = T(S(M))$. In addition, a suitable surface should also possess what is known as "local control": namely the property that local changes of the mesh cause only local changes of the surface. In the case of polynomial surfaces, one also wishes, in order to prevent unforeseen wiggles, that the degree be as low as possible.

There are various ways of limiting the problem. The constituent curves may be specified to be of a certain type. Often they are chosen to be Bézier splines, or alternatively, the topology of the mesh may be restricted in various ways . For example, "faces" of the mesh are often restricted to be triangular or rectangular. Other ways of limiting the problem are to modify one's expectations of a suitable interpolating surface: G^1 or G^2 continuity for instance is often deemed sufficient.

Wavelets, Images and Surface Fitting
P. J. Laurent, A. Le Méhauté, and L. L. Schumaker (eds.), pp. 501–510.

ISBN 1-56881-040-7.

The Coon's patch [5,6] spanned by a rectangular array of four arbitrary C^1 curves is a C^1 surface which agrees with given cross-boundary derivatives and mixed second-partials at vertices of the array. Given a rectangular mesh of curves together with all cross boundary derivatives and mixed partial information, an interpolating surface for the mesh is then constructed patch by patch.

The task of assigning mixed partials turned into a thorny problem which was addressed by Gregory [9]. Chiyokura and Kimura [4] adapted these results to the handling of Bézier style patches. The Chiyokura-Kimura and Gregory patches possess the common characteristic that polynomial boundary curves give rise to a rational surface patches. Although primarily applicable to rectangular patches, the methods are generalized to include triangular and pentagonal patches.

Gordon [7,8] took a different approach. Instead of building surfaces as collections of patches, he developed a method of extending univariate interpolation schemes, such as Lagrange interpolation, to bivariate schemes that could be used for interpolating a rectangular mesh as a single surface. The method, however, shares the difficulty associated with all univariate polynomial schemes - namely that the degree of the result increases as a function of the data being interpolated. In our case the degree of the surface does not increase as a function of the size of the mesh.

From another point of view, solutions were developed in terms of tensor product Bézier patches [1]. These were applicable to the case of a rectangular mesh. They were later generalized in a number of ways as in [13,14] to various classes of irregular meshes. Triangular Bézier patches were introduced by Farin [10,11]. Mann et al. [12] provide a recent survey of triangular methods in general; in all cases, curves of the mesh are Bézier curves.

In this paper the problem is specialized to the case of a rectangular mesh. Constituent curves, however, may be arbitrary. Results of the paper provide an explicit representation of an interpolating surface. The surface is affine invariant and possesses local control. The degree of global continuity depends only on the continuity of the mesh and of the blending functions used in its construction. Furthermore, if the mesh is piecewise polynomial, the total degree of the resulting surface is a linear function of the degree of the polynomial segments in the mesh and the desired degree of continuity. The results depend on a technique by which surfaces which interpolate the mesh locally are blended to obtain a surface which interpolates globally. Roots of this technique may be found in Catmull and Rom [2].

In Section 2 a general theorem is presented which, together with its corollary, is applied in Section 3 to the construction of a suitable interpolating surface. Results regarding the construction appear in Section 4. In conclusion, Section 5 sets forth some ideas for further research.

§2. A Theorem from Point-Set Topology

The following elementary theorem encapsulates the problem of constructing a function which interpolates globally from a family of functions which interpolate locally.

Theorem 2.1. *Let A be a subspace of a Hausdorff space X, let $\mathcal{B} = \{B_\alpha : \alpha \in I\}$ be a partition of unity on X, and $f : A \to Y$ a continuous function to a topological vector space over $\mathbb{R}$. If for every $\alpha \in I$, there exists a fuction $F_\alpha : X \to Y$ which extends $f|_{support(B_\alpha)\cap A}$, then $g : X \to Y$ defined by $g = \sum_{\alpha\in I} F_\alpha B_\alpha$ extends f.*

Proof: Since $\mathcal{B}$ forms a partition of unity on X, for every $x \in X$ the sum $g(x) = \sum_{\alpha\in I} F_\alpha(x)B_\alpha(x)$ is well defined. Furthermore, since for $x \in A$ and $\alpha \in I$, $F_\alpha(x) = f(x)$, it follows that $x \in A$ implies

$$g(x) = \sum_{\alpha\in I} F_\alpha(x) \cdot B_\alpha(x) = \sum_{\alpha\in I} f(x) \cdot B_\alpha(x) = f(x) \cdot \sum_{\alpha\in I} B_\alpha(x) = f(x). \quad \blacksquare$$

Each of the functions F_α is a global function which agrees locally with f, namely on the portion of the support of B_α that lies in A. The function g as constructed is then a globally defined function that agrees on the subset A with the function f. In the case when derivatives may be defined, we have the following Corollary.

Corollary 2. *If F_α has continuity of order r and if B_α has continuity of order $k-1$, then g has continuity $min\{k-1, r\}$.*

§3. Application

The problem of interpolating a rectangular curved mesh by a suitable surface as discussed in the introduction may be restated as follows. Given strictly increasing sequences $U = \{u_i : 0 \le i \le m\}$ and $V = \{v_j : 0 \le j \le n\}$, let

$$A = \left(\bigcup_{i=0}^{m}(\{u_i\} \times [v_0, v_n])\right) \bigcup \left(\bigcup_{j=0}^{n}([u_0, u_m] \times \{v_j\})\right)$$

and $X = [u_0, u_m] \times [v_0, v_m]$. We then ask the question, given a function $f : A \to \mathbb{R}^3$ can we find an extension $g : X \to \mathbb{R}^3$ such that the resulting surface $g(X)$ is sufficiently nice. In particular, can we insist that g have the property that local changes to f induce only local changes to g? Also if f has continuity of order r, what can be said of the continuity of g ?

3.1. The Tilde Construction

Considering the theorem of the previous section, choose a partition of unity on X to be the set $\mathcal{B}$ of bivariate B-spline functions $B_{i,j}$ of degree k, $0 \le i \le m-k-1$ and $0 \le j \le n-k-1$, defined by the sequences U and V which have support in $[u_i, u_{i+k+1}] \times [v_j, v_{j+k+1}]$.

It turns out that the set $\mathcal{B}$ forms a partition of unity only on the set $\tilde{X} = [u_k, u_{m-k}] \times [v_k, v_{n-k}]$. Therefore the theorem of Section 2 may not be applied directly. However, setting $\tilde{A} = \tilde{X} \cap A$ and $\tilde{f} = f|_{\tilde{A}}$, we may try to apply the theorem in this context to obtain an extension $\tilde{g} : \tilde{X} \to \mathbb{R}^3$ of $\tilde{f}$. First however it is necessary to find appropriate local interpolating functions, $\tilde{F}_{i,j} : \tilde{X} \to \mathbb{R}^3$, to match corresponding functions $B_{i,j}$.

The idea used here is to consider bivariate Lagrange interpolation of local parts of the mesh. In detail, for $0 \le i \le m-k-1$ and $0 \le j \le n-k-1$ define the $(i,j)^{th}$ portion of the mesh to be

$$A_{i,j} = \Big(\bigcup_{r=0}^{k+1} (\{u_{i+r}\} \times [v_0, v_n])\Big) \bigcup \Big(\bigcup_{s=0}^{k+1} ([u_0, u_m] \times \{v_{j+s}\})\Big)$$

and define $F_{i,j} : X \to \mathbb{R}^3$ to be the bivariate Lagrange interpolation of the mesh $A_{i,j}$ In detail, for $i \le r \le i+k+1$ set

$$L_{i,r}(u) = \frac{\prod_{p=0 p\neq r}^{k+1}(u - u_{i+p})}{\prod_{p=0 p\neq r}^{k+1}(u_r - u_{i+p})} \tag{3.1}$$

and similarly, for $j \le s \le j+k+1$ define $L_{j,s}$.

Next, for $0 \le i \le m-k-1$ and $0 \le j \le n-k-1$ and $(u,v) \in [u_0, u_m] \times [v_0, v_n]$, define

$$F_i^1(u,v) = \sum_{r=0}^{k+1} f(u_{i+r}, v) L_{i,r}(u), \tag{3.2}$$

$$F_j^2(u,v) = \sum_{s=0}^{k+1} f(u, v_{j+s}) L_{j,s}(v), \tag{3.3}$$

and

$$F_{i,j}^3(u,v) = \sum_{s=0}^{k+1} \sum_{r=0}^{k+1} f(u_{i+r}, v_{j+s}) L_{i,r}(u) L_{j,s}(v). \tag{3.4}$$

Setting

$$F_{i,j} = F_i^1 + F_j^2 - F_{i,j}^3, \tag{3.5}$$

the local interpolating functions are chosen to be the functions $F_{i,j}|_{\tilde{X}}$.

According to our construction, $F_{i,j}|_{\tilde{X}}$ extends $\tilde{f}|_{support(B_{i,j})} \cap \tilde{A}$. Abusing notation and writing $F_{i,j}$ for $F_{i,j}|_{\tilde{X}}$, it follows that $\tilde{g} : \tilde{X} \to \mathbb{R}^3$ defined by

$$\tilde{g} = \sum_{j=0}^{n-k-1} \sum_{i=0}^{m-k-1} F_{i,j} B_{i,j} \tag{3.6}$$

is an extension of $\tilde{f} \,.\, \tilde{A} \to \mathbb{R}^3$.

3.2. Remark

We have not yet solved the problem. Given the function $\tilde{g} : \tilde{X} \to \mathbb{R}^3$ which we have just constructed and which interpolates $\tilde{f} : \tilde{A} \to \mathbb{R}^3$, we are now left with the task of extending $\tilde{g}$ to a function $g : X \to \mathbb{R}^3$ which extends $f : A \to \mathbb{R}^3$.

This may be accomplished in a variety of ways. Two such methods are described in detail below. Both methods rely on first extending the given knot sequences and expanding the set of blending functions so that a partition of unity may be defined on the entire domain $X = [u_0, u_m] \times [v_0, v_m]$. The first uses a technique analogous to repeating boundary control vertices commonly used to force B-spline segments begin or end at chosen points [1].

The second method proceeds by constructing a new mesh $\hat{f} : \hat{A} \to \mathbb{R}^3$ where $\hat{A}$ is obtained from A by adding a border of width k and the function $\hat{f}$ is an extension of f. Applying the construction just described gives a function $\tilde{\hat{g}} : X \to \mathbb{R}^3$ that extends f. We therefore set $g = \tilde{\hat{g}}$. This solution therefore depends on the choice of extension $\hat{f} : \hat{A} \to \mathbb{R}^3$. There is of course a variety of ways in which $\hat{f}$ may be chosen. A simple method for choosing $\hat{f}$ is shown in Section 3.3.2 below.

3.3. Choosing Extensions $\tilde{g}$

Given $U = \{u_i : 0 \le i \le m\}$ and $V = \{v_j : 0 \le j \le n\}$. Construct $\hat{U}$ by adding points $u_{-k} \le u_{-k+1} < \cdots < u_{-1} < u_0$ and $u_m < u_{m+1} < u_{m+2} \le \cdots < u_{m+k}$ to the beginning and end of U, respectively. Likewise, construct $\hat{V}$ from V . Consider the B-spline functions of degree k defined by the sequences $\hat{U}$ and $\hat{V}$ for $-k \le i \le m$ and $-k \le j \le n$.

3.3.1. Repeating Interpolation Functions

On a rectangle $[u_p, u_{p+1}] \times [v_q, v_{q+1}]$, for $k \le p \le m - k - 1$ and $k \le q \le n - k - 1$, $\tilde{g}$ takes the following form:

$$\tilde{g}(u, v) = \sum_{i=p-k}^{p} \sum_{j=q-k}^{q} F_{i,j}(u, v) B_{i,j}(u, v)$$

For rectangles in the border, namely those for which $0 \le p < k$, $m - k \le p < m$, $0 \le q < k$, or $n - k \le q < n$, we alter the expression appearing in (3.6) by repeating certain of the control functions $F_{i,j}$. In particular, if $g_1 : X \to \mathbb{R}^3$ is the desired extension, we require g_1 restricted to such rectangles $[u_p, u_{p+1}] \times [v_q, v_{q+1}]$ to satisfy the defining characteristic that

$$g(u, v) = \sum_{i=p-k}^{p} \sum_{j=q-k}^{q} G_{i,j}(u, v) B_{i,j}(u, v),$$

where $G_{i,j} = F_{i,j}$ except in the following cases:

$$G_{i,j} = \begin{cases} F_{0,j}, & i < 0,\ 0 < j \leq n-k-1 \\ F_{0,n-k-1}, & i < 0,\ j > n-k-1 \\ F_{i,n-k-1}, & 0 \leq i \leq m-k-1,\ j > n-k-1 \\ F_{m-k-1,n-k-1}, & i > m-k-1,\ j > n-k-1 \\ F_{m-k-1,j}, & i > m-k-1,\ 0 < j \leq n-k-1 \\ F_{m-l-1,0}, & i > m-k-1,\ j < 0 \\ F_{i,0}, & 0 \leq i \leq m-k-1,\ j < 0 \\ F_{0,0}, & i < 0,\ j < 0. \end{cases} \tag{3.7}$$

3.3.2. Expanding the Mesh

Setting

$$\hat{A} = \Big(\bigcup_{i=-k}^{m+k} (\{u_i\} \times [v_{-k}, v_{n+k}])\Big) \bigcup \Big(\bigcup_{j=-k}^{n+k} ([u_{-k}, u_{m+k}] \times \{v_j\})\Big),$$

and $X = [u_{-k}, u_{m+k}] \times [v_{-k}, v_{m+k}]$, let $r : \hat{A} \to A$ be the retraction obtained by:

- setting $r(\mathbf{x}) = \mathbf{x}$ for $\mathbf{x} \in A$
- uniformly contracting the border $\hat{A} \setminus A$ onto the boundary of A: for $0 < i \leq k$ and $0 \leq t \leq 1$,

$$\begin{aligned} r\big((1-t)(u_{-i}, v_{-i}) + t(u_{-i}, v_{n+i})\big) &= (1-t)(u_0, v_0) + t(u_0, v_n), \\ r\big((1-t)(u_{m+i}, v_{-i}) + t(u_{m+i}, v_{n+i})\big) &= (1-t)(u_0, v_0) + t(u_m, v_n), \\ r\big((1-t)(u_{-i}, v_{-i}) + t(u_{m+i}, v_{-i})\big) &= (1-t)(u_0, v_0) + t(u_m, v_0), \\ r\big((1-t)(u_{-i}, v_{n+i}) + t(u_{m+i}, v_{n+i})\big) &= (1-t)(u_0, v_n) + t(u_m, v_n). \end{aligned}$$

We then take $\hat{f} = f \circ r$. Applying the tilde construction to the function $\hat{f}$, we obtain an extension $\tilde{\hat{g}} : \tilde{\hat{X}} \to \mathbb{R}^3$. Since $\tilde{\hat{X}} = X$, we define $g_2 = \tilde{\hat{g}}$, to be the desired extension of $\tilde{g}$.

3.4. Comments

In the case that the degree k of the B-spline functions is greater than either the dimensions m or n of the mesh, the tilde construction clearly does not make sense, for the reason that the border that must be subtracted to obtain $\tilde{A}$ and $\tilde{X}$ is greater than the mesh itself. Hence the tilde construction assumes $k < \min(m, n)$. However, in the case that $k \geq \min(m, n)$, the mesh may first be expanded as described above obtaining $\hat{f} : \hat{A} \to \mathbb{R}^3$. The tilde construction can then be applied to obtain the desired interpolation function.

§4. Results

Characteristics of surfaces created with our construction are listed below. An examination of the construction shows that the surface is constructed as an algebraic combination of elements of the mesh, the locally interpolating bivariate Lagrange functions, and the B-spline blending functions. Properties of these constituent elements are reflected in characteristics of the surface.

Suppose the mesh $f : A \to \mathbb{R}^3$ is C^r continuous in the sense that partial derivatives $\frac{\partial^k f}{\partial u^k}$ and $\frac{\partial^k f}{\partial v^k}$ exist and are continuous for $1 \leq k \leq r$ at points (u, v_j) , respectively (u_i, v), where $u_0 \leq u \leq u_m$, $v_0 \leq v \leq v_n$, $0 \leq j \leq n$, and $0 \leq i \leq m$. Since the B-spline functions $B_{i,j}$ of degree k are C^{k-1} continuous, examining lines (3.2) to (3.6), gives the following result.

Theorem 4.1. *If the mesh $f : A \to \mathbb{R}^3$ is C^r continuous, $\tilde{g} : \tilde{X} :\to \mathbb{R}^3$ has $C^{min(k-1,r)}$ continuity.*

If f is piecewise polynomial, then it follows that $\tilde{g}$ also is piecewise polynomial. In particular if f is piecewise polynomial of degree p, the Lagrange functions $L_{i,r}$ and $L_{j,s}$ of line (3.1) are each of degree $k+1$, and from (3.2)–(3.4) it follows that each of the functions $F_{i,j}$ and $\tilde{F}_{i,j}$ is of degree $\max(2k+2, p+k+1)$. Since the bivariate B-spline functions are of total degree $2k$, we have the following result.

Theorem 4.2. *If $f : A \to \mathbb{R}^3$ is piecewise polynomial of degree p, then $\tilde{g} : \tilde{X} \to \mathbb{R}^3$ is piecewise polynomial of total degree* $\max(2k+2, p+k+1)+2k$.

Example 4.3. From the above results if $k = 2$ and $p = 3$ then the extension $\tilde{g}$ is C^1 and of total degree 10. If $k = 3$ and $p = 3$, then $\tilde{g}$ is C^2 and of total degree 14.

If a portion of the mesh, say $f|_{\{u_i\}} \times [v_j, v_{j+1}]$, $k \leq i \leq m - k - 1$, is altered, examining the tilde construction, one sees that the local interpolating functions $F_{r,s}$ are affected only for $i - k \leq r \leq i$ and $j - k \leq j$. Thus, we have the following result regarding the extension $\tilde{g}$. In this sense we claim that the constructed surface has the characteristic of "local control".

Theorem 4.4. *If the mesh is altered on the segment $f|_{\{u_i\}} \times [v_j, v_{j+1}]$, $k \leq i \leq m-k-1$, the extension $\tilde{g}$ is altered in at most the region $[u_{i-k} u_{i+k}] \times [v_{j-k}, v_{j+k}] \cap \tilde{X}$.*

Finally we address the question of affine invariance.

Theorem 4.5. *Let $T : \mathbb{R}^3 \to \mathbb{R}^3$ be an affine transformation. If $\tilde{h} : \tilde{X} \to \mathbb{R}^3$ is the surface generated by the tilde construction applied to the mesh $T \circ f : A \to \mathbb{R}^3$, then $T \circ \tilde{g} = \tilde{h}$. In other words, the surface $\tilde{g}$ is affine invariant.*

Proof: For $0 \leq i \leq m - k - 1$ and $0 \leq j \leq n - k - 1$ let H_i^1, H_j^2, and $H_{i,j}^3$ be the functions obtained, as in equations (3.2) through (3.5), for the mesh $T \circ f : A \to \mathbb{R}^3$. For example, $(u, v) \in [u_0, u_m] \times [v_0, v_n]$, equation (3.2) becomes

$$H_i^1(u, v) = \sum_{r=0}^{k} T \circ f(u_{i+r,v}) L_{i,r}(u).$$

Expressing T as $S + b$ for $S : \mathbb{R}^3 \to \mathbb{R}^3$ a linear transformation and $b \in \mathbb{R}^3$, observe that

$$\begin{aligned}
H_i^1((u, v) &= \sum_{r=0}^{k} T \circ f(u_{i+r}, v) L_{i,r}(u) \\
&= \sum_{r=0}^{k} (S \circ f(u_{i+r}, v) + b) L_{i,r}(u) \\
&= \sum_{r=0}^{k} S \circ f(u_{i+r}, v) L_{i,r}(u) + \sum_{r=0}^{k} b L_{i,r}(u) \\
&= (\sum_{r=0}^{k} S \circ f(u_{i+r}, v) L_{i,r}(u)) + b \\
&= S(\sum_{r=0}^{k} f(u_{i+r}, v) L_{i,r}(u)) + b \\
&= S(F_i^1(u, v)) + b.
\end{aligned}$$

Similarly for $0 \le j \le n-k-1$ and $0 \le i \le m-k-1$, $H_j^2(u, v) = S(F_i^2(u, v)) + b$ and $H_{i,j}^3(u, v) = S(F_{i,j}^3(u, v)) + b$.

Thus, setting $H_{i,j} = H_i^1 + H_j^2 - H_{i,j}^3$ as in equation (3.5) , equation (3.6) becomes

$$\begin{aligned}
\tilde{h} &= \sum_{j=0}^{n-k-1} \sum_{i=0}^{m-k-1} H_{i,j} B_{i,j} \\
&= \sum_{j=0}^{n-k-1} \sum_{i=0}^{m-k-1} (H_i^1 + H_j^2 - H_{i,j}^3) B_{i,j} \\
&= \sum_{j=0}^{n-k-1} \sum_{i=0}^{m-k-1} (SF_i^1 + b) + (SF_j^2 + b) - (SF_{i,j}^3 + b) B_{i,j} \\
&= \sum_{j=0}^{n-k-1} \sum_{i=0}^{m-k-1} (SF_i^1 + SF_j^2 - SF_{i,j}^3) B_{i,j} + b \sum_{j=0}^{n-k-1} \sum_{i=0}^{m-k-1} B_{i,j} \\
&= S\Big(\sum_{j=0}^{n-k-1} \sum_{i=0}^{m-k-1} F_{i,j} B_{i,j} \Big) + b \\
&= T\Big(\sum_{j=0}^{n-k-1} \sum_{i=0}^{m-k-1} F_{i,j} B_{i,j} \Big) \\
&= T(\tilde{g}). \quad \blacksquare
\end{aligned}$$

It is not difficult to see that the results mentioned above may apply to either of the extensions $g : X \to \mathbb{R}^3$ of $\tilde{g} : \tilde{X} \to \mathbb{R}^3$ proposed in Sections 3.3.1 or 3.3.2. For instance in the case of Theorem 4.1, if g is constructed by the method of repeating "control functions" $F_{i,j}$ as in Section 3.3.1, the result follows immediately from an inspection of the definition g in line (3.7). On the other hand, for g constructed as in Section 3.3.2 by first expanding the original mesh, one need only observe that the new mesh is of the same continuity as the old.

§5. Conclusion

Using an elementary theorem from general topology, we have constructed a surface interpolating a rectangular curved mesh. The curves of the mesh may be completely arbitrary. The construction uses a technique by which local interpolating functions are blended to obtain a single global interpolating function. The resulting surface has the usual desirable characteristics of " affine invariance" and "local control" it also possesses the unusual characteristic that the degree of global continuity may be prescribed in advance.

In the described construction local interpolating functions are defined by extending univariate Lagrange interpolation to a bivariate interpolation scheme. If we were to have begun with a different univariate scheme, for instance some sort of spline interpolation scheme, then we would have arrived, in the same way, at a different bivariate scheme. The relative benefits of various different local interpolation methods should be examined.

The task of extending the methods to meshes that are not rectangular is of practical importance. This could be approached in a number of ways; for instance: 1) methods could be developed by which a single multi-sided patch can be inserted in a mesh that is principally rectangular, or 2) the techniques shown here could be extended to other regular subdivisions of the domain on which blending functions with local support can be defined. In particular triangular divisions should be relatively straightforward.

The techniques employ the idea that "control functions" may be blended in much the same way that we blend "control vertices". That is, instead of blending points in $\mathbb{R}^n$ we blend elements of some appropriate ring of functions. Further work is necessary to determine which of the usual tools for analysis of splines carry over to this setting. For instance, is it possible to extend results available to us on blossoming?

References

1. Bartels, R.H., Beatty, J.C., and Barsky, B.A., *An Introduction to Splines for Use in Computer Graphics & Geometric Modeling,* Morgan Kauffman, Los Altos, California, 1987.
2. Bézier, P.E., *Numerical Control – Mathematics and Applications,* John Wiley, London, 1987.

3. Catmull, E., and Rom, R., A class of local interpolating splines, in *Computer Aided Geometric Design*, R.E. Barnhill and R.F. Riesenfeld, eds., Academic Press, New York, (1974), 317–326.
4. Chiyokura, H. and Kimura, F., Design of solids with free-form surfaces, Computer Graphics ACM Vol. 17 No.3 (SIGGRAPH '83 Proceedings) Detroit, 1983, 289–298.
5. Coons, S., *Surfaces for Computer Aided Design*, Technical Report, MIT, 1964.
6. Coons, S., *Surfaces for Computer Aided Design of Space Forms*, Technical Report MIT, Project MAC, 1968.
7. Gordon, W., *Free-form Surface Interpolation through Curve Networks*, Technical Report GMR-921, GM Research Laboratories, 1969.
8. Gordon, W., Spline-blended surface interpolation through curve networks, J. Math Mechanics **18**, 10 (1969), 931–952.
9. Gregory, J.A., Smooth interpolation without twist constraints, in *Computer Aided Geometric Design*, R.E. Barnhill and R.F.Riesenfeld, eds., Academic Press, New York, (1974), 71–87.
10. Farin, G., A construction for the visual C^1 continuity of polynomial-surface patches, Computer Graphics and Image Processing **20** (1982), 272–282.
11. Farin, G., Designing C^1 surfaces consisting of triangular cubic patches, Computer Aided Design **14**, 5 (1982), 253-256.
12. Mann, S., Loop, C., Lounsbery, M., Meyers, D., Painter, J., DeRose, T., and Sloan, K., A survey of Parametric Scattered Data Fiting Using Triangular Interpolants, in *Curve and Surface Design*, H. Hagen, ed., SIAM, Philadelphia, (1992), 145–172.
13. Shirman, L.A., and Séquin, C.H., Local surface interpolation with Bézier patches, Computer Aided Design **4** (1988), 279–295.
14. Van Wijk, J.J., Bicubic patches for approximating non-rectangular control-point meshes, Computer Aided Geometric Design **3** (1986), 1–13.

Marshall Walker
Department of Computer Science and Mathematics
Atkinson College, York University
North York M3J 1P3
Ontario, CANADA
mwalker@sol.yorku.ca

Constrained Surface Fitting Using Powell-Sabin Splines

K. Willemans and P. Dierckx

Abstract. Convexity and boundary conditions for Powell-Sabin splines are derived and algorithms are presented for fitting a constrained Powell-Sabin spline to a set of scattered data. The Powell-Sabin splines are defined in terms of basis functions (B-splines) with local support. Using their Bernstein-Bézier representation leads to efficient calculations with these B-splines. Also special attention is paid to the elimination of redundant conditions. Numerical examples are supplied to illustrate the usefulness of surface fitting with constrained Powell-Sabin splines.

§1. Introduction

We study the problem of smoothing a discrete set of data values corresponding to points arbitrarily scattered in a bounded area. Hereby data reduction is an important issue and the approximation $s(x, y)$ should satisfy some additional conditions such as convexity constraints, boundary conditions, etc.

Although tensor product splines have already proven their usefulness while dealing with smoothing problems [4], in some cases they are not very appropriate. As an alternative, Powell-Sabin splines on conforming triangulations are considered. These piecewise quadratics with C^1-continuity have already been successfully used [3] for surface fitting problems. It was shown that a local basis is fairly easy to construct. In this paper we will show how additional convexity constraints and boundary conditions can be imposed.

In Section 2, Powell-Sabin splines are briefly discussed. Sections 3 and 4 deal with convexity and boundary conditions for Powell-Sabin splines. In Section 5 we present two subroutine packages developed for constrained surface fitting problems. We go into further detail on the practical aspects of the implementation and comment on some results obtained with the subroutine packages. Finally, in Section 6, some concluding remarks are given.

Wavelets, Images and Surface Fitting
P. J. Laurent, A. Le Méhauté, and L. L. Schumaker (eds.), pp. 511–520.

ISBN 1-56881-040-7.

§2. Powell-Sabin (PS) Splines

Definition 1. *Let D be a simply connected subset of $\mathbb{R}^2$ with polygonal boundary. A collection of non-degenerate triangles $\Delta = \{T_i, i = 1, \ldots, t\}$ is a conforming triangulation of D with vertices $V_i, i = 1, \ldots, n$ provided that*

a) $D = \bigcup_{i=1}^{t} T_i$,

b) *any pair of triangles T_i and T_j $(i \neq j)$ has disjoint interiors,*

c) *any triangle T_i contains no vertex of any other triangle T_j different from its own three vertices.*

Definition 2. *The linear space of piecewise polynomials of order d and smoothness r, $-1 \leq r < d$, associated with Δ is defined as*

$$S_d^r(\Delta) = \left\{ s(x,y) : \; s \in C^r(D) \text{ and } s_{|T_i} \in \Pi_d, \quad i = 1, \ldots, t \right\},$$

where Π_d is the space of bivariate polynomials of total degree d.

The dimension of $S_d^r(\Delta)$ can not be expressed in a general formula depending on the number of vertices, edges and triangles of Δ, as it is influenced by both the topology and the exact geometry of Δ (see e.g., [12,13,6]). Powell and Sabin therefore studied the space $S_2^1(\Delta^*)$, where Δ^* is obtained by subdividing each triangle T_i of Δ into six subtriangles as is explained in [11]. We call such a refined triangulation a *PS-refinement.* For any set of data $(\alpha_i, \beta_i, \gamma_i), i = 1, \ldots, n$ precisely one $s(x,y)$ in $S_2^1(\Delta^*)$ satisfies

$$s(V_i) = \alpha_i, \quad \frac{\partial s}{\partial x}(V_i) = \beta_i, \quad \frac{\partial s}{\partial y}(V_i) = \gamma_i, \qquad i = 1, \ldots, n \tag{1}$$

and thus the dimension of $S_2^1(\Delta^*)$ equals $3n$. Dierckx *et al.* [3] constructed a local basis for $S_2^1(\Delta^*)$ by considering for each vertex $V_l, l = 1, \ldots, n$ of Δ three linearly independent triplets $(\alpha_l^j, \beta_l^j, \gamma_l^j), j = 1,2,3$ and denoting the solution of (1) with all $(\alpha_i, \beta_i, \gamma_i) = (0,0,0)$ except for $i = l$ and $(\alpha_l, \beta_l, \gamma_l) = (\alpha_l^j, \beta_l^j, \gamma_l^j)$ as $B_l^j(x,y), j = 1,2,3$. Any spline $s(x,y)$ in $S_2^1(\Delta^*)$ then has a unique representation

$$s(x,y) = \sum_{i=1}^{n} \sum_{j=1}^{3} c_{i,j} B_i^j(x,y). \tag{2}$$

We refer to the $B_i^j(x,y)$ as B-splines with respect to the vertex V_i.

§3. Convex Powell-Sabin Splines

3.1. Convexity conditions for a quadratic polynomial

Consider a quadratic on a triangle T with vertices V_1, V_2, V_3 in its Bernstein-Bézier form

$$b(\tau) = \sum_{i+j+k=2} b_{i,j,k} B^2_{i,j,k}(\tau) \tag{3}$$

with respect to T, where the $B^2_{i,j,k}(\tau)$ are the usual Bernstein basis functions. Chang and Davis [1] and Chang and Feng [2] have proven that $b(\tau)$ is convex on T iff

$$\Delta_a + \Delta_b \geq 0, \quad \Delta_b + \Delta_c \geq 0, \quad \Delta_c + \Delta_a \geq 0, \tag{4a}$$

$$\Delta_a \Delta_b + \Delta_b \Delta_c + \Delta_c \Delta_a \geq 0, \tag{4b}$$

with $\Delta_a = b_{2,0,0} + b_{0,1,1} - b_{1,1,0} - b_{1,0,1}$, $\Delta_b = b_{0,2,0} + b_{1,0,1} - b_{0,1,1} - b_{1,1,0}$ and $\Delta_c = b_{0,0,2} + b_{1,1,0} - b_{1,0,1} - b_{0,1,1}$. In fact only two out of the three inequalities (4a) need to be combined with (4b) as the remaining one will then automatically hold.

3.2. Convexity conditions for Powell-Sabin splines

Given a conforming triangulation $\Delta = \{T_i, i = 1, \ldots, t\}$ on D with vertices $V_i, i = 1, \ldots, n$ and a corresponding PS-refinement Δ^*, we look for conditions on the B-spline coefficients $c_{i,j}$ in (2) in order for $s(x, y)$ to be convex on D. It is easily shown (see *e.g.*, [7], in a more general context) that the convexity of $s(x, y)$ on every triangle of Δ^* is necessary and sufficient for $s(x, y)$ to be convex on D. The restriction of (2) to a triangle T of Δ^* can be expressed by a single quadratic polynomial. So it is clear that for every T_i of Δ any two of the inequalities (4a) along with (4b) have to be imposed on each of the six subtriangles of T_i in Δ^*. The resulting eighteen conditions however are not linearly independent (a linear inequality corresponding to an edge shared by two subtriangles might even be considered twice). In [14] we have shown how the redundant linear restrictions can be eliminated. We retained a set of six conditions (4a) and six conditions (4b) on every triangle T_i of Δ and it was easily proven then that the retained conditions were theoretically still sufficient in order for $s(x, y)$ to be convex on T_i. The choice of the retained linear conditions was based on practical considerations such as the computational simplicity of the coefficients of the $c_{i,j}$ and the minimization of the number of $c_{i,j}$ involved in each restriction.

So far, a total of $6t$ linear and $6t$ non-linear restrictions is to be satisfied by the $c_{i,j}$ so that $s(x, y)$ is convex on D. Yet, for a triangulation Δ with more than one triangle ($t > 1$), these restrictions are not linearly independent either. In [14] we developed an algorithm in order to eliminate as many superfluous conditions as possible. For a properly numbered Δ, *i.e.*, no triangle T_i has three neighbour triangles T_j, T_k, T_l with the maximum of j, k and l smaller than i, the algorithm will reduce the total number of retained linear constraints to the minimum, *i.e.*, $6t - 2e_{in}$, e_{in} being the number of edges shared by different triangles of Δ. In general, $6t - 2e_{in} + t_{ip}$ linear conditions will be retained, with t_{ip} the number of triangles of Δ having three neighbour triangles which reference number is lower than the reference number of the enclosed triangle (see [14] for more details).

§4. Powell-Sabin Splines with Boundary Conditions

4.1. Boundary conditions for a quadratic polynomial

Let us reconsider the quadratic polynomial $b(\tau)$ of (3). We will now derive conditions for the Bézier ordinates $b_{i,j,k}$ in order for $b(\tau)$ to have zero function value and/or zero normal derivative on an edge of T. In the sequel, only the edge V_1–V_2 will be considered; analogous derivations for the other edges of T are then straightforward.

Zero function value on V_1–V_2. It is easily shown that $b(\tau) = b_{2,0,0}\tau_1^2 + b_{0,2,0}\tau_2^2 + b_{0,0,2}\tau_3^2 + 2b_{1,1,0}\tau_1\tau_2 + 2b_{0,1,1}\tau_2\tau_3 + 2b_{1,0,1}\tau_1\tau_3$ is identically zero on V_1–V_2 ($\tau_3 = 0$) iff

$$b_{2,0,0} = b_{1,1,0} = b_{0,2,0} = 0. \tag{6}$$

Zero normal derivative on V_1–V_2. A normal direction α with respect to V_1–V_2 is found by considering the intersection point P of the edge V_1–V_2 and the perpendicular on V_1–V_2 through V_3 and setting $\alpha = P - V_3 = (\alpha_1, 1-\alpha_1, 0) - (0,0,1) = (\alpha_1, 1-\alpha_1, -1)$ with $\alpha_1 = \frac{(V_2-V_3).(V_2-V_1)}{\|V_2-V_1\|^2}$. The directional derivative of $b(\tau)$ with respect to α [5],

$$\begin{aligned} D_\alpha b(\tau) = 2\,[&\tau_1(\alpha_1 b_{2,0,0} + (1-\alpha_1)b_{1,1,0} - b_{1,0,1}) \\ &+ \tau_2(\alpha_1 b_{1,1,0} + (1-\alpha_1)b_{0,2,0} - b_{0,1,1}) \\ &+ \tau_3(\alpha_1 b_{1,0,1} + (1-\alpha_1)b_{0,1,1} - b_{0,0,2})] \end{aligned}$$

is then identically zero on V_1–V_2 ($\tau_3 = 0$) iff

$$\begin{aligned} \alpha_1(b_{2,0,0} - b_{1,1,0}) + b_{1,1,0} - b_{1,0,1} &= 0, \\ (1-\alpha_1)(b_{0,2,0} - b_{1,1,0}) + b_{1,1,0} - b_{0,1,1} &= 0. \end{aligned} \tag{7}$$

Zero function value and zero normal derivative on V_1–V_2. Combining both equations (6) and (7) now results in

$$b_{2,0,0} = b_{1,1,0} = b_{0,2,0} = b_{1,0,1} = b_{0,1,1} = 0. \tag{8}$$

4.2. Boundary conditions for Powell-Sabin splines

Let again be given a conforming triangulation $\Delta = \{T_i, i = 1, \ldots, t\}$ on D with vertices $V_i, i = 1, \ldots, n$ and corresponding PS-refinement Δ^*. We are now interested in finding conditions for the B-spline coefficients $c_{i,j}$ in (2) so that $s(x,y)$ has zero function value and/or zero normal derivative on (part of) the boundary ∂D of D. In case there are only boundary conditions on part of

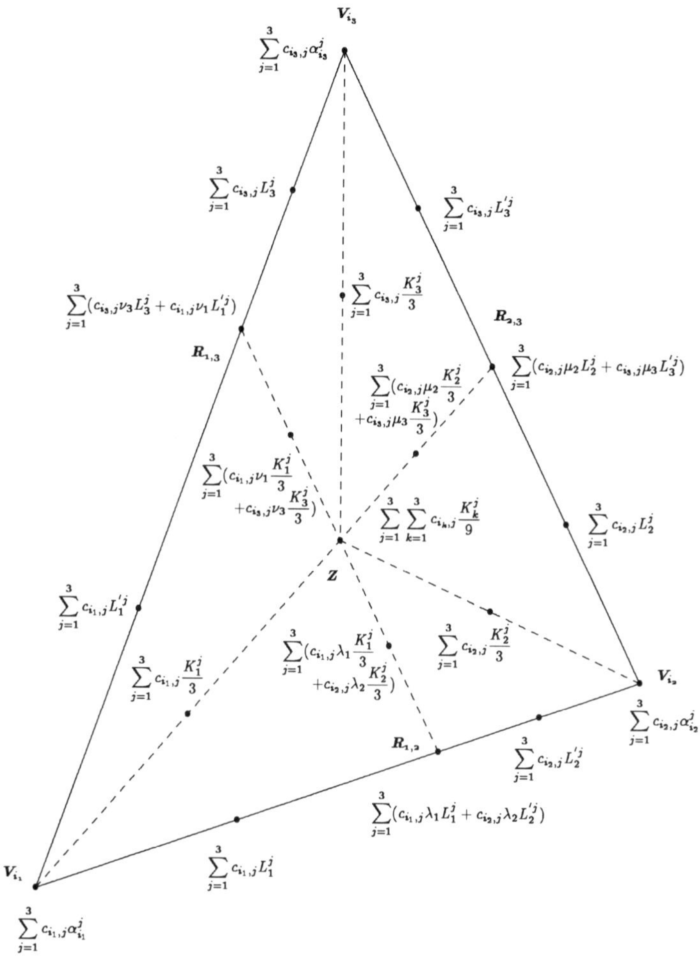

Fig. 1. The restriction of a PS-spline to a triangle.

∂D, we assume that the end points of this part coincide with some vertices of the triangulation Δ. It is then straightforward to see that it is necessary and sufficient to impose the boundary conditions on each triangle of Δ^* which has an edge in common with (the considered part of) ∂D in order for $s(x, y)$ to satisfy the boundary conditions.

Consider now a triangle T_i of Δ with vertices $V_{i_1}, V_{i_2}, V_{i_3}$ having cartesian coordinates $(x_{i_1}, y_{i_1}), (x_{i_2}, y_{i_2}), (x_{i_3}, y_{i_3})$ and a spline (2) of which the

Bernstein-Bézier form of the restriction to T_i is found in Figure 1. It shows the restriction of a PS-spline to a triangle with $Z = (\frac{1}{3}, \frac{1}{3}, \frac{1}{3})$, $R_{1,2} = (\lambda_1, \lambda_2, 0)$, $R_{2,3} = (0, \mu_2, \mu_3)$, $R_{1,3} = (\nu_1, 0, \nu_3)$ and $\tilde{\beta}_1^j = \beta_{i_1}^j(x_{i_2} - x_{i_1}) + \gamma_{i_1}^j(y_{i_2} - y_{i_1})$, $\tilde{\gamma}_1^j = \beta_{i_1}^j(x_{i_3} - x_{i_1}) + \gamma_{i_1}^j(y_{i_3} - y_{i_1})$, $K_1^j = 3\alpha_{i_1}^j + \frac{\tilde{\beta}_1^j + \tilde{\gamma}_1^j}{2}$, $L_1^j = \alpha_{i_1}^j + \lambda_2 \frac{\tilde{\beta}_1^j}{2}$, $L_1^{'j} = \alpha_{i_1}^j + \nu_3 \frac{\tilde{\gamma}_1^j}{2}$, $j = 1, 2, 3$ and $K_k^j, L_k^j, L_k^{'j}$, $k = 2, 3, j = 1, 2, 3$ analogously defined.

Suppose that T_i has the edge V_{i_1}–V_{i_2} along (the considered part of) ∂D. From what precedes, it is clear that equations (6), (7) or (8) have to hold on both subtriangle $(V_{i_1}, R_{1,2}, Z)$ with respect to the edge V_{i_1}–$R_{1,2}$ and subtriangle $(R_{1,2}, V_{i_2}, Z)$ with respect to the edge $R_{1,2}$–V_{i_2}.

Zero function value on V_{i_1}–V_{i_2}. Conditions (6) for the edges V_{i_1}–$R_{1,2}$ and $R_{1,2}$–V_{i_2} are now written as

$$\text{a)} \sum_{j=1}^{3} c_{i_1,j}\alpha_{i_1}^j = 0, \qquad \text{c)} \sum_{j=1}^{3} c_{i_2,j}L_2^{'j} = 0,$$

$$\text{b)} \sum_{j=1}^{3} c_{i_1,j}L_1^j = 0, \qquad \text{d)} \sum_{j=1}^{3} c_{i_2,j}\alpha_{i_2}^j = 0.$$

Zero normal derivative on V_{i_1}–V_{i_2}. Applying conditions (7) to the edges V_{i_1}–$R_{1,2}$ and $R_{1,2}$–V_{i_2} leads to

$$\text{a)} \sum_{j=1}^{3} c_{i_1,j}[\delta\alpha_1^j + (1-\delta)L_1^j - \frac{K_1^j}{3}] = 0,$$

$$\text{b)} \sum_{j=1}^{3} \{c_{i_1,j}[(\delta\lambda_2 + \lambda_1)L_1^j - \frac{\lambda_1 K_1^j}{3}] + c_{i_2,j}[(1-\delta)\lambda_2 L_2^{'j} - \frac{\lambda_2 K_2^j}{3}]\} = 0,$$

$$\text{c)} \sum_{j=1}^{3} c_{i_2,j}[(1-\epsilon)\alpha_2^j + \epsilon L_2^{'j} - \frac{K_2^j}{3}] = 0,$$

with $\delta = \frac{(R_{1,2}-V_{i_3}).(R_{1,2}-V_{i_1})}{\|R_{1,2}-V_{i_1}\|^2}$ and $\epsilon = \frac{(V_{i_2}-V_{i_3}).(V_{i_2}-R_{1,2})}{\|V_{i_2}-R_{1,2}\|^2}$.

Zero function value and zero normal derivative on V_{i_1}–V_{i_2}. We have proven [15] that equations (8) applied to the edges V_{i_1}–$R_{1,2}$ and $R_{1,2}$–V_{i_2} are identical to

$$c_{i_1,j} = c_{i_2,j} = 0, \qquad j = 1, 2, 3$$

if T_i is a non-degenerate triangle, $\lambda_1 \neq 0$, $\lambda_2 \neq 0$ and the $(\alpha_{i_1}^j, \beta_{i_1}^j, \gamma_{i_1}^j)$ and $(\alpha_{i_2}^j, \beta_{i_2}^j, \gamma_{i_2}^j)$ $j = 1, 2, 3$, are linearly independent, which indeed is presumed.

§5. Constrained Surface Fitting with Powell-Sabin Splines

Let Δ be a conforming triangulation of a subset D of the plane and let Δ^* be the corresponding PS-refinement. Given data points (X_r, Y_r) in D with corresponding data values Z_r and weights $W_r, r = 1, \ldots, m$ we search for a spline $s(x, y)$ in $S_2^1(\Delta^*)$ with representation (2) minimizing

$$\sum_{r=1}^{m}\left[W_r[Z_r-\sum_{i=1}^{n}\sum_{j=1}^{3}c_{i,j}B_i^j(X_r,Y_r)]\right]^2$$

and so that the $c_{i,j}$ satisfy inequality or equality constraints as discussed in Sections 3.2 and 4.2.

We have developed two Fortran subroutine packages for solving the surface fitting problems mentioned above. In addition to the set of data points (X_r, Y_r) and the corresponding data values Z_r and weights $W_r, r = 1, \ldots, m$, the user also has to define the triangulation Δ. The programmes then compute a basis for $S_2^1(\Delta^*)$ according to the approach of Dierckx *et al.* [3] (see also Section 2).

As for the convex surface fitting package, an appropriate selection of the linear convexity constraints is made by applying the algorithm mentioned in Section 3.2. Subsequently, the non-linear constrained minimization problem is solved. A suitable routine that deals with this kind of problem is routine E04VDF in the NAG Fortran Library [8] which indeed has been used in our implementation.

As concerns the subroutine package for surface fitting with boundary conditions, the user also has to specify the boundary conditions (if any) on each edge of Δ along the boundary of D. The corresponding linear constraints are then formed as explained in Section 4.2. Finally, a linear system matching the least-squares minimization problem with linear equality constraints is solved using SYMMLQ ([9,10]), an algorithm for solving large sparse systems of linear equations.

We now briefly report on some results obtained with our subroutine packages. Using a random number generator we generated 120 data points (X_r, Y_r) scattered uniformly in the region $[0,1] \times [0,1]$ and a set of normally distributed stochastic variates e_r with expected value 0 and standard deviation 0.05; corresponding data values $Z_r = f(X_r, Y_r) + e_r$ and weights $W_r = 1, r = 1, \ldots, 120$ were considered in order to find a convex PS-spline approximation for $f(x,y) = x^3 + 5(y-0.6)^2 + 1$ on $[0,1] \times [0,1]$. In Figure 2a the data points (*), the triangulation Δ ($n = 9, t = 8$) (—), the PS-refinement Δ^* (- -) and the contour map (...) of the unconstrained least-squares PS-spline are shown. For reasons of comparison, in Figure 2b the same was done for the convex least-squares PS-spline. The quality of fit of the convex approximation may be judged as satisfactory, whereas this is not the case with the unconstrained approximation.

The same data were then reconsidered, this time however to find a PS-spline with a normal derivative identically zero on the edge of the domain where $x = 0$. Figure 3 again shows the superior quality of the constrained approximation.

§6. Conclusions

We presented surface fitting algorithms using constrained splines on triangulations. As for the considered spline space and the corresponding suitable set

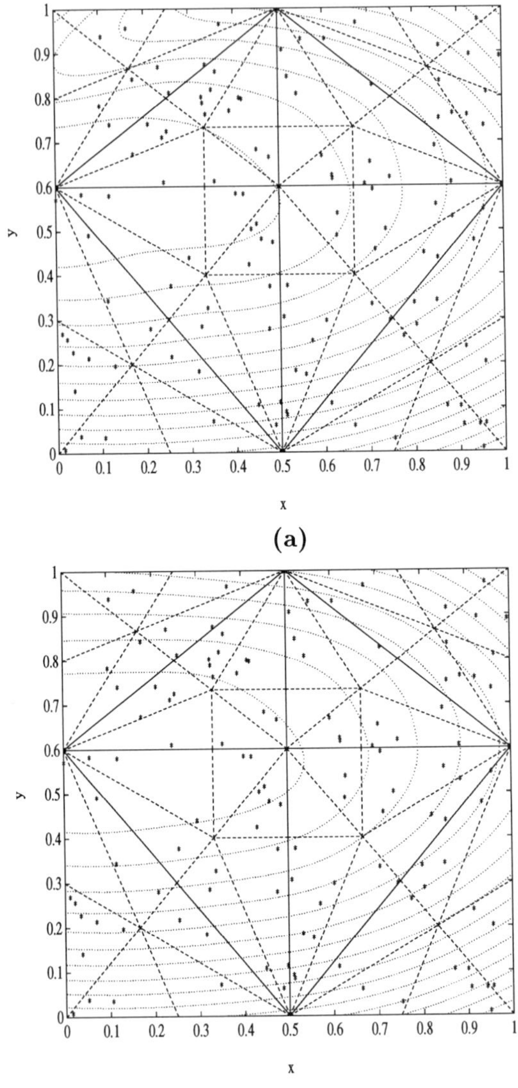

Figure 2. Approximation of $f(x, y) = x^3 + 5(y - 0.6)^2 + 1$. **a.** Least-squares PS-spline; **b.** Convex least-squares PS-spline.

of basis functions, we adopted the approach of Dierckx *et al.* [3] in view of the interesting characteristics of the PS-splines and the local support property of the B-splines. We also used the Bernstein-Bézier representation of the basis splines because it facilitates the manipulation of these splines and improves the computational efficiency. From the graphical results in Section 5 it is clear that one can successfully benefit from the advance knowledge on the

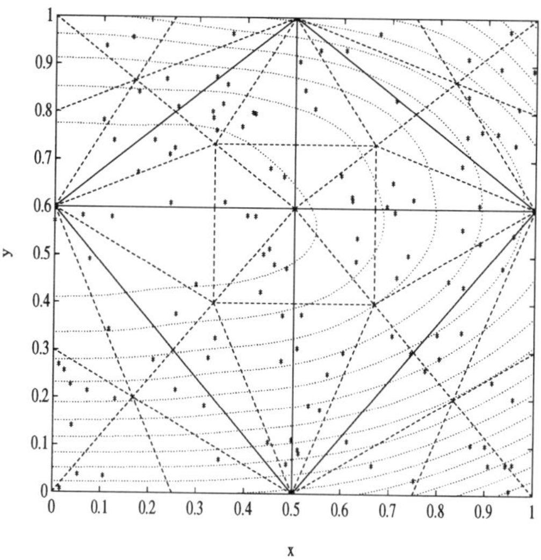

Figure 3. Least-squares PS-spline approximation for $f(x,y)$ with boundary constraints.

behaviour of the function underlying the data.

References

1. Chang, G., and P. J. Davis, The Convexity of Bernstein Polynomials over Triangles, J. Approx. Th. **40** (1984), 11–28.
2. Chang, G., and Y. Feng, An improved condition for the convexity of Bernstein-Bézier surfaces over triangles, Comp. Aided Geom. Design **1** (1984), 279–283.
3. Dierckx, P., S. Van Leemput, and T. Vermeire, Algorithms for surface fitting using Powell-Sabin splines, IMA Journal of numerical analysis **12** (1992), 271–299.
4. Dierckx P., Curve and Surface Fitting with Splines, *Monographs on Numerical Analysis*, Oxford University Press, 1993.
5. Farin, G. Triangular Bernstein-Bézier patches, Computer-Aided Geom. Design **3** (1986), 83–127.
6. Gmelig-Meyling, R. H. J., and P. R. Pfluger, On the dimension of the spline space $S_2^1(\Delta)$ in special cases, in *Multivariate Approximation Theory III*, W. Schempp & K. Zeller (eds.), Birkhäuser, Basel, 1985, 180–190.
7. Grandine, T. A. On convexity of piecewise polynomial functions on triangulations, Comp. Aided Geom. Design **6** (1989), 181–187.
8. The NAG Fortran Library Manual - Mark 13 (1988).
9. Paige, C. C., and M. A. Saunders, Solution of sparse indefinite systems of linear equations, SIAM J. Numer. Anal. **12** (1975), 617–629.
10. Paige, C. C., and M. A. Saunders, Private communication (1992).

11. Powell, M. J. D., and M. A. Sabin, Piecewise quadratic approximations on triangles, ACM Trans. Math. Software **3** (1977), 316–325.
12. Schumaker, L. L., On the dimension of spaces of piecewise polynomials in two variables, in *Multivariate Approximation Theory*, W. Schempp & K. Zeller (eds.), Birkhäuser, Basel, 1979, 396–412.
13. Schumaker, L. L., Bounds on the dimension of spaces of multivariate piecewise polynomials, Rocky Mt. J. Math. **14** (1984), 251–264.
14. Willemans, K., and P. Dierckx, Surface fitting using convex Powell-Sabin splines, J. Comp. Appl. Math. (1994), accepted.
15. Willemans, K., and P. Dierckx, Surface fitting with boundary conditions by means of Powell-Sabin splines, Report TW 199, Dept. Comp. Science, K.U. Leuven, 1993.

Karin Willemans and Paul Dierckx
Katholieke Universiteit Leuven
Departement of Computer Science
Celestijnenlaan 200 A
B-3001 Heverlee, BELGIUM
karinw@cs.kuleuven.ac.be

B-spline Patches for Surface Reconstruction in Computer Vision

ChangSheng Zhao and Roger Mohr

Abstract. This paper considers the problem of 3D surface reconstruction using B-spline patches in computer vision. We propose a direct method of global surface reconstruction which is based on regularized uniform bicubic B-spline patches. This reconstruction is achieved by observing the motion of occluding contours, i.e., where the view lines graze the surface. It has been shown that reconstruction of such a 3D surface is possible when the camera motion is known, except for instances when the surface is fully concave parts and no view line can be tangential to the surface. Experimental results are presented for real data.

§1. Introduction

The problem of representing and reconstructing 3D surfaces has received an enormous amount of attention in computer vision research over the past decade. The interest arises at least in part because robust surface reconstruction would have applicability to a wide variety of fields namely, design automation, manufacturing automation, terrain mapping, vehicule guidance, archeology, restoration of works of art, surveillance, intelligent robots, etc. But aside from any practical applications, the problem has a great deal of scientific and mathematical interest.

From the perspective of classical differential geometry, there are three representations for a surface of an object: explicit, implicit, and parametric. In this paper, we use the uniform bicubic B-spline patches to give a parametric representation of the surface [1].

Our aim is to reconstruct the 3D surface of an object from the observation of occluding contours [2, 5] together with a knowledge of the motion of the camera mounted on the robot. We would like to show that the reconstruction of such a 3D surface is possible when the camera motion is known, with the exception of those fully concave parts where no view line can be tangential to the surface. We assume that the camera is modeled as a pinhole. We suppose that the object is smooth, specifically, that its surface is at least C^2.

Wavelets, Images and Surface Fitting
P. J. Laurent, A. Le Méhauté, and L. L. Schumaker (eds.), pp. 521–528.

ISBN 1-56881-040-7.

In the first part of this paper we will introduce the main properties of uniform bicubic B-spline patches, and give the notions of the occluding contour and epipolar parametrization of a surface. The second part describes mathematically our method for B-spline patches reconstruction. In the last part, we present results on real data and discuss their accuracy.

§2. Definitions and Notations

B-spline curves are piecewise polynomial functions. These curves are defined by a sequence of control points. In applications, cubic B-spline curves are the most frequently used. Especially, uniform cubic B-spline curves which play a important role. Uniform bicubic B-spline patches are a natural and straightforward generalization of uniform cubic B-spline curves. They can be expressed by a double summation:

$$\vec{X}(s,t) = \sum_{i=-3}^{m-1} \sum_{j=-3}^{n-1} (\alpha_{ij} B_i(s) B_j(t), \beta_{ij} B_i(s) B_j(t), \gamma_{ij} B_i(s) B_j(t)) \tag{1}$$

where $s, t \in I \times J$ and $I, J \subseteq \mathbb{R}$. $(\alpha_{ij}, \beta_{ij}, \gamma_{ij})$ are called *control points.* The products $B_i(s)B_j(t)$ are the usual *tensor-product* B-splines, where B_i, B_j are simply the *basis functions* of uniform cubic B-spline curves, as defined in [3]. They have the two following interesting properties: local support and invariance under affine transforms.

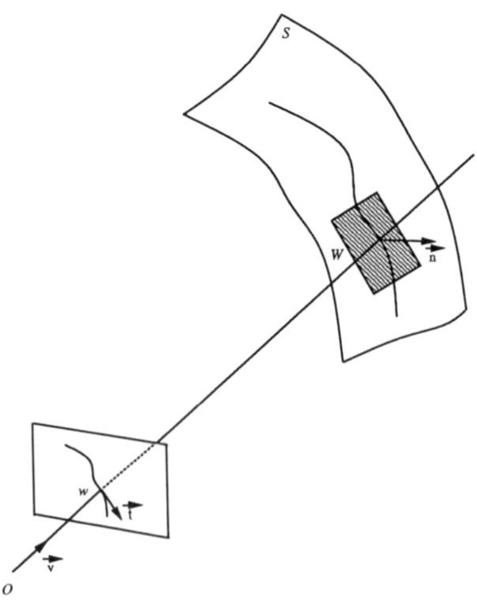

Fig. 1. Plane projection of the occluding contour w of a surface S..

Now we consider an object surface S. The camera model is assumed to be the pinhole, its optical center is at O. As shown in Figure 1, the *occluding contour* w of a surface S, viewed from a point O, is defined to be the projection

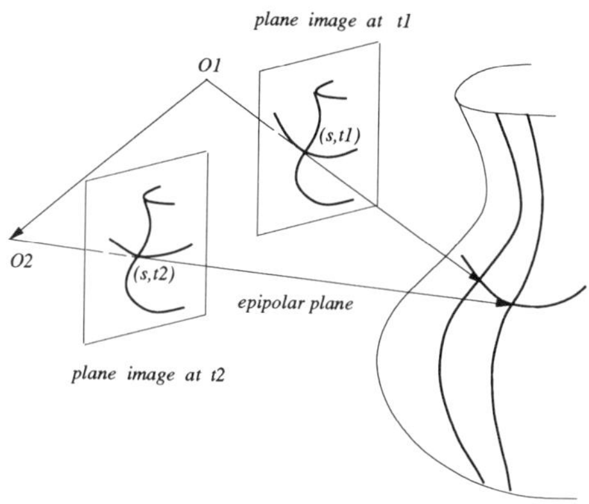

Fig. 2. Epipolar parametrisation.

of the set of points W for which $\vec{n} \cdot \vec{OW} = 0$, where $\vec{n}$ is the unit normal vector to the tangent plane at W.

We suppose that we observe a family of the occluding contours of an object in a sequence of images when the camera moves in space The perspective projection in the retina of these 3D curves over time describes a *spatio-temporal* surface. The parametrisation of the *spatio-temporal* surface and the parametrisation of the object surface are directly linked. We choose two parameters t and s: t is a temporal parameter, i.e., the parameter t is linked to the motion of the camera, and s is a spatial parameter. It is imposed by the arc length of the occluding contour (see Figure 2). This choice we made is very close to the epipolar parametrisation described by Cipolla [2].

§3. Mathematical Model

In this section, we will establish our mathematical model for the reconstruction surface. We suppose that the parametric representation of a surface is defined by the formula (1): $\vec{X}(s,t) = (X_1(s;t), X_2(s,t), X_3(s,t))$.

Therefore B-spline patches reconstruction requires the estimation of $3 \times (m+3) \times (n+3)$ unknowns α_{ij}, β_{ij}, γ_{ij}, $i = -3, ..., m-1$, $j = -3, ..., n-1$ in the formula (1).

We suppose that a point w with the coordinates (x, y) is the projection of a point W on the surface (see Figure 3) with the homogeneous coordinates $[X_1(s,t), X_2(s,t), X_3(s,t), 1]$. Let $M = (m_{pr})_{p=1,...,3,r=1,...,4}$ be the known projection matrix.

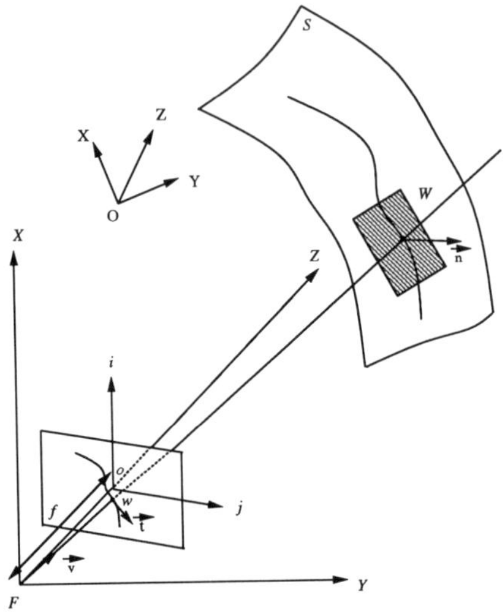

Fig. 3. Property of an occluding contour.

We have for homogeneous coordinates:

$$
\begin{aligned}
x &= \frac{m_{11}X_1(s,t) + m_{12}X_2(s,t) + m_{13}X_3(s,t) + m_{14}}{m_{31}X_1(s,t) + m_{32}X_2(s,t) + m_{33}X_3(s,t) + m_{34}}, \\
y &= \frac{m_{21}X_1(s,t) + m_{22}X_2(s,t) + m_{23}X_3(s,t) + m_{24}}{m_{31}X_1(s,t) + m_{32}X_2(s,t) + m_{33}X_3(s,t) + m_{34}}.
\end{aligned} \tag{2}
$$

Trivially, the formulas (2) can be rewritten after substituting (1) and multiplying by the appropriate denominators as

$$P_1(x, s, t, \alpha_{ij}, \beta_{ij}, \gamma_{ij}, m_{pr}) = 0, \quad P_2(y, s, t, \alpha_{ij}, \beta_{ij}, \gamma_{ij}, m_{pr}) = 0. \tag{3}$$

It is very easy to see that the two equations (3) are linear with respect to the unknowns $\alpha_{ij}, \beta_{ij}, \gamma_{ij}$.

Since w is a point of the occluding contour (see Figure 3), the optical ray $\vec{v}$ determined by FW is then tangent to the object surface S. The normal vector $\vec{n}$ to the tangent plane at W is given by

$$\frac{\partial \vec{X}(s,t)}{\partial s} \times \frac{\partial \vec{X}(s,t)}{\partial t}. \tag{4}$$

According to the definition of an occluding contour, we have the following relation:

$$\vec{n} \cdot \vec{v} = \left(\frac{\partial \vec{X}(s,t)}{\partial s} \times \frac{\partial \vec{X}(s,t)}{\partial t}\right) \cdot \vec{v} = 0. \tag{5}$$

Trivially, formula (5) can be rewritten after substituting (1) and (4) as

$$P_3(x, y, s, t, \alpha_{ij}, \beta_{ij}, \gamma_{ij}, m_{pr}) = 0. \tag{6}$$

Unfortunately, (6) is not linear with respect to the unknowns α_{ij}, β_{ij}, γ_{ij}.

Finally, we can regroup these equations (3) and (6) into the following system of three equations:

$$\begin{aligned} &P_1(x, s, t, \alpha_{ij}, \beta_{ij}, \gamma_{ij}, m_{pr}) = 0 \\ &P_2(y, s, t, \alpha_{ij}, \beta_{ij}, \gamma_{ij}, m_{pr}) = 0 \\ &P_3(x, y, s, t, \alpha_{ij}, \beta_{ij}, \gamma_{ij}, m_{pr}) = 0. \end{aligned} \tag{7}$$

We now suppose that we have k images and l points on each image; this leads us to $3 \times k \times l$ equations in $3 \times (m+3) \times (n+3)$ unknowns. So if k and l are large enough, we have a following redundant system of equations:

$$\begin{aligned} &q_1(x_l^k, s_l^k, t_l^k, \alpha_{ij}, \beta_{ij}, \gamma_{ij}, m_{pr}^k) = 0 \\ &q_2(y_l^k, s_l^k, t_l^k, \alpha_{ij}, \beta_{ij}, \gamma_{ij}, m_{pr}^k) = 0 \\ &q_3(x_l^k, y_l^k, s_l^k, t_l^k, \alpha_{ij}, \beta_{ij}, \gamma_{ij}, m_{pr}^k) = 0. \end{aligned} \tag{8}$$

Now the most direct procedure is to try to solve this system of nonlinear equations. Since the system is an overdetermined one, we can hope to solve it by standard least squares technique. The problem can be formulated as minimization over

$$E = \sum_{k=1} \sum_{l=1} [\lambda(q_1^2 + q_2^2) + \mu q_3^2] \tag{9}$$

where $\lambda, \mu \in \mathbb{R}$ are two weight parameters.

Since there is some noise, i.e., errors or approximations may occur at the contour detection stage, etc in the images. The above method does not allow the form of our surface to be easily adjusted. We therefore introduce a regularization term on the surface. Finally, the regularized B-spline patches functional can be written [6] as

$$\begin{aligned} E = &\sum_{k=1} \sum_{l=1} [\lambda(q_1^2 + q_2^2) + \mu q_3^2] \\ &+ \tau \int_I \int_J [(\frac{\partial^4 X_1}{\partial^2 s \partial^2 t})^2 + (\frac{\partial^4 X_2}{\partial^2 s \partial^2 t})^2 + (\frac{\partial^4 X_3}{\partial^2 s \partial^2 t})^2] ds dt \end{aligned} \tag{10}$$

where $\tau \in \mathbb{R}$ is called the tensor of B-spline patches.

Currently, two standard numerical least squares methods, quasi-Newton and Levenberg-Marquardt are implemented in our system. The starting values of the system are provided by the well-known method of triangulation. Practical experiments show that the Levenberg-Marquardt algorithm works better than the quasi-Newton algorithm.

§4. Experimental Results

All our experiments were conducted with a Pulnix 765 camera, a 18 *mm* lens and FG150 imaging technology grab board. The camera model is assumed to be a perfect pinhole one, as the problem of camera calibration is very difficult. In practice, the accuracy of the projection matrix (translation and rotation) does not reach the required precision. We therefore want to use stable points in the scene for computing a reliable projection matrix. We have built a transparent cube and with a carafe inside (see Figure 4). The polyhedral pieces of paper (white and black) have been added in the scene. This has been investigated by Mohr [4]. The contour points were obtained by a standard gradient based edge detector, and these contours were approximated with cubic B-splines [3]. The corners of the polyhedral pieces of paper in the scene were detected. Then these corners were tracked by the well-known correlation operator. Finally, we chose an absolute coordinate frame attached to the cube. The accurate projection matrices of the sequence were thus computed [4].

Fig. 4. One image of the sequence of our carafe.

Figure 4 shows one of a 10 images sequence of a carafe. Figure 5 shows results of the reconstructed B-spline patches of the carafe using a sequence of 10 images.

The parameters in this process are: the number of subdivisions of the interval on the axis s is $m = 3$, the number of subdivisions of the interval on the axis t is $n = 1$, the number of the images in the sequence is $k = 10$, the number of the extract points on an occluding contour is $l = 70$. The number of equations in our redundant system is then $3 \times l \times k = 2100$, the number of unknowns is $3 \times (m + 3) \times (n + 3) = 72$, and the number of the control points of the reconstructed B-spline patches is 24. The necessary number of iterations of Levenberg-Marquardt's algorithm is only 3. In particular, the total time for 3 iterations is 99 seconds on a Sparc10.

We remark that the quality of the reconstruction surface in Figure 5 is qualitatively good. Our final goal is to reconstruct the whole surface when the camera makes a tour of the carafe. Figure 6 shows the graph of Gaussian curvature for the estimated B-spline patches of our carafe.

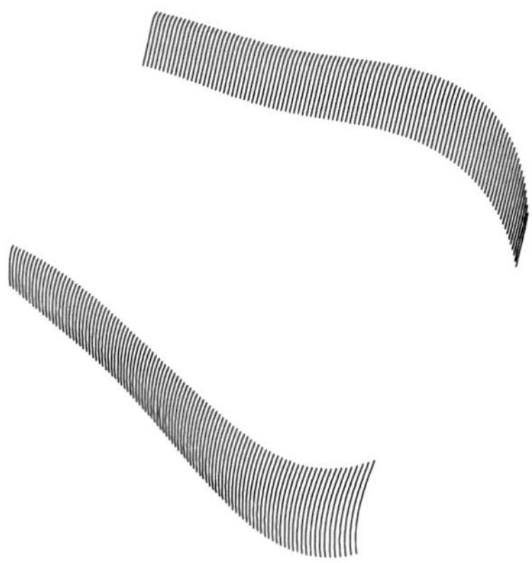

Fig. 5. A view of two reconstructed B-spline patches of the carafe.

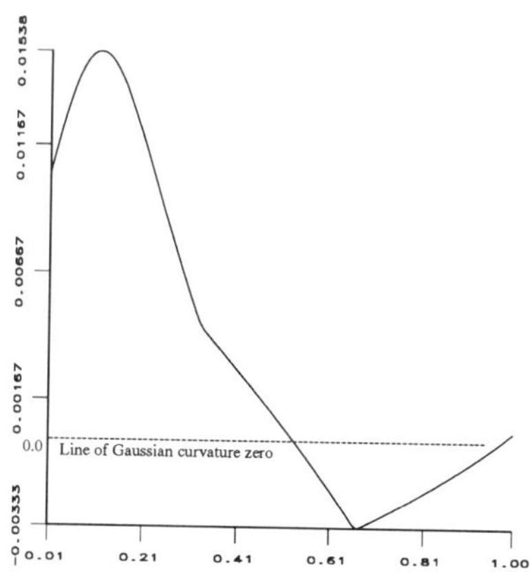

Fig. 6. Gaussian curvature of the reconstructed B-spline patches of the carafe.

In order to verify the quality of our reconstructed carafe, we compare the real diameter and the measured diameter. Measures of diameter of our carafe are performed with a pair of sliding callipers. The results are in general accurate to $\pm$ 5 mm.

§5. Discussion

We have shown that the reconstruction of a 3D surfaces is possible when the camera motion is known. The reconstruction process provided good results. The iterative algorithm of Levenberg-Marquard works well, convergence can generally be obtained in 2 to 10 iterations. Once the patches are estimated, the problem of fusing the different patches obtained with different motions still remains. This can be overcome by going back to a discrete representation with a grid of 3D points sampled on the surface patches, as net fusion is thus much easier. This step is presently being completed.

Acknowledgements. This work is partially supported by the French projet Orasis within the GDR-PRC "Communication Homme-Machine".

References

1. Bartels R. H., J. C. Beatty, and B. A. Barsky, *An Introdruction to Splines for Use in Computer Graphics and Geometric Modeling*, Morgan Kaufman, 1987.
2. Cipolla R., Active visual inference of qualitative geometry. Technical report OUEL 1795/89, University Oxford, Department of engineering science, Park Road, Oxford OX1 3PJ, U.K., July, 1989.
3. Laurent P. J., Courbes ouvertes ou fermées par B-splines régularisées. Technical report RR 652-M-, IMAG, Grenoble, France, march 1987.
4. Mohr R., L. Quan, F. Veillon, and B. Boufama. Relative 3D reconstruction using multiples uncalibrated images. Technical report RT 84-I-IMAG LIFIA 12, Lifia-Irimag, 1992.
5. Vaillant R., and O. Faugeras. Using extremal boundaries for 3-D objet modeling. *IEEE Transactions on PAMI*, 14(2):157–173, February 1992.
6. Zhao C. S., R. Mohr, and L. Quan, Global surface reconstruction through regularized B-spline patches, in *Geometric Methods in Computer Vision II, SPIE's 1993 International Symposium on Optical Instrumentation and Applied Science*, July, 1993, 134–145.

ChangSheng Zhao, Roger Mohr
LIFIA-INRIA
46, avenue Félix Viallet
38031 Grenoble cedex, FRANCE
zhao@lifia.imag.fr, mohr@lifia.imag.fr